DICTIONARY OF BIOCHEMISTRY AND MOLECULAR BIOLOGY

DICTIONARY OF BIOCHEMISTRY AND MOLECULAR BIOLOGY

Second Edition

J. STENESH

Professor of Chemistry
Western Michigan University

WILEY

A WILEY-INTERSCIENCE PUBLICATION

JOHN WILEY & SONS

New York / Chichester / Brisbane / Toronto / Singapore

Library of Congress Cataloging in Publication Data:

Stenesh, J., 1927–
    Dictionary of biochemistry and molecular biology / J. Stenesh. —
    2nd ed.
        p.   cm.
    Rev. ed. of: Dictionary of biochemistry, 1975.
    "A Wiley-Interscience publication."
    Bibliography: p.
    ISBN 0–471–84089–0
    1. Biochemistry—Dictionaries. 2. Molecular biology—
    —Dictionaries.   I. Stenesh, J., 1927–   Dictionary of biochemistry.
    II. Title.
    QP512.S73   1989
    574.19'2'0321—dc19   88-36561
                                                88-38561
                                                CIP

Printed in the United States of America

10 9 8 7 6 5 4 3 2 1

R
572.03
A825d2

239528

PREFACE

This dictionary, first published in 1975, was written to provide scientists and students in the life sciences with a reference work on the terminology of biochemistry and molecular biology. The expansion of knowledge in these areas created the need for an extensive revision of the first edition. All of the original entries were checked and reworked, if necessary, in view of new information. This second edition contains approximately 16,000 entries, of which some 4,000 are new, representing an increase of about 30% over that of the first edition. The source material consulted for revision of existing terms and for addition of new terms consisted of over 300 textbooks and reference books of various kinds and of over 600 journal articles from the research literature, all of which have been published since 1975. All told, the dictionary entries are drawn from over 500 books and 1,000 articles, including the recommendations of the Commission on Biochemical Nomenclature of the International Union of Pure and Applied Chemistry and the International Union of Biochemistry. Throughout, an effort has been made to include terms recently introduced into the biochemical literature and to exclude obsolete ones, except for a few of historical interest.

The terminology of biochemistry has a number of characteristics that influenced the selection of entries. One of these is the extensive use of terms from other sciences, since biochemistry, by its very nature, draws heavily on allied sciences. For this reason, many terms from such sciences as chemistry, immunology, genetics, virology, biophysics, and microbiology have been included in the dictionary. A second characteristic is the widespread use of abbreviations, both standard and nonstandard. Many of these are included to aid the reader of biochemical literature and to provide for the likelihood that some of the nonstandard abbreviations will become standard ones in the future. A third characteristic is the extensive use of synonymous expressions, frequently differing from each other only by minor variations. Since the synonymous nature of one expression to another may not always be apparent to the reader, principal synonymous expressions are included and cross-referenced. A fourth characteristic is the widespread use of jargon, especially in the area of molecular biology. While some of these terms may subsequently drop out of usage, others will end up becoming part of the standard terminology. For this reason, a large number of such expressions that are currently used in biochemistry and molecular biology have been included in this dictionary.

This second edition differs from the first in two important aspects. One change involves the names of specific compounds and other substances. The number of such entries included in the dictionary has been substantially enlarged. At the same time, however, no attempt was made to be exhaustive in this respect.

The second change involves the scope of the definitions. While the concise nature of the definitions of the first edition has by and large been maintained, an effort has been made to provide some additional information when this was considered useful. Thus, many terms, both original and new ones, have been defined in a slightly expanded fashion. In some cases, even lengthier definitions were deemed desirable. This was the case, for example, for many of the physical-chemical techniques, hypotheses, theories, and models used in modern biochemistry, for which a brief definition would fail to convey the essence of the term to the reader and would fail to distinguish it clearly from other, related terms. In all cases, however, a comprehensive, encyclopedic treatment was purposely avoided.

I would like to thank Dr. Mary Conway, Margery Carazzone, and Diana Cisek, my editors at Wiley, for their cooperation and helpful suggestions; Michele McCarville, Connie Gray, and Linda Thayer for their typing of the manuscript; and my wife, Mabel, and my sons, Ilan and Oron, for their understanding and support during the prolonged and time-consuming work on this book.

J. STENESH

Kalamazoo, Michigan
May 1989

EXPLANATORY NOTES

Arrangement of Entries The entries are arranged in alphabetical order, letter by letter; thus "acidimetry" precedes "acid number," and "waterfall sequence" precedes "water hydrate model." Identical alphabetical listings are entered so that lowercase letters precede capital ones and subscripts precede superscripts.

Chemical prefixes, in either abbreviated or unabbreviated form, are disregarded in alphabetizing when they are used in the ordinary sense of denoting structure of organic compounds. These include ortho-, meta-, para-, alpha-, beta-, gamma-, delta-, *cis-, trans-, N-, O-,* and *S-*. Such prefixes are, however, included in alphabetizing when they form integral parts of entries and are used in ways other than for the indication of structure of organic compounds, as in "alpha helix," "beta configuration," and "N-terminal." The prefixes mono-, di-, tri-, tetra-, and poly-, which form integral parts of entries, are included in alphabetizing, as in "monoglyceride" and "tetrahydrofolic acid."

All numbers are disregarded in alphabetizing; these include numbers denoting chemical structure, as in "glucose-6-phosphate dehydrogenase" and "5-HT," and numbers used for other purposes, as in "factor IV" and "S-100 fraction."

The letters D and L, denoting configuration, are omitted from names of terms as entered and are usually omitted from the definitions themselves.

Form of Entries All entries are direct entries so that, for example, "first law of cancer biochemistry" is entered as such and not as "cancer biochemistry, first law of." The entries are generally in the singular, with the plural indicated only when considered necessary. When several parts of speech of a term are in use, the term is generally entered in the noun form, and other parts of speech are entered only to the extent deemed useful. The different meanings of a term are numbered, chemical formulas are generally omitted, and the spelling is American.

Cross References Four types of cross-references are used in this dictionary; they are indicated by the use of *see, aka, see also,* all in italics, and by the use of words in small capital letters. The word *see* is used either in a directive sense, as in "coat—*see* spore coat; viral coat" and "hereditary code—*see* genetic code," or to indicate that the term is defined within the definition of another, separately entered term, as

in "E'_0—*see* standard electrode potential" and "MIH—*see* melanocyte-stimulating hormone regulatory hormone." The abbreviation *aka* (also known as) is used at the end of a definition to indicate expressions that are synonymous to the entry; principal synonymous expressions are entered separately in the text. The phrase *see also* is used at the end of a definition where it is considered useful to point out to the reader comparable, contrasting, or other kinds of related entries. Small capital letters are used to indicate an expression that is synonymous to the entry and that is defined in its alphabetical place in the book. Thus, the definition of the entry "amphiphilic" by the word "AMPHIPATHIC," and the definition of the entry "pentose oxidation cycle" by the term "HEXOSE MONOPHOSPHATE SHUNT" indicate that the terms in small capital letters are expressions that are synonymous to the entries and that are themselves defined in their appropriate alphabetical places in the text.

Abbreviations and Symbols The following standard abbreviations and symbols are used in this dictionary:

A	ampere
Å	angstrom unit
abbr	abbreviation
adj	adjective
adv	adverb
aka	also known as
atm	atmosphere
°C	degree Celsius
cal	calorie
cc	cubic centimeter
cd	candela
cm	centimeter
cps	cycles per second
deg	degree
dm	decimeter
e.g.	for example
esu	electrostatic unit
g	gram
i.e.	that is
J	joule
kcal	kilocalorie
kg	kilogram
L	liter
lb	pound
lm	lumen

DICTIONARY OF BIOCHEMISTRY AND MOLECULAR BIOLOGY

A

a 1. Subscript denoting the more active form of an interconvertible enzyme. 2. Atto.

A 1. Adenine. 2. Adenosine. 3. Absorbance. 4. Angstrom unit. 5. Mass number. 6. Alanine. 7. Helmholtz free energy. 8. Ampere.

2,5-A TWO-FIVE A.

Å Angstrom unit

AA 1. Amino acid. 2. Atomic absorption.

AA-AMP Aminoacyl adenylate.

AAN Amino acid nitrogen.

AAS Atomic absorption spectrophotometry.

AA-tRNA Aminoacyl-tRNA.

AA-tRNA^{AA} Aminoacyl transfer RNA; the prefix AA denotes the aminoacyl group attached to the transfer RNA (tRNA) molecule, while the superscript AA denotes the amino acid for which the transfer RNA is specific.

AAV Adenovirus-associated virus.

Ab Antibody.

ABA Abscisic acid.

A band A transverse dark band that is seen in electron microscope preparations of myofibrils from striated muscle and that consists of thick and thin filaments.

Abbe refractometer A refractometer for the direct measurement of the refractive index of a solution. A few drops of liquid are placed between two prisms in a water-thermostated compartment and light is then passed through the prisms into a telescope, attached to a measuring scale.

ABC Antigen binding capacity.

a × b × c code An early version of the genetic code according to which there exist, respectively, *a, b,* and *c* distinguishable and nonequivalent bases for each of the three positions of the codon, so that the product *a × b × c* is equal to the number of categories into which the triplet codons are divided. The original *a × b × c* code was thought to be a 4 × 3 × 2 code.

ABC excinuclease An enzyme, present in *E. coli,* that mediates both the incision and excision steps of the excision repair of DNA. The enzyme is composed of three subunits and appears to recognize helical distortions in DNA, such as those produced by ultraviolet irradiation or alkylating agents.

aberration *See* chromosomal aberration.

abetalipoproteinemia A genetically inherited metabolic defect in humans that is characterized by the absence of low-density lipoproteins.

abiogenesis 1. The formation of a substance other than by a living organism. 2. The doctrine that living organisms can come from nonliving matter; spontaneous generation.

abiogenetic Of, or pertaining to, abiogenesis.

abiogenic Of, or pertaining to, abiogenesis.

abiological Of, or pertaining to, nonliving matter.

abiosis The absence of life.

abiotic Of, or pertaining to, abiosis.

ablation The breakup and wearing of a solid surface by impact with particles or radiation; the etching of the surface of a biological tissue by exposure to ultraviolet lasers is an example.

ABM paper Aminobenzyloxy methylcellulose paper, used in the study of nucleic acids. When this paper is chemically activated, it binds single-stranded nucleic acid covalently.

abnormal hemoglobin A hemoglobin that differs from normal hemoglobin in its amino acid sequence.

ABO blood group system A human blood group system in which there are two antigens, denoted A and B, that give rise to four serum groups, denoted A, B, AB, and O. The antigens are mucopeptides and contain a mucopolysaccharide that is identical in both antigens except for its nonreducing end. The serum groups A, B, AB, and O are characterized, respectively, by having red blood cells that carry A antigens, B antigens, both A and B antigens, and neither A nor B antigens.

abortive complex 1. NONPRODUCTIVE COMPLEX. 2. A ternary, dead-end complex; an inactive complex, consisting of enzyme, substrate, and product.

abortive infection A viral infection that either does not lead to the formation of viral particles or leads to the formation of noninfectious viral particles.

abortive initiation An initiation of transcription that is terminated after only a few nucleotides have been polymerized. In this case, the 5′-fragment synthesized (consisting of $_{ppp}$A and one or more additional nucleotides) dissociates from the promoter so that the initiation process must start again. Abortive

1

initiation may occur if a needed nucleotide is missing as a result of other factors.

abortive transduction   Bacterial transduction in which the DNA from the donor cell is introduced into the recipient cell, but fails to become integrated into the chromosome of the recipient bacterium.

ABP   Androgen-binding protein.

abrin   A plant protein in the seeds of *Abrus precatorius* that is toxic to animals and humans and that has antitumor activity; it inhibits protein synthesis in eukaryotes by inhibiting the binding of aminoacyl-tRNA to ribosomes.

abscisic acid   A widely occurring sesquiterpene plant hormone that is antagonistic to many other plant hormones; it inhibits growth, seed germination, bud formation, and leaf senescence. *Abbr* ABA. *Aka* abscisin, dormin.

abscissa   The horizontal axis, or *x*-axis, in a plane rectangular coordinate system.

absolute alcohol   Anhydrous ethyl alcohol.

absolute configuration   The actual spatial arrangement of the atoms about the asymmetric carbon atoms in a molecule.

absolute counting   The counting of radiation that includes every disintegration that occurs in the sample; such counts are expressed as disintegrations per minute.

absolute defective mutant   A cell or an organism that exhibits its mutant phenotypic behavior under all conditions. *See also* conditional mutant.

absolute deviation   The numerical difference, regardless of sign, between an experimental value and a given value; the latter may be a constant, a sample value, or a mean.

absolute error   The absolute deviation of an experimental value from the true, or the best, value of the quantity being measured.

absolute oil   *See* essential oil.

absolute plating efficiency   The percentage of cells that give rise to colonies when a given number of cells are plated on a nutrient medium.

absolute reaction rates   *See* theory of absolute reaction rates.

absolute specificity   The extreme selectivity of an enzyme that allows it to catalyze only the reaction with a single substrate in the case of a monomolecular reaction, or the reaction with a single pair of substrates in the case of a bimolecular reaction. *Aka* absolute group specificity.

absolute temperature scale   A temperature scale on which the zero point is the absolute zero, and the degrees, denoted K (no degree sign), match those of the Celsius scale. *Aka* Kelvin temperature scale.

absolute zero   The zero point on the absolute temperature scale $(-273.2°C)$; the theoretical temperature at which all atomic motion ceases.

absorb   To engage in the process of absorption.

absorbance   A measure of the light absorbed by a solution that is equal to $\log I_o/I$, where I_o is the intensity of the incident light, and I is the intensity of the transmitted light. *Sym* A. *Aka* optical density.

absorbance index   ABSORPTIVITY.

absorbance unit   The amount of absorbing material contained in 1 mL of a solution that has an absorbance of 1.0 when measured with an optical path length of 1.0 cm.

absorbancy   Variant spelling of absorbance.

absorbate   A substance that is absorbed by another substance.

absorbed antiserum   An antiserum from which antibodies have been removed by the addition of soluble antigens.

absorbed dose   *See* radiation absorbed dose.

absorbent   1. *n* A substance that absorbs another substance. 2. *adj* Having the capacity to absorb.

absorber   A material used to absorb radioactive radiation.

absorptiometer   1. An instrument for measuring the amount of gas absorbed by a liquid. 2. A device for measuring the thickness of a layer of liquid between parallel glass plates. 3. COLORIMETER.

absorption   1. The uptake of one substance by another substance. 2. The passage of materials across a biological membrane. 3. The process by which all or part of the energy of incident radiation (includes heat, electromagnetic, and radioactive radiation) is transferred to the matter through which it passes. 4. The removal of antibodies from a mixture by the addition of soluble antigens, or the removal of soluble antigens from a mixture by the addition of antibodies.

absorption band   A portion of the electromagnetic spectrum in which a molecule absorbs radiant energy.

absorption cell   CUVETTE.

absorption coefficient   1. ABSORPTIVITY. 2. BUNSEN ABSORPTION COEFFICIENT. 3. The rate of change in the intensity of a beam of radiation as it passes through matter.

absorption cross section   The product of the probability that a photon passing through a molecule will be absorbed by that molecule and the average cross-sectional area of the molecule; the absorption cross section s is related to the molar absorptivity ϵ by $s = 3.8 \times 10^{-21} \epsilon$.

absorption optical system   An optical system that focuses ultraviolet light passing through a solution in such a fashion that a photograph is obtained in which the darkening of the photographic film depends on the amount of light transmitted by the solution. A boundary in the solution appears as a transition between a lighter and a darker region, and measurements are made on the film by means of a densitometer tracing. The optical system is used in the analytical ultracentrifuge.

absorption ratio The ratio of the concentration of a compound in solution to its absorptivity.

absorption spectrum A plot of the absorption of electromagnetic radiation by a molecule as a function of either the frequency or the wavelength of the radiation.

absorptive lipemia The transient increase in the concentration of lipids in the blood that follows the ingestion of fat.

absorptivity The proportionality constant ϵ in Beer's law, $A = \epsilon lc$, where A is the absorbance, l is the length of the light path, and c is the concentration.

abstraction The removal of either an atom or an electron from a compound.

abundance The average number of molecules of a specific mRNA type in a given cell. The abundance (A) is given by $A = NRf/M$, where N is Avogadro's number, R is the RNA content of the cell in grams, f is the fraction of the specific mRNA relative to the total RNA content of the cell, and M is the molecular weight of the specific mRNA in daltons. *Aka* representation.

Ac Acetyl group.

acanthocyte A cell that has numerous projecting spines or "thorns."

acanthocytosis 1. A condition characterized by blood that contains spherical erythrocytes that have numerous projecting spines. 2. ABETALIPOPROTEINEMIA.

acanthosome A membranous vesicle that appears in fibroblasts, isolated from the skin of hairless mice that have been subjected to chronic UV irradiation.

ACAT Acyl-CoA:cholesterol transferase; the enzyme that forms cholesteryl esters from cholesterol.

acatalasemia ACATALASIA.

acatalasia A genetically inherited metabolic defect in humans that is due to a deficiency of the enzyme catalase.

acceleration A stage in carcinogenesis in which, according to the Busch theory, an accelerator protein is synthesized which functions in accelerating the production of cancer RNA from cancer DNA.

accelerator An instrument for imparting high kinetic energy to subatomic particles by means of electric and magnetic fields.

accelerator globulin PROACCELERIN.

accelerator protein *See* acceleration.

accelerin The activated form of proaccelerin that converts prothrombin to thrombin during blood clotting.

acceptor 1. A protein that is activated by a hormone receptor and that directly mediates the action of a rate-limiting enzyme. Hormone action thus involves the following stages: (a) the hormone binds to a receptor which undergoes a conformational change; (b) the hormone–receptor complex interacts with an acceptor molecule to form a hormone–receptor–acceptor complex; (c) formation of the latter complex activates the acceptor; (d) the activated acceptor molecule mediates the activity of a rate-limiting enzyme. 2. The atom that receives a hydrogen in the formation of a hydrogen bond.

acceptor control The dependence of the respiratory rate of mitochondria on the ADP concentration. *See also* loose coupling; tight coupling.

acceptor-control ratio The rate of respiration, in terms of oxygen uptake per unit time, in the presence of ADP, divided by the rate in the absence of ADP; measured either in the intact cell or in isolated mitochondria.

acceptor end The trinucleotide CCA at the 3′-end of tRNA. The amino acid becomes esterfied to the 2′- or 3′-position of the terminal adenine nucleotide in this sequence.

acceptor junction *See* splicing junctions.

acceptor protein ACCEPTOR (1).

acceptor RNA TRANSFER RNA.

acceptor site AMINOACYL SITE.

acceptor splicing site *See* splicing junctions.

acceptor stem *See* arm.

accessible surface That part of the van der Waals surface of a protein that is defined by the center of a suitable probe, generally a water molecule having a radius of 1.4 Å. The accessible surface (A_s) for a small protein of molecular weight M can be approximated by the relation $A_s = 11.12 \times M^{2/3}$. For a large protein, with conspicuous domains, A_s becomes directly proportional to the molecular weight.

accessory factor A protein in blood clotting that, when activated proteolytically, serves to enhance the rate of proteolytic activation of some other blood clotting factor.

accessory pigment A photosynthetic pigment, such as a carotenoid or a phycobilin, that functions in conjunction with a primary photosynthetic pigment.

AcCoA Acetyl coenzyme A.

accumulation theory A theory of aging according to which aging is due to the accumulation of either a deleterious or a toxic substance.

accumulator organism An organism capable of

absorbing and retaining large amounts of specific chemical elements.

accuracy The nearness of an experimental value to either the true, or the best, value of the quantity being measured.

ACD solution Acid–citrate–dextrose solution.

acellular Not composed of cells.

ACES *N*-(2-Acetamido)-2-aminoethanesulfonic acid; used for the preparation of biological buffers in the pH range of 6.1 to 7.5. *See also* biological buffers.

acetal A compound derived from an aldehyde and two alcohol molecules by splitting out a molecule of water.

acetate hypothesis The hypothesis that a multitude of complex substances may be formed naturally as a result of modification of the linear chains formed by repeated head-to-tail condensation of acetic acid residues; typical modifications are cyclization, oxidation, and alkylation.

acetate-replacing factor LIPOIC ACID.

acetate thiokinase A fatty acid thiokinase that catalyzes the activation of fatty acids having two or three carbon atoms to fatty acyl coenzyme A.

acetification The spoilage of beverages, such as wine and beer, due to the aerobic oxidation of ethyl alcohol to acetic acid by microorganisms.

acetoacetic acid A ketoacid that can be formed from acetyl coenzyme A and that is one of the ketone bodies.

acetogenin One of a large number of compounds that are formally equivalent to head-to-tail condensation products of acetic acid residues. Acetogenins are biosynthesized by means of a multienzyme complex via condensations of acetyl coenzyme A molecules or other derivatives of coenzyme A. Acetogenins are responsible for many of the brilliant colors that occur in nature. Major subgroups include flavonoids, tetracyclines, and macrolide antibiotics. *Aka* polyketide.

acetoin 2-Keto-3-hydroxybutane; a compound that can be formed by air oxidation of butylene glycol in the course of butylene glycol fermentation.

acetoin fermentation BUTYLENE GLYCOL FERMENTATION.

acetone A ketone that can be formed from acetyl coenzyme A and that is one of the ketone bodies.

acetone body KETONE BODY

acetone-butanol fermentation The fermentation of glucose that is characteristic of some *Clostridium* species and which, at first, yields acetic acid and butyric acid, but after the pH drops, yields acetone and butanol as major end products. *Aka* solvent fermentation.

acetonemia 1. The presence of excessive amounts of acetone in the blood. 2. The presence of excessive amounts of ketone bodies in the blood.

acetone powder A preparation of one or more proteins that is produced by removal of acetone by vacuum filtration from an acetone extract of a tissue; used in the course of isolating and purifying an enzyme or other protein.

acetonuria 1. The presence of excessive amounts of acetone in the urine. 2. The presence of excessive amounts of ketone bodies in the urine.

acetonyl-SCoA An inhibitory analog of acetyl coenzyme A; the compound CH_3—CO—CH_2—SCoA.

acetylation An acylation reaction in which an acetyl radical CH_3CO—is introduced into an organic compound.

acetylcholine The acetylated form of choline; the hydrolysis of acetylcholine to choline and acetic acid is catalyzed by acetylcholinesterase and is a key reaction in the transmission of the nerve impulse. *Abbr* ACh.

acetylcholinesterase The enzyme that catalyzes the hydrolysis of acetylcholine to choline and acetic acid during the transmission of a nerve impulse. *Abbr* AChE. *Aka* true cholinesterase; choline esterase I; specific cholinesterase. *See also* cholinesterase.

acetyl CoA Acetyl coenzyme A.

acetyl-CoA carboxylase A multienzyme system that catalyzes the ATP-requiring biosynthesis of malonyl–SCoA from acetyl-SCoA and HCO_3^-. The enzyme from *E. coli* and plants consists of three components: (a) biotin carboxyl carrier protein (BCCP or BCP); a protein that contains two identical subunits, each of which has one molecule of biotin linked covalently to the ϵ-NH_2 group of a lysine residue; (b) biotin carboxylase (BC); an enzyme having two identical subunits; (c) transcarboxylase (TC or carboxyl transferase); a tetrameric enzyme containing two pairs of non-identical subunits.

acetyl coenzyme A The acetylated form of coenzyme A; a key intermediate in the citric acid cycle, in fatty acid oxidation, in fatty acid synthesis, and in other metabolic reactions. Variously abbreviated as acetyl-SCoA, acetyl-CoA, CoASAc, AcSCoA, and AcCoA.

acetyl coenzyme A carboxylase *See* acetyl-CoA carboxylase.

acetylene 1. The hydrocarbon CH≡CH. 2. ALKYne.

acetyl group The acyl group of acetic acid; the radical CH_3CO—. *Abbr* Ac, OAc.

N-acetylmuramic acid A compound derived from acetic acid, glucosamine, and lactic acid that is a major building block of bacterial cell walls.

N-acetylneuraminic acid A compound derived from acetic acid, mannosamine, and pyruvic acid that is a major building block of animal cell coats. *Abbr* NANA; NAcneu; NeuAc.

acetyl number A measure of the number of hydroxyl groups in a fat; equal to the number of milligrams of potassium hydroxide required to neutralize the acetic acid in 1 gram of acetylated fat. *Aka* acetyl value.

acetylornithine cycle A cyclic set of reactions in bacteria and plants that constitutes a major pathway for the synthesis of ornithine from glutamic acid and *N*-acetylornithine.

acetyl-SCoA Acetyl coenzyme A.

N-acetylserine The acetylated form of serine that is believed to function in the initiation of translation in mammalian systems, much as *N*-formylmethionine functions in the initiation of translation in bacterial systems.

acetyltransferase An enzyme that catalyzes the transfer of an acetyl group from acetyl coenzyme A to another compound.

AcG Accelerator globulin.

Ac globulin Accelerator globulin.

ACh Acetylcholine.

A chain 1. The shorter of the two polypeptide chains of insulin, containing 21 amino acids and one intrachain disulfide bond. 2. The heavy chain (H chain) of the immunoglobulins.

AChE Acetylcholinesterase.

achiral Not chiral.

achirotopic Not chirotopic.

achlorophyllous Lacking chlorophyll.

achromic Devoid of color.

achromic point A stage in the hydrolysis of starch at which the addition of iodine fails to produce a blue color.

achromotrichia factor *p*-AMINOBENZOIC ACID.

achromycin *See* tetracyclines.

acid *See* Bronsted acid; Lewis acid.

acidaminuria AMINOACIDURIA.

acid anhydride A compound containing two acyl groups bound to an oxygen atom. The compound is referred to as either a simple or a mixed anhydride depending on whether the two acyl groups are identical or different. In biochemistry, both simple and mixed anhydrides frequently contain the phosphoryl group.

acid–base balance The reactions and factors involved in maintaining a constant internal environment in the body with respect to the buffer systems and the pH of the various fluid compartments.

acid–base catalysis *See* general and specific acid–base catalysis.

acid–base indicator *See* indicator.

acid–base titration A titration in which either acid or base is added to a solution, and the titration is followed by means of pH measurements or by means of indicators.

acid–citrate–dextrose solution An aqueous solution of citric acid, sodium citrate, and dextrose, that is used as an anticoagulant in the collection and storage of blood.

acidemia A condition characterized by an increase in the hydrogen-ion concentration of the blood.

acid–fast Descriptive of the lipid-rich cell walls of some bacteria that resist decolorization by mineral acids after having been stained with basic aniline dyes.

acid hematin A hematin formed from hemoglobin by treatment with acid below pH 3.

acid hydrolase A hydrolytic enzyme that has an acidic optimum pH.

acidic 1. Of, or pertaining to, an acid. 2. Of, or pertaining to, a solution having a pH less than 7.0.

acidic amino acid An amino acid that has one amino and two carboxyl groups; an amino acid that has a net negative charge at neutral pH.

acidic dye An anionic dye that binds to, and stains, positively charged macromolecules. *Aka* acidic stain.

acidic food A food that is rich in phosphorus, sulfur, and chlorine and that leaves an acidic residue when subjected to combustion.

acidification of urine The process whereby the glomerular filtrate of the kidney that has an approximate pH of 7.4 is converted to urine that has a lower pH and may have a pH as low as 4.8.

acidimetry 1. The chemical analysis of solutions by means of titrations, the end points of which are recognized by a change in the hydrogen-ion concentration. 2. A determination of the amount of an acid by titration against a standard alkaline solution.

acidity constant ACID DISSOCIATION CONSTANT.

acid mucopolysaccharides GLYCOSAMINO-GLYCANS.

acid number The number of milligrams of potassium hydroxide required to neutralize the free fatty acids in 1g of fat. *Aka* acid value.

acidolysis Hydrolysis by means of an acid.

acidophil A cell that stains with an acidic dye.

acidosis A deviation from the normal acid–base balance in the body that is due to a disturbance which, by itself and in the absence of compensatory mechanisms, would tend to lower the pH of the blood. The actual change in pH depends on whether and to what extent the disturbance is compensated for. The disturbances and the compensatory mechanisms are considered primarily with respect to their effect on the bicarbonate/carbonic acid ratio of blood plasma. *See also* metabolic acidosis; primary acidosis; etc.

acidosome A nonlysosomal vesicle that functions in the acidification of digestive phagocytic vacuoles in *Paramecium.*

acidotic Of, or pertaining to, acidosis.

acid pH A pH value below 7.0.

acid phosphatase A phosphatase, the optimum pH of which is below 7.0.

acid plant A plant that accumulates organic acids in its leaves; these acids form ammonium salts.

acid rain The environmental phenomenon in which sulfur dioxide and nitrogen oxides, expelled into the air by industrial combustion, react with rainwater to produce dilute solutions of sulfuric and nitric acids. Acid rain leads to acidification of streams and lakes and depletion or loss of their fish life.

aciduria A condition characterized by the excretion of an excessively acidic urine.

aconitase The iron-containing enzyme that catalyzes the interconversion of citrate and isocitrate in the citric acid cycle. The reaction proceeds via the enzyme-bound intermediate *cis*-aconitate (a tricarboxylic acid). *Aka* aconitate hydratase.

aconitate hydratase ACONITASE.

cis-aconitic acid *See* aconitase.

acoustical phonon *See* phonon.

ACP 1. Acyl carrier protein. 2. Acid phosphatase.

ac polarography Alternating-current polarography; a polarographic method in which a small alternating potential is superimposed on the normal, direct-current applied potential, and the ac component of the resulting current is measured.

acquired antibody An antibody produced by an immune reaction as distinct from one occurring naturally.

acquired hemolytic anemia An autoimmune disease in which individuals form antibodies to their own red blood cells.

acquired immunity The immunity established in an animal organism during its lifetime.

acquired immunodeficiency syndrome *See* AIDS.

acquired tolerance The immunological tolerance produced in an animal organism by the injection of antigen into it; acquired tolerance persists only as long as the antigen remains in the organism.

acridine dye A planar heterocyclic molecule used to stain DNA and RNA. Acridine dyes are basic dyes that become intercalated into the nucleic acid molecule; they are mutagenic, since their intercalation produces insertions or deletions.

acridine orange An acridine dye that functions both as a fluorochrome for staining nucleic acids and as a mutagen, producing insertions or deletions.

acriflavin An acridine dye that leads to frame shift mutations.

acrolein test A qualitative test for glycerol, based on the dehydration and oxidation of glycerol to acrolein by heating with potassium bisulfate.

acromegaly A condition characterized by overgrowth of skeletal structures due to the excessive production of growth hormone.

acronym A word formed from the initial letters of other words; the words LASER and LET are two examples.

acrosome A cap-like structure, beneath the cell membrane, at the head of a spermatozoon; it serves to digest the egg coatings to permit fertilization.

acrosome reaction The release of the contents of an acrosome by exocytosis upon contact of a sperm with an egg.

acrylamide *See* polyacrylamide gel.

AcSCoA Acetyl coenzyme A.

ACTH Adrenocorticotropic hormone.

ACTH family A group of peptide hormones, including ACTH, lipotropin, and melanotropin, that are derived from a common precursor. The opioids β-endorphin and γ-endorphin are also derived from the same precursor which is known as prepro-opiomelanocortin. *Aka* ACTH/endorphin peptides.

actidione CYCLOHEXIMIDE.

actin A major protein component of the myofilaments of striated muscle and the principal constituent of the thin filaments of muscle and of the microfilaments of the cytoskeleton. *See also* F-actin; G-actin.

actin filament A thin filament of striated muscle that consists largely of actin and that is linked to thick filaments by means of cross-bridges which protrude from them; a myofilament. The polymerization of actin monomers to form filaments proceeds with polarity. The plus, or barbed, end of the filament is the fast-assembly end which requires a lower critical concentration of monomer (the concentration at which addition of monomer just balances dissociation); the minus, or pointed, end is the slow-assembly end which requires a higher critical concentration of monomer. *See also* microfilament; treadmilling.

actin-fragmenting protein One of a number of proteins, such as villin and gelsolin, that bind to actin filaments and sever them. These are generally calcium-dependent proteins and they are thought to bind so strongly to the actin filaments that the latter are broken at the binding sites.

actinin A minor protein component of striated muscle, believed to be part of the thin filaments and to be concentrated in both the Z line and the I band. Two components, denoted α-and β-actinin, have been identified: α-actinin links actin filaments together to form a random, three-dimensional network; β-actinin tends to reduce the length of an F-actin strand and may serve to determine the length of actin filaments.

actinometer A device for the determination of absorbed light by means of a photochemical reaction of known quantum yield.

actinometry A method of chemical analysis by means of an actinometer.

actinomyces A genus of gram-positive bacteria that belongs to the family of Actinomycetaceae (order Actinomycetales or Actinomycetes). Actinomyces are rods or branched filaments and are anaerobes with varying degrees of aerotolerance.

actinomycin D An antibiotic, produced by *Streptomyces chrysomallus*, that inhibits the transcription of DNA to RNA by binding to DNA and that also has immunosuppresive activity. *Aka* actinomycin C1.

action potential The membrane potential of a stimulated membrane, produced by the ion flux across the membrane, when its permeability is changed upon stimulation.

action spectrum A plot of a quantitative biological or chemical response as a function of the wavelength of the radiation producing the response; the death of bacteria, the occurrence of mutations, the occurrence of fluorescence, and photosynthetic efficiency are examples of responses.

activated *See also* active.

activated alumina Alumina that has been thoroughly dried.

activated carbon A porous material, consisting primarily of carbon, that is prepared by the destructive distillation of plants; used for adsorption of gases and decolorization of solutions.

activated complex theory THEORY OF ABSOLUTE REACTION RATES.

activated form *See* active form.

activated macrophage A macrophage that has been stimulated, generally by a lymphokine, to increase in its size, in its number of enzyme molecules, and in its phagocytic activity.

activating enzyme 1. FATTY ACID THIOKINASE. 2. AMINOACYL-tRNA SYNTHETASE.

activation 1. The conversion of a compound to a more reactive form; the change of an amino acid to aminoacyl transfer RNA, the change of a fatty acid to fatty acyl coenzyme A, and the change of an inactive enzyme precursor to the active enzyme are some examples. 2. The increase in the extent, and/or the rate, of an enzymatic reaction. 3. The drying of chromatographic supports. 4. The first stage in the conversion of a spore to a vegetative cell; this stage can frequently be produced by heat or aging and is believed to involve damage to an outer layer of the spore. 5. The conformational change of a receptor upon the binding of a hormone.

activation analysis A method for the qualitative and quantitative analysis of the chemical elements in a sample; based on identification and determination of the radionuclides formed when the sample is bombarded with neutrons or other particles.

activation energy The difference in energy between that of the activated complex and that of the reactants; the energy that must be supplied to the reactants before they can undergo transformation to products. *Sym* E_a; E_A.

activation hormone An insect hormone that controls the secretion of the corpora allata, the paired glands that synthesize the juvenile hormone in insect larvae. The activation hormone is a polypeptide, produced in the brain.

activation stage That part of the blood-clotting process that consists of the formation of active thrombin.

activator A metal ion that serves as a cofactor for an enzyme.

activator constant The equilibrium constant for the reaction $EA \rightleftharpoons E + A$, where E is an enzyme and A is an activator.

activator protein 1. CALMODULIN. 2. *See* Britten–Davidson model.

activator RNA *See* Britten–Davidson model.

active acetaldehyde An acetaldehyde molecule attached to thiamine pyrophosphate; α-hydroxyethylthiamine pyrophosphate.

active acetate ACETYL COENZYME A.

active acetyl 1. ACETYL COENZYME A. 2. Acetyl lipoic acid.

active acyl 1. An acyl coenzyme A. 2. An acyl lipoic acid.

active adenosyl ADENOSINE-5′-TRIPHOSPHATE.

active adenylate ADENOSINE-5′-TRIPHOSPHATE.

active aldehyde An aldehyde molecule attached to thiamine pyrophosphate; α-hydroxyalkylthiamine pyrophosphate.

active aldehyde theory The theory according to which the nonenzymatic browning of foods is due to reactions involving very active aldehydes that are formed by the dehydration of sugars.

active amino acid 1. An amino acid linked to the phosphate group of AMP; an aminoacyladenylate. 2. An amino acid linked to the hydroxyl group of ribose in the terminal adenosine nucleotide in transfer RNA; an aminoacyl–tRNA. 3. A Schiff base of an amino acid as that formed in transamination.

active ammonia 1. CARBAMOYL PHOSPHATE. 2. GLUTAMINE.

active anaphylaxis The anaphylactic reaction produced in an animal organism as a result of the injection of antigen.

active carbohydrate 1. A UDP-sugar. 2. A GDP-sugar. 3. An ADP-sugar.

active carbon dioxide CARBOXYBIOTIN.

active carboxylic acid A reactive derivative of a carboxylic acid that is capable of reactions which the free acid does not undergo. Biochemically important active carboxylic acids are acid anhydrides and thioesters.

active center ACTIVE SITE.

active concentration ACTIVE TRANSPORT.

active enzyme centrifugation A method that permits the hydrodynamic study of an enzyme-substrate complex; involves layering a small amount of an enzyme solution over a substrate solution and then centrifuging. While the enzyme layer sediments, one observes spectroscopically either the appearance of a product or the disappearance of a substrate. When carried out in the analytical ultracentrifuge, the method permits a determination of the sedimentation or diffusion coefficient of the actual active enzyme molecule. *Abbr* AEC.

active fatty acid A fatty acid linked to coenzyme A; a fatty acyl-SCoA. These thioesters are high-energy compounds.

active form 1. That derivative of a metabolite that can serve as a high-energy compound and/or as a compound that initiates a reaction or a series of reactions. 2. That form of a macromolecule that possesses biological activity.

active formaldehyde ACTIVE FORMYL.

active formate 1. ACTIVE FORMYL. 2. ACTIVE FORMIMINO.

active formimino A formimino group $NH=CH-$ attached to tetrahydrofolic acid.

active formyl A formyl group $O=CH-$ attached to tetrahydrofolic acid.

active fructose FRUCTOSE-1,6-BISPHOSPHATE.

active glucose 1. UDP-GLUCOSE. 2. ADP-GLUCOSE.

active glycolaldehyde A glycolaldehyde group $CH_2OH-CO-$ attached to thiamine pyrophosphate; α, β-dihydroxyethyl thiamine pyrophosphate.

active hydroxyethyl ACTIVE ACETALDEHYDE.

active hydroxymethyl 5,10-Methylene tetrahydrofolic acid.

active immunity The immunity acquired by an animal organism as a result of the injection of antigens into it.

active iodine That form of iodine, possibly an iodinium ion I^+, which reacts with tyrosine to form iodotyrosines in the thyroid gland.

active mediated transport An active transport that requires one or more transport agents.

active methionine S-ADENOSYLMETHIONINE.

active methyl 1. 5-Methyltetrahydrofolic acid. 2. S-ADENOSYLMETHIONINE.

active one-carbon unit A one-carbon fragment linked to tetrahydrofolic acid.

active oxygen The form of oxygen as it is used in reactions catalyzed by monooxygenases; the oxygen linked to the enzyme–copper complex of dopamine β-monooxygenase is an example.

active patch ANTIGEN BINDING SITE.

active phosphate 1. ADENOSINE-5'-TRIPHOSPHATE. 2. GUANOSINE-5'-TRIPHOSPHATE.

active phospholipid A cytidine-5'-diphosphate derivative of either a phospholipid or a component of phospholipids.

active pyrophosphate ADENOSINE-5'-TRIPHOSPHATE.

active pyruvate α-Hydroxyethylthiamine pyrophosphate; the compound formed by the reaction of pyruvate with enzyme-bound thiamine pyrophosphate. Active pyruvate is the first intermediate formed in the pyruvate dehydrogenase reaction whereby pyruvate is converted to acetyl-SCoA.

active site 1. That portion of the enzyme molecule that interacts with, and binds, the substrate, thereby forming an enzyme–substrate complex. 2. That portion of the antibody molecule that interacts with, and binds, the antigen, thereby forming an antigen–antibody complex.

active site-directed irreversible inhibitor An artificially designed inhibitor for the irreversible inhibition of a given enzyme. The inhibitor is a trifunctional molecule that contains (a) a functional group that can bind to the active site of the enzyme, (b) a nonpolar fragment that can attach to a nonpolar region just outside the active site, and (c) a group, such as sulfonyl chloride, that can alkylate a functional group of the enzyme just outside the nonpolar region. The first functional group serves to direct the inhibitor to the active site of the enzyme; the nonpolar fragment serves to align the inhibitor so that the alkylating group is brought into contact with a susceptible group on the enzyme; and the third functional group then leads to an alkylation reaction that results in the irreversible inhibition of the enzyme. *See also* affinity labeling.

active succinate Succinic acid linked to coenzyme A; succinyl-SCoA.

active sulfate 1. The compound 3'-phosphoadenosine-5'-phosphosulfate that serves as a sulfating agent in the esterification of sulfate with alcoholic and phenolic hydroxyl groups. *Abbr* PAPS. 2. The compound adenosine-5'-phosphosulfate that serves as an intermediate in the synthesis of 3'-phosphoadenosine-5'-phosphosulfate and that can be reduced directly to sulfite in *Desulfovibrio desulfuricans*. *Aka* adenylyl sulfate.

active translocation ACTIVE TRANSPORT.

active transport The movement of a solute across a biological membrane such that the movement is directed upward in a concentration gradient (i.e., against the gradient) and requires the expenditure of energy. When the energy is supplied by the simultaneous hydrolysis of ATP (ATPase activity), or some other high-energy compound, on the surface of the transport agent, the process is known as primary active transport or pump. When the energy is supplied by coupling the active transport to the simultaneous movement of a second substance down its concentration gradient, the process is known as secondary active transport. The second substance may be moving in the same direction as the first (symport) or in the opposite direction (antiport).

activity 1. A measure of the effective concentration of an enzyme, drug, hormone, or other substance, and by extension, the substance the effectiveness of which is being measured. 2. The product of the molar concentration of an ionic solute and its activity coefficient. Activity represents an effective concentration, reflecting solute–solute interactions, and must be used in place of molar concentrations for nonideal solutions.

activity coefficient The ratio of the activity of an ion to its molar concentration; the logarithm of the activity coefficient is equal to $-0.5Z^2\sqrt{I}$, where Z is the charge of the ion and I is the ionic strength. *See also* mean ionic activity coefficient.

actomere A subcellular organelle, believed to initiate the assembly of actin filaments in some sperm cells.

actomyosin The complex formed between myosin and actin, either as extracted from muscle or as prepared from the purified components.

acumentin A protein in macrophages that binds to the minus (pointed, slow-assembly) end of actin filaments.

acute disease A disease that has a rapid onset and is of short duration (days or weeks), terminating either in recovery or in death.

acute porphyria A porphyria that is of short duration and that is characterized by the excretion of excessive amounts of uroporphyrin III, coproporphyrin III, and porphobilinogen.

acute serum A serum obtained soon after the onset of a disease. *Aka* acute phase serum.

acute test A toxicity test that is performed on laboratory animals and that requires only a single dose of a chemical, administered in a single application.

acute transfection A brief infection of cells with foreign DNA.

acyclic ALIPHATIC.

acyclovir 9[2-Hydroxyethoxy)methyl]guanine; an antiviral drug that is particularly effective in the treatment of genital herpes. The antiviral activity of this drug is initiated when it is phosphorylated, a reaction catalyzed by the enzyme thymidine kinase.

acylated tRNA A transfer RNA molecule to which an amino acid is linked; an aminoacyl-tRNA molecule; a charged tRNA molecule.

acylation The introduction of an acyl radical RCO— into an organic compound.

acyl carrier protein A small protein that is a component of the fatty acid synthetase system; it carries a phosphopantetheine group, which contains an SH-group and which is esterified via its phosphate to a serine hydroxyl in the protein. All of the acyl intermediates in fatty acid biosynthesis are covalently linked to the SH-group of phosphopantetheine in the acyl carrier protein much as the acyl intermediates in β-oxidation of fatty acids are linked to the SH-group of phosphopantetheine in coenzyme A. *Abbr* ACP.

acyl-CoA synthetase THIOKINASE.

acyl-enzyme intermediate One of a group of structures formed transiently between an enzyme and its substrate during covalent catalysis; two examples are

where E represents the enzyme.

acylglycerol An ester of glycerol and one to three molecules of a fatty acid; a neutral fat. Depending on the number of fatty acid molecules esterified, the product is called mono-, di-, or triacylglycerol. *Aka* glyceride.

acyl group The radical RCO— that is derived from an organic acid by removal of the OH from the carboxyl group.

acyl-SCoA Acyl coenzyme A.

acyltransferase An enzyme that catalyzes the transfer of an acyl group from acyl coenzyme A to another compound.

AD Alzheimer's disease.

ADA *N*-(2-Acetamido)iminodiacetic acid; used for the preparation of biological buffers in the pH range of 6.0 to 7.2. *See also* biological buffers.

Adair equation A general equation for the binding of a ligand to a macromolecule; refers to the case where there are from 1 to *n* identical binding sites for a specific ligand per macromolecule and where the binding is independent (no interaction between the binding sites).

Adamkiewicz reaction The production of a violet color upon treatment of a solution containing protein with acetic acid and sulfuric acid.

Adam's catalyst Platinum oxide, a catalyst for hydrogenation reactions.

ada protein The protein product of the ada gene which is responsible for control of the adaptive response in *E. coli*; it participates mechanistically in the repair of damaged DNA and also regulates the expression of a number of genes whose products function in DNA repair. *See also* adaptive response.

adaptation DESENSITIZATION (3).

adapter hypothesis The hypothesis, proposed by Crick in 1958, that an amino acid is joined to a specific adapter molecule during protein synthesis. The adapter serves to carry the amino acid to the ribosome and becomes bound to the codon of the amino acid in the messenger RNA which is attached to the ribosome. In this fashion the adapter, now known to be transfer RNA, assures the insertion of the amino acid into its proper place in the growing polypeptide chain.

adapter RNA TRANSFER RNA.

adaptive enzyme INDUCIBLE ENZYME.

adaptive response A set of induced processes in *E. coli* that involve repair of damage made to DNA by methylating and ethylating agents. The lesions repaired by these processes include purine bases alkylated at ring nitrogens or at exocyclic oxygens, pyrimidine bases alkylated at exocyclic oxygens, and phosphotriesters. The regulation of the adaptive response is independent of the SOS regulatory network and is controlled by the ada protein.

adaptor A short, synthetic fragment of DNA that contains a restriction site and that is used in recombinant DNA research to join one molecule, having blunt ends, to a second molecule, having cohesive ends. When the resultant molecule is cleaved by a restriction enzyme, two DNA molecules are obtained that have mutually complementary cohesive ends.

adaptor RNA Variant spelling of adapter RNA.

ADCC Antibody-dependent cellular cytotoxicity.

Addison's disease The pathological condition resulting from adrenal insufficiency and characterized by general weakness, loss of appetite, gastrointestinal disturbances, and weight loss.

addition polymer CHAIN-GROWTH POLYMER.

addition reaction A chemical reaction in which there is an increase in the number of groups attached to carbon atoms so that the molecule becomes more saturated.

adduct The product formed by the chemical addition of one substance to another.

adductor muscle CATCH MUSCLE.

ade Adenine.

adenine The purine 6-aminopurine that occurs in both RNA and DNA. *Abbr* A; Ade.

adenine nucleotide barrier ATRACTYLOSIDE BARRIER.

adenohypophyseal Of, or pertaining to, the anterior lobe of the pituitary gland.

adenohypophysis The anterior lobe of the pituitary gland which produces the adrenocorticotropic, gonadotropic, lipotropic, somatotropic, and thyrotropic hormones.

adenoma A tumor of epithelial tissue that is generally benign and in which the cells form glands or glandlike structures.

adenosine The ribonucleoside of adenine. Adenosine mono-, di-, and triphosphate are abbreviated respectively, as AMP, ADP, and ATP. The abbreviations refer to the 5'-nucleoside phosphates unless otherwise indicated. *Abbr* Ado; A.

adenosine-3',5'-cyclic monophosphate A cyclic nucleotide, commonly called cyclic AMP, that is formed from ATP in a reaction catalyzed by the enzyme adenyl cyclase. Cyclic AMP functions as a second messenger and mediates the effect of a large number of hormones. The hormones interact with the adenyl cyclase system in the cell membrane, and the intracellular cyclic AMP then interacts with specific enzymes or other intracellular components. *Abbr* cAMP. *Aka* cyclic adenylic acid.

adenosine deaminase *See* Taka diastase.

adenosine diphosphate The high-energy compound, adenosine-5'-diphosphate, that can undergo hydrolysis to adenosine-5'-monophosphate and inorganic phosphate. *Abbr* ADP.

adenosine diphosphate glucose ADP-GLUCOSE.

adenosine monophosphate The nucleotide, adenosine-5′-monophosphate, that can be formed by hydrolysis of either of the high-energy compounds, ATP or ADP. *Abbr* AMP.

adenosine-5′-phosphosulfate *See* active sulfate (2).

adenosine triphosphatase One of a group of enzymes that catalyze the hydrolysis of ATP either to ADP and inorganic phosphate or to AMP and pyrophosphate. The enzymes are widely distributed in biological membranes and are named according to the cation(s) required for their activation. *Abbr* ATPase. *See also* Na$^+$, K$^+$-ATPase; H$^+$-ATPase.

adenosine triphosphate The high-energy compound, adenosine-5′-triphosphate, that functions in many biochemical systems. It can be hydrolyzed to either adenosine-5′-monophosphate or adenosine-5′-diphosphate; the hydrolysis reaction is accompanied by the release of a large amount of free energy which is used to drive a variety of metabolic reactions. *Abbr* ATP.

S-adenosylmethionine A high-energy compound that is derived from ATP and methionine and that functions as a biological methylating agent. *Abbr* SAM.

adenovirus A naked, icosahedral virus that contains double-stranded DNA. Adenoviruses infect mammals, often leading to respiratory infections; some are oncogenic.

adenovirus-associated virus A small, naked, icosahedral virus that contains single-stranded DNA and that is found in association with adenoviruses; a subclass of parvoviruses.

adenylate A compound consisting of adenylic acid that is esterified through its phosphate group to another molecule.

adenylate charge hypothesis *See* energy charge.

adenylate control hypothesis The hypothesis that cellular metabolism is regulated by feedback effects that are a function of the relative amounts of AMP, ADP, and ATP in the cell. *See also* energy charge.

adenylate cyclase *See* adenyl cyclase.

adenylate kinase The enzyme that catalyzes the interconversion between two molecules of ADP and one molecule each of ATP and AMP. *Aka* myokinase.

adenylate pool The total intracellular concentration of AMP, ADP, and ATP.

adenyl cyclase The enzyme that catalyzes the formation of cyclic AMP from ATP by the splitting out of pyrophosphate.

adenylic acid The ribonucleotide of adenine.

adenylylation The transfer of a 5′-AMP group (5′-adenylyl group) from ATP; used specifically for the reaction catalyzed by the enzyme glutamine synthetase adenylyl-transferase. In this reaction, a 5′-AMP group is transferred to form a phospho-diester bond with the phenolic hydroxyl group of a specific tyrosine residue in each of the 12 subunits of the enzyme glutamine synthetase. The progressive adenylylation of glutamine synthetase leads to its progressive inactivation and this forms part of the complex regulation of the activity of this enzyme.

adenylyl sulfate *See* active sulfate (2).

adermine VITAMIN B$_6$.

ADH 1. ALCOHOL DEHYDROGENASE. 2. ANTIDIURETIC HORMONE.

adhesion plaque *See* vinculin.

adhesion protein One of a group of proteins, such as fibronectin, collagen, and fibrinogen, that are present in the extracellular matrix and that function in cell adhesion, cell migration, and cell differentiation.

adhesive protein ADHESION PROTEIN.

adiabatic process A proces conducted without either a gain or a loss of heat; a process conducted in an isolated system.

adiabatic system A thermodynamic system that is thermally insulated from its surroundings.

adipocyte A fat cell; a cell of adipose tissue.

adipokinetic hormone LIPOTROPIN.

adipose tissue Lipid tissue; fat deposits in an organism. *Aka* depot fat. *See also* brown fat; white fat.

adiposis A condition characterized by excessive accumulation of fat in the body; the accumulation may be local or general.

adiposity OBESITY.

adipsin A serine protease, present in the blood, that is synthesized and secreted by adipose cells. Some genetic and some acquired obesity syndromes are associated with reduced expression of adipsin mRNA and with reduced concentration of circulating adipsin.

adjuvant A substance that increases the immune response of an animal to an antigen when injected together with the antigen.

adjuvanticity The capacity of a substance to function as an adjuvant.

ad libitum Referring to the feeding of experimental animals where the animals are allowed to eat without any imposed restrictions. *Abbr* ad lib.

admix To mix one substance with another.

admixture 1. A mixture. 2. The act of mixing.

A DNA *See* DNA forms.

Ado Adenosine.

AdoMet *S*-Adenosylmethionine.

adoptive immunity The immunity acquired by an animal organism when it is injected with lymphocytes from another organism; the immunity acquired through an adoptive transfer.

adoptive tolerance The immunological tolerance acquired by an animal organism when it is injected with lymphocytes from another organism; the tolerance acquired through an adoptive transfer.

adoptive transfer The transfer of an immune function from one organism to another that is brought about by the transfer of cells that are immunologically competent or active.

ADP 1. Adenosine diphosphate. 2. Adenosine-5′-diphosphate.

ADP–ATP translocation ATP–ADP CARRIER.

ADPG ADP-glucose.

ADP-glucose A nucleoside diphosphate sugar that is the donor of a glucose residue in the biosynthesis of starch in plants and in the biosynthesis of $\alpha(1 \rightarrow 4)$ glucans in bacteria. *Abbr* ADPG.

ADP-ribosylation The reaction whereby an ADP-ribose moiety is linked covalently to another compound. The cleavage of NAD^+ by cholera toxin and the subsequent attachment of the ADP-ribose moiety from NAD^+ to an arginine residue of a G protein, thereby inhibiting the latter's GTPase activity, is an example, Diphtheria toxin ADP-ribosylates elongation factor eEF_2 (translocase) in a similar manner.

ADR Adrenaline.

adrenal cortex That part of the adrenal gland, derived from mesodermal tissue, which secretes the adrenal cortical hormones.

adrenal cortical hormone A steroid hormone secreted by the adrenal cortex. Major adrenal cortical hormones are the glucocorticoids, cortisol and corticosterone, and the mineralocorticoid, aldosterone; minor adrenal cortical hormones are the sex hormones.

adrenal cortical steroid A steroid produced by the adrenal cortex. Many of these steroids are hormones, such as the glucocorticoids, mineralocorticoids, and sex hormones; some, such as cholesterol, are not hormones.

adrenal corticosteroid ADRENAL CORTICAL STEROID.

adrenalectomy The surgical removal of an adrenal gland.

adrenal gland The endocrine gland located near the kidney and composed of two parts, a medulla that secretes epinephrine and norepinephrine, and a cortex that secretes the adrenal cortical hormones.

adrenaline EPINEPHRINE.

adrenaline tolerance test A test used in the diagnosis of glycogen storage disease type I; the test is based on measuring the level of blood glucose as a function of time following the injection of an individual with adrenaline.

adrenalism A condition resulting from insufficient function of the adrenal glands.

adrenal medulla That part of the adrenal gland, derived from ectodermal tissue, which secretes the hormones epinephrine and norepinephrine.

adrenal virilism The appearance of male secondary sexual characteristics in a female as a result of excessive secretion of androgens by the adrenal cortex.

adrenergic Of, or pertaining to, nerve fibers that release epinephrine and norepinephrine at the nerve endings.

adrenergic receptor A tissue receptor that mediates the action of catecholamines. Adrenergic receptors are classified as α- and β-receptors, based on their relative response to the synthetic agonist isoproterenol: α-receptors are more sensitive to adrenaline than they are to isoproterenol, while β-receptors are more sensitive to isoproterenol than they are to adrenaline. Some of the physiological processes mediated by these receptors are the following: α-receptors—increased liver glycogenolysis, increased gluconeogenesis, and relaxation of intestinal smooth muscles; β-receptors—increased muscle glycogenolysis, increased liver gluconeogenesis and glycogenolysis, increased mobilization of depot fat, and increased heart rate and contractility. In addition to α- and β-receptors, which are widely distributed, there are dopamine adrenergic receptors which are largely confined to renal and mesenteral vasculature and to certain regions of the central nervous system.

adrenocortical steroid ADRENAL CORTICAL STEROID.

adrenocorticoid ADRENAL CORTICAL STEROID.

adrenocorticotrophin Variant spelling of adrenocorticotropin.

adrenocorticotropic hormone A polypeptide hormone of 39 amino acids that stimulates the synthesis and secretion of adrenal cortical hormones by the adrenal cortex. The adrenocorticotropic hormone is secreted by the anterior lobe of the pituitary gland. *Var sp* adrenocorticotrophic hormone. *Abbr* ACTH.

adrenocorticotropin ADRENOCORTICOTROPIC HORMONE.

adrenodoxin A nonheme, iron–sulfur protein that functions in nonphosphorylating electron transport systems such as the cytochrome P_{450}-mediated side chain cleavage of cholesterol.

adrenoleukodystrophy A genetically inherited metabolic defect in humans that is characterized by an unusual accumulation of very long-chain saturated fatty acids (VLCFA). These are normally present in small amounts in the diet and are also synthesized within the body. In an unknown manner, these fatty acids result in demyelination which leads to loss of voluntary motion and death. *Abbr* ALD.

adsorb To attract and hold a substance to the surface of another substance.

adsorbate A substance that is adsorbed to the surface of another substance from either a solution or a gas phase.

adsorbed antiserum An antiserum from which antibodies have been removed by the addition of particulate antigens.

adsorbent 1. *n* A substance that adsorbs another substance from either a solution or a gas phase. 2. *adj* Having the capacity to adsorb.

adsorption 1. The adhesion of molecules to surfaces of solids. 2. The removal of antibodies from a mixture by the addition of particulate antigens, or the removal of particulate antigens from a mixture by the addition of antibodies. 3. The attachment of phage particles to a bacterial cell.

adsorption chromatography A chromatographic technique in which molecules are separated on the basis of their adsorption properties. The stationary phase is a solid adsorbent, generally in the form of a column; the mobile phase is either an aqueous or an organic solution. The rate of movement of the molecules through the column depends on the degree of their adsorption to the solid adsorbent.

adsorption coefficient A constant, under defined conditions, that relates the elution of a substance from a chromatographic column to the weight of adsorbent.

adsorption isotherm A plot of the fractional saturation (or of some other property related to ligand binding to a macromolecule) as a function of the ligand concentration at constant temperature.

adsorptive endocytosis LIGAND-INDUCED ENDO-CYTOSIS.

adult hemoglobin The major form of hemoglobin in normal adults that is designated HbA; a minor form is designated HbA_2.

adult-onset diabetes *See* diabetes.

adult rickets OSTEOMALACIA.

advanced glycosylation end product One of a group of substances, derived from Amadori products by dehydration, rearrangement, and combination with other molecules. Many of these substances are able to cross-link adjacent proteins. *Abbr* AGE.

AEC Active enzyme centrifugation.

AE-cellulose Aminoethylcellulose, an anion exchanger.

aequorin A bioluminescent protein from jellyfish (*Aequorea* sp.) that is used for the assay of calcium in serum and subcellular organelles.

aerial mycelium That portion of a fungal mycelium that projects above the surface of the medium and frequently bears either reproductive cells or spores.

aerobe *See* facultative aerobe; obligate aerobe.

aerobic 1. In the presence of oxygen; in an environment or an atmosphere containing oxygen. 2. Requiring the presence of molecular oxygen for growth. 3. Capable of using molecular oxygen for growth. *See also* oxybiontic.

aerobic glycolysis The group of cellular reactions, occurring in the presence of oxygen, whereby glucose is converted to pyruvic acid. *See also* glycolysis.

aerobic respiration RESPIRATION (3).

aerobiosis Life under aerobic conditions.

aerobiotic Of, or pertaining to, aerobiosis.

aerogel A gel in which removal of the dispersing agent (the solvent) does not lead to shrinkage and an unswollen state, but rather results in a rigid structure.

aerogenic Of, or pertaining to, an organism that forms gas (as well as other metabolic by-products) from particular substrates.

aerosol A colloidal dispersion of liquid droplets or solid particles in a gas.

aerosporin POLYMYXIN.

aerotaxis A form of chemotaxis in which the chemical gradient is due to oxygen.

aerotolerant 1. Of, or pertaining to, an anaerobic organism that can survive, but not grow, in the presence of oxygen. 2. Of, or pertaining to, an anaerobic organism that can grow at suboptimal rates in the presence of oxygen. *See also* microaerophilic.

afferent 1. Leading or conveying toward a cell or an organ. 2. Of, or pertaining to, the stages involved in activating the immune system. *See also* efferent.

afferent inhibition The prevention of transplantation immunity through the binding of antibodies from the recipient animal to antigens in the transplant; as a result, the transplant antigens are unable to reach and/or to stimulate the antibody-forming cells in the recipient animal.

affinity 1. The capacity of an enzyme to bind substrate; generally measured by the affinity constant. 2. The capacity of an antibody to bind either antigens or haptens; frequently measured by the average intrinsic association constant for the binding reaction.

affinity chromatography A column chromatographic technique based on the specific affinity between a molecule to be isolated (such as a protein or an enzyme) and a molecule that it can bind (a ligand). The ligand may be a small molecule or a macromolecule, and its binding to the molecule of interest may involve biochemical or immunochemical reactions. The ligand is linked covalently to an insoluble support (sepharose, agarose, cellulose, etc.) without destroying its activity and specificity. Frequently, a spacer is inserted between the ligand and the matrix to avoid steric hindrance when the ligand binds the molecule of interest. When a mixture of molecules is passed through the column, the covalently linked ligands will bind specifically the molecule of interest. Elution of the latter is achieved by changing the conditions to such in which binding does not occur. Two examples are the use of DNA-cellulose for the isolation of DNA-dependent DNA polymerase, and the use of agarose-antibody preparations for the isolation of antigens. *See also* magnetic affinity chromatography.

affinity constant The reciprocal of the dissociation constant for the complex PL in the reversible system $P + L \rightleftharpoons PL$ where P is usually a protein and L is a ligand such as a substrate, an inhibitor, or an activator. The association or binding constant for a specific ligand to a macromolecule. *See also* association constant.

affinity electrophoresis Electrophoresis on a carrier that contains an immobilized ligand, capable of specific interaction with some component(s) of the mixture to be separated.

affinity elution A chromatographic technique in which compounds are adsorbed nonspecifically to a column and the compound of interest is then eluted specifically through its binding to a ligand in the eluting solvent.

affinity labeling A method for the specific labeling of the active site of an enzyme, antibody, or other protein. A reagent A–X that can bind specifically, reversibly, and noncovalently to the active site through its A group is first allowed to bind to the active site. The reagent is then linked covalently through its chemically reactive group X to an amino acid at or close to the active site. *See also* active site-directed irreversible inhibitor.

affinity partitioning A phase-partitioning technique, used for the isolation and purification of proteins, in which a polymeric ligand, having specific affinity for a given protein, is used. If the polymeric ligand partitions itself predominantly into one phase, then the corresponding protein is also shifted into that phase.

affinity ratio The ratio of the substrate constant for one reaction to the substrate constant for a second reaction that is

catalyzed by the same enzyme but involves a different substrate.

affinoelectrophoresis AFFINITY ELECTROPHORESIS.

affinophore A macromolecular polyelectrolyte bearing affinity ligands for a specific protein. When a mixture of proteins is electrophoresed in the presence of an affinophore, the protein having an affinity for the ligand will form a complex with the affinophore; as a result, the apparent electrophoretic mobility of the protein will be altered. If the protein is sufficiently accelerated, it can be separated from the other proteins.

affinophoresis The electrophoretic separation of proteins by means of affinophores.

afibrinogenemia A genetically inherited metabolic defect in humans that is characterized either by the complete absence of fibrinogen or by the presence of a defective fibrinogen.

aflatoxin A toxic and carcinogenic compound produced by fungi; a coumarin derivative that belongs to the group of mycotoxins. Aflatoxin has been found in a number of foodstuffs and is believed to inhibit RNA synthesis.

A form *See* DNA forms.

AFP Alpha-fetoprotein.

Ag 1. Antigen. 2. Silver.

agammaglobulinemia A genetically inherited metabolic defect in humans that is characterized by the complete absence of immunoglobins. *See also* hypogamma-globulinemia.

agar An acidic polysaccharide extracted from certain seaweeds; used as a solidifying agent of culture media in microbiology and as a support medium for zone electrophoresis.

agar diffusion method A method of determining the sensitivity of a micro-organism to an antimicrobial drug; based on measuring the zone of growth inhibition when the drug is placed in a cylinder, a hole, or a filter paper disk on a petri plate that has been seeded with the microorganism.

agar gel electrophoresis Zone electrophoresis in which the supporting medium consists of a gel prepared from agar.

agarose A sulfate-free, neutral fraction of agar; a linear galactan hydrocolloid that is used in gel filtration, electrophoresis, and immunodiffusion.

agar plate count A plate count in which the solid nutrient medium contains agar.

age The length of time that a preparation of cells or a subcellular fraction has been stored.

AGE Advanced glycosylation end product.

Agent Orange A herbicide used in the Vietman War. *See also* dioxin.

age pigment An insoluble pigment granule that accumulates in certain animal tissues upon aging; believed to be a lipid-protein complex resulting from crosslinking of protein with compounds formed by peroxidation of lipids. The pigment is brown colored and exhibits green-yellow fluorescence when activated with long wavelength ultraviolet light. *Aka* ceroid pigment; lipofuscin; senility pigment.

agglutinating antibody AGGLUTININ.

agglutination The clumping of bacterial and other cells that is brought about by an antigen–antibody reaction between the particulate antigens on the cell surface and added antibodies.

agglutinin An antibody that can bind to particulate antigens on the surface of cells to produce an agglutination reaction.

agglutinogen A surface antigen of bacterial and other cells that can induce the formation of agglutinins and can bind to them to produce an agglutination reaction.

aggregate 1. MULTIENZYME SYSTEM. 2. METABOLON (2).

aggregate anaphylaxis An anaphylactic shock that is produced by a single injection of antigen.

aggressin A substance that is produced by a microorganism and that, though not necessarily toxic by itself, promotes the invasiveness of the microorganism in the host; the enzymes hyaluronidase and collagenase are two examples.

aglucone The noncarbohydrate portion of a glucoside.

agonist A molecule, such as a drug, an enzyme activator, or a hormone, that enhances the activity of another molecule or receptor site. A hormone that binds to a receptor in a productive manner, triggering the normal response, is an example. *See also* decamethonium; full agonist; partial agonist.

agranulocyte A white blood cell (leukocyte) that contains few, if any, granules in the cytoplasm.

A/G ratio Albumin/globulin ratio.

Agrobacterium tumefaciens *See* crown gall tumor.

agrobactin A linear siderophore of the phenol–catechol type found in *Agrobacterium tumefaciens*.

AHF Antihemophilic factor.

AHG 1. Antihemophilic globulin. 2. Anti-human globulin.

AIA Anti-immunoglobulin antibodies.

AICAR 5-Aminoimidazole-4-carboxamide ribonucleotide; an intermediate in the biosynthesis of purines.

AICF Autoimmune complement fixation.

AIDS Abbreviation for acquired immunodeficiency syndrome; a severe viral disease, caused by a retrovirus. The virus destroys T lymphocytes of the immune system and infects cells within the central nervous system.

The syndrome first occurred among homosexuals and users of intravenous drugs (1981) but has since spread throughout the world. Most infections occur through sexual transmission, use of contaminated needles, and as a result of infected mothers passing the virus to newborns.

AIDS virus One of a group of retroviruses implicated as the cause of acquired immunodeficiency syndrome (AIDS). Various virus isolates appear to be closely related members of the same virus group. They have been designated LAV (lymphadenopathy-associated virus), HTLV-III (human T-cell lymphotropic virus type III), IDAV (immunodeficiency-associated virus), and ARV (AIDS-associated retrovirus). Two compound designations, HTLV-III/LAV and LAV/HTLV-III have also been used. It has been proposed that the AIDS retroviruses be officially designated as human immunodeficiency viruses, abbreviated as HIV. *See also* antigenic drift.

AIP Aldosterone-induced proteins.

air dose The dose of radiation delivered to a specified point in air.

air peak The gas chromatographic peak that is produced when a small amount of air is injected with the sample into the chromatographic column.

Akabori hypothesis The hypothesis that the origin of proteins is based on the polymerization of non-amino acid building blocks to form polyglycine and on the subsequent replacement of the α-hydrogens in polyglycine by various R groups in secondary reactions.

Akabori reaction The formation of an alkamine by the reaction of an aldehyde with the amino group of an amino acid.

Al Aluminum.

Ala 1. Alanine. 2. Alanyl.

alanine An aliphatic nonpolar amino acid; α-alanine occurs in proteins and β-alanine occurs in the peptides anserine and carnosine. *Abbr* Ala; A.

alarmone A signal molecule in bacteria that has a regulatory effect on metabolism by exerting control on many biochemical reactions at once. The action of an alarmone is similar to that of a hormone in multicellular organisms. In bacteria, such regulation may come into play in response to environmental stresses. As an example, amino acid starvation results in the accumulation of the compounds known as magic spots. These are believed to function as alarmones, leading to cessation of protein synthesis and cessation of transcription of rRNA genes.

alarm reaction GENERAL ADAPTATION SYNDROME.

albinism A genetically inherited metabolic

defect in humans that is characterized by the lack of skin pigmentation and that is due to a deficiency of the enzyme tyrosinase.

albino A person or an animal that is deficient in skin pigmentation.

albomycin An iron-containing antibiotic, produced by *Actinomyces subtropicus*; a cyclic polypeptide that contains cytosine. The compound is either similar, or identical, to grisein.

albumin A water-soluble, globular, and simple protein that is not precipitated by ammonium sulfate at 50% saturation.

albumin/globulin ratio The ratio of the concentration of serum albumin to that of serum globulin. *Abbr* A/G ratio.

albuminimeter An apparatus for determining protein in biological fluids on the basis of the volume of the precipitated protein.

albuminuria The presence of excessive amounts of protein, mainly albumin, in the urine.

Albustix test A rapid, semiquantitative test for protein in urine by means of paper strips impregnated with buffer and indicator. *See also* protein error.

alcapton Variant spelling of alkapton.

alcaptonuria Variant spelling of alkaptonuria.

alcohol 1. An alkyl compound containing a hydroxyl group. The alcohol is designated as a primary, a secondary, or a tertiary alcohol depending on whether the hydroxyl group is attached to a carbon atom that is linked to one, two, or three other carbon atoms. 2. Ethyl alcohol; ethanol.

alcohol dehydrogenase A pyridine-linked dehydrogenase that catalyzes the oxidation of ethanol to acetaldehyde.

alcoholic fermentation The group of reactions, characteristic of yeast, whereby glucose is fermented to ethyl alcohol.

alcoholic hydroxyl group A hydroxyl group attached to an aliphatic carbon chain.

alcoholic steroid STEROL.

alcoholysis The cleavage of a covalent bond of an acid derivative by reaction with an alcohol ROH so that one of the products combines with the H of the alcohol and the other product combines with the OR group of the alcohol.

ALD Adrenoleukodystrophy.

aldaric acid A dicarboxylic sugar acid of an aldose in which both the aldehyde group and the primary alcohol group have been oxidized to carboxyl groups.

aldehyde An organic compound that contains an aldehyde group.

aldehyde group The carbonyl group attached to one carbon and one hydrogen atom; the grouping —CHO.

aldehyde indicator SCHIFF'S REAGENT.

aldimine An organic compound that has the general formula R—CH=NH.

alditol A derived carbohydrate in which the aldehyde group of an aldose has been reduced to an alcohol group.

aldo- 1. Combining form meaning aldose. 2. Combining form meaning aldehyde.

aldofuranose An aldose in furanose form.

aldolase 1. An aldehyde lyase. 2. The enzyme of glycolysis that catalyzes the interconversion of fructose-1,6-bisphosphate to dihydroxyacetone phosphate and glyceraldehyde-3-phosphate.

aldol condensation An addition reaction of two ketones, or two aldehydes, or an aldehyde and a ketone.

aldonic acid A monocarboxylic sugar acid of an aldose in which the aldehyde group has been oxidized to a carboxyl group.

aldopyranose An aldose in pyranose form.

aldose A monosaccharide, or its derivative, that has an aldehyde group.

aldosterone The major mineralocorticoid in humans.

aldosterone-induced proteins The group of proteins whose synthesis is stimulated by mineralocorticoids; they mediate the effects of the mineralocorticoids on water and electrolyte balance. *Abbr* AIP.

aldosteronism A pathological condition characterized by the excessive production and secretion of aldosterone.

alexin COMPLEMENT.

ALG Antilymphocyte globulin.

alga (*pl* algae.) A chlorophyll-containing, photosynthetic protist; algae are unicellular or multicellular, are generally aquatic, and are either eukaryotic or prokaryotic.

algal Of, or pertaining to, algae.

algicide A chemical compound that selectively kills algae; used to inhibit the growth of algae in swimming pools and water reservoirs.

alginic acid An algal polysaccharide of mannuronic acid.

algorithm 1. A computational method or a set of rules for obtaining the solution of all problems of a specified type in a finite number of operations; a fixed sequence of formulas and/or algebraic and/or logical steps for calculations of a given problem. 2. A defined process consisting of a number of fixed step-by-step procedures for accomplishing a given result in a finite number of steps. *See also* heuristic process; stochastic process.

alicyclic Designating a compound derived from a saturated cyclic hydrocarbon.

alien addition monosomic The genome of a species that contains, in addition to the normal complement of chromosomes, a single chromosome from another species.

aliesterase CARBOXYLESTERASE.

alimentary 1. Of, or pertaining to, food or nutrition. 2. Nutritious.

alimentary canal DIGESTIVE TRACT.

alimentary glycosuria The temporary increase in the level of glucose in the urine that follows a meal rich in carbohydrates.

aliphatic Of, or pertaining to, an organic compound that has an open chain structure. *Aka* acyclic.

aliquot 1. A part of a whole that divides the whole without a remainder; thus 4 mL, but not 7 mL, is an aliquot of 12 mL. 2. Any part or fraction of a whole.

alkalemia A condition characterized by a decrease in the hydrogen-ion concentration of the blood.

alkali A base, specifically one of an alkali metal.

alkali disease One of a number of animal poisonings of either plant or mineral origin.

alkali metal An element of group IA in the periodic table that consists of the elements lithium (Li), sodium (Na), potassium (K), rubidium (Rb), cesium (Cs), and francium (Fr).

alkalimetry 1. The chemical analysis of solutions by means of titrations, the end points of which are recognized by a change in the hydrogen-ion concentration. 2. A determination of the amount of a base by titration against a standard acid solution.

alkaline BASIC.

alkaline earth An element of group IIA in the periodic table that consists of the elements beryllium (Be), magnesium (Mg), calcium (Ca), strontium (Sr), barium (Ba), and radium (Ra).

alkaline hematin A hematin formed from hemoglobin by treatment with alkali above pH 11.

alkaline hydrolase A hydrolytic enzyme that has a basic optimum pH.

alkaline pH A pH value above 7.0.

alkaline phosphatase A phosphatase, the optimum pH of which is above 7.0.

alkaline reserve The plasma bicarbonate concentration that is determined either from the carbon dioxide combining power of plasma or from the direct titration of plasma. *Aka* alkali reserve.

alkaline rigor The increase in pH upon death that occurs in some species of fish where death was preceded by struggling.

alkaline tide The increase in the pH of the blood and of the urine that occurs shortly after a meal; thought to be due to the withdrawal of chlorides from the blood for the formation of hydrochloric acid in the stomach.

alkaloids A group of basic, nitrogenous organic compounds which occur primarily in plants. Alkaloids are generally heterocyclic compounds of complex structure and almost invariably have intense pharmacological activity. Major classes of alkaloids, and the precursors from which they are biosynthesized, are the following: indole (tryptophan), isoquinoline (phenylalanine or tyrosine), piperidine (acetate and lysine), pyrrolidine (acetate and ornithine), pyrrolizidine (ornithine), quinolizidine (lysine), Rutaceae (anthranilic acid), terpene (mevalonic acid), and tropane (acetate and ornithine).

alkalophilic Of, or pertaining to, bacteria that grow at high external pH values.

alkalosis A deviation from the normal acid–base balance in the body that is due to a disturbance which, by itself and in the absence of compensatory factors, would tend to raise the pH of the blood. The actual change in pH depends on whether and to what extent the disturbance is compensated for. The disturbances and the compensatory mechanisms are considered primarily with respect to their effect on the bicarbonate/carbonic acid ratio of blood plasma. *See also* metabolic alkalosis; primary alkalosis; etc.

alkalotic Of, or pertaining to, alkalosis.

alkane A saturated aliphatic hydrocarbon.

alkapton HOMOGENTISIC ACID.

alkaptonuria A genetically inherited metabolic defect in humans that is characterized by the urinary excretion of black melanin pigments formed from homogentisic acid (alkapton); the defect is due to a deficiency of the enzyme homogentisic acid oxidase which functions in the metabolism of phenylalanine and tyrosine.

alkene An unsaturated aliphatic hydrocarbon that contains one or more double bonds.

alkenyl group The radical derived from an alkene, or from a derivative of an alkene, by removal of a hydrogen atom.

alkylating agent One of a group of compounds, including the nitrogen and sulfur mustards, that alkylates specific sites of biologically important molecules such as DNA and protein. Alkylating agents are frequently carcinogenic, mutagenic, and immunosuppresive; they are classified as mono-, bi-, and polyfunctional depending on the number of reactive groups per molecule of alkylating agent.

alkylation The introduction of an alkyl group into an organic compound.

alkyl group The radical derived from an alkane, or from a derivative of an alkane, by the removal of a hydrogen atom.

alkyne An unsaturated aliphatic hydrocarbon that contains one or more triple bonds.

alkynyl group The radical derived from an alkyne, or from a derivative of an alkyne, by the removal of a hydrogen atom.

allantoic acid The carboxylic acid that is the

end product of purine catabolism in some teleost fishes.

allantoin The heterocyclic compound that is the end product of purine catabolism in mammals, other than primates, and in some reptiles.

allantoinase The enzyme that catalyzes the hydrolysis of allantoin to allantoic acid.

allatum hormone An insect hormone that affects differentiation after molting and that is required for vitellogenesis in the adult female. *See also* juvenile hormone.

allele A specific form of a gene; one of several possible mutational forms of a gene.

allelic Of, or pertaining to, an allele.

allelic allotype ALLOTYPE.

allelic complementation INTRAGENIC COMPLEMENTATION.

allelic exclusion The phenomenon that, in any immunoglobulin producing cell, only one set of immunoglobulin genes (there are two sets per cell, one from each parent) will be expressed. The mechanism whereby expression of the other set of allelic genes is excluded is currently unknown.

allelism test COMPLEMENTATION TEST.

allelochemical A compound, produced by a microorganism or a plant, that is toxic to a microorganism or a plant of a different species. In the case of plants, such compounds may be exuded from living roots, leaves, or fruits, or they may be leached out from decaying plant tissue as a result of microbial action. *Aka* allelopathic chemical. *See also* allomone, pheromone.

allelomorph ALLELE.

allelopathy The phenomenon of plants or microorganisms producing substances that are toxic to plants or microorganisms of different species; the production of allelochemicals.

allelotype The frequency of alleles in a breeding population.

allelozyme One of a number of enzymes that catalyze the same reaction but are specified by different alleles within a group of closely related species.

Allen correction A method of correcting absorbance measurements for the absorbance due to interfering substances. The absorbance is measured at the peak wavelength and at two other wavelengths, generally equidistant from the peak. A baseline is drawn by connecting the measurements on either side of the peak, and the absorbance at the peak is corrected by subtracting the baseline value at the peak. The correction assumes that the absorbance change is linear between the three points.

Allen's test A modification of Fehling's test for glucose in urine; the urine is added to boiling Fehling's solution and turbidity develops as the solution is cooled.

allergen An antigen that produces an allergic response.

allergic contact dermatitis An inflammation of the skin that is due to an allergic response brought about by exposure of the skin to a chemical sensitizer.

allergic response The formation and the reactions of antibodies that occur when a sensitized animal is exposed to an allergen.

allergy A hypersensitive reaction to intrinsically harmless antigens, most of which are environmental. Allergy is manifested principally in the gastrointestinal tract, the skin, and the respiratory tract.

allo– 1. Combining form meaning other or dissimilar. 2. Combining form referring to an isomeric form such as an enantiomer of a compound that has more than one pair of enantiomers, or the more stable form of two geometrical isomers. 3. Combining form referring to a dissimilar genome.

alloantibody An antibody produced in response to the administration of an alloantigen.

alloantigen An antigen that produces an immune response when administered to a genetically different individual of the same species.

allogeneic Referring to genetically dissimilar individuals of the same species.

allogeneic disease GRAFT-VERSUS-HOST REACTION.

allogeneic inhibition The destruction of cells that is apparently nonimmunological and that is brought about by contact with genetically different cells or with extracts from such cells.

allogenic Variant spelling of allogeneic.

allograft A transplant from one individual to a genetically dissimilar individual of the same species.

allograft reaction The immune reaction whereby an allograft is rejected.

allolactose A variant form of the disaccharide lactose in which the two monosaccharides, galactose and glucose, are joined via an $\alpha(1 \rightarrow 5)$ glycosidic bond.

allomerism The variation in the chemical composition of substances that have the same crystalline form.

allomerization The oxidation of chlorophyll by air in the presence of alkali.

allometry The relation between the rate of growth of a part of an organism and the rate of growth of another part or of the organism as a whole.

allomone A compound that is produced by one organism and influences the behavior of a second organism from another species, resulting in some benefit only to the producer of the compound. In contrast, if only the recipient organism benefits, the compound is known as a kairomone; if both organisms benefit, it is known as a synomone. *See also* allelochemical; pheromone.

allomorphism The variation in the crystalline form of substances that have the same chemical composition.

allophenic Descriptive of a phenotype that is not due to the mutant makeup of the cells showing the phenotype.

allophycocyanin A red accessory pigment of algal chloroplasts that consists of a protein conjugated to a phycobilin.

alloplex interaction The interaction that takes place when a disordered protein molecule undergoes refolding upon contact with another protein molecule.

allopurinol An analogue of hypoxanthine, used to treat individuals suffering from hyperuricemia as a result of gout and other conditions.

all-or-none Descriptive of a reaction or a response that occurs either to its fullest extent or does not occur at all. The highly cooperative, thermal denaturation of DNA and the dose response of an animal to a drug are two examples.

all-or-none model CONCERTED MODEL.

allosteric Pertaining to two or more topologically distinct sites on the same protein molecule.

allosteric activation The activation of an allosteric enzyme by a positive effector.

allosteric coefficient A mathematical term that is a measure of the allosteric nature of an enzyme, based on the concerted model. It is equal to the ratio of the tensed and relaxed forms ($[T_0]/[R_0]$), multiplied by the ratio of an inhibition term to an activation term. An increase in the inhibition term (for example, by an increase in inhibitor concentration) will cause the binding function to become more sigmoidal; an increase in the activation term (for example, by an increase in activator concentration) will cause the binding function to become more hyperbolic. If the allosteric coefficient is equal to zero, then the binding function reverts to a normal hyperbolic curve. *See also* allosteric constant.

allosteric constant The equilibrium constant for the transition from the relaxed to the tensed form in the concerted model ($[T_0]/[R_0]$). *See also* allosteric coefficient.

allosteric effector *See* effector.

allosteric enzyme A regulatory enzyme that has the capacity of having its catalytic activity modified through the binding of one or more metabolites; it is generally an oligomeric protein that readily undergoes conformational

changes. An allosteric enzyme has two or more topologically distinct sites, either interacting catalytic (active) sites or interacting catalytic and regulatory (allosteric) sites. As a result of such interactions, the enzyme frequently exhibits sigmoidal, rather than hyperbolic, kinetics. The metabolites that bind to the regulatory sites may be activating or inhibiting and are called effectors (modifiers, modulators). *See also* concerted model; sequential model.

allosteric inhibition The inhibition of an allosteric enzyme by a negative effector.

allosteric interactions The interactions of an allosteric enzyme or nonenzyme protein with allosteric effectors.

allosteric protein A protein that has two or more topologically distinct binding sites such that the binding of ligands (effectors) to these sites alters the properties of the protein.

allosteric site REGULATORY SITE.

allosteric transition The conformational change of an allosteric enzyme or of an allosteric protein as a result of its interaction with an effector.

allosterism The phenomenon of allosteric interactions.

allostery ALLOSTERISM.

allotopic Of, or pertaining to, allotopy.

allotopy The phenomenon of a substance, such as an enzyme, possessing different properties when it exists in a particulate or in a soluble form.

allotropy The phenomenon of an element existing in different forms in the same phase; the different crystal forms of phosphorus and the molecular forms of oxygen and ozone are two examples.

allotype One of a group of different antigenic determinants of a given serum protein or immunoglobulin that occur in different individuals of the same species; such proteins are under the control of one genetic locus but are produced by different alleles of the same gene. *Aka* allelic allotype. *See also* idiotype; isotype.

allotype suppression The suppression of the expression of an immunoglobulin allotype in an individual that is brought about by the administration of antibodies against the allotype.

alloxan diabetes An experimentally produced diabetes in which the level of insulin in an animal is lowered through preferential destruction of the insulin-producing cells of the pancreas by the administration of the pyrimidine drug alloxan.

allozyme One of a group of enzymes that are produced by alleles of the same gene.

all-trans-retinal The isomeric form of retinal

that is produced by light from the 11-*cis* isomer.

allylic Next to a double bond; the allyl group has the structure $CH_2=CH—CH_2—$.

allysine A derivative of lysine in which the ϵ-amino group has been converted to an aldehyde group; allysine undergoes an aldol condensation with hydroxyallysine during the cross-linking of collagen chains.

alpha 1. Denoting the first carbon atom next to the carbon atom that carries the principal functional group of the molecule. 2. Denoting a specific configuration of the substituents at the anomeric carbon in ring structures of carbohydrates. 3. Denoting observed rotation (α) and specific rotation ($[\alpha]$) in optical rotation. *Sym* α.

alpha adrenergic receptor *See* adrenergic receptor.

alpha amanitin An amatoxin that is a potent inhibitor of eukaryotic RNA polymerase II and, to a lesser degree, of RNA polymerase III.

alpha amino acid *See* amino acid.

alpha amylase *See* amylase.

alpha amylose AMYLOSE.

alpha blocker An antagonist (inhibitor) of alpha adrenergic receptors. Some alpha blockers are used to treat migraine, diabetic gangrene, and spastic vascular disease. *Aka* alpha adrenergic blocker.

alpha bungarotoxin A small basic protein that is the snake venom poison of snakes from the genus *Bungarus*; a neurotoxin that binds noncovalently to nicotinic receptors of acetylcholine. It blocks the binding of acetylcholine at the postsynaptic cell and prevents depolarization of the postsynaptic membrane. It is referred to as an antagonist of the cholinergic system.

alpha chain 1. The heavy chain of the IgA immunoglobulins. 2. One of the two types of polypeptide chains present in adult hemoglobin.

alpha decay The radioactive disintegration of an atomic nucleus that results in emission of an alpha particle.

alpha error TYPE I ERROR.

alpha fetoprotein *See* carcinoembryonic antigen.

alpha fraction 1. HIGH-DENSITY LIPOPROTEIN. 2. VERY HIGH-DENSITY LIPOPROTEIN.

alpha helix A coil- or spring-like configuration of protein molecules that occurs particularly in globular proteins. In this configuration, the polypeptide chain is held together by means of intrachain hydrogen bonds between the >CO and >NH groups of perptide bonds in such a fashion that there are 3.6 amino acid residues per turn of the helix, that the rise per residue is 1.5 Å, and that the pitch of the helix is 5.4 Å; each >CO group is hydrogen-bonded to the >NH group of the fourth residue ahead of it in the chain. The helix may be left- or right-handed depending on whether it is twisted in the manner of a left- or right-handed screw. The right-handed alpha helix is the configuration most commonly encountered in proteins.

alpha keratin The helical form of keratin in which the polypeptide chains are in the alpha-helical configuration.

alpha lactalbumin A heat-stable protein in the milk of mammals; a component (B protein) of the enzyme lactose synthetase.

alpha lipoprotein HIGH-DENSITY LIPOPROTEIN.

alpha method DEAN AND WEBB METHOD.

alpha orientation The orientation of atoms or groups that are attached below the plane of the steroid molecule.

alpha oxidation An oxidative pathway of fatty acids in which they are oxidized at the alpha carbon and degraded one carbon at a time. The carbon is removed as CO_2 by decarboxylation and the residual fatty acid molecule is converted to an aldehyde; occurs in germinating plant seeds.

alpha particle 1. A subatomic particle consisting of two protons and two neutrons; the alpha particle is identical to the nucleus of the helium atom and is frequently emitted by radioactive isotopes. 2. A cluster of glycogen granules in the liver; the granules are referred to as beta particles.

alpha plateau The low-potential portion of the characteristic curve of a proportional radiation detector at which the count rate is almost independent of the applied voltage, and at which the potential is of sufficient magnitude to detect alpha particles.

alpha radiation A radiation consisting of alpha particles.

alpha ray A beam of alpha particles.

alpha receptor *See* adrenergic receptor.

alpha threshold The lowest potential at which alpha particles can be detected with a proportional radiation detector.

alpha tocopherol *See* vitamin E.

ALS Antilymphocyte serum.

alteration enzyme A phage T4 enzyme that is injected with the phage DNA into the host bacterium; it modifies the host RNA polymerase and, thereby, inhibits the initiation of host RNA synthesis.

alternate-site model A model proposed to explain the anticooperative effects of certain oligomeric enzymes. It is similar to the flip-flop model and is based on the notion that a chemical event or binding that occurs on one subunit can facilitate the release of product from another subunit.

alternation of generations The phenomenon, in the life cycle of certain organisms, in which a mature haploid individual alternates with a mature diploid individual; exhibited by some fungi, algae, and plants.

alternative pathway *See* complement.

alum A double sulfate salt of aluminum and either a monovalent metal or an ammonium ion.

alumina Aluminum oxide; an adsorbent used in column chromatography

alumina gel A gel prepared from ammonium sulfate and aluminum sulfate; used in the purification of proteins by adsorption chromatography.

aluminum adjuvant An aluminum compound, such as aluminum hydroxide, aluminum phosphate, or alum, that functions as an adjuvant in alum precipitation.

alum precipitated toxoid A toxoid precipitated with an aluminum adjuvant. *Abbr* APT.

alum precipitation An immunochemical technique in which soluble antigens are mixed with aluminum adjuvants to form a precipitate. When injected into an animal, the precipitate forms a depot from which the antigen is slowly released.

Alu sequences A set of some 300,000 copies of base sequences, consisting of about 300 base pairs each, that occurs in human DNA. The name is derived from the fact that each unit contains a tetranucleotide that can be cleaved by the restriction enzyme Alu I. The Alu sequences are scattered throughout the genome and account for about 5% of the total human DNA. They may constitute transposable elements.

alveolar Of, or pertaining to, alveoli.

alveolus (*pl* alveoli). One of a large number of air cells in the lung through which the gas exchange of respiration takes place.

Alzheimer's disease An age-related, progressive, neurodegenerative disease in humans that is characterized by gradual loss of memory, reasoning, orientation, and judgment. One of the hallmarks of the disease is the formation of numerous neuritic plaques in the brain which consist of degenerating axons and neurites surrounding an amyloid core. At least some forms of the disease are due to a specific genetic defect of chromosome 21.

am Abbreviation for amber mutation.

Amadori product A compound formed by the nonenzymatic reaction between the aldehyde group of glucose and the amino group of a protein.

Amadori rearrangement The isomerization of *N*-substituted aldosylamines into *N*-substituted 1-amino-1-deoxy-2-ketoses; occurs in the Maillard reaction, in the reaction of carbohydrates with phenylhydrazine, and in the biosynthesis of pteridines.

α-amanitin *See* alpha amanitin.

amatoxin One of a group of bicyclic octapeptides that are toxic components of the poisonous mushroom *Amanita phalloides*. An important member of the group is α-amanitin which inhibits eukaryotic RNA polymerase II. *See also* phallotoxin.

amaurotic familial idiocy TAY–SACHS DISEASE.

amber codon The codon UAG, one of the three termination codons.

Amberlite Trademark for a group of ion-exchange resins.

amber mutant A conditional lethal mutant that contains an amber codon in a gene with a vital function.

amber mutation A mutation in which a codon is mutated to the amber codon, thereby causing the premature termination of the synthesis of a polypeptide chain. *Abbr* am.

amber suppression The suppression of an amber codon.

ambient conditions The conditions, such as temperature and pressure, of the surrounding environment.

ambiguity The occurrence of mistakes in protein synthesis, particularly in in vitro systems, such as the incorporation of one amino acid in response to a codon for a different amino acid.

ambiguous codon A codon that can lead to the incorporation of more than one amino acid.

ambiquitous enzyme An enzyme whose distribution between soluble and particulate forms varies with the metabolic state of the cell.

ambiquity The property of an enzyme that can exist either by being bound to a structure or by being free in solution.

ambivalent codon A codon that is expressed in some mutants as a result of suppression but that is not expressed in other mutants; a nonsense codon; a termination codon.

ambivalent mutation NONSENSE MUTATION.

amboceptor A term introduced by Ehrlich to describe hemolysin, an antibody that possesses two different binding sites, one for the antigen and one for complement; currently used to describe an antibody to a surface antigen of erythrocytes. The combination of erythrocytes with homologous amboceptors results in sensitized erythrocytes that can function as detectors of complement.

amelogenin A protein in dental enamel.

Ames test A bacterial bioassay for detecting mutagenic compounds that was developed by Bruce N. Ames in 1974. Since many chemical carcinogens are also mutagenic, the Ames test is also used as a screen for the potential carcinogenicity of chemical compounds. The test involves growing cells of *Salmonella typhimurium* that are unable to grow in the absence of histidine (due to a mutation in a gene involved in the biosynthesis of histidine) in the presence of the test mutagen. This results in many new mutations, some of which are revertants of the original mutation and can now synthesize histidine and grow in the absence of exogenous histidine. The number of revertants formed is scored at various test mutagen concentrations and this permits the construction of a dose–response curve.

amethopterin A folic acid analogue that inhibits the enzyme dihydrofolate reductase and that is used in the treatment of leukemia.

amidation The introduction of an amide group into an organic compound.

amide group The radical —$CONH_2$, derived from an acid by replacement of the OH of the carboxyl group with an amino group.

amidinotransferase The enzyme that catalyzes the transamidination reaction in which a guanido group is transferred from arginine to glycine.

amido black 10B A dye used for the spectrophotometric determination of proteins. This method is not affected by most of the reagents that interfere with the Lowry method. *Aka* amidoschwarz 10B; buffalo black NBR; naphthol blue black.

amination The introduction of an amino group into an organic compound.

amine A basic organic compound derived from ammonia by substitution of one more organic radicals for the hydrogens. The amine is designated as a primary, a secondary, or a tertiary amine depending on whether one, two, or three organic radicals have been substituted for the hydrogen atoms in ammonia.

aminoacetic acid GLYCINE.

amino acid An organic compound that contains both a basic amino group and an acidic carboxyl group. The alpha amino acids, in which the amino group is attached to the alpha carbon, are the building blocks of peptides and proteins. The amino acids are commonly classified either as (a) neutral, basic, or acidic, or as (b) nonpolar, polar and uncharged, or polar and charged; the presence or absence of a charge on the amino acid refers to that at pH 7.0

amino acid accepting RNA TRANSFER RNA.

amino acid activating enzyme AMINOACYL-tRNA SYNTHETASE.

amino acid activation A set of two reactions, catalyzed by an aminoacyl-tRNA synthetase, whereby an amino acid becomes covalently linked first to AMP and then to a specific tRNA molecule.

amino acid analysis The analytical determination of both the relative amounts and the types of the amino acids in a peptide or in a protein.

amino acid analyzer An instrument for the automated amino acid analysis of peptide and protein hydrolysates. The amino acids are separated by ion-exchange chromatography and are quantitatively determined by colorimetry.

amino acid arm The acceptor stem in the clover leaf model of tRNA to which the amino acid is covalently linked; the segment contains both the 5'- and 3'-ends of the tRNA. *See also* arm.

amino acid attachment site The site, on a tRNA molecule, to which the amino acid becomes covalently linked; the 2'- or 3'-hydroxyl group of the terminal adenosine nucleotide, at the 3'-end of the tRNA molecule, to which the amino acid becomes esterified.

amino acid composition The makeup of a peptide or a protein in terms of both the relative amounts and the types of its constituent amino acids; generally expressed in terms of mole percent.

amino acid incorporation The in vivo or in vitro reactions whereby amino acids become constituents of proteins as a result of protein synthesis.

amino acid nitrogen The nitrogen of the amino acids in serum. *Abbr* AAN.

aminoacidopathy A genetically inherited metabolic defect in humans that involves amino acid metabolism.

amino acid oxidase An enzyme that catalyzes the oxidative deamination of amino acids. An L-amino acid oxidase is specific for L-amino acids and is a flavoprotein having FMN as a prosthetic group; a D-amino acid oxidase is specific for D-amino acids and is a flavoprotein having FAD as a prosthetic group.

amino acid replacement The substitution of one amino acid for another at a position in a polypeptide chain as a result of a mutation in the corresponding codon. *See also* conservative substitution; radical substitution.

amino acid residue That portion of an amino acid that is present in a peptide or a polypeptide; the amino acid minus the atoms that are removed from it in the process of linking it to other amino acids by means of peptide bonds. Depending on its position in the peptide or in the polypeptide chain, the amino acid loses a hydrogen atom, a hydroxyl group, or a molecule of water as it becomes linked to the other amino acids.

amino acid sequence The linear order of the amino acids as they occur in a peptide or in a protein; the amino acid sequence is conventionally written with the N-terminal amino acid on the left and with the C-terminal amino acid on the right.

amino acid sequencer *See* sequenator.

amino acid side chain The atoms of the amino acid molecule exclusive of the alpha carbon atom and its hydrogen atom, the alpha amino group, and the alpha carboxyl group.

amino acid starvation *See* starvation (2).

amino acid substitution AMINO ACID REPLACEMENT.

amino acid:tRNA ligase AMINOACYL-tRNA SYNTHETASE.

aminoaciduria The presence of excessive amounts of amino acids in the urine.

aminoacyl- Combining form denoting an amino acid that is esterified through its carboxyl group to another molecule.

aminoacyl adenylate An amino acid that has been esterified through its carboxyl group to the phosphate group of AMP; an intermediate in the activation of an amino acid to the aminoacyl-tRNA. *Abbr* AA-AMP.

aminoacyl site The site on the ribosome at which the incoming aminoacyl-tRNA is bound during protein synthesis; the A-site.

aminoacyl-tRNA An amino acid that has been esterified through its carboxyl group to the 2'- or 3'-hydroxyl group of the terminal adenosine at the 3'-end of a transfer RNA molecule; aminoacyl-tRNA is the form in which an amino acid is transported to the ribosomes for protein synthesis. *Abbr* AA-tRNA; AA-tRNA^{AA}. *Aka* aminoacylated-tRNA.

aminoacyl-tRNA site AMINOACYL SITE.

aminoacyl-tRNA synthetase The enzyme that catalyzes the coupled reactions of amino acid activation whereby an amino acid is first attached to AMP to form an aminoacyl adenylate, and is then attached to a transfer RNA molecule to form an aminoacyl-tRNA molecule.

aminoadipic pathway A biosynthetic pathway of lysine that proceeds by way of α-aminoadipic acid and occurs in fungi.

p-aminobenzoic acid A component of folic acid that is generally classified with the B vitamins, since it is a growth factor (provitamin) for some bacteria. It has no vitamin activity in humans because humans lack the ability to synthesize folic acid from it. *Abbr* PAB; PABA.

γ-aminobutyrate bypass A reaction sequence for the conversion of α-ketoglutaric acid to succinic acid that differs from the normal sequence in the citric acid cycle and occurs in brain tissue. *Aka* GABA shunt.

γ-aminobutyric acid A fatty acid derivative that functions in the metabolism of brain. *Aka* 4-aminobutyric acid.

aminoethyl cellulose An anion exchanger.

aminoglycoside antibiotics A diverse group of antibiotics, isolated from various bacterial species, that are valuable agents for the treatment of infectious diseases. Their primary growth-inhibiting action is due to their specific attachment to ribosomes and subsequent disruption of the translation mechanism of the microbial cell, leading to inhibition of protein synthesis at one or more steps. The group includes such antibiotics as kanamycin, neomycin, and streptomycin. *Aka* aminoglycosides; aminoglycoside–aminocyclitol antibiotics.

amino group The radical –NH$_2$.

p-aminohippuric acid A compound used for renal clearance tests. *Abbr* PAH.

δ-aminolevulinic acid A key intermediate in the biosynthesis of porphyrins in which two molecules of δ-aminolevulinic acid condense to form the pyrrole porphobilinogen. *Abbr* DALA.

aminopeptidase An exopeptidase that catalyzes the sequential hydrolysis of amino acids in a polypeptide chain from the N-terminal.

amino precursor uptake and decarboxylation *See* APUD theory.

aminopterin A folic acid analogue that inhibits the enzyme dihydrofolate reductase and that is used in the treatment of leukemia.

2-aminopurine A purine analogue that is incorporated into nucleic acids and thereby produces transitions. *Abbr* AP.

p-aminosalicylic acid An analogue of p-aminobenzoic acid that is used in the treatment of tuberculosis. *Abbr* PAS.

amino sugar A monosaccharide in which one or more hydroxyl groups have been replaced by an amino group.

amino terminal N-TERMINAL.

aminotransferase TRANSAMINASE.

amitosis A form of nuclear division in which a spindle is not formed, the nuclear membrane persists throughout the division, and the nucleus divides by constriction; occurs in the macronuclei of ciliates and in certain fungi.

ammonia A colorless gas that is the major form in which nitrogen is utilizable by living cells. Ammonia is the first compound formed in biological nitrogen fixation and is also the end product of purine catabolism in some marine invertebrates and in crustaceans.

ammonia fixation A group of three reactions, one or more of which occur in every organism, whereby ammonia is converted to glutamic acid, glutamine, or carbamoyl phosphate.

ammonification The formation of ammonia by the degradation of organic compounds.

ammonification of nitrate *See* nitrate respiration.

ammonium plant ACID PLANT.

ammonium sulfate fractionation A fractional precipitation by means of ammonium sulfate that is used in the purification of enzymes and other proteins.

ammonolysis The cleavage of a covalent bond of an acid derivative by reaction with ammonia so that one of the products combines with the hydrogen atom and the other combines with the amino group of ammonia.

ammonotelic organism An organism, such as a teleost fish, which excretes the nitrogen from amino acid and purine catabolism primarily in the form of ammonia.

amniocentesis A procedure for the sampling and testing of the amniotic fluid during pregnancy; permits a determination of the sex of the embryo and a detection of various genetic diseases.

amnion 1. The fluid-filled sac within which the embryos of reptiles, birds, and mammals develop. The wall of the sac consists of two layers of epithelium; the inner one is called amnion, the outer one chorion. 2. The inner layer of epithelium of the amniotic sac.

amniotic fluid The fluid that fills the membranous sac enclosing the embryo.

amorph A mutant allele that has little or no effect on the expression of a trait compared to the effect that the wild-type allele has.

amorphous 1. Noncrystalline; devoid of a regular shape and a molecular lattice structure. 2. Lacking a definite form or organization; descriptive of nonhelical regions in macromolecules.

AMP 1. Adenosine monophosphate (adenylic acid). 2. Adenosine-5'-monophosphate (5'-adenylic acid). 3. Avian myeloblastosis virus.

ampere A unit of electrical current intensity; equal to the constant current that, when passed through a standard aqueous solution of silver nitrate, deposits silver at the rate of 0.001118 g/s. *Sym* A.

amperometric titration A titration in which either the titrant or the substance being titrated is electroactive and the limiting current is plotted as a function of added titrant.

amphetamine The drug, 1-phenyl-2-amino-propane, that stimulates the central nervous system and inhibits sleep.

amphibaric Descriptive of a pharmacologically active substance that can either lower or raise the blood pressure depending on its dose or concentration.

amphibiotic Descriptive of an organism that can behave either as a symbiont or as a parasite with respect to a given host.

amphibolic pathway The metabolic pathway composed of the reactions of the citric acid cycle and some of the reactions of glycolysis. The pathway occupies a central position in metabolism, since it can be used either catabolically for the oxidation of metabolites to carbon dioxide and water, or anabolically for the synthesis and interconversion of metabolites. *Aka* central metabolic pathway.

amphipathic Descrptive of a molecule that has both polar (hydrophilic) and nonpolar (hydrophobic) groups.

amphiphilic AMPHIPATHIC.

amphiprotic Descriptive of a compound that can either gain or lose protons; synonymous with amphoteric if acids are defined as proton donors and bases as proton acceptors.

amphiprotic solvent A nonaqueous solvent that can act either as a proton donor or as a proton acceptor with respect to the solute.

amphitrophic Descriptive of an organism that can live photosynthetically in the light and chemotrophically in the dark.

ampholyte An amphoteric electrolyte.

amphoteric Descriptive of a compound that has at least one group that can act as an acid and one group that can act as a base; a compound that can act either as a proton donor or as a proton acceptor. Synonymous with amphiprotic if acids are defined as proton donors and bases as proton acceptors.

amphotropic virus A virus that can replicate either in cells from its host species or in cells from another species. *See also* ecotropic virus.

ampicillin A semisynthetic derivative of penicillin that is more effective against gram-negative bacteria than other derivatives of penicillin.

AMP kinase ADENYLATE KINASE.

amplicon A defective virus vector; a defective viral genome such as those derived from Herpes simplex virus.

amplification *See* cascade mechanism; enzyme amplification; gas amplification; gene amplification; plasmid amplification.

amplifier enzyme A membrane-located enzyme that mediates that action of a hormone in a multiple cascade mechanism. The hormone–receptor complex activates the amplifier enzyme by means of G proteins, and the amplifier enzyme then activates a second messenger molecule which initiates the cascade mechanism.

amplifier T cells A group of T cells that amplify the response and the proliferation of cytotoxic T cells. *Aka* amplifier T lymphocytes.

amplitude 1. The maximum response of an interconvertible enzyme that can be achieved with saturating concentrations of effectors for a given converter enzyme. 2. The maximum displacement of an oscillation, a vibration, or a wave.

ampoule A small glass container with a thin extended portion that is readily sealed by heating. *Var sp* ampule.

amu Atomic mass unit.

amyelination The failure to form myelin.

A myeloma protein An abnormal immunoglobulin of the IgA type that is produced by individuals suffering from multiple myeloma.

amygdalin A β-cyanogenic glycoside, similar in structure to laetrile but containing an additional glucose residue. It occurs naturally in the kernels or seeds of most fruits. Apricot kernels contain the enzymes β-glucosidase and oxynitrilase (mandelonitrile lyase). The former enzyme catalyzes the hydrolysis of amygdalin to two molecules of glucose and one molecule of mandelonitrile; the latter enzyme catalyzes the hydrolysis of mandelonitrile to cyanide (HCN) and benzaldehyde. *See also* laetrile; vitamin B_{17}.

amylase An enzyme that catalyzes the hydrolysis of starch at $\alpha(1 \rightarrow 4)$ glycosidic bonds. Alpha amylase is an endoamylase that catalyzes random hydrolysis; beta amylase is an exoamylase that catalyzes the sequential removal of glucose residues, commencing at the nonreducing end of the starch molecule.

amylo- Combining form meaning starch.

amyloclastic AMYLOLYTIC.

amyloclastic method A method of assaying for the enzyme amylase by determining the amount of unhydrolyzed starch that remains after incubation of the starch with the enzyme.

amylodextrin SOLUBLE STARCH.

amyloid One of a number of fibrous proteins that give a starch-like reaction with iodine and that are deposited in blood vessels and other tissues under certain pathological conditions. Amyloid fibers consist of stacks of pleated sheets and are highly resistant to degradation. One type of amyloid, deposited in the brain, is believed to represent the waste product of patients afflicted with Alzheimer's disease.

amyloidosis A pathological condition characterized by the formation of amyloid deposits.

amylolysis The hydrolysis of starch.

amylolytic Of, or pertaining to, amylolysis.

amylometric method A method of assaying for the enzyme amylase by determining the amount of starch that is hydrolyzed during incubation of the starch with the enzyme.

amylopectin The form of starch that is composed of branched chains of glucose units which are joined by means of $\alpha(1 \rightarrow 4)$ and $\alpha(1 \rightarrow 6)$ glycosidic bonds.

amylopectinosis GLYCOGEN STORAGE DISEASE TYPE IV.

amyloplast A starch-storing plastid.

amylopsin The α-amylase present in the pancreatic juice.

amylose The form of starch that is composed of long, unbranched chains of glucose units which are joined by means of $\alpha(1 \rightarrow 4)$ glycosidic bonds. *Aka* α-amylose.

amylose synthetase The enzyme that catalyzes the synthesis of amylose from ADP-glucose.

amytal The barbiturate drug, 5-ethyl-5-isoamylbarbituric acid, that inhibits the electron transport system between the flavoproteins and coenzyme Q.

anabiosis ANHYDROBIOSIS (2).

anabolic Of, or pertaining to, anabolism.

anabolic steroid A steroid that stimulates muscle grwoth and muscle strength; androgens and certain androgen derivatives have this effect.

anabolism 1. The phase of intermediary metabolism that encompasses the biosynthetic and energy-requiring reactions whereby cell components are produced. 2. The cellular assimilation of macromolecules and complex substances from low molecular weight precursors.

anacidity 1. A lack of acidity, particularly the lack of gastric hydrochloric acid. 2. The pathological condition due to a lack of gastric hydrochloric acid.

anaerobe *See* facultative anaerobe; obligate anaerobe.

anaerobic 1. In the absence of oxygen; in an environment or an atmosphere devoid of oxygen. 2. Not requiring the presence of molecular oxygen for growth. 3. Not capable of using molecular oxygen for growth. *See also* anoxybiontic.

anaerobic-aerotolerant MICROAEROPHILIC.

anaerobic fermentation *See* fermentation (2).

anaerobic glycolysis The group of cellular reactions, that do not require oxygen, whereby glucose is converted to lactic acid.

anaerobic respiration The energy-yielding metabolic breakdown of organic compounds in an organism that proceeds in the absence of molecular oxygen and with the use of inorganic compounds, such as nitrate or sulfate, as oxidizing agents. *See also* fermentation (2).

anaerobiosis Life under anaerobic conditions.

anaerobiotic Of, or pertaining to, anaerobiosis.

anaerogenic Of, or pertaining to, an organism that does not produce gas from a particular substrate.

analbuminemia A genetically inherited metabolic defect in humans that is characterized by an impaired synthesis of serum albumin.

analgesia The relief of pain without loss of consciousness.

analgesic 1. *n* An agent that brings about analgesia. 2. *adj* Of, or pertaining to, analgesia.

analog computer A computer that receives information in the form of continuous variables, such as temperature, pressure, and flow, and that processes the information by translating each variable into an analogous or a related mechanical or electrical variable, such as voltage.

analogous Having a similar function and a similar, but not identical, structure.

analogous enzyme variants Enzyme variants that differ significantly in their molecular structures and catalytic properties.

analogue A compound that is structurally similar to another compound and that is used for such purposes as the determination of structural prerequisites of enzyme substrates, the competitive inhibition of specific enzymatic and other reactions, and the synthesis of altered macromolecules. *Var sp* analog.

analysis of covariance A statistical analysis for determining the variability in the principal variable that is due to variability in some other variable; consists of the combined application of linear regression and analysis of variance techniques.

analysis of variance A statistical analysis for segregating the sources of variability in measurements, as in determining the extent to which the variability in sets of observations is due to differences between the sets and the extent to which it is due to random variations. An analysis of the total variability of a set of data into components which can be attributed to different sources of variation. *Abbr* ANOVA.

analyte The ion or compound that is being measured (determined) in a given analytical procedure.

analytical biochemistry A branch of biochemistry that deals with the qualitative and quantitative determination of substances in living systems.

analytical method A method, such as ultracentrifugation, eletrophoresis, or chromatography, that requires relatively small amounts of sample and that is used primarily for the identification and characterization of specific substances. *See also* preparative method.

analytical ultracentrifuge A high-speed centrifuge, equipped with one or more optical systems, that is used for measurements of sedimentation coefficients and molecular weights as well as for a variety of studies of macromolecules. The centrifuge is capable of generating speeds of approximately 60,000 rpm and centrifugal forces of approximately $500,000 \times g$. The optical systems used in conjunction with the analytical ultracentrifuge are a schlieren optical system, an absorption optical system, and an interferometric optical system.

analyzer The nicol prism in a polarimeter that is used for determining the rotation of the plane-polarized light. *See* also polarizer.

anamnestic response SECONDARY IMMUNE RESPONSE.

anaphase The third stage in mitosis during which the chromosomes move to opposite poles.

anaphoresis 1. The movement of charged particles toward the anode. 2. ELECTROPHORESIS.

anaphylactic response The immune reactions of anaphylaxis.

anaphylactic shock A severe and generalized form of anaphylaxis that is characterized by violent cardiac and respiratory symptoms and that may be produced by the injection of a substance to which an individual is either allergic or sensitized.

anaphylactoid reaction A condition that resembles an anaphylactic shock but that is not caused by an immunological reaction.

anaphylatoxin A pharmacologically active substance, apparently a polypeptide fragment of complement, that can cause the release of histamine from mast cells in anaphylaxis.

anaphylaxis An immediate-type hypersensitivity in which the first administration of an antigen to an animal is harmless, but the second administration leads to an intense secondary immune response accompanied by pathological reactions; involves the combination of antigens with homologous, mast cell-bound IgE (reaginic) antibodies. *See also* active anaphylaxis; passive anaphylaxis; reverse passive anaphylaxis.

anaplasia The loss by a cell of its characteristic structure accompanied by its reversion to a more primitive, embryonic type.

anaplastic Of, or pertaining to, anaplasia.

anaplerosis *See* anaplerotic reaction.

anaplerotic reaction A reaction whereby a metabolic intermediate is replenished; this is generally achieved through the insertion of either a one-carbon fragment, in the form of carbon dioxide, or a two-carbon fragment, in the form of acetyl coenzyme A, into the appropriate metabolic reaction.

anatoxin TOXOID.

anchimeric assistance The facilitation, by one part of a substrate molecule, of an enzyme reaction that occurs at a different part of the same substrate molecule. Thus, different parts of the same substrate molecule participate both in catalysis and in the actual chemical reaction.

anchorage dependence The difference between the extent of cellular transformation that is produced by an oncogenic virus, such as polyoma virus, with cells that are planted in agar and with cells thar are suspended in a viscous medium.

anchorage-independent growth The ability of transformed (tumorigenic) cells to grow progressively while suspended in a semisolid medium. This property generally distinguishes tumorigenic from normal (nontumorigenic) cells.

anchorin ANKYRIN.

Andersen's disease GLYCOGEN STORAGE DISEASE TYPE IV.

androgen 1. A 19-carbon steroid that is a male sex hormone or one of its metabolites. 2. Any 19-carbon steroid. *See also* male sex hormone.

androgen-binding protein A protein that is secreted by cells in the testes in response to follicle-stimulating hormone. The protein binds androgens and is believed to function in establishing high local concentrations of testosterone. *Abbr* ABP.

androstane The parent ring system of the androgens.

androsterone A major metabolite of testosterone that has weak androgenic activity and that belongs to the group of ketosteroids.

anemia A condition in which the number of red blood cells, the volume of red blood cells, or the hemoglobin content of the blood are below normal levels. *See also* hemolytic anemia; hypochromic anemia; pernicious anemia; sickle cell anemia.

anemic Of, or pertaining to, anemia.

anergy The total absence of an allergic response in an animal under conditions that would otherwise be expected to lead to such a response.

anesthetic drug A drug that induces either a local, or a total, loss of sensation in the body.

aneuploid state The chromosome state in which there is a loss or a gain of single chromosomes, and the chromosome number is not an exact multiple of the basic number in the genome. *Aka* aneuploidy.

aneurin THIAMINE.

aneurysm 1. A blood-containing tumor connected directly with the lumen of an artery. 2. A circumscribed dilation of an artery.

ANF 1. Antinuclear factor. 2. Atrial natriuretic factor.

angel dust Phencyclidine [1–(1–phenylcyclohexyl)piperidine; PCP]; a compound that was first introduced as a general anesthetic and that is now used as an animal tranquilizer. It is frequently abused as a drug by adolescents. Phencyclidine inhibits cholinergic activity, increases brain dopamine activity in rats, and induces psychoses in humans that are similar to schizophrenia.

angiogenesis The formation of new blood capillaries. It is now believed that, once a solid tumor take has occurred, every increase in tumor size must be preceded by an increase in new capillaries that converge upon the tumor.

angiogenic factors A group of naturally occurring substances that promote angiogenesis; includes a number of polypeptides, such as acidic and basic fibroblast growth factors, angiogenin, transforming growth factors α and β, and some lipids.

angiogenin A small protein, isolated from human tumor cells (adenocarcinoma) grown in culture, that induces new blood vessels to grow in living tissue; it is believed to be produced in healthy, nontumor tissue as well and has 35% sequence homology with pancreatic ribonuclease.

angioma A tumor consisting chiefly of blood or lymphatic vessels.

angiotensin I The inactive decapeptide precursor of angiotensin II; it is cleaved off from angiotensinogen in a reaction catalyzed by the enzyme renin.

angiotensin II The active octapeptide formed from angiotensin I by hydrolytic removal of two amino acids in a reaction catalyzed by the serum converting enzyme; a powerful hypertensive agent.

angiotensinogen The hepatic globulin from which the decapeptide angiotensin I is cleaved off in a reaction catalyzed by the enzyme renin.

angiotonin ANGIOTENSIN.

angle rotor A centrifuge rotor in which the tubes containing solution are held at a fixed angle. Such rotors are used for the preparative fractionation of macromolecules and their efficiency is due to he fact that convection is superimposed upon sedimentation in the tube. *Aka* angle head.

angle strain A strain in a ring structure that is

due to expansion or compression of bond angles.

angstrom unit A unit of length equal to 10^{-8} cm and used in describing atomic and molecular dimensions. *Sym* AU; Å; A. *Aka* angstrom.

angular methyl group A methyl group attached to the perhydrocyclopentanophenanthrene ring system of steroids.

angular resolved photoelectron spectroscopy A technique for the study of surfaces in which photons are allowed to strike a surface, leading to the ejection of photoelectrons from molecules adsorbed to the surface. Measurements of the angles and intensities of these photoelectron emissions allows a determination of the orientation of the adsorbed molecules. *Abbr* ARPES.

angular velocity The velocity of rotation expressed in terms of the central angle, in radians, transversed per unit time.

anhaptoglobinemia A genetically inherited metabolic defect in humans that is due to a lack of haptoglobin in the blood.

anhydride *See* acid anhydride.

anhydrobiosis 1. Life in the absence of water. 2. A state of suspended animation shown by some organisms in which they can sustain the removal of all, or almost all, of their cellular water and return to normal living conditions when resupplied with water.

anhydrous Devoid of water.

animal cephalin PHOSPHATIDYL SERINE.

animal charcoal BONEBLACK.

animal hormone *See* hormone.

animal protein factor VITAMIN B_{12}.

animal saponin A sulfur-containing steroid glycoside that has properties of a plant saponin but is isolated from a marine invertebrate.

animal starch GLYCOGEN.

animal toxin A toxin of animal origin, such as that in snake venom.

animal virus A virus that infects animal cells and multiplies in them. *See also* virus.

anion A negatively charged ion.

anion exchanger A positively charged ion-exchange resin that binds anions.

anion gap A measure for evaluating chemical disturbances of acid–base balance, particularly those of metabolic acidosis; defined as the difference between the concentration, in blood, of the major cation (Na^+) and the sum of the concentrations of the major anions (Cl^-, HCO_3^-), with all values expressed in terms of milliequivalents per liter of serum. Thus, the anion gap is given by $[Na^+] - ([Cl^-] + [HCO_3^-])$.

anionic detergent A surface-active agent in which the surface-active part of the molecule carries a negative charge. *Aka* anionic surface-active agent.

anion respiration The phenomenon whereby exposure of plant tissues to salt solutions frequently leads to an increase in respiration which appears to be proportional to the rate of anion absorption by the plant.

anion-transport protein An integral protein in the red blood cell membrane that spans the entire width of the membrane; a glycoprotein that has a large part of the molecule protruding on the cytoplasmic side and the carbohydrate chains protruding on the extracellular side of the membrane. The protein functions in the transport of anions across the membrane. *Aka* band 3.

anisotropic Of, or pertaining to, anisotropy.

anisotropic band A BAND.

anisotropy The variation in the physical properties of a substance as a function of the direction in which these properties are measured. *Aka* anisotropism.

ankyrin A peripheral protein of the red blood cell membrane that links spectrin molecules to anion-transport proteins. *Var sp* anchorin. *Aka* syndein.

annealing 1. The renaturation of heat-denatured proteins or heat-denatured nucleic acids by slow cooling. 2. The formation of hybrid nucleic acid molecules, containing paired strands from different sources, by slow cooling of a mixture of denatured nucleic acids. 3. The tempering of glass in glass blowing by slow cooling. *See also* reannealing.

annular Ring-shaped.

annulation reaction A chemical reaction that involves building a new ring onto a molecule.

anode The electrode by which electrons leave the solution of an electrolyte and toward which the anions move in solution. With respect to properties in solution, the anode is a positive electrode; with respect to the external flow of electrons, the anode is a negative electrode.

anodic 1. Of, or pertaining to, the anode. 2. Descriptive of a component that moves toward the anode in electrophoresis.

anomalous dispersion An optical rotatory dispersion that cannot be expressed by a simple, one-term Drude equation; such a dispersion is generally expressed $[m'] = a_o \lambda_o^2 / (\lambda^2 - \lambda_o^2) + b_o \lambda_o^4 / (\lambda^2 - \lambda_o^2)^2$, where $[m']$ is the reduced mean residue rotation, λ is the wavelength, and a_o, b_o, and λ_o are constants.

anomalous osmosis The electroosmotic flow of water through a charged membrane that is caused by a potential gradient across the membrane. The anomalous osmosis is said to be positive when the water moves from a dilute to a concentrated solution and is said to be negative when the flow of water is in the opposite direction.

anomer One of two isomeric carbohydrates (designated α and β) that differ from each other only in the configuration about the anomeric carbon of the ring structure. The α-isomer has the hydrogen at the anomeric carbon above (and the β-isomer has it below) the plane of the ring in a Haworth projection.

anomeric carbon The carbon atom of the carbonyl group in a carbohydrate.

anomeric effect The stereochemical effect in carbohydrate chemistry in which the interaction between the oxygen of the monosaccharide ring and the substituent (—OR; —O—CO—R; or halogen) at the anomeric carbon is such as to favor the maximum separation between the oxygen and the substituent; as a result, the axial substituent, or α-anomer, is favored over the equatorial substituent, or β-anomer. The molecule having an equatorial anomeric substituent is less stable than the one having an axial substituent.

ANOVA Acronym for analysis of variance. *Aka* ANOVAR.

anovar Acronym for analysis of variance.

anoxia HYPOXIA.

anoxybiontic Not capable of using atmospheric (molecular) oxygen for growth. *Aka* anoxybiotic. *See also* anaerobic (2,3).

anserine A dipeptide of β-alanine and methyl histidine that occurs in vertebrate muscle.

antagonism The phenomenon in which the action of one agent is counteracted by the action of another agent that is present at the same time.

antagonist A molecule, such as a drug, an enzyme inhibitor, or a hormone, that diminishes or prevents the action of another molecule or receptor site. *See also* α-bungarotoxin.

ante-iso fatty acid A fatty acid that is branched at the carbon atom preceding the penultimate carbon atom at the hydrocarbon end of the molecule.

antenna molecules Molecules that are not photochemically active and merely serve in the capacity of a large antenna, passing the excitation energy in photosynthesis from one molecule to another until it is trapped by the photochemically active molecules in the reaction center. Antenna molecules constitute the bulk of the photosynthetic pigment molecules. *Aka* antenna chlorophyll.

ante-penultimate carbon The third carbon atom from the end of a chain.

anterior 1. In front of, or in the front part of, a structure. 2. Before, in relation to time.

anthesin FLOWERING HORMONE.

anthocyanidin The aglycone of an anthocyanin.

anthocyanins Water-soluble plant pigments that occur largely in the form of glycosides of an anthocyanidin. Anthocyanins are bioflavonoids. *See also* bioflavonoid.

anthranilic acid *See* chorismic acid.

anthrone reaction A colorimetric reaction for carbohydrates, particularly hexoses, that is based on the production of a green color on treatment of the sample with anthrone.

anthropic principle The principle according to which the presence of life on earth may explain some of the conditions associated with life. It is usually argued that life arose on the earth because circumstances, such as a moderate temperature, were conducive to its existence. According to the anthropic principle, the argument is reversed; it is postulated that the presence of life on earth explains why the latter has a moderate temperature.

anti 1. Referring to a nucleoside conformation in which the base has been rotated around the sugar, using the C—N glycosidic bond as a pivot, so that the sugar is in direct opposition to the base. This represents a sterically less hindered conformation than the syn conformation; in polynucleotides, it leads to the bulky portions of the bases being pointed away from the sugar–phosphate backbone of the chain. 2. Referring to a trans configuration for certain compounds containing double bonds, such as the oximes which contain the group C=N—OH. 3. Referring to the position occupied by two radicals of a stereoisomer in which the radicals are farther apart as opposed to the syn position in which they are closer together. *See also* syn.

antiacrodynia factor VITAMIN B_6.

antiadrenergic *See* alpha blocker; beta blocker.

antianemia factor VITAMIN B_{12}.

antiantibody An antibody produced in response to an antigenic determinant of an antibody molecule.

antiauxin A compound that functions as a competitive inhibitor of auxin.

antibacterial agent *See* bactericide; bacteristat.

antiberiberi factor VITAMIN B_1.

antibiosis The association of two organisms in which one produces a substance, such as an antibiotic, or a condition that is harmful to the other.

antibiotic Originally, defined as a compound produced by a microorganism that inhibits the reproduction or causes the destruction of other microorganisms. Now more generally defined as a compound produced by a microorganism or a plant, or a close chemical derivative of such a compound, that is toxic to microorganisms from a number of other species. *See also* under individual antibiotics and classes of antibiotics, such as streptomycin and macrolide antibiotic.

anti-black-tongue factor NICOTINIC ACID.

antibody A glycoprotein of the globulin type that is formed in an animal organism in response to the administration of an antigen and that is capable of combining specifically with that antigen. *Abbr* Ab. *See also* immunoglobulin.

antibody-binding fraction FAB FRAGMENT.

antibody combining site ANTIGEN BINDING SITE.

antibody-dependent cellular cytotoxicity Cell-mediated cytotoxicity that requires prior binding of antibodies to target cells. *Abbr* ADCC. *Aka* antibody-dependent, cell-mediated cytotoxicity. *See also* killer cells.

antibody diversity ANTIBODY HETEROGENEITY.

antibody-excess zone A zone in the precipitin curve of the antigen–antibody reaction in which the amount of antibody precipitated increases with increasing amounts of antigen.

antibody fixation The binding of antibodies to cell receptors in immediate-type hypersensitivity.

antibody formation *See* theory of antibody formation.

antibody heterogeneity The state of a given preparation of antibodies in which the antibodies differ with respect to size, structure, charge, or other properties.

antibody-mediated hypersensitivity IMMEDIATE-TYPE HYPERSENSITIVITY.

antibody response IMMUNE RESPONSE.

antibody specificity *See* specificity (2).

antibody titer The highest dilution of an antiserum that will produce detectable precipitation or agglutination when reacted with antigens.

antibody valence The number of antigen binding sites, of which there are at least two, per antibody molecule.

antibonding orbital A molecular orbital in which there is a node of electron density between the bonding atomic nuclei, resulting in a weakening of the bond between the nuclei. Antibonding orbitals are generally of higher energy than sigma (σ) and pi (π) orbitals and are designated sigma star (σ^*) and pi star (π^*).

anticancer compound A compound that arrests or reverses the growth of a malignant tumor.

anticarcinogenesis The inhibition of the action of one carcinogen by the simultaneous administration of a second carcinogen.

antichaotropic agent A substance that decreases the solubility of hydrophobic (nonpolar) molecules; generally, a small, singly charged ion such as fluoride, or multiply charged ions such as citrate, phosphate, or sulfate.

anticholinesterase An inhibitor of the enzyme cholinesterase.

antichymotrypsin An inhibitor of the enzyme chymotrypsin.

anticlinal At right angles to the surface or the circumference; radial.

anticoagulant A substance that prevents the clotting of blood. Most anticoagulants function by binding calcium ions; these include oxalates, citrates, and ethylenediaminetetraacetic acid (EDTA). Another anticoagulant is heparin; it acts by combining with antithrombin, an inhibitor of the enzyme thrombin.

anticode *See* genetic anticode.

anticoding strand That strand of double-stranded DNA that is transcribed into RNA; the strand that serves as a template for transcription; the sense strand.

anticodon A sequence of three nucleotides in tRNA that, in the process of protein synthesis, binds to a specific codon in mRNA by complementary base pairing.

anticodon arm *See* arm.

anticodon deaminase An enzyme that catalyzes the deamination of adenine to hypoxanthine whenever the adenine occurs at the first position (5'-end) of the anticodon in the unmodified tRNA transcript.

anticodon loop *See* arm.

anticollagenase An inhibitor of the enzyme collagenase.

anticompetitive inhibition UNCOMPETITIVE INHIBITION.

anticomplementary Referring to a treatment or an agent that either removes or inactivates a component of complement.

anticomplement fluorescent antibody technique A fluorescent antibody technique in which an antigen–antibody complex is reacted with complement and the entire aggregate is then stained by means of fluorescent antibodies to complement.

anticooperativity NEGATIVE COOPERATIVITY.

antidepressant A stimulatory drug that reduces fatigue, appetite, and sleeping time. Antidepressants are amines (such as amphetamine and ephedrine) that are believed to function as competitive inhibitors of monoamine oxidase. This leads to an accumulation of catecholamines, the natural substrates of the enzyme, and results in stimulatory effects.

antidermatosis vitamin PANTOTHENIC ACID.

antidiuresis A decrease in the excretion of urine.

antidiuretic 1. *n* An agent that decreases the excretion of urine. 2. *adj* Of, or pertaining to, antidiuresis.

antidiuretic hormone VASOPRESSIN.

antidiuretin VASOPRESSIN.

antidotal agent ANTIDOTE.

antidotal therapy Therapy by means of antidotes.

antidote An agent that limits or reverses the effect of a poison.

anti-egg-white-injury factor BIOTIN.

antienzyme An antibody to an enzyme.

anti-fatty-liver factor LIPOCAIC.

antifoam A chemical substance added to liquid cultures of microorganisms to minimize foam formation during growth.

antifolate An antimetabolite of folic acid or of a derivative of folic acid. *Aka* antifolic acid agent.

antifreeze protein An unusual, extracellular glycoprotein that is found in the blood of some arctic and antarctic fish species. It contains a repeating sequence of alanine-alanine-threonine and has a disaccharide unit of D-galactosyl-*N*-acetyl-D-galactosamine attached to each threonine residue. The antifreeze protein depresses the freezing point of water, apparently because it inhibits the formation of ice crystals.

antigen A substance, frequently a protein, that can stimulate an animal organism to produce antibodies and that can combine specifically with the antibodies thus produced; called also complete antigen as distinct from a hapten. *Abbr* Ag.

antigen–antibody complex The generally insoluble molecular aggregate that is formed by the specific interaction of antigens and antibodies.

antigen-antibody lattice *See* lattice theory.

antigen-antibody reaction PRECIPITIN REACTION.

antigen binding capacity The total antibody concentration in an antiserum based on a determination of the amount of antigen bound by a given volume of the antiserum. *Abbr* ABC.

antigen-binding fragment FAB FRAGMENT.

antigen binding site One of at least two sites on the antibody molecule to which a complementary portion of an antigen, the antigenic determinant, becomes bound in the course of an antigen–antibody interaction; the active site of an antibody. *Aka antibody combining site.*

antigen-excess zone A zone in the precipitin curve of the antigen–antibody reaction in which the amount of antibody precipitated decreases with increasing amounts of antigen.

antigenic competition The decreases in the immune response to one antigen that is produced by the administration of a second antigen.

antigenic conversion 1. The appearance of one or more specific antigens on cells that have been infected by a virus. *See also* conversion. 2. The expression of new cell surface antigens, and the cessation of the expression of other cell surface antigens, that is brought about by antibodies; a switch in gene activities. *Aka* serotype transformation.

antigenic deletion The cellular loss of antigenic determinants, or the masking of existing cellular antigenic determinants.

antigenic determinant That portion of the antigen molecule that is responsible for the specificity of the antigen in an antigen–antibody reaction and that combines with the antigen binding site to which it is complementary.

antigenic drift A change in the specificity of viral antigens as a function of time. This occurs, for example, in the case of influenza and AIDS viruses. As the human population becomes immune to infection by existing viral strains, there is an increased tendency for natural selection of other, slightly different, strains that can evade the human immune response. As a result, strains of slightly different antigenicity become established.

antigenic gain The cellular acquisition of new antigenic determinants, or the unmasking of existing cellular antigenic determinants.

antigenicity The capacity of an antigen to stimulate the formation of specific antibodies.

antigenic modulation The suppression of cell-surface antigens in the presence of homologous antibodies.

antigenic sin *See* doctrine of original antigenic sin.

antigen-presenting cell A cell that carries a foreign antigen which is then recognized by a helper T cell.

antigen template theory An instructive theory of antibody formation according to which antigens taken up by a cell serve as templates for the synthesis of antibodies by that cell. The antigens are considered to bind to ribosomes or to mRNA, thereby modifying translation so that antibodies are formed, the combining sites of which are complementary to the antigenic determinants of the bound antigens.

antigen tolerance IMMUNOLOGICAL TOLERANCE.

antigen valence The number of antigenic determinants per antigen molecule; an antigen molecule may have one valence with respect to one antibody and have a different valence with respect to another antibody.

antigibberellin A compound that binds to the same active site as the plant hormone gibberellin; a competitive inhibitor of gibberellin.

antiglobulin An antibody formed against the antigenic determinant of a serum globulin molecule, usually an immunoglobulin.

antiglobulin consumption test A consumption test in which the binding of immunoglobulin with antigen is measured by subsequent consumption of added antiglobulin.

antiglobulin method INDIRECT FLUORESCENT ANTIBODY TECHNIQUE.

antiglobulin test COOMBS' TEST.

anti-gray-hair factor *p*-AMINOBENZOIC ACID.

antihemophilic factor An accessory protein that participates in the activation of Factor X in the intrinsic pathway of blood clotting. Controversy exists as to whether the antihemophilic factor is identical to Factor VIII or whether there are two distinct protein components that together define Factor VIII. *Abbr* AHF. *Aka* antihemophilic factor A; antihemophilic globulin; platelet cofactor I.

antihemophilic factor B CHRISTMAS FACTOR.

antihemophilic factor C. PLASMA THROMBO-PLASTIN ANTECEDENT.

antihemophilic globulin ANTIHEMOPHILIC FAC-TOR.

antihemorrhagic vitamin VITAMIN K.

antihistamine A drug that blocks the action of histamine and that is used in the treatment of immediate-type hypersensitivity.

antihormone 1. a substance that decrease or prevents the action of a hormone; a hormone antagonist. 2. An antibody to a hormone.

anti-idiotype antibody An antibody formed in response to an idiotypic marker; an antibody against an idiotype. *See also* idiotype.

anti-immunoglobulin antibodies Antibodies that are produced in an animal in response to the administration of foreign antibodies.

anti-infective vitamin VITAMIN A.

anti-insulin A compound, such as a sex hormone or a corticosteroid, that decreases the activity of insulin.

antilepton *See* elementary particles.

antilipotropic Descriptive of a substance that has the capacity of diverting methyl groups from the synthesis of choline.

antilogarithm The antilogarithm of X is that number the logarithm of which is X. *Abbr* antilog.

antilymphocyte globulin The globulin fraction of antilymphocyte serum. *Abbr* ALG.

antilymphocyte serum A serum that contains antibodies to lymphocytes and that is used as an immunosuppressive agent. *Abbr* ALS.

antimalarial 1. *n* A drug used to prevent or treat malaria. 2. *adj* Preventing or curing malaria.

antimer ENANTIOMER.

antimetabolite A compound that competitively inhibits a specific enzymatic or other reaction in metabolism because of its similarity in structure to the natural metabolite that participates in the reaction. *See also* competitive inhibitor.

antimicrobial spectrum The types of micro-organisms against which an antimicrobial drug is effective. *See also* sensitivity spectrum.

antimitotic agent A compound that inhibits mitosis.

antimorph 1. ENANTIOMER. 2. A mutant gene that has an effect opposite that of its corresponding wild-type gene.

antimutagen A substance that counteracts the action of a mutagen by decreasing the rate of induced, and occasionally of spontaneous, mutations.

antimycin A An antibiotic, produced by *Streptomyces griseus*, that inhibits the electron transport system between cytochromes b and c_1.

antineuritic factor VITAMIN B_1.

antinuclear factor An antibody against a constituent of the cell nucleus. *Abbr* ANF. *Aka* antinuclear antibody.

antioxidant A substance, generally an organic compound, that is more readily oxidized than a second substance and hence can retard or inhibit the autoxidation of the second substance when added to it.

antiparallel chains 1. Two peptide chains running in opposite directions, with the one progressing from the C-terminal to the N-terminal, and the other progressing in the opposite direction 2. ANTIPARALLEL STRANDS.

antiparallel spin The spin of two particles in opposite directions.

antiparallel strands Two polynucleotide strands running in opposite directions, with the one progressing from the 3'-terminal to the 5'-terminal, and the other progressing in the opposite direction.

antiparticle *See* elementary particles.

antipellagra factor NICOTINIC ACID.

antipeptic ulcer factor VITAMIN U.

antipernicious anemia factor VITAMIN B_{12}.

antipode OPTICAL ANTIPODE.

antipolarity The decrease that may occur in the synthesis of an enzyme if the enzyme is specified by a gene that precedes another gene that has undergone a polar mutation.

antiport The linked transport in opposite directions of two substances across a membrane. *See also* symport; uniport.

antipromoter A substance that counteracts the action of a promoter in carcinogenesis. It may act at the initiation stage by detoxifying a carcinogen or it may act at the promotion stage by, for example, inhibiting a protease that helps a tumor invade neighboring tissue. Several dietary factors are suspected of having antipromoter activity. These include dietary fiber, vitamin A (or its precursor, beta carotene), vitamin C, vitamin E, the trace element selenium, and indoles, flavones, and isothiocyanates derived from cruciferous vegetables (such as broccoli, cabbage, and cauliflower, which belong to the Cruciferae family of plants).

antipurine A purine analogue; a purine or a purine nucleoside that is structurally similar to the natural compound and that acts as an antimetabolite in nucleic acid metabolism.

antipyrimidine A pyrimidine analogue; a pyrimidine or a pyrimidine nucleoside that is structurally similar to the natural compound and that acts as an antimetabolite in nucleic acid metabolism.

antiquark *See* elementary particles.

antirachitic vitamin VITAMIN D.

antirepressor INDUCER.

antiscorbutic factor VITAMIN C.

antisense strand CODING STRAND.

antisepsis The destruction and the prevention of growth of microorganisms causing disease, decay, or putrefaction.

antiseptic Of, or pertaining to, antisepsis.

antiserum A serum that contains antibodies and that has been obtained from an animal organism subsequent to its immunization with an antigen.

anti-sigma factor *See* antispecificity factor.

antispecificity factor A protein that antagonizes the action of a specificity factor and thereby prevents the initiation of transcription of a certain section of DNA. The protein, synthesized during the infection of *E. coli* by T4 phage, is an example; it prevents recognition of the promoter region by the sigma subunit of RNA polymerase and is, therefore, also known as an anti-sigma factor.

antisterility factor VITAMIN E.

antitermination A transcriptional control mechanism that applies to intercistronic termination sites present in some operons. In the presence of antitermination factors, RNA polymerase is able to bypass the termination site of a given proximal gene and continue transcription of the adjacent, distal gene.

antitermination factor A protein that functions in antitermination; it enables RNA polymerase to ignore the signals for the termination of transcription at specific sites along the DNA molecule. *Aka* antiterminator.

antithrombin A protein in blood that reacts with, and inhibits, thrombin, factor X, and factor VII to form inert enzyme–substrate intermediates.

antithyroid agent An agent that inhibits thyroid function by affecting the synthesis, release, or utilization of thyroxine or triiodothyronine.

antithyroid compound A goitrogenic compound; a goitrogen.

antitoxin An antibody to a toxin; an antibody capable of neutralizing a toxin.

antitranscription terminator ANTITERMINATION FACTOR.

antitrypsin An inhibitor of the enzyme trypsin that is also active against other proteases. Important antitrypsins are the soybean trypsin inhibitor, isolated from soybeans, and the α-1-antitrypsin (α-1-AT or α-AT), isolated from human serum.

antitumor antibiotic An antibiotic that arrests or reverses the growth of a malignant tumor.

antitumor antimetabolite An antimetabolite that arrests or reverses the growth of a malignant tumor.

antitumor enzyme An enzyme that inhibits tumor growth. Such enzymes either stimulate the degradation of amino acids that cannot be synthesized by tumor cells or stimulate reactions that lead to an inhibition of the synthesis of tumor-specific DNA.

antitumor protein A protein that inhibits tumor growth. The plant proteins abrin and ricin, which inhibit protein synthesis, and the bacterial protein neocarcinostatin (isolated from *Streptomyces*), which inhibits DNA synthesis, are examples. Neocarcinostatin also has antibiotic-like activity against gram-positive bacteria.

antiviral agent A synthetic compound that inhibits viral replication; a drug, designed to selectively inhibit viral-encoded or virus-induced enzymes without affecting the enzymes normally involved in similar biochemical processes of the host cell. *See also* acyclovir.

antiviral protein A protein, induced by interferon, that binds to ribosomes and inhibits the translation either of viral RNA or of the mRNA which is derived from viral DNA. *Abbr* AVP.

antivitamin A structural analogue of a vitamin; a vitamin antagonist that acts as a competitive inhibitor of a vitamin.

antixerophthalmic factor VITAMIN A.

antizyme A noncompetitive protein inhibitor of the enzyme ornithine decarboxylase which is a key enzyme in the biosynthesis of polyamines. Antizyme is induced by the addition of the polyamines spermine, spermidine, and putrescine to a variety of eukaryotic cell lines and rat liver preparations.

anucleate Lacking a nucleus.

anucleolate Lacking a nucleolus.

anucleolate mutation A mutation that produces a cell lacking the nucleolus organizer.

anuresis The failure or the inability to void urine that is formed but which is retained in the urinary bladder.

anuria The lacking or the defective excretion of urine due to a failure in the function of the kidneys.

AP Aminopurine.

APA Apurinic acid.

apamine A polypeptide of 18 amino acids that is the toxin of bee venom.

AP endonuclease An enzyme that cleaves phosphodiester bonds next to an AP site. Two types are known, cleaving the DNA at the 5′-end and at the 3′-end, respectively, of the AP site. Removal of the sugar phosphate moiety of the AP site, plus removal of several adjacent nucleotides, prepares the way for DNA repair by means of DNA polymerase and DNA ligase. *See also* uracil-N-glycosylase.

aperiodic polymer A polymer consisting of nonidentical repeating units.

aphasic lethal A mutation that becomes expressed randomly during the development of an organism and leads to its death.

apholate An aziridine mutagen that is used for the sterilization of insects.

aphorism of Clausius A combined statement of the first and second laws of thermodynamics: The energy of the universe is constant, but its entropy is increasing to a maximum.

aphthovirus A virus that belongs to a subgroup of picornaviruses which includes foot-and-mouth disease virus.

aplasia 1. The defective development of an organ or a tissue. 2. The absence of an organ or a tissue from the body.

apnea *See* drug-induced apnea.

apo- Combining form denoting the protein portion of a conjugated protein.

apocrine gland A gland that contributes a part of its cytoplasm to its secretion.

apoenzyme The protein portion of a conjugated enzyme.

apoferritin The protein component of ferritin.

apoinducer A protein that becomes bound to DNA and stimulates transcription by RNA polymerase.

apolar NONPOLAR.

apoplast The extracellular compartment of a plant consisting of all the cell walls, the xylem tubes, and the water contained in both. *See also* symplast.

apoprotein The protein component of a conjugated protein; used among others, to designate the lipid-free protein components of the various lipoprotein fractions.

apoprotein D CHOLESTEROL ESTER TRANSFER PROTEIN.

aporepressor A protein that is the product of a regulator gene and that, when combined with a corepressor, forms an active repressor which binds to the operator and inhibits transcription.

a posteriori Proved by induction; of, or pertaining to, (a) reasoning that derives principles from observed facts; (b) knowledge that cannot be acquired except by experience; (c) tests in which comparisons are unplanned and are made only after experimental results are obtained.

apparent competitive inhibition A competitive inhibition of enzyme activity in which the presence of the inhibitor on the enzyme affects the affinity of the enzyme for the substrate.

apparent equilibrium constant 1. An equilibrium constant based on molar concentrations rather than on activities. 2. An equilibrium constant calculated for a fixed pH for a reaction in which protons are produced; $K_{eq}/[H^+] \cdot Sym$ K'.

apparent specific volume The change in volume per gram of solute when a known weight of solute is added to a known volume of solvent.

approach to sedimentation equilibrium ARCHIBALD METHOD.

a priori Marked by deduction; of, or pertaining to, (a) consequences from definitions formed or principles assumed; (b) knowledge that can be acquired by reason alone; (c) tests in which comparisons are made on theoretical grounds before experimental results are obtained.

A protein 1. A protein (attachment protein) of single-stranded RNA phages. *Aka* maturation protein. 2. An experimentally produced oligomer of protein coat subunits of tobacco mosaic virus. 3. A protein subunit of the enzyme tryptophan synthetase. 4. A component of the enzyme lactose synthetase.

aprotic solvent A nonaqueous solvent that acts neither as a proton acceptor nor as a proton donor with respect to the solute; a nonhydroxylic solvent such as benzene, ether, and acetone.

APS Adenosine-5′-phosphosulfate.

AP site An apurinic or apyrimidinic site; that is, a site in a polynucleotide strand from which a purine or a pyrimidine has been removed.

APT Alum-precipitated toxoid.

aptitude The physiological state of a lysogenic bacterium that enables it, upon induction, to produce infectious phage particles.

APUD theory Acronym for amino precursor uptake and decarboxylation theory. A theory according to which the ectopic production of small peptide hormones by certain tumors correlates with the presence of cells (APUD cells) that are derived from common ancestral cells. These APUD cells can be identified by a set of shared biochemical properties including the ability to take up and store certain amine precursors (such as DOPA and 5-hydroxytryptophan) and to decarboxylate them to bioactive amines (such as catecholamines and serotonin). These amines are related in an unknown way to the secretion of ectopic hormones.

apurinic acid A DNA molecule from which the purines have been removed by mild acid hydrolysis. *Abbr* APA.

apyrase The enzyme that catalyzes the hydrolysis of ATP to AMP and two molecules of orthophosphate.

apyrimidinic acid A DNA molecule from which the pyrimidines have been removed by treatment with hydrazine.

aq Aqueous.

aquametry The quantitative determination of water.

aquated ion AQUO-ION.

aquation The formation of aquo-ions.

aqueous Of, or pertaining to, water. *Abbr* aq.

aqueous humor The clear fluid that fills the anterior chamber of the eye.

aqueous micelle *See* micelle.

aqueous phase separator centrifugation An interface centrifugation in which particles are selectively transferred from a crude aqueous solution to a short isodensity column; used for the purification of viruses.

aqueous solution A solution having water as its principal solvent.

aquo-ion A complex ion containing one or more water molecules.

AR Analytical reagent; denotes a chemical reagent for which either the actual or the maximum permissible concentrations of impurities are known.

Ara Arabinose.

araBAD A cluster of three genes that code for arabinose degradation and that form part of the arabinose operon.

araban A high-molecular weight, branched polysaccharide of arabinose that occurs in the hemicellulose of plants.

arabinan A homopolysaccharide of arabinose.

arabinose An aldose having five carbon atoms.

araC Cytosine arabinoside.

arachidic acid A saturated fatty acid that contains 20 carbon atoms.

arachidonic acid A 20-carbon noncyclic, unsaturated fatty acid that is the precursor of leukotrienes, prostaglandins, and thromboxanes; its systematic name is 5,8,11,14-eicosatetraenoic acid.

arachidonic acid cascade The series of enzymatic reactions that leads from arachidonic acid to the prostaglandins, thromboxanes, prostacyclin, and leukotrienes.

Araldite Trademark for a plastic used in electron microscopy for the embedding of tissues.

arbovirus An enveloped virus containing single-stranded RNA. Arboviruses multiply in both vertebrates and arthropods, with the arthropods generally serving as vectors.

arc The line of antigen–antibody precipitate obtained in immunodiffusion and in immunoelectrophoresis.

archaebacteria A class of unusual bacteria that, phylogenetically, are neither prokaryotes nor eukaryotes. They have some characteristics of prokaryotes (such as absence of a nucleus and cell organelles), some characteristics of eukaryotes (such as initiation of protein synthesis with methionine and ribosome insensitivity to chloramphenicol), and some characteristics that are unique to them (such as composition of the cell wall and the types of membrane lipids). Accordingly, the archaebacteria are believed to represent a third primary kingdom such that three lines of descent lead from a common ancestor (progenote) to archaebacteria, prokaryotes, and eukaryotes, respectively. Archaebacteria include thermoacidophiles, extreme halophiles, and methanogens and may represent some of the earliest forms of living cells. *Var sp* archaeobacteria.

Archeozoic era One of the two subdivisions of the Precambrian era; the earliest period of geologic time and an era that is devoid of fossil remains. It extended over a period of about 600 million years and ended about 1.6 billion years ago.

Archibald method A centrifugal method for determining molecular weights and assessing size homogeneity of macromolecules that is generally performed in the analytical ultracentrifuge, using relatively low speeds of rotation. The method is based on applying sedimentation equilibrium criteria to both the meniscus and the bottom positions of the cell, and on measuring the curvature of the gradient curve at those positions.

arene An alkyl-substituted benzene.

Arg 1. Arginine. 2. Arginyl.

argentaffin cells Specialized cells that secrete gastrointestinal hormones and that have an affinity for silver stains.

argentation chromatography A chromatographic technique based on the rapid formation of loose complexes between silver ions in an adsorbent and the π electrons of double and triple bonds in the molecules being separated; the greater the extent of unsaturation in a molecule, the greater is the extent of complex formation, and hence the slower is the rate of chromatographic migration. The method is used particularly for the separation of lipids.

arginase The enzyme that catalyzes the hydrolysis of arginine to urea and ornithine in the urea cycle.

arginine An aliphatic, basic, and polar alpha

amino acid that contains the guanido group. *Abbr* Arg; R.

arginine cycle UREA CYCLE.

arginine dihydrolase pathway An energy-generating pathway in some microorganisms; involves the conversion of arginine to ornithine and carbamoyl phosphate and use of the latter for the synthesis of ATP.

argininemia HYPERARGININEMIA.

arginine-rich histone An older term for histones H3 and H4.

arginine vasopressin A vasopressin molecule in which the eighth amino acid residue has been replaced by an arginine residue. *Abbr* AVP.

arginine vasotocin A vasotocin molecule in which the eighth amino acid residue has been replaced by an arginine residue.

argininosuccinic aciduria A genetically inherited metabolic defect in humans that is associated with mental retardation and that is characterized by a high blood concentration and a large renal excretion of argininosuccinic acid; due to a deficiency of the enzyme argininosuccinate lyase (argininosuccinase). *Aka* argininosuccinic acidemia.

argininosuccinuria ARGININOSCUCCINIC ACIDURIA.

argon detector An ionization detector, employed in gas chromatography, in which argon is used to ionize the organic compounds being separated; useful for trace analysis of steroids, fatty acids, and related compounds of relatively high molecular weights.

arithmetic mean MEAN.

arithmetic growth LINEAR GROWTH.

arm A segment of the cloverleaf model of tRNA. It consists of a helical, hydrogen-bonded section (stem) and a section that is not hydrogen-bonded (loop). There are 4 or 5 arms per molecule: (a) amino acid arm; consists only of a stem, called the acceptor stem, to which the amino acid becomes covalently linked; (b) anticodon arm, which contains the anticodon; (c) dihydro U arm (DHU arm, D arm) which contains dihydrouracil; (d) TψC arm (pseudo-U arm, T arm) which contains pseudouridine (ψ); (e) variable arm (extra arm, S region) which consists only of a loop of variable length.

Aroclor Trademark for a mixture of polychlorinated biphenyls.

aromatase A microsomal enzyme system which catalyzes the conversion of testosterone to 17-β-estradiol by means of three successive hydroxylations. In this reaction, testosterone functions as a prohormone.

aromatic Of, or pertaining to, a carbocyclic organic compound that contains the benzene nucleus.

aromatic amino acid An amino acid that contains the benzene ring.

ARPES Angular resolved photoelectron spectroscopy.

Arrhenius activation energy *See* Arrhenius equation.

Arrhenius complex A catalyst–reactant complex for which the reversion to free catalyst plus reactant proceeds at a much faster rate than the conversion to free catalyst plus product. *See also* van't Hoff complex.

Arrhenius equation An equation relating the rate constant k of a reaction to the absolute temperature T; specifically $\ln k = \ln A - E/RT$, where R is the gas constant, E is the activation energy of the reaction, and A is a constant known as the frequency factor, preexponential factor, or Arrhenius factor.

Arrhenius factor *See* Arrhenius equation.

Arrhenius plot A plot of the logarithm of the rate constant of a reaction versus the reciprocal of the absolute temperature; used for determining the activation energy of the reaction. *See also* Arrhenius equation.

arrow poison One of a group of natural toxins used to coat arrows, spears, and blowpipe darts. Among the compounds used are ouabain, other cardiac glycosides, and curare alkaloids.

arsenic An element that is essential to humans and animals. Symbol, As; atomic number, 33; atomic weight, 74.9216; oxidation states, +3, +5; most abundant isotope, ^{75}As; a radioactive isotope, ^{76}As, half-life, 26.5 h, radiation emitted, beta particles.

arsenical An arsenic-containing compound; many have antimicrobial activity and some have been used as therapeutic agents.

arsenolysis The cleavage of a covalent bond of an acid derivative by reaction with arsenic acid H_3AsO_4 so that one of the products combines with the H and the other product combines with the H_2AsO_4 group of the arsenic acid.

artefact Variant spelling of artifact.

arterial Of, or pertaining to, arteries.

arteriole A small blood vessel; a small artery.

arteriosclerosis The hardening and calcification of the arteries.

artery A blood vessel that transports blood from the heart to the tissues.

arthropod-borne virus ARBOVIRUS.

Arthus reaction An allergic reaction characterized by skin inflammation in response to repeated subcutaneous injections of antigen; similar reactions can also be produced in tissues other than skin. The Arthus reaction may be active, passive, or reverse passive; *see* active, passive, and reverse passive anaphylaxis for a definition of these terms.

artifact Any structure or substance that is not representative of the in vivo state of the specimen or of the makeup of the original sample but which is, instead, a result of the isolation procedure, the handling, or other factors.

artificial induction The induction of a prophage by a change in the conditions of the bacterial culture such that the immunity substance is either inactivated or not synthesized.

artificial kidney HEMODIALYZER.

artificial nitrogen fixation A synthetic reaction, such as the Haber process, that converts atmospheric nitrogen to ammonia.

artificial pH gradient A pH gradient formed by the layering of two or more buffers of different pH values; used originally for isoelectric focusing, but, since it changes with time upon the application of an electric field, is useful only for experiments of short duration.

aryl group An organic radical derived from an aromatic compound by loss of a hydrogen atom.

ascending boundary The electrophoretic boundary that moves upward in one of the arms of a Tiselius electrophoresis cell. *See also* Tiselius apparatus.

ascending chromatography A chromatographic technique in which the mobile phase moves upward along the support.

Aschheim–Zondek test A test for pregnancy based on the injection of urine, voided during the early stages of pregnancy and containing human chorionic gonadotropin, into immature female mice or rats. A positive test is indicated by ripening and rupture of the ovarian follicles.

ascites The abnormal accumulation of serous fluid in the abdominal cavity.

ascitic Of, or pertaining to, ascites.

Ascoli test 1. RING TEST. 2. A precipitin test for anthrax antigens.

ascorbic acid VITAMIN C.

ascus (*pl* asci) A microscopic, sac-like, structure that encases the spores of some fungi.

-ase Combining form denoting an enzyme.

asepsis 1. The prevention of access of microorganisms causing disease, decay, or putrefaction to the site of a potential infection 2. A state of sterility.

aseptic Of, or pertaining to, asepsis.

A-site AMINOACYL SITE.

A-site–P-site model The model of translation according to which a ribosome possesses two binding sites; the aminoacyl-, or A-, site binds the incoming aminoacyl-tRNA and the peptidyl-, or P-, site binds the peptidyl-tRNA subsequent to the addition of each new amino acid to the growing polypeptide chain.

Asn 1. Asparagine. 2. Asparaginyl.

Asp 1. Aspartic acid. 2. Aspartyl.

asparagine An aliphatic, polar alpha amino acid that is the amide of aspartic acid. *Abbr* Asn; $AspNH_2$; N.

aspartame A dipeptide, L-aspartyl-L-phenylalanine methyl ester, used as an artificial sweetener; called Nutrasweet when it is an ingredient in a product and Equal when it is sold as a sugar substitute.

aspartate transcarbamoylase A regulatory enzyme that catalyzes the first step in the biosynthesis of the pyrimidines in which *N*-carbamoyl aspartic acid is formed from carbamoyl phosphate and aspartic acid. The enzyme consists of two classes of subunits referred to as catalytic and regulatory subunits. *Var sp* aspartate transcarbamylase; *Abbr* ATCase.

aspartic acid An aliphatic, acidic, and polar alpha amino acid. *Abbr* Asp; D.

aspartic semialdehyde A derivative of aspartic acid in which only one of the two carboxyl groups has been converted to an aldehyde group.

aspirin Acetylsalicylic acid, an analgesic.

AspNH₂ Asparagine.

asporogenous mutant A mutant that is unable to form spores. *Aka* asporogenic mutant.

assay A measurement of either the concentration or the activity of a substance. *See also* enzyme assay.

assembly *See* self-assembly.

assimilation The conversion of nutrients by an organism into intra- and extracellular compounds utilized by that organism.

assimilation time The average time required for the reduction of one molecule of carbon dioxide by a molecule of chlorophyll when a plant is exposed to bright light.

assimilation transfer The penetration of a protein, either partially or totally, into a membrane such that the protein remains lodged in, and becomes a functional component of, the membrane; may involve cotranslational or post-translational transfer.

assimilatory reduction The process in plants whereby sulfate and nitrate, after reduction, are assimilated into cellular organic compounds. *See also* dissimilatory reduction.

association colloid A surface-active agent that tends to aggregate and to form micelles in solution.

association constant 1. The equilibrium constant for the formation of a more complex compound from simpler components, as the association of a proton and an anion to form an acid. 2. The equilibrium constant for the formation of a complex containing one or more macromolecules, as the binding of an inhibitor to an enzyme, or the binding of an antigen to an antibody. *See also* affinity constant.

associative recognition A mechanism of T lymphocyte activation in which there is simultaneous binding to the cell of an antigen in association with another structure; the latter is normally a cell surface alloantigen coded for by the major histocompatibility complex. *See also* dual recognition.

Astrup method A method for determining the acid–base balance of blood by measuring the pH of blood, collected to avoid CO_2 loss, and then remeasuring the pH after equilibrating the blood with a rather high partial pressure of CO_2.

Asx The sum of aspartic acid and asparagine when the amide content is either unknown or unspecified.

asymmetric 1. Lacking symmetry; unsymmetric. 2. Descriptive of a molecule that is totally lacking in symmetry as contrasted with a disymmetric one. 3. Descriptive of an elongated macromolecule, the shape of which differs significantly from that of a sphere.

asymmetric carbon A carbon atom to which are attached four different substituents; a chiral carbon. *See also* chirality.

asymmetric center An asymmetric carbon atom or one of several identical asymmetric carbon atoms in a molecule.

asymmetric strand-transfer model A model for genetic recombination that is similar to the Holliday model and that is based on an initial asymmetric pairing of DNA strands.

asymmetric synthesis The synthesis of only one of two optical isomers; generally the case for enzymatic, but not for nonenzymatic, reactions.

asymmetric transcription The normal mode of in vivo RNA synthesis in which only one of the two complementary DNA strands of any double-helical segment is being transcribed.

asymmetric unit The smallest part of a structure which, when operated on by symmetry elements, will reproduce the complete structure. The asymmetric unit is equal to, or smaller than, the unit cell.

asymmetry effect The decrease in the electrophoretic mobility of a protein that is brought about by the cloud of counterions. The protein is believed to move along an irregular path and at every change in position, the counterion cloud is temporarily left behind, thereby setting up an opposing

electric field which leads to a decrease in the mobility of the protein. *See also* electrophoretic effect.

asymmetry factor SHAPE FACTOR.

asymmetry potential 1. The potential across two permselective membranes (M_1 and M_2) which are separated by a concentrated polyelectrolyte solution (B), and which have another, but identical, electrolyte solution (A) on the other side of each membrane (i.e., A-M_1-B-M_2-A). 2. The potential of a membrane electrode, such as the glass electrode, that arises from slight imperfections in the membrane.

asymmetry ratio The sum of the concentrations of adenine and thymine divided by the sum of the concentrations of cytosine and guanine for a given DNA; the concentrations are expressed in terms of mole percent.

asynchronous growth The growth of cells that are randomly distributed with respect to their stage in cell division.

asynchronous muscle A muscle that yields a number of contractions for every motor nerve impulse that it receives.

ATA Aurintricarboxylic acid.

atactic polymer A polymer in which the R groups of the monomer are randomly distributed on both sides of the plane that contains the main chain.

ATCase Aspartate transcarbamoylase.

AT content *See* GC content.

A + T/G + C ratio ASYMMETRY RATIO.

atherogenesis The development of atherosclerosis; the formation of atheromas.

atherogenic Having the capacity to initiate, or to increase, the process of atherogenesis.

atheroma 1. A lipid-containing deposit in arteries undergoing hardening. 2. ATHEROSCLEROSIS.

atheromatous Of, or pertaining to, atheromas.

atherosclerosis A disease of the arteries, characterized by a gradual accumulation of cholesterol, cholesterol esters, collagen, elastic fibers, and proteoglycans in the arterial wall. Cholesterol and cholesterol esters are major components of atherosclerotic lesions (plaques). An increased level of plasma cholesterol and an increase in the major cholesterol-carrying lipoprotein (LDL) are associated with an increased risk of atherosclerosis. A primary cause of atherosclerosis appears to be a deficiency of LDL membrane receptors. As a result, LDL particles are not removed efficiently from the blood and, therefore, have increased chance of invading the lining of the arteries and participating in plaque formation.

athymic mice NUDE MICE.

Atmungsferment CYTOCHROME OXIDASE.

atom The smallest part of an element that is capable of undergoing a chemical reaction and that is chemically indestructible and indivisible; a structural unit of matter that remains unchanged in chemical reactions except for the loss or gain of electrons.

atom beam scattering A technique for the study of surfaces in which a beam of low-energy noble gas atoms is used to probe the surface. The noble gas atoms, striking the surface, are diffracted. Measurements of the angles of diffraction provide information on the structure and periodicity of ordered surface layers.

atomic absorption spectrophotometry A sensitive analytical method for the spectrophotometric determination of elements; based on a measurement of the radiation absorbed by unexcited, nonionized, and ground-state atoms which are produced when compounds are dissociated into atoms by means of a flame. *Abbr* AAS.

atomic mass The mass of a neutral atom expressed in terms of atomic mass units.

atomic mass unit One-twelfth of the mass of the carbon isotope $^{12}_{6}C$, and equal to 1.661×10^{-24} g; prior to 1961 the atomic mass unit was defined as one-sixteenth of the mass of the oxygen isotope $^{16}_{8}O$. *Abbr* amu. *See also* atomic weight unit; dalton.

atomic number The number of protons in the nucleus of an atom which is also equal to the number of orbital electrons surrounding the nucleus of the neutral atom. *Sym* Z.

atomic orbital The orbital of an electron about the nucleus of an atom.

atomic radius The distance between the nucleus and the outermost electron shell of an atom. *See also* van der Waals radius.

atomic weight The average weight for the neutral atoms of an element, existing as a mixture of isotopes identical to that found in nature; expressed in atomic mass units.

atomic weight unit One-twelfth of the mass of the carbon isotope $^{12}_{6}C$, and equal to 1.661×10^{-24}g; identical to the atomic mass unit. Prior to 1961 the atomic weight unit was defined as one-sixteenth of the average mass of the oxygen isotopes weighted in the same ratio as they occur in nature. *Abbr* awu.

atomizer A spraying device for breaking up a solution into fine droplets.

atom percent excess A measure of the concentration of a stable isotope expressed in terms of the excess, in percent of atoms, of the isotope over its natural abundance.

atom smasher ACCELERATOR.

atopic Of, or pertaining to, atopy.

atopic reagin REAGIN.

atopy An immediate-type hypersensitivity,

such as asthma or hay fever, that is due to the production of reagins and that tends to occur as an inherited tendency.

ATP 1. Adenosine triphosphate. 2. Adenosine-5'-triphosphate.

ATP–ADP carrier The inner mitochondrial membrane system that couples the outward translocation of ATP (formed via oxidative phosphorylation) to the inward movement of ADP (formed by ATP hydrolysis in the cytosol); an electrogenic pump and active transport system that is inhibited by bongkrekic acid and atractyloside.

ATP-ADP cycle The sum of the reactions by which (a) ADP is converted to ATP by means of the energy derived from food in the course of catabolism, and (b) ATP is hydrolyzed to ADP with the release of energy which is used to drive the energy-requiring reactions of an organism.

AT pair An adenine–thymine base pair.

ATPase Adenosine triphosphatase.

ATPase-linked pump PRIMARY ACTIVE TRANSPORT.

ATP regenerating system An enzymatic system for the synthesis of ATP from ADP that is used in cell-free amino acid incorporation experiments for replenishing the ATP that is used up in the course of amino acid activation. The system consists either of phosphocreatine and creatine kinase or of phosphoenolpyruvate and pyruvate kinase.

ATP synthase COMPLEX V (F_1F_0-ATPase). *See also* H^+-ATPase.

atractyloside A toxic glycoside that inhibits the ATP–ADP carrier system of the inner mitochondrial membrane.

atractyloside barrier The block to adenine nucleotide transport across the inner mitochondrial membrane that is produced by atractyloside.

atrial natriuretic factor A peptide hormone, secreted by the heart, that lowers blood pressure and acts as a diuretic. *Abbr* ANF.

atrophy The reduction in size of a tissue or an organ as a result of nutritional deficiencies and/or decreased functional activity.

atropine A tropane alkaloid that inhibits muscarinic receptors and acts like decamethonium.

attachment site *See* prophage attachment site.

attenuate 1. To decrease the virulence of a bacterium or a virus. 2. To decrease the intensity of a beam of radiation by passage of the beam through matter. 3. To terminate transcription prematurely.

attenuation A proposed mechanism of control in some bacterial operons which results in premature termination of transcription and which is based on the fact that, in bacteria, transcription and translation can and do proceed simultaneously. Attenuation involves a provisional stop signal (attenuator), located in the DNA segment that corresponds to the leader sequence of mRNA. During attenuation, the ribosome becomes stalled in the attenuator region in the mRNA leader. Depending on the metabolic conditions, the attenuator either stops transcription at that point or allows read-through to the structural gene part of the mRNA and synthesis of the appropriate protein.

attenuator A nucleotide sequence in DNA that can lead to premature termination of transcription. *Aka* attenuator region. *See also* attenuation.

atto- Combining form meaning 10^{-18} and used with metric units of measurement. *Sym* a.

att site ATTACHMENT SITE.

A-type particles One of two types of particles (A and B) seen in electron microscope preparations of mouse mammary tumors. The A particles are intracellular particles from which viral particles (B particles) develop. *See also* B-type particles.

atypical insulin INSULIN-LIKE ACTIVITY.

AU Angstrom unit.

Auger effect The process whereby an orbital electron passes from an excited to a lower energy level and thereby produces an x ray that collides with, and ejects, an orbital electron (Auger electron); subsequent to the ejection of the electron, the x ray escapes from the sphere of influence of the atom.

Auger electron *See* Auger effect.

aureomycin *See* tetracyclines.

AU-rich DNA DNA-LIKE RNA.

aurintricarboxylic acid An inhibitor of the initiation of protein synthesis in both prokaryotic and eukaryotic systems; it inhibits the binding of mRNA to the small ribosomal subunit. *Abbr* ATA.

aurosome A gold-containing, electron-dense lysosomal organelle; it is artificially induced in cultured animal cells that have been administered gold complexes.

Australia antigen An antigen in the serum of individuals suffering from type B viral hepatitis; the antigen shows up as a virus-like particle in electron micrographs and was originally detected in the serum of an Australian aborigine. It constitutes part of the Dane particle which is believed to be the virus responsible for type B hepatitis. *See also* hepatitis.

autacoids A group of substances that occur naturally in the body, resemble hormones or drugs in their activity on cells, and act on restricted areas within the body. Histamine, serotonin, angiotensin, and the prostaglandins

are some examples. Hormones and chalones (2) are sometimes referred to as excitatory and inhibitory (restraining) autacoids, respectively. *Aka* local hormones.

autarky Self-sufficiency as opposed to parasitic existence.

autoallergic disease AUTOIMMUNE DISEASE.

autoallergy Allergy to autoantigens.

autoanalyzer An instrument for the automated analysis of elements and compounds. The instrument consists of automated devices that replace such manual steps as the pipetting of reagents, the preparation of protein-free filtrates, and the heating of solutions for the development of colors.

autoantibody An antibody that is formed in an individual in response to antigens of the same individual.

autoantigen An antigen that is a normal constituent of an individual and that has the capacity of producing an immune response against itself in the same individual.

autocatalysis The phenomenon in which the product of a reaction serves as a catalyst for the reaction that forms the product so that the velocity of the reaction keeps increasing with time.

autocatalytic Of, or pertaining to, autocatalysis.

autocatalytic induction A self-accelerating enzyme induction as that occurring when (a) an inducer induces both an enzyme and a transport system that actively transports the inducer, or (b) an enzyme is induced by the product of the reaction that it catalyzes.

autochthonous Originating or formed in the place where found; said of a disease or a tumor that originated in that part of the body in which it is found.

autocide A substance produced by bacteria, that is bactericidal to the producing and closely related strains, but that has no activity against other bacteria.

autoclave An instrument for sterilizing materials in an airtight chamber by the use of steam at high pressure.

autocoid Variant spelling of autacoid.

autocoupling hapten A reactive low molecular weight compound, such as a diazonium salt or an acid anhydride, that, when injected into an animal, will combine with tissue antigens to form hapten–antigen compounds which then lead to the formation of antibodies.

autocrine hormone A substance, such as interleukin 2, that acts on the same cells that released it.

autocrine hypothesis A theory of carcinogenesis according to which cell transformation is linked to growth factors. Most normal cells are considered to be controlled by growth factors produced by other cells. A cancerous (transformed) cell, on the other hand, is considered to be one that somehow develops the ability both to make growth factors and to respond to them (have receptors for them).

autocytolysis AUTOLYSIS.

autofluorescence The fluorescence of tissues that is due to molecules naturally present in the tissues and that is unrelated to the treatment of the tissues with fluorochromes. *See also* intrinsic fluorescence.

autogenous AUTOLOGOUS (1).

autogenous control A genetic regulatory mechanism in which a gene is regulated by its own product. In positive autogenous control, the product enhances the activity of the gene, and in negative autogenous control, the product inhibits the activity of the gene. An example of autogenous control would be the coding of the repressor protein in enzyme induction by a gene that is part of the operon itself; thus, the repressor regulates the rate of transcription of its own mRNA. *Aka* autogenous regulation; autoregulation.

autogenous vaccine A vaccine made from the microorganisms that have infected an individual and then used to reinoculate the same individual.

autograft A transplant from one site to another in the same individual.

autoimmune complement fixation A complement fixation test in which both the antigens and the antibodies are derived from the same individual. *Abbr* AICF.

autoimmune disease A disease produced by an autoimmune response.

autoimmune response 1. The formation of antibodies to autoantigens. 2. The allergic response produced by the injection of auto-antibodies.

autoimmunity The immune reactions in an individual in which both the antigens and the antibodies are derived from that individual.

autoinduction The induction of a drug-metabolizing enzyme by the chronic administration of that drug.

autointerference The decrease in viral multiplication in animal cells that are infected with a large dose of virus particles as compared to the extent of viral multiplication in cells that are infected with a small dose. The effect is due to the appreciable concentration of interferon that is produced when the cells are infected with the larger viral dose.

autointoxication theory The theory that psychoses and other mental diseases are caused by the endogenous production of toxins.

autologous 1. Derived from the same organism. 2. Designating a transplant from one site to another in the same individual.

autolysate The suspension of broken cells obtained upon autolysis. *Var sp* autolyzate.

autolysin An autolytic enzyme; an endogenous enzyme that functions in autolysis.

autolysis The self-destruction of a cell as a result of the action of its own hydrolytic enzymes.

autolysosome An organelle formed by fusion of an autophagosome with teleolysosomes.

autolytic Of, or pertaining to, autolysis.

automatic buret A buret that is connected to a reservoir of liquid and that is automatically refilled after delivery of a given volume of liquid; used for the repeated delivery of a fixed volume of liquid.

automutagen A compound that is produced by the metabolic reactions of an organism and that is mutagenic for the same organism.

autonomous Existing and functioning independently; said of a tumor cell that is free of host control, or of an episome that replicates independently of the chromosome.

autonomous controlling element A controlling element that appears to have functions corresponding to both receptor elements and regulator elements. *See also* controlling element.

autooxidation Variant spelling of autoxidation.

autophagic Self-digesting.

autophagic vacuole A secondary lysosome that contains intracellular membranes or organelles, such as mitochondria or secretory vesicles.

autophagolysosome A subcellular organelle formed by fusion of an autophagic vacuole with a primary lysosome.

autophagosome A vacuole into which cytoplasmic components of uncertain identity have been sequestered; a phagosome formed by autophagy. Such organelles have been observed in rat liver cells.

autophenic Descriptive of a phenotype that is due to the mutant makeup of the cells showing the phenotype.

autoprothrombin I PROCONVERTIN.

autoprothrombin II CHRISTMAS FACTOR.

autoprothrombin III STUART FACTOR.

autoradiogram The photographic record of a chromatogram that contains radioactively labeled compounds; prepared by exposing a sensitive photographic film to the radioactive radiation by placing it in contact with the chromatogram.

autoradiograph The photographic record obtained in autoradiography.

autoradiographic efficiency The number of activated silver grains produced in a photographic emulsion, in contact with a labeled tissue section, per 100 radioactive disintegrations occurring in that tissue section during the exposure interval.

autoradiography A technique for studying the location of radioactive isotopes in macromolecules and in larger structures. The material to be studied is labeled with a radioactive isotope and is placed in contact, in the dark, with a photographic film or a photographic emulsion; the latent image produced by the radioactive radiation is subsequently developed.

autoretardation The phenomenon whereby the rate of an endonuclease-catalyzed hydrolysis of a nucleic acid decreases as the high molecular weight nucleic acid is broken down to smaller fragments.

autoregulation AUTOGENOUS CONTROL.

autosome Any chromosome that is not a sex (X or Y) chromosome.

autosynthetic cell A cell-like structure obtained by combining lipid and protein extracts of the brain.

autotroph A cell or an organism that uses carbon dioxide as its sole carbon source, and that synthesizes all of its carbon-containing biomolecules from carbon dioxide and other small inorganic molecules.

autoxidation The oxidation of a compound by air alone.

auxanographic method A method for identifying growth factors of microorganisms; involves inoculating the surface of a solid basal or minimal medium with the organism to be tested, followed by discrete application of various carbon sources, vitamins, or other substances.

auxesis The growth in size that results from an increase in cell volume without an increase in cell number, that is, in the absence of cell division.

auxiliary enzyme An enzyme that is added to a reaction mixture to assay a second enzyme that is not readily assayed directly. The auxiliary enzyme converts the product, formed by the second enzyme, quantitatively to a substance that can be determined by some analytical technique. The entire system is known as a coupled assay.

auxiliary pigment ACCESSORY PIGMENT.

auxiliary protein of DNA polymerase δ CYCLIN.

auxin One of a group of plant hormones, such as indoleacetic acid, that promote cell enlargement, chromosomal DNA synthesis, and growth along the longitudinal axis of a plant.

auxocarcinogen An auxiliary group of atoms in the molecule of a chemical carcinogen that influences the activity of the carcinogenophore.

auxochrome A group of atoms that, when attached to a molecule containing a chromophore, intensifies the color of the chromophore.

auxotroph A mutant microorganism that has a block in a metabolic pathway as a result of either the lack of an enzyme or the presence of a defective enzyme. Such mutants require for their growth either the product of the blocked enzymatic reaction or other metabolites that are not required by the wild-type organism.

avalanche *See* Townsend avalanche.

average MEAN.

average affinity AVERAGE INTRINSIC ASSOCIATION CONSTANT.

average binding number The average number of ligands bound per molecule of protein; equal to the number of moles of bound ligand divided by the number of moles of total protein. *Sym v.*

average deviation The average of the absolute deviations for a set of measurements regardless of the signs of the individual deviations.

average intrinsic association constant The value of the association constant for the binding of a given antigen by the corresponding antibodies that is determined for the case when one-half of all the antibody sites are occupied by the antigen. *Sym* K_0. *See also* heterogeneity index.

average life The average length of time that a radioactive atom exists before it disintegrates; the average life is equal to the reciprocal of the decay constant.

average molecular weight The value of the molecular weight that is determined for a sample consisting of a mixture of molecules. The type of average molecular weight obtained depends on the physical method used in studying the molecules. *See also* number average molecular weight; weight average molecular weight; Z-average molecular weight; viscosity average molecular weight.

average radius of gyration The value of the radius of gyration that is based on all the different conformations that a molecule may assume; specifically, $R_G = (\bar{R}^2)^{\frac{1}{2}}$, where R_G is the average radius of gyration and $\bar{R}^2$ is the average of the squares of all possible radii of gyration.

avian Of, or pertaining to, birds.

avian leukosis virus An RNA-containing virus that produces leukemia in chickens and that belongs to the group of leukoviruses.

avian myeloblastosis virus An oncogenic RNA virus.

avian polypeptide PANCREATIC POLYPEPTIDE.

avian sarcoma virus ROUS SARCOMA VIRUS.

avidin A protein in raw egg white that

combines tightly with the vitamin biotin; when fed to an animal, avidin can produce symptoms of biotin deficiency.

avidity 1. The tendency of an antibody to bind antigen; measured by the rate of the binding reaction. 2. AFFINITY (2).

avirulent Not virulent.

A virus *See* oncornavirus; hepatitis.

avitaminosis HYPOVITAMINOSIS.

Avogadro's number 1. The number of molecules in a gram-molecular weight of a compound; denoted by the symbol N and equal to 6.023×10^{23}. 2. The number of atoms in a gram-atomic weight of an element; denoted by the symbol N and equal to 6.023×10^{23}.

AVP 1. Antiviral protein. 2. Arginine vasopressin.

awu Atomic weight unit.

axenic Descriptive of the condition in which an organism of a given species grows in the complete absence of organisms from other species. The growth of a pure culture and of a germ-free animal are examples of axenic growth.

axenic animal GERM-FREE ANIMAL.

axerophthal VITAMIN A ALDEHYDE.

axerophthol VITAMIN A.

axial bond A bond in a molecule having a ring structure that is at right angles to the plane of the ring.

axial ratio The ratio, for an ellipsoid of revolution, of the axis of revolution to the axis perpendicular to it. The axial ratio is an indication of the overall asymmetry of a macromolecule, since a macromolecule in solution is approximated by an equivalent ellipsoid of revolution.

axial substituent A substituent attached to an axial bond.

axis of rotational symmetry An axis of symmetry such that rotation of a body about it will yield one or more structures that are identical to the structure before rotation. The axis is an n-fold axis and is denoted C_n if the identical structure is produced by a rotation of $360/n$ degrees, or $1/n$ of a turn.

axis of symmetry An imaginary axis through a symmetrical body.

axon The long process of a nerve cell that generally conducts impulses away from the nerve cell body.

axoneme A bundle of microtubules in cilia and flagella of unicellular eukaryotic organisms.

axoplasm The cytoplasm of an axon.

5-azacytidine A pyrimidine analogue that, when incorporated into newly synthesized DNA, can turn on certain genes and lead to the synthesis of the corresponding proteins.

8-azaguanine A purine analogue that retards the growth of some cancers.

azaserine An antibiotic, either produced by a species of *Streptomyces* or prepared synthetically, that inhibits purine biosynthesis and leads to chromosomal aberrations.

azathioprene An immunosuppressive drug.

6-azauracil A pyrimidine analogue; consists of uracil in which a carbon atom of the ring has been replaced by a nitrogen atom.

azeotrope A mixture of two or more liquids that has a constant boiling point and that is distilled without a change in composition. The boiling point of the mixture is below the boiling points of the component liquids.

azide The grouping $N{=}N{=}N:^-(N_3{}^-)$; azide is an inhibitor of the electron transport system where it prevents the reduction of the oxidized a_3 component of cytochrome oxidase. Sodium azide is an antimicrobial agent used as a preservative of laboratory reagents.

azide group The grouping $-N{=}\overset{+}{N}{=}\overset{-}{N}$.

aziridine mutagen A chemical mutagen that contains the aziridinyl group (shown below) and that functions as an alkylating agent.

$$-N\begin{matrix} CH_2 \\ | \\ CH_2 \end{matrix}$$

azo compound A compound containing the azo group.

azo dye A dye that contains the azo group.

azo-dye protein AZOPROTEIN.

azoFd Azoferredoxin; *see* nitrogenase.

azofer *See* nitrogenase.

azofermo *See* nitrogenase.

azoferredoxin *See* nitrogenase.

azo group The grouping $-N{=}N-$.

azoic Without life, particularly in reference to geologic periods antedating life on earth.

azoprotein A modified protein in which a tyrosine residue has been coupled to an aromatic diazo compound.

azotobacter A genus of nonsymbiotic, nitrogen-fixing bacteria that is found in soil and water.

azotoflavin A flavodoxin that functions as an electron carrier in nitrogen fixation in *Azotobacter vinelandii*.

azure B A metachromatic dye used in cytochemistry.

azurin 1. A low molecular weight, blue, copper-containing, bacterial protein, believed to function in respiration as an electron carrier in an electron transport chain. 2. A solution of copper sulfate and ammonium hydroxide used in agriculture as an antifungal agent.

B

b Subscript denoting the less active form of an interconvertible enzyme.

B 1. The sum of aspartic acid and asparagine when the amide content is either unknown or unspecified. 2. Boron. 3. 5-Bromouridine.

bacillus (*pl* bacilli). 1. A bacterium having a cylindrically or rod-shaped cell; bacilli represent one of the three major forms of bacteria. *See also* coccus; spirillum. 2. A bacterium belonging to the genus *Bacillus*.

bacitracin A cyclic peptide antibiotic produced by *Bacillus subtilis* and *Bacillus licheniformis*; it inhibits peptidoglycan biosynthesis.

backbone The chain structure of a polymer from which the side groups, and/or the side chains, project.

background 1. The counts of radioactivity registered by a radiation detector in the absence of a radioactive sample; such counts are caused by cosmic radiation, instrument noise, radioactive contamination, and other factors. 2. The appearance of a chromatogram, an electropherogram, or an electron micrograph in areas that are devoid of sample substances.

background constitutive synthesis The occasional transcription of genes in a repressed operon; brought about by the temporary dissociation of the repressor from the DNA which permits the binding of RNA polymerase to the promoter and the initiation of transcription. *Aka* basal synthesis; sneak synthesis.

back mutation REVERSION (1).

backscattering *See* backward scattering.

backside displacement S_N2 MECHANISM.

back titration The titration of the reagent left after an excess of the reagent has been added to the solution and has been allowed to react with it.

backward flow The flow of the solvent of a solution of macromolecules that occurs in a closed vessel and that is in a direction opposite to the direction of movement of the macromolecules.

backward flow interface centrifugation Interface centrifugation, used for cells, bacteria, and viruses, in which the interface is displaced according to the hydrodynamic volume of the particles.

backward scattering The scattering of radiation in the direction of the beam of radiation and toward the source of the radiation.

bacteremia The presence of viable bacteria in the blood.

bacteria *See* bacterium.

bacterial Of, or pertaining to, bacteria.

bacterial ferredoxin *See* iron–sulfur protein.

bacterial nucleus NUCLEOID.

bacterial toxin *See* endotoxin; exotoxin.

bacterial-type ferredoxin *See* iron–sulfur protein.

bacterial virus BACTERIOPHAGE.

bactericidal agent BACTERICIDE.

bactericide An agent that kills bacteria.

bacterifacture The production of desired materials by bacteria such as the production of specific proteins following the insertion of recombinant DNA into bacteria.

bacterin A vaccine consisting of dead pathogenic bacteria.

bacteriochlorin A specific porphyrin structure; 7,8,17,18-tetrahydroporphyrin.

bacteriochlorophyll The chlorophyll of photosynthetic bacteria; it differs from chlorophyll *a* of plants in having one of the pyrrole rings in a more reduced form and having a vinyl group replaced by an acetyl group.

bacteriocidal Variant spelling of bactericidal.

bacteriocin A protein produced by one bacterium that is toxic for another bacterium; a bacteriocin differs from an antibiotic in being a protein, having a narrower microbial spectrum, and generally being much more potent. The synthesis of bacteriocins is controlled by plasmids.

bacteriocin factor BACTERIOCINOGEN.

bacteriocinogen The plasmid that controls the formation of a bacteriocin.

bacteriology The science that deals with studies of bacteria.

bacteriolysin An antibody capable of leading to the dissolution of bacterial cells in the presence of complement.

bacteriolysis The lysis of bacterial cells.

bacteriolytic Of, or pertaining to, bacteriolysis.

bacteriophage A virus that infects bacteria and multiplies in them. *Aka* bacterial virus; phage. *See also* virus.

bacteriophage packaging The introduction of phage DNA into a bacterial cell for replication and encapsidation to form infective phage particles; used specifically for the introduction of phage lambda DNA into *E. coli* cells.

bacteriorhodopsin The sole protein component

of the purple membrane of *Halobacterium halobium*; so called because it contains a retinal prosthetic group much as the rhodopsin of animals. Bacteriorhodopsin is a large protein that spans the membrane seven times, with each segment being a helix, and that functions as a light-driven transmembrane proton pump. It is an inside-out protein, having polar amino acids in the interior of the helix and nonpolar ones on the outside. *See also* purple membrane.

bacteriostatic agent An agent that prevents the growth of bacteria without killing the cells.

bacteriotropic index OPSONIC INDEX.

bacteriotropin IMMUNE OPSONIN.

bacteristasis The prevention of bacterial growth without killing the bacterial cells.

bacteristat BACTERIOSTATIC AGENT.

bacteristatic Variant spelling of bacteriostatic.

bacterium (*pl* bacteria). A minute, unicellular prokaryotic organism that is classified as a lower protist; bacteria occur in soil, water, and air and as symbionts, parasites, or pathogens of humans and other animals, plants, and other microorganisms.

bacteroid A bacterium-like cell; used, for example, to describe morphologically differentiated, nonmotile, nonviable derivative cells of *Rhizobium* in root nodules of leguminous plants.

bactogen CHEMOSTAT.

bactoprenol A long-chain, lipid-soluble, membrane-bound polyisoprenyl alcohol that consist of 11 isoprene units. It functions as a carrier in the biosynthesis of some bacterial polymers such as peptidoglycan and lipopolysaccharides, presumably by facilitating transfer of sugar nucleotides across the cell membrane.

baker's yeast One of several strains of yeast belonging to the species *Saccharomyces cerevisiae* and capable of rapid fermentation in dough under low oxygen tension.

BAL British anti-Lewisite.

balanced fermentation Fermentation in which there is no net oxidation or reduction of a substrate. Involves splitting of the substrate into two fragments such that electrons can be transferred from one to the other with a sufficiently negative free energy change to drive ATP synthesis.

balanced growth The growth of cells such that all of the cellular components increase by the same factor and that the overall composition of the cells remains constant. A given increase in the concentration of protein, DNA, RNA, lipid, or some other macromolecule is coincident with the same increase in the mass and number of cells.

balance study A study of the overall metabolism of a substance in an organism that is based on a determination of the amount of the substance ingested and of the amount of the same substance, or of its metabolic products, excreted; such a study indicates whether there is a net gain or loss of the substance by the organism.

BALB/c mice An inbred strain of white mice that develop myeloma following intraperitoneal injection of mineral oil, complete Freund's adjuvant, or other substances.

Balbiani ring A loop-like structure in polytene chromosomes that is formed by extremely large chromosomal puffs.

bal 31 exonuclease An exonuclease from *Brevibacterium albidum* that catalyzes the successive removal of nucleotides from both the 3'- and 5'-ends of DNA strands.

ball and stick model A molecular model in which the atoms are represented by spheres and the bonds by sticks; the bond lengths and the atomic radii are fixed, and the bond angles are correctly indicated for each atom.

Balzer freeze fracture apparatus An apparatus used for preparing electron microscope specimens for freeze etching or freeze fracturing; permits sectioning of the frozen specimen in a vacuum.

band 1. A zone of macromolecules, such as the zone obtained in density gradient sedimentation, zone electrophoresis, isoelectric focusing, chromatography, or similar techniques. 2. ABSORPTION BAND.

band 3 ANION-TRANSPORT PROTEIN.

bandpass The range of wavelengths of a radiation that passes through a filter or a similar device.

band sedimentation A type of density gradient centrifugation in which a sample, in a solution of low density, is layered over a denser solution (in the absence of a preformed density gradient) and then subjected to centrifugation. Diffusion of solute from the denser to the lighter solution leads to establishment of a self-generated density gradient through which the sample sediments.

bandwidth 1. The width of an absorption band. 2. A range of wavelengths.

Bang method A method for determining glucose in urine by means of alkaline copper thiocyanate.

bangosome LIPOSOME.

BAP 1. 6-Benzylaminopurine. 2. Bacterial alkaline phosphatase.

barbed end *See* actin filament.

barbital Diethylbarbituric acid; a barbiturate.

barbiturates A large group of derivatives of barbituric acid (2,4,6-trioxypyrimidine; pyrimidinetrione) that are used as sleep-producing drugs. *See also* narcotic drugs.

bar chart A chart used to represent a frequency distribution or a time series;

consists of vertically placed rectangles that are of equal width and adjacent to each other on a common baseline. The heights of the rectangles are proportional to the frequencies (or values) they represent. *Aka* block diagram. *See also* histogram.

bare lipid membrane BLACK LIPID MEMBRANE.

Barfoed's test A colorimetric test for distinguishing mono- from disaccharides by means of a solution of cupric acetate in dilute acetic acid.

barium An element that is essential to a few species of organisms. Symbol, Ba; atomic number, 56; atomic weight, 137.34; oxidation state, +2; most abundant isotope, ^{138}Ba, a radioactive isotope, ^{133}Ba, half-life, 7.2 years, radiation emitted, gamma rays.

barn A unit area of the atomic nucleus equal to 10^{-24} cm^2 and used as a measure of the capture cross section of the nucleus.

baroreceptor A receptor in the central nervous system that responds to changes in blood pressure.

Barr body An inactive, condensed X chromosome in the nuclei of somatic cells of female mammals.

barrier layer cell PHOTOVOLTAIC CELL.

baryon Any hadron with half-integer spin; that is, any hadron made of exactly 3 quarks. The proton and neutron are baryons. *See also* elementary particles.

basal body KINETOSOME.

basal enzyme An inducible enzyme that is produced in small amounts in the absence of an inducer.

basal granule KINETOSOME.

basal lamina (*pl* basal laminae). A thin layer of specialized extracellular matrix that contains collagen IV and the glycoprotein laminin as major components; a subdivision of the basement membrane. Basal laminae underlie epithelial cell sheets and tubes and surround muscle cells, fat cells, and nerve cells. They function in regulating passage of macromolecules, tissue regeneration, and synapse construction. *Aka* basement lamina.

basal level The low level of concentration at which a basal enzyme is produced by a cell.

basal medium A medium that supports the growth of a range of nutritionally undemanding chemoorganotrophs.

basal metabolic rate The rate of basal metabolism under the following standardized conditions: physical rest, but not sleep; an ambient temperature that does not require energy expenditure for physiological temperature regulation; and a postabsorptive state following a 12-h fast. The basal metabolic rate may be determined from the energy value of the food intake and the

excreted waste products, or from the heat produced by the organism, or from the oxygen consumption by the organism. *Abbr* BMR.

basal metabolism The level of metabolic activity that is required by an animal for the maintenance of vital functions such as respiration, heart contraction, and kidney and liver function; the maintenance of nonvital functions such as muscular work and digestion is excluded. *See also* basal metabolic rate.

basal synthesis BACKGROUND CONSTITUTIVE SYNTHESIS.

base 1. Purine. 2. Pyrimidine. 3. Bronsted base. 4. Lewis base. 5. The fixed number, such as 10 or *e* used in logarithms.

base analogue A purine or pyrimidine that is similar in its chemical structure to the normal base occurring in nucleic acids and that may be incorporated into the nucleic acid in place of the normal base.

base composition The relative amounts of the various purines and pyrimidines in a nucleic acid; generally expressed in terms of mole percent.

base excision repair *See* excision repair.

base line The line in a chromatogram, an ultracentrifuge pattern, or a similar tracing that corresponds to the solvent rather than to the solution.

basement lamina BASAL LAMINA.

basement membrane The fragile, thin layer that underlies the epithelium and is devoid of cells; an extracellular matrix, composed of collagens, glycoproteins, and proteoglycans. The basement membrane is a composite structure, which includes the basal lamina, and forms an interface between epithelial cells and the underlying connective tissue. Basement membranes are involved in the regulation of cellular growth, cell adhesion, and cell differentiation.

base pair 1. A pair of hydrogen-bonded bases; a purine and a pyrimidine that either link two separate polynucleotide strands as in double-helical DNA or link parts of the same polynucleotide strand as in the cloverleaf model of transfer RNA. 2. A unit of length in nucleic acid molecules that is equal to one base pair. *Abbr* bp.

base pairing The complementary binding of the bases in a nucleic acid by means of hydrogen bonding. The binding may be between two strands of a double-stranded molecule or between parts of a single-stranded molecule folded back upon itself.

base-pairing rules The requirements that in a double-helical nucleic acid structure adenine must form a base pair with thymine (or uracil) and cytosine must form a base pair with guanine, and vice versa.

base-pair ratio　ASYMMETRY RATIO.

base-pair substitution　A transition (1) or a transversion. *Aka* base-pair switch.

basepiece　*See* supermolecule.

base ratio　One of three concentration ratios for the bases in a nucleic acid that are generally expressed in terms of mole percent: adenine/thymine (or uracil); guanine/cytosine; and purines/pyrimidines.

base sequence　The linear order of the purine and pyrimidine bases, or of their nucleotides, as they occur in a nucleic acid strand.

base specificity　The selectivity of a nuclease that accounts for its reaction being limited to specific purine or pyrimidine sites in the nucleic acid.

base stacking　The arrangement of the base pairs in parallel planes in the interior of a double-helical nucleic acid structure.

base substitution　The replacement of one base for another in either a nucleotide or a nucleic acid.

base unit　*See* SI.

basic　1. Of, or pertaining to, a base. 2. Of, or pertaining to, a solution having a pH greater than 7.0.

BASIC　Acronym for Beginner's All Purpose Symbolic Instruction Code; the most widely used high-level language for small computers.

basic amino acid　An amino acid that has two amino groups and one carboxyl group; an amino acid that has a net positive charge at neutral pH.

basic dye　A cationic dye that binds to, and stains, negatively charged macromolecules. *Aka* basic stain.

basic food　A food that is rich in sodium, potassium, magnesium, and calcium and that leaves a basic residue when subjected to combustion.

basicity constant　BASE DISSOCIATION CONSTANT.

basic number　The lowest haploid chromosome number in a polyploid series; the monoploid number.

basic orange　ACRIDINE ORANGE.

basic set　The smallest number of functional proteins and/or functional genes that could have constituted a primitive cell (protocell).

basophil　A cell that stains with a basic dye. *See also* polymorphonuclear leukocyte.

batch adsorption　A technique for adsorbing a solute from a solution by stirring the solution together with an adsorbent to form a slurry; subsequently, the solution is separated from the adsorbent by decantation, filtration, or centrifugation.

batch culture　A culture grown in a given volume of medium. *Aka* closed culture. *See also* continuous culture.

batch elution　*See* stepwise development.

bathochromic group　A group of atoms that, when attached to a compound, shifts the adsorption of light by the compound to longer wavelengths.

bathochromic shift　A shift in the absorption spectrum of a compound toward longer wavelengths.

bathorhodopsin　A structurally altered form of rhodopsin that is produced after the exposure of rhodopsin to light and prior to its conversion to lumirhodopsin. *Aka* pre-lumirhodopsin.

baud　A measure of data transmission speed by a computer; the number of bits of information that can pass a given point in one second (bits per second, bps).

bay region theory　A theory of chemical carcinogenesis involving a specific structural element (a diol epoxide) of certain poly-nuclear aromatic hydrocarbons.

BC　Biotin carboxylase.

BCCP　Biotin carboxyl carrier protein.

B cell　A bursa-derived lymphocyte. In birds, hemopoietic stem cells (bone marrow cells) migrate to the bursa of Fabricius where they differentiate into bursa lymphocytes which then migrate to peripheral lymphoid tissues to become B cells. Mammals lack a bursa of Fabricius and hemopoietic stem cells develop into lymphocytes in the hemopoietic tissues themselves and then migrate to peripheral lymphoid tissues to become B cells. B cells function in humoral immune responses that are due to circulating antibodies in the blood and antibodies secreted onto mucous surfaces.

BCG　Bacillus Calmette–Guérin; an attenuated strain of *Mycobacterium tuberculosis* used as a vaccine for protection against tuberculosis and leprosy.

B chain　1. The longer of the two polypeptide chains of insulin that contains 30 amino acids and that is linked to the other chain by two disulfide bonds. 2. The light (L) chain of the immunoglobulins.

BChl　Bacteriochlorophyll.

B complex　*See* vitamin B complex.

BCP　Biotin carboxyl carrier protein.

BD-cellulose　Benzoylated diethylaminoethyl-cellulose; a chromatographic support.

B DNA　*See* DNA forms.

Beckmann thermometer　A thermometer with a large bulb and a fine bore that is used for the measurement of small differences in temperature.

bed　The solid support of the column in column chromatography.

Beer–Lambert law　A composite of Beer's law

and Lambert's law. It relates the absorbance of a solution to its concentration and to the length of the light path through the solution; specifically, $A = \epsilon lc$, where A is the absorbance, l is the length of the light path, c is the concentration, and ϵ is the absorptivity, When l is in centimeters and c is in moles per liter, then ϵ is the molar absorptivity. *Aka* Beer's law.

Beer's law 1. BEER–LAMBERT LAW. 2. The law that forms part of the Beer–Lambert law. It states that the intensity of monochromatic light passing through an absorbing medium decreases exponentially with increasing concentrations of the absorbing material.

beet sugar Sucrose that is isolated from beets.

behenic acid A saturated fatty acid that contains 22 carbon atoms.

BEI Butanol-extractable iodine; the iodine that can be extracted from serum by butanol.

Belling's hypothesis A forerunner of the copy-choice hypothesis of the mechanism of genetic recombination. It assumes that genes are replicated first and then intergenic connections are made between the newly synthesized genes; there is no breakage and reunion of parental chromosomes.

bell-shaped curve A symmetrical curve, usually of a continuous frequency distribution, having the overall shape of a vertical section through a bell.

belt desmosome A cell junction that consists of a continuous band around the interacting epithelial cells. *See also* cell junction.

Bence–Jones protein An abnormal immunoglobulin that consists only of light chains, generally in the form of dimers, and that is produced by individuals suffering from plasma cell tumors; Bence–Jones protein has unusual thermosolubility properties, is relatively homogeneous, and is formed in large amounts in patients with multiple myeloma.

bending vibration DEFORMATION VIBRATION.

Benedict's reagent An aqueous solution of copper sulfate, sodium citrate, and sodium carbonate that is used in Benedict's test for reducing sugars.

benign neoplasm A harmless, localized, and nonmetastasizing tumor.

Benson model An older model of a biological membrane according to which the proteins are largely globular and are located in the interior of the membrane, and the lipids are intercalated into the folds of the protein chains, with the polar portions of the lipid molecules at the exterior surfaces of the membrane.

bentonite A clay that consists principally of montmorillonite (aluminum–magnesium–silicate) and that is used as an inhibitor of nucleases.

Benzedrine Trade name for amphetamine.

benzidine test A test for blood that is based on the formation of a blue compound, benzidine blue, upon treatment of the sample with glacial acetic acid and hydrogen peroxide.

benzo(a)pyrene A polycyclic hydrocarbon, produced in the incomplete combustion of fossil fuels. It is inactive by itself, but is converted to highly reactive derivatives, which are mutagenic and carcinogenic, by enzymatic action within an organism.

6-benzylaminopurine A synthetic cytokinin.

benzylpenicillin *See* penicillin.

β-benzyme *See* beta benzyme.

beriberi The disease caused by a deficiency of the vitamin thiamine. Dry beriberi is characterized by rapid weight loss, muscle wasting, and muscular weakness; wet beriberi is characterized, additionally, by generalized edema.

Berkefeld filter A filter, made of diatomaceous earth, that retains bacteria and that is used for the sterilization of solutions.

Bernoulli distribution BINOMIAL DISTRIBUTION.

Berthelot reaction A colorimetric test for ammonia in urine that is based on the production of blue indophenol upon treatment of urine with phenol and sodium hypochlorite.

BES N,N-Bis(2-hydroxyethyl)-2-aminoethanesulfonic acid; used for the preparation of biological buffers in the pH range of 6.4 to 7.8. *See also* biological buffers.

best fit *See* goodness of fit; method of least squares.

beta 1. Denoting the second carbon atom next to the carbon atom that carries the principal functional group of the molecule. 2. Denoting a specific configuration of the substituents at the anomeric carbon in the ring structure of carbohydrates. 3. Denoting buffer value. *Sym* β.

beta adrenergic blocker *See* beta blocker.

beta adrenergic receptor *See* adrenergic receptor.

beta amylase *See* amylase.

beta barrel A cylindrical, barrel-shaped structure in proteins, formed by the twisting of polypeptide chains in the parallel or antiparallel pleated sheet configuration.

beta bend A tight fold in the polypeptide chain that sharply reverses the direction in which the chain runs; generally formed by having the carbonyl group of one amino acid residue form a hydrogen bond with the amino group of another amino acid residue, three positions farther along the polypeptide chain. Thus, the polypeptide chain is sharply folded back upon itself. Beta bends are largely responsible for giving globular proteins their spherical structure and frequently involve glycine and proline as part of the complement of the four consecutive amino acids involved in the folding. *Aka* beta turn; hairpin turn; reverse turn.

beta benzyme   An artificial, chymotrypsin-like molecule that contains no amino acids and that has a molecular weight of 1365; the catalytic site of this synthetic molecule is derived from β-cyclodextrin.

beta blocker   An antagonist (inhibitor) of β-adrenergic receptors. Agents that block these receptors have various effects, among them a decrease of the rate and force of heart contractions. *Aka* β-adrenergic blocker.

beta carotene   A carotene that is a precursor of vitamin A and that is cleaved in animals to yield two molecules of vitamin A per molecule of beta carotene. *See also* anti-promoter.

beta chain   One of the two types of polypeptide chains occurring in adult hemoglobin.

beta conformation   The structure of a polypeptide chain that is almost fully extended into a zigzag, rather than a helical conformation; the structure of a polypeptide chain as it occurs in pleated sheets.

beta decay   The radioactive disintegration of an atomic nucleus that results in the emission of a beta particle.

beta emitter   A radioactive nuclide that decays by emission of a beta particle. A beta emitter is considered to be soft or hard depending on whether the emitted beta particles are of low energy and have a short penetration range, or whether they are of high energy and have a long penetration range.

beta error   TYPE II ERROR.

beta fraction   1. LOW-DENSITY LIPOPROTEIN. 2. VERY LOW-DENSITY LIPOPROTEIN.

beta function   *See* Scheraga–Mandelkern equation.

beta galactosidase   An enzyme that catalyzes the hydrolysis of β-galactosides including lactose. It is an inducible enzyme in *E. coli*, coded for by a gene in the lac operon.

beta galactoside permease   One of the three enzymes coded for by the lac operon in *E. coli*; it controls the rate of entry of β-galactosides from the medium into the cells.

beta glucosidase   *See* amygdalin.

beta glucuronidase   *See* laetrile.

betaine   1. *N,N,N*-Trimethyl glycine; a compound that serves as a methyl group donor and occurs in plant and animal tissues. 2. Any dipolar compound in which the positive and negative charges are not adjacent.

beta keratin   The extended form of keratin, obtainable by stretching alpha keratin, in which the polypeptide chains are in the parallel pleated sheet configuration.

beta lactam antibiotics   Antibiotics in which the key structural feature is the presence of a beta lactam, a four-membered ring in which a carbonyl group and a nitrogen are joined in an amide linkage. Includes the penicillins and cephalosporins. Beta lactam antibiotics inhibit enzymes that are important for bacterial cell wall synthesis and thus are effective against actively growing bacteria. They have remarkably few side effects.

beta lactamase   PENICILLINASE.

beta lactoglobulin   The major whey protein in cow's milk; the first pure protein for which the complete amino acid composition was determined (1947).

beta lipoprotein   LOW-DENSITY LIPOPROTEIN.

beta meander   A conformation of antiparallel beta sheets in which the beta conformation segments are connected by relatively tight beta bends and all connections are equivalent.

beta method   RAMON METHOD.

beta orientation   The orientation of atoms or groups that are attached above the plane of the steroid molecule.

beta oxidation   The oxidation of fatty acids in metabolism through successive cycles of reactions, with each operation of the cycle leading to a shortening of the fatty acid by a two-carbon fragment that is removed in the form of acetyl coenzyme A.

beta particle   1. An electron originating in the atomic nucleus and emitted frequently by radioactive isotopes. Beta particles are considered to be soft or hard depending on whether they are of low energy and have a short penetration range, or whether they are of high energy and have a long penetration range. 2. A glycogen granule in the liver.

beta plateau   The high-potential portion of the characteristic curve of a proportional radiation detector at which the count rate is almost independent of the applied voltage and at which the potential is of sufficient magnitude to detect beta particles.

beta ray   A beam of beta particles.

beta-ray spectrometer   An instrument for the analysis of either the energy spectrum or the momentum spectrum of beta rays.

beta receptor   *See* adrenergic receptor.

beta sheet   PLEATED SHEET.

beta structure   *See* pleated sheet.

beta threshold   The lowest potential at which beta particles can be detected with a proportional radiation detector.

betatron   An accelerator, designed to impart high kinetic energy to electrons by means of electromagnetic induction.

beta turn   BETA BEND.

BeV   One billion (10^9) electronvolts.

bevatron   An accelerator designed to impart high kinetic energy to protons.

B form   *See* DNA forms.

BGG   Bovine gamma globulin.

bi-   1. Combining form meaning two or twice. 2. Referring to two kinetically important substrates and/or products of an enzymatic reaction; thus a uni bi reaction is a reaction with one substrate and two products.

Bial's reaction   ORCINOL REACTION.

biamperometric titration   An amperometric titration using two like electrodes.

biantennary   *See* complex oligosaccharides; high-mannose oligosaccharides.

bias   DETERMINATE ERROR.

bicine   *N,N*-Bis(hydroxyethyl)glycine; used for the preparation of biological buffers in the pH range of 7.6–9.0. *See also* biological buffers.

bidentate   Designating a ligand that is chelated to a metal ion by means of two donor atoms.

bidirectional replication   DNA replication proceeding in two directions; two replicating forks moving in opposite directions as in formation of a theta structure.

bifluorescence   The variation in the apparent intensity of plane-polarized fluorescence by the rotation of an analyzer through which the fluorescence is observed.

bifunctional antibody   An antibody that has two combining sites for antigen (two antigen binding sites); a divalent antibody.

bifunctional catalyst   A catalyst that can provide both an acidic and a basic catalytic function.

bifunctional feedback   A feedback mechanism that affords control in two directions so that the input of a system is affected both by an increase and by a decrease of the output. An example is a system in which the pH will be adjusted if the pH either rises above or falls below the normal value.

bifunctional reagent   A compound that has two reactive functional groups and that can interact either with two groups of one protein or with one group each of two different proteins.

big ACTH   One of several, high molecular weight forms of ACTH that have been identified in human pituitaries, plasma, and tumors that secrete ACTH ectopically.

big gastrin   *See* gastrin.

big insulin   PROINSULIN.

big T   T ANTIGEN.

bilayer   A layer that is two molecules thick. *See also* lipid bilayer.

bilayer lipid membrane   BLACK LIPID MEMBRANE.

bilayer model   UNIT MEMBRANE HYPOTHESIS.

bile   The secretion of the liver that aids in the digestion of fats by emulsifying them and that serves to excrete bile pigments, heavy metals, and other waste products of metabolism. *See also* bile salt; digestive juice.

bile acid   A 24-carbon steroid that occurs in the bile in the form of a bile salt. Cholic acid and chenodeoxycholic acid are sometimes referred to as primary bile acids; they may be hydroxylated by bacteria in the digestive tract to yield the secondary bile acids, deoxycholic acid and lithocholic acid, respectively.

bileaflet membrane   LIPID BILAYER.

bile alcohol   One of a group of polyhydroxylated steroids, derived structurally from cholestane, that occur as sulfate esters in the bile of lower vertebrates; they function like the bile acids in higher organisms.

bile pigment   A degradation product of the heme portion of hemoglobin and other heme proteins; the linear tetrapyrroles, bilirubin and biliverdin, are two examples. *See also* plant bile pigments.

bile salt   A surface-active agent in the bile that aids in the emulsification of fats during digestion and that consists of a bile acid coupled to either glycine or taurine.

biliary   Of, or pertaining to, the bile.

bilin   A colored bile pigment, such as urobilin or stercobilin, that is formed by the oxidation of a colorless bilinogen pigment.

bilinogen   A colorless bile pigment, such as urobilinogen or stercobilinogen, that forms a colored bilin pigment upon oxidation.

biliprotein   *See* phycobiliprotein.

bilirubin   A red-brown bile pigment that may have an antioxidant function. Bilirubin is formed mainly from the hemoglobin of aged erythrocytes in the spleen, the bone marrow, and the liver; it is transported as a complex with serum albumin and is excreted in the bile as a conjugate of glucuronic acid. *See also* direct-acting bilirubin; indirect-acting bilirubin.

bilirubin diglucuronide   A conjugated, soluble form of bilirubin that is formed in the liver by the esterification of two molecules of glucuronic acid to two propionic acid residues of bilirubin. *See also* direct-acting bilirubin.

bilirubinemia   HYPERBILIRUBINEMIA.

biliverdin   A green bile pigment that is reduced by NADPH to bilirubin.

bimolecular lamellar lipid membrane   BLACK LIPID MEMBRANE.

bimolecular layer   BILAYER.

bimolecular leaflet   BILAYER.

bimolecular lipid membrane   BLACK LIPID MEMBRANE.

bimolecular reaction   A chemical reaction in which either two molecules (or other entities) of a single reactant, or one molecule each of two reactants, interact to form products.

binal symmetry   Symmetry in which there are two types of symmetry elements.

binary   Consisting of two parts.

binary complex mechanism PING-PONG MECHANISM.

binary digit BIT.

binary fission Asexual division in which a cell divides into two approximately equal parts.

binary number system A numbering system in which all numerical values are expressed in terms of only two digits, 0 and 1.

binder A substance, such as calcium sulfate, that is mixed with a thin-layer chromatographic adsorbent to increase the mechanical strength of the adsorbent layer.

binding assay 1. Any method for measuring protein–ligand interaction as in the binding of sugars to periplasmic proteins and in the binding of cyclic AMP to protein kinase. 2. A method for measuring the binding of aminoacyl-tRNA to ribosomes in response to oligoribonucleotides of defined sequences and in the absence of peptide bond formation; based on using labeled aminoacyl-tRNA, collecting the aminoacyl-tRNA–ribosome–oligoribonucleotide complex on Millipore filters, and counting the radioactivity in this complex. *Aka* ribosome binding technique.

binding constant ASSOCIATION CONSTANT.

binding factor ^ protein factor required for the binding of aminoacyl-tRNA to ribosomes.

binding number *See* average binding number.

binding protein One of a number of soluble proteins that specifically, and reversibly, bind a number of substances, including amino acids, sugars, inorganic ions, and vitamins. Binding proteins have been found in the periplasmic space of gram-negative bacteria and in some yeasts; they appear to be devoid of enzymatic activity and are believed to function in the transport of the bound substance.

binding protein transport system *See* binding protein.

binding site The structural part of a macromolecule that directly participates in its specific combination with a ligand. Binding sites are said to be interacting or independent depending on whether the binding of one ligand to one site does, or does not, affect the binding of other ligands to other sites on the same molecule. *See also* active site.

binomial coefficient The coefficient of any term that results from the binomial expansion of $(a + b)^n$.

binomial distribution A frequency distribution in which the frequencies have the values of binomial coefficients. It represents a distribution of the number of successes in n trials when the probability of a success remains constant from trial to trial and the trials are independent. *Aka* Bernoulli distribution.

binomial nomenclature A system for naming species of plants and animals in which the name consists of two parts, the first designating the genus, and the second designating the species.

bio- 1. Combining form meaning life. 2. Combining form meaning a biological or a biochemical system.

bioassay The measurement of either the activity or the amount of a substance that is based on the use of living cells or living organisms; measurements of infectivity, antibody formation, weight gain, and bacterial growth are examples.

bioautograph The record obtained when a bioassay is used in conjunction with a chromatographic procedure, as in the case where a paper chromatogram is placed in contact with a bacterial culture on a solid medium.

bioautography A modification of paper chromatography in which the growth of bacteria is used as an indicator for locating growth factors and antibiotics on a paper chromatogram. The method involves placing a paper chromatogram, containing growth factors or antibiotics, in contact with an agar surface, seeded with the indicator organisms.

bioblast MITOCHONDRION.

biochemical Of, or pertaining to, biochemistry.

biochemical coupling hypothesis *See* chemical coupling hypothesis.

biochemical deletion hypothesis *See* deletion hypothesis.

biochemical energetics The free energy relations of biochemical reactions.

biochemical engineering *See* biotechnology; recombinant DNA technology.

biochemical evolution The evolutionary processes concerned with the formation of biomolecules, cells, cellular structures, metabolic pathways, and other attributes of living cells. *See also* biological evolution; chemical evolution.

biochemical fossil *See* chemical fossil.

biochemical genetics MOLECULAR GENETICS.

biochemical inflexibility of tumors The phenomenon that the control mechanisms of many enzyme systems, such as those of enzyme induction, appear to be frozen in tumor cells.

biochemical lesion A biochemical alteration, such as the inactivation of an enzyme, that leads to a pathological condition; may be caused by a mutagen, a carcinogen, or other factors.

biochemically deficient mutant AUXOTROPH.

biochemical marker A mutable site on a chromosome that, when mutated, leads to a specific enzyme defect which can be detected by biochemical means.

biochemical mutant AUXOTROPH.

biochemical oxygen demand The rate at which the oxygen in water is consumed by microorganisms for the oxidation of organic compounds to simple inorganic molecules. *Abbr* BOD.

biochemical predestination theory The theory that the development of the living cell is determined by, and follows naturally from, the physical–chemical properties of the simplest starting compounds.

biochemical similarity principle The assumption that the biochemical compounds and processes known to occur ubiquitously in contemporary living systems and to be essential for life, also occurred at the early stages of the development of life and were essential to the origin of life.

biochemistry The science that deals with the chemistry of living systems and their components; it deals with such areas as chemical composition, metabolism, nutrition, energetics, enzyme function, transfer of genetic information, membrane properties, cellular organization, and molecular diseases of living organisms.

biochrome A naturally occurring coloring matter in a plant, an animal, or a microorganism; a pigment.

biochronometry The science that deals with the temporal organization and the time-keeping mechanisms of biological systems.

biocide A chemical substance used either to kill or to arrest the growth of living organisms, particularly microorganisms; bactericides, fungicides, and pesticides are examples.

biocytin ϵ-Biotinyllysine; a compound formed by linking biotin through its carboxyl group to the ϵ-amino group of a lysine molecule. Biotin is believed to be similarly linked to a lysine residue in those enzymes that require biotin as a coenzyme.

biod The dry biomass per unit area of the earth's surface.

biodegradable Descriptive of a substance that can be decomposed by the enzyme systems of bacteria or other organisms. Biodegradable detergents are those containing unbranched hydrocarbon chains; branched-chain compounds resist degradation.

bioelements Those elements that are required by, and occur in, living organisms; this includes about 40 elements of which six (C, O, H, N, S, and P) account for about 90% of the mass of living matter.

bioenergetics BIOCHEMICAL ENERGETICS.

biofeedback A process whereby an individual is trained to gain some control over autonomic (involuntary) body functions such as blood pressure, muscle tension, and brain wave activity. The individual is provided with visual or auditory information about these processes and learns to consciously control them. The technique is based on the principle that a desired response is learned when received information (feedback) indicates that a specific thought complex or action has produced the desired response. An example of biofeedback would be the amplification of the electronic output of a muscle in response to a visual or other signal and the use of that signal to increase the output of the muscle as in physical therapy.

bioflavonoids A group of compounds, originally designated as vitamin P (permeability vitamin) or vitamin C_2 (synergist of vitamin C). Some of the flavonoids exhibit biological activities, including reduction of capillary fragility (vitamin P activity) and protection of biologically important compounds through antioxidant activity (vitamin C_2 activity), but none has been shown to be essential for humans or to cause deficiency syndromes. Hence, they are now considered to be pharmacological rather than nutritional agents. Bioflavonoids constitute a very large group of colored, phenolic pigments that are found in all higher plants and that are related to the parent compound flavone; they are responsible for most of the red, pink, and purple colors of higher plants. Some of the subgroups of the bioflavonoids include flavones, flavanones, flavonols, and anthocyanins. *See also* antipromoter.

Biogel Trademark for a group of polyacrylamide and agarose gels used in gel filtration.

biogenesis 1. The synthesis of a substance in a living organism; biosynthesis. 2. The doctrine that living things can some only from other living things.

biogenetic Of, or pertaining to, biogenesis.

biogenetic law RECAPITULATION THEORY.

biogenic 1. Produced by the action of living organisms. 2. Essential to life.

biogenic amine An amine that is produced by a living organism, particularly a physiologically important amine such as epinephrine, norepinephrine, serotonin, or histamine; an amino acid decarboxylation product or its derivatives.

biogenic amine hypothesis The hypothesis that depression results from depletion of neurotransmitter amines in the areas of the brain involved in sleep, arousal, appetite, sex drive, and psychomotor activity.

biogeochemistry The science that deals with the interaction of living organisms with the mineral environment of the earth's crust.

biolith A geological sediment, such as peat or humus, that consists of residues from organic matter.

biological assay BIOASSAY.

biological buffers A group of buffers, such as Good's buffers, that are especially suited for research of biochemical systems in the physiological pH range; Tris, HEPPS, and MES are three examples. A biological buffer should, ideally, meet all of the following specifications: (a) have a pK_a between 6 and 8; (b) be very water soluble and not move across a biological membrane; (c) be available in a high degree of purity; (d) have a pH that is minimally influenced by the concentration, temperature, ionic composition, and salt effects of the medium; (e) be nontoxic and noninhibitory in biological systems; (f) be stable to enzymatic and nonenzymatic hydrolysis; (g) form only soluble complexes with cations; (h) be devoid of light-absorbing capacity in the visible and ultraviolet regions. *See also* specific buffers.

biological chemistry BIOCHEMISTRY.

biological clock The periodicity of either a biological function or a biochemical reaction.

biological evolution The gradual development of living organisms to their present state; a process that followed the stage of chemical evolution and extended over a period of about 3 billion years. *See also* Oparin hypothesis.

biological half-life *See* half-life (2).

biological nitrogen fixation The conversion of atmospheric nitrogen to ammonia by living organisms.

biological oxidation–reduction 1. The oxidation–reduction reactions in aerobic cellular respiration whereby nutrients are oxidized in the citric acid cycle, electrons are transported by the electron transport system, and ATP is synthesized through oxidative phosphorylation. 2. Any oxidation–reduction reaction occurring in a living system.

biological oxygen demand BIOCHEMICAL OXYGEN DEMAND.

biological response modifier An agent that stimulates the body's own defenses to fight off disease; cytokines, such as interferon, interleukin, and tumor necrosis factor, are examples. *Abbr* BRM.

biological rhythm BIOLOGICAL CLOCK.

biological value The relative nutritional value of a protein that is based on the amino acid composition of the protein, the digestibility of the protein, and the availability of the digested products. The biological value has been defined in terms of (a) the fraction of absorbed protein nitrogen that is retained in the body; (b) the growth rate of young animals as a function of the dietary level of the protein; (c) the minimal protein concentration required to establish nitrogen balance in adults; and (d) the change in the concentration of essential amino acids in the serum as a function of the dietary level of the protein. *See also* net protein utilization.

biology The science that deals with living things; includes botany, zoology, embryology, genetics, morphology, and allied sciences.

bioluminescence The production of visible light by a living organism. *See also* luciferase.

biolysis Lysis by biological means.

biomacromolecule BIOPOLYMER.

biomass The mass of an organism or a group of organisms; variously used in reference to wet mass, dry mass, or mass per unit area. *See also* biod.

biomembrane A biological membrane. *See* fluid mosaic model.

biometry The science that deals with the application of statistics to biological systems.

biomolecule A molecule, such as a protein or a lipid, that occurs in a living system.

biomonomer A monomer, such as an amino acid or a nucleotide, that occurs in a living system.

bionics The application of structural and functional principles, discovered in living organisms, to man-made objects, especially those involving electronics; the development of robots, the construction of prosthetic devices, and the miniaturization of components are some examples.

bionomics ECOLOGY.

bioorthogonal code An early version of the genetic code based on 24 codons, each containing six nucleotides, such that each codon could undergo two base substitutions and still be recognized as being related to its original form.

biophysics The science that deals with the physics of living systems and their components.

biopolymer A polymer, such as a protein or a nucleic acid, that occurs in a living system.

biopsy The removal and microscopic examination of tissue or other material from the body for purposes of diagnosis.

biopterin A derivative of pteridine, the reduced form of which, dihydrobiopterin, serves as a coenzyme for hydroxylation reactions of amino acids.

bioregulation The ability of a living organism to adjust its processes in response to external and internal influences.

biorhythm BIOLOGICAL CLOCK.

bios A term previously used to denote a

growth-promoting substance for yeast; bios I referred to inositol and bios II (or II B) referred to biotin.

biosis LIFE.

biosphere The regions of and around the earth that support life; includes the oceans, the upper portion of the land masses, and the lower portion of the atmosphere.

biosterol VITAMIN A.

biosynthesis The formation of a substance in a living system.

biosynthetic Of, or pertaining to, biosynthesis.

biosynthetic pathway An anabolic pathway that leads to the synthesis of a biomolecule.

biotechnology The various industrial processes that involve the use of biological systems; the collection of microbial and other biochemical processes carried out on an industrial scale. It includes, but is not limited to, the industrial aspects of genetic engineering. Other areas of biotechnology deal with fermentation technology (antibiotics), hybridoma technology (monoclonal antibodies), and agricultural technology (plant and animal transformations). *See also* genetic engineering; recombinant DNA technology.

biotest BIOASSAY.

biotic Of, or pertaining to, life.

biotin A vitamin of the vitamin B complex that functions as a coenzyme in carboxylation–decarboxylation reactions.

biotin carboxylase ACETYL-COA CARBOXYLASE.

biotin carboxyl carrier protein *See* acetyl-CoA carboxylase.

biotinyllysine BIOCYTIN.

biotope 1. The environment of an organism or a group of organisms. 2. The location of a particular parasitic organism within the body. 3. The spatial distribution of the biomass in a cross section of a river, lake, and so on.

biotransformation The metabolic reactions in an organism whereby a foreign chemical, introduced into the organism, is either converted to a different compound or conjugated to a metabolite of the organism.

biotype SEROTYPE.

bireactant reaction A bimolecular reaction in which two different reactants interact to form products.

birefringence The phenomenon of a substance possessing two refractive indices depending on the direction along which light passes through the substance.

bis Prefix indicating two phosphate groups attached to a molecule at two different positions. Thus, fructose-1,6-bisphosphate is now preferred to fructose-1,6-diphosphate.

bispecific antibody HETEROSPECIFIC ANTIBODY.

1,3-bisphosphoglycerate A high-energy compound, the dephosphorylation of which to 3-phosphoglycerate leads to the synthesis of ATP from ADP in the second stage of glycolysis. *Aka* 1,3-diphosphoglycerate.

2,3-bisphosphoglycerate A compound that is present at high concentrations in the red blood cell and that binds to tetrameric hemoglobin to form a 1:1 complex. One molecule of 2,3-bisphosphoglycerate binds in the central cavity of the hemoglobin molecule and greatly reduces the affinity of hemoglobin for oxygen; the binding favors the dissociation of oxygen from oxyhemoglobin. The compound serves as an allosteric effector and regulates the oxygen-binding affinity of hemoglobin in relation to the partial pressure of oxygen in the lungs. *Abbr* DPG. *Aka* 2,3-diphosphoglycerate.

bisubstrate reaction An enzymatic reaction in which two substrates participate.

bit 1. A binary digit (0 or 1); a single character in a notation using two characters. 2. The smallest item of useful information that a computer can handle and that is expressed in terms of binary choices such as "yes" and "no". It is conveyed by an electrical impulse and is based on the fact that every microswitch in a computer can only have one of two positions, on (1) or off (0).

bitter acids A group of bitter-tasting, chemically labile compounds that occur in hops; they are monoacylphloroglucides.

bitter peptides A group of bitter-tasting peptides that sometimes are associated with the spoiling of foodstuffs; it is claimed that bitterness of a peptide correlates with its average hydrophobicity.

bitter principles A diverse group of bitter-tasting compounds, isolated from some plants, that are used as bitter spices to increase appetite and promote digestion.

Bittner's mouse milk factor MOUSE MAMMARY TUMOR VIRUS.

biuret A compound formed by the condensation of two molecules of urea.

biuret reaction A colorimetric reaction for the qualitative and quantitative determination of proteins; based on the production of a purple color upon treatment of biuret, peptides, proteins, or related compounds with copper sulfate in an alkaline solution.

Bjerrum formation function KLOTZ PLOT.

BJP Bence–Jones protein.

black lipid membrane An artificially prepared bimolecular lipid membrane that consists of either naturally occurring or synthetic lipids; usually formed in an annular space connecting two compartments filled with an aqueous medium that contains the lipids. The name black lipid membrane is due to the fact that, as the membrane forms, light reflected from it changes as a result of interference so that the membrane eventually turns grayish black. *Aka* black membrane.

blank A mixture of reagents that excludes the sample but that is identical to the mixture of reagents with which the sample is treated and that is carried through the same procedures as the sample. *See also* control.

blast cell A cell with a poorly differentiated, but RNA-rich, cytoplasm that actively synthesizes DNA.

blastoma A tumor consisting of immature and undifferentiated cells.

blast transformation TRANSFORMATION (3).

bleaching 1. Loss of chlorophyll as brought about, for example, by growth in the absence of light. 2. Change of a chromophore from a colored to a colorless form. 3. Whitening of material by oxidation–reduction reactions such as those involving hydrogen peroxide and chlorine.

blender A small appliance used to mix, homogenize, or disintegrate liquids and/or solids.

blender experiment HERSHEY–CHASE EXPERIMENT.

blind test A test of a substance or a procedure in which an independent observer records the results without knowing either the identity of the samples or the expected results. *See also* double-blind technique.

Blinks effect CHROMATIC TRANSIENT.

BLM 1. Bare lipid membrane. 2. Bilayer lipid membrane. 3. Bimolecular lamellar lipid membrane. 4. Bimolecular lipid membrane. 5. Black lipid membrane.

Block *See* metabolic block; genetic block.

block copolymer A copolymer in which different blocks of identical monomer units alternate with each other, as in the polymer —A—A—A—B—B—B—A—A—A—.

block diagram BAR CHART.

block electrophoresis An electrophoretic technique, used primarily for preparative procedures, in which the supporting medium is in the form of blocks. The material (starch, cellulose, Sephadex, etc.) is made up as a thick slurry, and is poured into a rectangular container. Excess liquid is then removed and samples are applied into slots, cut into the blocks.

blocking antibody 1. A protective antibody that prevents anaphylaxis by combining with the allergen; it is formed during desensitization and is primarily of the IgG type. 2. An incomplete antibody that actively inhibits agglutination.

blocking group PROTECTING GROUP.

block nucleotide A synthetic oligonucleotide that has the structure of a block copolymer. *Aka* block oligonucleotide.

blood The fluid that circulates through the cardiovascular system and that is composed of a fluid fraction, plasma, and a cellular fraction consisting of erythrocytes, leukocytes, and thrombocytes.

blood–brain barrier The slow penetration into the brain of substances that are transported by the blood as compared to their more rapid penetration into most other tissues. It serves to provide a nearly constant environment for brain function.

blood–cerebrospinal fluid barrier BLOOD–BRAIN BARRIER.

blood clot The insoluble network of polymerized fibrin molecules and trapped cells that is formed upon the solidification of blood, most commonly in the course of external bleeding.

blood clotting The two groups of complex biochemical reactions by which a blood clot is formed. The first group of reactions leads to the formation of the active enzyme thrombin from its inactive precursor, prothrombin; the second group of reactions leads to the thrombin-catalyzed conversion of fibrinogen to the fibrin clot. *Aka* blood coagulation. *See also* extrinsic pathway; intrinsic pathway.

blood count A determination of the number of red and white blood cells per cubic millimeter of blood.

blood group One of the classes of individuals that belong to a given blood group system.

blood group antigen BLOOD GROUP SUBSTANCE.

blood group chimera A chimera containing two different blood types.

blood grouping BLOOD TYPING.

blood group substance A genetically determined particulate isoantigen that is attached to the surface of red blood cells and that may be attached to the surface of other cells; related blood group substances represent alternative antigens, specified by allelic genes.

blood group system A classification of individuals into groups on the basis of their possession, or nonpossession, of specific blood group substances. Some of the blood group systems are known as the ABO, Lewis, M–N–S–s, and Rh systems.

blood plasma *See* plasma.

blood platelet *See* platelet.

blood serum *See* serum.

blood sugar 1. The glucose in the blood. 2. The trehalose in the hemolymph of insects.

blood typing The identification of the blood group substances of an individual.

blood urea nitrogen The nitrogen of the urea in serum. *Abbr* BUN.

blood vessel An artery, a vein, or a capillary through which the blood circulates.

Bloom's disease A genetically inherited metabolic defect in humans that is characterized by chromosomal aberrations and decreased immunity; it is due to defective DNA repair.

blot hybridization *See* northern blotting; Southern blotting.

blotting Any one of a number of techniques whereby chromatographically or electrophoretically separated DNA, RNA, or protein molecules can be transferred from the support medium, such as a gel, to another medium such as filter paper or a membrane matrix. The transfer can be achieved by capillary action (Southern blotting, northern blotting, western blotting) or by electrophoresis (electroblotting). *Aka* transfer. *See also* protein blotting.

blue copper protein *See* blue protein.

blue dextran A polymeric material with an average molecular weight of 2×10^6 that is prepared from dextran and that is used for the calibration of gel filtration columns from which it is excluded because of its size.

blue-green algae CYANOBACTERIA.

blue-green bacteria CYANOBACTERIA.

blue phase *See* liquid crystal.

blue protein 1. Azurin (1). 2. Plastocyanin.

blue shift HYPSOCHROMIC SHIFT.

blunt end ligation The joining of restriction fragments, terminating in blunt ends, by means of DNA ligase.

blunt ends *See* restriction enzyme.

B lymphocyte B CELL.

BMR Basal metabolic rate.

BND-cellulose Benzoylated–naphthoylated diethylaminoethyl cellulose; a chromatographic support. Also abbreviated BNC.

boat conformation The arrangement of atoms that resembles the outline of a boat and that is the less stable conformation of the two possible ones for cyclohexane and other six-membered ring systems.

tert–BOC-amino acid An amino acid in which the amino group has been protected by attachment of a tertiary butoxycarbonyl group; used in peptide synthesis by the solid phase (Merrifield) method. *Abbr* t-BOC-AA.

BOD Biochemical oxygen demand.

body A structural domain of prokaryotic 30S ribosomal subunits.

Bohr coefficient The change in the logarithm of the partial pressure of oxygen with change in pH at constant oxyhemoglobin concentration (usually at a constant oxygen saturation of 0.5); $\Delta \log P_{O_2}/\Delta pH$.

Bohr effect The decrease in the oxygen affinity of hemoglobin that is produced either by a decrease in the pH or by an increase in the partial pressure of carbon dioxide. The effect is due to changes in the pK values of ionizing groups in hemoglobin upon oxygenation and deoxygenation of the molecule. The effect above pH 6, when oxyhemoglobin is more negatively charged than deoxyhemoglobin, is known as the alkaline Bohr effect; the effect below pH 6, when deoxyhemoglobin is more negatively charged than oxyhemoglobin, is known as the acid Bohr effect.

Bohr magneton The magnetic dipole moment of a spinning electron.

boiled enzyme test A determination of the sensitivity of a reaction to heat; if the reaction is enzyme-catalyzed, it will generally be sensitive to heat as a result of the thermal denaturation of the enzyme.

boivin antigen A lipopolysaccharide–protein complex that can be extracted from the outer membrane of gram-negative bacteria with trichloroacetic acid.

bolometer A temperature transducer used for the measurement of minute quantities of radiant heat.

Boltzmann constant The gas constant divided by Avogadro's number. *Sym k*.

Boltzmann distribution The most probable distribution of a large number of molecules or particles in a nonuniform field of force at or near equilibrium; the centrifugal force in an ultracentrifuge, and the potential acting on an ion in electrophoresis are examples of such fields of force. The distribution can be expressed as $n_1 = n_2 \, e^{-(E_1 - E_2)/kT}$, where n_1 and n_2 are the number of molecules or particles in two locations or in two energy states, E_1 and E_2 are the respective energies, k is the Boltzmann constant, and T is the absolute temperature.

bombesin A substance in the gastrointestinal tract that, when injected into animals, stimulates gastric acid secretion and inhibits intestinal motility.

bond 1. The linkage between two atoms in a molecule. 2. The linkage between two atoms, groups of ions, or molecules, or between combinations of these.

bond angle The angle between any two bonds by which an atom is linked to other atoms in the molecule.

bond energy The energy required to break a chemical bond.

bonding orbital A molecular orbitral in which there is no node of electron density between the bonding atomic nuclei, resulting in a strengthening of the bond between the nuclei. Bonding orbitals are designated sigma (σ) and pi (π) and are generally of lower energy than the antibonding orbitals sigma star (σ^*) and pi star (π^*).

bond length The length of the bond between two atoms; equal to the distance between the centers of the nuclei of the two bonded atoms.

bond radius One-half of the bond length.

bond strength BOND ENERGY.

boneblack An impure charcoal prepared from bone and used as a decolorizing adsorbent.

bone mineral HYDROXYAPATITE.

bone remodeling *See* remodeling.

bongkrekik acid A fungal antibiotic that inhibits the ATP–ADP carrier of mitochondria.

bookmark hypothesis The hypothesis that peptides containing aromatic amino acids serve in vivo as a means of recognizing DNA sequences by having their aromatic amino acid residues become partially inserted between base pairs at certain intercalating sites in the DNA; the different intercalating sites are referred to as pages in a book, and the intercalating peptides are referred to as bookmarks.

booster dose A dose of antigen, particularly in the form of a vaccine, that is administered to an individual after a priming dose with the intent of stimulating the production of large amounts of antibodies.

booster response SECONDARY RESPONSE.

boron An element that is essential to a wide variety of plants. Symbol, B; atomic number, 5; atomic weight, 10.811; oxidation state, +3; most abundant isotope, ^{11}B.

bottromycin One of a group of peptide antibiotics, produced by *Streptomyces*, that inhibit bacterial protein synthesis.

botulism A severe, and often fatal, form of food poisoning due to the toxins of *Clostridium botulinum* which block neural transmission by inhibiting the synthesis or the release of acetylcholine at presynaptic sites.

Bouguer's law LAMBERT'S LAW.

boulevard peripherique GERL.

boundary A transition zone, either between solvent and solution or between two different solutions. In analytical ultracentrifugation, a boundary is produced when molecules are sedimented through the solution; in diffusion and electrophoresis experiments, using a Tiselius apparatus, a boundary is produced by the layering of the solvent over the solution.

boundary sedimentation Sedimentation in which a boundary is formed, as in sedimentation velocity.

bound enzyme INSOLUBLE ENZYME.

bound insulin INSULIN-LIKE ACTIVITY.

Boveri's theory of cancer *See* chromosome theory of cancer.

bovine Of, or pertaining to, cattle.

Bowman–Birk inhibitor A soybean trypsin inhibitor.

box A sequence of nucleotides that occurs repeatedly as a transcription or regulatory signal. *See also* CAAT box, GC box, Hogness box, Pribnow box, TATA box.

bp 1. Base pair; a unit of length in nucleic acid molecules equal to one base pair. *See also* base pair. 2. Boiling point; also abbreviated b.pt.

BPA Bovine plasma albumin.

B protein 1. A protein subunit of the enzyme tryptophan synthetase. 2. A component of the enzyme lactose synthetase.

bps Bits per second. A measure of data transmission speed by a computer; the number of bits of information that pass a given point in one second. *Aka* baud.

Bradford method A spectrophotometric method of high sensitivity for the determination of proteins; based on the shift in the adsorption maximum of the dye Coomassie brilliant blue upon binding to protein.

Bradshaw test A test for the presence of Bence–Jones protein in urine that is performed by layering dilute, acidified urine over concentrated hydrochloric acid.

bradykinin *See* kinin.

Bragg angle The angle of incidence, which equals the angle of reflection, in the Bragg equation.

Bragg curve A plot of specific ionization as a function of either distance or energy.

Bragg equation An equation relating the angle at which either light rays or x rays are reflected from a crystal to the spacing of the atomic planes in the crystal; specifically, $2d \sin \theta = n \lambda$, where θ is both the angle of incidence and the angle of reflection, λ is the wavelength of the radiation, d is the spacing between reflecting planes, and n is an integer.

Bragg law BRAGG EQUATION.

Bragg peak The peak of a Bragg curve.

Bragg scattering The scattering of radiation by a crystal that is described by the Bragg equation.

brain barrier system BLOOD–BRAIN BARRIER.

brain hormone PROTHORACICOTROPIC HORMONE.

brain sparing The phenomenon that, during starvation, the loss of bulk matter from the brain is smaller than that from other organs.

branched-chain ketoaciduria MAPLE SYRUP URINE DISEASE.

branched fatty acid A fatty acid having a

branched chain. Branched fatty acids occur in many different tissues and organisms and are frequently associated with specific physiological functions or structures. *See also* ante-iso fatty acid; iso fatty acid.

branched metabolic pathway A sequence of metabolic reactions that diverges and that can give rise to two or more different end products.

branched polymer A polymer that consists of a main chain to which side chains are attached.

branching The simultaneous decay of a given type of radioactive atoms by two different modes; a fraction of the atoms decays by one mode and the remaining fraction decays by the other mode.

branching enzyme A enzyme that catalyzes the synthesis of branch points in a polymer.

branch migration The process whereby a strand segment, that is base-paired with one strand (A) of double-helical DNA, is displaced by a second strand segment that is also able to base pair with the same DNA strand (A). The process is based on the breathing of DNA and results in a branch point of the DNA being displaced sequentially along the DNA molecule.

branch point 1. The point in the main chain of a polymer at which either an additional molecule or a second chain is attached. 2. The point at which a sequence of metabolic reactions diverges so that it can give rise to two or more different end products.

Bray's solution A dioxane-based liquid scintillator designed for the detection of weak beta-emitting isotopes such as ^{14}C or ^{3}H in biological samples.

breakage and reunion model The model of genetic recombination according to which parts of the parental chromosomes are exchanged as a result of physical breakage of the chromosomes and reunion of the broken fragments. *Aka* breakage hypothesis; break and exchange model.

breakage–reunion enzyme One of a group of enzymes that catalyze the breaking and rejoining of DNA molecules and that act on internal segments, rather than on termini, of the DNA.

breakdown potential The potential at which a Geiger–Mueller tube begins to produce a continuous discharge.

breathing 1. The opening up of double-stranded regions in DNA to become single-stranded bubbles; a random and transient breaking and reforming of hydrogen bonds between the two strands. The process is considered to be necessary to allow for the interaction of specific proteins with the DNA. 2. A local unfolding of the polypeptide chain

in a protein as that which occurs during deuterium exchange.

breeding true The production, by homozygotes, of offsprings that have a phenotype that is identical to that of the parents.

Brei HOMOGENATE.

bremsstrahlung An electromagnetic radiation that is produced when high-energy beta particles are decelerated in the electrostatic fields of atomic nuclei.

brewer's yeast One of several strains of yeast belonging to the species *Saccharomyces cerevisiae* and capable of producing and tolerating high levels of ethanol.

bridge complex One of a number of ternary complexes formed between an enzyme (E), a metal ion (M), and a substrate (S). Major types of bridge complexes are substrate bridge complexes (E—S—M), metal bridge complexes (E—M—S), and enzyme bridge complexes (M—E—S).

bridge migration BRANCH MIGRATION.

bridging atom 1. An atom that connects two groups in a molecule, such as the oxygen atom that connects two phosphoryl groups in ATP. 2. The metal ion in a metal bridge complex. *Aka* bridge atom.

Briggs–Haldane treatment The treatment of enzyme kinetics that is based on the assumptions that (a) a steady state attains for the enzyme substrate complex, and (b) the velocity of the reaction is an initial velocity, proportional to the concentration of enzyme–substrate complex, so that the reverse reaction from products to enzyme–substrate complex can be neglected. The resulting rate equation has the form $v = V[S]/(K_m + [S])$. where v is the initial velocity of the reaction, V is the maximum velocity, $[S]$ is the substrate concentration, and K_m is the Michaelis constant. This rate equation is known as the Michaelis–Menten equation. *See also* Michaelis–Menten treatment.

Briggsian logarithm *See* logarithm.

Brij One of a number of polyoxyethylene ethers of higher aliphatic alcohols that are used as nonionic detergents for the solubilization of membrane fractions.

brilliant cresyl blue A basic dye used in cytochemistry.

British anti-Lewisite The sulfhydryl compound, 2,3-dimercaptopropanol, developed during World War II as a detoxicant for certain war poisons and used subsequently as a chelating agent for heavy metal ions. *Abbr* BAL.

British thermal unit The amount of heat required to raise the temperature of 1 lb of water by 1 °F (from 63 to 64 °F). *Abbr* BTU.

Britten–Davidson model A model of gene regulation in eukaryotes that is somewhat similar to the operon (Jacob and Monod) model in prokaryotes. The model postulates two levels of positive control, requiring two classes of regulatory proteins. The first control level consists of the interaction of a primary regulatory protein, a sensor protein, first with its specific effector, and then with a sensor gene that controls the transcription of a set of integrator genes. Activation of the sensor gene results in the production of an activator RNA. The second control level consists of the synthesis of secondary regulatory proteins, activator proteins, that are coded for by the activator RNA. The activator proteins then interact with specific receptor sequences (receptor sites) to promote the expression of structural genes.

brittle diabetes A disease in which lack of glucose tolerance and sufficiency of insulin activity vary in an unpredictable manner so that glucose and ketone bodies may be present in the urine at one time, and acceptable levels of insulin activity may be present at another time.

BRM Biological response modifier.

broad-beta lipoprotein FLOATING BETA LIPOPROTEIN.

broad-spectrum antibiotic An antibiotic, such as chloramphenicol or a tetracycline, that has a wide range of antibacterial activity.

Brockmann scale A scale for classifying various grades of alumina for adsorption chromatography; based on the adsorption of a series of dyes by the alumina.

Brodie's solution A manometer fluid containing NaCl, sodium choleate, and Evan's blue; it has a density of 1.033 g/mL so that a column 10,000 mm high is equivalent to a pressure of 1 atm.

bromatology Food science.

bromine An element that is essential to a wide variety of species. Symbol, Br; atomic number, 35; atomic weight, 79.909; oxidation states, -1, $+1$, $+5$; most abundant isotope, ^{79}Br; a radioactive isotope, ^{82}Br, half-life, 35.4 hours, radiation emitted, beta particles and gamma rays.

5-bromodeoxyuridine A thymidine analogue used as an antiviral drug. *Abbr* BUDR. *Aka* 5-bromouracildeoxyriboside.

5-bromouracil A mutagenic pyrimidine analogue that is readily incorporated into DNA in place of thymine to produce transitions. *Abbr* BU.

Bronsted acid A molecule or an ion that acts as a proton donor.

Bronsted base A molecule or an ion that acts as a proton acceptor.

Bronsted catalysis law A quantitative expression of the fact that general acid or base catalysis of a given reaction is primarily a function of the acid or base strength of the catalyst. This may be stated as $\log k = a \log K + b$, where k is the rate constant of the reaction, a and b are constants, and K is either the dissociation constant in the case of an acid catalyst or the reciprocal of the dissociation constant of the conjugate acid in the case of a base catalyst. *Aka* Bronsted equation.

Bronsted–Lowry theory The theory that describes acids as proton donors and bases as proton acceptors.

Bronsted plot A plot of the logarithm of the rate constant versus the negative logarithm of the dissociation constant. *See also* Bronsted catalysis law.

bronzed diabetes A disease characterized by the presence of excessive amounts of glucose in the urine as a result of the deposition of iron in the pancreas, liver, and other organs. *See also* hemochromatosis.

broth A complex liquid medium for growing bacteria; commonly refers to nutrient broth or any medium based on it.

brown adipose tissue BROWN FAT.

brown fat A special type of adipose tissue that serves to generate heat from fat oxidation. The fat cells are characterized by the presence of numerous fat droplets. The tissue is brown because it contains many mitochondria which, in turn, are rich in red-brown cytochromes. These specialized mitochondria dissipate the energy from electron transport as heat to maintain body temperature, especially in young and hibernating animals. *See also* white fat.

Brownian motion The random, thermal motion of solute molecules that is due to their continual bombardment by molecules of the solvent. *Aka* Brownian movement.

browning reactions A group of complex reactions, both enzymatic and nonezymatic, that occur in various foods upon processing and/or storage; the enzymatic reactions are thought to involve the oxidation of phenolic compounds, and the nonenzymatic reactions are thought to involve caramelization, decomposition of ascorbic acid, and the Maillard reaction.

BrUrd 5-Bromouridine.

brush border membrane The plasma membrane of certain epithelial cells that is characterized by the presence of many microvilli, frequently packed together like the bristles of a brush. The brush border membrane of the epithelial cells of the small intestine plays an important role in digestion; it contains many hydrolytic enzymes in its microvilli.

BSA Bovine serum albumin.

BSV Bushy stunt virus.

BTU British thermal unit.

B-type particles One of two types of particles (A and B) seen in electron microscope preparations of mouse mammary tumors; now referred to as the mouse (or murine) mammary tumor virus (MMTV) which belongs to the group of leukoviruses. B-type particles differ from C-type particles in that they have an intracellular precursor from which they develop (A-type particles) and have an eccentrically situated genome in the virion. *Aka* B-type virus.

B-type virus *See* B-type particles.

BU 5-Bromouracil.

bubble structure 1. EYE STRUCTURE. 2. D-LOOP.

budding 1. A form of reproduction in which a daughter cell is produced in yeast and other fungi. 2. An outfolding of the host cell wall through which some phages that have infected the cell release their progeny.

BUDR 5-Bromodeoxyuridine.

bufadienolide A steroid that is a 24-carbon homologue of a cardenolide and occurs as a cardiac glycoside in plants and toads.

bufanolide The fully saturated system of the lactones that occur in toad poison.

buffalo black NBR AMIDO BLACK 10B.

buffer A solution containing a mixture of a weak acid and its conjugate weak base that is capable of resisting substantial changes in pH upon the addition of small amounts of acidic or basic substances.

buffer capacity 1. The number of equivalents of either protons or hydroxyl ions that is required to change the pH of 1 L of a 1 M buffer by one pH unit; equal to $(1/m)(dn/d\text{pH})$, where m is the number of moles of buffer, and $d\text{pH}$ is the change in pH produced by the additon of dn equivalents of either protons or hydroxyl ions. 2. BUFFER VALUE.

buffer index BUFFER VALUE.

buffer molarity The total concentration of both buffer components. Thus, a 0.1 M $H_2PO_4^-$/HPO_4^{2-} buffer refers to one in which the sum of the concentrations of the $H_2PO_4^-$ and HPO_4^{2-} forms is 0.1 M.

buffer value The number of equivalents of either protons or hydroxyl ions that is required to change the pH of a buffer by one pH unit; equal to $dn/d\text{pH}$, where $d\text{pH}$ is the change in pH produced by the addition of dn equivalents of either protons or hydroxyl ions.

buffy coat The thin layer of white blood cells that forms at the surface of the packed layer of red blood cells when unclotted blood is centrifuged.

building block A molecule that serves as a structural unit in a biopolymer; a biomonomer.

bulbogastrone A substance, released by duodenal cells, that inhibits gastric secretion.

bulk element An element, such as carbon, hydrogen, oxygen, nitrogen, calcium, magnesium, sodium, potassium, phosphorus, and sulfur, that is an essential nutrient for an organism and that is required in relatively large amounts (of the order of grams/day) for humans and animals. *See also* trace elements; macronutrients; micronutrients.

BUN Blood urea nitrogen.

bundle sheath cells The inner layer of cells in the leaves of C_4 plants; site of the reactions of the Calvin cycle which follow preliminary fixation of CO_2 that occurs in the mesophyll cells.

bundling proteins One of a number of proteins that bind to the sides of actin filaments and cause their association into bundles.

α-bungarotoxin *See* alpha bungarotoxin.

Bunsen absorption coefficient The number of gas volumes that are dissolved by one volume of liquid when the liquid is equilibrated with the gas under 1 atm of pressure; the gas volume is calculated as that volume occupied by the gas at 0°C and 1 atm of pressure. *Aka* Bunsen solubility coefficient.

Bunsen–Roscoe law RECIPROCITY.

Bunsen solubility coefficient BUNSEN ABSORPTION COEFFICIENT.

buoyancy factor The term $1 - \bar{v}\rho$ that appears in equations pertaining to hydrodynamic methods of studying macromolecules, where $\bar{v}$ is the partial specific volume of the solute and ρ is the density of the solution.

buoyant density The density of a molecule as determined by density gradient sedimentation equilibrium.

buret A graduated tube with stopcock used for delivery of known volumes of liquid as in a titration. *Var sp* burette.

buried residue MASKED RESIDUE.

Burkitt's lymphoma A lymphoma, afflicting children in Africa, from which the Epstein–Barr virus has been isolated.

Burnet's theory CLONAL SELECTION THEORY.

bursa of Fabricius A sac-like structure in birds where B lymphocytes mature into antibody-secreting plasma cells.

bursicon A insect polypeptide hormone that appears after molting and that promotes the tanning and hardening of cuticle.

burst 1. The rapid pre-steady-state release of the first product in a ping-pong mechanism. 2. The explosion of a phage-infected bacterial cell that is accompanied by the release of phage particles into the medium. *See also* respiratory burst.

burst size The average number of phage particles released per infected bacterial cell; equal to the ratio of the final titer to the initial titer in a phage multiplication cycle.

burst titration A titration procedure for the active site of an enzyme; based on measuring the increase in absorbance due to the release of a chromogenic product from the stoichiometric reaction between a suitable substrate and the active site of the enzyme. *See also* reverse burst titration.

Busch theory A theory of carcinogenesis that postulates three stages of disease referred to as initiation, promotion, and acceleration. According to this theory, a portion of the DNA is responsible for carcinogenesis and is inhibited in normal cells by combination with a suppressor protein. The carcinogen binds to the cancer DNA–suppressor complex in the initiation stage, releases the DNA in the promotion stage, and allows it to form cancer RNA in the acceleration stage.

bushy stunt virus *See* tomato bushy stunt virus.

busulfan A mutagenic alkylating agent.

butanediol fermentation BUTYLENE GLYCOL FERMENTATION.

butanol fermentation *See* acetone–butanol fermentation.

butterfly mode A mode of DNA replication observed in some animal viruses and plasmids; so called because a partially replicated molecule resembles the shape of a butterfly. The unreplicated portion is supercoiled and the replicated portion is untwisted in the form of two loops.

butyl alcohol fermentation *See* acetone–butanol fermentation.

butylene glycol fermentation The fermentation of glucose, characteristic of *Aerobacter aerogenes* and related forms, that yields primarily butylene glycol and ethanol, and secondarily a number of other products.

butyric–butylic fermentation The fermentation of glucose that yields butyric acid, *n*-butanol, acetone, and isopropanol in varying proportions.

B virus *See* oncornavirus; B-type particle; hepatitis.

B vitamin *See* vitamin B complex.

bypass *See* metabolic bypass.

by-product A minor product in a chemical reaction.

byte A sequence of bits, usually 8, that is treated as a unit by a computer and that stores a unit of information.

bz The benzoyl group; the acyl group derived from benzoic acid; the grouping

bzl The benzyl group; the grouping $C_6H_5^-$ derived from benzene.

C

c 1. Concentration. 2. Curie. 3. Centi.

C 1. Cytosine. 2. Cytidine. 3. Cysteine. 4. Carbon. 5. Degree Celsius (centigrade). 6. Complement. 7. Coulomb. 8. Heat capacity.

^{14}C A radioactive isotope of carbon that has a half-life of 5730 years and emits beta particles.

C23 Nucleolin.

Ca Calcium.

CAAT box A nucleotide sequence, present in many eukaryotic promoters at about 75 base pairs upstream from the site at which transcription starts. It has the consensus sequence GG(T/C)CAATCT in which T and C are equally frequent at the third position from the left.

cable properties The electrical characteristics of an axon that are involved in the passive spread of an electrical signal.

cachectin TUMOR NECROSIS FACTOR.

cachexia The malnutrition and wasting of bodily tissue that is produced by chronic diseases, such as the drain on host nutrients produced by the proliferation of cancer cells.

C$_4$ acid cycle HATCH–SLACK–KORTSCHAK PATHWAY.

CaCl$_2$ transformation *See* calcium chloride transformation.

cacodylic acid Dimethylarsinic acid [(CH$_3$)$_2$AsO$_2$H]. Cacodylate buffers are used in preparing fixatives for electron microscopy.

cadaverine A five-carbon polyamine that contains two amino groups; a biogenic amine formed by decarboxylation of lysine; 1,5-diaminopentane.

Ca^{2+}-dependent regulatory protein CALMODULIN.

cadmium An element that is essential to humans and animals. Symbol, Cd; atomic number, 48; atomic weight, 112.40; oxidation state, +2; most abundant isotope, ^{114}Cd; a radioactive isotope, ^{109}Cd, half-life, 453 days, radiation emitted, gamma rays.

CAF *See* calpain.

caffeine A purine alkaloid (1,3,7-trimethylxanthine) that occurs in coffee beans and tea leaves and that has a stimulatory effect on the central nervous system. Caffeine inhibits the phosphodiesterase that converts cyclic AMP to inactive 5'-AMP; this prolongs the adrenalin-producing effect of cyclic AMP.

cage A cavity or enclosed region in the solvent structure into which solute molecules can fit. *See also* clathrate.

caged ATP A derivative of ATP[P^3-1-(2-nitro)phenylethyladenosine] which, upon hydrolysis, yields ATP.

Cahn–Ingold–Pregold sequence rules RS SYSTEM.

Cairns experiment An experiment that provided evidence for the existence of one replicating fork per molecule of DNA undergoing replication. The experiment consisted of labeling the DNA of growing *E. coli* cells with tritiated thymine, isolating the DNA by mild procedures, and determining the distribution of label in the DNA by means of radioautography.

Cairns model The bidirectional mode of replication of double-stranded circular DNA in which the two replicating forks move in opposite directions as is the case for the replication of bacterial DNA.

Cairns molecule THETA STRUCTURE.

cal Small calorie.

Cal Large calorie; a kilocalorie, equal to 1000 small calories.

calcifediol Trivial name, proposed for 25-hydroxycholecalciferol.

calciferol ERGOCALCIFEROL.

calcifetriol Trivial name, proposed for 1,25-dihydroxycholecalciferol.

calcification The formation of calcium salt deposits in a tissue.

calcified Having undergone calcification.

calcimedin One of a group of Ca^{2+}-binding proteins that occur in several tissues and that differ from calmodulin in their isoelectric points, DEAE-cellulose binding characteristics, and heat stability.

calcineurin An inhibitory protein, composed of two subunits, that is located in nervous tissue and that binds both Ca^{2+} ions and calmodulin. The binding of calcineurin to calmodulin prevents the activation of several Ca^{2+}-dependent enzymes by calmodulin. *Aka* calmodulin-binding protein; inhibitory protein of cyclic nucleotide phosphodiesterase.

calcitonin A polypeptide hormone that lowers the level of calcium in the blood and that is secreted by both the thyroid and the parathyroid glands. *Abbr* CT.

calcium An element that is essential to all plants and animals. Symbol, Ca; atomic number, 20; atomic weight, 40.08; oxidation state, +2; most abundant isotope, ^{40}Ca; a radioactive isotope, ^{45}Ca, half-life, 165 days, radiation emitted, beta particles.

calcium-activated factor CALPAIN.

calcium-activated neutral proteinase CALPAIN.

calcium chloride transformation A common technique for cloning DNA that has been inserted into a plasmid; involves mixing recipient bacteria and the modified plasmid in a solution of cold $CaCl_2$. In such a solution, bacteria can take in DNA molecules and thus a plasmid is easily transferred from one strain to another.

calcium-dependent proteinase CALPAIN.

calcium-dependent regulatory protein CALMODULIN.

calcium phosphate gel A gel prepared from calcium chloride and trisodium phosphate and used in the purification of proteins by adsorption chromatography.

calcium pump The structure and/or the mechanism that medicates the active transport of calcium across a biological membrane. A primary active transport mechanism in which a Ca^{2+}-ATPase spans the membrane in an asymmetric fashion.

calculus (*pl* calculi). A hard aggregate or stone that is found in the body and that may consist chiefly of inorganic matter, as in the case of kidney stones, or of organic matter, as in the case of uric acid stones associated with gout.

caldesmon A calmodulin-binding protein that is unique in that it also binds F-actin. The former binding is Ca^{2+}-dependent, the latter is not. Formation of the two complexes is regulated by Ca^{2+} which functions as a flip-flop switch.

calelectrin One of a group of proteins that occur in the electric organ of *Torpedo marmorata* and in several mammalian tissues. They have synexin-like activity and become associated with the plasma membrane in a Ca^{2+}-dependent manner; they exist in various states of aggregation (from 32,500 to 67,000 daltons) and include such proteins as chromobindin, synhibin, and endonexin.

calibration The standardization and graduation of a measuring instrument.

calibration curve STANDARD CURVE.

C-alkaloids CURARE ALKALOIDS.

callose A linear homopolysaccharide that occurs in higher plants and that is composed of D-glucose units linked by means of $\beta(1 \rightarrow 3)$ glycosidic bonds.

callus A mass of relatively undifferentiated cells formed from a single plant cell in tissue culture.

calmodulin A calcium-binding protein in eukaryotic cells which mediates the control of a large number of enzymes by Ca^{2+}. The process generally involves two steps: a binding of Ca^{2+} to calmodulin, accompanied by a conformational change of the protein, followed by the binding of the calmodulin–Ca^{2+} complex to an enzyme, resulting in enhanced enzymatic activity. *Abbr* CAM; CaM.

calomel Mercurous chloride.

calomel electrode A reference electrode for pH measurements that contains mercury, mercurous chloride, and a saturated solution of potassium chloride.

caloric intake The caloric equivalent of the food ingested, calculated on the basis of the energy yield obtainable by complete oxidation of the food.

caloric value The quantity of heat, generally expressed in kilocalories per gram, that is released when a foodstuff is subjected to complete oxidation; the heat of combustion of a foodstuff. *Aka* calorific value.

calorie A measure of energy equal to the amount of heat required to raise the temperature of 1.0g of water by 1°C (from 14.5 to 15.5°C) at a pressure of 1 atm; 1 calorie = 4.184 J. *Abbr* cal. *Aka* small calorie *See also* empty calorie; large calorie.

calorific Of, or pertaining to, the production of heat.

calorigenesis The production of heat or energy in an organism; an increase in heat or energy production; an increase in oxygen consumption.

calorigenic Of, or pertaining to, calorigenesis.

calorigenic action SPECIFIC DYNAMIC ACTION.

calorimeter An instrument for measuring the heat that is either absorbed or released by a chemical reaction or by a group of chemical reactions.

calorimetry The measurement of the heat change in a chemical reaction or a group of chemical reactions, either in an in vitro system or in an intact organism.

calpain Calcium-dependent papain-like proteinase; one of a group of calcium-activated neutral proteinases. The calpains are sulfhydryl proteases that have about 33% sequence homology, around the catalytic SH-group, with the enzyme papain. *Aka* calcium-activated factor (CAF); calcium-activated neutral proteinase (CANP); calcium-dependent proteinase (CAP, CDP).

calpastatin A specific protein inhibitor of the calpains which has been isolated from a variety of mammalian and avian tissues; the molecular weight of the protein varies from 24,000 to 400,000, depending on the source and the method of extraction.

calsequestrin An acidic glycoprotein, rich in aspartic and glutamic acids, that is present in the sarcoplasmic reticulum where it binds more than 40 Ca^{2+} ions per molecule of protein. It serves to store Ca^{2+} and to release it upon muscle contraction.

Calvin cycle A cyclic set of reactions, occurring in chloroplasts, that results in the fixation of carbon dioxide and in its conversion to glucose by means of the ATP and the NADPH formed in the light reaction of photosynthesis. *Aka* Calvin–Bassham cycle; Calvin–Benson cycle.

calvinosome A prokaryotic compartment containing all of the enzymes of the Calvin cycle for autotrophic carbon dioxide fixation.

Calvin plant C_3 plant.

CaM Calmodulin.

CAM 1. Calmodulin. 2. Cell adhesion molecule. 3. Chloramphenicol.

camera lens A lens that focuses an image on a photographic plate.

cAMP Cyclic AMP.

Campbell model *See* insertion model.

cAMP–CRP The complex formed between cyclic AMP (cAMP) and cyclic AMP receptor protein (CRP).

cAMP-dependent protein kinase One of a group of enzymes in animal cells that mediate the effect of cyclic AMP; they catalyze the transfer of a phosphate group from ATP to a serine or a threonine residue of a protein in the target cell.

cancellous bone *See* lamellar bone.

cancer 1. A disease of multicellular organisms that is characterized by seemingly uncontrolled cellular growth and by the spreading within the organism of apparently abnormal forms of the organism's own cells; cancer cells thus show excessive multiplication, autonomy with respect to the host, and invasiveness (metastasis). 2. A malignant tumor.

cancer biochemistry The biochemistry of cancer cells. *See also* first law of cancer biochemistry; second law of cancer biochemistry.

cancer gene ONCOGENE.

cancer-inducing virus ONCOGENIC VIRUS.

cancerocidal Capable of killing cancer cells.

cancerogenesis CARCINOGENESIS.

cancer theory *See* theory of cancer.

cancroid 1. *n* A skin tumor of low malignancy. 2. *adj* Cancer-like.

candidate hormone A substance whose hormonal status is not yet clearly established; vasoactive intestinal peptide, enteroglucagon, bombesin, motilin, and urogastrone are some examples.

cane sugar Sucrose isolated from sugar cane.

canine Of, or pertaining to, dogs.

cannabinol *See* THC.

cannabis *See* hashish.

canonical sequence CONSENSUS SEQUENCE.

canonical structure Any one of the possible resonance structures of a compound.

CANP Calcium-activated neutral proteinase.

cap 1. The modified 5′-end of eukaryotic mRNA. The cap is introduced enzymatically shortly after initiation of mRNA synthesis and consists of 7-methylguanosine-5′-monophosphate linked 5′ → 5′ via a phosphate group to the terminal nucleotide of the mRNA [(7-MeG)-5′-ppp-terminal nucleoside]. In unicellular eukaryotes, such as yeast and slime molds, this is the predominant structure and it is designated as Cap 0. In multicellular eukaryotes (animals), the predominant structure has an additional methyl group at the 2′-position of the original terminal nucleotide; this structure is designated as Cap 1. In some cases, there is a further methylation at the 2′-position of the next nucleotide and this structure is designated as Cap 2. The cap is believed to be necessary for efficient protein synthesis by serving to protect the mRNA against degradation by nucleases. 2. A network of antigens and antibodies that forms on the surface of a cell and then redistributes itself to one region of the cell surface. *Aka* cell cap.

CAP 1. Catabolite activator protein. 2. Calcium-dependent proteinase. 3. Calf intestine alkaline phosphatase.

capacitance An electrical unit equal to the total charge that can be stored in a condenser divided by the potential difference across the plates.

capillarity The action by which the surface of a liquid, where it is in contact with a solid, is either elevated or depressed as a result of the relative attractions of the molecules of the liquid for each other and for the molecules of the solid.

capillary 1. *n* A tube or a vessel having a very small diameter. 2. *adj* Of, or pertaining to, a tube or a vessel having a very small diameter.

capillary action CAPILLARITY.

capillary attraction The force of adhesion between a solid and a liquid in capillarity.

capillary precipitin test RING TEST.

capillary viscometer An instrument for measuring the viscosity of a liquid from the time required for a given volume of the liquid to flow through a capillary. *See also* Ostwald viscometer.

capneic Descriptive of organisms that actually require increased carbon dioxide tensions, rather than reduced oxygen tensions, for growth.

capon test A bioassay for androgen activity that is based on the stimulation of comb growth in capons. Castration of a capon causes secondary sex characteristics to disappear and this can be counteracted by administration of androgens. A capon unit is defined as the amount of androgen that, when administered to a castrated capon, causes an increase of 20% in the surface area of the degenerated comb.

capped 5′-end The 5′-end of eukaryotic mRNA that carries a methylated cap.

capping 1. Addition of a methylated cap to eukaryotic mRNA. 2. Formation of a cell cap composed of antigen–antibody complexes. 3. The lateral movement of some membrane proteins to specific sites or zones in the membrane according to the fluid mosaic model.

capping proteins Proteins that bind selectively to one or the other end of actin filaments and microtubules which are subject to treadmilling.

capric acid A saturated fatty acid that contains 10 carbon atoms; the systematic name decanoic acid is preferred.

caprin The triacylglycerol (triglyceride) of capric acid.

caprine Of, or pertaining to, goats.

caprinized vaccine A vaccine that contains live organisms whose virulence has been decreased by serial passage through goats.

CAPS 3-(Cyclohexylamino)-1-propanesulfonic acid; used for the preparation of biological buffers in the pH range of 9.7 to 11.1 *See also* biological buffers.

capsid The protein coat, composed of capsomers, that surrounds the nucleic acid of a virus and determines the overall shape of the virus.

capsomer The morphological unit, one or more of which constitute the viral capsid. The capsomer, in turn, consists of one or more structural units, called protomers or monomers. A capsomer that consists of five structural units is known as a pentagonal capsomer, or pentamer, and a capsomer that consists of six structural units is known as a hexagonal capsomer, or hexamer. *Var sp* capsomere.

capsular polysaccharide A polysaccharide that is a component of a bacterial capsule and that is frequently antigenic.

capsule A loose gel- or slime-like structure that is rich in polysaccharides and that frequently coats the outer surface of a bacterial cell wall. Capsules may be divided into three categories: (a) macrocapsules, which are sufficiently thick to be detected by light microscopy; (b) microcapsules, which are too thin to be detected by light microscopy but can be detected by serological techniques; and (c) slime layers, which are diffuse secretions that adhere loosely to the cell wall and have no definite borders; they generally become dispersed in the medium when the organism is grown in liquid culutre. *See also* cell coat.

capsule swelling reaction QUELLUNG REACTION.

capture cross section The product of the probability that a particle impinging on an atomic nucleus will be captured by that nucleus and the cross-sectional area of the nucleus (in barns).

caramelization The browning of sugars when they are heated above their melting points.

carbamide UREA.

carbamino compound A compound formed by the reaction of carbon dioxide with a primary aliphatic amine.

carbaminohemoglobin The carbamino compound that is formed by the reaction of hemoglobin with carbon dioxide and that represents one of the forms in which carbon dioxide is transported by the blood.

carbamoyl group The acyl group of carbamic acid; the radical H_2NCO—. *Aka* carbamyl group.

carbamoyl phosphate The high-energy compound NH_2—COO—PO_3H_2 that functions in the urea cycle, ammonia fixation, and pyrimidine biosynthesis. *Aka* carbamylating agent; carbamyl phosphate.

carbanion A carbon anion; the species $R_3C:^-$ in which the carbon atom carries an unshared pair of electrons. A carbanion is formed by removal of a group attached to the carbon atom without removing the pair of bonding electrons.

carbene A neutral organic compound that contains a divalent carbon atom and that is formed by removal of two groups attached to one carbon atom together with one pair of the bonding electrons; the species

carbocation CARBONIUM ION.

carbocyclic Of, or pertaining to, an organic compound that has a ring structure consisting only of carbon atoms.

carbohydrase An enzyme that catalyzes the hydrolysis of glycosidic bonds in carbohydrates; a glycosidase.

carbohydrate An aldehyde or a ketone derivative of a polyhydroxy alcohol that is synthesized by living cells. Carbohydrates may be classified either on the basis of their size into mono-, oligo-, and polysaccharides, or on the basis of their functional group into aldehyde or ketone derivatives.

carbohydrate tolerance test *See* glucose tolerance test; galactose tolerance test.

carbolic acid PHENOL (1).

carboligase The enzyme that catalyzes the formation of acetoin from acetaldehyde and active acetaldehyde.

carbometer An instrument for measuring the carbon dioxide content of breath.

carbon An element that is essential to all plants and animals. Symbol, C; atomic number, 6; atomic weight, 12.01115; oxidation states, -4, $+2$, $+4$; most abundant isotope, ^{12}C; the stable isotope, ^{13}C; a radioactive isotope, ^{14}C, half-life, 5730 years, radiation emitted, beta particles.

carbonaceous Consisting in part, or entirely, of carbon.

carbon assimilation CARBON DIOXIDE FIXATION.

carbon chain A chain of covalently linked carbon atoms.

carbon clearance test *See* phagocytic index (2).

carbon clock The radioactive isotope of carbon, ^{14}C, that is used in radiocarbon dating for establishing the age of biological remains.

carbon cycle 1. The set of reactions whereby photosynthetic organisms reduce carbon dioxide to carbohydrates and heterotrophic organisms oxidize the carbohydrates back to carbon dioxide. 2. CALVIN CYCLE. 3. A series of thermonclear reactions in the sun believed to be responsible for the energy released by it.

carbon dating *See* radiocarbon dating.

carbon dioxide assimilation CARBON DIOXIDE FIXATION.

carbon dioxide capacity of plasma *See* carbon dioxide combining power of plasma.

carbon dioxide combining power of plasma The maximum amount of carbon dioxide that 100 mL of plasma can hold in the form of bicarbonate when the plasma is saturated with carbon dioxide at a tension corresponding to the tension of carbon dioxide in normal arterial blood.

carbon dioxide compensation point The concentration of carbon dioxide, at a given light intensity, at which the rate of CO_2 fixation by a plant (photosynthesis) is equal to the rate of photorespiration. *See also* light compensation point.

carbon dioxide fixation The photosynthetic conversion of carbon dioxide to carbohydrates. *See also* Calvin cycle.

carbon dioxide transport The carrying of carbon dioxide by the blood from the tissues to the lungs.

carbon fixation CARBON DIOXIDE FIXATION.

carbon-fixation cycle CALVIN CYCLE.

carbonic anhydrase The enzyme, located in the erythrocytes, that catalyzes the reversible decomposition of carbonic acid to carbon dioxide and water in the course of respiration.

carbonium ion A carbon cation; the species R_3C^+ that is formed by the removal of a group attached to a carbon atom together with the pair of bonding electrons. Thus, the

carbon has only six electrons in its valence shell.

carbon monoxide hemoglobin CARBOXYHEMO-GLOBIN.

carbon number EQUIVALENT CHAIN LENGTH.

carbon–oxygen cycle CARBON CYCLE (1).

carbon radical A radical formed from a compound by removal of a group attached to a carbon atom together with one of the two bonding electrons.

carbon reduction cycle CALVIN CYCLE.

carbon skeleton The structure of a molecule considered solely in terms of its carbon atoms.

carbonyl group The grouping that occurs in aldehydes and ketones:

Carborundum Trademark for silicon carbide (SiC); an abrasive.

carboxybiotin A biotin molecule to which a molecule of carbon dioxide has been attached.

carboxydismutase RIBULOSE-1,5-BISPHOSPHATE CARBOXYLASE.

carboxyhemoglobin A hemoglobin derivative in which the sixth coordination position of the iron is occupied by carbon monoxide. *Abbr* HbCO.

carboxylation The introduction of a molecule of carbon dioxide into an organic compound.

carboxylation phase The first stage of the Calvin cycle in which one molecule of ribulose-1,5-bisphosphate is converted to two molecules of 3-phosphoglyceric acid.

carboxyl carrier protein *See* biotin; biocytin.

carboxylesterase An enzyme of low specificity that catalyzes the hydrolysis of esters of carboxylic acids.

carboxyl group The radical —COOH of an organic acid.

carboxylic acid An organic compound containing the carboxyl group; an organic acid.

carboxyl terminal C-TERMINAL.

carboxyltransferase *See* acetyl-CoA carboxylase.

carboxylyase An enzyme that catalyzes a decarboxylation reaction.

carboxypeptidase An exopeptidase that catalyzes the hydrolysis of amino acids in a polypeptide chain from the C-terminal. Carboxypeptidase *A* catalyzes the hydrolysis of most amino acids and leads to the sequential degradation of the polypeptide chain from the C-terminal; carboxypeptidase *B* catalyzes only the hydrolysis of C-terminal lysine and C-terminal arginine.

carboxysome A polyhedral inclusion body, present in some blue-green algae and autotrophic bacteria, that contains the enzyme ribulose-1,5-bisphosphate carboxylase and that functions in the fixation of carbon dioxide by these organisms.

Carbowax Trademark for polyethylene glycol.

carcinoembryonic antigens A group of plasma glycoproteins that serve as an aid for the early recognition of certain tumors and that are present in the embryo but cannot be detected shortly after birth. They can be formed again later in life in response to malignant tumors. Examples are α-fetoprotein, embryonogenic colon antigen, and Regan isoenzyme. *Abbr* CEA.

carcinogen A physical or a chemical agent that is capable of producing cancer.

carcinogenesis The production of cancer.

carcinogenic Capable of producing cancer.

carcinogenic index A measure of the activity of a carcinogen; equal to $100A/B$, where A is the number of animals bearing a tumor divided by the number of animals living on the day of appearance of the first tumor, and B is the mean time in days of the appearance of tumors.

carcinogenicity The capacity to produce cancer.

carcinogenophore A grouping of atoms in a chemical carcinogen that is primarily responsible for the carcinogenic activity of the molecule. *See also* auxocarcinogen; K region; L region.

carcinoid A cancer-like tumor of the gastrointestinal tract that grows slowly and rarely metastasizes.

carcinoma A malignant tumor, derived from epithelial cells, that can occur in a variety of sites such as skin, breast, and liver.

carcinomatosis A condition in which multiple carcinomas develop simultaneously in different parts of the body as a result of the widespread dissemination of a carcinoma from a primary source.

carcinosis CARCINOMATOSIS.

carcinostasis The inhibition of tumor growth.

carcinostatic Of, or pertaining to, carcinostasis.

cardanolide The fully saturated lactone system of the cardenolides.

cardenolide A steroid that contains 23 carbon atoms and that is characterized by a 14-β-hydroxyl group and and α,β-unsaturated γ-lactone ring; cardenolides occur as cardiac glycosides in plants and insects.

cardiac Of, or pertaining to, the heart.

cardiac genins A collective term for the steroids present in cardiac glycosides; cardenolides and bufadienolides.

cardiac glycoside One of a group of steroid glycosides, such as oubain and digitalis, that act directly on the heart muscle, increasing the force of systolic contraction and thereby improving cardiac output; they also cause mild vasoconstriction and influence the $Na^+ - K^+$ transport across erythrocyte and other cell membranes.

cardiac muscle The involuntary, striated muscle of the heart.

cardiac puncture A technique for withdrawing blood from an animal by inserting a syringe directly into the heart.

cardiolipin The phospholipid hapten, diphosphatidyl glycerol, that is used as an antigen for reacting with the Wasserman antibody in the Wasserman test for syphilis; it accounts for more than 10% of the lipid of the inner mitochondrial membrane. *Abbr* CL.

cardiotonic Tending to increse the contraction of heart muscle.

cardiotonic steroid CARDIAC GLYCOSIDE.

cardiotoxin A substance that, in toxic doses, causes heart damage and may lead to heart stoppage. A cardiac glycoside is an example.

cardiovascular Of, or pertaining to, the heart and the blood vessels.

cardiovirus A virus that belongs to a subgroup of picornaviruses which includes encephalomyocarditis virus and Mengo virus.

cariogenic Promoting dental caries (tooth decay).

carnitine A compound that functions in beta oxidation by transporting fatty acyl groups across the inner mitochondrial membrane. It has been classified as a B vitamin by some investigators due to its requirement in the diets of some organisms and its solubility in water. *Sym* Vit B_t; Vit B_7.

carnitine barrier The limited ability of long-chain fatty acids to cross the inner mitochondrial membrane in the form of fatty acyl coenzyme A, as contrasted with their ability to cross the membrane in the form of fatty acyl carnitine.

carnosine A dipeptide of β-alanine and histidine occurring in vertebrate muscle.

carotene A hydrocarbon carotenoid. *See also* beta carotene.

carotenoid A polyisoprenoid that may be linear or cyclic and that consists of eight isoprene units (a tetraterpenoid) linked largely in a head-to-tail manner. Carotenoids are water-insoluble pigments that occur in plants and photosynthetic bacteria and that frequently function as accessory pigments in photosynthesis.

carrageenan A mixture of polysaccharides, obtained from red algae, that forms a gel similar to that produced by agar.

carrier 1. An element or a compound that is added to a sample as an aid in the chemical manipulation of the same, but labeled, element or compound that is present in the sample. 2. Any related or unrelated substance that is added to a sample as an aid in the chemical manipulation of another substance that is present in the sample. 3. A transport agent, generally a protein or an enzyme, that combines with a substance and transports it either across a biological membrane or within a biological fluid. 4. CARRIER GAS.

carrier ampholyte The ampholyte that forms the pH gradient in isoelectric focusing, as distinct from the sample ampholyte which is fractionated.

carrier culture A culture, infected with a virus, that nevertheless maintains the multiplication of both the cells and the virus particles. Such a culture can be obtained by making most of the cells of a culture resistant to viral infection as a result of the interferon produced by a small portion of virus-infected cells present in the same culture.

carrier displacement chromatography Displacement chromatography in which either related or unrelated substances are added to the mixture being chromatographed as an aid in the chemical manipulation of the separated components.

carrier-facilitated diffusion MEDIATED TRANSPORT.

carrier-free Descriptive of a radioactive nuclide that is essentially free of its stable isotopes.

carrier gas The inert gas that functions as the mobile phase in gas chromatography.

carrier protein 1. A protein that functions as a transport agent. 2. The protein to which a hapten may be conjugated in vitro or to which it may become conjugated in vivo.

carrier state infection A viral infection in which only a small portion of the cells are infected.

Carr–Price reaction A colorimetric reaction for the determination of vitamin A that is based on the production of a blue color upon treatment of the sample with antimony trichloride.

cartilage Connective tissue that consists largely of collagen and chondroitin sulfate.

casamino acids An acid hydrolysate of casein used for bacteriological media.

cascade mechanism 1. The sequence of successive activation reactions that constitute the process of blood clotting and that achieve a continuous amplification of the initial event due to the fact that the concentrations of the components in the blood increase step by step

from the initiating factor to fibrinogen. 2. A sequence of successive activation reactions pertaining to either enzymes or hormones, such as the reaction sequences for the activation of phosphorylase, adrenocorticotropin, and complement.

casein A phosphoprotein that is the principal protein in milk.

cassette One of a group of eukaryotic DNA sequences of related loci that can be substituted for a genetically identical or different sequence by transposition. The expression of mating types in yeast is believed to involve the transposition of such cassette sequences.

castanospermine An alkaloid from the nuts of the Australian tree *Castanospermum australe* that is toxic to animals eating these nuts; a competitive inhibitor of α-glucosidases and an inhibitor of glycoprotein processing.

CAT 1. Computer of average transients; an instrument, used in conjunction with NMR spectroscopy, that permits the accumulation and averaging of spectral data and that increases the signal to noise ratio. 2. Computerized axial tomography.

catabolic Of, or pertaining to, catabolism.

catabolic deletion hypothesis An early formulation of the deletion hypothesis of cancer, according to which loss of one or more key catabolic enzymes through a deletion mutation in DNA was thought to increase the availability of building blocks for polymers and thereby permit continued cell growth. *Aka* catabolic enzyme deletion hypothesis.

catabolism 1. The phase of intermediary metabolism that encompasses the degradative and energy-yielding reactions whereby nutrients are metabolized. 2. The cellular breakdown of complex substances and macromolecules to low molecular weight compounds.

catabolite Any metabolic intermediate produced in the catabolism of food molecules.

catabolite activator protein CYCLIC AMP RECEPTOR PROTEIN.

catabolite inactivation The irreversible inactivation of some enzymes, involved in the catabolism of sugars, that is brought about by the addition of glucose to the medium. This occurs in yeast, but not in *E. coli*, and is so called to distinguish the process from catabolite repression.

catabolite repression The inhibition of the synthesis of a number of enzymes, involved in the catabolism of sugars, that is produced by glucose or closely related compounds such as glucose-6-phosphate and fructose. The effect is due to the inhibition of adenyl cyclase by glucose and a resultant decrease in the level of cyclic AMP (cAMP). The latter must complex with the cyclic AMP receptor protein (CRP) and the complex binds to the promoters of operons subject to catabolite repression. In the absence of cAMP–CRP binding, RNA polymerase cannot bind to the promoter, and transcription of the genes coding for some of the enzymes of sugar catabolism is repressed. Catabolite repression occurs in bacteria and may also occur in yeast.

catabolite-sensitive operon One of several inducible operons that control the synthesis of enzymes, involved in the metabolism of various sugars, and that cannot be induced if glucose is present. Each of these operons is regulated by the cyclic AMP–cyclic AMP receptor protein complex.

catalase The hemoprotein enzyme that catalyzes the decomposition of hydrogen peroxide to oxygen and water.

catalysis The change in the rate of a chemical reaction, generally an increase, that is brought about by the action of a catalyst. *See also* acid–base catalysis; covalent catalysis, etc.

catalysome An organelle, found in adipose tissue, that functions in lipid metabolism; it might represent a specialized type of mitochondrion.

catalyst A substance that changes the rate of a chemical reaction, generally increasing it. A catalyst remains either unchanged during the reaction or is regenerated in its original form at the end of the reaction. A catalyst functions by changing the activation energy of the rate-determining step, by affecting the orientation of molecules in collison, or by making possible another pathway or mechanism that has a different activation energy.

catalytic Of, or pertaining to, a catalyst or catalysis.

catalytic amount The amount of substance that is used in a chemical reaction for catalytic purposes and that is much smaller than the stoichiometric amount of either a reactant or a product; a catalyst, a primer, and a sparker are all used in catalytic amounts.

catalytic antibody An antibody that has the potential of behaving like an enzyme and thus catalyzes specific reactions. Such antibodies function by stabilizing the transition states of selected chemical reactions. One approach to designing such antibodies has been to use a compound, resembling the transition state of the reaction, as an antigen for generating monoclonal antibodies.

catalytic center 1. CATALYTIC SITE. 2. A region within the catalytic site.

catalytic center activity A measure of enzymatic activity that is equal to the number of molecules of substrate transformed into products per minute per catalytic center of the enzyme. *See also* molar activity.

catalytic constant *See* k_{cat}.

catalytic exchange method A method for randomly labeling a compound with tritium by dissolving the compound in a tritiated hydroxylic solvent in the presence of metal catalysts and either acid or base.

catalytic rate constant *See* k_{cat}.

catalytic reduction method A method for labeling a compound with tritium in which the compound is dissolved in a nonhydroxylic solvent and double bonds in the compound are reduced by exposure of the solution to tritium gas.

catalytic site The active site of an enzyme, specifically the active site of an allosteric enzyme as distinct from its regulatory site.

catalytic subunit The subunit of the regulatory enzyme aspartate transcarbamoylase that has enzymatic activity but does not bind the negative effector CTP. *See also* regulatory subunit.

catalyze To change the rate of a chemical reaction through catalysis.

cataphoresis 1. The movement of charged particles toward the cathode. 2. ELECTROPHORESIS.

catatoxic Descriptive of a substance or a condition that causes an animal to defend itself vigorously against an irritation, a toxin, or other factors.

catatoxic steroid A steroid that protects an animal against a drug by stimulating the activity of drug-metabolizing enzymes.

catch muscle A muscle, occurring in mollusks, that can remain locked in a contracted form for long periods of time.

catecholamine A dihydroxyphenylalkylamine derived from tyrosine, such as dopa, dopamine, epinephrine, or norepinephrine; they are amine-containing derivatives of catechol (1,2-dihydroxybenzene). Catecholamines affect blood vessels, intermediary metabolism, and nerve transmission.

catecholestrogen A compound formed by enzymatic hydroxylation of an estrogen; catecholestrogens are devoid of estrogenic activity and are structurally related to the catecholamines.

catechol-O-methyl transferase An enzyme that functions in the metabolism of epinephrine and norepinephrine and that utilizes *S*-adenosyl-L-methionine as a methyl group donor. *Abbr* COMT.

catemer CONCATEMER.

catenane A structure consisting of two

interlocking, circular, double-stranded, DNA molecules; catenation and decatenation are catalyzed by DNA gyrase. *Aka* catenate; catenated dimer.

catenase A collective term for an enzyme of either the endo or the exo type that catalyzes the cleavage of a polymeric chain; ribonuclease, lysozyme, and carboxypeptidase are examples.

catenated Interlocked like the links in a chain; said of interlocking, circular, double-stranded DNA molecules.

catharometer Variant spelling of katharometer.

cathepsin One of a group of intracellular proteolytic enzymes that occur in most animal tissues.

cathode The electrode by which electrons enter a solution of electrolytes and toward which the cations move in solution. With respect to properties in solution, the cathode is a negative electrode; with respect to the external flow of electrons, the cathode is a positive electrode.

cathodic 1. Of, or pertaining to, the cathode. 2. Descriptive of a component that moves toward the cathode in electrophoresis.

cation A positively charged ion.

cation exchanger A negatively charged ion-exchange resin that binds cations.

cationic detergent A surface-active agent in which the surface-active part of the molecule carries a positive charge. *Aka* cationic surface-active agent.

cavitand A synthetic organic compound that contains an enforced, rigid cavity of dimensions at least equal to those of the smaller ions, atoms, or molecules. The shape of the cavity can vary and is determined by the organic chemical structure of the molecule.

cavitation 1. The phenomenon in a flowing liquid that entails formation of vapor bubbles in a low-pressure area and collapse of these bubbles in a subsequent high-pressure area. 2. The forcing of an inert gas, such as nitrogen, into cells under high pressure and the subsequent release of the pressure, resulting in an "explosion" of the cell membrane and cell lysis.

CBG Cortisol-binding globulin.

CBN Commission on Biochemical Nomenclature of the International Union of Pure and Applied Chemistry and the International Union of Biochemistry.

CBZ-amino acid An amino acid in which the amino group has been protected by attachment of a carbobenzoxy group; used in peptide synthesis by the solid phase (Merrifield) method. *Abbr* CBZ-AA.

cc Cubic centimeter.

CCA-end ACCEPTOR END.

CCA-enzyme The enzyme that catalyzes addition of the segment 3'-CCA to the 3'-end (acceptor stem) of tRNA; a nucleotidyltransferase.

cccDNA Covalently closed circular DNA.

CCD Countercurrent distribution.

CCF Crystal-induced chemotactic factor; a chemotactic factor of the complement system. A peptide, produced by polymorphonuclear leukocytes upon phagocytosis of crystalline substances such as monosodium urate.

C-chain CONNECTING PEPTIDE.

CCK Cholecystokinin.

cd Candela; a unit of luminous intensity.

CD Circular dichroism.

^{14}C dating *See* radiocarbon dating.

cDNA Complementary DNA; a molecule of DNA that is complementary to a molecule of RNA. The DNA synthesized by the enzyme RNA-dependent DNA polymerase (reverse transcriptase) is cDNA. Complementary DNA may be single-stranded or double-stranded. *Aka* copy DNA.

C DNA *See* DNA forms.

cDNA clone A double-stranded DNA segment that is complementary to an RNA molecule and that is carried in a cloning vector.

cDNA library A clone library that differs from a gene library in that it contains only transcribed DNA sequences (exons) and no nontranscribed sequences (introns, spacer DNA). It is established by making complementary DNA from a population of cytoplasmic mRNA molecules, using the enzyme RNA-dependent DNA polymerase (reverse transcriptase), converting the single-stranded cDNA to double-stranded DNA, and cloning the latter as in the establishment of a gene library.

CDP 1. Cytidine diphosphate. 2. Cytidine-5'-diphosphate. 3. Calpain.

CDPC CDP-choline.

CDP-choline A cytidine diphosphate derivative of choline that serves as a donor of a choline residue for the synthesis of certain phosphoglycerides.

CDP-sugar Cytidine diphosphate sugar; an activated form of carbohydrates. CDP-glucose functions in the biosynthesis of cellulose in some plants and CDP-ribitol functions in the biosynthesis of bacterial cell walls.

CDR Calcium-dependent regulatory protein; calmodulin.

CEA Carcinoembryonic antigen.

Celite Trademark for a preparation of diatomaceous earth.

cell 1. The fundamental unit of living organisms; a structure that is capable of independent reproduction and that consists of cytoplasm and a nucleus, or a nuclear zone, surrounded by a cell membrane. 2. An electrical device capable of converting chemical energy into electrical energy, or vice versa; consists of two half-cells, each of which is characterized by a half-reaction. 3. A small container, such as that which holds a solution subjected to centrifugation in an analytical ultracentrifuge.

cell adhesion CELL AFFINITY.

cell adhesion molecule A neuronal cell surface glycoprotein that helps to mediate the cohesive interactions between developing neurities. *Abbr* CAM.

cell affinity The property of eukaryotic cells of a given type to adhere to each other but not to cells of a different type. *Aka* cell adhesion.

cell-associated virus The virus that is released into the medium upon the disruption of infected cells which have previously been washed to remove extracellularly adsorbed virus.

cell blotting One of a number of techniques in which cells are blotted on nitrocellulose paper and then reacted with a dye, a protein, or some other substance.

cell body PERIKARYON.

cell cap CAP (2).

cell cloning *See* cloning.

cell coat The covering of the outer surface of many eukaryotic cells that is rich in glycoproteins and mucopolysaccharides; it plays a role in contact inhibition. *See also* capsule.

cell count *See* total cell count; viable cell count.

cell culture The in vitro growth of either single cells or groups of cells that are not organized into tissues.

cell cycle The sequence of events occurring in a eukaryotic cell from one mitotic division to the next. The cell cycle is commonly divided into four periods: a mitotic phase (M), a DNA synthesis phase (S), and two gap phases (G) which separate the mitotic and synthesis phases. The sequence of phases is M, G_1, S, G_2; the G_2 phase is then followed by another mitotic phase, and so on.

cell-detaching factor PENTON.

cell differentiation *See* differentiation.

cell disruption The breakage of cells.

cell division The process whereby a parent cell divides, giving rise to two daughter cells.

cell envelope CELL MEMBRANE.

cell factor A protein that is produced by cells of solid tumors, occurring in humans and animals, and that may be produced in very small quantities by normal cells. The cell factor is an arginine-specific protease that acts

as a plasminogen activator and thereby leads to proteolysis of fibrin.

cell fractionation The separation of subcellular components; entails breakage of cells by such techniques as lysis, ultrasonication, or grinding, followed by removal of unbroken cells and cell debris (typically by ultracentrifugation) and fractionation of the remaining cell-free extract by such techniques as ultracentrifugation, electrophoresis, precipitation, or chromatography.

cell-free amino acid incorporating system A reconstituted cell-free system for the in vitro study of protein synthesis. It generally consists of ribosomes, messenger RNA (natural or synthetic), transfer RNA, enzymes, amino acids, ATP, GTP, an ATP regenerating system, buffer, and other inorganic and organic compounds.

cell-free extract A cytoplasmic extract of cells, prepared by rupturing the cells and removing unbroken cells and cell debris, commonly by centrifugation.

cell-free protein synthesis *See* cell-free amino acid incorporating system.

cell-free system A system composed of subcellular fractions and/or cell-free extracts, but devoid of intact cells.

cell fusion HYBRIDIZATION (3).

cell hybridization HYBRIDIZATION (3).

cell interaction genes A number of genes in the major histocompatibility complex of the mouse that affect the ability of various cellular components of the immune system to mount an effective immune response.

cell junction A specialized intercellular region formed by the interaction of two eukaryotic cell membranes. Cell junctions are grouped into three functional categories: (a) adhering junctions, which hold cells together mechanically (spot desmosomes, belt desmosomes, and hemidesmosomes); (b) impermeable junctions, which hold cells together and seal them so that molecules cannot leak in between them (tight junctions and septate junctions); and (c) communicating junctions, which mediate the passage of small molecules from one cell to another (gap junctions and chemical synapses).

cell line A heterogeneous group of cells derived by the first subculturing (transfer), or at any stage during the serial subculturing, of a primary culture. *See also* established cell line.

cell-mediated immunity CELLULAR IMMUNITY.

cell-mediated lympholysis An in vitro test for cellular immunity in which activated T lymphocytes are used to destroy target cells by direct contact.

cell membrane The membrane, composed of

lipids and proteins, that surrounds a cell; in eukaryotic cells the cell membrane is frequently covered by a cell coat, and in prokaryotic and plant cells it is covered by a cell wall.

cellobiose A disaccharide of glucose in which the gluose molecules are linked by means of a $\beta(1 \rightarrow 4)$ glycosidic bond; the repeating unit in cellulose.

cellogel Gelatinized cellulose acetate; an electrophoretic support.

Cellophane Trade name for transparent sheets of regenerated cellulose.

Cellosolve Trademark for ethylene glycol mono-ethyl ether.

cell plate The structure that is formed between the two daughter nuclei of a dividing plant cell during mitosis and that is the precursor of the cell wall.

cell renewal system A steady-state normal cell population in an animal in which there is a rapid cell loss that is offset by a rapid replacement of cells.

cell sap CYTOSOL.

cell strain A group of cells of limited transferability that have been derived from either a primary culture or an established cell line by selection and cloning of cells that have specific properties or markers.

cell theory The theory, proposed by Schleiden and Schwann in 1838, that all animals and plants are composed of cells and products of cells, that cells are the structural and functional units of an organism, and that an organism grows and reproduces by cell division.

cellular Of, or pertaining to, cells.

cellular immunity Immunity that is due to cell-bound antibodies. in contrast to humoral immunity. Cellular immunity involves immune responses against invading microorganisms, including fungi, parasites, intracellular viruses, cancer cells, and foreign tissues. It is responsible for such reactions as allograft rejection and delayed-type hypersensitivity and is associated with T lymphocytes. *Aka* cellular immune response; cell-mediated immune response.

cellular oncogene PROTOONCOGENE.

cellular respiration *See* respiration (1).

cellular retinol-binding protein A protein, found in many tissues of the rat, that is smaller than the plasma retinol-binding protein and that binds vitamin A but does not bind transthyretin. *Abbr* cRBP.

cellulase An enzyme that catalyzes the hydrolysis of cellulose.

cellulifugal In a direction away from the center of the cell.

cellulolytic organism An organism that has the

ability to digest (hydrolyze) cellulose because it contains a cellulase that can cleave the $\beta(1 \rightarrow 4)$ glycosidic bonds in cellulose.

cellulose A straight chain polysaccharide composed of glucose molecules linked by means of $\beta(1 \rightarrow 4)$ glycosidic bonds; the major structural material in the plant world.

cellulose acetate electrophoresis Zone electrophoresis in which a cellulose acetate sheet is used as the supporting medium.

cellulytic Capable of causing cell lysis.

cell wall The rigid structure that is external to the cell membrane and that encloses prokaryotic and plant cells; the cell wall of prokaryotic cells consists primarily of peptidoglycan and that of plant cells consists primarily of cellulose.

CELO virus Chick embryo lethal orphan virus; an adenovirus.

Celsius temperature scale A temperature scale on which the freezing and boiling points of water at 1 atm of pressure are set at 0 and 100, respectively, and the interval between these two points is divided into 100 degrees. *Aka* centigrade temperature scale.

cementum The calcified covering of dentine at the submerged portion of a tooth.

Cenozoic era The most recent geologic time period that began about 63 million years ago and that is characterized by the development of mammals.

center of symmetry The central point of a symmetrical body about which like faces are arranged in opposite pairs.

centi- Combining form meaning one-hundredth and used with metric units of measurement. *Sym* c.

centigrade temperature scale CELSIUS TEMPERATURE SCALE.

centile PERCENTILE.

central complex An intermediate in an enzyme-catalyzed reaction that cannot participate in a bimolecular reaction with substrate or product because all binding sites are occupied; a transitory complex that can only undergo a unimolecular reaction with release of a substrate or a product. *See also* transitory complex.

central dogma A description of the basic functional relations between DNA, RNA, and protein. The central dogma states that DNA serves as the template for its own replication and for the transcription of RNA, and that RNA serves as the template for translation into protein. Hence, the flow of genetic information is $\widehat{}$DNA $\rightarrow$ RNA $\rightarrow$ protein. In view of our current knowledge of reverse transcriptase and RNA replication, the central dogma must now be schematically represented as $\widehat{}$DNA $\rightleftharpoons$ $\widehat{}$RNA $\rightarrow$ protein. *Aka*

Central dogma of molecular biology. *See also* reverse transcriptase.

central ion The ion, in the Debye–Hueckel theory, that is surrounded by the ion atmosphere in which there is a statistical preference for ions of opposite charge (counterions).

central metabolic pathway AMPHIBOLIC PATHWAY.

central nervous system That part of the nervous system of vertebrates that consists of the brain, the spinal cord, and the nerves originating therefrom. *Abbr* CNS.

centrifugal elutriation A centrifugal separation technique that is based on the sedimentation of particles through a liquid which flows in a direction opposite to the direction of particle sedimentation. A specially designed rotor is used so that particles that sediment more slowly than the flow velocity of the liquid are washed out from the rotor. The separation process is controlled by varying the flow velocity of the liquid and by varying the rotation rate of the rotor.

centrifugal field The space within which a centrifugal force is of sufficient strength that its effect can be detected.

centrifugal force The force exerted on a rotating particle and directed away from the center of rotation; the force increases with increasing distance from the center of rotation.

centrifugal partition chromatography A liquid chromatographic technique for separating complex mixtures of chemical substances in the absence of a solid support; involves the simultaneous application of countercurrent distribution and centrifugation. Separation columns are connected in series with column cartridges, which are arranged in a circle around the rotor of a centrifuge with their longitudinal axes parallel to the direction of the applied centrifugal force. *Abbr* CPC.

centrifugation The process of subjecting either a solution or a suspension to a centrifugal force to separate the components of the solution or the suspension; used for the collection of precipitates, the separation of phases, and the sedimentation of macromolecules. Separation of the components is based on differences in their size, shape, and density.

centrifuge An instrument capable of generating centrifugal forces by the rotation of a rotor; the rotor holds tubes filled with the solution that is being subjected to centrifugation.

centrifuge cell *See* cell (3).

centrifuge head CENTRIFUGE ROTOR.

centrifuge rotor *See* rotor.

centrifuge tube The container, constructed of glass, metal, or plastic, that holds the solution that is subjected to centrifugation.

centriole The central granule in the centrosome; a self-replicating organelle that consists of nine groups of microtubules, arranged in the form of a hollow cylinder.

centripetal force The force that is exerted on a rotating particle and that is directed toward the center of rotation.

centromere The junction between the two arms of a chromosome to which the spindle fibers attach during mitosis.

centrosome A macromolecular complex, consisting of two centrioles, satellite bodies, and differentiated cytoplasm, that is responsible for the organization of the mitotic spindle during nuclear division. *Aka* centrosphere.

cephalic Of, or pertaining to, the head.

cephalin 1. PHOSPHATIDYL ETHANOLAMINE. 2. PHOSPHATIDYL SERINE.

cephalin–cholesterol flocculation test A liver function test that is based on the formation of a flocculant precipitate when serum from individuals with one of several forms of hepatitis is treated with a cephalin–cholesterol suspension.

cephalosporin An antibiotic, produced by the mold *Cephalosporium*, that resembles penicillin in its action. *See also* beta lactam antibiotics.

cer Ceramide.

ceramidase An enzyme that cleaves the bond between the fatty acid and sphingenine in ceramides.

ceramide An *N*-acylated sphingoid; an *N*-acyl fatty acid-substituted compound formed from sphingenine, its homologues, its isomers, or its derivatives. *Abbr* cer. See *also* glycosyl ceramide.

ceramide glycoside *See* glycosyl ceramide.

ceramide lactoside *See* cytolipin.

cercidosome A specialized organelle in trypanosomes for terminal oxidative metabolism.

cerebral Of, or pertaining to, the brain.

cerebrocuprein A copper-containing protein present in the brain.

cerebroside A monoglycosyl derivative of a ceramide that generally contains either glucose or galactose and that is abundant in the myelin sheath of nerves and in brain tissue; a 1-β-glycosylceramide; a ceramide monosaccharide.

cerebrospinal fluid The fluid that circulates through the subarachnoid spaces of the brain and the spinal cord. *Abbr* CSF.

Cerenkov radiation A radiation consisting of photons and produced when high-energy beta particles pass through either a solid or a liquid medium at speeds greater than that of light in the same medium.

ceride A wax that is an ester of a long-chain fatty acid and a higher aliphatic alcohol.

ceroid A lipid granule that may be formed in an animal, particularly in the liver, as a result of either the injection of oils rich in unsaturated fatty acids or the ingestion of various experimental diets.

ceroid pigment AGE PIGMENT.

cerotic acid A saturated fatty acid that contains 26 carbon atoms.

ceruloplasmin A serum globulin that binds eight atoms of copper per molecule and that serves to transport copper in the blood; an enzyme that is also called ferroxidase I because of its ability to catalyze the oxidation of ferrous iron to the ferric state.

cesium chloride gradient centrifugation *See* density gradient sedimentation equilibrium.

Cetavlon Trade name for a cationic detergent and bacteriostatic agent; cetyl trimethyl ammonium bromide (CTAB).

cevitaminic acid ASCORBIC ACID.

CF 1. Citrovorum factor. 2. Complement fixation.

CF$_1$ Chloroplast coupling factor.

CF$_0$–CF$_1$ complex A chloroplast system that has very similar properties to the mitochondrial F$_0$F$_1$-ATPase.

C form *See* DNA forms.

CFT Complement fixation test.

CG Chorionic gonadotropin.

CGA Catabolite gene activator protein.

C genes Genes coding for segments of the constant regions of immunoglobulin molecules.

cGMP Cyclic GMP.

cgs units The units of measurement that are based on the centimeter–gram–second system. *See also* SI.

c$_H$ The constant region of the heavy chains of the immunoglobulins. *See also* c$_L$.

Ch Choline.

chain 1. A group of like atoms linked together in succession. 2. A group of repeating units linked together in succession to form a polymer.

chain conformation The combined secondary and tertiary structure of either a polypeptide chain or a polynucleotide strand.

chain elongation *See* elongation.

chain-growth polymer A polymer formed by the addition of monomers to the growing chain through the breaking of double bonds in the monomers. *Aka* addition polymer.

chain initiation *See* initiation.

chain isomer One of two or more isomers that differ from each other in the manner in which

the side chains are attached to the main chain and in the lengths of the chains.

chain length *See* double chain length; triple chain length.

chain propagation 1. CHAIN ELONGATION. 2. The second stage in a chain reaction.

chain reaction 1. A series of chemical reactions characterized by initiation, propagation, and termination steps. The reactions of the propagation step are such that each produces a product that can serve as a reactant for a subsequent reaction and the last reaction regenerates a reactant for the first reaction; in this fashion the entire sequence of reactions in the propagation step can be repeated over and over. 2. An autocatalytic reaction, particularly a nuclear one, in which the products react with the rectants to produce more products.

chain termination *See* termination.

chain termination codon *See* termination codon.

chain termination mutation A mutation in which a normal codon, that specifies an amino acid, is altered to one of the three termination codons; a nonsense mutation.

chain terminator A compound that stops the extension of a DNA strand during replication; 2′,3′-dideoxynucleoside triphosphates, which are analogues of normal 2′-deoxynucleoside triphosphates, are an example.

chain terminator method SANGER–COULSON METHOD.

chair conformation The arrangement of atoms that resembles the outline of a chair and that is the more stable of two possible conformations for cyclohexane and other six-membered ring systems.

chalcones A group of yellow-orange bioflavonoids, derived from the parent compound chalcone. They occur primarily as glycosides and serve as intermediates in bioflavonoid biosynthesis.

challenge A dose of antigen, particularly the second or a subsequent dose, that is injected into an animal for the purpose of provoking an immune response.

challenge virus The virus that is introduced into a host subsequent to, or simultaneously with, the introduction of an interfering virus.

chalone 1. A chemical substance, such as a hypothalamic hormone, that acts like a hormone in having a target-specific effect but that is not secreted by an endocrine gland. 2. An endogenous, tissue-specific, but specifies nonspecific inhibitor of cell proliferation; a peptide, or other compound, that is a growth inhibitor. Chalones are of potential use in the control of neoplasia.

chance variable VARIATE.

channel 1. The interval between the settings of

the two discriminators in a scintillation detector that defines the range of pulse intensities that will be recorded by the system. *See also* differential counting. 2. An opening in a biological membrane through which transport of solutes may occur. A channel is generally presumed to be a water-filled passage, lined by hydrophilic groups of integral membrane proteins. *See also* ionophore.

channeling The control of a biosynthetic pathway brought about by having an assembly of enzymes such that metabolic intermediates pass directly from one enzyme to the next. In this way, metabolites are channeled along the pathway with restricted opportunity for their diffusion into the medium or their entry into the general intracellular metabolic pool. Nucleotide biosynthesis appears to involve such a control system. *Aka* processivity.

channel protein A protein that mediates passive transport across a biological membrane by forming an aqueous channel through which solutes of appropriate size and charge can diffuse.

channels ratio method A method of correcting for quenching in liquid scintillation counting by using two channels to measure the average energies of pulses of beta particles both before, and after, quenching.

chaotic oscillations *See* oscillating reactions.

chaotropic agent A substance that enhances the partitioning of nonpolar molecules from a nonaqueous to an aqueous phase as a result of the disruptive effect that the substance has on the structure of water. Chaotropic agents tend to solubilize hydrophobic (nonpolar) molecules; they are generally ions, such as thiocyanate (SCN^-), perchlorate (ClO_4^-), and trichloroacetate (CCl_3COO^-), that have a large radius, a single negative charge, and a low charge density; they are used to solubilize membrane-bound proteins, to alter the secondary and tertiary structure of proteins and nucleic acids, and to increase the solubility of small molecules. *See also* antichaotropic agent.

chaotropic series An arrangement of ions in the order of their effectiveness as chaotropic agents.

CHAPS 3-[(3-Cholamidopropyl) dimethylammonio]-1-propanesulfonate; a nondenaturing, zwitterionic detergent used for membrane biochemistry.

CHAPSO 3-[(3-Cholamidoproppyl) dimethylammonio]-2-hydroxyl-1-propanesulfonate; an unusual detergent that has a high critical micelle concentration (8 mM) and that does not denature membrane bound proteins.

characteristic The whole-number part of a logarithm.

characteristic curve A plot of the potential applied to a radiation detector versus the count rate.

characteristic ratio A quantity that describes the variation of a real polymer from a random flight chain; equal to the ratio of the square of the average end-to-end distance of the polymer divided by the square of the end-to-end distance of the random flight chain.

Chargaff's rules A set of two quantitative rules that express the base composition of double-stranded, Watson–Crick-type DNA: (1) [A] = [T]; (2) [G] = [C], where the brackets indicate concentrations of the bases in mole percent and any minor bases, if present, are included in the appropriate concentration terms. Three corollaries follow from these rules: (a) [A]/[T] = [G]/[C] = 1; (b) Σ purines = Σ pyrimidines; (c) [A] + [C] = [G] + [T], or Σ 6-aminobases = Σ 6-ketobases according to the former numbering system of purines and pyrimidines.

charge density The net electrical charge of a particle per unit surface area of the particle.

charged polar amino acid A polar amino acid that carries a charge in the intracellular pH range of about 6 to 7.

charged tRNA A transfer RNA molecule to which an amino acid has been covalently linked; an aminoacyl-tRNA molecule.

charge effect *See* primary charge effect; secondary charge effect.

charge fluctuation interactions 1. KIRKWOOD–SHUMAKER INTERACTIONS. 2. Any interaction between molecules and/or atoms that is due to fluctuating charges; Kirkwood–Shumaker interactions and London dispersion forces are two examples.

charge relay system A series of hydrogen bonds between amino acid side chains that is present at the active site of chymotrypsin and other serine proteases. It is responsible for the high degree of a nucleophilicity of the hydroxyl group of the serine residue at the active site. The series of hydrogen bonds permits a flow of electrons from a negatively charged carboxyl group of aspartic acid to the oxygen of the hydroxyl group of the serine residue, thus making the hydroxyl oxygen highly nucleophilic.

charge reversal spectrum A series of mono- and polyvalent ions in which the ions are arranged in the order of their effective concentrations for reversing the charge of an oppositely charged molecule to which the ion can bind.

charge transfer complex A complex that may be formed in oxidation–reduction reactions when, as a result of the electron transfer, the electron donor becomes positively charged and the electron acceptor becomes negatively charged so that the two are held together in a complex by electrostatic attraction.

charge transfer relay system CHARGE RELAY SYSTEM.

charging The covalent attachment of an amino acid to a transfer RNA molecule to form an aminoacyl-tRNA molecule.

Charon bacteriophage One of a group of bacteriophage derivatives, prepared by modifying phage lambda, that are used as cloning vectors in recombinant DNA technology. They are named for the boatman of Greek mythology who ferried passengers across the River Styx to the underworld.

chase 1. The effective stoppage of the incorporation of either an isotope or a labeled compound into a substance by the addition of large amounts of either the nonradioactive element or the unlabeled compound; used particularly to stop incorporation following a pulse. 2. The amount of nonradioactive element or unlabeled compound used to stop the incorporation of an isotope or a labeled compound.

chaulmoogric acid An unsaturated, cyclic fatty acid that occurs in plants and that contains 18 carbon atoms.

Chauvenet's criterion A criterion for deciding whether or not to reject a measurement that differs greatly from other, identical measurements of the same sample. The criterion states that the measurement should be rejected if the probability of its occurrence is equal to, or less than, $N/2$ where N is the total number of measurements.

ChE Cholinesterase.

chelate The ring structure formed by the reaction of two or more groups on a ligand with a metal ion. *Aka* chelate compound.

chelating agent A compound that can form a chelate with a metal ion; ethylene-diaminetetraacetic acid (EDTA) is an example.

chelation The formation of a chelate.

chelator CHELATING AGENT.

cheluviation The downward movement of chelate complexes in the soil.

chemical 1. *n* COMPOUND. 2. *adj* Of, or pertaining to, chemistry.

chemical bond *See* bond.

chemical coupling hypothesis A hypothesis of the coupling of ATP synthesis to operation of the electron transport system in oxidative phosphorylation. According to this hypothesis, the transport of electrons leads to the formation of high-energy phosphorylated intermediates which are then used to

phosphorylate ADP to ATP in coupled reactions using the high-energy compounds as common intermediates.

chemical element *See* element.

chemical equilibrium (*pl* chemical equilibria) The state of a chemical reaction in which there is no more change in the concentrations of the reactants and the products, and the free energy is at a minimum; the rate of the forward reaction is equal to the rate of the reverse reaction so that a small change in one direction is balanced by a small change in the opposite direction. *See also* steady state.

chemical evolution The gradual development of the structure and function of biomolecules which includes the synthesis of primitive molecules, their condensation to form primitive polymers, and the self-assembly of these polymers to form large molecular aggregates; a process preceding the stage of biological evolution and extending over a period of about 2 billion years. *See also* Oparin's hypothesis.

chemical fossil A fossil that contains one or more types of organic compounds that were part of the original animal or plant.

chemical interference The interference that occurs in atomic absorption spectrophotometry if chemical compounds react with the sample and prevent its dissociation into free atoms.

chemical ionization mass spectrometry A mass spectrometric technique that requires volatilization of the sample, generally in the form of derivatives, prior to ionization by collisions with ions of a reagent gas. The technique usually yields higher intensities of molecular ions and fragment ions in the high mass range than those obtained with electron impact mass spectrometry. *Abbr* CI-MS.

chemical kinetics The branch of chemistry that deals with the rate behavior of chemical reactions.

chemical messenger HORMONE.

chemical potential The partial molar free energy of a substance.

chemical quenching The quenching that occurs in liquid scintillation counting if some of the energy of the radiation is absorbed either by the sample itself or by other substances in the solution.

chemical race CHEMOVAR.

chemical reaction A reaction in which there are changes in the orbital electrons of the reacting atoms as distinct from a nuclear reaction in which there are changes in the atomic nuclei.

chemical score A measure of the nutritional

quality of a protein based on a comparison of its amino acid composition with that of egg, which has a nearly ideal balance of essential amino acids. The amount of each amino acid in the protein is expressed as a percentage of the amount of the same amino acid in egg; the lowest value, or score, is given by the essential amino acid that is limiting for growth and is a measure of the nutritional quality of the protein.

chemical shift The shift in the position of a peak in nuclear magnetic resonance relative to the position of the peak produced by a standard nucleus; it is equal to the difference between the applied magnetic field strength required to produce absorption of energy in a nucleus and the magnetic field strength predicted by the gyromagnetic ratio for an identical nucleus. The chemical shift is due to the fact that each nucleus is in a different part of the molecule and hence experiences a different field, determined by its environment. By convention, chemical shifts are measured from a reference point which is the signal obtained with tetramethylsilane $[(CH_3)_4Si;$ TMS]; this signal is arbitrarily set as the zero point. Chemical shifts are recorded in terms of an arbitrary scale, called the delta scale; one delta unit is equal to 1 ppm of the spectrometer frequency.

chemical synapse A coupling between two neurons that is mediated by chemical means as opposed to one that is mediated by electrical means (electrical synapse). *See also* synapse.

chemical taxonomy The deduction of taxonomic relationships between organisms that is based on an analysis of the distribution, composition, and structure of certain natural products found in these organisms.

chemical thermodynamics The branch of thermodynamics that deals with chemical compounds and chemical reactions.

chemical transmitter A compound, such as acetylcholine, that mediates the transmission of a nerve impulse from one nerve cell to another.

chemiluminescence The production of visible light as a result of a chemical reaction.

chemiosmotic coupling hypothesis A hypothesis of the coupling of ATP synthesis to operation of the electron transport system in oxidative phosphorylation. According to this hypothesis (proposed by Mitchell), the transport of electrons generates an energy-rich proton gradient across the mitochondrial membrane and the proton motive force associated with this gradient then drives the phosphorylation of ADP to ATP.

chemisorption Sorption that requires strong chemical forces, such as those operative in the formation of chemical bonds.

chemistry The science that deals with the composition, structure, properties, and transformations of substances.

chemoautotroph 1. A chemotrophic autotroph. 2. LITHOTROPH.

chemoencephalography The study of the metabolic patterns of the brain as a function of the behavioral experiences of the individual.

chemoheterotroph 1. A chemotrophic heterotroph. 2. ORGANOTROPH.

chemokinesis The random migration of cells, brought about by a specific substance, in the absence of a concentration gradient.

chemolithotroph An organism or a cell that utilizes for its growth (1) oxidation–reduction reactions as a source of energy, (2) inorganic compounds as electron donors for these oxidation–reduction reactions, and (3) carbon dioxide as its source of carbon atoms.

chemoorganotroph An organism or a cell that utilizes for its growth oxidation–reduction reactions as a source of energy, and organic compounds both as electron donors for these oxidation–reduction reactions and as a source of carbon atoms.

chemoreceptor A receptor that is stimulated by chemical compounds.

chemostat An apparatus for maintaining bacteria in the exponential phase of growth over prolonged periods of time. This is achieved by the continuous addition of fresh medium, which is balanced by the continuous removal of the overflow, so that the volume of the growing culture remains constant.

chemosynthetic Chemotrophic.

chemosynthetic organism 1. CHEMOLITHOTROPH. 2. CHEMOORGANOTROPH.

chemotactic Of, or pertaining to, chemotaxis.

chemotactic hormone A hormone that has a chemotactic effect such as a steroid that causes aggregation of amoeba and slime molds.

chemotaxin A substance that is derived from complement and that induces leukocytes to move from an area of lower to one of higher chemotaxin concentration.

chemotaxis A taxis in which the stimulus is a chemical compound and cells or organisms move along a concentration gradient. Such directed migration is believed to play a role in the localization of immune effector cells at inflammation sites, the movement of phagocytic cells toward various attractants, and the secretion of lysosomal enzymes. *See also* lymphocyte-derived chemotactic factors.

chemotaxonomy The classification of organisms on the basis of the distribution and/or the composition of chemical substances in these organisms; the use of DNA base composition data for the taxonomy of bacteria is an example.

chemotherapeutic agent A chemical that interferes with the growth of either microorganisms or cancer cells at concentrations at which it is tolerated by the host cells.

chemotherapy The treatment of a disease by means of chemotherapeutic agents.

chemotroph An organism or a cell that uses oxidation–reduction reactions as a source of energy.

chemotropism A tropism in which the stimulus is a chemical compound.

chemovar A plant that, when grown in one locality, contains one or more different chemical substances as compared to those it contains when grown in a different locality.

chemurgy A branch of applied chemistry dealing with the industrial use of chemicals derived from farm produce.

chenic acid CHENODEOXYCHOLIC ACID.

chenodeoxycholic acid A bile acid that has two hydroxyl groups and that is the major component of the bile of hens, geese, and other fowl; it occurs in small amounts in the bile of other animals and humans. *Aka* chenic acid.

CHES 2-(*N*-Cyclohexylamino)ethanesulfonic acid; used for the preparation of biological buffers in the pH range of 8.6 to 10.0. *See also* biological buffers.

chiasma (*pl* chiasmata) The cytological manifestation of crossing over; the visible connection between chromatids during meiosis.

chick antidermatitis factor PANTOTHENIC ACID.

chickenpox VARICELLA.

chicle *See* gutta.

chimera An individual composed of two or more, genetically different, types of cells, derived from genetically different zygotes. *See also* blood group chimera; mosaic; radiation chimera.

chimeric DNA A recombinant DNA molecule that carries unrelated genes from different species.

chimeric protein 1. A fused protein in which the two linked proteins are derived from two different organisms. 2. A multifunctional protein prepared by artificial manipulation of a native protein.

Chinese restaurant syndrome A condition characterized by severe headaches, numbness, palpitation, and other symptoms of neurological disturbance resulting from increased levels of glutamic acid; can be brought about by ingestion of Chinese food to which

large quantities of monosodium glutamate are generally added as seasoning. *Aka* Kwok's disease.

chip A generic term for an integrated circuit; a single package (a slice of silicon; a wafer) holding hundreds of thousands of microscopic components and used in computers.

chirality The necessary condition that allows for a discrimination between two enantiomers; the right- or left-handedness of an asymmetric molecule or of an asymmetric object. Due to their chirality, enantiomeric molecules cannot be brought into coincidence with each other by rotation about axes of symmetry, by reflection in planes of symmetry, or by a combination of these maneuvers. An asymmetric carbon is a chiral carbon.

chiroptical properties Optical properties (such as circular dichroism and optical rotatory dispersion) that relate to specific stereochemical and electronic structural features of a compound (such as chirality).

chirotopic Describing an atom residing in a chiral environment; both an asymmetric carbon atom and the atoms attached to it are considered to be chirotopic. Thus, in the compound CHBrClF, all of the atoms and the spaces between them are chirotopic because the entire molecule is chiral. Chirotopicity deals with the local geometry of compounds as opposed to stereogenicity which deals with stereoisomerism.

chi sequence A segment of eight bases (GCTGGTGG) in the DNA of *E. coli* that acts as a hot spot in genetic recombination involving the rec A protein; the segment occurs at about every 10 kb in the genome.

chi-squared distribution A distribution that may be regarded as that of the sum of squares of independent variates from a normal population. The distribution is of importance for inferences concerning population variances or standard deviations.

chi-squared test A test of significance based upon chi-squared statistics; permits a comparison of the goodness of fit of a set of observed values with theoretically expected ones.

chi structure A structure that resembles the Greek letter chi (χ) and that is formed in genetic recombination during crossover between two double-stranded DNA molecules.

chitin A homopolysaccharide of *N*-acetyl-D-glucosamine that is a major constituent of the hard, horny exoskeleton of insects and crustaceans.

chitosamine GLUCOSAMINE.

chitosome An organelle in fungi that contains

the enzyme chitin synthetase and that functions in the synthesis of chitin microfibrils.

Chl Chlorophyll.

chloragosome A cytoplasmic particle of unknown composition that is found in modified peritoneal cells of the earthworm.

chlorambucil A mutagenic, alkylating agent.

chloramphenicol A broad-spectrum antibiotic, produced by *Streptomyces venezuelae*, that inhibits protein synthesis by attaching to the 70S prokaryotic ribosome and inhibiting peptidyl transferase, thereby preventing peptide bond formation. Chloramphenicol also inhibits the peptidyl transferase of eukaryotic mitochondrial (not cytoplasmic) ribosomes and has immunosuppresive activity. *Abbr* CM; CAM.

chloramphenicol particle A ribosomal subparticle isolated from bacteria in which protein synthesis has been inhibited by chloramphenicol. *Abbr* CM particle.

chlorella A genus of unicellular, nonmotile, green algae used for the study of photosynthesis.

chloremia HYPERCHLOREMIA.

chlorenchyma Plant tissue that contains chloroplasts.

chlorhemin crystals TEICHMANN'S CRYSTALS.

chloride shift The movement of chloride and hydroxyl ions across the erythrocyte membrane as a result of the movement of bicarbonate ions in the opposite direction; this exchange of ions occurs at both the tissue and the lung level, but the relative directions of ion movement are reversed at the two levels.

chlorin The parent compound of the chlorophylls; dihydroporphyrin.

chlorine An element that is essential to humans and several classes of animals and plants. Symbol, Cl; atomic number, 17; atomic weight, 35.453; oxidation states, -1, $+1$, $+5$, $+7$; most abundant isotope, ^{35}Cl, half-life, 3×10^5 years, radiation emitted, beta particles and positrons.

Chlorobium chlorophyll A chlorophyll occurring in some sulfur bacteria.

Chlorobium vesicle CHLOROSOME.

chlorocruorin A hemoglobin-like, respiratory pigment of invertebrates that has a molecular weight of 3.5×10^6 and that contains 190 heme groups per molecule.

p-chloromercuribenzoic acid A reagent that reacts with the sulfhydryl groups of proteins. *Abbr* PCMB.

chloromycetin CHLORAMPHENICOL.

chlorophyll The green pigment that occurs in plants and that functions in photosynthesis by absorbing the radiant energy of the sun. The chlorophylls are a group of closely related pigments, structurally related to the porphyrins, but containing magnesium instead of iron. Major chlorophylls of land plants are chlorophyll a and b, that of some marine organisms is chlorophyll c, and that of photosynthetic bacteria is bacteriochlorophyll.

chlorophyllide A molecule of chlorophyll from which the phytol side chain has been removed by hydrolysis.

chlorophyll unit PHOTOSYNTHETIC UNIT.

chloroplast A chlorophyll-containing chromoplast that is the site of photosynthesis in green plants.

chloroplast coupling factor A protein factor in chloroplasts that is analogous to the mitochondrial coupling factor F_1.

chlorosis The yellowing of normally green plant components that is due to a failure of chlorophyll development.

chlorosome The light-harvesting structure in green, anaerobic bacteria of the families *Chlorobiaceae* and *Chloroflexaceae*. Chlorosomes are ovoid, bag-like structures, closely associated with the cytoplasmic membrane, that contain the bulk of the antenna bacteriochlorophyll; they do not contain the photosynthetic reaction centers which are located in the cytoplasmic membrane. *Aka* chlorobium vesicle.

cholagogue A substance that aids in the solubilization of cholesterol.

cholecalciferol A compound that has vitamin D activity and that is obtained by ultraviolet irradiation of 7-dehydrocholesterol; designated vitamin D_3.

cholecystokinin A polypeptide hormone, secreted by the duodenum, that stimulates the secretion of digestive enzymes by the pancreas and that stimulates the contraction of the gall bladder. *Abbr* CCK.

cholecystokinin–pancreozymin CHOLECYSTOKININ.

choleic acid A specific complex formed between a steroid, particularly a bile acid, and fatty acids, hydrocarbons, or other organic compounds.

cholelithiasis A disease, characterized by the formation of concretions (calculi) in the biliary tract that consist chiefly of cholesterol.

choleragen A toxin of the cholera bacillus that affects the plasma membrane of intestinal cells.

cholera toxin An endotoxin, produced by *Vibrio cholerae*, that has enzymatic activity; it catalyzes the transfer of ADP-ribose from intracellular NAD^+ to a G protein which, in turn, functions as a regulatory protein of adenyl cyclase. *See also* ADP-ribosylation.

cholestane The parent ring system of the sterols.

cholestanol A minor sterol of the animal body; 5,6-dihydrocholesterol.

cholesteremia CHOLESTEROLEMIA.

cholesteric structure One of the three specific structures of a thermotropic liquid crystal. *See also* liquid crystal.

cholesterol The principal sterol of vertebrates and a precursor of bile acids and steroid hormones; it is synthesized entirely from acetyl coenzyme A by condensation reactions of isoprene units.

cholesterol desmolase *See* desmolase.

cholesterolemia The presence of excessive amounts of cholesterol in the blood.

cholesterol ester An ester formed from cholesterol and a fatty acid. Cholesterol esters are very hydrophobic. *Aka* cholesteryl ester.

cholesterol ester transfer protein A protein, believed to catalyze the transfer of cholesterol esters from high-density lipoproteins to very low-density or low-density lipoproteins.

cholesterol intoxication theory A theory according to which atherogenesis results from either the ingestion of high-cholesterol fat or the deficiency of certain vitamins.

cholesterolosis A condition that is characterized by the formation of cholesterol deposits in various organs and tissues, and that is caused by a disturbance in lipid metabolism.

cholesterol oxidase DESMOLASE.

cholesterosis CHOLESTEROLOSIS.

cholesteryl ester CHOLESTEROL ESTER.

cholic acid The most abundant bile acid in human bile; it has three hydroxyl groups.

choline A methyl group donor that occurs in some phospholipids and in acetylcholine. It is generally classified with the B vitamins, since it is required in the diet under certain conditions, but it is not a typical vitamin and has no known coenzyme function. *Abbr* Ch.

choline acetyltransferase The enzyme that catalyzes the reaction in which choline and acetyl coenzyme A are converted to acetylcholine and coenzyme A.

cholinergic Of, or pertaining to, nerve fibers that release acetylcholine at the nerve endings.

cholinergic agonist *See* decamethonium.

cholinergic antagonist *See* alpha bungarotoxin.

cholinesterase The enzyme that catalyzes the hydrolysis of acetylcholine and a variety of other choline esters and that is present in various tissues other than the nervous system. *Aka* cholinesterase II; nonspecific cholinesterase; pseudocholinesterase. *See also* acetylcholinesterase.

cholinolytic Descriptive of a pharmacological substance that blocks the action of acetylcholine.

cholinomimetic Descriptive of a pharmacological substance that imitates the action of acetylcholine.

chondriogene 1. A plasmagene that is attached to a mitochondrion. 2. A gene of mitochondrial DNA.

chondrioid MESOSOME.

chondriome 1. A collective term for all of the mitochondria of a cell. 2. A composite intracellular structure involving several organelles, one of which is a mitochondrion; the complex of kinetoplast and mitochondrion found in trypanosomes is an example.

chondriosome 1. MITOCHONDRION. 2. Any mitochondrial macromolecular complex involved in biosynthetic reactions.

chondrocyte A cartilage cell; a specialized cell of connective tissue that produces collagen and proteoglycans.

chondroitin A glycosaminoglycan (mucopolysaccharide) composed of D-glucuronic acid and N-acetyl-D-galactosamine.

chondroitin sulfate The sulfate ester of chondroitin and a major constituent of bone and cartilage. *See also* glycosaminoglycan.

chondrome The genetic information contained within the mitochondria of a cell.

chondronectin A factor, distinct from fibronectin, which mediates the attachment of chondrocytes to collagen.

chopper A rotating wheel with alternate silvered and cut out sections that is placed in the light path of a spectrophotometer, thereby allowing the light beam to pass alternately through the sample solution and through the reference solution.

chorioallantoic membrane The membrane, used in the assay of viruses, that surrounds the embryo of the chicken and other birds.

choriogonadotropin HUMAN CHORIONIC GONADOTROPIN.

choriomammotropin PLACENTAL LACTOGEN.

chorion *See* amnion.

chorionic Of, or pertaining to, the placenta.

chorionic gonadotropin *See* human chorionic gonadotropin.

chorionic somatomammotropin PLACENTAL LACTOGEN.

chorismic acid An important intermediate in the biosynthesis of aromatic compounds. It can be converted to anthranilic acid or prephenic acid; the former is a precursor of tryptophan and the latter is a precursor of both phenylalanine and tyrosine.

Chou–Fassman method A statistical method for predicting the secondary structure of a protein from its amino acid sequence. The method uses the known three-dimensional structures of soluble proteins to calculate the frequency of occurrence of amino acid residues in specific secondary structures, such as the alpha helix, the beta sheet, and the beta turn. On the basis of this information, and the known amino acid sequence of a given protein, one can then predict the type of secondary structure that a given segment of the polypeptide chain is likely to have. *Aka* Chou and Fassman rules.

Christmas disease　HEMOPHILIA B.

Christmas factor　The factor in the intrinsic pathway of blood clotting that activates factor X.

ChRNA　Chromosomal RNA.

chromaffin granule　A subcellular organelle in the adrenal medulla that synthesizes, stores, and releases the catecholamines epinephrine and norepinephrine.

chroman　A redox lipid such as tocopherol.

chromatic transient　A sudden short-lived increase or decrease in the rate of a photosynthetic reaction when the wavelength, but not the effective intensity, of the incident light is suddenly changed.

chromatid　One of the two strands that result from the duplication of a chromosome and that are held together by a centromere; the chromatids become separate chromosomes upon division of the centromere.

chromatin　The nuclear material of the chromosomes in higher organisms that consists principally of DNA and histones. *See also* euchromatin; heterochromatin.

chromatin body　NUCLEOID (1).

chromatofocusing　A separation method for proteins that is similar to isoelectric focusing but requires no specialized equipment; it is carried out on an ion exchange column. Like isoelectric focusing, chromatofocusing uses a stabilizing medium containing a pH gradient but in this case, a titration reaction, rather than an electric field, is employed to create the gradient. Gradient stability presents no problem since the gradient is defined by the rate of titration which can be controlled.

chromatogram　The visual record of a chromatographic separation, either in the form of the chromatographic support itself, or in the form of a tracing thereof.

chromatographic　Of, or pertaining to, chromatography.

chromatographic resolution　*See* resolution (3).

chromatographic spray　An atomizer used in the spraying of chromatograms for detecting the separated sample spots.

chromatographic support　*See* support.

chromatography　The separation of complex mixtures of molecules that is based on the repetitive distribution of the molecules between a mobile and a stationary phase. The mobile phase may be either a liquid or a gas, and the stationary phase may be either a solid or a solid coated with a liquid. The distribution of the molecules between the two phases is determined by one or more of four basic processes, namely adsorption, gel filtration, ion exchange, and partitioning. The operation of these processes, coupled to the movement of the mobile phase, results in a differential migration of the molecules along the stationary phase. *See also* adsorption chromatography; gel filtration chromatography; ion exchange chromatography; partition chromatography.

chromatophore　A bacteriochlorophyll-containing chromoplast of photosynthetic bacteria.

chromatophoresis　A separation technique for complex protein mixtures that involves the sequential use of chromatography (HPLC) and electrophoresis (PAGE).

chromatophorotropic hormone　MELANOCYTE-STIMULATING HORMONE.

chromatopile　A stack of filter paper disks that have the same diameter and that are compressed within a chromatographic column which is used for preparative-scale separations.

chromatoplate　The plate, covered with a support, that is used in thin-layer chromatography.

chromatosome　A macromolecular complex consisting of one molecule of H1 or H5 histone, two molecules each of histones H2A, H2B, H3, and H4, and a DNA segment of about 160 base pairs. *See also* nucleosome.

chromium　An element that is essential to humans and animals and that is a component of the glucose tolerance factor. Symbol, Cr; atomic number, 24; atomic weight, 51.996; oxidation states, +2, +3, +6; most abundant isotope, ^{52}Cr; a radioactive isotope, ^{51}Cr, half-life, 27.8 days, radiation emitted, gamma rays.

chromobindin　*See* calelectrin.

chromocenter　A heterochromatic structure formed by the aggregation of polytene chromosomes in the fruit fly, *Drosophila*.

chromogen　1. The colorless precursor of a pigment. 2. The colorless parent compound of a dye.

chromogenic　1. Producing a pigment or a color. 2. Of, or pertaining to, a chromogen.

chromogranin　A soluble protein in chromaffin granules.

chromoisomer　One of two or more isomers that differ from each other in their color.

chromomere A thickening along a eukaryotic chromosome that results from the local coiling of the chromosome threads.

chromonema (*pl* chromonemata). One of the coiled threads in a eukaryotic chromosome.

chromoneme The thread of DNA in bacterial cells and in viruses.

chromophobe A cell that does not stain readily.

chromophore The group of atoms in a compound that is capable of absorbing light and that is responsible for the color of the compound.

chromoplast A pigment-containing plastid, such as a chloroplast or a chromatophore, that functions in photosynthesis.

chromoprotein A conjugated protein in which the nonprotein portion is a pigment or some other chromophoric material.

chromosomal aberration An abnormality in a chromosome that results from the deletion, the duplication, or the rearrangement of the genetic material; the abnormality is referred to as intrachromosomal or interchromosomal depending on whether it is the result of changes in one or in two chromosomes. *Aka* chromosomal mutation.

chromosomal puff *See* chromosome puff.

chromosomal RNA RNA that is associated with the chromosome and that is distinct from RNA involved in protein synthesis (messenger, transfer, and ribosomal RNA); primer RNA is an example.

chromosome 1. A structure in the nucleus of eukaryotic cells that consists of one or more large double-helical DNA molecules that are associated with RNA and histones; the DNA of the chromosome contains the genes and functions in the storage and in the transmission of the genetic information of the organism. 2. The nuclear DNA of eukaryotic cells, the DNA of prokaryotic cells, or the DNA of viruses. 3. The RNA of viruses. *See also* genophore.

chromosome break A break in the structure of a chromosome as that produced by some carcinogenic alkylating agents.

chromosome jumping A type of chromosome walking in which use is made of chromosomal aberrations to shift the walking to another position on either the same or a different chromosome. *See also* chromosome walking.

chromosome map 1. GENETIC MAP. 2. CYTOGENETIC MAP.

chromosome puff A localized swelling in a polytene chromosome that represents a region of active RNA or DNA synthesis. Very large chromosome puffs form loop-like structures known as Balbiani rings.

chromosome rearrangement A chromosomal aberration in which chromosomal segments are rearranged by inversion and translocation.

chromosome scaffold *See* scaffold.

chromosome set The entire group of chromosomes representing the genome of an organism.

chromosome substitution Replacement of one or more chromosomes by homologous or homoeologous chromosomes from a different source.

chromosome theory of cancer A theory, proposed by Boveri in 1912, according to which cancer is due to the presence of abnormal chromosomes in the cells as a result of irregularities in mitosis.

chromosome walking A technique for the sequential identification and isolation of DNA clones representing regions larger than, and adjacent to, a given DNA region. A specific gene that has been cloned is used as a probe to screen a gene library for all the DNA fragments containing the marker gene. The fragment, containing a nucleotide sequence farthest removed from the marker gene, is then cloned. This represents a single walking step. The procedure is then repeated, using the new fragment as a new probe, and so on. Thus, step by step, nucleotide sequences farther and farther removed from the original marker gene, are identified and isolated.

chromosorb A chromatographic adsorbent prepared by the fusion of diatomaceous earth either with, or without, sodium carbonate.

chromotrope A substance that can appear in two or more different colors depending on the extent to which it is covered with a metachromatic dye.

chronic disease A disease that persists for a relatively long period of time (months or years), terminating either in recovery or in death.

chronic exposure Prolonged exposure to radiation; used to describe experimental conditions in which an organism is exposed either to a continuous low-level of radiation or to a fractionated dose.

chronic toxicity test A toxicity test performed on laboratory animals that requires the administration of a chemical at least once daily for periods of 1 to 2 years.

chronometric method A method of assaying for the enzyme amylase by measuring the time that is required for the complete hydrolysis of all the starch in the reaction mixture.

chronon A hypothetical linear sequence of DNA, the transcription of which takes about a day, that is believed to be related to circadian rhythms.

chronopotentiometry An electroanalytical method for studying electrolysis reactions by measuring the potential, as a function of time, at a microelectrode on which is impressed a small constant current.

chronotropic effect An effect on the rate of rhythmic movements, especially those of the heart.

C'H$_{50}$ unit UNIT OF COMPLEMENT.

chyle The lymphatic fluid that is taken up by the lymph vessels from the intestine during digestion and that is characterized by its high content of fat globules.

chylomicron (*pl* chylomicrons; chylomicra) A colloidal fat globule that occurs in blood and lymph and that serves to transport fat from the intestine. Chylomicrons represent the plasma lipoprotein fraction of lowest density (less than 0.95 g/mL) and contain about 2% protein, 7% phospholipid, 8% cholesterol and cholesterol esters, and 83% triglycerides (triacylglycerols). Chylomicrons have a molecular weight of about 10^9 to 10^{10}, a flotation coefficient above 400 S, and are classified as the omega fraction on the basis of electrophoresis. *See also* lipoprotein.

chymase A chymotrypsin-like, mast cell proteinase, that is present in fairly large concentrations in skeletal muscle, lung, and skin, and that is believed to be involved in inflammatory reactions.

chyme The semifluid mass of partially digested material that is passed from the stomach into the intestine.

chymodenin A basic polypeptide, isolated from duodenal mucosa, that stimulates pancreatic secretion of chymotrypsinogen.

chymosin A proteolytic enzyme from the fourth stomach of the calf that has properties similar to those of pepsin. It leads to milk coagulation by hydrolyzing casein to give paracasein which then reacts with calcium to yield the insoluble curd.

chymotropic pigment A pigment dissolved in the vacuole of a plant cell.

chymotrypsin An endopeptidase that catalyzes the hydrolysis of peptide bonds, principally those in which the carbonyl group is contributed by tryptophan, phenylalanine, or tyrosine.

chymotrypsinogen The inactive precursor of chymotrypsin.

chymotryptic Of, or pertaining to, chymotrypsin.

chymotryptic peptides The peptides obtained by the digestion of a protein with the enzyme chymotrypsin.

Ci Curie.

CIE Crossed immunoelectrophoresis.

CIEP Counterimmunoelectrophoresis.

CIG Cold insoluble globulin.

ciliary Of, or pertaining to, cilia.

cilium (*pl* cilia). A thread-like cellular extension that functions in the locomotion of bacteria and unicellular eukaryotic organisms; cilia are more numerous and shorter than flagella.

CI-MS Chemical ionization mass spectrometry.

cinchona alkaloids *See* quinoline alkaloids.

circadian rhythm A biological clock that has a period of approximately 24 hours.

circular birefringence The birefringence produced by left and right circularly polarized light.

circular chromatography A paper chromatographic technique in which the material is allowed to migrate radially; may be carried out by the use of a circular piece of filter paper, to the center of which the sample is applied, and from which a sector is cut out and dipped into the solvent.

circular covalent Descriptive of a circular polynucleotide strand in which the 3'- and 5'-termini are linked together covalently.

circular dichroism The dichroism that results from the differences, at a given wavelength, between the extinction coefficients of left and right circularly polarized light when such light is passed through a solution containing a chromophore. Circular dichroism depends on the asymmetry of the electric charge distribution around the chromophore and may either be an intrinsic property of the molecule or be induced in the molecule as a result of perturbations in the surroundings. Circular dichroism is used in the study of the secondary structure of macromolecules. *Abbr* CD. *See also* magnetic circular dichroism.

circular DNA A DNA molecule that has a closed ring-type structure, not necessarily that of a geometric circle.

circular genetic map The genetic map of a closed, ring-type chromosome as that of *E. coli* and other bacteria. *Aka* circular linkage map.

circularly polarized light Light in which the electric field vectors, at any point on the axis along which the light is being propagated, rotate in a plane perpendicular to that axis. The light is referred to as right or left circularly polarized light, depending on whether the electric field vector rotates in a clockwise or in a counterclockwise direction as seen when looking toward the light source.

circular noncovalent Descriptive of a circular polynucleotide strand in which the 3'- and 5'-termini are held together by noncovalent bonds.

circular permutation The formation of different linear segments when the same circle is opened at one or more different points; a principle that is invoked in relating a circular genetic map to the genetic structure of linear DNA molecules.

circulating water bath A waver bath equipped with a pump so that the water can be pumped from the bath to some apparatus and returned from there to the bath.

circulation The movement of a liquid through a circuit, such as the movement of blood or lymph through the blood or lymphatic vessels, respectively.

circulative virus PERSISTENT VIRUS.

circulin One of a group of peptide antibiotics, produced by *Bacillus circulans*, that are related in structure and function to polymyxin.

cirrhosis An inflammatory disease of the liver that is characterized by the replacement of liver cells with fat and fibrous tissue.

cis 1. Referring to the configuration of a geometrical isomer in which two groups, attached to two carbon atoms linked by a double bond, lie on the same side with respect to the plane of the double bond. 2. Referring to two mutations, particularly those of pseudoalleles, that lie on the same chromosome.

cis-acting locus A region on a DNA molecule that affects the activity of genes that are located on the same molecule. *See also* trans-acting locus.

cis-dominant Descriptive of a genetic region that affects the activity of one or more adjacent regions in the same chromosome.

cis effect The influence of one gene on the expression of another gene that is located on the same chromosome.

cis interactions Noncovalent interactions between adjacent segments, located on the same polypeptide chain of an immunoglobulin molecule.

cis isomer *See* cis (1).

cisplatin A coordination compound, *cis*-diamminedichloroplatinum, that is highly cytotoxic and that has been used as an effective antitumor agent.

cisterna (*pl* cisternae) A sac in a cell or in an organism that serves as a reservoir.

cis–trans isomers *See* cis (1); trans (1).

cis–trans test A complementation test of pseudoalleles.

cistron The unit of genetic function; a section of the chromosome that codes for a single polypeptide chain; a structural gene.

citrate-activated thrombin The material obtained by the dissociation of thrombin when thrombin is dissolved in 25% sodium citrate solution.

citrate cleavage enzyme The cytoplasmic enzyme, ATP citrate lyase, which cleaves citrate to acetyl-SCoA and oxaloacetate. This reaction provides acetyl-SCoA for the cytoplasmic reactions of fatty acid biosynthesis.

citrate cycle CITRIC ACID CYCLE.

citrate lyase CITRATE CLEAVAGE ENZYME.

citrate–pyruvate cycle A cyclic set of reactions, involving the sequence citrate–oxaloacetate–pyruvate, that functions in fatty acid synthesis.

citrate synthase The enzyme that catalyzes the first reaction of the citric acid cycle in which acetyl coenzyme A condenses with oxaloacetic acid to yield citric acid and coenzyme A.

citric acid The symmetrical tricarboxylic acid that is formed in the first reaction of the citric acid cycle in which acetyl coenzyme A condenses with oxaloacetic acid.

citric acid cycle The cyclic set of reactions that constitutes the core of the central metabolic pathway in most living cells. The citric acid cycle is initiated by the condensation of acetyl coenzyme A with oxaloacetic acid which yields citric acid and coenzyme A. One turn of the citric acid cycle, in conjunction with the operation of the electron transport system and oxidative phosphorylation, achieves the equivalent of the oxidation of one molecule of acetic acid to carbon dioxide and water and the synthesis of 12 molecules of ATP. *Abbr* TCA cycle.

citroens A group of complex information-transforming reproducing objects that evolve by natural selection; a family of "living" organisms, terrestrial or nonterrestrial, that are believed to have been the forerunners of intelligent life.

citrogenase CITRATE SYNTHASE.

citrovorum factor Folinic acid; a growth factor for *Leuconostoc citrovorum*. *Abbr* CF.

citrulline A nonprotein alpha amino acid that is an intermediate in the urea cycle.

citrullinuria A genetically inherited metabolic defect in humans that is associated with mental retardation and that is characterized by high concentrations of citrulline in the blood and in the urine; due to a deficiency of the enzyme argininosuccinate synthetase. *Aka* citrullinemia.

CK Creatine kinase.

Cl Chlorine.

c_L The constant region of the light chains of the immunoglobulins, *See also* c_H.

CL Cardiolipin.

cladogenesis Branching evolution; dendritic evolution.

cladogram PHYLOGENETIC TREE.

Clark electrode An oxygen electrode that usually consists of a platinum cathode and a silver anode, both immersed in the same solution of concentrated KCl, and separated from the test solution by a membrane. Oxygen is reduced at the cathode to water and the current thus produced is proportional to the oxygen activity in the solution.

classical pathway *See* complement.

classical sedimentation equilibrium SEDIMENTATION EQUILIBRIUM.

classical thermodynamics EQUILIBRIUM THERMODYNAMICS.

classification The systematic arrangement of organisms into groups based on the natural relations between the organisms. The groups, proceeding from the largest to the smallest, are kingdom, phylum, class, order, family, genus, and species.

class II MHC antigens I-REGION-ASSOCIATED ANTIGENS.

class switching The changeover by a lymphocyte from the synthesis of antibodies of one class (such as IgM) to the synthesis of antibodies of a different class (such as IgG or IgA). *Aka* heavy chain class switching.

clastic reaction *See* phosphoroclastic reaction; thioclastic reaction.

clastogen A compound that produces chromosomal abnormalities.

clathrate An inclusion complex produced by trapping molecules of one kind in the lattice network formed by molecules of a second kind; frequently refers to the stable complex produced by trapping nonpolar solute molecules in a cage formed by water molecules. *Aka* clathrate compound; clathrate crystal.

clathrin The scaffolding protein that covers the surface of coated pits. Three molecules of clathrin (MW 180,000) associate with three molecules of smaller polypeptides to form a three-legged protein complex, called a triskelion; a network of the latter covers the coated pits.

Clausius' law APHORISM OF CLAUSIUS.

clearance A measure of the efficiency of the kidney in removing a substance from the blood; specifically, $C = UV/P$, where C is the clearance in milliliters of plasma per minute, U and P are the concentrations of the substance in the urine and in the plasma, and V is the flow of urine in milliliters per minute. The clearance is known as either a maximum or a standard clearance depending on whether the flow of urine is more or less than 2 mL/min.

clearance factor LIPOPROTEIN LIPASE.

clearing factor LIPOPROTEIN LIPASE.

clear plaque A plaque that is produced when all of the cells in the area of the plaque are lysed.

cleavage map RESTRICTION MAP.

Cleland's convention A simplified representation of multisubstrate enzyme systems in which the different enzyme forms are written from left to right below a horizontal line and reactants (A,B,C,...) and products (P,Q,R,...) are indicated by vertical arrows pointing toward the line (for reactants) or away from it (for products). The terms uni, bi, ter, and so on are used to designate the number of substrates (reactants) and products; thus a uni bi reaction is one having one substrate and two products. *Aka* Cleland's notation.

Cleland's reagents The compounds dithioerythritol (DTE) and dithiothreitol (DTT), that are used for the protection of sulfhydryl groups against oxidation to disulfides and for the reduction of disulfides to sulfhydryl groups.

Cleland's rules A set of rules for predicting the type of inhibition of an enzyme-catalyzed reaction, or the type of interaction between an enzyme and a cosubstrate or other substance, from an inspection of experimental kinetic data. The rules apply to a steady-state mechanism that either contains no random sequences or contains random sequences in rapid equilibrium. The two fundamental rules may be stated as follows: (a) The ordinate intercept of a double reciprocal plot is affected by a substance that associates reversibly with an enzyme form other than the one with which the variable substrate combines. (b) The slope of a double reciprocal plot is affected by a substance that associates with an enzyme form that is the same as, or is connected by a series of reversible steps to, the enzyme with which the variable substrate combines.

climacteric rise The increase to a maximum in the respiration of ripening fruits that may occur either before or after the removal of the fruit from the plant, depending on the fruit and on the harvesting procedure.

clinical centrifuge A centrifuge, generally considered to be a small table model, that is capable of generating speeds of approximately 3000 rpm and centrifugal forces of approximately $2000 \times g$.

clinical chemistry A branch of chemistry that deals with the qualitative and quantitative determination of chemical substances in humans, particularly of substances related to medicine.

Clinistix Trademark for a group of paper strips impregnated with chemicals and used for semiquantitative determinations of components in urine and/or blood.

CLIP Corticotropin-like intermediate lobe peptide.

C$_{55}$ lipid carrier BACTOPRENOL.

clock-timing A method of timing used in scintillation counters in which the timing device is not turned off during the interval that is required for the electronic processing of a pulse. *See also* live-timing.

clonal selection theory A selective theory of antibody formation according to which an antigen selects a particular cell clone from among a large number of lymphoid cell clones and then stimulates these cells to proliferate and to synthesize antibodies. Each cell clone is believed to be different and to contain a unique set of genes for specific immunoglobulins so that each clone can synthesize either antibodies having only one type of specificity or, at most, antibodies having a few types of specificity.

clone 1. A group of genetically identical cells or organisms, derived from a single cell or organism by asexual reproduction; proceeds via binary fission in prokaryotes and via mitosis in eukaryotes. 2. Multiple copies of identical DNA sequences produced by recombinant DNA technology.

cloned DNA PASSENGER.

cloned line A cell line consisting of a single clone.

clone library A large collection of bacterial or viral recombinant DNA clones that contain many of the DNA sequences from a single organism. Two general kinds of clone libraries are gene libraries and cDNA libraries.

cloning 1. Molecular cloning. The production of many identical copies of a gene, replicated from a single gene introduced into a host cell. The phrase "to clone a particular gene" is common usage to mean "to form a vector containing a particular gene." *See also* recombinant DNA technology. 2. Cell cloning. The production of a group of genetically identical cells from a single cell; the production of cells that synthesize a specific antibody is an example. *See also* monoclonal antibodies; hybridoma.

cloning host HOST (3).

cloning vector VECTOR (3).

cloning vehicle VECTOR (3).

clonotype 1. The phenotype of a clone of cells. 2. The homogeneous product of a clone of cells.

closed chain RING.

closed circle CIRCULAR DNA.

closed circuit system A system for measurements of indirect calorimetry in which the oxygen consumption, but not the carbon dioxide production, is determined.

closed culture BATCH CULTURE.

closed-promoter complex The initial conformation in transcription in which RNA polymerase has become bound to the promoter but the two DNA strands have not yet become locally unwound.

closed reading frame A segment in mRNA that cannot be translated into an amino acid sequence because it contains one or more termination codons.

closed system A thermodynamic system that can exchange energy, but not matter, with its surroundings.

close packing Descriptive of a structure in which nonbonded atoms are surrounded by other nonbonded atoms in such a way that the distances between the atoms are equal, to the extent possible, to the sum of their van der Waals contact radii.

Clostridium A genus of spore-forming, chemoorganotrophic bacilli that is widespread in soil, mud, and the intestinal tract of humans and animals; includes the organisms causing botulism, gangrene, and tetanus.

closure transformation The transformation of a micellar membrane from one having large spaces between the micelles ("open") to one having small spaces between them ("closed").

clot *See* blood clot.

clot-promoting factor HAGEMAN FACTOR.

clot retraction The shrinking of a blood clot that occurs upon standing and that is accompanied by the expressing of the serum.

clotting time The time, in minutes, required for blood to clot when it is exposed to air.

cloud chamber A chamber that contains a supersaturated atmosphere and that is used for observing the tracks produced by ionizing particles; the ions formed by an ionizing particle serve as nuclei for the formation of fog droplets which indicate the path taken by the ionizing particle.

cloverleaf model A model for the structure of transfer RNA that resembles a cloverleaf and that is based upon the folding of the transfer RNA strand back upon itself so as to permit the formation of a maximum number of intrachain hydrogen bonds; the structure contains four (or five) hydrogen-bonded segments, referred to as arms, to which are attached three (or four) non-hydrogen-bonded segments, referred to as loops.

clupein A protamine isolated from herring; a protein containing 30 amino acid residues.

cluster 1. MULTIENZYME SYSTEM. 2. METABOLON (2).

cm Centimeter.

CM Chloramphenicol.

cmc Critical micelle concentration.

CM-cellulose *O*-(Carboxymethyl)cellulose; a cation exchanger.

CMP 1. Cytidine monophosphate (cytidylic acid). 2. Cytidine-5'-monophosphate (5'-cytidylic acid).

CM-particle Chloramphenicol particle.

CM-sephadex *O*-(Carboxymethyl)Sephadex; a cation exchanger which contains the grouping —$CH_2CO_2^-$, linked via ether bonds to the sephadex.

CMV Cytomegalovirus.

CNS Central nervous system.

Co Cobalt.

CoI Cozymase I.

CoII Cozymase II.

CoA Coenzyme A.

coacervate A polymer-rich phase or droplet that is formed by coacervation and that is believed by some to have been a forerunner of primitive cells.

coacervation The spontaneous separation of an aqueous solution of a highly hydrated polymer into two phases, one having a relatively high, and the other having a relatively low concentration of the polymer.

coagulase The enzyme, produced by *Staphylococcus*, that has thrombokinase-like activity and causes citrated, or oxalated, plasma to coagulate. *Aka* coagulating enzyme.

coagulation The formation of a clot as that formed in blood clotting or in the boiling of egg white; clots may be soft, semisolid, or solid.

coarctation An increase in the cross-linking, the hardening, and the shrinking of a membrane.

coarse control The control of biochemical systems that is achieved by the regulation of the amount of an enzyme, as in enzyme induction and enzyme repression.

CoASAc Acetyl coenzyme A.

CoASH Coenzyme A.

coat *See* spore coat; viral coat.

coated pit A cell membrane depression, so called because it is coated with a scaffolding protein (clathrin). Various polypeptide hormones, such as insulin and epidermal growth factor, show the phenomenon of down regulation which involves internalization and degradation of the hormone receptors via coated pits. Receptors may either be located in the coated pit and bind ligands there, or move toward the coated pit, after binding ligands elsewhere. In either case, the receptor–ligand complexes aggregate in the coated pits. Small vesicles, called receptosomes, bud off from the coated pits and entrap the receptor–ligand complexes. The receptosomes move into the cell (endocytosis), associate with structures known as GERL, and fuse with lysosomes which then degrade both the receptors and their ligands (the hormones). *Aka* coated vesicle.

coated vesicle COATED PIT.

coat protein A protein of the viral coat.

cobalamin A 5,6-dimethylbenzimidazole derivative of a cobamide. Depending on the substituent (R) at the 6th coordination position of the cobalt atom, these compounds are named as follows: R = —CN, cyanocobalamin (vitamin B_{12}); R = —OH, hydroxycobalamin (vitamin B_{12a}); R = —H_2O, aquacobalamin (vitamin B_{12b}); R = —NO_2, nitritocobalamin (vitamin B_{12c}); R = 5′-deoxyadenosyl, 5′-deoxyadenosyl cobalamin (coenzyme B_{12}); R = —CH_3, methylcobalamin (methyl B_{12}). *Var sp* cobalamine.

cobalt An element that is essential to several classes of animals and plants. Symbol, Co; atomic number, 27; atomic weight, 58.9332; oxidation states, +2, +3; most abundant isotope, ^{59}Co; a radioactive isotope, ^{60}Co, half-life, 5.26 years, radiation emitted, beta particles and gamma rays.

cobamide A coenzyme form of vitamin B_{12} in which the 6th coordination position of the cobalt atom is linked covalently to the 5′-carbon of 5′-deoxyadenosine. *Var sp* cobamid; *Aka* coenzyme B_{12}; 5′-deoxyadenosyl cobalamin.

coboglobin An artificially prepared hemoglobin or myoglobin molecule in which the iron atom has been replaced by a cobalt atom.

cobratoxin A snake venom poison from the cobra, *Naja naja*, that acts like α-bungarotoxin.

cocaine A tropane alkaloid and the main alkaloid of the coca plant, *Erythroxylan coca*, and related forms. A narcotic drug that has euphoric or hallucinogenic effects.

cocarboxylase THIAMINE PYROPHOSPHATE.

cocarcinogen An agent that enhances the effect of a carcinogen either by increasing the yield of tumors or by shortening the time required for a tumor to appear.

cocarcinogenesis The enhancement of the action of one carcinogen by the simultaneous administration of a second carcinogen.

coccus (*pl* cocci). A bacterium having a more or less spherically shaped cell; cocci represent one of the three major forms of bacteria. *See also* bacillus; spirillum.

cochromatography A chromatographic technique for establishing the identity of a compound by applying the compound, together with one or more known compounds, to a chromatographic support.

cocktail 1. The mixture of reagents required for a cell-free amino acid incorporating system; excludes the mRNA, ribosome, and enzyme fractions. 2. The solution of fluors used for liquid scintillation counting.

coconversion The concurrent conversion of two chromosomal sites during gene conversion.

codase AMINOACYL-tRNA SYNTHETASE.

code 1. *n* GENETIC CODE. 2. *v* To direct the incorporation of an amino acid in response to a codon.

codecarboxylase PYRIDOXAL PHOSPHATE.

codegenerate codon SYNONYM CODON.

codehydrogenase I NICOTINAMIDE ADENINE DINUCLEOTIDE.

codehydrogenase II NICOTINAMIDE ADENINE DINUCLEOTIDE PHOSPHATE.

codeine An opium alkaloid which is converted to morphine as the poppy ripens. A narcotic drug which is slightly analgesic, strongly inhibits coughing, and is fairly nondangerous regarding the development of addiction.

codeword CODON.

code word family A group of codons that code for either one or two amino acids and that differ only in their 3'-terminal base.

codeword triplet CODON.

coding DNA Sections of DNA that actually code for proteins or nontranslated RNAs such as tRNA and rRNA.

coding ratio The ratio of the number of nucleotides in an mRNA molecule to the number of amino acids in the polypeptide chain that is coded for by the mRNA; the number of nucleotides in a codon.

coding strand That strand of double-stranded DNA that has the same base sequence as that in mRNA except that thymine in DNA substitutes for uracil in RNA; the strand that does not serve as a template for transcription; the antisense strand.

coding triplet CODON.

codogenic strand ANTICODING STRAND.

codon The sequence of three adjacent nucleotides that occurs in mRNA and that functions as a coding unit for a specific amino acid in protein synthesis. The codon determines which amino acid will be incorporated into the protein at a particular position in the polypeptide chain. There are 64 codons, 61 of which code for amino acids and 3 of which serve as termination codons. Codons are written 5'-XYZ-3'.

codon dictionary *See* dictionary.

codon recognizing site ANTICODON.

coef Coefficient; used in the Cleland nomenclature of enzyme kinetics to indicate a factor, composed of one or more rate constants, by which the concentration of a component must be multiplied. Thus, (coef A)*A* may mean, for example, $(k_3 + k_5)$ [A], where [A] is the concentration of component A, and k_3 and k_5 are rate constants.

coefficient of coincidence COINCIDENCE (2).

coefficient of correlation *See* correlation coefficient.

coefficient of variation A measure of the relative variation of data with respect to the mean; equal to the ratio of the standard deviation to the mean. Often multiplied by 100 to express it as a percentage. Thus, CV = $(\sigma/\bar{X})100$, where CV is the coefficient of variation, σ is the standard deviation, and $\bar{X}$ is the observed mean.

coefficient of viscosity VISCOSITY.

coenocyte A multinucleate organism that lacks cell walls.

coenzyme The organic molecule that functions as a cofactor of an enzyme.

coenzyme I NICOTINAMIDE ADENINE DINUCLEOTIDE.

coenzyme II NICOTINAMIDE ADENINE DINUCLEOTIDE PHOSPHATE.

coenzyme A The coenzyme form of the vitamin pantothenic acid that functions in metabolism as a carrier of an acetyl or some other acyl group; the acyl group is linked to the sulfhydryl group of coenzyme A. *Abbr* CoASH; CoA.

coenzyme B$_{12}$ COBAMIDE.

coenzyme-coupled reactions A set of coupled reactions which are linked by means of a coenzyme; the coenzyme is a product of the first reaction and serves as a reactant for the second reaction.

coenzyme F FOLATE COENZYME.

coenzyme M 2,2'-Dithiodiethanesulfonic acid; an intermediate in the formation of methane by methanogenic bacteria.

coenzyme Q One of a group of benzoquinone derivatives that have an isoprenoid side chain of varying length and that function as electron carriers in the electron transport system. Coenzyme Q is structurally related to vitamin K and is not tightly bound or covalently linked to a protein. Instead, a small pool of this compound is in the lipid phase of the mitochondrial membrane and serves as a mobile carrier of electrons. *Abbr* CoQ.

coenzyme Q–cytochrome c reductase complex COMPLEX III.

coenzyme R BIOTIN.

CoF Coenzyme F.

cofactor The nonprotein component that may be required by an enzyme for its activity. The cofactor may be either a metal ion (activator) or an organic molecule (coenzyme) and it may be attached either loosely or tightly to the enzyme; a tightly attached cofactor is known as a prosthetic group.

cofactor-requiring mutant A phage mutant that requires a cofactor for its adsorption to the host cell.

cognates A transfer RNA molecule (cognate tRNA) and its corresponding aminoacyl-tRNA synthetase (cognate synthetase).

cognon The passive part of a two-component cell–cell interaction (agglutination) system that provides the site that is recognized;

analogous to the substrate in an enzyme–substrate complex, the antigen in an antibody–antigen complex, or the ligand in a receptor–ligand complex. *See also* cognor.

cognor The active recognizer part of a two-component cell–cell interaction (agglutination) system; analogous to the enzyme in an enzyme–substrate complex, the antibody in an antibody–antigen complex, and the receptor in a receptor–ligand interaction. *See also* cognon.

coherent light Light in which all of the waves are in phase.

coherin A heat-stable, hypophyseal polypeptide that stimulates the coordinate contractions of the intestine.

cohesive end One of two complementary, single-stranded segments at opposite ends of each of the two strands of a double-stranded nucleic acid molecule; the presence of these segments permits the joining of the ends of the molecule and its conversion to a double-stranded circular form. *Abbr* cos. *Aka* cohesive sites (termini). *See also* restriction enzymes.

cohesive end ligation The joining of restriction fragments, terminating in cohesive ends, by means of DNA ligase.

Cohn fraction One of a number of fractions of proteins that are precipitated from plasma when the plasma is treated with ethanol at low temperatures.

coiled coil SUPERHELIX.

coimmune Denoting two mutants of the same phage that do not differ in the gene that controls the synthesis of the immunity substance.

coincidence 1. The occurrence of radioactive events within a span of time that is too short to permit their resolution by a radiation counter. 2. The ratio of the observed number of double crossovers to the theoretical number of double crossovers.

coincidence circuit An electronic circuit that has two inputs but only one output and that produces an output pulse only if two input pulses arrive either simultaneously or within a known time interval; used in liquid scintillation counting to decrease the level of the dark current due to background counts.

coincidence correction The correction that is applied in radiation counting for coincidence loss.

coincidence counting The counting of pulses, produced by radioactive disintegrations, by means of a coincidence circuit.

coincidence loss The loss of register of pulses as a result of their occurring within too short an interval to permit their resolution by the electronic circuit.

coincidence time 1. The minimum length of time that must elapse between two events to permit them to be registered as two separate events. 2. The maximum length of time that may separate two pulses and still permit the registration of an output pulse by means of a coincidence circuit.

cointegrate formation REPLICON FUSION.

cointegrate structure The circular molecule, formed in plasmid fusion, that contains the two plasmids and two copies of the transposon. Formation of the cointegrate structure is believed to be a necessary step in transposition.

co-ion An ion that has a charge of the same sign as that of another ion; ions, used in ion-exchange chromatography, that have charges of the same sign as those of the ion-exchange resin, are considered to be co-ions.

coisogenic Descriptive of animals that are genetically identical except for one or two genetic loci that have been altered by mutation.

Colcemid Trade name for a colchicine derivative that is a mitotic poison.

colchicine An alkaloid that binds to tubulin and prevents its polymerization; an anti-mitotic drug that prevents formation of the mitotic spindle and blocks cells in mitosis.

cold Containing no radioactive isotopes.

cold agglutinin *See* cold hemagglutinin.

cold antibody An antibody that has a higher titer at lower temperatures.

cold-blooded POIKILOTHERMIC.

cold hemagglutinin A hemagglutinin that causes agglutination of red blood cells at lower temperatures but leads to their dispersion at higher temperatures.

cold-insoluble globulin A serum-derived protein that behaves like a cryoglobulin and that is closely related to the membrane protein fibronectin; it represents the plasma form of fibronectin. *Abbr* CIG. *See also* cryoglobulin.

cold-sensitive enzyme An enzyme that loses its activity and stability as the temperature is lowered; due to dissociation of the enzyme into inactive subunits as hydrophobic and/or electrostatic interactions become weakened as the temperature is decreased.

cold-sensitive mutant A mutant that has a higher minimum temperature of growth than the wild-type organism.

cold shock A sudden chilling.

cold-stable enzyme An enzyme that has an unusually low optimum temperature.

Col factor Colicin factor.

colicin factor A bacterial plasmid that allows the organism to produce colicins. *Abbr* Col factor.

colicinogen The bacteriocinogen of colicin; the colicin factor.

colicinogenic factor Colicin factor.

colicins A group of proteins, produced by certain strains of *Enterobacteriaceae*, that are bactericidal for certain other strains of the same family; bacteriocins produced by strains of *E. coli*.

coliform bacteria 1. Bacteria belonging to the genera *Escherichia* and *Aerobacter*. 2. A large and diverse group of bacteria that includes *E. coli* and bacteria related to it.

colinear code A code in which the sequence of the codons in mRNA corresponds to the sequence of the amino acids in the polypeptide chain that is coded for by that mRNA.

colinearity The concept that the sequence of the nucleotides in a gene corresponds to the sequence of the amino acids in the polypeptide chain that is specified by that gene. The concept is supported by studies of the enzyme tryptophan synthetase from *E. coli* and of the gene specifying this enzyme; it has been shown that the order and spacings of mutational changes in the gene correspond to the order and spacings of the amino acid substitutions in the enzyme.

colipase A protein, present in pancreatic juice, that stabilizes the interaction of lipase with fat droplets in the intestine.

coliphage A phage that infects the bacterium *E. coli*.

collagen A fibrous scleroprotein that is the major protein of connective tissue and the most abundant protein in higher animals. Collagen forms an unusual triple helix and has an unusual amino acid composition in which glycine, proline, and hydroxyproline together constitute about two-thirds of the total amino acid residues. The basic unit of collagen is tropocollagen. Several types of collagen occur: Type I—the major adult form, widespread, lowest carbohydrate content, low hydroxylation of lysine; type II—in cartilage, intermediate hydroxylation of lysine; Type III—in blood vessels and fetal skin, disulfide bonds, low hydroxylation of lysine; Type IV—in basement membranes, disulfide bonds, highest carbohydrate content, high content of hydroxylysine and hydroxyproline; Type V—in basement membranes, high content of carbohydrate and hydroxylysine.

collagenase An enzyme that catalyzes the hydrolysis of collagen; the only proteolytic enzyme capable of degrading native collagen to low molecular weight, soluble peptides.

collagen helix The unusual triple helix of collagen in which the polypeptide chains do not have the alpha helical configuration.

collateral sensitivity The increased sensitivity of an individual to an anticancer drug that results from the individual's resistance to a different anticancer drug.

colligative property A property of a solution, such as osmotic pressure, that depends on the number of solute particles per unit volume of solution and that does not depend on the size or shape of the particles.

collimating lens A lens that converts light striking it into a beam of parallel rays.

collimator A device, composed of either lenses or slits, that is used to convert incident radiation into a narrow beam of parallel rays.

collisional quenching The energy transfer from an excited molecule to another molecule that occurs when the two molecules approach each other to within the contact distance that they attain during molecular collisions.

collision theory The theory of chemical kinetics according to which the velocity of a chemical reaction is a direct function of the molecular collisions. The velocity depends on the frequency of these collisions and on the energy and the relative orientations of the colliding molecules.

colloid 1. A macromolecule or a particle in which at least one dimension has a length of 10^{-9} to 10^{-6} m. 2. THYROID COLLOID.

colloidal Of, or pertaining to, colloids.

colloidal dispersion A colloidal system that consists of a dispersed phase and a dispersion phase and that is thermodynamically unstable and not readily reconstituted after separation of the phases. *See also* colloidal solution; suspension.

colloidal electrolyte ASSOCIATION COLLOID.

colloidal solution A true solution that consists of colloidal macromolecules and solvent and that is thermodynamically stable and readily reconstituted after separation of the macromolecules from the solvent. *See also* colloidal dispersion; suspension.

colloid osmotic pressure The osmotic pressure of a colloidal system that is separated by a membrane that is impermeable to the colloidal particles but is permeable to crystalloids.

colon bacillus *See Escherichia coli*.

colony A group of contiguous cells that grow in or upon a solid medium and are derived from a single cell. A bacterial colony may be of smooth or rough morphology depending on whether or not the cells possess either a capsule or other surface components.

colony bank GENE LIBRARY.

colony hybridization An in situ hybridization technique in which bacterial colonies are transferred from an agar surface to a filter paper. The cells are then lysed on the paper, and the released DNA is fixed, and then hybridized with a labeled DNA or RNA probe. After washing the paper, the labeled DNA segments are located by autoradiography. The technique can also be used for phage plaques.

colony-stimulating factor One of a number of protein growth factors that are required for the proliferation of hematopoietic cells in tissue culture. *Abbr* CSF.

color Whimsical name for a kind of internal charge possessed by quarks. *See also* elementary particles.

colorimeter 1. An optical or a photoelectric instrument for measuring either color differences or color intensities; used for the quantitative determination of compounds in solution by colorimetry. 2. An instrument for the exact matching of two colored solutions.

colorimetry 1. A method of quantitative analysis in which a compound is determined by a comparison of the color produced by the reaction of a reagent with both standard and test solutions of the compound. 2. A method of quantitative analysis in which a compound is determined by the exact matching of the colors produced by the reaction of a reagent with both a standard and a test solution of the compound.

color quenching The quenching that occurs in liquid scintillation counting when some of the light that is emitted by the fluor is absorbed by colored components of the sample.

color vision The capacity to perceive colors that is due to the cones in the retina.

colostral milk COLOSTRUM.

colostrum The milk secreted during the first few days after parturition. *Aka* colostral milk.

col pasmid A plasmid that contains genes for the synthesis of colicins.

column A cylindrical tube that is filled with a chromatographic support and is used in column chromatography.

column chromatography A chromatographic technique in which the stationary phase consists of a porous solid contained in a cylindrical tube, and the mobile phase percolates through the solid; used primarily for adsorption, gel filtration, and ion-exchange chromatography.

coma A state of profound unconsciousness from which the individual cannot be aroused.

comb growth test CAPON TEST.

combination The selection of one or more of a set of distinct objects without regard to order. The number of possible combinations, each containing r objects, that can be formed from a collection of n distinct objects is $n!/(n - r)!r!$ and is denoted as $\binom{n}{r}$, $_nC_r$, C_r^n, or $C(n, r)$.

combination code An early version of the genetic code according to which the nucleotide sequence in mRNA was assumed to be random so that all possible sequence permutations of a given triplet could code for the same amino acid.

combination electrode An electrode that consists of a glass tube into which both a reference electrode and a glass electrode have been incorporated.

combinatorial association The association of heavy chains, of any type, with light chains, of any type, in a given population of immunoglobulin molecules.

combinatorial translocation The association of any variable region gene with any constant region gene, for a given immunoglobulin chain, within the same multigene family.

combined acidity The acidity of gastric juice that is due both to protein-bound hydrochloric acid and to acids other than free hydrochloric acid, such as lactic acid and butyric acid. The combined acidity is equal to the difference between the total titratable acidity of gastric juice and the acidity due to free hydrochloric acid.

combining site *See* antigen binding site.

cometesimal A body of matter formed from primordial dust; the consolidation of cometesimals is believed to have led to the formation of the planets close to the periphery of the solar system.

comicellization The solubilization of an insoluble or a slightly soluble compound through the formation of a mixed micelle that consists of the compound and of an amphipathic compound. The process occurs at concentrations of amphipathic compound that are considerably below its critical micelle concentration.

comma-less code A genetic code in which there are no signals to indicate either the beginning or the end of a codon; in such a code, the displacement of the starting point will lead to the reading of a different set of codons. *Aka* comma-free code.

command voltage The chosen voltage that maintains a fixed membrane potential in the voltage clamp technique.

commensalism A stable condition in which two organisms of different species live in close physical association and neither benefit nor harm accrues to either organism as a result of this association. The term is also used to refer to either symbiosis or parasitism.

comminuted Finely divided.

committed step A reaction that forms part of a sequence of reactions and that, once it takes place, ensures that all the subsequent reactions in the sequence will also take place. It generally proceeds with a large loss in free energy so that the step is essentially irreversible; it may produce a metabolite that has no other role than to serve as an intermediate in the biosynthesis of the end product of the reaction sequence. A committed step may be (a) the first reaction catalyzed by a multienzyme system; (b) the first reaction in a biosynthesis pathway; or (c) the reaction at a branch point in a biosynthetic pathway.

common intermediate principle The principle that two energetically coupled reactions must proceed by having a common intermediate that transfers the energy from one reaction to the other. *See also* pacemaker enzyme.

commonsense phenomenon PROXIMITY EFFECT.

comparative biochemistry A branch of biochemistry that deals with the nature, the origin, and the control of biochemical differences among organisms.

compartmentation The unequal distribution of a substance, such as a metabolite or an enzyme, within a cell or within an organism; may refer to the occurrence of the substance in particular structures or to its being a part of a given pool.

compensated acidosis An acidosis in which the pH of the blood remains constant due to the effect of mechanisms that counteract the decrease in pH produced initially.

compensated alkalosis An alkalosis in which the pH of the blood remains constant due to the effect of mechanisms that counteract the increase in pH produced initially.

compensation point The concentration of carbon dioxide below which, for a given organism, its uptake by photosynthesis is less than its output by respiration.

competence 1. The physiological state of a bacterial cell that enables it to undergo transformation. 2. The physiological state of a cell that enables it to either recognize an antigen or synthesize antibodies. 3. The physiological state of a part of an embryo that enables it to react to an inductor by determination and differentiation in a specific direction.

competent cell A cell possessing competence.

competitive inhibition The inhibition of the activity of an enzyme that is characterized by an increase in the apparent Michaelis constant (substrate concentration required for one-half the maximum velocity) and by an increase in the slope of a double reciprocal plot (1/velocity versus 1/substrate concentration) compared to those of the uninhibited reaction; the maximum velocity remains unchanged. *See also* degree of inhibition.

competitive inhibitor An inhibitor that produces competitive inhibition and that generally bears a structural similarity to the substrate of the inhibited enzyme. The competitive inhibitor competes with the substrate for, and binds to, the active site of the enzyme. *See also* degree of inhibition.

competitive protection The protection of biomolecules against damage from an ionizing

radiation that is provided by chemical substances (radical scavengers) which compete with the biomolecules for the harmful free radicals produced by the radiation. *See also* restitutive protection.

competitive protein-binding technique An assay for a hormone in body fluids that is similar to a radioimmunoassay except that the binding protein is not antibody but either a plasma or a cellular receptor site for the hormone.

competitive radioassay RADIOIMMUNOASSAY.

competitive radioligand assay RADIOIMMUNOASSAY.

complement A group of at least 9 serum proteins, found in the blood of all vertebrates, that are not immunoglobulins but participate in a variety of immunological reactions; so called, because it complements the action of antibodies in killing cells. Complement may be activated by two separate enzyme cascades, called the classical and alternate pathways. The former involves activation by antigen–antibody complexes containing certain immunoglobulins; the latter involves activation by cell wall polysaccharides of bacteria and yeast. When activated, complement has three functions: (1) it increases vascular permeability and vasodilation (anaphylatoxins); (b) it facilitates ingestion and destruction of antigens by phagocytes (phagocytosis, opsonization); (c) it lyses certain types of foreign cells, such as invading bacteria and erythrocytes from other species (immune hemolysis, immune adherence). *Sym* C; C′. *See also* properdin.

complement activation *See* complement binding reaction; complement fixation.

complemental air The volume of air that can be forcibly drawn into the lungs after the normal tidal air has been inspired.

complementarity The matching up and the mutual adaptation of surfaces in two interacting macromolecules. Complementarity plays a role in such processes as the binding of a substrate to an enzyme, the binding of an antigen to an antibody, and the binding of one nucleic acid strand to another.

complementary Of, or pertaining to, complementarity.

complementary base pairing The linking of bases in double-stranded DNA (via H bonds) according to the base-pairing rules.

complementary base sequence The base sequence in a nucleic acid strand that is related to the base sequence in another strand by the base-pairing rules; thus, the sequence A-T-G-C in a DNA strand is complementary to the sequence T-A-C-G in a second DNA strand and to the sequence U-A-C-G in an RNA strand.

complementary RNA *See* cDNA.

complementary genes Two genes that are similar in their phenotypic effect when they are present separately but which, when they are present together, interact to produce a different phenotypic effect; two nonallelic genes that complement each other to produce a single trait. *Aka* complementary factor.

complementary interaction The interaction of two genes that leads to phenotypic effects that are different from those produced by either one alone.

complementary RNA A synthetic RNA molecule, transcribed in vitro from a specific DNA molecule; may be radioactively labeled and used as a probe. *Abbr* cRNA.

complementary strand A polynucleotide chain that has a complementary base sequence to that in another chain.

complementation The interaction between two sets of either cellular or viral genes that occurs within the same cell and that permits the cell or the virus to function even though each set of genes carries a mutated and nonfunctional gene. *See also* intergenic complementation; intragenic complementation; in vitro complementation.

complementation group A group of mutants that carry mutations located within the same cistron.

complementation map A genetic map constructed on the basis of complementation experiments.

complementation test A test for determining whether the mutations of two mutant chromosomes occurred in the same gene so that complementation between the genes is possible; performed by introducing the two mutant chromosomes simultaneously into the same cell.

complement binding reaction The initial event that activates the complement system and that involves binding of the C1 component of complement (C1q) to the Fc portion of an immunoglobulin molecule (IgG or IgM).

complement fixation The activation of complement by either the classical pathway or the alternative pathway. *Abbr* CF. *See also* complement.

complement fixation inhibition test The inhibition of a complement fixation test, as that produced by the presence of certain haptens or antibodies.

complement fixation test A test for either antigen or antibody that is based on the binding of complement to the antigen–antibody complex and on the consequent disappearance of complement activity from a mixture of antigen, antibody, and complement. *Abbr* CFT.

complement-fixing antibody An immunoglobulin of the IgG or IgM type that binds to (fixes) complement.

complete antibody An antibody that is fully reactive and that gives the ordinary serologic reactions of precipitation and agglutination.

complete antigen *See* antigen.

complete medium A minimal medium that is fortified with yeast extract, casein hydrolysate, and the like to permit the growth of auxotrophs.

complete oxidation The oxidation of organic compounds such that carbon dioxide is the only carbon-containing product; the term may refer either to a single reaction or to a group of reactions.

complete protein A protein that contains all of the amino acids commonly found in proteins.

complete transduction Transduction in which the DNA from the donor bacterium becomes fully integrated into the chromosome of the recipient bacterium.

complete virion A fully assembled and infective virus particle; a mature virus.

complex 1. An aggregate of two or more molecules, particularly macromolecules, held together by noncovalent forces in a definable structural relation and as a result of specific interactions. The binding can result from any combination of hydrogen bonding, hydrophobic interactions, ionic interactions and van der Waals interactions. 2. The product formed by the interaction of a metal ion and ligands. *See also* complex ion.

complex I One of the four complexes derived from electron transport particles that, by itself, can catalyze the oxidation of NADH by coenzyme Q.

complex II One of the four complexes derived from electron transport particles that, by itself, can catalyze the oxidation of succinate by coenzyme Q.

complex III One of the four complexes derived from electron transport particles that, by itself, can catalyze the oxidation of reduced coenzyme Q by cytochrome *c*.

complex IV One of the four complexes derived from electron transport particles that, by itself, can catalyze the oxidation of reduced cytochrome *c* by molecular oxygen.

complex V F_0F_1-ATPase.

complex glycoproteins *See* GLYCOSYLATION.

complex hapten A high molecular weight hapten that constitutes a separate part of a complete antigen and that gives a visible precipitin reaction with the appropriate antibody.

complex ion The product that is formed by the interaction of a metal ion and ligands and that carries a charge. *See also* complex (2).

complexity A measure of the amount of nonrepetitive DNA in a given sample; defined as the combined length of all the unique DNA segments in the sample. It can be expressed in terms of the number of base pairs or in terms of some mass unit. The complexity increases in evolution from a primitive to an advanced species.

complex lipid 1. AMPHIPATHIC LIPID. 2. One of a group of diverse lipids. *See also* lipid.

complex locus A closely linked cluster of functionally related genes resulting in a plethora of apparently different phenotypes due to mutations in a single gene. Some of the loci in *Drosophila* and the locus for the human hemoglobin genes are complex loci.

complex medium A medium that contains a variety of both known and unknown chemical ingredients.

complex oligosaccharides A group of N-linked oligosaccharides that contain several different monosaccharides in addition to mannose, including sialic acid, fucose, galactose, and *N*-acetyl glucosamine. These oligosaccharides may have 2, 3, or 4 branches and are then designated as bi- tri-, or tetraantennary.

complexone IONOPHORE.

complex RNA The group of different mRNA molecules of which each one occurs in the form of only a few copies per cell.

complex virion A virus the morphology of which is more intricate than that of either an icosahedral or a helical virus.

component 1. An independently variable, and chemically distinct, substance. 2. An ingredient of a mixture.

component I *See* nitrogenase.

component II *See* nitrogenase.

composite transposon A larger and more complex transposable element in bacteria than an insertion sequence. Composite transposons are flanked by insertion sequences in either an inverted repeat or a direct repeat configuration. They contain genes unrelated to the insertion function, such as genes carrying antibiotic resistance or genes for sugar fermentation.

compositionism HOLISM.

compound A substance composed of two or more elements, such that the atoms of the elements are firmly linked together and are present in definite proportions.

compound lipid COMPLEX LIPID.

compound microscope A microscope having two or more lenses.

Compton effect The ejection of an orbital electron from an atom by the impingement on the atom of a high-energy photon, such as a photon of x rays or gamma rays. Part of the energy of the photon is used to eject the electron and to impart kinetic energy to it; the remainder of the energy is emitted as a photon having a lower energy and a longer wavelength than the impinging photon.

Compton recoil electron The electron ejected from an atom in the Compton effect.

Compton smear The continuous spectrum of the energies of the photons that are emitted in the Compton effect; a continuous spectrum is obtained, since any fraction of the impinging x-ray or gamma-ray energy can be dissipated in this fashion.

compulsory ordered mechanism ORDERED MECHANISM.

computer An automatic electronic system that can receive a large number of items of information, subject them to specific and often complex calculations, and provide the results either in direct form or in terms of control of other systems; a device that can receive and then follow instructions to manipulate information. *See also* analog computer; digital computer.

computer graphics The combination of various communication and graphic arts skills, computer equipment, and computer techniques that results in rapid and economical production of detailed drawings by a computer.

computer hardware *See* hardware.

computer interface The auxiliary equipment used in linking a computer to an apparatus or to an instrument.

computerized axial tomography A noninvasive x-ray technique for obtaining visualization of cross-sectional planes of living tissues at various depths. The image is produced by computer synthesis of x-ray transmission data obtained in many different directions through the given plane. *Abbr* CAT.

computer language A computer programing system; a set of conventions (symbols and terms) specifying how to tell a computer what to do; the items that, when entered into a computer, cause it to respond and to carry out specific operations.

computer network Two or more connected computers that have the ability to exchange information.

computer program A series of commands, instructions, or statements, put together in a way that tells a computer to do a specific thing or a series of things.

computer software *See* software.

COMT Catechol–O–methyl transferase.

comutation A mutation that occurs in the vicinity of, and simultaneously with, a selected mutation.

Con A Concanavalin A.

conalbumin OVOTRANSFERRIN.

concanavalin A A lectin, isolated from jack beans, that agglutinates red blood cells and stimulates T lymphocytes to undergo mitosis.

concatemer An oligomeric nucleic acid molecule in which complete genomes are held together in an end-to-end manner by either covalent or noncovalent bonds; occurs in the replication of some viral genomes. *Aka* catemer; concatenate; concatener.

concatenate CONCATEMER.

concatenation The formation of concatemers.

concatener CONCATEMER.

concave exponential gradient An exponential density gradient that is formed if the solution introduced into a mixing chamber of constant volume has a lower concentration than the solution initially present in the mixing chamber.

concentrated solution A solution that contains a large amount of solute.

concentration The amount of solute in a solution. *See also* formal solution; molal solution; molar solution; osmolal solution; osmolar solution; percent solution; ppb; ppm.

concentration equilibrium constant APPARENT EQUILIBRIUM CONSTANT (1).

concentration gradient The change of concentration with distance, as the change of concentration along a density gradient or across a membrane.

concentration of enzymatic activity A measure of the concentration of an enzyme in solution that is equal to the enzymatic activity divided by the volume of the solution; it is expressed in terms of katals per liter.

concentration work OSMOTIC WORK.

concentric cylinder viscometer An instrument for measuring the viscosity of a liquid by placing the liquid in the space between two concentric cylinders, rotating one cylinder at a constant speed, and measuring the torque exerted on the other cylinder.

concerted acid–base catalysis Catalysis that consists of the simultaneous action of both acidic and basic catalytic groups.

concerted catalysis Catalysis that results from the presence of more than one catalytic grouping in the active site of an enzyme.

concerted divalent inhibition The inhibition of an allosteric enzyme that is produced when two effectors are bound to the enzyme simultaneously, but that is not produced when either effector is bound to the enzyme alone.

concerted feedback inhibition The feedback inhibition of an enzyme that is produced when two or more end products are present simultaneously, but that is not produced when an end product is present alone.

concerted model A model for allosteric enzymes—proposed by Monod, Wyman, and Changeux—according to which the enzyme exists in two different conformational forms (a relaxed, R-form, and a tensed, T-form) that are in equilibrium with each other. The two forms differ in their capacity to bind substrate, positive effectors, and negative effectors, but the overall symmetry of the molecule is maintained throughout the various binding steps so that no two identical subunits are in different conformational states at any given time. The binding of an effector or of the substrate shifts the quilibrium from one form to another. *Abbr* MWC model. *Aka* concerted transition model.

concerted reaction A chemical reaction in which a new bond is formed at the same time as, and as a direct consequence of, the breaking of another bond.

concrete oil *See* essential oil.

condensate 1. The crystalline particles of DNA that are formed during an early stage in the maturation of T-even phages. 2. The liquid obtained by condensation of either a gas or a vapor.

condensation 1. The linking of two like, or two unlike, molecules with the elimination of either a molecule of water or some other small molecule. 2. POLYMERIZATION. 3. An early stage in the maturation of T-even phages during which the condensate is formed. 4. The transition of either a gas or a vapor to a liquid.

condensation polymer STEP-GROWTH POLYMER.

condensation principle CONDENSING PRINCIPLE.

condensed conformation A low-energy conformation of mitochondria that occurs in mitochondrial preparations containing an excess of ADP, and that is characterized by a mitochondrial matrix which is not squeezed together tightly and does not stain heavily. *See also* orthodox conformation.

condensing enzyme CITRATE SYNTHASE.

condensing principle A factor that aids in the aggregation of DNA to form a condensate during an early stage in the maturation of T-even phages.

condensing site PEPTIDYL SITE.

conditional lethal mutant A conditional mutant, the ability of which to grow depends on the conditions: it grows as a normal organism under permissive conditions but does not grow, and thus expresses its lethal mutation, under restrictive conditions.

conditional mutant A mutant whose behavior depends on the physical conditions: it shows normal, wild-type behavior under certain (permissive) conditions and abnormal, mutant behavior under other (restrictive) conditions. A temperature-sensitive mutant and a suppressor-sensitive mutant are two examples.

conditioned vitamin deficiency A disorder that is caused by an interference with the digestion, absorption, or utilization of a vitamin as distinct from one that is caused by a lack of the vitamin in the diet. *See also* secondary deficiency.

conductance The property of an electrical circuit that determines the rate at which electrical energy is converted into heat when a given potential is applied across the electrodes; equal to the reciprocal of the electrical resistance.

conduction 1. The act of conveying either matter of energy from one location to another. *See also* nerve impulse conduction. 2. The transfer to a recipient cell of a nonmobilizable plasmid via genetic recombination with a different, conjugative plasmid.

conductivity 1. The capacity to conduct either electricity or heat. 2. CONDUCTANCE.

conductometry A method of chemical analysis that is based on measurements of electrical conductivity.

cone A light receptor in the retina of vertebrates that functions in day and color vision.

cone threshold The lowest light intensity to which the cones are sensitive.

confidence interval An interval for which one can assert with a given probability (called degree of confidence or confidence coefficient) that it will contain the parameter it is intended to estimate. The parameter may be the mean, a standard deviation, a proportion, or any other estimate of a point. The endpoints of a confidence interval are referred to as upper and lower confidence limits.

confidence limits *See* confidence interval.

configuration A unique and fixed spatial arrangement of the atoms in a molecule such that the molecule may be isolated in this stereochemical form. The change from one configuration to another requires the breaking and forming of covalent bonds. *See also* conformation.

confluent growth The growth of bacterial cells on a solid medium such that the entire surface of the medium is covered by the cells.

confluent lysis The complete lysis, in a plaque assay, of the entire bacterial lawn.

conformation A spatial arrangement of the atoms in a molecule that results from the rotation of the atoms about single bonds without a change in the covalent structure of the molecule. Conformation thus refers to a family of structures and not to a single, isolatable stereochemical form. The change from one conformation to another does not require the breaking and forming of covalent bonds. *See also* configuration.

conformational analysis An analysis that attempts to deduce the most stable (least strained) conformation of a flexible molecule in solution; based on minimizing the potential energy of the molecule by considering bond lengths, bond angles, and other factors.

conformational coupling hypothesis A hypothesis of the coupling of ATP synthesis to the operation of the electron transport system in oxidative phosphorylation. According to this hypothesis, the transport of electrons leads to the formation of energy-rich conformations of mitochondrial membrane components, and the energy associated with the relaxation of these membrane components then drives the phosphorylation of ADP to ATP.

conformational isomer CONFORMER.

conformational map RAMACHANDRAN PLOT.

conformer One of two or more isomers that differ from each other in their conformation; any one of the various possible conformations of a molecule.

conformon 1. A quantized package of energy that, according to the conformational coupling hypothesis, is associated with a localized conformational change in the mitochondrial membrane. 2. A vibrational or electronic excitation of a macromolecule that is accompanied by a local deformation. 3. A collection of a small number of catalytic residues of enzymes or segments of nucleic acids that are arranged in space and time with appropriate force vectors so as to cause chemical transformations or physical changes of a substrate or a bound ligand. 4. A packet of energy and/or genetic information, stored as a transient localized conformational strain in a biological macromolecule.

congener One of a family of related chemical substances, such as derivatives of a compound or elements belonging to the same group in the periodic table.

congenic strains Strains of an organism that differ from each other only with respect to a small, restricted, region of the chromosome.

congenital Existing at birth.

congenital goiter FAMILIAL GOITER.

congenital hyperammonemia HYPERAMMONEMIA.

congenital parahemophilia PARAHEMOPHILIA.

congenital porphyria A genetically inherited metabolic defect in humans that is characterized by an overproduction of Type I porphyrins and the excretion of excessive amounts of uroporphyrins in the urine.

conglutination The agglutination of antigen–antibody–complement complexes by conglutinin.

conglutinin A protein that is present in normal

serum and that causes the agglutination of antigen–antibody–complement complexes; conglutinin is not an antibody. *See also* immunoconglutinin.

conglutinogen A site on the bound C3b component of complement that is modified enzymatically by KAF and then binds conglutinin during conglutination.

conidium (*pl* conidia). An asexual spore of certain fungi. Large and usually multinucleate conidia are known as macroconidia; small and usually uninucleate conidia are known as microconidia.

conjugate acid–base pair A Bronsted acid and its corresponding Bronsted base; a proton donor and the corresponding proton acceptor.

conjugated antigen An antigen consisting of a protein that is covalently linked to either a molecule or a group which contains an antigenic determinant.

conjugated double bonds *See* conjugation (2).

conjugated enzyme An enzyme that is a conjugated protein.

conjugated protein A protein that contains a nonprotein component in addition to the amino acids. The nonprotein component may be either a metal ion or an organic molecule such as a lipid, a carbohydrate, or a nucleic acid. The nonprotein component may be either loosely associated with the protein or bound to it tightly as a prosthetic group.

conjugate redox couple REDOX COUPLE.

conjugation 1. The covalent or noncovalent combination of a large molecule, such as a protein or a bile acid, with another molecule. 2. The alternating sequence of single and double bonds in a molecule. 3. The genetic recombination in bacteria and in other unicellular organisms that resembles sexual reproduction and that entails a transfer of DNA between two cells of opposite mating type which are associated side by side.

conjugative plasmid A plasmid that carries genes that determine the effective contact function for the transfer of the plasmid DNA from a donor to a recipient cell.

conjugon A genetic element, such as the fertility factor, that is required for bacterial conjugation.

connecting peptide A peptide of 30 amino acids that, together with 4 basic amino acids, serves to link together the A and B chains of insulin in the proinsulin molecule. *Abbr* C peptide.

connective tissue The extracelluar matrix and the cells found in it, such as fibroblasts, macrophages, and mast cells. Connective tissue is distributed throughout the body in cartilage, tendons, ligaments, and the matrix of bone; it underlies the skin, binds blood vessels, and binds cells in such tissues as liver and muscle. It consists of insoluble fibers,

formed by polymers of high molecular weight, embedded in a matrix called the ground substance.

connexon *See* gap junction.

consecutive reactions A series of two or more reactions in which the product of one reaction serves as a reactant for the next reaction.

cons electrophoresis ISOTACHOPHORESIS.

consensus sequence A basic sequence of nucleotides derived from a large set of observed similar sequences in a specific region of a nucleic acid molecule. The sequences of the Pribnow box and the Hogness box are examples. *Aka* canonical sequence.

conservation equation An equation that expresses the total concentration of a component in terms of all the various forms in which it occurs. For example, the total concentration of enzyme in a simple enzymatic reaction is equal to the concentration of the free enzyme plus the concentration of the enzyme in the form of the enzyme–substrate complex.

conservative amino acid replacement CONSERVATIVE SUBSTITUTION.

conservative recombination Genetic recombination, involving breakage and reunion of preexisting DNA strands, in the absence of DNA synthesis.

conservative replication A mode of replication for double-stranded DNA (now considered obsolete) in which the parental strands do not separate and in which the progeny consists of both original parental duplexes and of newly synthesized duplexes.

conservative substitution The replacement in a protein of one amino acid by another, chemically similar, amino acid, such as the replacement of a polar (nonpolar) amino acid by another polar (nonpolar) amino acid. A conservative substitution is generally expected to lead to either no change or only a small change in the properties of the protein. *See also* radical substitution.

conserved sequence A base sequence in a nucleic acid (or an amino acid sequence in a protein) that has changed only very slightly in the course of evolution.

conspecific Belonging to the same species.

constant region That part of the immunoglobulin molecule in which virtually no changes in the amino acid sequence are found when immunoglobulins from different sources are compared. The constant region comprises portions of both the light c_L and the heavy c_H chains and does not constitute part of the antigen binding site. *Aka* constant domain. *See also* variable region.

constituent concentration The concentration of a component that takes into account all of the forms in which the component occurs; the total concentration of a macromolecule that exists in two conformational states is an example.

constituent parameter The concentration average of a parameter, such as a sedimentation coefficient or a diffusion coefficient, for a component that exists in several forms in the solution.

constitutive enzyme An enzyme that is present in a given cell in nearly constant amounts regardless of the composition of either the tissue or the medium in which the cell is contained; an enzyme that is constantly produced regardless of the growth conditions.

constitutive expression The unregulated expression of an operon resulting from a constitutive mutation.

constitutive gene A gene whose activity depends only on the efficiency with which the promoter of the gene binds RNA polymerase.

constitutive mutation A mutation that results in extensive constitutive synthesis of an inducible enzyme in the absence of an inducer and that involves an alteration in either the operator or the regulator gene of the enzyme.

constitutive secretory cell A cell, such as a muscle or a liver cell, that secretes proteins as fast as they are synthesized inside the cell. Such cells do not have a large intracellular pool of proteins and the rate of protein secretion is affected by altering the rate of protein synthesis. *See also* regulated secretory cell.

constitutive synthesis The synthesis of a specific protein (or of its mRNA) at a nearly constant rate independent of the presence of any molecule that interacts with the protein (or with its mRNA).

constraint A limitation imposed on a set of data by external conditions. Thus, the requirement that a set of data have a mean of a given value, represents a constraint.

constructive interference *See* interference (1).

constructive metabolism ANABOLISM.

consumption test Any test in which the amount of antigen or antibody removed (consumed), in the form of an antigen–antibody precipitate, is being determined.

contact activation cofactor HIGH MOLECULAR WEIGHT KININOGEN.

contact dermatitis *See* allergic contact dermatitis.

contact factor HAGEMAN FACTOR.

contact guidance The guiding of migrating cells along a specific pathway in the extracellular matrix by contact with oriented surfaces or structures.

contact hypersensitivity The hypersensitivity that is brought about by exposure of the skin to a chemical substance. *See also* allergic contact dermatitis.

contact inhibition The inhibition of cell growth that occurs in tissue culture when cells of multicellular organisms come into contact with each other. Contact inhibition permits the growth of monocellular layers and prevents the disorderly piling up of cells. The loss of contact inhibition is one of the characteristics of a tumorigenic cell.

contact map A two-dimensional graphical representation of the structure of a protein; involves a plot of the distances between all pairs of alpha carbon atoms, either in the order of the amino acid sequence or as falling within a certain distance. *Aka* distance map.

contact skin sensitivity The capacity of an animal organism to respond to a percutaneous application of a chemical sensitizer.

contact surface Those parts of the van der Waals surface of a protein that make contact with the surface of an appropriate probe, generally considered to be a water molecule with a radius of 1.4 Å.

contamination 1. The mixing of an impurity, such as a heavy metal ion or a radioactive substance, with the sample. 2. An impurity that is present in the sample.

continuity equation An equation that expresses the conservation of mass during ultracentrifugation on the basis of sedimentation and diffusion.

continuous assay An assay in which the reaction mixture is analyzed continuously by some monitoring technique without interfering with the progress of the reaction by the removal of samples.

continuous cell line ESTABLISHED CELL LINE.

continuous culture A culture of cells that is maintained in a growing state over prolonged periods of time *Aka* open culture. *See also* batch culture; chemostat.

continuous density gradient A density gradient in which the density changes in an uninterrupted fashion, rather than in a stepwise fashion, from one end of the gradient to the other.

continuous development A chromatographic technique, used particularly with paper and thin-layer chromatography, in which the solvent is allowed to run continuously over the support.

continuous discharge region That portion of the characteristic curve of an ionization chamber in which, during gas amplification, there is a continuous discharge in the chamber so that it is no longer usable as a detector.

continuous distribution A set of experimental data in which the variable being measured can vary continuously and is expressed as a number having one or more decimal places; the weight gain per animal in a group of animals is an example.

continuous emission The emission of light over a range of wavelengths that is produced in flame photometry when nonionic materials are present in the sample.

continuous flow centrifugation A preparative-type centrifugation, used for collecting materials from large volumes of liquid, in which a liquid is fed continuously into a rotor, the sediment is accumulated, and the supernatant is continuously withdrawn.

continuous flow electrophoresis An electrophoretic technique in which the flow of liquid is in a vertical direction and the electric field is at right angles to the direction of liquid flow. The sample is applied continuously at the top of the apparatus and fractions are collected at the bottom, at various spacings along the supporting medium.

continuous flow isoelectric focusing An isoelectric focusing technique in which the flow of liquid is in a vertical direction and the electric field is at right angles to the direction of liquid flow. The sample is applied continuously at the top of the apparatus and fractions are collected at the bottom, at various spacings along the supporting medium.

continuous flow scintillation counter A liquid scintillation counter designed for the continuous monitoring of radioactivity and used for effluents from amino acid analyzers and gas chromatographs.

continuous flow technique RAPID FLOW TECHNIQUE.

continuous spectrum A spectrum in which either the absorption or the emission of radiation covers a range of wavelengths.

continuous variation *See* method of continuous variation.

contour length The length of an extended polymer as distinct from the end-to-end distance of the folded polymer.

contracted muscle A muscle that has been shortened by contraction.

contractile Capable of contraction.

contractile protein A protein, such as actin or myosin, that is a component of fibrous tissues and that is capable of producing changes in the lengths of its constituent fibrous elements.

contraction The shortening of a muscle. *See also* isometric contraction; isotonic contraction.

contributing structure CANONICAL STRUCTURE.

control 1. An experiment that serves as a standard of comparison for other experiments; the control is carried out exactly as the other experiments except that it differs from them in one variable, the significance of which can thereby be assessed. *See also* blank. 2. The regulation of a biochemical process. *See also* coarse control; fine control.

control analysis A theoretical and experimental approach applied to intact metabolizing systems for the purpose of obtaining quantitative measures of the relative importance of individual steps in the control of a given metabolic pathway.

controlled atmospheric storage GAS STORAGE.

controlling element A genetic element that becomes inserted in a gene and makes it an unstable, hypermutable gene. There are two types of controlling elements, receptor and regulator elements. The former causes inactivation of the target gene; the latter maintains the mutational instability of the target gene.

controlling gene A gene, such as a regulator gene, that can turn other genes on or off.

control strength A measure of flux changes in a metabolic pathway that arise from changes in the activity of an enzyme in the pathway. Specifically,

$$C_i = \frac{\partial \ln v_g}{\partial \ln v_i} = \frac{\partial v_g}{\partial v_i} \frac{v_i}{v_g}$$

where C_i is the control strength, v_g is the net flux through the pathway, and v_i is the net flux through a step in the pathway, catalyzed by enzyme E_i. If the substrates of the enzyme are present in large excess, the relation becomes

$$C_i = \frac{[E_i]}{v_g} \frac{\partial v_g}{\partial [E_i]}$$

convalescent serum The serum obtained during convalescence from a disease.

convection The bulk movement of fluid in which both solvent and solute move together and that is usually due to either density inversions caused by temperature variations or local concentration changes.

conventional animal An animal raised under ordinary conditions as distinct from one raised in a germ-free environment.

conventional sedimentation equilibrium SEDIMENTATION EQUILIBRIUM.

convergence theory of cancer GREENSTEIN HYPOTHESIS.

convergent evolution An evolutionary pattern in which the lines of development for more recent species come together as a result of independent development from earlier species; such a pattern can be depicted as two or more independent networks, arising from different origins.

conversion A change in the properties of the host bacterium, such as antigenic character or toxin production, that is brought about by the prophage of that bacterium. *See also* antigenic conversion.

conversion coefficient The fraction of gamma rays that produce Auger electrons.

conversion electron The electron emitted from an atom that is undergoing internal conversion.

conversion factor A number that converts one set of dimensions into another.

conversion period The theoretical time required for the conversion of all of the substrate to product by a given amount of enzyme.

conversion stage That part of the blood clotting process that consists of the conversion of fibrinogen to fibrin under the influence of thrombin.

converter enzyme An enzyme that catalyzes the interconversion of two forms of another enzyme; an enzyme that carries out the chemical modification of a regulatory enzyme. The enzymes synthase phosphorylase kinase (SPK) and phosphoprotein phosphatase (PP-1) that catalyze the interconversion of phosphorylase a and b are two examples.

convertin The activated form of proconvertin; one of the factors in the extrinsic pathway of blood coagulation.

converting enzyme See serum converting enzyme.

converting phage A phage that brings about conversion in its host cell.

convex exponential gradient An exponential density gradient that is formed if the solution introduced into a mixture chamber of constant volume has a higher concentration than the solution initially present in the mixing chamber.

Conway microdiffusion apparatus An apparatus for the microchemical analysis of a gas, such as ammonia or carbon dioxide, that can be liberated from a sample by treatment with specific reagents. The apparatus consists of two concentric plates, much like a modified petri dish; the central well contains the sample, and the outer space contains a solution for trapping the gas that will be liberated and that will diffuse away from the sample upon addition of the reagents.

Cooley's anemia THALASSEMIA.

Coomassie brilliant blue Trademark for a dye used for the detection of protein bands following electrophoresis. See also Bradford method.

Coombs' reagent An antiserum that contains antibodies to human immunoglobulins and that is prepared by injecting these immunoglobulins into rabbits.

Coombs' test A test for demonstrating incomplete antibodies against red blood cells; based on an agglutination reaction in which the incomplete antibodies bind simultaneously to red blood cell antigens and to antibodies against themselves. See also direct Coombs' test; indirect Coombs' test.

Coon's method INDIRECT FLUORESCENT ANTIBODY TECHNIQUE.

cooperative binding The binding of ligands to a macromolecule such that the binding of one ligand to one binding site affects the binding of subsequent ligands to other binding sites on the same molecule. The ligands and the binding sites may be of one kind each or they may be different. The binding sites in cooperative binding are known as interacting (binding) sites.

cooperative feedback inhibition The feedback inhibition of an enzyme that is produced by two or more end products such that the inhibition caused by a mixture of two end products present together is greater than that caused by either end product present alone at the same total specific concentration (i.e., the concentration relative to the inhibitor constant).

cooperative hydrogen bonding The interaction between neighboring hydrogen bonds in a molecule such that the energy required to form these bonds is smaller than the sum of the energies for the individual bonds, and the energy required to break these bonds is greater than the sum of the energies for the individual bonds.

cooperative interactions See cooperative binding; cooperativity.

cooperative kinetics The kinetics of cooperative binding reactions. See also sigmoid kinetics.

cooperativity 1. The interaction between either identical or different binding sites of a macromolecule so that the binding of a ligand to one site affects the binding of subsequent ligands to other sites on the same molecule. 2. The interaction between neighboring hydrogen bonds in either a protein or a nucleic acid. See also cooperative hydrogen bonding.

cooperativity coefficient HILL COEFFICIENT.

cooperativity index The radio of substrate concentrations required to achieve any two fractions of the maximum velocity of an enzymatic reaction. Thus, if a substrate concentration of $[S]_{0.9}$ yields 90% of V_{max} and a substrate concentration of $[S]_{0.1}$ yields 10% of V_{max}, then the cooperativity index is given by $[S]_{0.9}/[S]_{0.1}$.

cooperativity models See concerted model; sequential model; Ferdinand model; Rabin model; nearest-neighbor cooperative model.

coordinate covalent bond A covalent bond formed between two atoms and consisting of two electrons, both of which are donated by only one of the bonded atoms.

coordinated enzymes The enzymes that are controlled by genes of one operon and that are either induced in coordinate induction or repressed in coordinate repression.

coordinated enzyme synthesis The synthesis of coordinated enzymes.

coordinate induction Enzyme induction in which a single inducer brings about the synthesis of a number of inducible enzymes that catalyze a sequence of either consecutive, or related, reactions in which the inducer is generally the first substrate. The structural genes of the coordinated enzymes are contiguous and form part of one operon. *See also* sequential induction. The induction by lactose of the synthesis of the enzymes coded for by the lac operon is an example.

coordinate regulation *See* coordinate induction; coordinate repression.

coordinate repression Enzyme repression in which a single repressor brings about the decreased synthesis of a number of repressible enzymes that catalyze a sequence of either consecutive or related reactions in which the repressor is generally the last end product. The structural genes of the coordinated enzymes are contiguous and form part of one operon. The repression by histidine of the synthesis of the nine enzyme-catalyzed steps in histidine biosynthesis is an example.

coordination 1. The formation of a complex between a metal ion and ligands. 2. COORDINATED ENZYME SYNTHESIS.

coordination number The number of ligands that can be bound to a metal ion to form a complex.

coordination position The position in the space surrounding a metal ion that can be occupied by a ligand for coordination with the metal ion.

cop I Copolymer I; a synthetic copolymer of 4 amino acids that are prevalent in myelin basic protein. The compound is being investigated as a possible means of preventing the immune system in multiple sclerosis from attacking native myelin, thereby leading to de-myelination. *See also* myelin basic protein.

copia elements A group of transposable elements in *Drosophila* that have closely related base sequences and code for large (copious, hence the name copia) amounts of RNA. Copia elements contain about 5 kb and are terminated by a 267-bp sequence in direct repeats.

coplanar Lying in the same plane.

copolymer A polymer formed from two or more different types of monomers that polymerized together.

copolymer I *See* Cop I.

copper An element that is essential to all plants and animals. Symbol, Cu; atomic number, 29; atomic weight, 63.54; oxidation states, $+1$, $+2$; most abundant isotope, ^{63}Cu; a radioactive isotope, ^{64}Cu, half-life, 12.8 h, radiation emitted beta particles, positrons, and gamma rays.

copper proteins A group of metalloproteins, often blue in color, that contain copper. The copper is generally in the divalent form and can usually be removed by dialysis against complex-forming compounds. *See also* azurin; blue protein.

coprecipitating antibody An antibody that does not form an antigen–antibody precipitate by itself, but can be incorporated in an antigen–antibody precipitate under suitable conditions.

coproantibody An antibody present in feces.

coproporphyrin The urinary pigment that is derived from coproporphyrinogen. *Abbr* CP. *See also* porphyrin.

coproporphyrinogen An intermediate in the biosynthesis of heme that is derived from uroporphyrinogen. *Abbr* CPG.

coprostanol A sterol alcohol excreted in the feces and formed by reduction of cholesterol as a result of the action of intestinal bacteria; it represents a major form in which cholesterol is eliminated.

copy-choice hypothesis A hypothesis of the mechanism of genetic recombination according to which the recombinant DNA molecule is synthesized by a system that uses both of the parental DNA molecules as templates, but that copies them in an alternating fashion.

copy DNA COMPLEMENTARY DNA.

copyediting *See* proofreading function.

copy error A mistake in replication.

copy-error mutation A mutation that results from a mistake in replication.

copy number The number of plasmids per cell. A low copy number (stringent plasmid) refers to the case of one or a few plasmids per cell; a high copy number (relaxed plasmid) typically refers to the case of 10–100 plasmids per cell.

copy-splice mechanism A mechanism that describes the genetic control of the synthesis of immunoglobulin chains in terms of the germ line theory.

CoQ Coenzyme Q.

CoQH$_2$ Reduced coenzyme Q.

cor The bare 15-membered ring of the corrin ring system.

cord factor A glycolipid in the cell walls of certain strains of *Mycobacterium* that is toxic for some laboratory animals. The compound is a mycolic acid diester of trehalose (trehalose-6,6'-dimycolate) and the name derives from the fact that it can be isolated primarily from those bacterial strains that grow in long cord-like skeins.

cordycepin   3′-Deoxyadenosine; an inhibitor of the polyadenylation of RNA.

core   *See* spore core.

core DNA   The DNA of a nucleosome core particle.

core enzyme   1. The portion of the enzyme RNA polymerase that consists of an aggregate of four subunits and that possesses catalytic activity, but that requires the attachment of the sigma factor before it can recognize an initiation site of transcription. *Aka* core polymerase. 2. The smallest aggregate of DNA-dependent DNA polymerase III that has enzymatic activity.

core particle   A particle obtained from ribosomes by removal of some of the ribosomal proteins, known as split proteins. *Abbr* CP. *See also* intersome. 2. NUCLEOSOME.

core polymerase   CORE ENZYME (1).

corepressor   A small molecule that combines with an aporepressor to form an active repressor which binds to the operator and inhibits transcription in enzyme repression. The corepressor is generally either the end product of the enzymatic reaction or a compound that is structurally similar to the product.

core protein   1. A structural protein molecule that occurs in complex IV, one of the respiratory assemblies. 2. *See* proteoglycan aggregates.

core region   A sequence of three monosaccharides that is common to most N-linked oligosaccharides and that is linked to the amide group of asparagine in the glycosylated protein [Man β1–4 GlcNAc β1–4 GlcNAC β–Asn].

core sequence   The DNA base sequence in a prophage attachment site in which exchange occurs.

Cori coefficient   A measure of the rate of monosaccharide absorption by rat intestine that is expressed as the number of milligrams of monosaccharide absorbed per 100 g of rat per hour.

Cori cycle   The cyclic group of reactions whereby glycogen is broken down and re-synthesized. The sequence begins with the breaking down of muscle glycogen to lactic acid, which is carried by the blood to the liver where it is converted back to glycogen; the liver glycogen, in turn, is then broken down to glucose, which is carried by the blood to the muscle where it is reconverted to glycogen.

Cori ester   Glucose-1-phosphate.

Cori's disease   GLYCOGEN STORAGE DISEASE TYPE III.

corn sugar   GLUCOSE.

corpus allatum   An endocrine gland in insects that synthesizes and secretes the allatum hormone.

corpuscle   A small particle or body.

corpuscular   Of, or pertaining to, corpuscles.

corpus luteum   (*pl* corpora lutea). A yellow progesterone-secreting body that is formed in a ruptured follicle.

corrected absorbance   1. The absorbance of a solution that has been corrected for the absorbance of either a blank or a reference solution. 2. The absorbance of a solution that is obtained after applying an Allen correction.

correction   The replacement in DNA, via excision repair, of mismatched base pairs by complementary base pairs.

correlation   The extent to which two statistical variables vary together; the interdependence between two variables. It is measured by the correlation coefficient.

correlation coefficient   A measure of the correlation between two statistical variables; it varies from zero for no correlation to +1 or −1 for perfect positive or negative correlation. A correlation coefficient of +0.3 (−0.3) means that as one variable increases, the other will increases (decrease) 30% of the time in the long run. *Sym* r.

correlation time   A measure of time, used in resonance studies, that is equal to one-third of the relaxation time.

correndonuclease   A correctional endonuclease; an endonuclease that specifically acts on damaged DNA resulting in correctional pathways in vivo. Correndonucleases that act on DNA having single or multiple base modifications are known, respectively, as Type I and Type II correndonucleases.

correxonuclease   A correctional exonuclease; an exonuclease that functions in the repair of damaged DNA.

corrin   The basic ring structure of vitamin B_{12} in which a cobalt atom is chelated.

corrinoid   Any compound containing the corrin ring system.

cortex   *See* adrenal cortex; spore cortex.

cortical bone   *See* lamellar bone.

corticoid   ADRENAL CORTICAL STEROID.

corticoliberin   CORTICOTROPIN RELEASING HORMONE.

corticosteroid   ADRENAL CORTICAL STEROID.

corticosteroid-binding globulin   TRANSCORTIN.

corticosteroid-binding protein   TRANSCORTIN.

corticosterone   A glucocorticoid that is biosynthesized from progesterone.

corticotrophin   Variant spelling of corticotropin.

corticotropin   ADRENOCORTICOTROPIC HORMONE.

corticotropin-like intermediate lobe peptide   A polypeptide consisting of residues 18–39 of ACTH and formed from it in the intermediate lobe of the pituitary gland; its physiological role is not yet established. *Abbr* CLIP.

corticotropin releasing hormone The hypothalamic hormone that controls the secretion of corticotropin. *Var sp* corticotrophin releasing hormone. *Abbr* CRH. *Aka* corticotropin releasing factor (CRF); corticoliberin.

cortin 1. ADRENAL CORTICAL STEROID. 2. An acetone extract of the adrenal cortex.

cortisol The major glucocorticoid in humans that occurs in the blood bound to the protein transcortin, is biosynthesized from progesterone, and has strong antiinflammatory activity.

cortisol-binding globulin TRANSCORTIN.

cortisol-binding protein TRANSCORTIN.

cortisone A glucocorticoid that is biosynthesized from cortisol. When administered, it is converted in vivo back to cortisol which largely accounts for its strong antiinflammatory activity.

cortoic acid A carboxylic acid formed by oxidation of cortisol, and some of its metabolites, at carbon 21.

COS cells Monkey cells that have been transformed by Simian virus 40 DNA that contains a defective origin of viral replication.

cosmic rays The high-energy ionizing radiation that originates outside the earth's atmosphere and that consists primarily of protons and other nuclei.

cosmid A cos site carrying plasmid; a plasmid vector that carries the cos sites of phage lambda DNA. This allows the plasmids to be packaged into phage particles for efficient introduction into bacteria. The cos site also guards against breakage of the plasmids and increases the probability of selecting a recombinant plasmid carrying foreign DNA. Lastly, cosmids are useful for cloning very large DNA inserts such as those of eukaryotic DNA.

cosmochemistry The study of the chemical composition of, and changes in, the universe.

cos sites The nucleotide sequences of the cohesive ends (termini) of a phage DNA molecule. When these become linked to form a double-stranded circular molecule, the hydrogen-bonded region of the cohesive ends is designated as cos.

cosubstrate A compound that acts somewhat in the capacity of a substrate during an enzymatic reaction, such as a dissociable coenzyme molecule. A cosubstrate participates stoichiometrically in the reaction and is consumed as the substrate is consumed. The NAD^+ and $NADP^+$ of pyridine-linked dehydrogenases are examples.

COSY Correlated spectroscopy; a two-dimensional nuclear magnetic resonance technique.

cot curve The curve obtained by plotting the data of a reassociation kinetics experiment. Since the reassociation of DNA is a bimolecular, second-order reaction, it follows that $C/C_0 = 1/(1 + kC_0t)$ where k is the second-order rate constant (L mol^{-1}s^{-1}), t is the time (s), C_0 is the initial concentration of single-stranded DNA (moles of nucleotide per liter), and C is the concentration of single-stranded DNA remaining in the reaction mixture at time t (moles of nucleotide per liter). The cot curve is obtained by plotting the fraction of single-stranded DNA remaining (C/C_0) as a function of log (C_0t), that is, the logarithm of the product of the initial concentration and the elapsed time. The cot curve is an S-shaped curve. *See also* reassociation kinetics.

cotransduction The simultaneous transduction of two or more genes that lie on the same segment of the bacterial DNA that is being transduced.

cotransformation The simultaneous transformation of two or more genes that lie on the same segment of DNA that is being transformed.

cotranslational transport The translocation of a protein across a biological membrane that is coupled to the synthesis of the protein; the translocation is in process before the synthesis of the polypeptide chain is completed. *Aka* cotranslational transfer. *See also* signal hypothesis; post-translational transport.

cotransport The simultaneous transport of two different substances across a biological membrane; symport.

Cotton effect The change in sign of the optical rotation in the neighborhood of an absorption band. The effect is due to the circular dichroism of the left and right circularly polarized light components of plane-polarized light. A Cotton effect is referred to as positive or negative depending on whether the optical rotation, with increasing wavelength, passes first through a minimum value and then through a maximum value, or vice versa.

cot value The value of C_0t in a cot curve; the product of the initial concentration of single-stranded DNA and the time allowed for reassociation. The cot value, when the reaction has proceeded to half completion (the point at which 50% of the DNA has reassociated or annealed, that is, $C/C_0 = 0.5$) is designated as cot$_{1/2}$ or (cot)$_{1/2}$. This point is also known as half reaction time and is equal to $1/k$, where k is the second-order rate constant.

cotyledon The first leaf that develops in the embryo of a seed plant.

Couette viscometer A concentric cylinder viscometer in which the outer cylinder is rotated at a constant speed and the viscosity is determined from the torque exerted on the inner cylinder.

coulomb A quantity of electricity equal to a current of 1 A flowing for 1 s.

Coulomb effect ION–ION INTERACTION.

Coulombic interactions The electrostatic interactions that can be described by Coulomb's law.

Coulomb's law An expression for the electrostatic force F between two point charges; specifically, $F = Q_1Q_2/Dr^2$, where Q_1 and Q_2 are the two point charges, r is the distance between the charges, and D is the dielectric constant of the medium. The force is repulsive if the charges have the same sign, and the force is attractive if the charges have opposite signs. The energy of such interactions is proportional to the reciprocal of r.

coulometer An instrument for measuring the quantity of electricity.

coulometry A method of chemical analysis that is based on measurements of the quantity of electricity, in coulombs, which is associated with a quantitative electrode reaction. In constant-current coulometry, the current is kept at a constant level so that the elapsed time is proportional to the total number of coulombs consumed; in constant-potential coulometry, the potential is kept at a constant level and the quantity of electricity consumed is measured with a coulometer.

Coulter counter A particle counter used for counting blood cells and bacteria.

counter 1. An instrument for indicating, and frequently recording, radioactive radiation; it may include a detector, sample changer, scaler, and printer. Some counters, such as ionization chambers and Geiger–Mueller counters, use the ionization of a gas to measure the radiation; other counters, such as liquid and solid scintillation counters, use the scintillations produced by fluors to measure the radiation. 2. Any instrument for counting, such as a cell counter, a drop counter, etc.

countercurrent diffusion multiplier system A system for the production of hypertonic urine by certain nephrons in the kidney.

countercurrent distribution A multistep separation procedure that is based on solubility differences of compounds in two immiscible liquid phases. The compounds are partitioned repeatedly between the two immiscible phases as they "move" along a large number of partition tubes. *Abbr* CCD. *Aka* countercurrent extraction.

countercurrent multiplication mechanism COUNTERCURRENT DIFFUSION MULTIPLIER SYSTEM.

counter double current distribution A variation of countercurrent distribution in which the sample is injected continuously into the apparatus and the unwanted components are removed at both ends of the apparatus.

counterelectrophoresis COUNTERIMMUNOELECTRO-PHORESIS.

counterflow The movement of a substance from side A of a membrane to side B, after equilibrium has been established between the two sides, in response to the addition of a structurally related substance to side B. In this process, the substrate moves against its own concentration gradient. The occurrence of counterflow is taken as evidence for the existence of a single carrier which moves both of the substances.

counterflow centrifugation CENTRIFUGAL ELUTRI-ATION.

counterimmunoelectrophoresis A variant of immunoelectrophoresis, performed in a gel. Antigens and antibodies are placed in two wells, close together, and caused to migrate toward one another under the influence of an electric field; when the two components meet and interact, a band of precipitin is formed.

counterion An ion that has a charge of opposite sign to that of another ion. Ions having opposite charges to, and surrounding, either a macromolecule or the central ion in the Debye–Hueckel theory, are examples.

counterselective marker The gene that prevents growth of the desired organism in bacterial conjugation.

counterstain The staining of either a tissue or a culture with a dye that follows a previous staining with another dye.

counterstreaming centrifugation CENTRIFUGAL ELUTRIATION.

counting efficiency The ratio of the number of registered radioactive counts to the number of actual radioactive disintegrations that occurred during the same time; generally multiplied by 100 to give percent efficiency.

counting loss COINCIDENCE LOSS.

counting plateau That portion of the characteristic curve of an ionization chamber that is almost independent of the applied voltage.

counts A measure of radioactivity that represents the fraction of the radioactive disintegrations that are detected and registered by means of a counter.

coupled assay *See* auxiliary enzyme.

coupled layer chromatography A thin-layer chromatography technique in which a chromatoplate is used, the two halves of which are covered with two different, but adjacent, layers of a chromatographic support.

coupled neutral pump A coupled pump in which the movement of one ion across the membrane must be linked to the movement of another ion, of equal valence, in the opposite direction.

coupled pump A pump for the transport of one solute across a membrane that also drives the transport of a second solute across the same membrane in the opposite direction and in such a fashion that the transport of the second solute is physically dependent on the pump.

coupled reactions An endergonic and an exergonic reaction that are linked energetically; the endergonic reaction is driven by the exergonic reaction which occurs simultaneously and which shares a common intermediate with the endergonic reaction, such that the overall free energy change for the coupled reactions is negative. The ultimate coupling requirement is that the free energy change for each step in the mechanism (usually $\Delta G'$, at pH 7.0) must be ≤ 0. *Aka* energetically coupled reactions; energy coupling.

coupled transcription–translation The process, characteristic of prokaryotes, in which transcription and translation proceed simultaneously; the mRNA is being translated into protein before transcription of DNA into the mRNA has been completed.

coupled transport A transport system in which the movement of one solute across the membrane must be linked to the movement of a second solute across the same membrane but in the opposite direction.

coupling 1. The linking of aerobic respiration, specifically the operation of the electron transport system, to the synthesis of ATP. 2. The tendency of linked genes to be inherited together on the same chromosome. 3. CHANNELING.

coupling constant The separation between any two bands of multiple peaks in nuclear magnetic resonance; it is proportional to the magnitude of the spin–spin coupling. *Sym* J.

coupling factors A group of proteins that are required for the coupling of ATP synthesis to the operation of the electron transport system either in mitochondrial oxidative phosphorylation or in chloroplast photosynthesis. The mitochondrial coupling factor 1 is now called F_1-ATPase. *See also* F_0F_1-ATPase.

coupling inhibition UNCOMPETITIVE INHIBITION.

covalent bond A bond formed between two atoms and consisting of one or more shared pairs of electrons such that one electron in a

pair is donated by each of the two bonded atoms. *See also* coordinate covalent bond.

covalent catalysis Catalysis that requires the formation of a covalent enzyme–substrate intermediate.

covalent chromatography A column chromatographic technique in which a chemical reagent is linked covalently to the solid support. When a sample is passed through the column, the reagent reacts with, and binds covalently, the substance of interest. An additional chemical reaction then releases the substance from the support and permits its elution from the column, thereby restoring the initial form of the support.

covalent circle *See* circular covalent; covalently closed circle.

covalent enzyme–substrate complex ENZYME–SUBSTRATE COMPOUND.

covalent extension The initiation of DNA replication in which the leading strand is covalently attached to a parental strand as in the rolling circle replication.

covalent intermediate 1. A substance formed during covalent catalysis such as the intermediate formed in the transaminase reaction. 2. A covalently linked, high-energy intermediate that, according to the chemical coupling hypothesis, functions in the coupling aspect of oxidative phosphorylation.

covalent labeling AFFINITY LABELING.

covalently circular *See* circular covalent.

covalently closed circle A circular, double-stranded, DNA molecule in which each single strand is an unbroken, uninterrupted circle.

covalently modified enzyme A regulatory enzyme that has the capacity of having its catalytic activity modified through chemical alteration of the molecule which, in turn, is catalyzed by other enzymes. The enzyme-catalyzed phosphorylation and dephosphorylation of the enzyme phosphorylase is an example.

covalent orbital An orbital that functions in the bonding of a low-spin complex.

covalent structure analysis The determination of the covalent bonds that describe the arrangement of monomers in a macromolecule; the bonds that describe the amino acid sequence and the location of disulfide bonds in a protein, or those that describe the nucleotide sequence in a nucleic acid are examples.

covariance The average product of the deviations from the respective means for all pairs of values for the variables X and Y; the average of $(X - \bar{X})(Y - \bar{Y})$ for all pairs of values of X and Y, where $\bar{X}$ and $\bar{Y}$ are the means for the X values and Y values, respectively.

covariance analysis *See* analysis of covariance.

covariation PLEIOTROPISM.

covarion A concomitantly variable codon; the number of codons for a protein that are free to fix mutations at a point in time; it is a property of a single nucleotide sequence. *See also* varion.

covirus A virus, such as some plant viruses, that consists of two or more different viral particles that must be present together for the initiation of infection. *See also* segmented genome.

covolume The difference between the volume of a compound in solution and that given by the sum of its atomic volumes. The additional volume results from the intermolecular forces that set a lower limit to the distance of approach between molecules in a liquid. *Aka* excluded volume.

coxsackievirus A virus that belongs to the enterovirus subgroup of picornaviruses and that is similar in physical parameters to the polio virus.

cozymase An early designation of a heat-stable fraction, consisting chiefly of ATP, ADP, AMP, and NAD^+, that was isolated from yeast and participated in the reactions of alcoholic fermentation. Subsequently, cozymase I was used to denote NAD^+ and cozymase II was used to denote $NADP^+$.

CP 1. Coproporphyrin. 2. Core particle. 3. Chemically pure.

C₃ pathway CALVIN CYCLE.

C₄ pathway HATCH–SLACK–KORTSCHAK PATHWAY.

CPC Centrifugal partition chromatography.

cpDNA Chloroplast DNA.

C peptide CONNECTING PEPTIDE.

CPG Coproporphyrinogen.

CPK Creatine phosphokinase.

CPK model Cory–Pauling–Koltun model; a space-filling molecular model.

C₃ plants Plants that use the reaction of ribulose–bisphosphate carboxylase as the first CO_2 fixation step; so called because the CO_2 is incorporated into a 3-carbon compound (3-phosphoglycerate.)

C₄ plants Plants in which the C₃ pathway (Calvin cycle) of CO_2 fixation is preceded by additional steps; so called because these additional steps involve a preliminary fixation of CO_2 into a 4-carbon compound (oxaloacetate). These plants grow in hot, arid climates and have an increased efficiency of CO_2 fixation over that of C₃ plants.

cpm Counts per minute; the number of radioactive counts per minute.

C-protein A minor muscle protein; it is tightly bound to myosin and its function is unknown. *See also* protein C.

Crabtree effect The inhibition of oxygen consumption in cellular respiration that is produced by increasing concentrations of glucose. *See also* Pasteur effect.

cRBP Cellular retinol-binding protein.

C-reactive protein A protein that reacts with the pneumococcal type C polysaccharide and that is present in plasma during some bacterial infections.

creatine A nitrogenous compound, the phosphorylated form of which, phosphocreatine, is a high-energy compound that serves as a free energy storage compound in muscle.

creatine kinase The enzyme that catalyzes the reversible reaction whereby ATP and creatine react to form ADP and creatine phosphate. *Abbr* CK. *Aka* creatine phosphokinase.

creatine phosphate *See* phosphocreatine.

creatine phosphokinase CREATINE KINASE.

creatinine A cyclic compound, formed from creatine, that represents one of the major forms in which nitrogen is excreted in the urine.

creatinine coefficient The number of milligrams of creatinine excreted per 24 hours per kilogram of body weight.

creatinuria The presence of excessive amounts of creatine in the urine.

creationism A movement that advocates the inclusion of the biblical story of creation in school curricula and its presentation as an alternative explanation to that provided by the theory of evolution. Efforts to find support for the biblical story have been termed creation science.

C region Constant region.

crenation The shrinking of red blood cells that occurs when they are placed in a hypertonic solution.

cretinism A condition of arrested growth and impaired mental development brought about by thyroid deficiency that is present at birth.

CRF Corticotropin releasing factor; *see* corticotropin releasing hormone.

CRH Corticotropin releasing hormone.

Crick strand The DNA strand of Watson–Crick-type DNA that is not transcribed in vivo; the antisense strand. *Abbr* C strand.

Crigler–Najjar syndrome A genetically inherited metabolic defect in humans that is characterized by defective bilirubin metabolism and by jaundice, and that is due to a deficiency of the enzyme uridine diphosphate glucuronosyl transferase.

crinophagy A variant of autophagy in which a secretory vesicle is removed by fusion with a lysosome and the complex is subsequently degraded.

crisis 1. A sudden change in the course of an

acute disease. 2. The state of a primary cell culture, following a number of cell divisions, during which most of the secondary cells die and disintegrate. If the few remaining cells are maintained in culture, they will ultimately begin to grow, giving rise to an established cell line, the cells of which can grow and divide indefinitely.

crista (*pl* cristae). An extended infolding of the inner mitochondrial membrane.

cristael Of, or pertaining to, cristae.

critical concentration *See* actin filament; critical micelle concentration.

critical group CRITICAL PAIR.

critical micelle concentration The concentration of a surface-active compound above which the formation of micelles by this compound becomes appreciable. *Abbr* cmc. *See also second critical concentration.*

critical pair Two compounds that are not readily separable from each other by their partitioning between two liquid phases, as in countercurrent distribution.

critical point The point at which the temperature is the critical temperature and the pressure is the critical pressure.

critical point drying A method for drying specimens for electron microscopy. It involves replacing the water in the specimen by a liquid having a lower critical temperature (such as ethanol or liquid CO_2) and then raising the temperature above this critical temperature. In theory, this allows removal of the liquid phase without shrinkage in the specimen since at the critical point the surface tension of a liquid becomes zero.

critical pressure The minimum pressure that must be applied to a gas, at the critical temperature, to liquefy it.

critical temperature The temperature above which a gas cannot be liquefied by pressure alone; at that temperature the properties of the liquid and of its saturated vapor become indistinguishable.

CRM Cross-reacting material.

cRNA Complementary RNA.

CRO protein A small, basic, regulatory protein, made by phage lambda, that binds to a specific base sequence in the phage DNA.

cross *See* genetic cross.

cross-absorption The absorption of either antigens or antibodies by means of the corresponding cross-reacting antibodies or cross-reacting antigens.

cross-agglutination test CROSS-MATCHING.

crossbreeding OUTBREEDING.

cross-bridge A short projection from the thick filament of striated muscle; cross-bridges are regularly spaced and link the thick filaments to the thin filaments.

cross-β-conformation The structure of a polypeptide chain in the β-conformation that loops back upon itself to form an intramolecular pleated sheet; the loops constitute crossover connections between the β-conformation segments. A variety of such structures are possible.

crossed affino-immunoelectrophoresis CROSSED IMMUNOAFFINOELECTROPHORESIS.

crossed hydrophobic interaction immunoelectrophoresis An analytical technique that combines the principles of hydrophobic interaction chromatography and crossed immunoelectrophoresis.

crossed immunoaffinoelectrophoresis A variation of crossed immunoelectrophoresis in which a lectin is either incorporated into the first-dimension gel or incorporated into a gel that is used as an intermediate gel between the first-dimension gel and the antibody-containing gel.

crossed immunoelectrophoresis An electrophoretic technique in which antigens are first separated by one-dimensional gel electrophoresis; the antigens are then separated in the second dimension by gel electrophoresis, using a gel that contains antibodies and applying an electric field at right angles to the direction of the first separation. *Abbr* CIE.

cross-electrophoresis An electrophoretic technique designed to determine whether two charged substances interact; involves an analysis of the crossing point obtained by having the two substances move across each other in the form of two slanted lines. A pattern of ✕ indicates no interaction, while a pattern of ⅄ indicates interaction.

cross-feeding The phenomenon of two organisms that can grow only in the vicinity of each other or in each other's medium, since each is dependent on the other for an essential growth factor. *See also* syntrophy.

cross-hybridization The molecular hybridization of a nucleic acid probe to a nucleotide segment that is not completely complementary. *See also* hybridization (1).

cross-induction The induction of a prophage in a nonirradiated, lysogenic F^- bacterium in response to compounds transferred to the bacterium by conjugation with an ultraviolet-irradiated F^+ cell.

cross-infection The infection of a bacterium by two or more different phage mutants.

crossing over The process whereby genetic material is exchanged between homologous chromosomes, leading to gene combinations that are different from those in the parental chromosomes. *See also* breakage and reunion model.

crossing-paper electrophoresis CROSS-ELECTRO-PHORESIS.

cross-linker One of a number of proteins that bind to the sides of actin filaments and act as spot welds to cross-link these filaments. *Aka* gelation protein.

cross-linking The formation of covalent bonds between chains of polymeric molecules.

cross-matching A serologic procedure for blood typing in which erythrocytes from a donor of unknown blood type are mixed with the serum of recipients of known blood types.

cross of isocline The cross-like pattern observed in flow birefringence; the arms of the cross appear dark on a light background.

crossover 1. The chromosome resulting from crossing over. 2. The individual resulting from crossing over.

crossover connection *See* cross-β-conformation.

crossover method A method for studying a sequence of oxidation–reduction reactions from the changes produced in the sequence upon the addition of inhibitors. *See also* crossover theorem.

crossover point 1. The step in a sequence of oxidation–reduction reactions that is being inhibited by the addition of an inhibitor. *See also* crossover theorem. 2. The step in a sequence of metabolic reactions at which the metabolic flux is altered with a resultant change in the concentrations of the remaining reactants. The crossover point is referred to as positive or negative depending on whether the metabolic flux is increased or decreased at the particular step. A positive crossover point results in a decrease of the steady-state levels of the intermediates preceding the crossover point and in an increase of the steady-state levels of the intermediates following the crossover point. These concentration changes are reversed for a negative crossover point.

crossover region That section of a chromosome that lies between two specified marker genes.

crossover theorem The principle that perturbation of a metabolic pathway at a given enzymatic step will lead to opposite changes in metabolite concentrations before and after the particular step. Thus, if a specific enzyme inhibitor is added, the concentrations of the substrates of the enzyme will increase while the concentrations of the products of the enzyme reaction will decrease. Likewise, addition of an inhibitor to a series of oxidation–reduction reactions, such as the electron transport system, will cause the components on the reduced side of the inhibited reaction to become more reduced, and will cause those on the oxidized side to become more oxidized.

crossover unit A crossover value of 1% between a pair of linked genes.

cross-partition A phase partition technique that permits a determination of the isoelectric point of subcellular particles; involves a measurement of partitioning as a function of pH, using two different salt media. A plot of partition coefficient (or percentage of particles) in one phase as a function of pH yields two curves that cross each other at a pH corresponding to the isoelectric point of the particles.

cross-reacting antibody An antibody that can combine with antigens that are specific for stimulating the production of different antibodies.

cross-reacting antigen An antigen that can combine with antibodies that are produced in response to different antigens.

cross-reacting material A defective protein that is produced by a mutant and that is antigenically similar to the protein produced by the normal wild-type gene. *Abbr* CRM.

cross-reaction The reaction of an antigen with an antibody that is produced in response to a different antigen; the reaction occurs because of structural similarities between the antigenic determinants of the different antigens. *Aka* reaction of partial identity.

cross-reactivation The restoration of the activity of a mutant virus, which carries a lethal mutation as a result of previous exposure to a mutagen, by the simultaneous infection of a host cell with both the mutant virus and with one or more active viruses. The mutant virus is activated by a genetic exchange that leads to a replacement of its damaged DNA. *See also* multiplicity reactivation.

cross-resistance The resistance of a microorganism to a specific antibiotic that is associated with its resistance to a chemically related antibiotic.

cross-sensitization The immunological sensitization of an organism with an antigen that is different from the antigen that will be used subsequently to trigger an anaphylactic response.

cross-term diffusion coefficient The diffusion coefficient that a component has when it diffuses in the presence of another component; used in the treatment of diffusion data in a system showing interaction of flows. *See also* main diffusion coefficient.

cross tolerance The immunological tolerance against one antigen that is produced by the administration of a different, but cross-reacting antigen.

crotonase Enoyl CoA-hydratase; the enzyme that catalyzes the hydration reaction of trans unsaturated fatty acyl-CoA in the β-oxidation of fatty acids. So called because one enzyme is most active with crotonyl-CoA $(CH_3CH_2=CH_2COCoA)$.

crown ether A cyclic polyether; synthetic crown ethers have been used as hosts for the study of host–guest systems.

crown gall tumor A tumor that may develop on the stems or roots of a wide variety of plants following infection by the soil bacterium *Agrobacterium tumefaciens*. This transformation of normal to malignant cells is a natural form of genetic engineering and results from the transmission of a tumor-inducing plasmid (Ti plasmid) from the bacterium to the plant cell. During tumor induction, a small DNA fragment of the Ti plasmid (called transferred DNA, or T-DNA) becomes integrated into the plant cell chromosome and alters the hormonally regulated cell division of the plant cell.

CRP Cyclic AMP-receptor protein.

CRS Codon-recognizing site.

cruciferous *See* antipromoter.

cruciform DNA *See* foldback DNA.

crude extract A preparation, derived from biological material, that has not been extensively purified; used particularly to describe a preparation of either homogenized tissue or broken cells from which unbroken cells and cell debris have been removed, commonly by centrifugation.

cryo- Combining form meaning cold or freezing.

cryoenzymology The study of enzymes at low temperatures.

cryogenic Of, or pertaining to, low temperatures.

cryoglobulin An immunoglobulin that precipitates, gels, or crystallizes upon cooling of either a serum or a solution containing the globulin. Cryoglobulins are frequently found in inflammatory illnesses and in multiple myeloma. *See also* cold-insoluble globulin.

cryoprecipitagogue A substance that induces the formation of a cryoprecipitate.

cryoprecipitate A precipitate formed in the cold such as that of cryoglobulins.

cryoscope An instrument for the determination of freezing points.

cryoscopic method The determination of either molecular weight or osmotic pressure from the depression of the freezing point of a solvent that is produced by the addition of solute.

cryostat An apparatus for producing and maintaining a controlled low-temperature environment.

cryosublimation The process whereby water is sublimed from a frozen sample and is collected in a cold trap; cryosublimation refers to the collection of the water, while lyophilization refers to the collection of the dry residue.

cryptate A cyclic polyether containing two nitrogen atoms in the ring.

cryptic DNA A DNA of unknown function.

crypticity 1. The phenomenon that a particulate enzyme has different properties than the same enzyme in soluble form; solubilization of the particulate enzyme requires its detachment from the solid matrix to which it was attached. 2. The phenomenon that an intact cell is unable to use a metabolite because of a deficiency in the transport system that moves the metabolite across the cell membrane; disruption of such a cell permits the utilization of the metabolite by components of the cell. *See also* latency.

cryptic mutant 1. A cell that can synthesize an inducible enzyme but cannot synthesize a component of the transport system required to move the substrate of that enzyme across the cell membrane. 2. A cell that lacks one or more components of a membrane transport system so that a particular substrate cannot enter the cell and cannot be utilized even though the cell may possess the necessary complement of metabolic enzymes.

cryptic plasmid A plasmid to which no phenotypes (such as toxin production, ability to cause plant tumors, etc.) can be attributed.

cryptic prophage A phage that has suffered a deletion of some of its genes while it was being integrated as a prophage.

cryptic satellite DNA Satellite DNA that cannot be separated from the bulk DNA by density gradient centrifugation but can be separated by other means (for example, by its more rapid reannealing as a result of its high content of repetitive DNA).

cryptoactive Descriptive of triglycerides that show negligible optical rotation despite the fact that they contain an asymmetric center which results from the attachment of different acyl groups to carbon atoms 1 and 3 of the glycerol.

cryptobiosis Latent life, as that in a spore; the absence of detectable metabolism.

cryptogenic phage A phage that can give rise to a cryptic prophage when subjected to ultraviolet curing.

cryptogram A shorthand presentation of viral properties that consists of four pairs of symbols which indicate the following: type of the nucleic acid/strandedness of the nucleic acid; molecular weight of the nucleic acid/percentage of the nucleic acid in infective particles; outline of the viral particle/outline of the nucleocapsid; kinds of hosts infected/kinds of vectors.

crystal A solid of definite form that is characterized by geometrically arranged, external plane surfaces as a result of a regularly repeated, internal arrangement of the atoms.

crystal field splitting The separation of the degenerate *d* orbitals of a metal ion into orbitals having different energies that is produced by the ligands in a metal ion–ligand complex.

crystal field theory A description of the way in which the *d* orbitals of a metal are deformed by the electrons of a ligand in a metal ion–ligand complex. According to this theory (so called since it was developed to explain the spectra of transition metal impurities in crystals) the ligands in a transition-metal complex are treated as point charges.

crystal lattice The three-dimensional arrangement of the atoms in a crystal.

crystallin The major structural protein of the lens of the eye. A water-soluble protein that occurs in a number of forms not all of which occur in all species. Mammalian crystallins are classified into three major groups, designated α, β, and γ. The first two are oligomeric proteins that occur as various aggregates and cover a large range of molecular weights; γ-crystallins are monomeric proteins having a molecular weight of less than 28,000.

crystalline Of, or pertaining to, crystals.

crystallizable fragment FC FRAGMENT.

crystallization The transition of a substance from the molten, the liquid, or the gaseous state to the crystalline state.

crystallographic model A molecular model, such as a ball and stick or a framework model, in which the bond lengths and the bond angles are clearly indicated.

crystalloid A noncolloidal low molecular weight substance.

crystal protein One of a group of globular proteins that form crystalline inclusions in bacterial cells; they are widespread among *Bacillus* species. The formation of crystal proteins coincides with spore morphogenesis and may be related to it.

crystal violet A basic dye used in cytochemistry.

Cs Cesium.

CS Chorionic somatomammotropin; *see* placental lactogen.

CSF 1. Cerebrospinal fluid. 2. Colony-stimulating factor.

CSM Corn–soya–milk; a protein-rich baby food (20% protein) made in the United States from 68% precooked corn, 25% defatted soya flour, and 5% skim milk powder, with added vitamins B_1, B_2, B_6, and B_{12}, nicotinic acid, pantothenic acid, folic acid, vitamins A, D, and E, and $CaCO_3$. The mixture is used as a protein supplement in regions where either a low-protein diet or malnutrition is prevalent.

c-src gene A gene that is present in normal cells of various vertebrates and that is closely related to src, the oncogene of Rous sarcoma virus. The c-src gene codes for a protein (designated $_{pp}60$ c-src) that has similar properties

to those of the protein coded for by the src gene.

C strand Crick strand.

C substances A group of serologically distinct carbohydrates only one of which may occur in a given strain of *Streptococcus*; used as a basis for the identification and classification of streptococci.

CT Calcitonin.

CTAB *See* quat.

C-terminal The end of a peptide or a polypeptide chain that carries the amino acid that has a free alpha carboxyl group; in representing amino acid sequences, the C-terminal is conventionally placed on the right side. *Aka* C-terminus.

CTP 1. Cytidine triphosphate. 2. Cytidine-5'-triphosphate.

CTSH Chorionic thyroid stimulating hormone.

C-type particles Particles first seen in neoplastic mouse tissue and now known to be oncogenic RNA viruses belonging to the group of leukoviruses. The C-type particles differ from the B-type particles in that they appear to have intracellular precursors (A-type particles) and have a centrally situated genome in the virion. *Aka* C-type virus. *See also* B-type particles.

C-type virus *See* C-type particles; oncornavirus.

Cu Copper.

cubic symmetry Descriptive of a body that has at least four threefold axes of rotational symmetry; includes a perfect cube and point groups that are tetrahedral, octahedral, and icosahedral.

cultivar A variety or a strain of a plant that is produced by humans and that is maintained by cultivation.

cultivation The deliberate propagation of cells or organisms by means of a suitable culture.

culture A population of either microbial cells or tissue cells that grow in or on a nutrient medium.

cumulative feedback inhibition The inhibition of an enzyme that is produced when the enzyme is inhibited separately and independently by two or more end products. When one end product is present, there is a partial inhibition of the enzyme; when two or more end products are present, the inhibition is cumulative.

C_1 unit *See* active one-carbon unit.

cuprammonium rayon Cellulose that has been regenerated from a solution of cuprammonium hydroxide.

cuproprotein A conjugated protein containing copper as a prosthetic group.

curare A plant extract, used as an Indian arrow poison, that contains alkaloids that block the transmission of nerve impulses at the neuromuscular junction; a neurotoxin.

curie 1. A unit of radioactivity equal to 3.7×10^{10} disintegrations per second. 2. The quantity of radioactive nuclide that contains 3.7×10^{10} disintegrations per second. *Sym* Ci; c.

curing Removal of either a prophage or a plasmid from a bacterium. The former involves conversion of a lysogenic bacterium to a sensitive bacterium that may, upon subsequent infection, be either lysogenized or lysed; commonly achieved either by exposing the bacterial cells to radiation (radiation curing) or by superinfecting them with phage (superinfection curing). Plasmid curing can be achieved by, for example, treating some plasmid-containing cells with intercalating acridine dyes.

CURL 1. Endosome. 2. Compartment of uncoupling receptor and ligand; a vesicle that fuses with an endosome and functions in the receptor-mediated endocytosis of LDL. The CURL has a low internal pH (about 5.0) which induces the dissociation of LDL from its receptor.

curve fitting 1. The process of describing (approximating) an observed frequency distribution by means of a mathematically specified distribution. 2. The process of fitting a curve to points representing paired data.

curvilinear Consisting of, or bounded by, curved lines.

curvilinear correlation A nonlinear relation between two or more variables.

curvilinear regression *See* regression line.

Cushing's disease A disease characterized by an overproduction of adrencorticotropin and caused by either overactivity or a tumor of the adrenal cortex.

cut A break in both strands of a double-stranded nucleic acid. *See also* nick.

cut and patch repair EXCISION REPAIR.

cutaneous Of, or pertaining to, the skin.

cutaneous anaphylaxis The anaphylactic reaction that is produced in an animal organism by intradermal injections; cutaneous anaphylaxis can be of the active, the passive, or the reverse passive type.

cuvette A small container for a liquid sample that is to be subjected to optical measurements. Typical cuvettes used in spectrophotometry are rectangular containers, constructed of either Pyrex or quartz, that have a light path of 1 cm. Cuvettes selected to have a specific tolerance with respect to their light transmitting properties are referred to as matched cuvettes. *Var sp* cuvet.

CV 1. Coefficient of variation. 2. Cyclic voltammetry.

C value The mass of DNA (expressed, for ex-

ample, in picograms per cell) in the haploid genome of a species.

C value paradox The phenomenon that, frequently, C values do not correlate with the evolutionary complexity of species; they are large in some small organisms. This is presumably due to the fact that sizeable portions of the DNA do not code for proteins and either have other regulatory functions or are functionless.

C virus *See* oncornavirus; C-type particle.

cyanide The radical CN^- that is a strong poison due to its inhibition of the enzyme cytochrome oxidase at the terminal step of the electron transport system.

cyanobacteria A heterogeneous group of prokaryotic photosynthetic organisms that contain chlorophyll, evolve oxygen, and use water as the electron donor; many are also able to fix atmospheric nitrogen. Cyanobacteria were originally classified as plants (and referred to as blue-green algae) on the basis of their capacity for photosynthesis, but are now considered to be bacteria. *Aka* blue-green algae; cyanophyta; cyanophyceae.

cyanocobalamin *See* cobalamin.

cyanogen bromide reaction The hydrolysis by cyanogen bromide of those peptide bonds in which the carbonyl function is contributed by methionine.

cyanogenic glycoside A glycoside that is a plant toxin containing a residue of hydrocyanic acid.

cyanoguanidine DICYANAMIDE.

cyanohemoglobin A derivative of hemoglobin in which the sixth coordination position of the iron atom is occupied by cyanide.

cyanophage A virus whose host is a blue-green alga (cyanobacterium). All cyanophages appear to contain double-stranded DNA and most are virulent, producing host cell lysis.

cyanophyceae CYANOBACTERIA.

cyanophycin granule A storage granule of cyanobacteria that contains a copolymer of arginine and aspartic acid.

cyanophyta CYANOBACTERIA.

cyanopsin A visual pigment in freshwater fish that consists of opsin and retinal$_2$ and that has an absorption maximum at 620 nm.

cyanosis The bluish coloration of the skin that is caused by insufficient oxygenation of the blood.

cyanosome A phycobilisome of cyanobacteria.

cybernetics The comparative study of the automatic control system formed by the nervous system and the brain on the one hand and that formed by mechanical–electrical communication systems and devices (computers, thermostats, etc.) on the other hand.

cybotactic Of, or pertaining to, cybotaxis.

cybotaxis The spatial arrangement of solute molecules in a liquid, particularly of long-chain molecules, such that there is an equilibrium between molecules that have crystal-like orientations and molecules that have random orientations.

cybrid A hybrid formed by the fusion of cytoplasts of one parent line with cells of the other parent.

cyclamate A synthetic sweetener that is 30 times as sweet as sugar; sodium cyclohexylsulfamate. The compound is carcinogenic in animals and has now been banned from prepared foods.

cycle 1. A closed sequence of metabolic reactions, such as the citric acid cycle, in which an end product serves as a reactant for the initiation of the sequence, and in which most of the intermediates serve likewise as both reactants and products. 2. A closed sequence of large-scale processes, such as the nitrogen cycle, that describes the nutritional interdependence of plants, animals, and microorganisms.

cyclic 1. Of, or pertaining to, a cycle; circular. 2. Of, or pertaining to, a ring.

cyclic adenylic acid ADENOSINE-3′,5′-CYCLIC MONOPHOSPHATE.

cyclic AMP ADENOSINE-3′,5′-CYCLIC MONOPHOSPHATE.

cyclic AMP receptor protein A protein in *E. coli* that binds to, and is activated by, cyclic AMP and that is necessary for the efficient transcription of certain operons which are subject to catabolite repression. *Abbr* CAP.

cyclic electron flow The movement of electrons that is limited to photosystem I of chloroplasts and to its associated electron carriers; cyclic electron flow can lead to the synthesis of ATP but does not lead to an accumulation of NADPH.

cyclic GMP GUANOSINE-3′,5′-CYCLIC MONOPHOSPHATE.

cyclic metabolic pathway CYCLE (1).

cyclic peptide A peptide that consists of a closed chain and that is devoid of a free alpha amino group and a free alpha carboxyl group.

cyclic permutation CIRCULAR PERMUTATION.

cyclic photophosphorylation The synthesis of ATP that is coupled to the cyclic electron flow of photosynthesis.

cyclic symmetry ROTATIONAL SYMMETRY.

cyclic voltammetry A technique for observing the redox state of a substance over a wide range of potentials. It consists of cycling the potential of an electrode, which is immersed in an unstirred solution, and measuring the resulting current at the electrode. A plot of current as a function of potential yields a voltammogram. Cyclic voltammetry can be used to follow both fast and slow reactions and is a powerful probe to monitor reactive redox species.

cyclin A stable, cell cycle-regulated, nuclear protein that is synthesized mainly in the S-phase of the cell cycle and is believed to be a key component in DNA replication and cell division. Cyclin is absent or present in very low amounts in nondividing cells, but is synthesized by proliferating cells of both normal and transformed origin. Moreover, the rate of cyclin synthesis correlates with the proliferative state of normal cultured cells and tissues. *Aka* PCNA; auxiliary protein of DNA polymerase δ.

cyclitol A cyclic polyhydroxy alcohol.

cyclitol antibiotic An antibiotic that contains a cyclitol; the aminoglycoside antibiotics are an example.

cyclization The formation of a ring.

cycloaddition reaction A chemical reaction that involves the addition of one rectant to another to form a cyclic product.

cyclodepsipeptide *See* depsipeptide antibiotics.

cyclodextrins A class of naturally occurring macrocyclic polymers of glucopyranose. Cyclodextrins have interior cavities that make them useful as hosts in host–guest systems. Many inorganic and organic compounds have been shown to bind within cyclodextrin cavities.

cycloheximide An antibiotic, produced by *Streptomyces griseus*, that blocks protein synthesis by inhibiting peptidyl transferase in eukaryotic systems. *Aka* actidione.

cyclol hypothesis An early, and now discarded, hypothesis of protein structure. According to this hypothesis, all proteins had a unit architecture, composed of interlocking hexagons, with each hexagon (cyclol) formed by linking together two amino acid residues.

cyclooxygenase A key enzyme in the biosynthesis of prostaglandins; it catalyzes the first reaction of the cyclooxygenase pathway whereby arachidonic acid is oxidized to endoperoxide which is then converted to prostaglandins, prostacyclin, and thromboxanes.

cyclophilin A protein, present in the cytoplasm of T lymphocytes and in the brain, that has strong binding affinity for cyclosporin.

cyclophorase system A mitochondrial preparation from either kidney or liver that catalyzes the reactions of the citric acid cycle and of beta oxidation.

cyclophosphamide An immunosuppressive drug.

cycloserine An antibiotic, produced by *Streptomyces orchidaceus*, that inhibits the biosynthesis of the peptidoglycan component of bacterial cell walls; it is used in the treatment of tuberculosis.

cyclosporin One of a group of cyclic peptides produced by certain fungi; cyclosporin A is an immunosuppressive agent used to control the rejection of transplanted tissues.

cyclotron An accelerator designed to impart high kinetic energy to particles, such as protons and deuterons, by means of an oscillating electric field and a fixed magnetic field; the particles move in a circular path.

cyclum An automatic system that permits various column chromatographic separations to be repeated precisely a large number of times. Parameters can be adjusted during the operation and the system is applicable to both analytical and preparative separations.

Cyd Cytidine.

cylindrical axis of symmetry An axis of rotational symmetry such that $n = \infty$. *Sym* C_∞. *See also* axis of rotational symmetry.

cymograph Variant spelling of kymograph.

Cys 1. Cysteine. 2. Cysteinyl.

CySH 1. Cysteine. 2. Cysteinyl.

CySO$_3$H Cysteic acid. 2. Cysteyl.

CYSSOR Cysteine-specific scission by organic reagents; one of a group of reagents that produce cleavage at the amino terminal side of cysteine in peptides containing cysteine residues.

cystathionine A nonprotein alpha amino acid that is an intermediate in the mammalian biosynthesis of cysteine from methionine.

cystathioninuria A genetically inherited metabolic defect in humans that is characterized by an accumulation of cystathionine in the plasma and by mental retardation. It is due to a deficiency of the enzyme γ-cystathionase (cystathionine-γ-lyase) which catalyzes the hydrolysis of cystathionine to cysteine and α-ketobutyrate.

cystatins A group of proteins that act as competitive inhibitors of cysteine peptidases. Type I cystatins (sometimes called stefins) contain about 100 amio acid residues, no disulfide bonds, and no carbohydrates; type II cystatins contain about 115 amino acid residues and two disulfide bonds; type III cystatins, the kininogens, contain about 355 amino acid residues, a number of disulfide bonds, and are glycosylated.

cysteamine β-MERCAPTOETHYLAMINE.

cysteic acid A sulfonic acid that is obtained by oxidation of the sulfhydryl group of cysteine to –SO$_3$H. *Sym* CySO$_3$H.

cysteine An aliphatic, polar alpha amino acid that contains a sulfhydryl group. *Abbr* Cys; CySH; C.

cysteine peptidases A group of proteolytic enzymes (endopeptidases or exopeptidases) that contain a sulfhydryl group of cysteine in their active sites.

cystic fibrosis A hereditary disease in humans that is characterized by the functional failure of mucus-secreting glands and the resultant presence of an unusual glycoprotein in the mucus that causes the mucus to have an abnormal viscosity. A generalized dysfunction of the exocrine glands that leads to a progressive blocking of various ducts. Frequently, the exocrine glands of the pancreas are also affected, leading to a deficiency of pancreatic enzymes.

cystine The dimer formed from two cysteine residues, linked by means of a disulfide bond.

cystinosis A pathological condition, characterized by the deposition of cystine in the lysosomes of many cells; may result from defects in glutathione metabolism and may lead to crystal formations in the kidney.

cystinuria A genetically inherited metabolic defect in humans that is characterized by the presence of excessive amounts of cystine, lysine, and arginine in the urine.

cyt 1. Cytosine. 2. Cytochrome.

cytidine The ribonucleoside of cytosine. Cytidine mono-, di-, and triphosphate are abbreviated, respectively, as CMP, CDP, and CTP. The abbreviations refer to the 5′-nucleoside phosphates unless otherwise indicated. *Abbr* Cyd; C.

cytidylic acid The ribonucleotide of cytosine.

cyto- Prefix meaning cell.

cytochalasins A group of metabolites, excreted by various species of molds, that paralyze many types of vertebrate cell movement (locomotion, phagocytosis, cytokinesis, etc.) by binding specifically to one end of actin filaments.

cytochemistry The science that deals with the chemistry of cellular components.

cytochrome A hemoprotein that contains an iron–porphyrin complex as a prosthetic group. Cytochromes function as electron carriers by virtue of the reversible valence change that the heme iron can undergo; they are classified into four groups, designated *a*, *b*, *c*, and *d*.

cytochrome a A cytochrome in which the heme prosthetic group contains a formyl side chain; a cytochrome that contains heme A. *Aka* type *a* cytochrome; class *a* cytochrome.

cytochrome b A cytochrome that contains protoheme or a related heme (without a formyl group) as its prosthetic group and in which the prosthetic group is not bound covalently to the protein. *Aka* type *b* cytochrome; class *b* cytochrome.

cytochrome c A cytochrome in which there are covalent linkages between the side chains of the heme and the protein. *Aka* type *c* cytochrome; class *c* cytochrome.

cytochrome c′ RHP CYTOCHROME.

cytochrome c oxidase complex COMPLEX IV.

cytochrome c:oxygen oxidoreductase COMPLEX IV.

cytochrome d A cytochrome with a tetrapyrrolic chelate of iron as a prosthetic group in which the degree of conjugation of double bonds is less than that in porphyrin; dihydroporphyrin (chlorin) is an example. *Aka* type *d* cytochrome; class *d* cytochrome.

cytochrome oxidase The enzyme that catalyzes the terminal reaction in the electron transport system in which molecular oxygen is reduced to water.

cytochrome P$_{450}$ One of a class of enzymes that are heme proteins in which the iron of the heme is linked to the sulfur of an SH group of cysteine in the polypeptide chain. They form carbon monoxide complexes that have a major absorption band at 450 nm. The enzymes are widely distributed in animal tissues, plants, and microorganisms, and catalyze the monooxygenation of a vast variety of hydrophobic substances; they play an important role in the detoxification of drugs, mutagens, and carcinogens.

cytochromoid c RHP CYTOCHROME.

cytocidal Capable of killing cells.

cyto-differentiator A drug that renders a malignant cell benign instead of killing it.

cytoduction The physical or genetic transfer of cytoplasmic, specifically mitochondrial, genomes for the purpose of constructing new strains.

cytoflav An impure preparation of riboflavin from heart muscle.

cytogenetic disease A disease, such as mongolism, that is due to a chromosome abnormality.

cytogenetic map A genetic map that shows the location of the genes in a chromosome.

cytogenetics The science that deals with cellular changes that are related to hereditary phenomena; combines the methods of both cytology and genetics.

cytohemin Heme A; the prosthetic group of cytochrome a.

cytokines A group of substances, formed by an animal in response to infection. They are similar to hormones in their function; they are produced in one cell and stimulate a response in another cell. They are biological response modifiers (BRM) and include such substances as interferon, interleukin, and tumor necrosis factor. *See also* lymphokines.

cytokinesin A plant growth-regulating substance that affects cell division and that apparently acts synergistically with auxins.

cytokinesis The division of the cytoplasm that follows both mitosis and meiosis.

cytokinins A group of N-substituted derivatives of adenine that function as plant hormones; they promote cell division and stimulate plant metabolism, particularly RNA and protein synthesis.

cytolipin A cytoside; cytolipin H, or ceramide lactoside, can function as a hapten under certain conditions; cytolipin K is probably identical with globoside.

cytological hybridization IN SITU HYBRIDIZATION.

cytological map A representation of the location of genes at specific sites on the chromosome, especially of genes at sites on polytene or human mitotic chromosomes.

cytology The branch of biology that deals with the origin, the structure, the function, and the life history of cells.

cytolysin An antibody that can lead to the lysis of cells.

cytolysis The lysis of cells.

cytolysosome A membrane-bound cytoplasmic region which is formed during intracellular digestion and which is subsequently digested.

cytolytic Of, or pertaining to, cytolysis.

cytomegalovirus A virus of the herpes virus group that can cause diseases of the salivary glands and may cause degeneration of the central nervous system in the newborn. *Abbr* CMV.

cytomembrane A bacterial membranous system or structure that occurs in the cytoplasm and that may or may not be continuous with the cell membrane; mesosomes and chlorosomes are two examples.

cytopathic Causing either injury or disease to cells.

cytophilic antibody An antibody that can adhere specifically to macrophages.

cytophotometry A technique for the quantitative determination of substances by the combined use of microscopy and spectrophotometry; based on measurements of the light absorbed by cellular components that either have or have not been treated with specific stains.

cytoplasm The protoplasm of a cell, excluding the nucleus or the nuclear zone.

cytoplasmic gene A nonnuclear gene, such as a gene of mitochondria or chloroplasts.

cytoplasmic inheritance The non-Mendelian hereditary transmission that depends on replicating cytoplasmic organelles such as mitochondria, viruses, and plastids, rather than on nuclear genes.

cytoplasmic membrane CELL MEMBRANE.

cytoplasmon The total extranuclear genetic information of a eukaryotic cell excluding that in the mitochondria and the plastids.

cytoplast The intact cytoplasmic contents, as distinguished from the nuclear contents, of a cell; the structural and functional unit of a eukaryotic cell that is formed by a network of cytoskeletal proteins to which the cytoplasmic organelles are linked.

cytoribosome A cytoplasmic ribosome, as distinct from a nuclear or a mitochondrial ribosome.

cytosegresome An intracellular, membrane-bound, vacuole that is formed during intracellular self-digestion (autophagy) and that has enclosed some of the cell's own constituents.

cytoside A diglycosyl derivative of a ceramide that contains only simple sugars.

cytosine The pyrimidine 2-oxy-4-amino-pyrimidine that occurs in both RNA and DNA. *Abbr* C; Cyt.

cytosine arabinoside An antitumor agent. The compound, by itself, is inactive but, after intracellular conversion to the nucleoside triphosphate, functions as a competitive inhibitor (with respect to dCTP) of DNA polymerase. *Abbr* ara c.

-cytosis Combining form meaning an incrase in the number of cells.

cytoskeleton The filamentous, flexible, and dynamic framework in the cytoplasm of eukaryotic cells. Its major components are microtubules, microfilaments, and intermediate filaments; these are interconnected by the microtrabecular network. The cytoskeleton gives cells their characteristic shape and is responsible for changes in shape as in locomotion, cell division, and phagocytosis. It provides attachment sites for organelles and provides for communication between different parts of the cell.

cytosol The soluble portion of the cytoplasm that includes dissolved solutes but that excludes the particulate matter. *Aka* cell sap.

cytosome 1. A cytoplasmic organelle (vacuole) of unknown function frequently found in eukaryotic organisms. 2. CYTOPLASM.

cytostatic agent An agent that suppresses cell multiplication and cell growth.

cytotaxin A compound that has the capacity of promoting chemotaxis.

cytotaxis The attraction or repulsion of cells for one another that leads to the ordering and arranging of new cell structures under the influence of preexisting ones.

cytotoxic Causing cell death.

cytotoxic anaphylaxis The anaphylactic reaction that takes place when an animal organism is injected with antibodies which are specific for cell surfaces.

cytotoxic antibody An antibody that damages antigen-bearing cells, particularly in the presence of complement.

cytotoxic T cells A group of T cells that specifically kill foreign or virus-infected vertebrate cells. *Aka* cytotoxic T lymphocytes.

cytotropic anaphylaxis The anaphylactic reaction that is mediated by cytotropic antibodies.

cytotropic antibody An antibody that binds to target cells, particularly mast cells, thereby sensitizing the animal for anaphylaxis.

D

d 1. Deoxy. 2. Dextrorotatory. 3. Deci. 4. Dalton.

D 1. D configuration 2. Deuterium. 3. Dielectric constant. 4. Translational diffusion coefficient. 5. Aspartic acid. 6. Dihydrouridine. 7. Dalton.

2,4-D 2,4-Dichlorophenoxyacetic acid; a synthetic auxin.

D_{10} Decimal reduction time.

$D^0_{20,w}$ Standard diffusion coefficient.

da Deca.

Da Dalton.

DABITC 4-Dimethylaminoazobenzene-4'-isothiocyanate. A reagent that reacts with the alpha amino groups of amino acids and that can be used in the Edman degradation of proteins; it yields highly fluorescent derivatives that are easily identified by thin layer chromatography.

dactinomycin ACTINOMYCIN D.

DALA δ-Aminolevulinic acid.

dalton A unit of mass equal to one-twelfth the mass of one atom of the carbon isotope ^{12}C; equal to 1.661×10^{-24} g and identical to the officially defined atomic mass unit (amu). *Sym* Da, D, d.

dampening of charge effects The decrease of various electrostatic effects in solution, resulting from the presence of charged groups on macromolecules, that is brought about by an increase in the ionic strength of the solution. The presence of large amounts of other ions tends to minimize electrostatic interactions between the charged macromolecules.

Dane particle A particle that contains the Australia antigen, and that is now known to represent the infectious virion, responsible for type B hepatitis. *See also* hepatitis.

Danielli–Davson model An early model for the structure of biological membranes which postulated a lipid bilayer, with lipid material separating the two layers, and with proteins located on the outside of the lipid bilayer. *See also* unit membrane hypothesis; fluid mosaic model.

Danielli–Davson–Robertson model UNIT MEMBRANE HYPOTHESIS.

dansyl amino acid An amino acid derivative formed by the reaction of 1-dimethylaminonaphthalene-5-sulfonyl chloride with the alpha amino group of an amino acid. Dansyl amino acids are fluorescent and are used for the measurement of amino acids by fluorometry.

dansylation The introduction of a dansyl group into an organic compound.

Danysz phenomenon The phenomenon that the extent of formation and dissociation of an antigen–antibody comlex depends on whether the antigen is added all at once, or in small doses.

DAP pathway Diaminopimelate pathway.

dapsone *See* sulfone.

dark adaptation The time required for the rods in the retina of an animal, previously placed in bright light, to become fully responsive to dim light.

dark current The current that is obtained in a photoelectric instrument, such as a spectrophotometer or a scintillation counter, in the absence of radiation and that is caused by thermionic emissions of the photomultiplier tube.

dark field illumination The illumination of a specimen with rays directed from the side so that only scattered light enters the microscope lenses.

dark field microscope A microscope in which dark field illumination is used so that objects appear bright on a dark background as a result of the light scattered by the objects.

dark reaction The photosynthetic reaction or reaction sequence that can occur in the absence of light; the fixation of carbon dioxide by plant chloroplasts is an example.

dark reactivation The enzymatic repair of DNA, previously damaged by exposure to a mutagen, that does not require light; applies specifically to repair of thymine dimers, produced by exposure of DNA to ultraviolet light. Thymine dimers can be repaired by means of three distinct mechanisms: (a) excision repair; (b) recombination repair; and (c) SOS repair. *See also* photoreactivation.

dark recovery DARK REACTIVATION.

dark repair DARK REACTIVATION.

D arm *See* arm.

Darwinian evolution The cumulative changes, including mutation and selection, that occur through successive generations in organisms that are related by descent. *See also* natural selection.

Darwinian selection NATURAL SELECTION.

d-assay The construction of a thermal denaturation profile by measurements of the system at various ambient conditions; the assay measures the extent of the transition. *See also* i-assay.

data bank A computer library; an electronic storehouse of information.

data base A collection of related data that can be retrieved by a computer.

dative bond COORDINATE COVALENT BOND.

dative covalent bond COORDINATE COVALENT BOND.

datum (*pl* data). An experimental finding; a fact; a measurement.

dauermodification An environmentally induced, phenotypic change that appears to be inherited and survives through one or more generations but that eventually disappears.

daughter 1. One of the two cells formed from a parent cell by cell division. 2. The DNA molecule or the chromosome formed from either parental molecules or chromosomes by replication. 3. The nuclide formed from a parent nuclide by radioactive decay.

daughter-strand gap repair RECOMBINATION REPAIR.

Davie and Ratnoff theory A theory of blood clotting that is based on a cascade mechanism.

Davis U-tube A U-tube that contains a porous filter in its lower portion so that bacterial cultures may be placed in either one or both arms of the tube. The filter prevents the passage of bacteria but allows the passage of small, diffusible substances.

(dA)$_x$ Deoxyadenylic acid; a homopolynucleotide strand of deoxyriboadenylic acid containing about x residues.

day vision The capacity to see in bright light; due to the cones in the retina.

DBM paper Diazobenzyloxymethyl paper; an impregnated paper, used in some blotting experiments, that binds single-stranded DNA, RNA, and proteins. These polymers become linked covalently via the diazonium groups in the paper.

DCC; DCCD Dicyclohexylcarbodiimide.

DCMU Dichlorophenyldimethylurea.

D-configuration The relative configuration of a molecule that is based upon its stereochemical relation to D-glyceraldehyde.

DDS *See* sulfone.

DDT Dichlorodiphenyltrichloroethane; 1,1,1-trichloro-2,2-bis (*p*-chlorophenyl)ethane; an insecticide.

deacylase An enzyme that catalyzes a deacylation reaction.

deacylated tRNA A transfer RNA molecule from which either a previously bound aminoacyl group or a previously bound peptidyl group has been removed.

deacylation 1. The removal of an acyl group from a compound. 2. The formation of acetoacetyl coenzyme A and coenzyme A from two molecules of acetyl coenzyme A.

deadaptation The changes that occur in an inducible system in the time interval between the point at which the inducer is removed and the point at which synthesis of the inducible enzyme stops.

dead-end complex A complex of enzyme and substrate, or enzyme and product, that is catalytically inactive and, therefore, ties up the enzyme in a useless form. As an example, in linear noncompetitive inhibition, the ternary ESI (enzyme–substrate–inhibitor) complex does not break down to yield products; it is, therefore, a dead-end complex.

dead-end inhibitor A competitive inhibitor that forms an enzyme–inhibitor complex that cannot react further and that cannot lead to the formation of products until the inhibitor dissociates from the complex.

dead time 1. COINCIDENCE TIME. 2. The period of time between effective mixing and the start of observations in a rapid-flow or stopped-flow experiment.

dead-time loss COINCIDENCE LOSS.

dead vaccine KILLED VACCINE.

dead volume 1. VOID VOLUME. 2. Any section of a chromatographic flow system between the inlet or injection port and the detector outlet in which solute and mobile phase are mixed but are not passing over the stationary phase.

DEAE-cellulose *O*-(Diethylaminoethyl)cellulose; an anion exchanger.

DEAE-Sephadex *O*-(Diethylaminoethyl) Sephadex; an anion exchanger that contains the grouping $-C_2H_4N^+(C_2H_5)_2H$ linked via ether bonds to the Sephadex.

deamination The removal of an amino group from an organic compound.

Dean and Webb method A method for determining the equivalence zone of a precipitin reaction by mixing a constant amount of antiserum with varying dilutions of antigen and taking the tube in which recipitation occurs most rapidly to be indicative of the equivalence zone. *See also* method of optimal proportions; Ramon method.

death phase The phase of growth of a bacterial culture that follows the stationary phase and during which there is a decrease in the number (or the mass) of the cells.

debranching enzyme An enzyme that catalyzes the hydrolysis of branch points in a polymer.

de Broglie wavelength The wavelength associated with a moving particle; it is given by the de Broglie relationship $\lambda = h/mv$, where λ is the de Broglie wavelength, h is Planck's constant, m is the mass of the moving particle (electron, proton, etc.), and v is the velocity of the particle.

debye A unit of dipole moment; equal to the dipole moment of two charges, of 4.8×10^{-10} esu each, that are separated by 1 Å.

Debye–Hueckel limiting law *See* mean ionic activity coefficient.

Debye–Hueckel theory The theory that relates the activity coefficients of ions in solution to the electrostatic interactions between the ions and to the diameter of the ion atmosphere around each ion.

Debye length The thickness of the ion atmosphere surrounding a central ion according to the Debye–Hueckel theory; the most probable distance from the central ion to a counterion. *Aka* Debye radius.

deca- Combining form meaning 10 and used with metric units of measurement. *Var sp* deka. *Sym* da.

decade scaler A scaler that produces one output pulse for every 10 input pulses.

decamer An oligomer that consists of 10 monomers.

decamethonium A 16-carbon compound containing two quaternary nitrogen atoms; a divalent cation which binds noncovalently to muscarinic receptors of acetylcholine. It blocks the binding of acetylcholine at the postsynaptic cell and locks the ion channel of the receptor in the open state, thereby leading to constant depolarization of the postsynaptic membrane. Referred to as an agonist of the cholinergic system since it mimics the action of acetylcholine.

decant To pour off the liquid layer that is above sedimented material or above a precipitate.

decapitate To cut off the head.

decapsidate To remove the viral capsid.

decarboxylase CARBOXY-LYASE.

decarboxylation The removal or the loss of a molecule of carbon dioxide from the carboxyl group of an organic compound.

decatenation The conversion of a catenane to two separate circular molecules.

decay 1. The decomposition of organic matter through the action of microorganisms. 2. RADIOACTIVE DECAY.

decay chain RADIOACTIVE SERIES.

decay constant The fraction of radioactive atoms that are decaying per unit time; the constant λ in the equation $N = N_o e^{-\lambda t}$, where N is the number of radioactive atoms present at time t, N_o is the number present at time zero, and e is the base of natural logarithms. The term $e^{-\lambda t}$ is known as the decay factor.

decay factor *See* decay constant.

decay series RADIOACTIVE SERIES.

deci- Combining form meaning one tenth and used with metric units of measurement. *Sym* d.

decile One of the 9 values of a statistical variable which divide the total frequency distribution into 9 equal parts. The first, second, ..., ninth decile are values at, or below which, fall the lowest 10, 20, ..., 90% of a set of data.

decimal reduction time The time required, at a given temperature, to heat inactivate (kill) 90% of a given population of viable bacterial cells or spores. *See also* F value.

deciphering of the code The experimental determination of the nucleotide composition and the nucleotide sequence of the codons of the genetic code.

decline phase DEATH PHASE.

decoding site AMINOACYL SITE.

decomplementation Any process that removes complement activity from a serum; treatment of the serum with heat or antigen–antibody complexes are two examples.

decomposition The breakup of a chemical substance into two or more simpler substances.

decontamination The removal of a contamination particularly a radioactive one.

dedifferentiation The loss of differentiation, as that due to anaplasia.

deduction A conclusion arrived at by reasoning from generals to particulars.

deep groove MAJOR GROOVE.

defective lysogenic strain A strain of mutant bacterial cells that have incorporated a prophage that cannot replicate upon induction and that cannot give rise to intact phage particles.

defective organism AUXOTROPH.

defective prophage A prophage that cannot replicate upon induction and that cannot give rise to intact phage particles.

defective virus A virus that cannot synthesize one or more of its structural proteins and that can form intact particles in the host cell only when it is in the presence of a helper virus. *See also* deficient virus.

defibrinated blood Blood from which fibrin has been removed.

deficiency DELETION.

deficiency disease A disease that results from the deficiency of a nutrient, as that resulting from either a vitamin or a mineral deficiency.

deficiency mutant AUXOTROPH.

deficient virus A virus that cannot synthesize one or more of its functional proteins so that its nucleic acid cannot be replicated autonomously. *See also* defective virus.

defined medium A synthetic medium, containing only known chemical ingredients at known concentrations.

deformation vibration The vibration of a molecule in which there is a change in a bond angle.

deformylase   A prokaryotic enzyme that catalyzes the removal of the formyl group from *N*-formylmethionine at the N-terminal of the polypeptide chain in protein biosynthesis.

deg   Degree.

degassing   The removal of air bubbles from a chromatographic support, particularly a gel; readily done by placing a suspension of the support in a suction flask and connecting the latter to a water aspirator.

degeneracy   1. The existence of two or more synonym codons for a given amino acid. The degeneracy is termed complete or partial depending on whether all, or only some, of all the possible codons code for amino acids; the degeneracy is termed regular if it follows certain rules as distinct from one that is entirely random. 2. The existence of two or more atomic or molecular levels of equal energy; thus an atom may have two or more orbitals of equal energy, and a molecule may exist in two or more conformational states of equal energy.

degenerate   Possessing degeneracy.

degenerate codon   SYNONYM CODON.

degradation   The gradual and stepwise breakdown of a macromolecule to smaller fragments that proceeds by the breaking of covalent bonds.

degree   *See* absolute temperature scale; Celsius temperature scale; Fahrenheit temperature scale.

degree of inhibition   The extent of inhibition of an enzyme; specifically, $i = (v_o - v_i)/v_o$ where i is the degree of inhibition, v_o and v_i are the initial velocities of the uninhibited and inhibited reaction, respectively. Noncompetitive inhibition exists if i is unaffected by the substrate concentration; competitive inhibition exists if i is decreased as the substrate concentration is increased; and uncompetitive inhibition exists if i is increased as the substrate concentration is increased.

degree of ionization   The concentration of ions divided by the total concentration; for a weak acid ($HA \rightleftharpoons H^+ + A^-$), the degree of ionization is given by $[A^-]/([HA] + [A^-])$. The degree of hydrolysis and the degree of dissociation are defined in a similar manner.

degree of polymerization   The number of monomers in a polymer.

degrees of freedom   1. The total number of variables that can be varied arbitrarily without causing the disappearance of a phase. *See also* phase rule. 2. The total number of coordinates along which the velocity of a molecule has either a translational, or a rotational, component. 3. The total number of items,

such as observations, deviations, and means, that can vary independently in the light of restrictions imposed on the calculations. *Abbr* df; D/F.

degrowth   The decrease in the mass of an organism that occurs at the end of a period of growth.

dehydrase   DEHYDRATASE.

dehydratase   An enzyme that catalyzes a dehydration reaction.

dehydrated food   Food from which water has been removed.

dehydration   1. The removal of water from a compound. 2. The loss of body water.

dehydroascorbic acid   The oxidized form of ascorbic acid.

dehydrogenase   An enzyme that catalyzes the removal of hydrogen from a substrate using a compound other than molecular oxygen as an acceptor.

dehydrogenase-type mechanism   The mechanism of an enzyme reaction that resembles that of pyridine-linked dehydrogenases. For the latter enzymes, NAD^+ and $NADP^+$ function as cosubstrates, and the mechanism of the reaction is generally an ordered sequential one.

dehydrogenation   The removal of hydrogen from an organic compound.

deinhibitor   A substance that counteracts or eliminates the effect of an inhibitor.

deionized water   Water from which ions have been removed; usually done by passing water through an ion-exchange resin, particularly a mixed-bed demineralizer that removes both anions and cations.

deionizer   A device for removing ions from water or from other fluids, usually by means of ion-exchange resins.

deka   Variant spelling of deca.

delayed-type hypersensitivity   An allergic response that occurs a few hours or a few days after the administration of an antigen. The response depends on the sensitization of certain cells, especially lymphocytes, rather than on circulating antibodies.

deletion   1. A point mutation in either RNA or DNA in which one or more nucleotides are removed from a polynucleotide strand. In double-stranded nucleic acid this also leads to the removal of the complementary nucleotide from the second strand so that an entire base pair is ultimately deleted from the nucleic acid. 2. A chromosomal aberration in which there is a loss of genetic material from the chromosome. The portion lost may be either a nucleotide or a larger fragment, consisting of one or several genes.

deletion hypothesis   A hypothesis according to which cancer is due to the loss of one or more specific enzymes or other proteins. *See also* catabolic deletion hypothesis.

deletion loop The loop formed in a DNA strand that is hybridized with a complementary DNA strand that has a deletion; a region of noncomplementarity in only one strand. *See also* substitution loop.

deletion mapping A method for locating the position of a gene on a genetic map by means of overlapping deletions.

deletion method A method for isolating specific mRNA molecules by hybridizing them with DNA molecules that contain deletions.

deletion-substitution particle A transducing bacteriophage in which deleted phage genes have been replaced by bacterial genes.

deliberate immunization The purposeful introduction of antigens, antibodies, or lymphocytes into an organism to either stimulate the production of antibodies by the organism or provide the organism with antibodies.

delipidation The removal of lipid from a biological sample.

delipidized protein A lipoprotein from which part or all of the lipid has been removed.

deliquescence The uptake of moisture from the air by a substance under ordinary conditions of temperature and pressure.

delocalization of electrons The distribution of electrons over a number of nuclei, rather than their being located between two specific nuclei. Electron delocalization leads to lower energy orbitals and greater stability.

delocalized bond A bond involving more than two atoms.

delocalized orbital A molecular orbital that is spread over two or more bonding atoms or even over an entire molecule.

delta 1. Denoting a small difference between two values. *Sym* Δ. 2. Denoting the fourth carbon atom from the carbon atom that carries the principal functional group of the molecule. *Sym* δ.

delta chain 1. The heavy chain of the IgD immunoglobulins. 2. One of the polypeptide chains of a minor hemoglobin component in normal human adults.

delta ray A beam of high-energy secondary electrons that have energies above 100 eV.

delta scale A scale used for nuclear magnetic resonance measurements. One delta (δ) unit is equal to 1 ppm of the spectrometer frequency. Thus, if an instrument is operated at 60 MHz, one delta unit equals 60 Hz. *See also* chemical shift.

demineralizer DEIONIZER.

demyelination *See* multiple sclerosis.

denaturant DENATURING AGENT.

denaturation Any change in the native conformation of a protein or of a nucleic acid

other than the breaking of the primary chemical bonds that join either the amino acids or the nucleotides in the chain. Denaturation may involve the breaking of noncovalent bonds such as hydrogen bonds, and the breaking of covalent bonds such as disulfide bonds; it may be partial or complete, reversible or irreversible. Denaturation leads to changes in one or more of the characteristic chemical, biological, or physical properties of the protein or of the nucleic acid.

denaturation loop *See* denaturation mapping.

denaturation mapping An electron microscopic method for locating AT-rich regions in double-stranded DNA; involves heating the DNA to a point at which melting is just detected, and then stabilizing the denaturation loops by treatment with formaldehyde.

denatured Having undergone denaturation.

denaturing agent A physical or a chemical agent that can bring about denaturation.

dendrite A short, and usually branched, process of a nerve cell that conducts impulses to the cell body.

dendritic evolution An evolutionary pattern that, when diagramed, resembles a tree. Such phylogenetic trees are characteristic of the evolutionary patterns of animal species.

dengue viruses A virus group of four distinct serotypes that belong to the family of *Flaviviridae*. They are transmitted to humans by mosquitoes (*Aedes aegypti*) and usually cause a benign syndrome (dengue fever), characterized by fever, headache, and joint and muscle pains. In cases when the syndrome becomes severe it is associated with hemorrhage and increased vascular permeability (dengue hemorrhagic fever, DHF, or dengue shock syndrome, DSS).

Denhart's solution A solution that contains ficoll, polyvinylpyrrolidone, and bovine serum album; used to treat filters containing bound nucleic acids to prevent the binding of single-stranded DNA probes.

denitrification The formation of molecular nitrogen from nitrate by way of nitrite. *See also* nitrate respiration.

denitrifying bacteria The bacteria that carry out denitrification.

de novo initiation The initiation of DNA replication in which the leading strand is started afresh, being joined to an RNA primer; involves formation of a D-loop.

de novo synthesis The synthesis of a molecule, particularly a macromolecule, from simple precursors as distinct from its formation by way of anabolism, catabolism, or modification of other macromolecules. Thus, the de novo synthesis of a protein refers to its synthesis from the amino acid level and not to (a) its synthesis from the peptide or polypeptide level, (b) its formation by the addition of amino acids to other proteins, or (c) its formation by the breakdown of other proteins. *See also* salvage pathway.

densitometer 1. An instrument for measuring either absorbed or reflected light in materials other than solutions. Densitometers are used for the scanning of chromatograms and electropherograms, and for measuring the blackening of photographic films. 2. An instrument for measuring the density or the specific gravity of a gas or a liquid.

density 1. Weight per unit volume. 2. The degree of blackening of a photographic film.

density-dependent growth CONTACT INHIBITION.

density gradient The change of density with distance; refers particularly to a solution in which there is such a change from one end of the tube or the cell that holds the solution to the other. A density gradient may be set up in an ultracentrifuge cell by virtue of the variation of the centrifugal force with distance from the center of rotation, so that the density increases from the meniscus to the bottom of the cell; solutions of cesium chloride are commonly used for such experiments. A density gradient may also be set up in a certrifuge tube by layering solutions of different densities above each other so that the density increases from the top to the bottom of the tube; solutions of sucrose are commonly used for such experiments.

density gradient centrifugation The centrifugation of macromolecules through a density gradient for either preparative or analytical purposes. *See also* band sedimentation.

density gradient sedimentation equilibrium Density gradient centrifugation that is typically performed in an analytical ultracentrifuge and that permits the separation of macromolecules which differ only slightly in their densities. The technique involves centrifuging a concentrated salt solution that contains macromolecules, such as a cesium chloride solution, until the salt achieves its equilibrium distribution and thereby produces a density gradient in the cell; the macromolecules band in this density gradient at positions where their densities equal those of the gradient. Also referred to as isopycnic ("having the same density") gradient centrifugation since separation of the molecules is based on their differences in buoyant density.

density gradient sedimentation velocity Density gradient centrifugation that is typically performed in a preparative ultracentrifuge and that is used for the fractionation of macromolecules, coupled with a variety of assay techniques for detection of the various com-ponents. The technique involves layering a solution of macromolecules on top of a preformed density gradient and centrifuging the gradient in a swinging bucket rotor. The macromolecules sediment through the gradient as bands and are separated on the basis of their differences in sedimentation rates.

density gradient zonal centrifugation DENSITY GRADIENT SEDIMENTATION VELOCITY.

density inhibition CONTACT INHIBITION.

-dentate Combining form indicating the number of groups in a molecule that coordinate with a metal ion to form a complex; used with mono-, bi-, tri-, tetra-, etc.

dentine The major calcified tissue of teeth; it is covered by either enamel or cementum depending on whether the portion of the tooth is exposed or submerged.

deoxy- 1. Combining form indicating a compound that contains 2-deoxy-D-ribose. 2. Combining form indicating a compound that contains less oxygen than the parent compound.

deoxyadenosine The deoxyribonucleoside of adenine.

5′-deoxyadenosylcobalamin COBAMIDE.

deoxyadenylic acid The deoxyribonucleotide of adenine.

deoxycholic acid A bile acid that is derived from cholic acid by the loss of an oxygen atom. *Abbr* DOC.

deoxycorticosterone A mineralocorticoid that regulates the excretion and retention of minerals by the kidney, particularly with respect to sodium and potassium. *Abbr* DOC.

deoxycytidine The deoxyribonucleoside of cytosine.

deoxycytidylic acid The deoxyribonucleotide of cytosine.

deoxyguanosine The deoxyribonucleoside of guanine.

deoxyguanylic acid The deoxyribonucleotide of guanine.

deoxyhemoglobin Hemoglobin uncombined with molecular oxygen.

deoxymyoglobin Myoglobin uncombined with molecular oxygen.

deoxynojirimycin An antibiotic, synthesized by some *Bacillus* species, that is a potent inhibitor of α-glucosidases; it is a glucose analogue (5-amino-1,5-dideoxy-D-glucose) in which an NH group substitutes for the oxygen atom in the pyranose ring. Deoxynojirimycin is the reduced form of nojirimycin and does not carry a hydroxyl group at the anomeric carbon.

deoxyribofuranose Deoxyribose that has a 5-membered ring structure resembling that of the compound furan.

deoxyribonuclease An endonuclease that

catalyzes the hydrolysis of DNA. *Abbr* DNase; DNAase.

deoxyribonuclease I A deoxyribonuclease that catalyzes the hydrolysis of DNA to mono- and oligonucleotides consisting of, or terminating in, a 5′-nucleotide. *Abbr* DNase I; DNAase I.

deoxyribonuclease II A deoxyribonuclease that catalyzes the hydrolysis of DNA to mono- and oligonucleotides consisting of, or terminating in, a 3′-nucleotide. *Abbr* DNase II; DNAase II.

deoxyribonucleic acid The nucleic acid (*abbr* DNA) that constitutes the genetic material in most organisms and that is composed of the genes; together with histones it makes up the chromosomes of higher organisms. DNA is a polynucleotide that is characterized by its content of 2-deoxy-D-ribose and the pyrimidines cytosine and thymine. *See also* DNA forms; Watson–Crick model.

deoxyribonucleoprotein A conjugated protein that contains DNA as the nonprotein portion. *Abbr* DNP.

deoxyribonucleoside A nucleoside of 2-deoxy-D-ribose.

deoxyribonucleotide A nucleotide of 2-deoxy-D-ribose.

deoxyribose The five-carbon aldose, 2-deoxy-D-ribose, that is the carbohydrate component of deoxyribonucleic acid. *Abbr* dRib; deRib.

deoxyribose nucleic acid DEOXYRIBONUCLEIC ACID.

deoxyriboside A glycoside of deoxyribose.

deoxyribotide A deoxyribonucleotide.

deoxysugar A monosaccharide in which one or more hydroxyl groups have been replaced by hydrogen atoms.

deoxythymidine THYMIDINE.

deoxythmidylic acid THYMIDYLIC ACID.

deoxyuridine The deoxyribonucleoside of uracil.

deoxyuridylic acid The deoxyribonucleotide of uracil.

depancreatize To surgically remove the pancreas.

dependent form The phosphorylated form of the enzyme glycogen synthase that is a regulatory enzyme for which glucose-6-phosphate is a positive effector. *Abbr* D-form.

dependent variable A quantity that is a mathematical function of one or more independent variables; the value of a dependent variable is fixed once the values for the related independent variables are chosen.

depolarization The elimination of polarization, as that occurring in a muscle or a nerve membrane upon electrical stimulation. A decrease in membrane potential; the membrane potential becomes less negative than it is in the normal resting state.

depolarization fluorescence *See* fluorescence depolarization.

depolymerization The degradation of a polymer to oligomers and/or monomers.

depolymerizing enzyme An enzyme that catalyzes the hydrolysis of a biopolymer to oligomers and/or monomers.

depot fat The fat that is stored in an organism. *Aka* adipose tissue.

deproteinization The removal of protein from a biological sample.

depside A natural or synthetic ester formed by condensation of phenol carboxylic acids; depsides occur in lichens and tannins.

depsipeptide antibiotics A group of peptide-like antibiotics, produced by *Fusaria* fungi. They consist of alternating amino acid and hydroxy acid residues, with the residues being linked by alternating peptide and ester bonds. Depsipeptide antibiotics are frequently cyclic and are then referred to as cyclodepsipeptides, peptolides, or enniatins. Cyclic depsipeptides act as ionophores.

depurination The removal of purines from a nucleic acid.

depyrimidination The removal of pyrimidines from a nucleic acid.

derepression Any modification that eliminates the repression of a gene and permits the synthesis of the gene product. Possible modifications include a decrease in the repressor concentration produced by starving the organism of a required nutrient, a reaction of the inducer with the repressor, a mutation of the regulator gene, or a mutation of the operator gene.

deRib Deoxyribose.

derivative A compound, usually an organic one, that is obtained by modification of a parent compound as a result of one or more chemical reactions.

derivative spectroscopy A method for analyzing spectroscopic measurements by plotting the first-, second-, or higher-order derivatives of a spectrum with respect to the wavelength.

derivatize To synthesize a derivative.

derived carbohydrate A derivative of a simple sugar, such as a sugar acid or an amino sugar.

derived lipid A lipid obtained by hydrolysis of a naturally occurring lipid.

derived protein A product obtained by treatment of a protein with heat, acid, base, enzymes, or other agents. Primary derived proteins, such as proteins and metaproteins, are proteins that have been altered only slightly; secondary derived proteins, such as proteoses and peptones, are proteins that have been altered more extensively.

dermal Of, or pertaining to, the skin, especially the true skin.

dermatan sulfate A heterogeneous glycosaminoglycan that contains disaccharide repeat units consisting of N-acetyl-D-galactosamine and D-glucuronic acid and disaccharide repeat units consisting of N-acetyl-D-galactosamine and L-iduronic acid. The uronic acids are present with variable degrees of sulfation. Dermatan sulfate is found in skin, blood vessels, and heart valves.

dermotropic virus A virus, the target organ of which is the skin.

des Prefix, describing a specific lack; an example is a protein or a peptide from which the N-terminal amino acid has been removed, such as des-his-glucagon or des-asp-angiotensin I.

DES Diethylstilbestrol.

desalanine insulin Insulin from which the alanine residue at the carboxyl end of the B chain has been removed by the action of trypsin.

desalting The removal of inorganic salt ions from a sample; techniques used include electrodialysis, ion-exchange chromatography, gel filtration, and electrophoresis.

desaspidin A toxic substance occurring in some ferns that is an uncoupler of both oxidative and photosynthetic phosphorylation.

desaturase An enzyme that catalyzes a desaturation reaction.

desaturation A reaction, or a reaction sequence, whereby a saturated compound is converted to an unsaturated one; the introduction of double (or triple) bonds into a molecule.

descending boundary The electrophoretic boundary that moves downward in one of the arms of a Tiselius electrophoresis cell. *See also* Tiselius apparatus.

descending chromatography A chromatographic technique in which the mobile phase moves downward along the support.

desensitization 1. The modification of an allosteric enzyme by either mutation or chemical means that results in an enzyme that has retained its catalytic activity but has lost the capacity to respond to effectors. 2. The attempt to minimize the response of an individual suffering from immediate-type hypersensitivity upon subsequent exposure to an allergen. Common methods include either the repeated injection of small doses of the allergen to form protective blocking antibodies, or the depletion of the individual's tissue stores of histamine and serotonin. 3. The loss of the ability of target cells to respond to a signaling ligand (such as a hormone); often occurs after prolonged exposure of the cells to the ligand and may be due to a decrease in the number or activity of cell-surface receptors (resulting from an inactivation of these receptors) or due to other factors. Desensitization may be homologous or heterologous. Two major mechanisms of desensitization are down regulation and uncoupling. *Aka* adaptation; refractoriness, tolerance.

desensitized enzyme An allosteric enzyme that has been so altered by either mutation or chemical modification that, while it is still catalytically active, it no longer responds to an effector.

desert *See* NMR desert.

deshielded nucleus An atomic nucleus in a molecule that is surrounded by a relatively smaller electron density than another nucleus. In nuclear magnetic resonance, such a nucleus will absorb radio frequencies of lower energy (downfield) than the other nucleus.

desiccant A substance that absorbs water and that is used to dry air or another substance.

desiccate To dry by means of a desiccant.

desiccator A closed container that holds a desiccant and that is evacuated and used for the removal of moisture from a substance and for maintaining the substance in the dry state.

desmin A protein that is closely related to vimentin and that, together with vimentin, occurs in intermediate filaments of muscle.

desmoenzyme PARTICULATE ENZYME.

desmolase The enzyme complex, composed of a mixed function oxidase and cytochrome P_{450}, that catalyzes the removal of the side chain from cholesterol; cholesterol is thereby converted to pregnenolone, a precursor of the steroid hormones.

desmosine An unusual amino acid that is formed enzymatically from four lysine residues; it is found only in elastin where it cross-links the polypeptide chains. While desmosine could theoretically cross-link four different polypeptide chains, current models of elastin suggest that it cross-links only two chains.

desmosome An intercellular junction; a thickened zone in the cell membranes of adjoining eukaryotic cells; believed to function in cell adhesion. *See also* cell junction.

desmotubule *See* plasmodesmata.

desorb To remove adsorbed molecules from the surface of a solid.

desorption 1. The removal of adsorbed molecules from the surface of a solid. 2. ELUTION. 3. Anyone of a variety of loosely related techniques that result in the release of ions from surfaces; used in conjunction with mass spectrometry.

desoxy- DEOXY-.

destain To remove the excess dye that has not been utilized in staining the materials of interest.

destructive interference   *See* interference (1).

destructive metabolism   CATABOLISM.

desulfurase   *See* desulfurication.

desulfuration   DESULFURIZATION.

desulfuricants   Anaerobic bacteria of the genera *Desulfovibrio* and *Desulfotomaculum* that utilize sulfate respiration, yielding hydrogen sulfide. The most important species of these bacteria is *Desulfovibrio desulfuricans*.

desulfurication   1. The anaerobic degradation of sulfur-containing organic compounds to inorganic sulfur. These processes include the removal of SH groups from proteins to yield H_2S, a reaction catalyzed by enzymes known as desulfurases. 2. The formation of H_2S by desulfuricants.

desulfurization   The removal of sulfur from a compound. Cysteine, for example, can be desulfurized to yield alanine by treating it with Raney nickel. *Aka* desulfuration.

detachment   PROPHAGE EXCISION.

detailed balancing   PRINCIPLE OF MICROSCOPIC REVERSIBILITY.

detector   1. A device for detecting the presence of the organic compounds that come off a column in gas chromatography. 2. A device for detecting the presence of radioactive radiation that is given off by a sample in a radiation counter.

detergent   A synthetic, or a naturally occurring, surface-active agent. Detergents are used for cleaning and, in cell fractionations, to lyse cells and solubilize membranes. Detergents may be grouped into three classes: ionic detergents (such as sodium dodecyl sulfate), nonionic detergents (such as Triton), and bile salts.

detergent builder   Any additive to a detergent (surfactant) that makes the latter more effective. An example is phosphate derivatives, such as tripolyphosphate, that chelate metal ions present in hard water. As a result, a detergent, containing phosphate, will be more effective in hard water than a detergent without phosphate.

detergent degradation   *See* biodegradable.

determinant   1. EFFECTOR. 2. ANTIGENIC DETERMINANT. 3. A square array of numbers, called elements, symbolizing certain mathematical operations. The number of rows (or column is called the order of the determinant. For example. the determinant $\begin{vmatrix} a & b \\ c & d \end{vmatrix}$ is a second-order determinant and has the value of ($a \times d - b \times c$).

determinate error   An error in a measurement that can be accounted for and that can be avoided, at least in principle; an error due to methodology, instruments, etc.

determination   The establishment of a specific course of development by a part of an embryo that will be pursued regardless of subsequent situations.

deterpenation   The removal of terpenes from essential oils.

detoxification   The enzymatic reactions in an organism whereby foreign compounds, produced within the organism or introduced into it, are converted to less harmful forms and to more readily excretable products; the foreign compounds are either chemically altered or conjugated to normally occurring metabolites of the organism. *Aka* detoxication.

deuridylic acid   An RNA molecule from which uracil residues have been removed by treatment with hydroxylamine.

deuterated   Labeled with deuterium.

deuteri-   Proposed prefix for the ^2H isotope. *See also* proti-, proto-; deutero-.

deuterium   The stable, heavy isotope of hydrogen that contains one proton and one neutron in the nucleus. *Sym* D.

deuterium exchange   A technique for studying the conformation of a protein by measuring the rate of exchange of the hydrogen (or deuterium) atoms in the protein with the deuterium (or hydrogen) atoms in the medium. The hydrogen (or deuterium) atoms that are in direct contact with the solvent exchange more rapidly than those that are located in the interior of the molecule or those that participate in hydrogen bonding.

deutero-   1. Combining form indicating a compound that contains deuterium. 2. Proposed prefix for the ^2H isotope. *See also* deuteri-.

deuteron   The deuterium nucleus that consists of one proton and one neutron.

developer   A chemical reducing agent that converts exposed silver halide grains to metallic silver and thereby renders an image visible on a photographic film.

development   1. The process whereby a mixture, which has been applied to a chromatographic support, is separated into individual components by treatment with a mobile phase. 2. The process whereby an image is rendered visible on a photographic film by means of a developer. 3. The series of orderly changes by which a mature cell, tissue, organ, organ system, or organism comes into existence.

deviation   The difference between a measured value and a reference value, usually the mean.

devolution   Retrograde evolution; degeneration.

Devoret test   A bioassay for detecting carcinogenic compounds; based upon the induction of lysis of phage λ in lysogenic *E. coli* cells.

dex   Dextrorotatory.

dextran   One of a group of branched-chain polysaccharides of D-glucose found in yeast and bacteria. They serve as storage materials and as components of bacterial capsules.

dextrin One of a group of polysaccharides of intermediate chain length that are formed during the hydrolysis of starch.

dextrogenic amylase ALPHA AMYLASE.

dextrorotatory Having the property of rotating the plane of plane-polarized light to the right, or clockwise, as one looks toward the light source. *Abbr* dex; d.

dextrose D-Glucose; a dextrorotatory monosaccharide.

df Degrees of freedom; also abreviated D/F.

DF Dissociation factor.

D-form Dependent form.

DFP Diisopropylfluorophosphate.

DFP peptide A peptide that contains a serine residue that has been linked to diisopropylfluorophosphate; obtained by treatment of a protein with diisopropylfluorophosphate, followed by partial hydrolysis. DFP peptides provide information about the amino acid sequence near the active site of those enzymes that possess a serine residue at or near the active site.

D genes Genes coding for segments of the hypervariable regions of immunoglobulin molecules; so called because they add to the diversity of antibodies. *Aka* diversity genes.

DHF Dihydrofolic acid.

DHU Dihydrouridine; the nucleoside of dihydrouracil.

DHU arm The base-paired segment in the cloverleaf model of transfer RNA to which the loop, containing dihydrouracil, is attached. *See also* arm.

di- Combining form meaning two or twice.

diabetes *See* diabetes innocens; diabetes insipidus; diabetes mellitus; renal diabetes; starvation diabetes; steroid diabetes; brittle diabetes; bronzed diabetes.

diabetes innocens RENAL GLUCOSURIA.

diabetes insipidus A disease caused by vasopressin insufficiency and characterized by the excretion of large volumes of hypotonic urine.

diabetes mellitus A complex disease characterized by derangements of carbohydrate, lipid, and protein metabolism; the primary symptom is the presence of excessive amounts of glucose in the blood. Two types are recognized clinically: juvenile onset (or insulindependent) diabetes and adult onset (insulinindependent or maturity onset) diabetes. The former usually appears in childhood and is due to a deficiency of insulin which may be caused by inadequate proinsulin production by the pancreas, by an accelerated destruction of insulin, or by insulin antagonists and inhibitors. Uncontrolled, it is characterized by

hyperglycemia, hyperlipoproteinemia (VLDL and chylomicra), and ketoacidosis. Adult onset diabetes usually occurs in middle-aged individuals whose insulin is present at near normal or even elevated levels. The defect in these individuals may be at the level of insulin receptors located on the cell membrane. Obesity appears to be a major contributing factor to the development of this disease. It is characterized by hyperglycemia and hyperlipoproteinemia (VLDL) but not by ketoacidosis.

diabetogenic Having a tendency or capacity to enhance diabetes.

diabetogenic hormone The hormone hydrocortisone or one of the other 11-oxysteroids that are secreted by the adrenal cortex and that stimulate gluconeogenesis.

diacylglycerol An acylglycerol (glyceride) formed by the esterification of one molecule of glycerol with two fatty acid molecules.

diacytosis The discharge of an empty pinocytotic vesicle from a cell after it has transported its contents into the cell.

diafiltration A modification of ultrafiltration in which there is repeated or continuous dilution with fresh solvent in conjunction with the filtration; as a result, the rate of desalting is increased.

diagonal method A two-dimensional electrophoretic or chromatographic technique for mixtures of peptides in which both dimensions are carried out under identical conditions and a chemical reaction is performed on the paper or gel between the two separations. Peptides that have not been modified by the chemical treatment will behave in the second dimension exactly as they did during the first and, as a result, will form a diagonal pattern. Peptides that have been modified by the chemical treatment will not fall on the line and can be identified. The method can be used, for example, to identify the position of disulfide bonds in a protein. In this case, a partial hydrolysate of the protein, after the first dimension, is exposed to formic acid vapors which oxidize disulfide bonds to cysteic acid groups.

dialysate 1. The solution containing the material that has diffused through a dialysis membrane. 2. The solution containing the material that has not diffused through a dialysis membrane; to avoid confusion, the term retentate has been proposed to describe the material retained by a semipermeable membrane. *Var sp* dialyzate.

dialysis The separation of macromolecules from ions and low molecular weight compounds by means of a semipermeable membrane that is impermeable to colloidal macromolecules but is freely permeable to crystalloids and water. *See also* forced dialysis; reverse dialysis.

dialysis equilibrium *See* equilibrium dialysis.

dialyzate DIALYSATE.

dialyze To process by means of dialysis.

dialyzer An apparatus for performing dialysis that consists of one or more compartments separated by membranes.

diamagnetic Descriptive of a substance that has paired electrons and has no magnetic dipole moment; when such a substance is placed in a magnetic field, a magnetic dipole is induced in the substance which opposes the applied field and tends to move the substance out of it.

diameters A measure of optical magnification; a measure of 20,000 diameters means that the diameter of the object has been magnified 20,000 times when viewed through a microscope.

diaminopimelate pathway A pathway for the biosynthesis of lysine that proceeds by way of diaminopimelic acid and occurs in bacteria and higher plants. *Abbr* DAP pathway.

2,6-diaminopurine A purine analogue that is a mutagen.

diamond code An early version of the genetic code according to which the R groups of the amino acids fit into the diamond-shaped pockets that are present in Watson–Crick type double-helical DNA.

diapause The period of rest, cessation of growth, and decreased level of metabolism that occurs in the life cycle of insects.

diaphorase One of a group of enzymes that catalyze the reduction of an artificial electron acceptor, such as a dye, ferricyanide, or a quinone, by either reduced nicotinamide adenine dinucleotide or by reduced nicotinamide adenine dinucleotide phosphate. Such enzymes were originally thought to function in the reduction of metabolites in the electron transport system between NADH and the cytochromes, but this need not be the case. One preparation of diaphorase has been shown to be identical with lipoamide dehydrogenase.

diasolysis The separation of solutes by diffusion across a membrane that is somewhat related to dialysis but in which the solubility of the solute in the membrane has a major influence on its diffusion rate.

diastase AMYLASE.

diastatic Of, or pertaining to, amylase.

diastatic index A measure of amylase activity that is equal to the number of milliliters of 0.1% (w/v) starch solution, the starch of which is hydrolyzed by the enzyme present in 1.0 mL of a sample at 37°C in 30 min.

diastereomer One of two or more optical isomers of a compound that are not enantiomers. *Aka* diastereoisomer.

diasteromorph DIASTEREOMER.

diastereotopic Descriptive of either atoms or groups of atoms in a molecule that bear a diastereomeric relation to each other. *See also* enantiotopic.

diatomaceous earth A material that is composed principally of the siliceous skeletons of diatoms and that is used as an aid in filtration and as an adsorbent in column chromatography.

diauxic Of, or pertaining to, diauxie.

diauxie The biphasic growth curve that is obtained when an organism is exposed to two substrates that are utilized by the organism in such a fashion that one of the substrates is metabolized by constitutive enzymes while the other substrate is metabolized by inducible enzymes. The substrate requiring the constitutive enzymes is utilized first and represses the induction of the enzymes required for the utilization of the second substrate. Only after the first substrate is used up, can the enzymes be induced by the second substrate which is then metabolized. *Var sp* diauxy.

diazoate The ion $R—N=N—O^-$ formed from a diazonium salt.

diazobenzyloxymethyl paper *See* DBM paper.

diazo compound A compound containing the diazo group and having the general formula $R_2C=N=N$.

diazo group The grouping $=\overset{+}{N}=\bar{N}$.

diazonium compound DIAZONIUM SALT.

diazonium group The grouping $—\overset{+}{N}≡N$.

diazonium salt The salt of a compound that contains a diazonium group; used specifically for a compound that is a true ionic salt and that is prepared by the reaction of nitrous acid with a primary aromatic amine.

diazotization The formation of a diazonium salt.

dibasic Descriptive of a compound that contains two hydrogen atoms replaceable by a metal (such as KH_2PO_4) or an acid that can furnish two hydrogen ions (such as H_2SO_4).

dicarboxylic acid cycle GLYOXYLATE CYCLE.

2,4-dichlorophenoxyacetic acid A plant hormone used as a weed killer.

2,6-dichlorophenoxyacetic acid An antiauxin.

dichlorophenyldimethyl urea A herbicide that blocks electron transport between photosystem I and II by preventing the oxidation of water to oxygen. *Abbr* DCMU. *Aka* Diuron.

dichroic Of, or pertaining to, dichroism.

dichroic ratio The absorbance of plane-polarized light by a polymer in a direction that is parallel to the axis of the polymer divided by the absorbance in a direction that is perpendicular to that axis.

dichroism The directional effect in the absorption of light that results from the relative orientations of the absorbing chromophoric groups and the direction of polarization of the light. Thus, the absorption of light may be limited either to atoms that vibrate in a specific direction or to a component of the light that is polarized in a specific direction.

dicoumarol Variant spelling of dicumarol.

dictionary *See* genetic code dictionary.

dictyosome 1. The Golgi apparatus of plant cells. 2. A vacuole formed by fusion of Golgi vesicles with a vacuolar membrane. 3. Golgi stack. 4. A structure in the motile spore of uniflagellate fungi which may be the site for anchoring the base of the flagellum.

dicumarol A polycyclic aromatic compound, formed from a natural constituent of clover, that is an uncoupler of oxidative phosphorylation. It is an analogue of vitamin K and is used clinically to reduce the blood clotting tendency of an individual.

dicyanamide A dimer of cyanamide, used as a condensing agent for amino acids.

dicyclohexylcarbodiimide A compound used as a condensing agent for either amino acids or nucleotides; in this reaction, water is split out between the two condensing molecules and then reacts with the reagent, converting it to dicyclohexylurea. Dicyclohexylcarbodiimide is also an inhibitor of ATPase. *Abbr* DCC.

2′,3′-dideoxynucleoside triphosphate *See* chain terminator.

dideoxyribonucleotide sequencing SANGER–COULSON METHOD.

dielectric An insulating material that does not conduct an electric current.

dielectric constant A measure of the polarizability of a medium that is equal to the ratio of the electrostatic force, given by Coulomb's law, in a vacuum to that in the medium. The dielectric constant increases with an increase in the dipole moment of the molecules in the medium. *Sym* D.

dielectric dispersion The variation of the dielectric constant as a function of the frequency of the electric field.

dielectric increment The rate of change of the dielectric constant as a function of the solute concentration.

dielectrophoresis The migration of dipolar molecules toward the region of maximum field strength when placed in an inhomogeneous electric field; a concentration gradient is established when an equilibrium is achieved between dielectrophoretic migration and difussion.

diesterase *See* phosphodiesterase.

diet 1. The eating pattern of an organism. 2. The daily food intake of an organism.

dietary deficiency Undernutrition that is due to the inadequate intake of one or more essential nutrients even though the diet as a whole may be quantitatively unrestricted.

dietary fiber That part of plant food that is not digested in the small intestine and, therefore, reaches the large intestine relatively intact; consists primarily of cellulose and other nondigestible cell-wall polymers such as pectins and lignins. Some of the components of dietary fiber are broken down to varying degrees by bacterial action in the large intestine. Dietary fiber aids motility of the intestine and is thought to have protective value against a number of pathological conditions. *See also* antipromoter.

diethylpyrocarbonate Ethoxyformic anhydride; diethyl oxydiformate. A ribonuclease inhibitor used in the extraction of nucleic acids.

diethylstilbestrol A synthetic compound that has high estrogenic activity; a diethyl derivative of 4,4′-dihydroxystilbene (stilbestrol). *Abbr* DES.

difference matrix A tabular representation of the amino acid differences in a given protein isolated from various organisms. Both the columns and rows of the table list the organisms and any number entered in the table represents the number of amino acid differences between any two organisms. The table can also be constructed using the number of base differences of the corresponding amino acid codons. Difference matrices are used to construct phylogenetic trees.

difference spectrophotometry A technique for measuring small changes in absorbance by determining the absorbance of one test solution directly against that of another test solution rather than against that of a reference solution. The two test solutions contain the same solute of interest, and the solute is present in both solutions at the same concentration but under slightly modified conditions. The technique may be used, for example, to measure the absorbance of a solution containing a denatured protein against that of a solution containing the native protein, or to measure the absorbance of a solution containing a protein plus ligand against that of solutions containing the protein and ligand separately.

differential boundary The boundary formed between two solutions that contain the same components but at different concentrations.

differential centrifugation The centrifugation of a solution at various speeds so that particles of different sizes can be separated and collected.

differential counting The counting of pulses in a scintillation counter that fall within a range determined by two preselected levels of intensity. The range of intensities covered is called a window or a channel, and the pulses are selected by means of a pulse height analyzer which consists of two discriminators that reject pulses which are above or below the selected range of intensities.

differential dialysis. The separation of molecules by dialysis through a membrane that has known and specific permeability properties.

differential discrimination The selection of pulses that takes place in differential counting.

differential flotation centrifugation Interface centrifugation, used for tissue culture cells, in which particles are separated according to their densities.

differential labeling A method for investigating the structure of the antigen binding site of an antibody. The antibody is first allowed to bind a hapten and is then reacted with an unlabeled reagent that binds covalently to regions of the antibody that are not protected by the bound hapten. After removal of the hapten, the antibody is reacted with a labeled compound that reacts with those amino acids that had previously been protected by the hapten.

differential medium A medium that aids in the identification of bacteria by revealing specific properties of the organism grown on or in it; different types of organisms may be distinguished by their different forms of growth.

differential method A method for determining the order of a reaction by varying the concentration of each reactant and measuring the resultant rate of the reaction.

differential refractometer A refractometer for measuring the difference between two refractive indices.

differential scanning calorimetry A variation of the differential thermal analysis method in which the temperature of both the sample and the reference are maintained either at an equal level or at a fixed differential throughout the analysis. The variation in heat flow to the sample that is required to maintain this level during a transition is then measured. *Abbr* DSC.

differential sedimentation coefficient The sedimentation coefficient that corresponds to the movement of a differential boundary.

differential thermal analysis An analytical technique for studying temperature-induced transitions in a sample. The sample and an inert reference are heated or cooled at the same rate, and the difference in temperature between them is recorded. This difference is either zero or constant until a point is reached where a thermal reaction occurs in the sample. The course of this reaction shows up as a

peak in a plot of differential temperature versus either time or temperature. The direction of the peak indicates whether the transition was endothermic or exothermic. *Abbr* DTA.

differentiating circuit An electronic circuit for counting the discontinuous ionizations and the discontinuous electrical currents that are produced in an ionization chamber. *See also* integrating circuit.

differentiation The process whereby the structures and functions of the cells of a developing organism are progressively changed, thereby giving rise to specialized cells and structures.

differentiation antigen A cell surface antigen that is expressed only while the embryo undergoes a specific type of differentiation.

diffraction The modification that radiation undergoes when it passes the edge of an opaque body, passes through small apertures or slits, or is reflected from ruled surfaces or from atomic planes in a crystal.

diffraction grating A device that serves to separate light into wavelengths and that is used as a monochromator in some spectrophotometers. A diffraction grating consists of a large number of small grooves that are cut into it at such angles that each groove behaves like a prism, so that light is either reflected from, or transmitted through, the grating.

diffraction spots The spots produced on a photographic film in x-ray diffraction; they are referred to as first, second, third order, etc., corresponding to values of of n equal to 1, 2, 3, etc., in the Bragg equation.

diffusate 1. The material that passes through a dialysis bag. 2. DIALYSATE.

diffusible Capable of diffusing.

diffusing factor HYALURONIDASE.

diffusion 1. The process whereby molecules, as a result of their random Brownian motion, either change their orientation, resulting in rotational diffusion, or move from a region of higher to one of lower concentration, resulting in translational diffusion. 2. TRANSLATIONAL DIFFUSION.

diffusion chamber A chamber with porous walls that allows the diffusion of dissolved substances but does not allow the passage of cells; may be used for studies of substances released by cells.

diffusion coefficient *See* rotational diffusion coefficient; translational diffusion coefficient.

diffusion constant DIFFUSION COEFFICIENT.

diffusion current The polarographic current caused by the diffusion of ions toward the electrode.

diffusion-limited reaction Descriptive of a reaction in which the rate of the reaction depends solely on the frequency of molecular encounters as a result of diffusion; as soon as two molecules collide, due to random movement by diffusion, they undergo reaction (ligand binding, enzyme–substrate complex formation, and so on). No additional adjustments of the structure of one or both of the colliding molecules are required for the reaction to proceed. *Aka* diffusion-controlled reaction.

diffusion potential The membrane potential that arises from the diffusion of ions across the membrane.

DIFP Diisopropylfluorophosphate.

digest 1. *n* The mixture of compounds obtained by enzymatic or chemical hydrolysis of macromolecules. 2. *v* To subject macromolecules to enzymatic or chemical hydrolysis either in vivo, as in the digestive tract, or in vitro.

digestion 1. The process whereby the macromolecules of food are hydrolyzed in the digestive tract to smaller molecules that are absorbed across the intestinal wall and pass into the circulation. 2. Any chemical or enzymatic hydrolysis of macromolecules.

digestive Of, or pertaining to, digestion.

digestive enzyme A hydrolytic enzyme that functions in the breakdown of nutrient macromolecules in the digestive tract.

digestive juice One of the four secretions (salivary, gastric, pancreatic, and intestinal juices) that contain the digestive enzymes. Frequently, bile is included among the digestive juices. While bile aids in the digestion of fats by emulsifying them, it does not contain digestive enzymes and hence, strictly speaking, is not a digestive juice. *Aka* digestive fluid.

digestive system The digestive tract together with the related organs.

digestive tract The passage in animals that serves for the digestion of food, the absorption of nutrients, and the elimination of waste products.

digestive vacuole A secondary lysosome formed by phagocytosis of large particles, such as bacteria.

digital computer A computer that receives information in the form of discrete and discontinuous data, usually in the form of bits, and processes it by performing a series of arithmetic and logical operations on the data.

digitalis A plant toxin that is active as a cardiac glycoside.

digitalis glycosides A group of cardiac glycosides that belong to the class of cardenolides and that occur in the leaves of the foxglove plant, *Digitalis*. These compounds are cardiotonic steroids which are used for treatment

of heart weakness and defective heart valves.

digitonin A steroid glycoside, derived from the purple foxglove plant, *Digitalis purpurea*. A saponin, that is used for the fractional precipitation and determination of steroids, for solubilization of the outer mitochondrial membrane, and for extraction of rhodopsin from the retina.

diglyceride DIACYLGLYCEROL.

dihedral angle TORSION ANGLE.

dihedral symmetry Descriptive of a body that has at least one twofold rotational axis perpendicular to another n-fold rotational axis; that is, a structure that has two different axes of symmetry. For a structure containing subunits, the oligomer will consist of a minimum of $2n$ subunits.

dihydrobiopterin *See* biopterin.

dihydrofolate reductase The enzyme that catalyzes the reduction of dihydrofolic acid to tetrahydrofolic acid and that is competitively inhibited by aminopterin and amethopterin.

dihydro-U arm *See* arm.

dihydrouracil The minor base 5,6-dihydrouracil that occurs in tRNA; its nucleoside is dihydrouridine and its nucleotide is dihydrouridylic acid. *Abbr* DiHU.

dihydroxyacetonephosphate One of the two triose phosphates formed by cleavage of fructose-1,6-bisphosphate in glycolysis.

dihydroxyacetonephosphate shuttle GLYCEROL PHOSPHATE SHUTTLE.

1,25-dihydroxycholecalciferol The active form of vitamin D in humans; a steroid derivative, made by the combined action of the skin, the liver, and the kidneys, that functions as a hormone. It stimulates the absorption of Ca^{2+} and inorganic phosphate across the intestinal wall, leads to an increase in the serum concentrations of Ca^{2+} and inorganic phosphate, and is required for the calcification of bone. *See also* vitamin D.

2,5-dihydroxyphenylacetic acid HOMOGENTISIC ACID.

dihydroxyphenylalanine DOPA.

diisopropylfluorophosphate A reagent that reacts with the hydroxyl group of serine in proteins. The occurrence of this reaction with the enzyme acetylcholinesterase accounts for the reagent being a component of nerve gas. *Abbr* DIFP; DPFP; DFP; DIPFP.

diisopropylphosphofluoridate DIISOPROPYLFLUOROPHOSPHATE.

dilatation An enlargement or expansion, as that of a volume. *Aka* dilation.

dilatometry The measurement of small volume changes, particularly those of liquids, that are produced by either chemical or physical reactions.

dilaudid A semisynthetic, narcotic drug made

by converting morphine to dihydromorphinone.

diluent The solvent or the solution that is added to another solution for purposes of dilution.

dilute solution A solution that contains a small amount of solute.

dilution 1. The lowering of the concentration of a solution. 2. A solution having a lower concentration than another solution.

dilution end point method A method for determining the antibody titer of an antiserum by titrating a given amount of antigen with various dilutions of the antiserum. The highest dilution of antiserum that produces a detectable precipitin reaction is taken as the antibody titer.

dilution quenching The quenching in liquid scintillation counting that is caused by the dilution of the fluor with the sample so that the probability of exciting the fluor is decreased.

dilution value of a buffer The change in the pH of a buffer that occurs when the buffer is diluted with an equal volume of water; the dilution value is positive or negative depending on whether the pH increases or decreases.

dimensional analysis A check for the validity of an equation that is made by ascertaining that all the separate terms have the same units.

dimer 1. An oligomer consisting of two monomers. 2. The condensation product of either two identical or two similar molecules.

dimerization The formation of a dimer.

dimethyl sulfate protection A method for identifying specific protein-binding regions in DNA; based on the principle that dimethyl sulfate cannot methylate the adenines and guanines in the protein-binding region of the DNA if these bases are in close contact with the bound protein. These sites of contact can be identified by determining the positions of nonmethylated adenines and guanines in the endonuclease protected region with or without bound protein. *See also* DNase protection; footprinting.

dimorphism The occurrence of two forms. *See also* polymorphism.

dim vision NIGHT VISION.

dinitrofluorobenzene SANGER REAGENT.

2,4-dinitrophenol An uncoupler of oxidative phosphorylation. *Abbr* DNP.

dinitrophenyl amino acid An intensely colored amino acid derivative formed by the reaction of 1-fluoro-2,4-dinitrobenzene with the free alpha amino group of amino acids, peptides, or proteins; used for end group analysis of N-terminal amino acids and for chromatographic detection and quantitative estimation of amino acids. *Abbr* DNP-amino acid. *See*

also Sanger reaction.

dinucleotide 1. A compound consisting of two nucleotides linked by means of a phosphodiester bond. 2. A compound, such as NAD^+ or FAD, that is structurally related to a compound formed from two nucleotides and that contains two phosphate groups.

dinucleotide fold A characteristic protein-binding structure that consists of two Rossman folds (two mononucleotide-binding domains).

diol lipid A lipid that is a derivative of a dihydroxyalcohol (rather than of the trihydroxyalcohol, glycerol). These compounds include derivatives of ethylene glycol, propanediol, butanediol, and pentanediol.

dionin A semisynthetic, narcotic drug made by converting morphine to ethylmorphine.

dioscin *See* Marker synthesis.

diose DISACCHARIDE.

diosgenin *See* Marker synthesis.

dioxin One of a family of compounds in which two benzene rings are linked by means of two oxygen atoms. The compounds are highly toxic in animals but their effect in humans is not as well established. Substitution of chlorine atoms for the hydrogen atoms on the rings produces chlorinated dioxins. One such compound, 2,3,7,8-tetrachlorodibenzo-*p*-dioxin (TCDD) is a by-product of the manufacture of trichlorophenol. The latter is used for the production of a herbicide (2,4,5-trichlorophenol, or 2,4,5-T, one of the ingredients of Agent Orange) and an antibacterial agent (hexachlorophene).

dioxygen Molecular oxygen; O_2.

dioxygenase OXYGENASE.

dipalmitoylphosphatidylcholine *See* lung surfactant.

dipeptide A peptide consisting of two amino acids linked via one peptide bond.

DIPF; DIPFP Diisopropylphosphofluoridate.

diphenylamine reaction A colorimetric reaction for deoxypentoses that is based on the production of a blue color upon treatment of the sample with an acidic solution of diphenylamine; used for the determination of DNA.

1,3-diphosphoglyceric acid *See* 1,3-bisphosphoglycerate.

2,3-diphosphoglyceric acid *See* 2,3-bisphosphoglycerate.

diphosphopyridine nucleotide NICOTINAMIDE ADENINE DINUCLEOTIDE.

diphosphoribulose carboxylase RIBULOSE-1,5-BISPHOSPHATE CARBOXYLASE.

diphosphothiamine THIAMINE PYROPHOSPHATE.

diphthamide An unusal amino acid, derived by post-translational modification of histidine, that is found in the eukaryotic elongation factor eEF$_2$ at the site at which the factor becomes modified as a result of ADP-ribosylation by diphtheria toxin.

diphtheria toxin A bacterial protein of high toxicity, secreted by *Cornybacterium diphtheriae*; an acidic, single-chain protein (MW 62,000) that inhibits eukaryotic protein synthesis by inactivating elongation factor EF-2 (translocase; eEF_2). *See also* ADP-ribosylation; diphthamide.

dipicolinic acid The compound, pyridine-2,6-dicarboxylic acid, that is present in large concentrations in spores and that is related to the heat resistance of the spores.

diplochromosome A chromosome resulting from an abnormal duplication in which the centromere fails to divide and the daughter chromosomes fail to move apart; the resulting structure contains four chromatids.

diploid state The chromosome state in which each of the various chromosomes, except the sex chromosome, is represented twice. *Aka* diploidy.

diplornavirus A term used by some to include viruses of the genera orbivirus and reovirus.

dipolar ion A molecule, the dipole of which is due to two or more ionized groups.

dipolar potential The membrane potential that arises from both the orientation of polar molecules at the surface of the membrane and the ionic double layer of the membrane.

dipole An atom or a molecule that possesses an asymmetric charge distribution so that the center of its positive charges does not coincide with the center of its negative charges. The charges may arise from ionizations and/or from polar bonds, and the degree of charge separation is measured by the dipole moment.

dipole–dipole interaction The attractive or repulsive electrical force between two molecules that have permanent dipoles; the energy of such interactions is proportional to r^{-6}, where r is the distance between the dipoles.

dipole flip A sudden change in the orientation of a dipole.

dipole-induced dipole interaction The attractive electrical force between a molecule that has a permanent dipole and the dipole that is induced by this molecule in a neighboring atom or molecule. The energy of such interactions is proportional to r^{-6}, where r is the distance between the permanent dipole and the induced dipole.

dipole interactions *See* van der Waals interactions.

dipole moment A measure of the tendency of an atom or a molecule to orient itself in an electric field as a result of the separation of its charges; equal to the product of one of the two separated charges and the distance between the two charges.

dipole strength A measure of absorption intensity in absorption spectroscopy that is analogous to the rotational strength in circular dichroism.

diprotic acid An acid containing two dissociable protons.

direct-acting bilirubin The water-soluble conjugated form of bilirubin, bilirubin diglucuronide, that gives an immediate reaction with diazotized sulfanilic acid to yield a diazo dye. *See also* indirect-acting bilirubin.

direct calorimetry A method for determining the basal metabolic rate of an animal from the heat produced by the animal when it is placed in an insulated chamber. *See also* indirect calorimetry.

direct Coombs' test A Coombs' test in which the red blood cells have been coated with antibody in vivo. *See also* indirect Coombs' test.

direct effect The change, such as an excitation or an ionization, that is produced in a molecule as a result of its direct interaction with radiation. *See also* indirect effect.

direct fluorescent antibody technique A fluorescent antibody technique in which either the antigen or the antibody of interest is reacted directly with the corresponding antibody or antigen that has been labeled with a fluorochrome. *See also* indirect fluorescent antibody technique.

directionality of replication *See* bidirectional replication; unidirectional replication.

direct isotope dilution analysis A technique for determining the amount of an unlabeled compound by the addition of a known amount of the same, but labeled compound. The mixture of labeled and unlabeled compounds is then isolated and its specific activity is determined.

direct mutagenesis *See* mutagenesis.

direct oxidation pathway HEXOSE MONOPHOSPHATE SHUNT.

direct plot A plot of Y as a function of X; *see also* reciprocal plot.

direct repeats Identical or closely related DNA sequences, occurring in the same molecule and in the same orientation (running in the same direction); they may be adjacent or they may be separated by another segment of DNA.

direct transfer CHANNELING.

disaccharidase An enzyme that catalyzes the hydrolysis of a disaccharide; sucrase, maltase, and lactase are examples.

disaccharide A carbohydrate composed of two monosaccharide units linked by means of a glycosidic bond. Disaccharides are divided into two groups on the basis of the glycosidic bond between the two units. One group (trehalose type) consists of nonreducing sugars such as sucrose and trehalose; the other group (maltose type) consists of reducing sugars such as maltose and lactose.

disassembly The stepwise removal of ribosomal proteins from ribosomes, as that achieved with concentrated salt solutions.

disassociation DISSOCIATION (2).

disc gel electrophoresis An electrophoretic technique in which discontinuities of pH, ionic strength, buffer composition, and gel concentration are purposely built into the gel system. Used particularly for a zone electrophoretic technique that permits high-resolution analysis of small samples and that is performed in a polyacrylamide gel; the gel consists of three portions which differ in their composition and in their pH and which are loaded in three stages into open-ended cylindrical glass tubes. The topmost layer is called the sample gel; the middle layer, in which proteins are stacked in the form of thin bands, is called the spacer or stacking gel; and the lower layer, in which the protein bands are separated, is called the running, resolving, or separation gel. *Var sp* disk gel electrophoresis.

Dische reaction DIPHENYLAMINE REACTION.

discontinuous assay An assay in which a reaction mixture is analyzed by some sampling technique; aliquots of the reaction mixture are withdrawn at certain time intervals, the reaction is stopped in the aliquots and they are then analyzed. *Aka* fixed-time assay.

discontinuous density gradient A density gradient in which the density changes in a stepwise fashion from one end of the gradient to the other.

discontinuous distribution A set of experimental data in which the variable being measured cannot vary continuously and is expressed as a whole number; the number of amino acids per protein in a group of proteins is an example.

discontinuous replication 1. The currently accepted mechanisms of DNA replication. 2. The replication of the lagging strand in the currently accepted mechanism of DNA replication. *See* DNA replication.

discriminator An electronic device that is capable of either rejecting or accepting a pulse depending on the intensity of the pulse. *See also* differential counting; integral counting.

discriminator hypothesis The hypothesis that there is one site that is identical in a given group of isoacceptor tRNA molecules, and that this site is a factor, though not necessarily the only one, in the recognition of the isoacceptor tRNA molecules by the corresponding aminoacyl-tRNA synthetase.

disease A disturbance in the structure and/or the function of an organ, a tissue, or an organism, as that produced by either a viral or a bacterial infection.

disease of regulation A disease that is characterized by the abnormality of an otherwise normal bodily function and that is caused by a disturbance of the equilibria that control the bodily function; high blood pressure is an example.

disequilibrium ratio A measure that describes how far from, or how close to, equilibrium a given reaction is; equal to the concentration term for the actual (not equilibrium) conditions divided by the equilibrium constant. Thus, for the reaction $A \rightleftharpoons D$, the disequilibrium ratio Γ is given by $\Gamma = ([D]/[A])/K_{eq}$.

disinfectant A chemical substance used for disinfection.

disinfection The destruction of viable and harmful bacteria, or the destruction of the toxins and the vectors of these bacteria.

disintegration The spontaneous transformation of a radioactive nuclide into another nuclide, generally with the emission of radioactive radiation.

disintegration constant DECAY CONSTANT.

disk A round and flat piece of magnetic coated material used to store computer data with greater density, speed, and reliability than is possible with cassette tape. Disks can be of rigid metal (hard) or of flexible plastic (floppy). *Aka* diskette.

dismutation reaction A chemical reaction in which a single compound serves as both an oxidizing and a reducing agent and gives rise to two or more compounds by gain and loss of electrons. An example is the conversion of two molecules of pyruvate plus a molecule of water to one molecule each of lactate, acetate, and carbon dioxide. The reaction catalyzed by superoxide dismutase is another example. *See also* disproportionation reaction.

dispensable amino acid NONESSENTIAL AMINO ACID.

dispensable enzyme NONESSENTIAL ENZYME.

dispensable gene NONESSENTIAL GENE.

dispersed phase The solute particles of a colloidal dispersion.

dispersion *See* dispersion medium; colloidal dispersion.

dispersion effect DISPERSION FORCES.

dispersion forces The attractive electrical forces between atoms and/or nonpolar molecules that result from the formation of small, transient, induced dipoles. The motions of the electrons in one atom or molecule lead to the formation of a small, instantaneous, and transient dipole that induces a corresponding dipole in a neighboring atom or molecule. The energy of such interactions is proportional to r^{-6}, where r is the distance between the instantaneous dipoles. *Aka* London dispersion forces.

dispersion medium The solvent of a colloidal dispersion.

dispersive replication A mode of replication for double-stranded DNA (now considered obsolete) in which the parental strands are broken into segments that are incorporated equally, together with newly formed segments, into the two daughter DNA molecules.

displacement analysis *See* displacement chromatography; displacement electrophoresis; radioimmunoassay.

displacement chromatography A chromatographic technique, useful for preparative separations, in which a compound is applied to a chromatographic column and is then displaced from the column by elution with a solution containing a second compound.

displacement development DISPLACEMENT CHROMATOGRAPHY.

displacement electrophoresis ISOTACHOPHORESIS.

displacement loop D-LOOP.

displacement reaction A chemical reaction in which a group is displaced from a carbon atom by the attack of a nucleophile. *See also* S_N1 mechanism; S_N2 mechanism.

displacer The eluent used in displacement analysis.

disproportionation reaction A chemical reaction in which a single compound gives rise to two or more different compounds; an example is the reaction $2AB \rightarrow AB_2 + A$, where AB is a copolymer. *See also* dismutation reaction.

disseminated infection A viral or a bacterial infection in which the infecting agent spreads from the site of entry to other parts of the body.

dissimilation CATABOLISM.

dissimilatory reduction 1. NITRATE RESPIRATION 2. SULFATE RESPIRATION.

dissociation 1. The breakdown of a compound into smaller components, such as the dissociation of an acid to a proton and an anion. 2. The breakdown of a molecular complex, particularly one of macromolecules, such as the dissociation of an enzyme–substrate complex. 3. The appearance of a novel bacterial colony as a result of mutation and selection.

dissociation constant The equilibrium constant for dissociation (1,2).

dissociation factor *See* ribosome dissociation factor.

dissolved carbon dioxide The equilibrium mixture of CO_2 and H_2CO_3 that exists in solution, particularly in blood, plasma, or serum.

dissymmetric Descriptive of a molecule that lacks overall molecular symmetry but that possesses elements of symmetry within the molecule.

dissymmetry constant FRICTIONAL RATIO.

dissymmetry of scattering The ratio of the intensity of light scattered by a solution at the angle θ to that scattered at the angle $(180 - \theta)$.

dissymmetry ratio 1. The ratio of the intensity of light scattered by a solution at an angle of $45°$ to that scattered at an angle of $135°$. 2. ASYMMETRY RATIO. 3. FRICTIONAL RATIO.

distal Remote from a particular location or from a point of attachment.

distance map CONTACT MAP.

distillate The liquid produced by the condensation of its vapor during distillation.

distillation The process whereby a liquid is purified by boiling and the vapors are condensed and collected.

distilled water Water that is purified and collected by distillation.

distorted bond model A model for the structure of water according to which the water structure is an altered ice lattice that is produced by the bending and the distorting, but not the breaking, of the hydrogen bonds in ice.

distribution coefficient PARTITION COEFFICIENT.

distribution equation One of a set of equations that describe the distribution of an enzyme among various possible forms; such forms may include EI or EP, where E is the enzyme, I is an inhibitor, and P is a product. Each equation provides an expression for a ratio such as $[EI]/[E_t]$ or $[EP]/[E_t]$ in terms of kinetic constants, equilibrium constants, and concentrations of components; $[E_t]$ is the total enzyme concentration.

distribution isotherm PARTITION COEFFICIENT.

distribution law PARTITION LAW.

distribution potential The membrane potential that arises from the unequal distribution of ions on both sides of the membrane.

distribution ratio PARTITION COEFFICIENT.

disulfide A compound containing a disulfide (S–S) bond.

disulfide bond The covalent bond formed between two sulfur atoms, particularly that formed in peptides and proteins between two sulfhydryl groups of two cysteine residues.

disulfide bridge DISULFIDE BOND.

disulfide interchange A chemical reaction in which there is an interchange between the groups attached to two or more disulfide bonds, as in the reaction R_1—S—S—R_2 + R_3—S—S—R_4 $\rightleftharpoons$ R_1—S—S—R_3 + R_2—S—S—R_4.

disulfide link DISULFIDE BOND.

DIT Diiodotyrosine.

dithioerythritol *See* Cleland's reagent.

dithiothreitol *See* Cleland's reagent.

diuresis An increase in the excretion of urine.

diuretic 1. *n* An agent that increases the excretion of urine. 2. *adj* Of, or pertaining to, diuresis.

diurnal 1. Pertaining to the daytime. 2. Occurring daily.

diuron Dichlorophenyldimethylurea.

divergence 1. The difference, in percent, between the nucleotide (base) sequences of two nucleic acid strands. 2. The difference, in percent, between the amino acid sequences of two polypeptide chains.

divergent evolution An evolutionary pattern in which the lines of development for more recent species branch out from earlier species; such a pattern can be depicted as one network branching out from a common origin.

divergent transcription Transcription in which sections of DNA that are being transcribed have opposite orientations relative to a central region.

diversity genes D GENES.

Dixon plot A graphical method for determining the inhibitor constant of either competitive or noncompetitive inhibition of an enzymatic reaction; consists of plotting the reciprocal of the velocity of the reaction as a function of the inhibitor concentration at a constant substrate concentration.

dizygotic twins Twins that are formed from two separate eggs which were fertilized by separate sperms in the same maternal organism.

D-loop A structure formed early in the replication of double-stranded (linear or circular) DNA. It consists of one double-stranded branch (one parental strand paired with the leading strand) and one single-stranded branch (the so-far unreplicated second parental strand). Since the leading strand displaces this unreplicated strand, the structure is known as a displacement loop, or D-loop.

D-loop synthesis The replication mechanism of double-stranded circular mitochondrial DNA. Synthesis of one strand begins at one origin and proceeds unidirectionally around the circle. Synthesis of the other strand begins from a second origin which becomes activated when the displacement loop passes it.

d,l-pair A pair of enantiomers.

DME Dropping mercury electrode.

DMSO Dimethyl sulfoxide.

DNA Deoxyribonucleic acid.

DNA–agar technique A technique for measuring the extent of hybridization between nucleic acid molecules. Nucleic acid fragments from one source are trapped in an agar gel and are allowed to hybridize with radioactively labeled nucleic acid fragments from a second source. The amount of radioactivity in the agar is then determined.

DNA-arrest mutant A phage mutant that initiates the synthesis of DNA in normal fashion but stops the synthesis soon thereafter.

DNAase Deoxyribonuclease.

DNAase protection method *See* DNase protection.

DNA binding protein SINGLE STRAND BINDING PROTEIN.

DNA–cellulose chromatography Affinity chromatography in which the column material consists of cellulose to which DNA molecules have been either adsorbed noncovalently or linked covalently.

DNA chimera CHIMERIC DNA.

DNA clone A DNA segment that has been inserted as a passenger into a plasmid or a viral vector, and that has been subsequently replicated along with the vector in a host to form many copies per cell.

DNA cloning RECOMBINANT DNA TECHNOLOGY.

DNA complexity *See* complexity.

DNA-delay mutant A phage mutant that initiates the synthesis of DNA after a delay of several minutes.

DNA-dependent (directed) DNA polymerase *See* DNA polymerase.

DNA-dependent (directed) RNA polymerase *See* RNA polymerase.

DNA dot blot *See* dot blot assay.

DNA-driven hybridization A hybridization reaction, used in reassociation kinetics, in which DNA is hybridized with radioactive RNA, and the DNA is present in large excess. Used in the construction of cot curves for determining the degree of repetitiveness of the nucleotide sequences in the DNA.

DNA duplex A double-stranded DNA molecule; a double helix of DNA.

DNA duplicase DNA POLYMERASE.

DNA forms The configurations of DNA that are a function of the relative humidity and the type of positive counterions present. The A, B, and C forms of DNA are right-handed, double-stranded helices which, in fibrous form, are stable at intermediate, high, and low relative humidities, respectively, and have 11, 10, and 9.3 base pairs per turn of double helix. The B form is considered to be the biologically most important form. DNA maintains its B form in solution but with slightly altered helix parameters (10.4 base pairs per turn instead of 10.0). The C form is very similar to the B form, while the A form differs from the B form in having the base pairs tilted at 20°, rather than being perpendicular, to the helix axis. At high salt concentrations, the B form can be converted to a left-handed helix in which the sugar–phosphate backbone follows a zigzag path. Hence, this DNA is referred to as Z-DNA. In contains 12 base pairs per turn of double helix and has only a single groove as opposed to B-DNA which has two (a major and a minor one). It is believed that left-handed helical regions of Z-type DNA exist in vivo and possibly have regulatory functions.

DNA glycosylase   One of a number of enzymes that cleave *N*-glycosidic bonds at specific sites in the DNA. The enzymes remove mutated or altered bases, thereby forming apurinic or apyrimidinic sites (AP sites); the reaction is a step in the course of DNA repair.

dna G protein   The product of the dna G gene in *E. coli*; the enzyme primase.

DNA groove   *See* major groove; minor groove.

DNA gyrase   A topoisomerase of Type II that functions in the discontinuous replication of DNA. It removes the positive superhelicity produced during replication by introducing negative twists ahead of the advancing replicating fork.

DNA helicase   *See* helicase.

DNA library   GENE LIBRARY.

DNA ligase   An enzyme that catalyzes the formation of a phosphodiester bond between a 3'-hydroxyl group and a 5'-phosphate group in DNA; the enzyme catalyzes the joining together of two single-stranded DNA segments which may be either parts of the same duplex or parts of different duplexes. The enzyme functions in DNA replication and in DNA repair by linking DNA fragments together.

DNA-like RNA   An RNA molecule that resembles DNA in its overall base composition and base ratios; the RNA is generally rich in adenine and uracil and has an adenine/uracil ratio of approximately 1.0. This RNA is not part of ribosomal and transfer RNA and much of it has a high molecular weight and a short half-life. *Abbr* dRNA; DRNA.

DNA-melting protein   SINGLE-STRAND BINDING PROTEIN.

DNA methylase   An enzyme that catalyzes the methylation of the bases in DNA; the methylation occurs subsequent to, rather than prior to, the incorporation of the bases into the polynucleotide strand.

DNA modification   POSTREPLICATIVE MODIFICATION.

DNA modification and restriction   *See* restriction-modification system.

DNA-negative mutant   A phage mutant that is unable to initiate the synthesis of DNA.

DNA nucleotidyl transferase   DNA POLYMERASE.

DNA packing ratio   *See* packing ratio.

DNA phage   A DNA-containing phage.

DNA polymerase   An enzyme that catalyzes the synthesis of DNA from the deoxyribonucleoside triphosphates, using either single- or double-stranded DNA as a template; referred to as DNA-dependent (directed) DNA polymerase to distinguish it from RNA-dependent (directed) DNA polymerase which uses an RNA template for the synthesis of DNA.

DNA polymerase I   A DNA-dependent DNA polymerase, originally thought to be the major enzyme in prokaryotic DNA replication. While it does play a role in DNA replication, its major function is now believed to be in the repair–synthesis of DNA. The enzyme catalyzes the synthesis of DNA in the 5' to 3' sense and has associated with it two other enzymatic activities, a proofreading function (3' → 5' exonuclease) and a 5' → 3' exonuclease. *Abbr* pol I. *Aka* Kornberg enzyme. *See also* proofreading function; 5' → 3' exonuclease.

DNA polymerase II   A DNA-dependent DNA polymerase that occurs in prokaryotes and that catalyzes the synthesis of DNA in the 5' to 3' sense; it has a proofreading function (3' → 5' exonuclease) but no 5' → 3' exonuclease activity. It is believed to function in DNA repair. *Abbr* pol II.

DNA polymerase III   A DNA-dependent DNA polymerase that occurs in prokaryotes and that catalyzes the synthesis of DNA in the 5' to 3 ' sense; it has both a proofreading function (3' → 5' exonuclease) and a 5' → 3' exonuclease activity. It is the major synthetic enzyme in the replication of DNA. *Abbr* pol III.

DNA polymerase α   A eukaryotic DNA-dependent DNA polymerase that catalyzes the synthesis of DNA in the 5' to 3' sense. It is located in the nucleus and is responsible for the replication of chromosomal DNA. It lacks both the 3' → 5' and the 5' → 3' exonuclease activities. *Abbr* pol α.

DNA polymerase β   A eukaryotic DNA-dependent DNA polymerase that catalyzes the synthesis of DNA in the 5' to 3' sense. It is located in the nucleus and its function is unknown. *Abbr* pol β.

DNA polymerase γ   A eukaryotic DNA-dependent DNA polymerase that catalyzes the synthesis of DNA in the 5' to 3' sense. It is located both in the nucleus and in mitochondria and is believed to function solely in the replication of mitochondrial DNA. *Abbr* pol γ.

DNA polymerase chain reaction   *See* polymerase chain reaction.

DNA polymorphism   *See* restriction fragment length polymorphism.

DNA primase   *See* primase.

DNA primer   A single strand of DNA onto

which are added deoxyribonucleotides by DNA polymerase I during DNA replication. *See also* primer.

DNA probe *See* probe.

DNA puff CHROMOSOME PUFF.

DNA-relaxing enzyme *See* topoisomerase.

DNA repair *See* cut and patch repair; excision repair; patch and cut repair; SOS repair.

DNA replicase system REPLISOME.

DNA replication The process whereby new DNA is synthesized from parental DNA. The major, currently accepted, mechanism is known as discontinuous (or, more precisely, semidiscontinuous) replication. It entails the local unwinding of the two strands of DNA by the action of helicase to form a Y-fork (replicating fork). Single-strand binding proteins (SSB) bind to the single strands and prevent their coming back together and annealing. Replication of both strands proceeds by the addition of complementary nucleotides in the 5′ to 3′ sense. One newly synthesized strand grows largely continuously; it grows in the direction of fork movement and is known as the leading strand. The other newly synthesized strand grows discontinuously and in the opposite direction; it is known as the lagging strand. Synthesis of the lagging strand entails binding of the primosome to the complementary parental strand, synthesis of RNA primers by the primase of the primosome, and synthesis of DNA fragments (Okazaki fragments), linked to the RNA primers, by DNA polymerase III. The Okazaki fragments are then extended, and the RNA primers excised, by the action of DNA polymerase I. Lastly, the extended Okazaki fragments are linked together by the action of DNA ligase. *See also* primosome; DNA gyrase; replication (1); conservative replication; dispersive replication; end-to-end conservative replication; semiconservative replication.

DNA restriction enzyme *See* restriction enzyme.

DNA–RNA hybrid A double helix composed of a single strand of DNA which is hydrogen-bonded to a single strand of a complementary RNA.

DNA–RNA virus *See* retrovirus.

DNase Deoxyribonuclease.

DNase protection A method for determining the size of a specific protein-binding region in DNA; based on the principle that such a region is protected by the bound protein against endonuclease action. The DNA is allowed to bind the specific protein and is then subjected to endonuclease action which degrades most of the DNA outside the protected region to mono-and dinucleotides; the section of DNA in close contact with the protein remains in-

tact. This fragment can then be isolated and characterized. The method can be used, for example, to study the binding of RNA polymerase to the promoter region of DNA. *See also* footprinting; dimethyl sulfate protection.

DNA sequencing *See* Maxam–Gilbert method; Sanger–Coulson method.

DNA splicing *See* recombinant DNA technology.

DNA swivelase *See* topoisomerase.

DNA synaptase An enzyme, isolated from *E. coli*, that catalyzes the fusion of double-stranded DNA molecules at a region of homology and that may play a role in genetic recombination.

DNA synthesizer An automated setup for the chemical synthesis of oligonucleotide segments of DNA. *See also* gene machine.

DNA topoisomerase *See* topoisomerase.

DNA transcriptase DNA-DEPENDENT RNA POLYMERASE.

DNA unwinding protein SINGLE-STRAND BINDING PROTEIN.

DNA vector VECTOR (3).

DNA virus A DNA-containing virus.

DNFB Dinitrofluorobenzene.

DNP 1. 2,4-Dinitrophenyl group. 2. Deoxyribonucleoprotein. 3. 2,4-Dinitrophenol.

DNP-amino acid Dinitrophenyl amino acid.

Dns-amino acid Dansyl amino acid.

dNTP Deoxynucleoside triphosphate.

DOC 1. Deoxycorticosterone. 2. Deoxycholic acid.

DOCA Deoxycorticosterone acetate.

docking protein A protein that serves to remove the block of protein synthesis exerted by the signal recognition protein (SRP) after the ribosome–SRP complex has become bound to the membrane of the rough endoplasmic reticulum. The docking protein functions in or on the endoplasmic reticulum. *Aka* SRP-receptor.

doctrine of monomorphism The belief that microorganisms show constancy with respect to their morphological form and their physiological function. *See also* doctrine of pleomorphism.

doctrine of original antigenic sin The phenomenon that antibodies formed in a secondary response that is elicited by an immunogen related to, but not identical with, the immunogen that elicited the primary response react more strongly with the primary, than with the secondary, immunogen.

doctrine of pleomorphism The belief that microorganisms have great capacity for variation with respect to their morphological form and their physiological function. *See also* doctrine of monomorphism.

doctrine of uniformitarianism The assumption that the chemical and physical laws have remained unchanged from the time of the formation of the earth, throughout the development of life, and up to the present time.

dol Dolichol.

dolichol One of a group of long-chain lipids, containing 16–22 isoprene units, that serve as carriers (in the form of dolichol phosphates) of carbohydrates in the biosynthesis of glycoproteins in animals. *Abbr* dol. *See also* glycosylation; undecaprenol.

dolichol phosphate A lipid carrier of oligosaccharides that functions in glycoprotein synthesis; an isoprenoid phosphate which, in vertebrates, contains 18–20 isoprene units. Dolichol phosphate transfers the oligosaccharide to protein in the biosynthesis of N-linked oligosaccharides. *Abbr* dol-p.

dol-p Dolichol phosphate.

domain 1. An independently folded, and relatively globular, region of a polypeptide chain; a region that is spatially isolated and that could be physically separated from other parts of the molecule by a suitable cut, or cuts, in the polypeptide chain. Domains may interact slightly or extensively with each other, they may be associated with specific functions, and they vary in size from about 40 to 400 amino acid residues. 2. A region of homology in an immunoglobulin molecule; anyone of the various segments of the light and heavy chains. 3. A region in the chromosome in which supercoiling is independent of that in other regions. 4. An extensive region of DNA which includes an expressed gene that is very sensitive to degradation by DNase I.

dominant 1. DOMINANT GENE 2. The trait produced by a dominant gene.

dominant gene A gene that is fully expressed in a heterozygote. A dominant gene may partially or entirely suppress the expression of another allelic gene (recessive gene). *Aka* dominant allele.

donation The transfer, to a recipient cell, of a nonconjugative plasmid via the effective contact provided by a different, conjugative plasmid.

Donnan equilibrium GIBBS–DONNAN EQUILIBRIUM.

Donnan potential The membrane potential that arises from the unequal distribution of ions which is produced by the establishment of a Gibbs–Donnan equilibrium.

Donnan ratio The ratio of the concentration of a diffusible cation on one side of the membrane to the concentration of the same ion on the other side of the membrane in a Gibbs–Donnan equilibrium. For an anion, the ratio of the two concentration terms is reversed. Thus, if r is the Donnan ratio and A and B denote the two sides of the membrane, then $r = [Na^+]_A/[Na^+]_B = [Cl^-]_B/[Cl^-]_A$.

Donnan term A measure of the ionic distributions attained in a Gibbs–Donnan equilibrium; equal to the difference between the sum of the concentrations of the diffusible ions on that side of the membrane that contains nondiffusible ions, and the sum of the concentrations of the diffusible ions on the other side.

donor The atom that donates a hydrogen in the formation of a hydrogen bond.

donor junction *See* splicing junctions.

donor site PEPTIDYL SITE.

donor splicing site *See* splicing junctions.

DOPA 3,4-Dihydroxyphenylalanine; an intermediate both in the conversion of tyrosine to melanin pigments and in the synthesis of epinephrine and norepinephrine from tyrosine; used in the treatment of Parkinson's disease.

dopamine 3,4-Dihydroxyphenylethylamine or hydroxytyramine; a neurotransmitter catecholamine formed by decarboxylation of L-dopa. A deficiency of dopamine is one of the characteristics of Parkinson's disease. Dopamine is also an intermediate in tyrosine metabolism.

dopamine adrenergic receptor *See* adrenergic receptor.

dormancy The inactive state of a spore.

dormant infection A bacterial infection that does not produce overt disease symptoms but in which the bacteria can be detected by the use of proper techniques. *See also* latent infection.

dormin ABSCISIC ACID.

Dorn effect SEDIMENTATION POTENTIAL.

dosage 1. A specified dose. 2. The administration of a dose.

dosage repetition The occurrence of a given gene in multiple copies per nucleus; due to the fact that the cell requires a large amount of the particular gene product per unit time, which a single copy could not produce. The genes coding for ribosomal and transfer RNA exhibit dosage repetition.

dose A measured quantity of a chemical, a microorganism, a virus, or radiation that is administered to an organism.

dose–action curve DOSE–RESPONSE CURVE.

dose–effect curve DOSE–RESPONSE CURVE.

dose fractionation The exposure of an object or an organism to small doses of radiation, administered at regular intervals.

dose meter DOSIMETER.

dose rate The dose of radiation received per unit time.

dose–response curve A curve that describes the relation between the dose that is administered to organisms and either the number or the percentage of organisms that show a response. Examples of kinds of doses are those of carcinogens, drugs, poisons, viruses, and radiation; examples of kinds of responses are tumor development, recovery from a disease, and death. *Abbr* DRC.

dosimeter A instrument for measuring the cumulative dose of radiation to which an organism or an object has been exposed. *Aka* dose meter.

dosimetry The measurement of doses of radiation, microorganisms, viruses, or chemicals.

dot blot assay A quantitative method for the estimation of specific nucleic acid sequences (DNA or RNA) by means of a blotting technique.

double beam in space spectrophotometer A spectrophotometer in which the original light beam is split into two beams that pass through two identical optical systems at the same time; one beam passes through the sample solution while the other passes through a reference solution.

double beam in time spectrophotometer A spectrophotometer in which the original light beam is made to pass, at any given time, either through the sample solution or through a reference solution; commonly achieved by means of a light chopper.

double-beam spectrophotometer A spectrophotometer in which a light beam passes through both the sample solution and a reference solution so that a direct measurement of the difference in absorbance between the two solutions can be made. *See also* double beam in space spectrophotometer; double beam in time spectrophotometer.

double-blind technique A technique for testing either a substance or a procedure in which neither the subject nor the experimeter know the identity of the samples and the expected results; used in the testing of drugs, vaccines, etc., particularly in conjunction with the use of placebos. *See also* blind test.

double bond A covalent bond that consists of two pairs of electrons, shared by two atoms.

double carbon dioxide fixation The conversion of two molecules of carbon dioxide to organic compounds by means of two different reactions, as that occurring in *Chromatium*.

double-chain length A crystalline form of glycerides in which two acyl groups form a unit structure.

double diffusion A technique of immunodiffusion in which both the antigen and the antibody are allowed to diffuse through a gel; can be used for either one-dimensional or two-dimensional diffusion.

double-displacement mechanism PING-PONG MECHANISM.

double helix *See* Watson–Crick model.

double infection The infection of a bacterial cell with two different phage particles.

double irradiation SPIN DECOUPLING.

double isotope dilution analysis A technique for determining the amount of a labeled compound of unknown specific activity by treating two aliquots of the labeled compound with different and known amounts of the same, but unlabeled, compound. The mixtures of labeled and unlabeled compounds are then isolated and their specific activities are measured.

double-label experiment An experiment based on the use of either one compound labeled with two different isotopes, or two compounds, each labeled with a different isotope.

double-label method HANDLE METHOD.

double layer *See* ionic double layer; bilayer.

double lysogeny The phenomenon in which a bacterium is infected twice with the same temperate phage, the two prophages of which are inserted in tandem into the host bacterial DNA.

double-reciprocal plot A plot of $1/Y$ versus $1/X$, where Y and X are two variables; a Lineweaver–Burk plot is an example.

double refraction BIREFRINGENCE.

double refraction of flow FLOW BIREFRINGENCE.

double resonance *See* electron–electron double resonance; electron–nuclear double resonance.

double-sector cell An analytical ultracentrifuge cell that is divided into two compartments; useful, for example, for simultaneous analysis of the sedimentation behavior of two solutions.

double-sieve mechanism A model proposed to explain the low frequency of misacylation (mischarging) of amino acids to tRNA in protein synthesis. According to this model, the active site of an aminoacyl-tRNA synthetase is too small for amino acids that are larger than the correct amino acid. Thus, the active site serves as a first sieve with the result that amino acids larger than the correct one are not activated. Additionally, the hydrolytic site of the enzyme is presumed to be too small for the correct amino acid. This site, therefore, serves as a second sieve with the result that amino acids smaller than the correct one are removed by hydrolysis.

double strand break CUT.

double-stranded Descriptive of a nucleic acid molecule that consists of two polynucleotide chains. *Abbr* ds.

doublet 1. A sequence of two adjacent nucleotides in a polynucleotide strand. 2. A double peak, as that obtained in nuclear magnetic resonance.

double-tailed test TWO-SIDED TEST.

doublet code A genetic code in which an amino acid is specified by two adjacent nucleotides in mRNA; thought to have been a forerunner of the present genetic code.

double transformation COTRANSFORMATION.

double-well technique An immunoelectrophoretic procedure for determining the common precipitin arcs produced by two mixtures.

doubling dilution Serial dilution such that the dilution in each tube is twice that in the preceding tube.

doubling dose The dose of ionizing radiation that results in a doubling of the spontaneous mutation rate for a given species.

doubling time The observed time required for a cell population to double in either the number of cells or the cell mass; it is equal to the generation time only if all the cells in the population are capable of doubling, have the same generation time, and do not undergo lysis. *Sym* t_D.

doubly lysogenic strain A bacterial strain in which the cells carry two prophages per genome.

doughnut model A model of the conformational changes produced in a cell membrane by the action of complement. According to this model, complement brings about the formation of a rigid, stable channel that stretches across the phospholipid bilayer of the membrane; the outside of this channel is hydrophobic, but the hollow inside core is hydrophilic and permits the free passage of water molecules and ions. *See also* leaky patch model.

Dowex Trademark for a group of ion-exchange resins.

downfield Describing a peak in nuclear magnetic resonance that is on the low magnetic field side of the spectrum. *See also* deshielded nucleus.

downhill reaction EXERGONIC REACTION.

down promoter mutation *See* promoter down mutation.

down regulation The decrease in receptor activity of cell membranes for a specific ligand that is brought about by incubating the cells, or stimulating them repeatedly, with the ligand; observed for many polypeptide hormones such as epidermal growth factor and insulin. Thus, incubation of cells with insulin leads to a decrease in the binding of insulin by the cells. Down regulation involves receptor internalization and degradation via coated pits

and receptosomes. *See also* coated pit; receptosome.

downshift The transfer of a bacterial culture from a growth medium in which growth is rapid to one in which growth is slow.

Down's syndrome MONGOLISM.

downstream Describing a location or a sequence of units in the direction in which a process occurs: (a) during transcription, the location or sequence of deoxyribonucleotides on the transcribed strand of DNA from the 3'- to the 5'-end; (b) during translation, the location or the sequence of ribonucleotides on the mRNA from the 5'- to the 3'-end; (c) during replication, the location in the direction of replicating fork movement; (d) in a polypeptide chain, the sequence of amino acids in the direction in which they are linked together, from the N- to the C-terminal; (e) in the electron transport system, the sequence of electron carriers in the direction in which the electrons flow, from metabolite to oxygen. *See also* upstream.

DPFP Diisopropylfluorophosphate.

DPG 2,3-Diphosphoglycerate.

dpm Disintegrations per minute; the number of radioactive disintegrations per minute.

DPN$^+$ Diphosphopyridine nucleotide. The preferred abbreviation is NAD$^+$.

DPNH Reduced diphosphopyridine nucleotide. The preferred abbreviation is NADH.

dps Disintegrations per second; the number of radioactive disintegrations per second.

DPT Diphosphothiamine.

drag effect SOLVENT DRAG.

Draper's law FIRST LAW OF PHOTOCHEMISTRY.

DRC Dose–response curve.

Dreiding model A framework model in which a bond length is proportional to the internuclear distance between a hydrogen atom and the respective atom; these bond lengths are based on a scale of 0.4 Å/1.0 cm.

dRib Deoxyribose.

Drierite A brand of anhydrous calcium sulfate used as a desiccant.

drift *See* genetic drift.

driving force The source of free energy and/or the conditions and factors that are responsible for causing a reaction to proceed in a given direction; frequently refers to the mechanism by which the free energy of activation of the transition state is lowered.

dRNA DNA-like RNA; also abbreviated DRNA.

droplet sedimentation The phenomenon that may occur in density gradient sedimentation velocity when a solution of macromolecules is layered over a density gradient that contains a highly diffusible low molecular weight solute;

the unequal diffusion of this solute and of the macromolecules may lead to a density inversion and to convection at the interface between the solution of the density gradient and that of the macromolecules.

dropping mercury electrode An electrode, used in polarography, in which mercury drops from a reservoir through a capillary into the solution that is being studied; the capillary is mounted vertically below the reservoir. *Abbr* DME.

Drosophila The genus of the fruit fly; an organism used in genetic research.

Drude equation An equation that describes the variation of optical rotation with wavelength; specifically $[m'] = a_0 \lambda_0^2/(\lambda^2 - \lambda_0^2)$, where $[m']$ is the reduced mean residue rotation, λ is the wavelength, and a_0 and λ_0 are constants.

Drude term The expression $a_0\lambda_0^2/(\lambda^2 - \lambda_0^2)$ that is part of the Drude equation and that is also the first term in the Moffit–Yang equation.

drug-detoxication enzyme DRUG-METABOLIZING ENZYME.

drug-induced apnea A genetically inherited metabolic defect in humans that is due to a deficiency of the enzyme pseudocholinesterase.

drug-metabolizing enzyme A mammalian enzyme that acts on drugs and other foreign chemicals but that is not known to act on any metabolite normally occurring within the same organism. *See also* normal enzyme.

drug resistance The relative resistance of mutant microorganisms to the action of drugs.

drug-resistance plasmid R PLASMID.

dry application A method of sample application in electrophoresis in which the sample is applied to a dry support, such as filter paper, and this is followed by wetting of the support with buffer.

dry box A moisture-free enclosure.

dry column chromatography A chromatographic technique in which an adsorbent is packed in a solvent-free, dry column. The mixture to be separated is placed on top of the column and this is followed by addition of a developing solvent. When the solvent reaches the bottom of the column, the separation is complete, and fractions can be removed from the column and collected by various techniques.

dry ice Solid carbon dioxide.

drying agent DESICCANT.

drying oil A highly unsaturated oil that tends to undergo autoxidation.

dry weight The weight of a sample from which the water has been removed.

ds Double-stranded.

DSC Differential scanning calorimetry.

dsDNA Double-stranded DNA.

D-site Donor site.

dsRNA Double-stranded RNA.

DTA Differential thermal analysis.

dTDP 1. Thymidine diphosphate. 2. Thymidine-5'-diphosphate; the 5'-diphosphate of 2'-deoxyribosyl thymine.

DTE *see* Cleland's reagents.

dTMP 1. Thymidine monophosphate; thymidylic acid. 2. Thymidine-5'-monophosphate; 5'-thymidylic acid; the 5'-phosphate of 2'-deoxyribosyl thymine.

DTNB 5,5'-Dithiobis (2-nitrobenzoic acid); Ellman's reagent.

DTT See Cleland's reagents.

dTTP 1. Thymidine triphosphate. 2. Thymidine-5'-triphosphate; the 5'-triphosphate of 2'-deoxyribosyl thymine.

dual-bed chromatography COUPLED-LAYER CHROMATOGRAPHY.

dual-effect mutant A mutant possessing a polar mutation; a polarity mutant.

dual recognition A model of associative recognition in which a T lymphocyte is assumed to have two receptor sites, both of which must simultaneously bind one of two different ligands to activate the cell. One of these ligands is an antigen, the other is a self molecule coded for by the major histocompatibility complex.

dual signal hypothesis *See* synarchy.

ductless gland ENDOCRINE GLAND.

Du Nouy ring tensiometer An instrument for measuring surface tension and interfacial tension by a determination of the force required to detach a platinum ring from a surface.

duocrinin The gastrointestinal hormone that controls the secretion from Brunner's gland, located in the duodenum.

duplex 1. *n* A double helix. 2. *adj* Double-helical; double-stranded.

duplex DNA The double helix of the Watson–Crick model of DNA.

duplicase DNA-DEPENDENT DNA POLYMERASE.

duplicate gene One of the multiple copies of a gene, produced by gene duplication.

duplication A chromosomal aberration in which a chromosome bears two identical groups, each composed of one or several genes.

Duponol Trademark for a group of detergents composed of sulfate esters of alcohols that are derived from long-chain fatty acids.

D value DECIMAL REDUCTION TIME.

dwarfism A condition of being undersized that results from the premature arrest of skeletal growth and that may be caused by insufficient secretion of growth hormone.

dyad symmetry The symmetry of a body that exists when identical structures are produced when the body is rotated by 180°; used to describe double-helical DNA regions containing palindromes. When such a segment is rotated by 180°, the same base sequence is obtained.

dye   A compound that strongly absorbs light in the visible region and that can be firmly attached to a surface as a result of chemical and/or physical interactions.

dye-sensitized photooxidation   The oxidation of a biologically important molecule that occurs in the absence of oxygen but in the presence of a photosensitizing dye and an appropriate electron and/or hydrogen acceptor. *See also* photosensitization.

dynamic capacitor electrometer   VIBRATING REED ELECTROMETER.

dynamic equilibrium   1. EQUILIBRIUM. 2. STEADY STATE.

dynamic osmometer   An osmometer, the operation of which is based on the application of hydrostatic pressure sufficient to just prevent osmosis from occurring.

dynamic viscosity   The viscosity of a liquid that has not been corrected for the density of the liquid. The unit of dynamic viscosity is the poise. *See also* kinematic viscosity.

dyne   A unit of force equal to the force that, when applied to a mass of one gram, will impart to it an acceleration of 1 cm/s^2.

dynein   A protein that forms the "arms" in the axoneme structure of cilia and flagella. Dynein is a very large protein (MW about 5×10^5) and an ATPase, requiring either Ca^{2+} or Mg^{2+} for activity.

dynorphin   An opioid (endorphin) that has great potency and contains the Leu-enkephalin amino acid sequence at its N-terminal; it is derived from prodynorphin.

dysfunction   An impairment of normal function; a malfunction.

dysgammaglobulinemia   A condition in which there is an imbalance in the relative amounts of the various types of immunoglobulins in an individual.

dysgenic   Genetically harmful or injurious; detrimental to the genetic qualities of a race or a breed. *See also* eugenic.

dysmyelination   The formation of a myelin that has a faulty structure.

dystrophy   Defective nutrition.

E

e 1. The base of natural logarithms; a constant, equal to 2.7183.... 2. Electron. 3. Eukaryotic, as in eIF (eukaryotic initiation factor).

E 1. Enzyme. 2. Energy. 3. Extinction. 4. Reduction potential. 5. Glutamic acid. 6. Equivalent. 7. Exa.

E_0 *See* standard electrode potential.

E^0 *See* standard electrode potential.

E'_0 *See* standard electrode potential.

$E_{1\,cm}^{1\%}$ The extinction coefficient of a 1% solution, the absorbance of which is measured in a cuvette having a light path of 1 cm.

E_a ; E_A Activation energy.

EAA Essential amino acids.

E_a antigens A group of alloantigens in mouse erythrocytes.

Eadie–Hofstee plot A single reciprocal plot of the Michaelis–Menten equation in which $v/[S]$ is plotted versus v; v is the velocity of the reaction and $[S]$ is the substrate concentration. *Aka* Eadie plot; Eadie–Scatchard plot.

EAG Electroantennogram.

early enzyme A virus-specific enzyme that is transcribed from an early gene.

early gene A viral gene that is transcribed early after the infection of a host cell by the virus. *Aka* early function gene.

early protein A virus-specific protein that is transcribed from an early gene; early proteins are typically viral enzymes, as distinct from late proteins which are structural proteins of the virus.

early RNA A virus-specific RNA that is synthesized by RNA polymerase shortly after the infection of the host cell by the virus.

earthy group A group of compounds that have high melting points and that are believed to have occurred in the original gas dust of the solar nebula. *See also* gaseous group; icy group.

EBV Epstein–Barr virus.

EB virus Epstein–Barr virus.

EC Electron capture.

E.C. Enzyme Commission; in the listing of an enzyme, the abbreviation is followed by four numbers indicating the classification of the enzyme according to main division, subclass, sub-subclass, and serial number in the sub-subclass.

EC_{50} Effective concentration; the concentration of a compound at which a specified effect is observed under the test conditions in a spe-

cified time in 50% of the organisms being tested.

ECC End carbon chain.

eccrine gland A sweat gland.

ecdysone A steroid hormone that stimulates molting of caterpillars, pupa formation, and emergence from the pupa in insects.

ECF Extracellular fluid.

ECG Electrocardiogram.

ECGF Endothelial cell growth factor. *See* fibroblast growth factor.

echovirus A virus that belongs to the enterovirus subgroup of picornaviruses and that is similar to the polio virus in its physical parameters; the name echovirus is derived from enteric cytopathogenic human orphan virus. Echoviruses occur in the respiratory and intestinal tracts of healthy people and are not associated with any human diseases.

ECL Equivalent chain length.

eclipse 1. The time interval between the infection of a bacterial cell by a phage and the appearance of intracellular infective phage particles. 2. The time interval in bacterial transformation during which the transforming DNA cannot be extracted from the cell in a form that retains its activity for transformation.

eclipsed antigen An antigen that is effectively common to both a parasite and its host. The antigen of the parasite is so similar to that of the host that it does not elicit an immune response in the host.

eclipsed conformation The conformation of a molecule in which, in a Newman projection, a large number of the atoms are either partially or completely concealed from view by other atoms. In such a conformation, interatomic distances are relatively small and interatomic interactions are maximized; as a result, an eclipsed conformation is less stable than a staggered one.

eclipsing strain TORSIONAL STRAIN.

eclosion hormone An insect polypeptide hormone that appears at the end of molting and aids in the shedding of the old cuticle.

E. coli *Escherichia coli.*

ecological system ECOSYSTEM.

ecology The study of the interrelations between organisms and the interrelations between organisms and their external environments.

economic species A desirable species; a species, such as humans, that uses chemicals to eliminate other, undesirable species.

economic toxicology The branch of toxicology that deals with human use of chemicals to selectively affect tissue function or to selectively eliminate uneconomic species.

ecosystem The system of interrelated organisms and nonliving components in a particular environment.

ecotropic virus A virus that can replicate in cells from its host species but cannot replicate in cells from other species. *See also* amphotropic virus.

ECTEOLA-cellulose Epichlorohydrin triethanolamine cellulose; an anion exchanger.

ectochemistry The chemistry of the surface of either a cell or an organism.

ectocrine ECTOHORMONE.

ectoderm The outermost of the three germ layers of an embryo from which the epidermis, nervous tissue, and sense organs develop.

ectoenzyme An enzyme that is an integral protein of the cell membrane but has its active site located on the outside surface of the cell so that its activity is directed outward.

ectohormone A chemical, such as a pheromone, that is produced by one organism and that exerts an effect on another organism.

-ectomy Combining form meaning surgical removal.

ectopic hormone syndrome The aberrant synthesis and secretion of polypeptide hormones by nonendocrine malignant tissues.

ectopic protein A protein produced by a neoplasm that is derived from a tissue that is not normally engaged in the synthesis of that protein.

ectopic tumor A tumor of some organ that synthesizes and secretes a polypeptide hormone, normally synthesized and secreted by an endocrine gland.

ectoplasm The layer of the cytoplasm near the periphery of the cell that is more rigid than that in the interior of the cell. *See also* endoplasm.

ectotherm A cold-blooded animal; a poikilothermic organism.

ED$_{50}$ Median effective dose.

eddy diffusion The irregularity in the diffusion of solute molecules that occurs in a porous chromatographic support. The phenomenon is due to the fact that (a) the pathlength for some solute molecules is either shorter or longer than that for the bulk of the molecules; and (b) the rate of solvent flow varies in different regions of the porous support.

eddy migration The irregularity in the electrophoretic mobility of solute molecules that occurs in a porous support. The phenomenon is due to the fact that (a) the pathlength for some solute molecules is either shorter or longer than that for the bulk of the molecules; and (b) the rate of solvent flow varies in different regions of the porous support.

edema An abnormal accumulation of interstitial fluids in the tissues so that they become puffy.

editing PROOFREADING.

editing function PROOFREADING FUNCTION.

Edman degradation A method for the stepwise removal of amino acids from peptides and proteins that allows a determination of both the N-terminal amino acid and the amino acid sequence. The method is based on the reaction of the Edman reagent, phenylisothiocyanate, with the free alpha amino group of amino acids, peptides, or proteins, and on the removal of the N-terminal amino acid in the form of a phenylthiohydantoin derivative. These steps are then carried out repetitively. *See also* subtractive Edman degradation.

Edmundson wheel A diagrammatic representation for assessing the stabilizing effect of nonpolar amino acid side chains on alpha-helical segments in a protein; prepared by projecting the helix onto a flat surface, looking down on the helix axis. Consecutive amino acid residues are depicted as segments of a circle; nonpolar residues can be distinguished from polar ones by shading or coloring the corresponding segments.

ED pathway Entner–Doudoroff pathway.

EDS Energy-dispersive spectrometry.

EDTA 1. Ethylenedinitrolotetraacetic acid. 2. Ethylenediaminetetraacetic acid.

eEF Eukaryotic elongation factor. *See* elongation factor.

EEG Electroencephalogram.

EELS Electron energy loss spectroscopy.

EF 1. Extrinsic factor. 2. Elongation factor.

EFA Essential fatty acids.

E-face *See* fracture faces.

effective concentration *See* EC$_{50}$.

effective contact The formation of a specific donor–recipient pair in the course of plasmid DNA transfer.

effective half-life The half-life of a radioactive isotope in a biological system that is equal to the product of the biological half-life and the radioactive half-life.

effective lethal phase The stage in the development of an organism carrying a lethal gene at which that gene generally brings about the death of the organism.

effective mean residue rotation REDUCED MEAN RESIDUE ROTATION.

effector A metabolite that, when bound to an allosteric enzyme, alters the catalytic activity of the enzyme. The effector generally alters either the Michaelis constant of the enzyme or the maximum velocity of the reaction. An effector that functions as an activator and leads to an increase in the binding of the substrate and other effectors is known as a positive effector; an effector that functions as an inhibitor and leads to a decrease in the binding of the substrate and other effectors is known as a negative effector. An effector may likewise bind to a nonenzymatic, allosteric protein and lead to a change in the properties of the protein. *See also* Britten–Davidson model.

effector cell A lymphocyte in peripheral lymphoid tissue that is actively engaged in making an immunological response; T effector cells carry out cell-mediated responses while B effector cells secrete antibodies.

effector sequence *See* Britten–Davidson model.

efferent 1. Leading or conveying away from a cell or an organ. 2. Of, or pertaining to, the stages involved in the response of the sensitized immune system. *See also* afferent.

efficiency of infection The ratio of the number of viral particles in a group of animals, cell cultures, or other test units to the number of infective test units produced.

efficiency of plating 1. The proportion of animal cells that give rise to colonies. 2. The ratio of the plaque count in a plaque assay under a given set of conditions to that under standard conditions. *Abbr* EOP.

efflorescence The giving up of water by a substance when it is exposed to air under ordinary conditions of temperature, pressure, and moisture.

effluent ELUATE.

efflux Outward flow, as that out of a cell.

EF-G Elongation factor G; translocase.

EF hand A characteristic structure that is frequently involved in ion-binding sites of a protein such as those for Ca^{2+}; it is composed of 6 to 8 oxygen atoms, derived from aspartic and glutamic acid side chains and from backbone carbonyl groups of the protein. These various oxygen atoms interact simultaneously with the ion.

EF-T Elongation factor T.

EF-Ts *See* elongation factor.

EF-Tu *See* elongation factor.

egest To discharge material, such as waste, from either a cell or a body.

EGF Epidermal growth factor.

EGF-Uro Epidermal growth factor-urogastrone.

egg white injury The condition of biotin deficiency in humans or animals that results from an excessive intake of raw egg white; the condition is due to the protein avidin that is present in egg white and that combines tightly with biotin.

egg white injury factor BIOTIN.

Ehlers–Danlos syndrome A group of clinical disorders, characterized by structural laxity or defects in connective tissue, resulting from defects in the structure of collagen. All of the disorders are genetically inherited metabolic defects; for some types, the specific lesions are known, for others they are unknown.

Ehrlich ascites tumor A tumor, derived from a mouse carcinoma, that grows in the peritoneal cavity and that has been kept alive in tissue culture.

Ehrlich reaction A colorimetric reaction for tryptophan and other compounds containing the indole ring; based on the production of a red color upon treatment of the sample with *p*-dimethylaminobenzaldehyde and concentrated hydrochloric acid.

Ehrlich's reagent 1. A reagent that contains *p*-dimethylaminobenzaldehyde and that gives a red color with porphobilinogen, δ-aminolevulinic acid, and related compounds. 2. A reagent that contains diazotized sulfanilic acid and that gives a blue color with bilirubin in the Van den Bergh reaction.

Ehrlich's receptor theory An early selective theory of antibody formation that was formulated by Ehrlich in 1900. The theory proposed that cells were covered with antibody-like receptors that contained haptophore side chains with which the antigens combined. The receptors were then thought to be liberated from the cells and to enter the blood stream as circulating antibodies. *Aka* Ehrlich's side chain theory.

EHTP Equivalent height of a theoretical plate.

EI Enzyme–inhibitor complex.

EIA Enzyme immunoassy.

eicosanoids Collective term for arachidonic acid and all of the compounds derived from it; includes prostaglandins, thromboxanes, prostacyclin, and leukotrienes. Eicosanoids are ubiquitous and have a variety of potent effects. *See also* prostanoids.

eIF Eukaryotic initiation factor. *See* initiation factor.

eIF-2 stimulating protein *See* hemin-controlled repressor.

EI-MS Electron ionization mass spectrometry.

einstein A mole of photons; Avogadro's number of photons.

Einstein law of photochemical equivalence FIRST LAW OF PHOTOCHEMISTRY.

Einstein–Sutherland equation An equation relating Brownian motion and diffusion; specifically, $D = RT/Nf$, where D is the diffusion coefficient, R is the gas constant, T is the absolute temperature, N is Avogadro's number, and f is the frictional coefficient.

EIS Enzyme–inhibitor–substrate complex.

EKG Electrocardiogram.

elaidinization The cis–trans isomerization of mono- or polysaturated fatty acids. The term is derived from the isomerization of oleic acid (cis) to elaidic acid (trans).

elaioplast A lipid-rich plastid.

elastase An enzyme that catalyzes the hydrolysis of the peptide bonds of elastin and other peptide bonds which are formed by neutral amino acids.

elastic collision A collision in which there is no loss of kinetic energy; the sum of the kinetic energies of the colliding particles before and after the collision is the same; the kinetic energy is conserved.

elastin A major scleroprotein of connective tissue, especially of the elastic tissue of tendons and arteries.

elastomer A natural or a synthetic polymer that possesses rubber-like properties.

ELDOR Electron–electron double resonance.

elective enrichment ENRICHMENT CULTURE.

elective theory SELECTIVE THEORY.

electric 1. Of, or pertaining to, electricity. 2. Pertaining to the motion, emission, and behavior of currents of free electrons in passive elements such as wires, resistors, and inductors. *Aka* electrical.

electrical synapse The direct electrical coupling between two neurons which occurs at a number of sites in the nervous systems of many different species.

electric birefringence The birefringence produced by molecules that have become oriented as a result of the application of an electric field. *Aka* electrical birefringence.

electric dichroism The dichroism that occurs when polarized light is absorbed by molecules, the orientation of which is affected by the direction of an applied electric field. *Aka* electrical dichroism.

electric dipole DIPOLE.

electric double layer IONIC DOUBLE LAYER.

electric field The space surrounding electrical charges in which a mechanical force will be exerted on a charge introduced into it.

electride A new class of crystalline materials, derived from cesium, potassium, and other alkali metals, in which a lattice of positively charged atoms, caged in neutral molecules, is held together by trapped clouds of free-floating electrons. The electronic and optical properties of electrides range from those of nonmetals to those of metals.

electroantennogram An electrophysiological measurement of pheromone activity that is based on the nerve impulse generated by isolated receptor organs. A standard solution of pheromone is blown over isolated antennae in a stream of air, and the resulting cell potential is measured with a recorder and a set of microelectrodes.

electroblotting *See* blotting.

electrochemical potential The free energy change for the transport of a charged solute either up or down a concentration gradient; equal to the sum of the free energy changes due to the concentration gradient and due to the electrical potential.

electrochemistry The branch of chemistry that deals with the interrelations and the transformations of chemical and electrical energy.

electrochromatogram ELECTROPHEROGRAM.

electrochromatography 1. Any electrophoretic procedure in which the separation of solute particles does not depend solely on the mobilities of the particles, but is also significantly affected by the sorptive interactions between the solute particles and the solid support. 2. ZONE ELECTROPHORESIS. 3. CONTINUOUS FLOW ELECTROPHORESIS.

electrochromatophoresis ZONE ELECTROPHORESIS.

electrochromic effect A change in absorbance that is induced by changes in the electric field. *Aka* electrochromism.

electrocortin ALDOSTERONE.

electrode A device, frequently a wire or a plate, by which an electric current passes into, or out of, an electric cell, a solution, an apparatus, or a body.

electrodecantation A technique for the separation and fractionation of proteins by means of electrodialysis. During this process, the protein molecules move toward the electrodes and tend to form zones of high protein concentration at those portions of the retaining membrane that are near the electrodes. These zones are gravitationally unstable and tend to move downward in the container, carrying the protein molecules along. Thus, in the absence of stirring, there is a gradual decrease in the concentration of protein molecules in the top layers of the solution, and a gradual increase in their concentration in the lower layers.

electrode couple HALF-REACTION.

electrode potential 1. A measure of the tendency of an oxidation–reduction half-reaction to occur by either loss or gain of electrons. The electrode potential measuring electron loss is known as an oxidation potential and that measuring electron gain is known as a reduction potential; the two are numerically equal, but of opposite sign, and are denoted E. The electrode potential at pH 7.0 is the biochemical electrode potential and is denoted E'. *See* *also* standard electrode potential. 2. The potential across a chemically reactive electrode that participates either in the formation of ions from atoms or in the formation of atoms from ions.

electrodialysis A technique for the removal of ions from a solution in which dialysis of the solution is carried out with the simultaneous application of an electric field across the dialysis bag.

electrodiffusion The spreading of a spot or a zone in electrophoresis as a result of diffusion; may occur when an electrophoretic component is present as a number of rapidly interconvertible forms that have different mobilities.

electroelution The elution of compounds, separated by electrophoresis on agarose or acrylamide gels, that is brought about by subjecting the gel, under appropriate conditions, to an electric field; the compounds move by electrophoresis through, and out of, the separation gel.

electroencephalogram The recording of the potential changes of the brain. *Abbr* EEG.

electroendosmosis ELECTROOSMOSIS.

electrofocusing ISOELECTRIC FOCUSING.

electrofusion A technique (developed by Zimmerman) for promoting the fusion of cells in tissue culture by electrical means. The method involves exposing the cells to a nonuniform alternating electric field which brings the cells into close contact, and then subjecting them to a short, direct, current pulse of high intensity. This pulse introduces a reversible breakdown in the area of membrane–membrane contact; micropores are opened and permeability properties of the membrane are altered so that the cytoplasms can mix and cell fusion may follow. The changes in cell permeability may be such as to allow DNA fragments the size of genes to enter the cell. This is referred to as gene transfer by electroporation. *Aka* Zimmerman method. *See also* electroporation.

electrogenic carrier A carrier (transport agent) that leads to an alteration in the preexisting membrane potential.

electrogenic pump A pump that generates a gradient of electrochemical potential in the course of its operation. This results from the fact that either (a) the movement of one ion across the membrane is not linked to the movement of another ion in the opposite direction, or (b) the charges of one ion moved in one direction do not balance the charges of another ion moved in the opposite direction.

electroimmunodiffusion ROCKET ELECTROPHORESIS.

electrokinetic phenomenon One of four phenomena that describe either the electrical forces produced by the relative motion of solids and liquids, or the relative motion of solids and liquids produced by electrical forces. The four phenomena are electrophoresis, electroosmosis, streaming potential, and sedimentation potential.

electrokinetic potential ZETA POTENTIAL.

electrolysis The decomposition of a substance by the action of an electric current.

electrolyte A substance that dissociates, partly or entirely, into two or more ions in water; solutions of electrolytes conduct an electric current.

electrolyte balance The reactions and factors involved in maintaining a constant internal environment in the body with respect to the distribution of electrolytes between the various fluid compartments.

electrolytic Of, or pertaining to, electrolytes or electrolysis.

electrolytic desalting The removal of ions from a solution by means of electrolysis.

electromagnetic spectrum The entire range of either the wavelengths or the frequencies of electromagnetic radiations, from the shortest cosmic rays to the longest radio waves.

electromechanochemical coupling hypothesis A modified form of the conformational coupling hypothesis. *Abbr* EMC hypothesis.

electrometer An instrument for measuring electrical potential, electrical charge, or electrical current.

electrometric titration A titration in which electromotive force is measured as a function of titrant added.

electromigration 1. ZONE ELECTROPHORESIS. 2. ISOTACHOPHORESIS.

electromotive force The directly measurable electrical energy that can be derived from an electrical cell composed of two half-cells. *Abbr* EMF.

electromyogram The recording of the potential changes of a muscle. *Abbr* EMG.

electromyography The study of the electrical properties of muscle by means of recordings of its action potential.

electron 1. The elementary particle of nature that has a charge of minus one and a mass of 0.000549 amu (9×10^{-28} g). 2. The elementary particle of nature that has a mass of 0.000549 amu and a charge of either plus one or minus one. *Sym* e. *See also* elementary particles.

electron acceptor A substance that is being reduced; an oxidant; an oxidizing agent.

electron affinity The tendency of a substance to accept electrons and to function as an oxidizing agent.

electron capture A mode of radioactive decay

in which an orbital electron is attracted to the nucleus of an atom and combines with a proton in that nucleus. The electron is generally derived from the K shell of the atom and its combination with the proton leads to the formation of a neutron and to the emission of energy in the form of x rays. *Abbr* EC.

electron carrier A substance that serves as a donor and acceptor of either electrons or electrons and protons in an electron transport system. *See also* electron transfer protein.

electron configuration The arrangement of the electrons about an atomic nucleus.

electron dense Descriptive of a substance, such as a heavy metal ion or concentrated macromolecular matter, that has the capacity to scatter impinging electrons and to prevent their passage through it.

electron-dense label An atom or a compound that is electron dense and that is frequently used as a marker in electron microscopy. *See also* ferritin-labeled antibody.

electron-density map The three-dimensional representation of the structure of a molecule that is based on x-ray diffraction data and that is prepared by superimposing layers that correspond to electron densities of various planes in the molecule.

electron diffraction The diffraction of a beam of electrons that results from the wavelike nature of the electrons.

electron donor A substance that is being oxidized; a reductant; a reducing agent.

electronegative 1. Describing the tendency of an atom or a group of atoms to gain electrons. 2. Having a negative charge; having an excess of electrons.

electron–electron double resonance A spin-label technique used for evaluating rotational correlation times; involves the induction of saturation at one point in the spectrum by a first microwave source, and an investigation of the effect of this saturation at other points in the spectrum by means of a second microwave source. Such saturation-transfer spectroscopy is a function of the resonant conditions of the system. *Abbr* ELDOR.

electron energy loss spectroscopy A technique for the study of surfaces in which the energy loss of low-energy electrons, reflected from a surface, is used to determine vibrational modes of molecules adsorbed on the surface. *Abbr* EELS.

electron equivalent A measure of reducing power that is equivalent to one electron.

electroneutral symport A symport mechanism which results in no net movement of charge across the membrane (no change in membrane potential), sometimes because the charges of the two transported species cancel each other.

electron-exchange resin A resin that contains groups that are capable of undergoing reversible oxidation and reduction; such resins can be used as insoluble oxidizing and reducing agents in column chromatography.

electron hole The energy level, in either the ground state or a deexcited state, of an atom or a molecule that has lost an electron and that has a great tendency to recapture an electron. *See also* hole.

electronic Pertaining to the motion, emission, and behavior of currents of free electrons in vacuum tubes, gas tubes, semiconductors, and superconductors.

electronic transition The transition of electrons of an atom or a molecule from one atomic orbital to another. Electronic transitions require large amounts of energy and can be induced by visible light, ultraviolet light, and x rays.

electron ionization mass spectrometry A mass spectrometric technique that requires volatilization of the sample (generally in the form of volatile derivatives) prior to ionization, and that is especially useful for ions smaller than those in the molecular region. The sample is ionized by means of a beam of electrons usually produced by an incandescent filament. *Abbr* EI-MS. *Aka* electron impact ionization mass spectrometry.

electron magnetic resonance *See* electron paramagnetic resonance; electron spin resonance.

electron micrograph The photographic record of a pattern observed in an electron microscope.

electron microscope A microscope in which beams of electrons are focused by means of magnetic fields onto a fluorescent screen or onto a photographic film. The electron microscope has great resolving power because of the short wavelengths that are associated with electrons. *Aka* transmission electron microscope.

electron microscope radioautography The use of electron microscopy in conjunction with radioautography.

electron–nuclear double resonance Resonance spectroscopy that combines features of both electron spin resonance and nuclear magnetic resonance; a technique in which electron spins and nuclear spins are irradiated simultaneously. The method is useful for determining whether particular types of nuclei are interacting with free radicals; that is, whether the unpaired electron of the free radical has appreciable density on magnetic nuclei. *Abbr* ENDOR.

electron pair bond COVALENT BOND.

electron paramagnetic resonance A method for

studying the interaction of unpaired electrons in a substance with the environment of these electrons. A substance containing unpaired electrons has a permanent magnetic moment (i.e., it is paramagnetic) as a result of the magnetic properties of its spinning electrons and will tend to orient itself in an applied magnetic field; the magnetic field of the unpaired electron may be either parallel or antiparallel to that of the applied field. A transition from one state to the other is associated with an energy difference and occurs when the applied field is of sufficient strength and the transition is accompanied by absorption of electromagnetic radiation; the relative magnitude of this radiation and that of the applied magnetic field are interpreted in terms of the electron's interaction with its environment. *Abbr* EPR. *See also* electron spin resonance.

electron pressure The tendency of a substance to donate electrons and to function as a reducing agent.

electron probe microanalysis A method for in situ chemical analysis of microvolumes in a specimen. After obtaining an image of the specimen by means of transmission or scanning electron microscopy, the electron beam is adjusted to a "probe" which is restricted to a microregion of the specimen. The x rays excited by this beam are then analyzed spectroscopically.

electron projection function A three-dimensional representation of molecular electron distributions obtained by integration of molecule electron densities.

electron sink An electronegative atom or a group of atoms that captures an electron from other components of the system.

electron spin echo spectroscopy A form of electron paramagnetic resonance in which a number of pulses are applied to the sample in succession and the spin echo, emitted by the sample, is recorded and analyzed. *Abbr* ESE.

electron spin resonance The common name for electron paramagnetic resonance when it is applied to compounds that are characterized by g values close to that of the free electron, such as most organic free radicals. *Abbr* ESR.

electron-stimulated desorption ion angular distribution A technique for the study of surfaces in which molecules, adsorbed to the surface, are ionized by electron impact. Measurement of the trajectories of the ions thus formed provides information about the orientation of the parent molecules on the surface. *Abbr* ESDIAD.

electron transfer chain ELECTRON TRANSPORT SYSTEM.

electron transfer flavoprotein An electron carrier that serves as a link between the reduced form of fatty acyl coenzyme A dehydrogenase and the electron transport system. An extra flavoprotein of the electron transport system that is involved in the oxidation of fatty acids. *Abbr* ETF.

electron transfer potential The free energy change per mole for an oxidation–reduction reaction.

electron transfer protein A protein that serves as a donor and acceptor of either electrons or electrons and protons in oxidation–reduction reactions. Six types of electron transfer proteins have been identified: flavoproteins, proteins containing reducible disulfide groups, cytochromes, iron–sulfur proteins, cuproproteins, and molybdoproteins.

electron transfer system ELECTRON TRANSPORT SYSTEM.

electron transport chain ELECTRON TRANSPORT SYSTEM.

electron transport particle 1. RESPIRATORY ASSEMBLY. 2. SUPERMOLECULE.

electron transport system 1. The group of biological oxidation–reduction substances that are present in mitochondria and that act sequentially in the transport of either electrons or electrons and protons. The electrons and protons are abstracted from metabolites in glycolysis, the citric acid cycle, beta oxidation, and other metabolic reactions. Each oxidation–reduction substance, or electron carrier, is capable of oxidizing a preceding one in the sequence and the oxidation proceeds from a metabolite to molecular oxygen as the ultimate oxidizing agent. Two modes of electron transport occur, which differ in the initial sequence of electron carriers. One consists of metabolite–NAD^+-FMN-FeS-CoQ-; the other consists of succinate-FAD-FeS-CoQ-. From CoQ on, the two modes utilize the same sequence of electron carriers, specifically, CoQ-Cyt b-FeS-Cyt c_1-Cyt c-Cyt oxidase-oxygen. The reduction potential of the electron carriers becomes progressively more positive from metabolite to oxygen. The free energy change corresponding to the potential difference between metabolite and oxygen is utilized for the synthesis of ATP from ADP, a reaction that is coupled to the electron transport system at various sites. 2. Any group of electron carriers such as that functioning in photosynthesis. *Abbr* ETS.

electron trap ELECTRON SINK.

electronvolt A unit of energy equal to the energy acquired by an electron in passing through a potential gradient of 1 V. *Sym* eV.

electroosmosis The movement of a charged liquid, relative to a fixed medium carrying the opposite charge, under the influence of an electric field; an electrokinetic phenomenon that is the reverse of streaming potential.

electropherogram Variant spelling of electrophoregram.

electrophile An atom or a group of atoms that is electron pair seeking.

electrophilic Of, or pertaining to, either an electrophile or a reaction in which an electrophile participates.

electrophilic catalysis Catalysis in which the catalyst abstracts a pair of electrons from a reactant.

electrophilic displacement A chemical reaction in which an electrophilic group attacks and displaces a susceptible group in a compound and then binds covalently to the compound at that site. *Aka* electrophilic substitution.

electrophoregram The record of a zone electrophoresis pattern, either in the form of the electrophoretic support itself or in the form of a tracing thereof.

electrophorese To cause a compound or a mixture of compounds to move by electrophoresis.

electrophoresis The movement of charged particles through a stationary liquid under the influence of an electric field. Electrophoresis is a powerful tool for the separation of particles and for both preparative and analytical studies of macromolecules. The particles are separated primarily on the basis of their charge and to a lesser extent on the basis of their size and shape. Moving boundary or free electrophoresis refers to electrophoresis performed in solution, while zone electrophoresis refers to electrophoresis performed in a porous medium.

electrophoretic Of, or pertaining to, electrophoresis.

electrophoretic carrier A carrier (transport agent) that transports an ion across a membrane under the driving force of a constant, imposed potential.

electrophoretic effect The decrease in the electrophoretic mobility of a protein that is brought about by the movement of hydrated counterions in the opposite direction; this migration of counterions constitutes, effectively, a flow of solvent in the opposite direction. *See also* asymmetry effect.

electrophoretic injection *See* iontophoresis.

electrophoretic mobility The velocity with which a charged particle moves in electrophoresis divided by the electric field strength that is applied across the support; generally expressed in units of $cm^2/(s)(V)$.

electrophoretic retardation ELECTROPHORETIC EFFECT.

electrophoretogram ELECTROPHOREGRAM.

electrophorogram Variant spelling of electrophoregram.

electroplax A flat plate in the electric organ of some fish that, when stacked, allows for the generation of a large potential difference.

electroporation A technique for introducing foreign macromolecules into cells; based on subjecting cells to a controlled electrical impulse which leads to reversible formation of nanometer-sized pores in the cell membrane, thus permitting transfection or cell fusion. The technique permits, for example, the introduction of DNA into animal cells and plant protoplasts. *See also* electrofusion.

electropositive 1. Describing the tendency of an atom or a group of atoms to lose electrons. 2. Having a positive charge; having a deficiency of electrons.

electrosorptive spreading The spreading of a spot or zone in electrophoresis that occurs if the moving particle is strongly adsorbed to the support.

electrostatic Pertaining to electrical charges that are not in motion.

electrostatic bond IONIC BOND.

electrostatic catalysis Catalysis in which charge distributions around the active site of an enzyme serve to stabilize the transition state and/or guide a polar substrate to its binding site.

electrostatic interactions The attractive and repulsive electrical forces between atoms, and/or groups of atoms, and/or molecules that are caused both by the presence of ionized species and by the electropositive and electronegative properties of the atoms. *See also* ion–ion interaction.

electrostriction The decrease in the apparent volume occupied by a charged molecule in water as compared to the volume occupied by an uncharged molecule that has the same empirical formula as the charged one. The difference in volume is due to the strong attraction of water molecules to, and their resultant compression and close packing around, the charged groups of the solute molecule.

electroviscous effect The dependence of the viscosity of a polymer on the electrical charge of the polymer.

element A fundamental form of matter that has special properties and that is not decomposable by ordinary chemical means; it consists of atoms that are of one type and that have the same atomic number.

elementary analysis The quantitative chemical analysis of the relative amounts of the different elements in a compound.

elementary forces A group of four fundamental

forces that are believed to govern the interactions of matter; these are gravity, electromagnetism, weak force, and strong force. These four forces vary in range and strength; the ranges are infinite, infinite, less than 10^{-16} cm, and less than 10^{-13} cm, respectively. All the forces are conveyed by force particles. The force particles for the forces listed are graviton, photon, boson, and gluon, respectively. *See also* elementary particles.

elementary particle 1. SUPERMOLECULE. 2. RESPIRATORY ASSEMBLY. 3. F_0F_1-ATPase. *See also* elementary particles.

elementary particles Fundamental, structureless, and indivisible particles of matter. According to current theory, there are 12 such particles which fall into two groups: 6 leptons and 6 quarks; the six members of each group are called flavors. The leptons include the electron, the muon, the tau, and three neutrinos (electron neutrino, muon neutrino, and tau neutrino). The first three leptons carry an identical charge of -1 but differ in their mass. The neutrinos are electrically neutral and two of them (electron neutrino and muon neutrino) are nearly massless. The quarks are designated up, down, charm, strange, top, and bottom. These particles carry fractional charges and have varying masses. Different combinations of quarks result in hadrons, which are subnuclear particles that are not elementary particles and that vary in mass, charge, spin, and other properties. Protons, neutrons, and mesons are examples of hadrons. For each elementary particle (lepton or quark) there is a corresponding antiparticle. Antiparticles have the same mass and spin as their respective particles but carry opposite values for other properties such as charge. Collision of a particle and an antiparticle leads to mutual annihilation. The antiparticle of the electron, for example, is the antielectron, or positron. *See also* elementary forces.

elementary step The simplest possible chemical reaction that involves the conversion of reactants to products without the formation of any, chemically stable, intermediate. An elementary step is a reaction that proceeds in a single step and one for which the rate equation can be deduced directly from the stoichiometry of the reaction. *Aka* elementary reaction.

elements of symmetry The centers, axes, and planes of symmetry of either a molecule or a body.

eleostearic acid An unsaturated fatty acid that contains 18 carbon atoms and three double bonds.

eliminase PECTATE LYASE.

elimination reaction A chemical reaction in

which there is a decrease in the number of groups attached to carbon atoms so that the molecule becomes more unsaturated.

ELISA Enzyme-linked immunosorbent assay.

ellipsoid of revolution A geometrical solid formed by the rotation of an ellipse about one of its axes. The ellipsoid of revolution is called oblate or prolate depending on whether the rotation occurred about the minor or the major axis of the ellipse. *See also* equivalent ellipsoid of revolution.

ellipsosome A compartment in the retinal cones of certain fish that contains a cytochrome-like pigment.

elliptically polarized light Light that is composed of left and right circularly polarized light components of unequal amplitude.

ellipticity 1. The angle whose tangent is the ratio of the minor axis to the major axis of an ellipse. 2. MOLAR ELLIPTICITY.

Ellman's reagent 5,5′-Dithiobis(2-nitrobenzoic acid); a reagent for the determination of sulfhydryl groups in proteins.

elongation The stage in the polymerization of amino acids during protein synthesis that covers all of the steps between the initiation and the termination of the polypeptide chain. *See also* elongation cycle.

elongation cycle The set of repetitive reactions that occur during the elongation stage of protein synthesis. These reactions are (a) the attachment of the incoming aminoacyl-tRNA to the aminoacyl site on the ribosome; (b) the peptidyl transferase-catalyzed formation of a peptide bond between the incoming amino acid and the growing polypeptide chain; and (c) the translocase-catalyzed shifting of the peptidyl-tRNA from the aminoacyl site to the peptidyl site on the ribosome with a simultaneous shift of the messenger RNA by one codon.

elongation factor One of at least two protein factors that function in the elongation of polypeptide chains during protein synthesis and that are designated as T and G (EF-T, EF-G) in prokaryotes and as eEF-1 and eEF-2 (EF-1, EF-2) in eukaryotes. Elongation factor T is required for the binding of aminoacyl-tRNA to the ribosome and is dissociable into a stable and an unstable subunit referred to as T_s and T_u (EF-Ts, EF-Tu); elongation factor G is the translocase. In eukaryotes, there are four elongation factors that roughly correspond in function to the prokaryotic factors as follows: EF-1α (EF-Tu), EF-1β (EF-Ts), EF-1γ, EF-2 (EF-G).

eluant 1. Variant spelling of eluent. 2. The solute that is being separated on, and eluted from, a chromatographic column.

eluate The liquid that is collected by elution

from a chromatographic column.

eluent The liquid that is used for the elution of substances from a chromatographic column.

eluent strength The solvent adsorption energy per unit surface area of a fully activated adsorbent.

eluotropic series A group of solvents arranged in the order of their relative eluting strength for a given chromatographic adsorbent.

elute To remove and collect a solute from the stationary phase in chromatography by means of elution with a solvent.

eluting agent ELUENT.

elution The process whereby a solute is removed and collected from the stationary phase in chromatography by passage of a solvent over the chromatographic support.

elution analysis ELUTION CHROMATOGRAPHY.

elution band A zone of separated sample components that is obtained in column chromatography.

elution centrifuge A centrifuge in which a cylindrical holder, containing a separation column, revolves about the center of the apparatus and simultaneously rotates about its own axis; both feed and return tubes for the solutions are led through the center of the holder.

elution chromatography A chromatographic technique in which the sample is applied to a column and the sample components are separated into bands or zones that can be eluted and collected. *Aka* elution development.

elution profile A plot of some property of a column eluate, such as absorbance or radioactivity, as a function of either the eluate fraction number or the cumulative volume of the collected eluate.

elution time RETENTION TIME.

elution volume The volume of eluate collected in column chromatography from the time of sample application to the time at which a given component is eluted at maximal concentration.

elutriation 1. The separation of suspended particles according to their size by washing, decantation, and settling. 2. CENTRIFUGAL ELUTRIATION.

EMB agar A solid medium used to distinguish between members of the enterobacteriaceae; contains two dyes (eosin and methylene blue), peptone, lactose, and agar.

Embden ester A mixture of D-glucose-6-phosphate and D-fructose-6-phosphate, both of which are intermediates in glycolysis.

Embden–Meyerhof–Parnas pathway GLYCOLYSIS.

Embden–Meyerhof pathway GLYCOLYSIS.

embolus (*pl* emboli). A blood clot, or some other mass of undissolved material, that is carried by the blood and forced from one blood vessel into a smaller one, thus causing an obstruction. *See also* thrombus.

embryo 1. The unborn or unhatched young vertebrate during the early stages of its development. The developing human is considered to be an embryo up to the beginning of the third month of pregnancy, and a fetus subsequently. 2. The rudimentary plant within a seed.

embryology The science that deals with the development of an organism.

embryonated egg An egg (usually a hen's or duck's egg) that contains a live embryo; used for preparing tissue cultures and viral vaccines.

embryonic induction MORPHOGENIC INDUCTION.

EMC hypothesis Electromechanochemical coupling hypothesis.

emergency hormone CATECHOLAMINE.

emergency hypothesis The hypothesis that the catecholamines constitute the principal regulatory mechanism of an animal in an emergency situation, and that this mechanism allows the animal to mobilize to meet physical or emotional challenges.

Emerson enhancement effect *See* enhancement effect.

EMF 1. Electromotive force. 2. Erythrocyte maturation factor.

EMG Electromyogram.

-emia 1. Combining form meaning the presence of a substance in blood 2. Combining form meaning the presence of excessive amounts of a substance in blood.

emission The discharge of energy in the form of radiation.

emission spectrum A plot of the emission of electromagnetic radiation by a molecule as a function of either the wavelength or the frequency of the radiation.

emit To discharge energy in the form of radiation.

emphore A protein that is not an enzyme, but that specifically binds a ligand and thereby involves the ligand in biological activity.

empirical formula The chemical formula of a compound that indicates the number and the types of the different atoms in the compound but not the manner in which they are linked together.

EMP pathway Embden–Meyerhof–Parnas pathway.

empty calorie Descriptive of a calorie that is unassociated with nutritive proteins, minerals, and vitamins; a source of energy (usually sugar) that is largely devoid of nutritious value. The calories in alcohol and many types of soft drinks are examples.

empty capsid The viral capsid without the nucleic acid that it normally encloses.

EMS Ethylmethane sulfonate.

emulsification The formation of an emulsion.

emulsifying agent A substance, such as a surface-active agent, that stabilizes an emulsion by coating the droplets of the dispersed phase, thereby preventing their coalescence. *Aka* emulsifier.

emulsion A colloidal dispersion of one liquid in another, immiscible or partially miscible, liquid.

emulsion fractionation A chromatographic technique in which the stationary phase is the surface of droplets and the mobile phase is the liquid that flows between the droplets. *See also* foam fractionation.

enamel The calcified covering of dentine at the exposed portion of a tooth.

enamine An organic compound that contains the grouping

$$\begin{array}{ccc} \diagdown & | & \diagup \\ N{-}C{=}C & \\ \diagup & & \diagdown \end{array}$$

enantiomer One of two optical isomers of a compound that are nonsuperimposable mirror images of each other.

enantiomorph ENANTIOMER.

enantiotopic Descriptive of either the atoms or the groups of atoms in a molecule that bear an enantiomeric relation to each other; the two identical substituents of a meso carbon atom are an example of enantiotopic groups. *See also* diasterotopic.

encapsidate To enclose the viral nucleic acid with a capsid. *Aka* encapside.

encapsulated Having a capsule.

encephalitis Inflammation of the brain.

encephalitogenic protein MYELIN BASIC PROTEIN.

encephalovirus ARBOVIRUS.

encode Code for; thus, a DNA gene is said to encode a specific mRNA or protein molecule.

end absorption The intense absorbance of saturated compounds at and below wavelengths of 200 nm.

end carbon chain The number of carbon atoms in a fatty acid methyl ester molecule from the terminal methyl group of the carbon chain to the center of the double bond nearest this methyl group. The length of the end carbon chain is important in determining the retention of fatty acid methyl esters in gas chromatography. *Abbr* ECC.

endemic Of, or pertaining to, a disease that has a low, and more or less constant, incidence in a particular locality or region.

endergonic reaction A chemical reaction that requires the input of energy; an uphill reaction with a positive free energy change.

end-group analysis The determination of both the type and the number of the terminal groups in a polymer; used as an assessment of purity and for calculations of minimum molecular weights. The Sanger reaction is used for end-group analysis of proteins and exhaustive methylation is used for end-group analysis of carbohydrates.

end labeling The attachment of a radioactive or nonradioactive chemical group to the end of a polymer; used specifically for the attachment of a ^{32}P-labeled group to the end of a polynucleotide strand.

endo- Combining form meaning within or internal.

endo conformation *See* puckered conformation.

endocrine gland A ductless gland of internal secretion that produces one or more hormones which are secreted directly into the circulation.

endocrine hormones Hormones produced by specialized cells, clustered together in endocrine glands, and secreted into the blood stream from these glands. Endocrine hormones act on cells distant from their site of release.

endocrine system The endocrine glands.

endocrinology The science that deals with the structure, the function, and the products of the endocrine glands and of other specialized secretory cells.

endocrinopathy A pathological condition resulting from endocrine dysfunction. Such disease states can be grouped into three major categories: (a) excessive production of a hormone; (b) insufficient production of a hormone; (c) decreased target-tissue sensitivity to a hormone.

endocytic vesicle ENDOSOME; RECEPTOSOME.

endocytosis The process whereby cells take up fluids and particles by pinching off of the plasma membrane. The uptake of large particles is known as phagocytosis and that of small particles, solutes, and fluids is known as pinocytosis.

endocytotic vesicle ENDOSOME; RECEPTOSOME.

endoderm The innermost of the three germ layers of an embryo from which the epithelial lining of the intestine and all of the outgrowths of the intestine develop.

endoenzyme 1. An enzyme that catalyzes a reaction involving internal parts of a polymer, rather than the ends of the polymer. 2. An enzyme that is released inside the cell.

endoergic reaction ENDERGONIC REACTION.

endogenote The genetic complement of the recipient cell that combines with a genetic fragment from a donor cell in bacterial transformation. *See also* exogenote; merozygote.

endogenous Originating within the organism.

endogenous metabolism The level of metabolism in the absence of added nutrients.

endogenous minimum MINIMUM PROTEIN (NITROGEN) REQUIREMENT.

endogenous opiates 1. OPIOIDS. 2. OPIOID PEPTIDES.

endogenous virus An inactive virus that has become integrated into the DNA of germ-line cells. Such viral genomes are transmitted from parent to offspring like other cellular genes. Endogenous viruses may often be activated by chemical, physical, or immunological mechanisms.

endohormone An unused term for a hormone that is active within the organism producing it, in contrast to an ectohormone.

endohydrolase *See* endonuclease; endopeptidase.

endolysin PHAGE LYSOZYME.

endomembrane system ENDOPLASMIC RETICULUM.

endometrium The mucous membrane that lines the uterus.

endomitosis A process in which the chromosomes in a nucleus replicate but the nucleus does not divide.

endonexin *See* calelectrin.

endonuclease An enzyme that catalyzes the hydrolysis of a polynucleotide strand at internal positions of the strand, rather than at the ends. *See also* restriction enzyme.

endopeptidase An enzyme that catalyzes the hydrolysis of a polypeptide chain at internal positions of the chain, rather than at the ends.

endoplasm The portion of the cytoplasm in the interior of the cell that is more fluid than that near the periphery of the cell. *See also* ectoplasm.

endoplasmic reticulum The cytoplasmic network that consists of cisternae, vesicles, and tubules, and that is frequently continuous with the cell membrane and the nuclear membrane. Part of the endoplasmic reticulum has ribosomes attached to it and serves as a transport system for the proteins synthesized on the adhering ribosomes. *Abbr* ER. *See also* RER; SER.

ENDOR Electron–nuclear double resonance.

endorphin A term that is derived from endogenous morphines and that, at first, was used to describe all of the opioid peptides. Currently, the term endorphin is applied only to those opioid peptides (such as α-, β-, and γ-endorphin) that are derived from the processing of proopiomelanocortin.

endoskeleton The internal skeleton of an organism, such as the bony skeleton of vertebrates.

endosmosis The osmotic movement of fluid toward the inside of a cell or a vessel.

endosome A vesicle, derived from the cell membrane, that can fuse with lysosomes; it is composed of many vacuolar and tubular elements and is probably identical to a receptosome. *Aka* CURL.

endosperm The nutritive tissue surrounding and nourishing the embryo in most flowering plants.

endospore A spore formed within a cell, as that formed in a bacterium.

endosymbiotic infection A viral infection in which the infected cells multiply for several generations even though they continue to release virus particles. The phenomenon applies particularly to animal viruses and is due to differences in the rates of cell and virus multiplication rather than to lysogeny.

endosymbiont hypothesis The hypothesis that eukaryotic cells started out in evolution without mitochondria or chloroplasts and later established endosymbiotic relationships with bacteria (that is, the bacteria penetrated the eukaryotic cells). These symbiotic bacteria are postulated to have developed into mitochondria and chloroplasts at a later stage in evolution.

endothelial cells The cells that form the lining of blood vessels.

endotherm A warm-blooded animal; a homoiothermic organism.

endothermic reaction A chemical reaction that requires the input of heat; a reaction with a positive enthalpy change.

endotoxin A toxic lipopolysaccharide that is released from the cell wall of gram-negative bacteria upon destruction of the cell by autolysis or by other means.

end plate The base plate of a T-even phage. *See also* T-even phage.

end point The experimental point in a titration that corresponds to the theoretical equivalence point.

end-point method A virus assay in which a fixed volume of a serially diluted virus sample is inoculated into a number of animals, cell cultures, or other test units.

end-point mutation A mutation that is expressed in an organism after a period of growth following exposure of the organisms to a mutagen.

end product The chemical compound that represents a final substance formed in a sequence of chemical reactions.

end-product inhibition FEEDBACK INHIBITION.

end product repressible Descriptive of a biosynthetic enzyme, the level of which is regulated in response to the presence or absence of the end product of the reaction.

end-product repression COORDINATE REPRESSION.

end-to-end conservative replication A mode of replication for double-stranded DNA (now considered obsolete) in which the parental molecule breaks into two halves so that each daughter molecule consists of one-half of the parental duplex and one-half of a newly synthesized duplex.

end-to-end distance The distance between the ends of a folded polymer as distinct from the contour length of the extended polymer. *See also* root-mean-square end-to-end distance.

end window counter A Geiger–Mueller counter in which either a thin window or a membrane separates the sample from the detector.

enediol An organic compound that contains two hydroxyl groups attached to the two carbon atoms of a double bond.

energetically coupled reactions *See* coupled reactions.

energized conformation ORTHODOX CONFORMATION.

energy The capacity to do work. *Sym* E.

energy barrier 1. The difference between the bond energy at which a molecule dissociates and the energy of the ground state of the molecule. 2. The difference between the ground state energy of the activated complex of a reaction and the sum of the ground state energies of the reactants. 3. A highly endergonic step in a metabolic pathway.

energy charge A measure of the availability of high-energy phosphate bonds in the ATP–ADP–AMP system. It is equal to the expression ($[ATP] + \frac{1}{2}[ADP]$)/($[AMP] + [ADP] + [ATP]$) and has a value of 1.0 if all of the adenine nucleotide is present as ATP, and a value of zero if all of the adenine nucleotide is present as AMP. The energy charge is the mole fraction of ATP or its equivalent in the total adenylate pool; ADP is considered to be half-charged since it contains only one high-energy bond while ATP contains two. *See also* phosphate potential.

energy coupling 1. The synthesis of ATP that is linked to the operation of the electron transport system. 2. COUPLED REACTIONS. 3. ENERGY TRANSDUCTION.

energy diagram 1. POTENTIAL ENERGY DIAGRAM. 2. The diagrammatic representation of the energy content of various states such as those of (a) reactants, activated complex, and products in a reaction, (b) the nuclear energy levels of an atom, and (c) the electronic energy levels of an atom.

energy dispersive spectrometry An analytical x-ray technique for the detection of elements in a sample; based on exciting the sample with a high-energy source and analyzing the emitted x-ray photons which are characteristic of the elements present in the sample.

energy fluence The energy carried by the photons in photon fluence; the energy fluence rate is the energy carried by the photons in photon fluence rate.

energy of activation *See* activation energy.

energy-poor compound LOW-ENERGY COMPOUND.

energy-regenerating system ATP REGENERATING SYSTEM.

energy-rich bond HIGH-ENERGY BOND.

energy-rich compound HIGH-ENERGY COMPOUND.

energy sink A molecule or a group of atoms in a molecule that readily accepts energy transferred to it from other components of the system.

energy transduction *See* transduction (2).

energy transfer The transfer of excitation energy from one chromophore or one molecule to another by a radiationless process. The energy may then be dissipated in a number of ways, such as through fluorescence. Energy transfer is strongly dependent on the distance between the chromophores and is useful for studying structural relations among groups of atoms in a macromolecule.

energy well *See* potential energy well.

engram A memory trace; a postulated protoplasmic change in neural tissue that accounts for the persistence of memory.

enhancement 1. The prolongation of the life of a transplant in a recipient by injection of killed tissue from the donor into the recipient prior to implantation of the transplant. 2. The increase in the rate of growth of a tumor that occurs in an animal that has been immunized with antigens of the tumor. 3. The increased binding of a monoclonal antibody to the epitope of a soluble antigen that is brought about by the binding of a second monoclonal antibody to a second epitope of the same antigen. The second antibody is termed the enhancing antibody.

enhancement effect The increase in photosynthetic efficiency (quantum yield) of chloroplasts that occurs when light of longer wavelengths (above 700 nm) is supplemented with light of shorter wavelengths (680 nm).

enhancer A base sequence in DNA that increases the rate of transcription of genes present on the same molecule but does not have promoter activity. An enhancer can be moved upstream, downstream, and to the other side of the promoter without significant loss of activity; it can also be spliced into the DNA of other cells. The distinguishing feature of an enhancer is the fact that it can be cut out from its site and reinserted in reverse orientation with respect to the promoter, without loss of activity.

enkephalin One of two pentapeptides that bind to opioid receptors in the brain; an opioid peptide. The enkephalins have the same amino acid sequence for the first four amino acids and differ only in their C-terminal amino acid. Met-enkephalin has the sequence Tyr-gly-gly-phe-met and Leu-enkephalin has the sequence Tyr-gly-gly-phe-leu. The enkephalin amino acid sequence also occurs in other compounds. Thus, Met-enkephalin occurs in β-lipotropin and Leu-enkephalin occurs in dynorphin.

enniatin *See* depsipeptide antibiotics.

enol The tautomer of a ketone in which the carbonyl group has been converted to an alcoholic hydroxy group that is attached to a double bond.

enolase The enzyme that catalyzes the formation of the high-energy compound, phosphoenolpyruvic acid, from 2-phosphoglyceric acid in glycolysis.

enol–keto tautomerism *See* keto–enol tautomerism.

enology The science of wine and wine making.

enriched food Food to which nutrients have been added after naturally occurring ones have been removed; enrichment does not replace all of the lost nutrients.

enriched medium MAXIMAL MEDIUM.

enrichment 1. PURIFICATION (2). 2. The selective growth of bacteria by means of an enrichment culture. 3. The increase in the concentration of a stable isotope above its natural abundance.

enrichment culture A culture used for the selection of specific bacterial strains from among a mixture; such a culture favors the growth of the desired bacteria under the conditions used.

enrichment medium A liquid selective medium of such a composition that it favors the growth of a specific bacterial strain over others in a mixed bacterial population.

entatic Pertaining to a chemical bond that is strained.

entatic site hypothesis The hypothesis that in reactions catalyzed by metalloenzymes, a metal ion–enzyme complex is formed prior to the formation of an enzyme–substrate or enzyme–inhibitor complex. The metal ion–enzyme complex is believed to be characterized by distorted geometries of the metal coordination sites so that the enzyme is in a state of tension, and the metal ion is in a state that approximates its transition state for the particular reaction. Such an activated metal ion–enzyme complex would serve to lower the activation energy of the reaction.

enteric Of, or pertaining to, the intestine.

enteric virus A virus, the target organ of which is the intestine.

enterobacteriaceae A family of gram-negative, facultatively anaerobic bacteria that are widespread as parasites or pathogens of humans, other animals, and plants; includes the intestinal bacteria *E. coli* and *Proteus vulgaris*.

enterobactin A cyclic siderophore of the phenol–catechol type that appears to be common to all enteric bacteria.

enterocrinine A gastrointestinal hormone, present in the intestinal mucosa, that controls the secretion of intestinal juice.

enterogastrones Substances produced in the gastrointestinal tract that inhibit the secretion of gastric acid and pepsin, and that inhibit gastric motility.

enteroglucagon A peptide or peptides, found in the small intestine, that cross-react with antiglucagon antisera; believed to contain the complete structure of glucagon and may be related to pancreatic precursor forms of glucagon. The actions of enteroglucagon are similar to, but less striking than, those of glucagon. *Aka* glucagon-like immunoreactivity.

enterohepatic circulation The circulatory system that connects the intestine and the liver by way of the bile and the portal blood. The system transports, for example, cholesterol and bile acids; these are excreted from the liver to the intestine by way of the bile and are reabsorbed from the intestine and returned to the liver by way of the portal blood.

enterokinase ENTEROPEPTIDASE.

enteropeptidase A proteolytic enzyme, secreted by the intestine, that catalyzes the activation of trypsinogen to trypsin. The enzyme was originally known as enterokinase.

enterotoxin An exotoxin which, when ingested or produced within the intestine (for example, by *Staphylococcus aureus*), is absorbed by the intestine and affects its function, inducing nausea, cramps, diarrhea, and vomiting.

enterovirus 1. A virus that belongs to a subgroup of picornaviruses which includes polio virus, coxsackievirus, and echovirus; enteroviruses infect the gastrointestinal tract of humans and some also cause acute respiratory infections. 2. ENTERIC VIRUS.

enthalpy Heat content; the thermodynamic function H in $H = E + PV$, where E is the internal energy of the system, P is the pressure exerted on the system, and V is the volume of the system.

enthalpy change The difference between the enthalpy of formation of the products and that of the reactants in a chemical reaction. *Sym* ΔH.

Entner–Doudoroff pathway An anaerobic catabolic pathway for the production of ATP from glucose that is found in some microorganisms, especially *Pseudomonas* species, which lack some of the glycolytic enzymes. The pathway entails the conversion of glucose to glyceraldehyde-3-phosphate by alternate reactions and the conversion of glyceraldehyde-3-phosphate to pyruvate by the regular reactions of glycolysis. *Abbr* ED pathway.

entoderm ENDODERM.

entomology The study of insects.

entropic doom The terminal equilibrium state of the universe at which the free energy is at a minimum and the entropy is at a maximum; this state is predicted on the basis of the first and second laws of thermodynamics which imply that the universe is progressing toward a state of increased randomness and disorder.

entropic strain The concept that a good substrate, when bound to an enzyme, is restricted in its rotation and is in a conformation that approximates its conformation in the activated complex. A poor substrate, on the other hand, is considered to be able to rotate more freely about critical bonds, so that it will achieve the correct conformation only occasionally and at random.

entropy The thermodynamic function that is a measure of that part of the energy of a system that cannot perform useful work; the degree of randomness or disorder of a system. *Sym* S. *See also* second law of thermodynamics.

entropy change The difference between the entropy of the products and that of the reactants in a chemical reaction. *Sym* ΔS.

entropy compensation The increase in entropy that may accompany the formation of an enzyme–substrate complex as a result of conformational changes of the enzyme.

entropy stabilization The stabilization of a structure due to an increase in entropy associated with the formation of the structure.

entropy trap A concept in enzymology according to which the binding of the substrate to the enzyme results in a loss of translational and rotational entropy of the substrate; the substrate becomes more ordered, less random, hence the decrease in entropy. This decrease helps to overcome the energy of activation barrier of the reaction.

entropy unit A unit of entropy equal to 1 cal/(deg)(mol). *Abbr* eu.

entry site AMINOACYL SITE.

enucleated Without a nucleus.

envelope 1. The two nuclear membranes of a eukaryotic cell. 2. The membrane surrounding some viruses and consisting of lipid, carbohydrate, and protein. 3. The bacterial cell membrane, cell wall, and capsule. 4. A specific conformation of the furanose ring of monosaccharides.

envelope conformation The conformation of furanoses in which the ring is not planar; four atoms of the ring, including the ring oxygen, are coplanar, but the fifth atom of the ring is outside this plane. *See also* twist conformation.

enveloped nucleocapsid A nucleocapsid surrounded by a membrane.

Enz Enzyme.

enzymatic Of, or pertaining to, enzymes.

enzymatic activity 1. The catalytic activity of an enzyme. 2. The rate of reaction of a substrate that may be attributed to catalysis by an enzyme and that is expressed in terms of katals.

enzymatic reversion DEADAPTATION.

enzyme A protein molecule, produced by living cells, that functions as a catalyst of biochemical reactions. The number and the types of reactions catalyzed by an enzyme are determined by the specificity of the enzyme. Enzymes are classified into the six main divisions of oxidoreductases, transferases, hydrolases, lyases, isomerases, and ligases. *Abbr* Enz; E. *See also* enzyme classification; ribozyme.

enzyme I A soluble bacterial enzyme that is part of the phosphotransferase system for the transport of sugars across the cell membrane. The enzyme catalyzes the reaction P-enolpyruvate + HPr $\rightarrow$ pyruvate + P-HPr, where HPr is a heat-stable, low molecular weight protein, and P designates phosphate.

enzyme II A membrane-bound bacterial enzyme that is part of the phosphotransferase system for the transport of sugars across the cell membrane. The enzyme catalyzes the reaction P-HPr + sugar $\rightarrow$ sugar-P + HPr, where HPr is a heat-stable, low molecular weight protein, and P designates phosphate. The enzyme is responsible for the specificity of the transport with respect to the sugar and functions in some systems in conjunction with another protein (factor III).

enzyme III A third enzyme required in the bacterial phosphotransferase system for the transport of some, but not all, sugars across the cell membrane.

enzyme adaptation ENZYME INDUCTION.

enzyme amplification A technique for enhancing the speed and sensitivity of enzyme assays; involves using the enzyme in a primary system to provide a trigger substance for a secondary system that can generate a large quantity of product. As an example, the enzyme of the primary system may generate NAD^+ which then feeds into a redox cycle (two coupled half-reactions) to generate a large amount of product. The technique is particularly useful for enhancing enzyme immunoassays, such as ELISA. In these systems, the immunoassay (primary system) is coupled to a secondary system that yields a large amount of colored product.

enzyme analogue SYNZYME.

enzyme assay The measurement of enzymatic activity that is based on a determination of either the rate or the extent of the formation of a product or the disappearance of a reactant.

enzyme-bridge complex *See* bridge complex.

enzyme cascade CASCADE MECHANISM.

enzyme classification The systematic arrangement and the naming of enzymes that is based on the 1972 recommendations of the Enzyme Commission of the International Union of Biochemistry. Each enzyme is denoted by a number composed of four figures. The first figure denotes one of the six main divisions: oxidoreductases, transferases, hydrolases, lyases, isomerases, ligases. The second figure denotes the subclass and the third figure denotes the sub-subclass. The last figure denotes the serial number of the enzyme in its sub-subclass. The enzyme number is preceded by the abbreviation E.C.

Enzyme Commission A special commission of the International Union of Biochemistry that made recommendations for the classification and naming of enzymes and for the definitions of the mathematical constants used in enzymology. The recommendation were first published in 1964 and were published in revised form in 1972, 1978, and 1984. *Abbr* E.C.

enzyme complex MULTIENZYME SYSTEM.

enzyme concentration *See* concentration of enzymatic activity.

enzyme conservation equation *See* conservation equation.

enzyme deletion hypothesis *See* deletion hypothesis.

enzyme electrode An electrode that incorporates an enzyme into its design and that is specific for measuring the concentration of a reactant or a product of the reaction catalyzed by the enzyme. The enzyme is frequently trapped within a gel matrix around the electrode or trapped within a liquid film in a cellophane membrane.

enzyme engineering *See* genetic engineering; enzyme therapy; biotechnology.

enzyme fractionation Protein fractionation applied to an enzyme preparation.

enzyme graph A graphical representation of an enzyme reaction in the form of a network; enzyme species are located at the nodes of the network and are connected by means of arrows. The latter indicate the direction of the reaction and are designated by the corresponding rate constants. *Aka* enzyme network.

enzyme immunoassay An immunoassay that in-

volves the use of an enzyme, specifically an assay in which antigens have been labeled with an enzyme. *Abbr* EIA. *See also* immunoenzymometric assay; enzyme amplification.

enzyme induction The process whereby an inducible enzyme is synthesized in response to an inducer. The inducer combines with a repressor and thereby prevents the blocking of an operator by the repressor. The operator controls the structural gene of the enzyme and the active, unblocked operator permits the transcription of that gene.

enzyme-inhibitor complex The complex that consists of an enzyme and an inhibitor that is bound either to the catalytic site of the enzyme or to some other site on the enzyme. *Abbr* EI.

enzyme interconversion *See* covalently modified enzyme.

enzyme intermediates ENZYME SPECIES.

enzyme kinetics The kinetics of enzyme-catalyzed reactions; includes derivations of rate equations and graphical analysis of experimental data for all types of enzyme reactions such as single or multiple substrate reactions, uninhibited or inhibited reactions, equilibrium- or steady-state systems.

enzyme labeling A method for locating antigens or antibodies in tissues; based on binding the antigen or the antibody to an enzyme and then determining the location of the enzyme in the tissues by making use of the known properties of the enzyme.

enzyme-linked immunosorbent assay An immunoenzymometric assay based on the use of antigens that have been adsorbed to a solid surface and antibodies that have been labeled with a specific enzyme; combines the virtues of solid phase technology and enzyme-labeled immunoreagents. The antigen–antibody complex is determined by means of an enzyme assay, involving incubation of the complex with an appropriate substrate of the enzyme. *Abbr* ELISA. *See also* enzyme amplification.

enzyme modulation *See* covalently modified enzyme.

enzyme multiplicity The occurrence of two or more forms of the same enzyme, all of which catalyze the same reaction.

enzyme multiplicity feedback inhibition Feedback inhibition in which two or more forms of an enzyme, all of which catalyze the same reaction, are inhibited to different degrees and by different end products.

enzyme network ENZYME GRAPH.

enzyme nomenclature *See* Enzyme Commission.

enzyme pH electrode An enzyme electrode that incorporates in its design a conventional glass electrode which is sensitive to hydrogen ions.

enzyme replacement therapy The treatment of a genetic disease by administration of the enzyme that is either missing entirely or present in defective form. Some approaches include delivery of the enzyme by coupling it to a carrier, use of an encapsulated enzyme, organ transplantation, and introduction of the required gene into recipient cells. *See also* enzyme therapy.

enzyme repression The process whereby the synthesis of a repressible enzyme is decreased in response to either a repressor or a repressor–corepressor complex. The repressor or the repressor–corepressor complex binds to and blocks an operator and thereby prevents the transcription of the structural gene of the enzyme which is controlled by that operator.

enzyme species All of the isomeric forms of an enzyme and all of the covalent and noncovalent complexes formed between an enzyme and a substrate and/or a product and/or an effector. *Aka* enzyme intermediates.

enzyme-specific electrode ENZYME ELECTRODE.

enzyme specificity *See* specificity (1).

enzyme–substrate complex The complex that consists of an enzyme and the substrate that is bound to the enzyme noncovalently at its catalytic site. *Abbr* ES.

enzyme–substrate compound The enzyme–substrate complex in which the substrate is covalently linked to the enzyme. *Aka* enzyme–substrate intermediate.

enzyme–substrate intermediate 1. The unstable intermediate, formed in some reactions, in which the substrate is linked covalently to the enzyme. *Aka* enzyme–substrate compound. 2. SUBSTITUTED ENZYME (1).

enzyme system MULTIENZYME SYSTEM.

enzyme therapy Any therapeutic approach that is based on the use of an enzyme; includes cancer chemotherapy, treatment of genetic diseases (see enzyme replacement therapy), dissolution of blood clots, and treatment of toxic reactions. *See also* gene therapy.

enzyme unit The amount of enzyme that, under defined conditions, will catalyze the transformation of 1 μmol of substrate per minute, or, where more than one bond of each substrate molecule is attacked, 1 μeq of the group concerned per minute. *Sym* U. *See also* katal.

enzyme variant *See* variant (1); multiple forms of an enzyme; isozyme.

enzymic ENZYMATIC.

enzymoblotting The blotting of enzymes to nitrocellulose paper and their subsequent detection by reactions with specific substrates. The method has been used for detection of proteinases and their zymogens. *See also* blotting.

enzymoelectrophoresis An electrophoretic technique for the detection and determination of isozymes by the combined use of electrophoresis and an enzyme assay. The technique is used specifically for the isozymes of lactate dehydrogenase which are first separated by gel electrophoresis and are then assayed by covering the gel with another gel that contains substrate and coenzyme; the paired gels are then scanned spectrophotometrically.

enzymology The study of enzymes and enzyme-catalyzed reactions.

enzymolysis Hydrolysis by means of enzymes.

enzymopathy A disturbance of enzyme function, including the genetic deficiency of a specific enzyme.

eobiogenesis The first instance of the formation of living matter from nonliving matter.

eobiont A primitive prototype of a living cell.

Eoff process The formation of glycerol from dihydroxyacetone phosphate by yeast under alkaline conditions.

EOP Efficiency of plating.

eosin Tetrabromofluorescein; a red dye and fluorochrome.

eosinophil *See* polymorphonuclear leukocyte.

eosome A fundamental molecule of the ribosome; a molecule of ribosomal RNA or a molecule of ribosomal protein.

EP Enzyme–product complex.

Epa antigens Mouse alloantigens that are specific to epidermal cells.

epidemic Of, or pertaining to, a disease that has widespread incidence that is significantly above the endemic level.

epidemic hepatitis *See* hepatitis.

epidemiology The study of diseases in populations.

epidermal growth factor A polypeptide mitogen (MW 6400) that stimulates the proliferation of epidermal and epithelial tissues. *Abbr* EGF. *Aka* epithelial growth factor. *See also* urogastrone.

epidermal growth factor-urogastrone Name given to human epidermal growth factor which appears to be identical to urogastrone.

epigenesis The concept that an organism develops through the appearance and diversification of structures and functions that are not present in the egg. *See also* preformation.

epigenetics The study of the mechanisms that result in the expression of the phenotypic effects of genes during differentiation and development.

epimer One of two optical isomers that differ from each other only in the configuration about one asymmetric carbon atom (the epimeric carbon).

epimerase An enzyme that catalyzes the interconversion between two optical isomers, each of which contains only one asymmetric center. *See also* racemase.

epimeric carbon The asymmetric carbon atom of two optical isomers (epimers) that differ from each other only in the configuration about this carbon atom.

epimerization The formation of one epimer from another.

epinephrine A catecholamine hormone that is secreted by the adrenal medulla. Epinephrine raises the level of blood sugar by increasing the breakdown of glycogen, stimulates the mobilization of free fatty acids, and has various effects on the cardiovascular system and on muscular tissue. *Aka* adrenaline.

epiphysis PINEAL GLAND.

episome A plasmid that can become integrated reversibly into the chromosome of the bacterial host; phage lambda and the F factor are two examples. A plasmid may behave as an episome (replicating with the bacterial chromosome) in one cell and as a regular plasmid (replicating independently of the bacterial chromosome) in another cell.

epistasis The interaction between nonallelic genes such that one gene interferes with, or prevents, the expression of the other gene.

epistatic gene A gene, the expression of which suppresses or reduces the expression of another, nonallelic gene.

epithelial body PARATHYROID GLAND.

epithelial cell A cell of the epithelium.

epithelial growth factor EPIDERMAL GROWTH FACTOR.

epithelioma A malignant tumor derived from epithelial cells.

epithelium The sheet of cells, consisting of one or more layers, that covers surfaces and lines tubes of animal tissue.

epitope ANTIGENIC DETERMINANT.

epitype A family of related epitopes.

EPO Erythropoietin.

epoxide A cyclic, 3-membered ether; a compound containing the structure below. Epoxides are highly reactive compounds as a result of the strain in the ring; they have been implicated as mutagens, cytotoxins, and teratogens.

$$-\underset{\diagdown\diagup}{\underset{O}{C-C}}-$$

EPPS N-(2-Hydroxyethyl)piperazine-N'-3-propanesulfonic acid; used for the preparation of biological buffers in the pH range of 7.3 to 8.7. *Aka* HEPPS. *See also* biological buffers.

EPR Electron paramagnetic resonance.

epsilon Denoting the fifth carbon atom from the carbon atom that carries the principal functional group of the molecule. *Sym* ϵ.

epsilon chain The heavy chain of the IgE immunoglobulins.

Epstein–Barr virus A herpesvirus that is the causal agent of infectious mononucleosis and that has been implicated in the etiology of Burkitt's lymphoma. *Abbr* EBV.

eq Equivalent.

Equal *See* aspartame.

equation of state An equation that relates the volume, pressure, temperature, and mass of a substance.

equator The line in an x-ray diffraction pattern that passes through the zero point and that is perpendicular to the axis of the fiber that is being studied; the direction that is perpendicular to the meridian and that bisects the film when the latter is considered to be wrapped cylindrically about the fiber axis. *Aka* zero layer line.

equatorial bond A bond in a molecule having a ring structure that is roughly in the plane of the ring.

equatorial reflection An x-ray diffraction spot that lies on the equator.

equatorial substituent A substituent attached to an equatorial bond.

equilibrium (*pl* equilibria) The state of a system in which no further change is occurring and in which the free energy is at a minimum. At equilibrium, the rate of the forward reaction is equal to the rate of the reverse reaction so that a small change in one direction is balanced by a small change in the opposite direction. *See also* steady state.

equilibrium banding DENSITY GRADIENT SEDIMENTATION EQUILIBRIUM.

equilibrium constant The constant K that is characteristic of a given chemical reaction at a specified temperature and that is based on the activities of all of the reactants and all of the products at equilibrium. The equilibrium constant is generally written using molar concentrations, rather than activities (thus, actually an apparent equilibrium constant); for the reaction $aA + bB \rightleftharpoons cC + dD$, the equilibrium constant is given by

$$\frac{[C]^c[D]^d}{[A]^a[B]^b}$$

Sym K; K_{eq} *See also* apparent equilibrium constant.

equilibrium dialysis A method for measuring the binding of low molecular weight ligands to macromolecules. The macromolecules are placed inside a dialysis bag, the ligands are placed outside the bag, and dialysis is allowed to proceed until equilibrium is established. From the known concentration of macromolecules, the expected distribution of ligands across the dialysis membrane in the absence of binding, and the initial and final concentrations of ligand on either side of the dialysis membrane, it is possible to calculate the average number of ligand molecules that are bound per macromolecule. The method permits a determination of the number of binding sites per macromolecule for a given ligand and a determination of the intrinsic association constant that governs that binding. *See also* forced dialysis.

equilibrium potential   The membrane potential that arises from the differences in the concentrations of ions across a membrane and that exactly balances the tendency of these ions to diffuse from the more concentrated to the more dilute solution.

equilibrium thermodynamics   The branch of thermodynamics that deals with changes between equilibrium states of closed systems; that change from one equilibrium state to another may be either a reversible or an irreversible process.

equimolar   Containing an equal number of moles and, hence, an equal number of molecules.

equine   Of, or pertaining to, horses.

equivalence point   1. The point in a titration where a chemically equivalent amount of a compound has been added to the compound being titrated. 2. EQUIVALENCE ZONE.

equivalence rules   BASE PAIRING RULES.

equivalence zone   A zone in the precipitin curve of the antigen–antibody reaction in which maximum precipitation of the antigen–antibody complex occurs.

equivalent   GRAM-EQUIVALENT WEIGHT.

equivalent chain length   The number of carbon atoms, generally a nonintegral number, that is derived for an organic compound on the basis of its retention time in gas chromatography. The number is obtained by interpolation of a semilogarithmic plot of retention time versus the number of carbon atoms in the chain for a series of saturated, straight-chain organic compounds. *Abbr* ECL.

equivalent ellipsoid of revolution   The ellipsoid of revolution that has the same volume as the actual hydrodynamic unit in the solution; the hydrodynamic unit consists of the macromolecule together with tightly bound solvent.

equivalent height of a theoretical plate   *See* height equivalent to a theoretical plate.

equivalent thickness   The thickness, in centimeters, of an absorbing material that is equivalent to a thickness of 1 cm of air with respect to absorption of alpha particles, multi-

plied by the density of the absorbing material in grams per cubic centimeter.

ER   Endoplasmic reticulum.

E_1 reaction   A unimolecular elimination reaction closely analogous to an S_N1 reaction.

E_2 reaction   A bimolecular elimination reaction closely analogous to an S_N2 reaction.

eRF   Eukaryotic release factor; *see* release factor.

erg   A unit of work that is equal to the work done when a force of 1 dyne acts through a distance of 1 cm; 1 erg $= 10^{-7}$ J.

ergastoplasm   ROUGH-SURFACED ENDOPLASMIC RETICULUM.

ergocalciferol   A compound that has vitamin D activity and that is obtained by ultraviolet irradiation of ergosterol. Previously called calciferol and denoted vitamin D_2.

ergosome   POLYSOME.

ergosterol   A sterol that yields ergocalciferol (calciferol) upon irradiation with ultraviolet light.

Ergot alkaloid   One of a group of indole alkaloids obtained from fungi of the genus *Claviceps*; includes lysergic acid derivatives and clavine derivatives. Ergot alkaloids stimulate smooth muscle contraction and are used to control hemorrhage after childbirth; some specific compounds are ergotamine, ergotoxine, and ergometrine.

erosion model   RELIC MODEL.

error   The difference between a measured value of a quantity and either the true or the expected value.

error catastrophe   A model for biological aging according to which there is a decrease in the fidelity of protein synthesis with aging. As a result, errors in protein synthesis increase and accumulate with time to a point where cell death ensues.

error curve   NORMAL DISTRIBUTION.

error function   NORMAL DISTRIBUTION.

error of the first kind   TYPE I ERROR.

error of the second kind   TYPE II ERROR.

error-prone repair   SOS REPAIR.

error theory   A theory of aging according to which aging is due to the occurrence of errors in the biosynthesis of proteins; the accumulation of the partially active, or the inactive, proteins that are produced in this fashion then leads to the death of cells. *See also* error catastrophe.

erythorbic acid   ISOASCORBIC ACID.

erythroblastosis fetalis   Hemolytic disease of the newborn; *see* Rh blood group system.

erythrocin   ERYTHROMYCIN.

erythro configuration   The configuration of a compound in which two asymmetric carbon atoms have identical substituents on the same side, as is the case in erythrose.

erythrocruorin A hemoglobin-like respiratory pigment of invertebrates that has a molecular weight of about 4×10^5 to 6.7×10^6 and contains 30 to 400 heme groups per molecule.

erythrocyte A mature red blood cell that is no longer engaged in the synthesis of hemoglobin and that derives its energy primarily from anaerobic glycolysis and from the phosphogluconate pathway.

erythrocyte ghost *See* ghost (1).

erythrocyte maturation factor VITAMIN B_{12}.

erythrocytosis POLYCYTHEMIA.

erythromycin A macrolide antibiotic, produced by *Streptomyces erythreus*, that binds to the 50S ribosomal subunit and prevents formation of the 70S ribosome in prokaryotic protein synthesis.

erythron The sum of the circulating erythrocytes and the total mass of erythropoietic cells from which erythrocytes are derived.

erythropoiesis The formation of red blood cells.

erythropoietic Of, or pertaining to, erythropoiesis.

erythropoietic porphyria A congenital porphyria that is caused by an excessive formation of heme precursors in the developing red blood cells of the bone marrow; due to a deficiency of the enzyme ferrochelatase.

erythropoietin A glycoprotein mitogen and hormone (MW 23,000) that is produced by the kidneys and that stimulates the formation of erythrocytes. *Abbr* EPO.

erythrose A four-carbon aldose that is an intermediate in the phosphogluconate pathway.

erythrosome An artificially prepared subcellular fraction consisting of red blood cell ghosts that are cross-linked with glutaraldehyde and coated with phospholipids.

ES Enzyme–substrate complex.

escape synthesis The constitutive synthesis of a group of enzymes, coded for by bacterial genes closely linked to the prophage genome, that results from the induction of prophage lambda in *E. coli*.

Escherichia coli A bacterium that is normally present in the intestine and that is widely used in biochemical and genetic research. *Abbr E. coli. Aka* colon bacillus.

ESDIAD Electron-stimulated desorption ion angular distribution.

ESE Electron spin-echo spectroscopy.

ESI Enzyme–substrate–inhibitor complex.

E site A postulated third site on the ribosome (in addition to the A and P sites) that is introduced into the 70S ribosome by the 50S subunit. The site is specific for deacylated tRNA and characterized by relatively weak and reversible binding; hence, it is regarded as an "exit" site.

ESP Abbreviation for eIF-2 stimulating protein; *see* hemin-controlled repressor.

ESR 1. Electron spin resonance. 2. Erythrocyte sedimentation rate.

essential amino acid An amino acid that is required by an organism for normal growth and functioning, but that cannot be synthesized by the organism; such an amino acid must be obtained through the diet. Leucine, isoleucine, lysine, methionine, phenylalanine, threonine, tryptophan, and valine are essential amino acids for positive nitrogen balance in adult humans. *Abbr* EAA.

essential amino acid index The chemical score of the essential amino acids of a protein. *Abbr* EAA index.

essential enzyme An enzyme without which a cell or an organism cannot grow or survive.

essential fatty acid A fatty acid that is required by an organism for normal growth and functioning, but that cannot be synthesized by the organism; such a fatty acid must be obtained through the diet. The polyunsaturated fatty acids, linoleic acid and linolenic acid, are essential fatty acids in humans. *Abbr* EFA.

essential gene A gene that codes for an essential enzyme.

essential hypertension Hypertension for which no cause can be found; hypertension without preexisting renal disease or other pathological conditions.

essential metabolite A metabolite required for the growth of cells or the growth of an organism.

essential nutrient A nutrient that is required by an organism for normal growth and functioning, but that cannot be synthesized by the organism; such a nutrient must be obtained through the diet.

essential oil An oil that is produced by a plant and that has a characteristic odor and flavor. Essential oils are rich in terpenes and in oxygenated compounds such as alcohols, aldehydes, ketones, acids, and esters; the oxygenated compounds are most responsible for the odors and flavors of the essential oils. Some essential oils, obtained by extraction with organic solvents, contain waxes and paraffins which make them insoluble in dilute alcohol. These are called concrete oils. Removal of the insoluble components from concrete oils yields absolute oils.

established cell line A cell line that appears to be capable of unlimited in vitro propagation. The cell culture can be subcultured indefinitely; the cells grow and divide indefinitely. *Aka* established cell strain. *See also* crisis (2).

established tissue culture A long-term tissue culture.

established tumor A tumor that has been transplanted and that has been allowed to grow in the new host.

ester An organic compound formed from an alcohol and an acid by splitting out a molecule of water; the water is formed from the H of the alcohol and from the OH of the acid.

esterase An enzyme that catalyzes the hydrolysis of an ester.

esterification The formation of an ester.

ester interchange An interesterification reaction in which two esters react with each other to produce two new esters.

esterolytic protease A proteolytic enzyme that catalyzes the hydrolysis of ester bonds in appropriate substrates at a faster rate than it catalyzes the hydrolysis of natural peptide bonds.

ester value The number of milligrams of potassium hydroxide that are required to saponify the esters in one gram of a fat or an oil; identical to the saponification number if the sample does not contain free fatty acids.

estolide An intermolecular lactone of hydroxy fatty acids.

17-β-estradiol A steroid hormone, secreted by the ovaries, that is a major sex hormone and that is responsible for regulation of feminine characteristics and for regulation of the menstrual–ovulatory cycle. *Aka* β-estradiol; estradiol.

estrane The parent ring system of the estrogens.

estrogen An 18-carbon steroid that is a female sex hormone or one of its metabolites; the major estrogens are estrone and 17-β-estradiol. *See also* female sex hormone.

estrone A major female sex hormone.

estrophilin An estrogen receptor protein in the cytosol of target tissues; also designated as estrophilin I in contrast to the estrogen–receptor complex which is designated as estrophilin II.

estrous cycle The sequence of endocrine-related events from the beginning of one estrus to the beginning of the next.

estrus 1. The period of reproductive activity in an animal. 2. ESTROUS CYCLE.

Et Ethyl group.

ETC Electron transport chain.

ETF Electron-transfer flavoprotein.

ether 1. An organic compound derived from an alcohol by replacing the hydrogen of the hydroxyl group with an organic radical. 2. Diethyl ether; ethyl ether.

ethereal sulfate One of a group of phenolic, sulfur-containing compounds that are excreted in the urine.

etherification The formation of an ether.

ethidium bromide A planar, fluorescent molecule that intercalates with DNA and thereby decreases its density. Covalently closed circles cannot bind as much ethidium bromide as other DNA forms (linear or nicked circles) and therefore show a smaller decrease in density. Consequently, they can be separated from the other DNA forms by density gradient centrifugation.

ethionine An amino acid analogue that can be incorporated into protein during protein synthesis.

ethology The scientific study of animal behavior, particularly, the study of the biological processes underlying animal behavior.

ethylenediaminetetraacetic acid ETHYLENEDINIT-ROLOTETRAACETIC ACID.

ethylenedinitrolotetraacetic acid A chelating agent. *Abbr* EDTA.

ethylenimine An aziridine mutagen.

N-ethylmaleimide A reagent for sulfhydryl groups in proteins. *Abbr* NEM.

ethylmethane sulfonate A mutagenic, alkylating agent that commonly reacts with guanine. In that reaction, an ethyl group is linked to guanine so that the latter hydrogen bonds with thymine, resulting in a transition-type mutation. *Abbr* EMS.

etiology The study of causes, specifically the causes of disease. British: aetiology.

ETP Electron transport particle.

ETS Electron transport system.

ETS particle Electron transport system particle; *see* electron transport particle.

eu 1. Enzyme unit; the abbreviation U is now preferred. 2. Entropy unit.

eubacteria A term used originally to denote "true" bacteria as opposed to other microorganisms; now used to designate all bacteria other than archaebacteria.

eucaryon The nucleus of a eukaryotic cell.

eucaryote Variant spelling of eukaryote.

euchromatin Chromatin, or chromosome regions, having normal staining properties that are characteristic of the bulk of the chromosomal complement. Euchromatin is relatively uncoiled and stains weakly during interphase but is condensed and stains strongly during metaphase.

euchrysine ACRIDINE ORANGE.

eugenic Related to, or aimed at improving, the genetic qualities of a race or a breed. *See also* dysgenic.

eugenics The science that deals with the improvement of the genetic qualities of a race or a breed through alteration of its genetic makeup.

Euglena A genus of unicellular photosynthetic flagellates that are used for genetic studies and that are classified as either green algae or protozoa.

euglobulin A water-insoluble globulin.

eukaryon Variant spelling of eucaryon.

eukaryote A higher organism (unicellular or multicellular) that consists of cells that possess a true nucleus; the nucleus is surrounded by a nuclear membrane and contains the genetic material within multiple chromosomes. *See also* prokaryote.

eukaryotic Of, or pertaining to, eukaryotes.

eumelanin One of a group of brown or black pigments that are widespread in skin, hair, and feathers of animals and in cuticles of insects; they are cross-linked polymers derived from tyrosine.

euploid state The state of a cell or an organism having a chromosome number that is an exact multiple of the haploid number. *Aka* euploidy.

euroxic Descriptive of organisms exhibiting a wide range of oxygen tolerance.

euthenics The science that deals with the improvement of the human race through control and alteration of its physical, biological, and social environments.

eutrophication The process whereby a body of water becomes deficient in oxygen.

eutrophic lake A shallow lake, having a depth of 10 m or less, that is murky due to dense growth of planktonic algae and that has a high rate of nutrient supply in relation to its volume of water. In such a lake, both plant biomass and productivity are high and the bottom layers of the lake frequently contain low concentrations of dissolved oxygen at the end of the summer. *See also* mesotrophic lake; oligotrophic lake.

eV Electron volt.

evagination An outpocketing.

Eve *See* mitochondrial Eve.

evenly labeled UNIFORMLY LABELED.

even-numbered fatty acid A fatty acid molecule that has an even number of carbon atoms; refers to the total number of carbon atoms, those in the hydrocarbon portion of the molecule plus that in the carboxyl group. *Aka* even-carbon fatty acid.

eversible Capable of being turned inside out.

eversion theory A theory of aging according to which aging is due to changes in the structures of macromolecules.

everted sac technique A technique for studying the absorption of substances across the intestinal wall by means of in vitro experiments performed on small segments of the intestine, turned inside out.

evocation The process whereby a morphogenetic effect is brought about by an evocator.

evocator A chemical substance that is morphogenetically active and that is emitted by an organizer.

evolution 1. DARWINIAN EVOLUTION. 2. The process of continuous change by which something complex develops from something simpler. 3. The liberation of a gas during a chemical reaction.

evolutionary clock The rate at which mutations accumulate in a specific gene.

evolutionary tree PHYLOGENETIC TREE.

evolve To undergo evolution.

exa- Combining form meaning 10^{18} and used with metric units of measurement. *Sym* E.

EXAFS Extended x-ray absorption fine structure.

exchange diffusion The passive transport of two solutes across a biological membrane such that one solute moves in one direction while an equimolar amount of the second solute moves in the opposite direction. A form of antiport; the movement of two substances, in opposite directions across a membrane as a result of simple diffusion.

excimer An excited dimer; a complex between an excited molecule and an identical, but unexcited, molecule.

excinuclease A nuclease that makes two nicks in a polynucleotide strand, thereby allowing excision of a specific segment.

excision 1. The enzymatic removal of a nucleotide segment, particularly a thymine dimer, from a nucleic acid strand. 2. Removal of the prophage from the host bacterial DNA (prophage excision); requires the action of two enzymes, integrase and excisionase.

excisionase An enzyme required for the excision of the prophage from the bacterial chromosome; it works in conjunction with the enzyme integrase. *Abbr* Xis. *Aka* integration host factor (IHF).

excision repair A repair mechanism for damaged DNA. First, the damaged nucleotide segment of a DNA strand is removed by two nuclease-catalyzed incisions on either side of the damaged segment. Second, the correct segment is synthesized by means of DNA polymerase with the second strand of the DNA serving as a template. Last, the newly synthesized segment is joined to the existing strand by means of DNA ligase. This mechanism occurs, for example, in *E. coli* and serves to remove and repair lesions containing thymine dimers. *Aka* cut and patch repair. *See also* patch and cut repair.

excitability The capacity of living matter to respond to an external stimulus.

excitation 1. The transition of an atom or a molecule from a lower to a higher energy level as that brought about by the raising of electrons from lower to higher energy orbitals. 2. The process whereby an external stimulus brings about changes in living matter, such as the changes in a muscle fiber that are initiated by a nerve impulse.

excitatory autacoid *See* autacoid.

excited state Any state of a nucleus, an atom, or a molecule that is of a higher energy level than that of the ground state.

exciton A quantum of excitation energy; a photon.

exciton transfer RESONANCE ENERGY TRANSFER.

excluded site binding An extreme case of anticooperative binding in which the binding of a ligand to one site excludes the binding of other ligands at the nearest adjacent sites. Such a binding model can account for the intercalation of a dye with DNA when a maximum of one dye molecule is intercalated per two base pairs (4 nucleotides); that is, the dye fills every other potential site in the DNA. *See also* nearest-neighbor cooperative model.

excluded volume 1. COVOLUME. 2. The volume of solvent that is required to elute a component in gel filtration when that component is neither adsorbed to, nor permeates into, the gel bed.

exclusion chromatography GEL FILTRATION CHROMATOGRAPHY.

exclusion diagram RAMACHANDRAN PLOT.

exclusion limit The molecular weight of the largest molecules of a particular shape that can be effectively fractionated by a specific gel in gel filtration. Molecules of similar shape but higher molecular weights will not penetrate the gel particles and will move right through the column without being fractionated.

exclusion limit chromatography GEL FILTRATION CHROMATOGRAPHY.

exclusion reaction The prevention of further phage infection of a phage-infected bacterium that is brought about by strengthening the cell envelope when the phage-infected cell is heated.

exergonic reaction A chemical reaction that is accompanied by a release of energy; a downhill reaction with a negative free energy change.

exhaustion theory A theory of aging according to which aging is due to the exhaustion of an essential nutrient.

exhaustive methylation Maximal methylation; in carbohydrate chemistry this refers to the formation of a methyl ether at every free hydroxyl group on the carbohydrate. The hydrolysis of an exhaustively methylated carbohydrate, followed by separation, identification, and quantification of the components, is used as an aid in determining the structure of the carbohydrate.

EX₁ mechanism A type of deuterium exchange in which the rate-limiting step is the opening up of the protein structure.

EX₂ mechanism A type of deuterium exchange in which the rate-limiting step is the exchange reaction between hydrogen and deuterium atoms.

exo- Combining form meaning outside or external.

exobiology The study of extraterrestrial life.

exocellular Descriptive of an enzyme or a structure that is external to the cell membrane but still attached to it.

exo conformation *See* puckered conformation.

exocrine gland A gland of external secretion that discharges its secretion by means of a duct.

exocyclic Pertaining to an atom or a bond that occurs in a ring compound but that does not constitute part of the ring structure itself.

exocytosis The process whereby fluids and particles are discharged from a cell; involves the endoplasmic reticulum, the Golgi apparatus, and secretory granules.

exoenzyme 1. An enzyme that catalyzes a reaction involving the end of a polymer rather than internal positions. 2. An enzyme that is released outside the cell.

exoergic reaction EXERGONIC REACTION.

exogenote The genetic fragment of the donor cell that is transferred to a recipient cell in bacterial transformation. *See also* endogenote; merozygote; exosome.

exogenous Originating outside the organism.

exogenous metabolism The level of metabolism due to added nutrients.

exogenous virus A virus that undergoes vegetative replication but is not subject to vertical transmission via the gametes.

exohydrolase *See* exonuclease; exopeptidase.

exon A coding sequence in the DNA of eukaryotic genes. Such sequences are transcribed into RNA and are subsequently translated into protein. The term is also used for the translatable RNA sequence. Exons and introns (nontranslated, intervening sequences) make up split genes.

exon shuffling The linking together of DNA fragments, which serve as exons, by means of intron-mediated recombination. The DNA fragments can be genes that, previously, specified different proteins or different segments of the same protein. The linked DNA fragments constitute a new gene that codes for a new protein. Exon shuffling, intron intrusion, and junctional sliding have been proposed to help explain the evolutionary diversification of genes.

exonuclease An enzyme that catalyzes the sequential hydrolysis of nucleotides from one end of the polynucleotide strand. *See also* 3′ → 5′ exonuclease; 5′ → 3′ exonuclease; proofreading function.

3′ → 5′ exonuclease PROOFREADING FUNCTION.

5′ → 3′ exonuclease An exonuclease activity associated with DNA polymerases I and III. The catalytic site for this enzyme is distinct from the polymerization center on the enzyme molecule. The 5′ → 3′ exonuclease activity removes nucleotides from the 5′-end of a polynucleotide strand; it thus moves in the same direction as the polymerase activity. The main function of this exonuclease activity is to remove the RNA primers in the discontinuous replication of DNA. *See also* nick translation.

exonuclease III An exonuclease from *E. coli* that catalyzes the removal of nucleotides from the 3′-ends of both strands in duplex DNA.

exonuclease IV An enzyme that catalyzes the degradation of single-stranded DNA to oligonucleotides; it initiates hydrolysis at both the 3′- and 5′- ends of the DNA.

exonuclease V A complex, multifunctional, enzyme in *E. coli* that catalyzes the ATP-dependent hydrolysis of both linear duplex and single-stranded DNA. The enzyme also has helicase activity.

exopeptidase An enzyme that catalyzes the sequential hydrolysis of amino acids from one end of the polypeptide chain.

exophthalmic goiter GRAVE'S DISEASE.

exophthalmos-producing substance An uncharacterized substance, thought to be a hormone that functions in the production of bulging eyes.

exorphins Opioid peptides present in partial enzymatic digests of proteins derived from food stuffs. These peptides are called exorphins because of their exogenous origin and their morphine-like activities.

exoskeleton A protective external covering, such as the scales of fish or the horny structure of insects and crustaceans.

exosmosis The osmotic movement of fluid toward the outside of a cell or a vessel.

exosome A genetic fragment that is transferred to a recipient cell in bacterial transformation and that is not readily integrated into the recipient chromosome but can remain unintegrated and can replicate, be transcribed, and express biochemical function in this state. *See also* endogenote.

exospore A spore formed by budding, as that generally formed by a fungus.

exosporium A loose outer layer that frequently covers a spore.

exothermic reaction A chemical reaction that is accompanied by a release of heat; a reaction with a negative enthalpy change.

exotoxin A toxic protein that is discharged from a bacterial cell. Exotoxins are generally produced by gram-positive bacteria and can be demonstrated in a bacterial culture in which no appreciable autolysis has occurred.

experimental error The error that results from the inability to reproduce precisely the experimental conditions in an otherwise carefully and accurately performed analysis.

expire To exhale.

explant The tissue or organ fragment that is excised and used to start an in vitro culture in tissue culture experiments.

exponential A function obtained by raising the constant e to a power; the exponential of x is e^x, where e is the base of natural logarithms and x is a variable.

exponential curve A curve that is described by an equation of the form $Y = ab^x$.

exponential decay The mode of radioactive decay that can be described by the equation $N = N_0 e^{-\lambda t}$, where N is the number of radioactive atoms present at time t, N_0 is the number of radioactive atoms originally present, e is the base of natural logarithms, and λ is the decay constant.

exponential density gradient A density gradient in which the density increases exponentially from one end of the gradient to the other. It is described by the equation $c = a + b\ e^{-dV}$, where V is the volume of the gradient, c is the concentration, e is the base of natural logarithms, and a, b, and d are constants. *See also* concave exponential gradient; convex exponential gradient.

exponential growth The growth of cells in which the number of cells (or the cell mass) increases exponentially and the growth at any time is proportional to the number of cells (or the cell mass) present.

exponential growth rate constant The reciprocal of the doubling time, expressed as the number of generations per hour.

exponential phase The phase of growth of a bacterial culture in which the number of cells (or the cell mass) increases exponentially so that a plot of the logarithm of the number of cells (or the cell mass) as a function of time yields a straight line.

exponential survival curve A survival curve that indicates an exponential loss of active units as a function of increasing dose. Such data will yield a straight line when they are replotted in terms of the logarithm of the surviving fraction as a function of the dose.

export *See* protein export.

exposure The rate of irradiation per unit area, perpendicular to the beam of radiation, and multiplied by the time interval of irradiation.

exposure dose The intensity of a dose that is based on the ionizations produced by the radiation in air; generally expressed in terms of roentgens. *See also* radiation absorbed dose.

expression vector A cloning vehicle that promotes the expression of a gene inserted into it; typically, a restriction fragment that carries a regulatory sequence for the particular gene.

expressor protein A protein, produced by a regulator gene, that is required for the expression of one or more other genes.

exsanguinate To drain of blood.

extant Living or existing at the present time, as opposed to extinct.

extended x-ray absorption fine structure A method for obtaining highly accurate structural information of biological samples. Involves measurements of the absorption of x rays by an amorphous solid or a solution as a function of wavelength at energies just above the absorption transition of a particular atom. Under these conditions, a series of rapid oscillations in absorbance is observed due to interference by neighboring atoms. An analysis of these oscillations provides information on the number, types, and distances of neighboring atoms. *Abbr* EXAFS.

extensin A protein, rich in hydroxyproline, that is attached to cellulose fibrils in plant cell walls.

extension peptides Polypeptide segments at both the amino and carboxyl termini of procollagen molecules that are removed prior to conversion of procollagen to collagen.

extensive property A property of a system, such as volume or energy, that can be defined only by specifying the precise amounts of all of the substances involved. *See also* intensive property.

external indicator An indicator that is outside the titration vessel and to which is added a drop of the liquid that is being titrated.

external monooxygenase A monooxygenase in which the cosubstrate that incorporates the second oxygen atom is not itself a product of the reaction.

external quenching The stoppage of secondary and subsequent ionizations in an ionization detector that is caused by a momentary reduction in the applied potential.

external respiration *See* respiration (1).

external-sample scintillation counter A scintillation counter in which radiation from an external source interacts with a solid fluor that is coupled to a photomultiplier.

external standard A standard that is treated separately from the sample.

external suppression INTERGENIC SUPPRESSION.

extinct Not living or existing at the present times, as opposed to extant.

extinction ABSORBANCE.

extinction angle The angle between the plane of polarization and the cross of isocline in flow birefringence.

extinction coefficient ABSORPTIVITY.

extinction coefficient $\epsilon(P)$ A molar absorptivity used for nucleotides and nucleic acids, the concentration of which is expressed in terms of moles of phosphorus per liter; specifically. $\epsilon(P) = A/cd$, where $\epsilon(P)$ is the extinction coefficient, A is the absorbance, c is the concentration in terms of gram-atoms of phosphorus per liter, and d is the length of the lightpath in centimeters.

extinction dilution The dilution of a bacterial or a viral sample to such an extent that the diluted preparation is no longer infectious.

extinction point The critical oxygen tension for a green plant below which aerobic fermentation occurs as part of the metabolism of the plant, and above which aerobic respiration occurs exclusively.

extra- Prefix meaning outside.

extra arm *See* arm.

extracellular enzyme EXOENZYME.

extracellular titer The titer of phage particles that is obtained after the infected cells are removed by centrifugation; a measure of the phage particles that have been released by the host cells. *See also* intracellular titer.

extrachromosomal inheritance CYTOPLASMIC INHERITANCE.

extracistronic suppression INTERGENIC SUPPRESSION.

extract 1. CRUDE EXTRACT. 2. A preparation that serves as a source of vitamins and coenzymes for microbiological media and that is prepared from meat, yeast, or other materials by destruction of the cells.

extraction The removal of a substance from a solid or a liquid mixture by dissolving it in a solvent.

extraction ratio The fraction of a substance that is reabsorbed by the kidneys from the glomerular filtrate.

extragenic suppression INTERGENIC SUPPRESSION.

extranuclear inheritance CYTOPLASMIC INHERITANCE.

extrapolation The process of extending a graph from a region containing experimental data to one that is devoid of data.

extrapolation number The extrapolated value for a multitarget survival curve that may or may not correspond to the actual number of sensitive targets per irradiated unit.

extremophile A bacterium that grows under extreme environmental conditions; a thermophile and a psychrophile are two examples.

extrinsic blood coagulation EXTRINSIC PATHWAY.

extrinsic Cotton effect A Cotton effect that is caused by a small molecule which is bound to the protein, and not by the protein itself.

extrinsic factor VITAMIN B_{12}.

extrinsic fluorescence Fluorescence that is caused by a small molecule which is bound to the protein, and not by the protein itself. *See also* autofluorescence.

extrinsic heterogeneity The heterogeneity of antibodies that results from extrinsic factors, such as the large number of antigenic determinants in any particular antigen.

extrinsic pathway The series of reactions in blood clotting that involve factors normally present in the circulation plus tissue factors, with the latter serving as initiators of the sequence of reactions. The pathway proceeds in the form of a cascade mechanism, with each stage involving protease activation of a zymogen to its enzymatically active form. Major steps of the pathway are as follows (Roman numerals indicate factor numbers; the subscript "a" indicates the active form of the factor):

The sequence from factor X_a on is common to both the extrinsic and the intrinsic pathways. The common names of the factors are listed under factor. *Aka* extrinsic system.

extrinsic proteins PERIPHERAL PROTEINS.

extrinsic system EXTRINSIC PATHWAY.

extrinsic thromboplastin THROMBOPLASTIN.

extrude To remove material by extrusion.

extrusion The process of expelling or thrusting out, as the expelling of an adsorbent from a chromatographic column.

extrusome A vacuole, excreted by parasitic protists, that contains substances used for penetration of the host cells.

eye structure 1. The DNA structure formed by two replicating forks moving in opposite directions (bidirectional replication) in linear double-stranded DNA, as is the case for eukaryotic DNA. 2. D-LOOP.

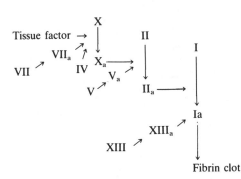

F

f 1. Frictional coefficient. 2. Femto.

f₀ The frictional coefficient of a sphere that has the same volume as the macromolecule being studied.

F 1. Degree Fahrenheit. 2. Fertility factor. 3. Faraday. 4. Farad. 5. Fluorine. 6. Fick unit. 7. Force. 8. Phenylalanine. 9. Folic acid. 10. Formality.

F⁺ A male, or donor, bacterial strain.

F⁻ A female, or recipient, bacterial strain.

F₀ *See* F_0F_1-ATPase.

F′ *See* F′ factor.

F₁ *See* F_0F_1-ATPase.

FA 1. Fatty acid. 2. Folic acid. 3. Filtrable agent.

Fab fragment The fragment of the IgG immunoglobulin molecule that is obtained by treating the molecule with the enzyme papain. The fragment consists of an intact light chain and the Fd fragment of one heavy chain, held together by means of a disulfide bond. Two Fab fragments are obtained per IgG molecule, and each fragment contains one antigen binding site.

Fab′ fragment The fragment of the IgG immunoglobulin molecule that is obtained by treating the molecule with the enzyme pepsin, followed by reduction. The fragment consists of an intact light chain and the Fd′ fragment of one heavy chain, held together by means of a disulfide bond. Two Fab′ fragments are obtained per IgG molecule, and each fragment contains one antigen binding site. An Fab′ fragment is slightly larger than an Fab fragment.

F(ab′)₂ fragment The fragment of the IgG immunoglobulin molecule that is obtained by treating the molecule with the enzyme pepsin without subsequent reduction. The fragment is a dimer of two Fab′ fragments, held together by means of two disulfide bonds. The F(ab′)₂ fragment contains both of the antigen binding sites of the molecule.

FAB-MS Fast atom bombardment mass spectrometry.

Fabry's disease A genetically inherited metabolic defect in humans that is associated with kidney failure and that is due to a deficiency of the enzyme trihexosyl ceramide α-galactosylhydrolase.

facilitated diffusion MEDIATED TRANSPORT.

facilitation The phenomenon that a second nerve impulse will often be transmitted through a synapse more effectively than the first; the phenomenon may be related to the nature of memory.

facilitator TRANSPORT AGENT.

F-actin The polymerized, fibrous form of actin that consists of a double helix of G-actin monomers.

factor A component that is not yet completely identified; frequently the term is retained after the factor has been fully identified.

factor I FIBRINOGEN.

factor II PHOTHROMBIN.

factor III 1. THROMBOPLASTIN. 2. A protein that functions in conjunction with enzyme II in some systems.

factor IV The calcium ions that function in blood clotting.

factor V PROACCELERIN.

factor VI Factor Va (accelerin) was once called factor VI but, according to current terminology, there is no factor VI among the blood clotting factors.

factor VII PROCONVERTIN.

factor VIII ANTIHEMOPHILIC FACTOR.

factor IX CHRISTMAS FACTOR.

factor X STUART FACTOR.

factor XI PLASMA THROMBOPLASTIN ANTECEDENT.

factor XII HAGEMAN FACTOR.

factor XIII FIBRIN STABILIZING FACTOR.

factor XIV Protein C; a zymogen which, when activated by thrombin, inactivates Factors V and VIII in the extrinsic and intrinsic pathways of blood clotting. Factor XIV is the inactive precursor of a serine protease.

factor F INITIATION FACTOR.

factor G TRANSLOCASE (2).

factorial A function of a positive integer n that is denoted $n!$ and that is equal to the product of all the integers between 1 and n; thus, $5! = 5 \times 4 \times 3 \times 2 \times 1 = 120$.

factor IF INITIATION FACTOR.

factor R RELEASE FACTOR.

factor T ELONGATION FACTOR T.

factor theory The theory according to which blood clotting proceeds by a cascade mechanism, involving a large number of components.

factor X 1. VITAMIN B_{12}. 2. BIOTIN.

factor Y PYRIDOXINE.

factory A cytoplasmic region in poxvirus-infected animal cells that is actively engaged in the replication of the viral DNA.

facultative Capable of living under more than one set of conditions.

facultative aerobe FACULTATIVE ANAEROBE.

facultative anaerobe An organism or a cell that can grow either in the absence or in the presence of molecular oxygen; the cell or the organism is aerobic but has the ability to grow anaerobically.

facultative water excretion The urinary excretion of water that is greater than that required for the elimination of waste products.

FAD Flavin adenine dinucleotide.

FADH$_2$ Reduced flavin adenine dinucleotide.

F agent FERTILITY FACTOR.

Fahrenheit temperature scale A temperature scale on which the freezing and boiling points of water at 1 atm of pressure are set at 32 and 212, respectively, and the interval between these two points is divided into 180 degrees.

fall curve The decrease in color intensity of a sample solution as a function of time; used in reference to determinations with an autoanalyzer.

falling sphere viscometer An instrument for measuring the viscosity of a solution from the time required for a sphere to fall through a known column height of the solution. *Aka* falling ball viscometer.

fallout The radioactive substances that are produced by nuclear explosions and that fall through the atmosphere onto the earth's surface.

familial Transmitted in families.

familial goiter A genetically inherited metabolic defect in humans that is characterized by an excessive loss of iodinated tyrosines from the thyroid gland and that is due to a deficiency of the enzyme iodotyrosine dehalogenase.

familial high-density lipoprotein deficiency TANGIER DISEASE.

familial hypercholesterolemia A genetically inherited metabolic defect in humans that is characterized by elevated levels of plasma cholesterol and by accelerated atherosclerosis; it is caused by a deficiency of functional low-density lipoprotein (LDL) receptors on cells and surfaces. Due to this deficiency, the entry of cholesterol into cells is impaired, intracellular feedback on cholesterol biosynthesis is lacking, and cholesterol synthesis in the liver and other tissues is excessive. The genetic defects are of three types: (a) lack of receptors for LDL and hence no binding of LDL; (b) decreased binding of LDL to the receptors; and (c) normal binding of LDL to the receptors but no internalization of LDL.

familial lysosomal lipase deficiency WOLMAN'S DISEASE.

familial methemoglobinemia METHEMOGLOBINEMIA.

family antigen An antigen that is common to a group of viruses that constitute a family.

Fanconi's anemia A genetically inherited metabolic defect in humans involving defective DNA repair and characterized by chromosomal aberrations and a decrease of all types of white blood cells.

Fanconi's syndrome A genetically inherited metabolic defect in humans that is characterized by an increased excretion of amino acids.

farad A unit of electrical capacitance, equal to the capacitance of a capacitor that requires 1 C to raise its potential by 1 V. *Abbr* fd.

faraday The quantity of electricity that is transported per gram-equivalent-weight of an ion; 96,500 C/eq; 23,060 cal/(V)(eq). *Sym* F.

Faraday effect The exhibition of optical rotatory power by an optically inactive substance that is placed in a magnetic field.

Farber's disease A genetically inherited metabolic defect in humans that is due to a deficiency of the enzyme ceramidase. A sphingolipidosis that is characterized by respiratory difficulty, arthritis, and painful movement. *Aka* Farber's lipogranulomatosis.

farnoquinone Vitamin K$_2$.

far-red drop The decrease in photosynthetic activity that occurs in the far red region (above 700 nm) despite the fact that the light is still absorbed. The decrease can be reversed by supplementing the long wavelength light with light of shorter wavelengths (enhancement effect).

Farr test A radioimmunoassay for determining the antigen binding capacity of an antiserum. The antiserum is treated with radioactively labeled antigen, and the labeled antigen–antibody complex is precipitated with 50% saturated ammonium sulfate; the radioactivity of the precipitate is then determined.

FAS system Fatty acid synthetase system.

fast-assembly end *See* actin filament.

fast atom bombardment mass spectrometry A mass spectrometric technique (similar to SIMS) that involves the transfer of kinetic energy from a beam of highly energetic atoms, such as argon or xenon atoms, to a matrix, such as glycerol, and then to the sample. The sample need not be volatile and does not require prior derivatization. Nonvolatile polar substances usually give intense peaks in the molecular ion region and a high intensity of fragment ions can be maintained for prolonged periods of time. *Abbr* FAB-MS.

fast component 1. The sample component that, in electrophoresis, moves the greatest distance in a given time. 2. The fraction of DNA that renatures first in reassociation kinetics; it consists of highly repetitive DNA.

fast hemoglobin A hemoglobin that, after electrophoresis, is located closer to the anode than is normal, adult hemoglobin.

fast protein, peptide, and polynucleotide liquid chromatography A modified form of HPLC in which high pressures are replaced by high flow rates. Separation techniques frequently involve ion exchange chromatography and are capable of both high capacity and high resolution. A variety of chromatographic support materials can be used. *Abbr* FPLC.

fast reaction A reaction, or a step in a reaction sequence, that has a very large rate constant and that proceeds very rapidly; special techniques are required for analysis of such reactions.

fast-sweep polarography A sensitive polarographic technique in which a hanging mercury drop electrode is used.

fast-twitch muscle WHITE MUSCLE.

fat 1. NEUTRAL FAT. 2. The oily and greasy material of adipose tissue.

fat body A structure that contains storage fat and that occurs in insects, reptiles, and amphibians.

fat-soluble A A fraction of fat-soluble vitamins prepared from egg yolk.

fat-soluble vitamin One of a diverse group of vitamins, including vitamins A, D, E, and K, that is soluble in organic solvents and insoluble in water.

fat solvent A nonpolar organic solvent, such as chloroform, that will extract lipids from tissues.

fat-splitting The hydrolysis of fats to free fatty acids that is produced by water at elevated temperatures.

fatty acid A long-chain carboxylic acid that occurs in lipids; may be branched or unbranched, and saturated or unsaturated.

fatty acid activating enzyme THIOKINASE.

fatty acid activation The conversion of a fatty acid to a fatty acyl coenzyme A ester which is the first step in the reactions of beta oxidation. The fatty acyl coenzyme A ester can be formed in a reaction catalyzed by a thiokinase or in a reaction catalyzed by a thiophorase.

fatty acid CoA ligase THIOKINASE.

fatty acid oxidation 1. BETA OXIDATION. 2. Any set of reactions whereby a fatty acid is oxidized in metabolism.

fatty acid synthetase system A cytoplasmic, multienzyme complex that catalyzes a cyclic set of reactions whereby a fatty acid is synthe-sized entirely from acetyl-SCoA; the complex consists of six enzymes and the acyl carrier protein. The synthesis proceeds by condensation of acetyl-acyl carrier protein (acetyl-ACP), formed from acetyl-SCoA, with successive molecules of malonyl-ACP, also formed from acetyl-SCoA. *Abbr* FAS system.

fatty acid thiokinase THIOKINASE.

fatty acyl group The acyl group of a fatty acid; the grouping

fatty alcohol One of a group of long-chain, aliphatic, monohydroxy alcohols that contain 10–20 carbon atoms and that occur as components of waxes.

fatty degeneration The degeneration of a tissue due to the formation of fat globules in the cytoplasm of the affected cells; the fat of these globules is derived from the cells themselves. *See also* fatty infiltration.

fatty infiltration The degeneration of a tissue due to the formation of fat globules in the cytoplasm of the affected cells; the fat of these globules is derived from outside the cells. *See also* fatty degeneration.

fatty liver A liver that is characterized by fatty degeneration and/or fatty infiltration and that may develop due to various conditions such as diabetes, chemical poisoning of the liver, or a deficiency of lipotropic agents.

favism A severe type of hemolytic anemia that occurs in individuals suffering from a hereditary glucose-6-phosphate dehydrogenase deficiency upon eating broad beans.

FB elements Feedback elements

FBP Fructose-1,6-bisphosphate.

FCCP Carbonylcyanide *p*-trifluoromethoxy phenylhydrazone; an uncoupling agent.

Fc fragment The fragment of the IgG immunoglobulin molecule that is obtained by treating the molecule with the enzyme papain. The fragment consists of two heavy chain fragments joined by means of two disulfide bonds.

Fc receptor A structure on the surface of many cells of the immune system to which the Fc fragment of the immunoglobulin molecule becomes bound.

fd Farad.

Fd Ferredoxin.

FDA Food and Drug Administration; an agency of the U.S. Public Health Service.

Fd fragment The fragment of the IgG immunoglobulin molecule that is obtained by treating the molecule with the enzyme papain, followed by reduction. The fragment consists of that portion of the heavy chain that is joined to an intact light chain in the Fab fragment. Two Fd fragments are obtained per molecule of IgG.

Fd′ fragment The fragment of the IgG immunoglobulin molecule that is obtained by treating the molecule with the enzyme pepsin, followed by reduction. The fragment consists of that portion of the heavy chain that is joined to an intact light chain in the Fab′ fragment. Two Fd′ fragments are obtained per molecule of IgG.

FD-MS Field desorption mass spectrometry.

FDNB 1-Fluoro-2,4-dinitrobenzene; the Sanger reagent.

FDP Fructose-1,6-diphosphate now designated fructose-1,6-bisphosphate.

F-duction SEXDUCTION.

Fe Iron.

FEBS Federation of European Biochemical Societies.

feedback 1. A process in which part of the output of a system is returned to the input of the same system in such a fashion that it affects the subsequent output by the system. *See also* negative feedback; positive feedback. 2. That part of the output of a system that is returned to the input in a feedback process.

feedback deletion hypothesis A modification of the deletion hypothesis of cancer according to which cancer results from the loss of a repression mechanism that restrains DNA synthesis in normal cells. The loss of the repression mechanism may be due to a number of factors such as lack of repressor synthesis or segregation of the repressor in a part of the cell.

feedback inhibition A negative feedback mechanism in which a distal product of an enzymatic reaction inhibits the activity of an enzyme that functions in the synthesis of this product. Feedback inhibition may be concerted, cumulative, or sequential. *Aka* poisoning.

feedback loop The cyclic system of components that participate in a feedback mechanism.

feedback repression A negative feedback mechanism in which a metabolite regulates the amount of an enzyme through enzyme repression.

feeder cell An irradiated cell that is metabolically active but that cannot multiply; such cells are used to provide growth factors for unirradiated cells.

feeder layer A layer of feeder cells that is used in tissue culture.

feeder pathway A metabolic pathway that provides metabolites for another, major pathway.

feed-forward activation The activation of an enzyme by means of an initial reactant of the reaction; the activation of aspartate transcarbamoylase by ATP and the activation of glycogen synthetase by glucose-6-phosphate are two examples.

Fehling's test A test for reducing sugars that is based on the reduction of cupric ions by a reducing sugar when an alkaline solution of the sugar is treated with copper sulfate.

Felderstruktur The structure of muscle fibers that is characterized by an incomplete separation of the thick myofibrils.

feline Of, or pertaining to, cats.

female hormone Any one of the three related compounds, estradiol, estrone, and estriol.

female protein VITELLOGENIN.

female sex hormone An estrogen that affects the estrous cycle, the reproductive cycle, and the development of secondary sex characteristics in the female; the female sex hormones are produced primarily by the ovaries and the placenta.

FeMo protein *See* nitrogenase.

femto- Combining form meaning 10^{-15} and used with metric units of measurement. *Sym* f.

Fenton chemistry The generation of powerful oxidizing species from the reduction of peroxides by metal ions; includes, but not limited to, reactions brought about by Fenton's reagent.

Fenton's reagent A ferrous ion-hydrogen peroxide mixture; a strong oxidizing agent that can oxidize most organic compounds.

F episome FERTILITY FACTOR.

Fe protein *See* nitrogenase.

Ferdinand model A model, proposed by Ferdinand, according to which cooperative interactions can be ascribed to kinetic considerations. Specifically, the model applies to a random single displacement mechanism of a bisubstrate enzyme reaction.

Ferguson equation An equation that relates the mobility of a protein in polyacrylamide gel electrophoresis to the concentration of the gel. Specifically, $\log R = -kT + \log R_0$, where R is the relative mobility of the protein, R_0 is the relative mobility of a small molecule that encounters no sieving effect, k is the retardation coefficient, and T is the total gel concentration (acrylamide + cross-linker).

Ferguson plot A plot of the relative electrophoretic mobility of a protein in gel electrophoresis as a function of the concentration of the gel. From such a plot, the parameters k and R_0 of the Ferguson equation can be determined.

ferment 1. *n* An early term for enzyme. 2. *v* To process by means of fermentation.

fermentation 1. The energy-yielding, metabolic

breakdown of organic compounds by microorganisms that generally proceeds under anaerobic conditions and with the evolution of gas. 2. The energy-yielding, metabolic breakdown of organic compounds in an organism that proceeds in the absence of molecular oxygen and with the use of organic compounds both as oxidizing agents and as oxidizable substrates. *See also* anaerobic respiration.

fermentor An apparatus for the growth of microorganisms in liquid media and for the study of microbial metabolism, including fermentation. *Var sp* fermenter.

ferredoxin One of a group of low molecular weight, nonheme, iron–sulfur proteins that have low reduction potentials and that serve as early electron acceptors in both photosynthesis and nitrogen fixation. *Abbr* Fd. *See also* spinach ferredoxin.

ferredoxin–NADP-oxidoreductase A flavoprotein that accepts electrons from ferredoxin and transfers them to NADP (thereby reducing it to NADPH) in the operation of photosystem I of chloroplasts. *Aka* ferredoxin–NADP reductase.

ferredoxin reducing substance An electron carrier, possibly a chlorophyll or a quinone, that is the immediate acceptor for the electrons from pigment P_{700} in photosystem I of chloroplasts; the compound transfers an electron to ferredoxin, thereby reducing it. *Abbr* FRS.

ferric Designating iron that has a valence of three; Fe^{3+}; Fe(III).

ferrichrome A cyclic hexapeptide, composed of three residues each of glycine and hydroxyornithine, that is complexed with a ferric atom and occurs in fungi; a siderophore of the hydroxamic acid type.

ferricytochrome A cytochrome in which the iron is in the ferric form.

ferriheme A heme in which the iron is in the ferric form.

ferrihemochrome A hemochrome in which the iron is in the ferric form.

ferrimycin A sideromycin produced by *Actinomyces.*

ferrioxamine A sideramine produced by *Streptomyces.*

ferriphore SIDEROPHORE.

ferriprotoporphyrin A protoheme in which the iron is in the ferric form.

ferritin A conjugated and electron-dense protein that is composed of a protein shell, apoferritin, which surrounds four discrete micelles of ferric hydroxide–phosphate salts. Ferritin functions in the absorption of iron through the intestinal mucosa and serves as a storage protein for iron in the liver, the spleen, and other animal tissues. *See also* ferritin-labeled antibody.

ferritin-labeled antibody An antibody to which ferritin has been linked covalently; the electron-dense ferritin provides a suitable label for localizing the antibody in electron microscope preparations. Antigens or other proteins may be labeled with ferritin in an analogous manner.

ferrochelatase The enzyme that catalyzes the final step in the biosynthesis of heme whereby an iron atom (Fe^{2+}) is inserted into the ring system of protoporphyrin IX. *Aka* heme synthetase.

ferrocytochrome A cytochrome in which the iron is in the ferrous form.

ferroflavoprotein A complex flavoprotein that contains iron in the form of a heme or in some other form, in addition to containing either FMN or FAD.

ferroheme A heme in which the iron is in the ferrous form.

ferrohemochrome A hemochrome in which the iron is in the ferrous form.

ferroprotoporphyrin A protoheme in which the iron is in the ferrous form; ferroprotoporphyrin IX is the prosthetic group of hemoglobin, myoglobin, catalase, peroxidase, and cytochrome *b*.

ferrous Designating iron that as a valence of two; Fe^{2+}; Fe(II).

ferrous wheel mechanism A postulated mechanism for the enzyme aconitase according to which a molecule of *cis*-aconitate is bound to the enzyme at three sites and is bound to the essential Fe(II) at the active site. The stereospecific addition of water to *cis*-aconitate, to form either citrate or isocitrate, is achieved by a partial rotation of this ferrous wheel.

ferroxidase A serum enzyme that catalyzes the oxidation of iron from the ferrous to the ferric form. Ferroxidase I is also known as ceruloplasmin; ferroxidase II appears to be the major serum component responsible for the oxidation of iron.

fertility factor The bacterial episome that enables a bacterium to function as a male in bacterial conjugation; the male bacterium serves as a donor of DNA and produces an F-pilus through which the DNA is transferred to the recipient female cell. *Sym* F.

fertility vitamin VITAMIN E.

FeS Iron–sulfur protein.

Fe-S protein Iron–sulfur protein.

fetal Of, or pertaining to, the fetus.

fetal hemoglobin The hemoglobin that occurs during the development of the fetus and that diminishes rapidly after birth; consists of two alpha and two gamma chains. *Sym* HbF.

α-fetoprotein *See* alpha fetoprotein.

fetuin A glycoprotein that constitutes about

50% of the plasma proteins in the bovine fetus; it is absent in the adult animal.

fetus The unborn or unhatched young vertebrate during the later stages of its development. The developing human is considered to be a fetus after the beginning of the third month of pregnancy and an embryo prior to that. *See also* embryo.

Feulgen reaction A staining reaction that is specific for DNA and that is based on converting the DNA to apurinic acid and treating the apurinic acid with Schiff's reagent for aldehydes.

fever blisters HERPES SIMPLEX.

f/f₀ Frictional ratio; f is the frictional coefficient of the molecule being studied and f_0 is the frictional coefficient of a sphere that has the same volume as the molecule.

FFA Free fatty acids.

F factor FERTILITY FACTOR.

F′ factor An augmented F factor; a fertility factor that has an additional chromosomal segment attached to it; it contains genes obtained from the bacterial chromosome in addition to the plasmid genes.

F₀F₁-ATPase A proton-translocating ATPase that functions in mitochondria, chloroplasts, and bacteria; consists of two oligomeric components, designated F_0 and F_1. The F_0 component is embedded in the lipid bilayer of the membrane and serves as a proton channel; the F_1 component carries the ATPase active site and leads to synthesis of ATP. It is also known as mitochondrial ATPase or coupling factor 1. *See also* H⁺-ATPase.

FFF Field flow fractionation.

(F₁ + F₂) fragment HEAVY MEROMYOSIN.

F₁ fragment The globular head portion of the myosin molecule.

F₂ fragment The central tail fragment of the myosin molecule.

F₃ fragment LIGHT MEROMYOSIN.

FGAR Formylglycinamide ribonucleotide; an intermediate in the biosynthesis of purines.

F-genote strain F′ STRAIN.

FGF Fibroblast growth factor.

FH₂ Dihydrofolic acid.

FH₄ Tetrahydrofolic acid.

FIA Fluoroimmunoassay.

fiber 1. A threadlike structure, as that of a muscle or a nerve, that consists of bundles of fibrils. 2. DIETARY FIBER.

fiber diagram The x-ray diffraction pattern that a fibrous material yields when it is analyzed by the rotating crystal method.

fiber optics probe A flexible probe consisting of fine glass or of fine plastic fibers that are optically aligned to transmit an image and/or to transmit light.

fibril A fine threadlike structure, bundles of

which constitute a fiber; the myofibrils of a muscle and the neurofibrils of a nerve are two examples.

fibril ghost A myofibril that has lost its myosin.

Fibrillenstruktur The structure of muscle fibers that is characterized by a relatively uniform distribution of well-separated myofibrils.

fibrils long spacing An artificially prepared assembly of collagen molecules that is characterized by having long periodicities of about 2500 Å when examined in the electron microscope; produced from neutral solutions of collagen in the presence of either glycoproteins or excess mucopolysaccharides. *Abbr* FLS. *Aka* fibrous long spacing. *See also* segment long spacing.

fibrin The protein that is formed from fibrinogen by the action of thrombin and that is polymerized to form the blood clot; the monomeric, soluble form is denoted fibrin-s, and the polymerized, insoluble form is denoted fibrin-i. *See also* hard clot; soft clot.

fibrinase FIBRIN-STABILIZING FACTOR.

fibrin monomer The monomeric fibrin that is formed from fibrinogen by the action of the enzyme thrombin.

fibrinogen The serum protein that gives rise to two fibrinopeptides and fibrin when it is acted upon by the enzyme thrombin during blood clotting; the fibrin is subsequently polymerized to form the blood clot.

fibrinoligane FACTOR XIII.

fibrinolysin PLASMIN.

fibrinolysis The dissolution of fibrin in blood clots, brought about by the hydrolytic action of plasmin.

fibrinopeptide One of two peptides, denoted A and B, that are removed from fibrinogen during its conversion to fibrin by the action of thrombin.

fibrin polymer The polymerized fibrin molecules that form the basis of the blood clot.

fibrin-stabilizing factor The zymogen of a plasma transglutaminase which covalently links fibrin monomers of the soft clot to form a hard clot during blood clotting; the enzyme catalyzes the joining of the γ-carboxyl groups of glutamic acid residues to the ε-amino groups of lysine residues. *Aka* fibrinase.

fibroblast A specialized cell of connective tissue which synthesizes fibrous proteins such as collagen. Human fibroblasts have a definite life span in culture that is somewhat related to the potential life expectancy of the individual from whom the cells are obtained.

fibroblast growth factors Two polypeptides (MW 13,400) that are endothelial cell growth factors (mitogens), bind heparin, and promote angiogenesis. *Abbr* FGF.

fibroin *See* silk fibroin.

fibroma A benign tumor derived from fibrous connective tissue.

fibronectin A large peripheral membrane protein that occurs at the surface of many animal cells; a major noncollagenous glycoprotein in the extracellular matrix and basement membranes. Fibronectin has strong adhesive properties and mediates the attachment of fibroblasts to collagen; it is believed to function in cell–cell interaction (contact inhibition) and cell–substratum adhesion. Fibronectin is either lost or greatly reduced in many cells upon malignant transformation. *Aka* LETS protein. *See also* cold-insoluble globulin.

fibrosome An artificial liposome-like vesicle, prepared by layering purified human plasma fibronectin molecules on an agar-coated substrate.

fibrous Consisting either lwggely, or entirely, of fibers.

fibrous actin F-ACTIN.

fibrous lamina A thick filamentous layer that reinforces the inner membrane of the nucleus of many cells.

fibrous long spacing FIBRILS LONG SPACING.

fibrous protein A protein in which the polypeptide chains are either extended or coiled in one dimension. The polypeptide chains in such a protein are held together largely by interchain hydrogen bonds and form sheets or fibers. Fibrous proteins serve principally as structural proteins. *See also* globular protein.

FICA technique Fluoroimmunocytoadherence; a method for the identification, enumeration, and isolation of antigen-binding cells. Based on column chromatographic fractionation of cell suspensions using antigen-coated Sephadex beads.

ficin An endopeptidase of broad specificity.

Fick's first law The law that relates the flow of material in translational diffusion to the concentration gradient and that may be expressed as $dm/dt = -DA(dc/dx)$, where dm is the mass transferred in time dt through a cross-sectional area A, D is the diffusion coefficient, and dc/dx is the concentration gradient. The negative sign indicates that the transfer of material occurs from a region of higher to one of lower concentration.

Fick's second law The law that relates the flow of material in translational diffusion to the concentration gradient and to the change of this gradient with time. The law may be expressed as a differential equation, and the equation can be solved for the concentration gradient; this yields a normal distribution which describes the diffusion pattern.

Fick unit A unit of the diffusion coefficient equal to 10^{-7} cm^2/s. *Sym* F.

ficoll A synthetic water-soluble copolymer of sucrose and epichlorohydrin that has a weight average molecular weight of about 400,000 and that is used for the preparation of density gradients.

FID Flame ionization detector.

fidelity The degree to which replication, transcription, or translation proceeds without errors.

fidelity group A group of four proteins (designated S4, S5, S11, and S12) in the 30S subunit of *E. coli* ribosomes that are involved in the streptomycin response of the ribosome and that affect the fidelity of translation.

field desorption mass spectrometry A mass spectrometric technique in which the sample does not require prior derivatization since it is being ionized concurrently with its desorption from an activated emitter wire in a high electric field; the heated sample is desorbed as an ion from the surface of one of the electrodes. The mass spectrum is dominated by ions representing the intact molecule; hence, the method is useful for molecular weight determinations. *Abbr* FD-MS.

field effect ELECTROSTATIC INTERACTIONS.

field flow fractionation A group of techniques for the analysis of high molecular weight polymers and particles in suspension. The essence of all such techniques involves the application of a force field (thermal, flow, electrical, or centrifugal) across a long, flat, ribbon-like channel containing a liquid sample. The polymers or particles become layered toward one wall, due to the applied force field. A liquid is then passed through the channel and the separated components are eluted in the manner of chromatography. The method has been termed "one-phase chromatography" since the applied field takes over the role played by the stationary phase in ordinary chromatography. *Abbr* FFF.

field inversion gel electrophoresis A technique of agarose gel electrophoresis in which the applied electric field is periodically inverted. This causes the molecules to make, on an alternating basis, large movements forward and small ones backward. Under these conditions, large DNA molecules can be separated from other ones of similar size that, in standard gel electrophoresis, all migrate at the same rate. *Abbr* FIGE. *See also* Reptation; *PFG*.

field ionization mass spectrometry A mass spectrometric technique that requires volatilization of the sample (generally in the form of derivatives) prior to ionization. The latter is achieved by placing the volatilized sample in a strong electric field between two electrodes. *Abbr* FI-MS.

field ion microscope A microscope for the analysis of surface structures in which the sample serves as the source of radiation, in the form of ions. As a result, no lenses are required. An imaging gas, usually an inert gas, is allowed to cover the sample which is a single crystal held in an electric field exceeding 10^8 V/cm. The enhanced potential of the atoms at the surface of the sample leads to ionization of the gas atoms. These ions then travel to a fluorescent screen and their emission pattern provides information about the arrangement of surface atoms in the sample. *See also* scanning tunneling microscope.

FIGE Field inversion gel electrophoresis.

fight hormones Epinephrine and norepinephrine.

figure eight An intermediate in genetic recombination that consists of two connected, double-stranded, circular DNA molecules.

figure of merit A measure of the ability of a liquid scintillation fluid to permit counting of materials that either are not miscible with the organic solvent system used or highly quench the system; defined as $(E)(H)$, where E is the efficiency of counting and H is the percentage of the total sample that is aqueous. The greater the figure of merit, the better the fluid fulfills the above requirements.

filaggrin One of a group of cationic structural proteins that associate specifically with intermediate filaments and not with other types of cytoskeletal proteins.

filament 1. A very thin, threadlike structure. 2. MYOFILAMENT.

filamentous phage *See* minute phage.

filamin A long, fibrous protein with a molecular weight of about 500,000. It consists of two subunits of identical size and links actin filaments together to form a random, filamentous, three-dimensional network. Filamin occurs in many cells, including smooth muscle cells and fibroblasts.

filiform Filamentous; threadlike.

film badge A photographic film holder that is worn by an individual and that is used for approximate measurements of the radiation to which the individual has been exposed.

film diffusion The rapid diffusion of ions in an ion-exchange resin such that the exchange rate is controlled by the speed at which the ions diffuse through the solution surrounding the resin particles.

filter 1. A porous material used for the collection of either a precipitate or suspended matter. 2. A light-absorbing material that transmits only selected wavelengths of light.

filterable agent An early designation of a phage that mediates transduction in *Salmonella*.

filterable virus An early term for virus.

filter affinity transfer A method for the in situ characterization of proteins in polyacrylamide gels. Based on coupling an affinity ligand of the protein to a chemically derivatized and activated cellulose filter paper which is then overlaid on the gel containing the separated proteins. Proteins that interact with the affinity ligand are transferred from the gel to the paper.

filter fluorometer A fluorometer in which filters are used to select the desired excitation and emission wavelengths.

filter hybridization The hybridization of nucleic acids that is carried out on a nitrocellulose filter; only hybrid, double-stranded, molecules remain on the filter after washing. *See also* liquid hybridization.

filter paper chromatography *See* paper chromatography.

filter photometer A photometer in which filters are used for the isolation of bandwidths.

filtrable Variant spelling of filterable.

filtrable calcium FREE CALCIUM.

filtrate The liquid that has passed through a filter.

filtration enrichment A method for the isolation of fungal auxotrophs by growing either wild-type or mutagenized fungal spores on a minimal medium. Since on this medium only normal spores will germinate and develop mycelia, they can be removed by filtration, and the remaining auxotrophic spores can then be supplied with an enriched medium to permit their germination and growth.

filtration fraction The fraction of plasma that is filtered through the glomeruli of the kidney and that is equal to the glomerular filtration rate divided by the renal plasma flow; frequently taken as being equal to the ratio of inulin clearance to *p*-aminohippuric acid clearance.

FIM Field ion microscopy.

fimbrin An actin-binding protein.

FI-MS Field ionization mass spectrometry.

fine control The control of biochemical systems that is achieved by regulating the activity of an enzyme, as in the case of allosteric enzymes.

fines Very small particles; a very finely divided solid.

fine structure The splitting of a spectral peak into a number of peaks, as that which occurs in nuclear magnetic resonance and in electron paramagnetic resonance.

fine-structure genetic mapping The determination of the relative positions of the mutable sites on a chromosome in terms of intervals of decreasing size and, ultimately, in terms of nucleotides.

finger nucleases Nucleases that are present in the skin of the fingers. Due to the presence of these enzymes, special care must be exercised when handling nucleic acid solutions.

fingerprint 1. A pattern of spots that is obtained when a partial hydrolysate of either a protein or a peptide is subjected to paper chromatography in one dimension and to paper electrophoresis in the second dimension. The term fingerprint is likewise applied to a similar two-dimensional map of a hydrolysate of either a nucleic acid or a nucleotide, as well as to maps obtained by modified procedures, involving other support media and/or other separation techniques. 2. The infrared absorption spectrum of a molecule.

fingerprint region The complex middle region of the infrared spectrum that is of most use for determinations of molecular structure because both group frequency bands and skeletal bands occur in this region; covers the wavelength range of about 7×10^{-4} to 1.6×10^{-3} cm.

firone Proposed term to designate a substance, of either intra- or extracellular origin, that can increase the probability of replicon misfiring. Tumor promoters could represent one class of firones. *See also* replicon misfiring.

first law of cancer biochemistry The principle, enunciated by Warburg in 1930, that cancer cells carry out glycolysis virtually universally, whether under aerobic or anaerobic conditions. The principle has been restated to mean that shifting of the metabolism of a cell to the anaerobic state is the main biochemical difference between a tumor and a normal cell.

first law of photochemistry The law that light must be absorbed by a molecule before that molecule can undergo a photochemical reaction. *Aka* Draper's law; Grotthus–Draper law.

first law of thermodynamics The principle of conservation of energy: the total energy of a system plus its surroundings is constant and is independent of any transformation that they may undergo; energy can neither be created nor destroyed by chemical means. In view of the interconversion of mass and energy, according to the theory of relativity, the modern version of the first law is as follows: the sum of energy and mass remains constant, but energy may change from one form to another, and energy may be converted to mass and vice versa.

first messenger The external signal that turns a multicyclic cascade system on or off; the chemical (such as acetylcholine, adrenalin, glucagon, and growth factors) that binds to a specific receptor on the external surface of the cell membrane. The binding sets off a series of transduction reactions by which the signal is transmitted to the inner surface of the membrane and leads to activation of an amplifier enzyme, located in the inner surface of the membrane, which then converts a precursor to a second messenger. The second messenger (such as cyclic AMP, diacyl glycerol, and inositol triphosphate) then acts on the first interconvertible enzyme of the multicyclic cascade system.

first-order kinetics The kinetics of a first-order reaction.

first-order reaction A chemical reaction in which the velocity of the reaction is proportional to the concentration of one reactant.

first-set rejection The sequence of events that leads to the rejection of an initial transplant by an unprimed individual.

first-step-transfer DNA A DNA fraction of T5 phage that is injected into the bacterial host during the first few minutes of infection and that controls the breakdown of the host DNA and the injection of the remainder of the phage DNA. *Abbr* FST-DNA.

Fischer formula 1. A straight-chain, planar projection of carbohydrates. *Aka* Fischer projection. 2. A simplified, planar formulation of porphyrins.

Fischer–MacDonald degradation A degradative procedure for aldoses whereby the sugar is converted to the next lower aldose; based on the removal of the anomeric carbon by treatment with ethyl mercaptan.

Fiske–SubbaRow method A colorimetric procedure for the determination of phosphate in biological materials that is based on the production of a blue color by treatment of the sample with ammonium molybdate and 1-amino-2-naphthol-4-sulfonic acid.

fission *See* nuclear fission; binary fission.

Fitzgerald factor HIGH MOLECULAR WEIGHT KININOGEN.

fixation 1. The chemical reactions whereby an atmospheric gas is converted to either an inorganic or an organic compound. *See also* carbon dioxide fixation; nitrogen fixation. 2. The preparation of tissues for cytological or histological study by converting cellular substances to insoluble components with as little alteration of the original biological structures as possible.

fixative A protein-denaturing solution that is used for the fixation of biological tissues.

fixed-angle rotor A centrifuge rotor in which the tubes are kept at a constant angle during centrifugation as opposed to a swinging bucket rotor where the tubes change their orientation during the centrifugation.

fixed time assay DISCONTINUOUS ASSAY.

fixed virus A virus that is obtained by passage through an organism or by cultivation in tissue culture. *See also* street virus.

fixer A chemical reagent that removes the unexposed and unreduced silver halide grains from a photographic film.

flagellar Of, or pertaining to, flagella.

flagellin The monomeric protein of flagella.

flagellum (*pl* flagella) A threadlike, cellular extension that functions in the locomotion of bacterial cells and of unicellular eukaryotic organisms; flagella are longer and less numerous than cilia.

flame emission spectrophotometer A spectrophotometer used for flame photometry.

flame ionization detector An ionization detector that is used in gas chromatography for the detection of organic compounds and that converts these compounds into ions by means of a flame. *Abbr* FID.

flame photometry The determination of elements in solution from the emission spectrum that is produced when the solution is sprayed into a flame. The electrons of the atoms are excited by the flame and emit the excitation energy as light when they fall back to lower energy levels.

flanking DNA Segments of DNA immediately adjacent to the two ends of a DNA region under consideration.

flash evaporator An apparatus for the removal of solvent from a solution; the solvent is evaporated from a thin film of solution, formed by the rotation of a round-bottom flask that is submerged in a water bath.

flash photolysis A technique for studying short-lived primary or subsequent photochemical intermediates; based on increasing the concentration of these intermediates by irradiating the sample with light pulses of short duration and great intensity. The light transmitted by the sample is then separated into its component wavelengths by means of a spectrograph and allowed to expose a photographic film.

flat-bed chromatography Chromatography, such as paper or thin-layer chromatography, in which a flat layer of chromatographic support is used.

flat-bed electrophoresis Electrophoresis, such as paper or thin-layer electrophoresis, in which a flat layer of electrophoretic support is used.

flat spectrum counting A counting method, used in liquid scintillation, that minimizes the effects of quenching.

Flaujeac factor HIGH MOLECULAR WEIGHT KININOGEN.

flavanone *See* bioflavonoid.

flavin adenine dinucleotide The flavin nucleotide, riboflavin adenosine diphosphate, which is a coenzyme form of the vitamin riboflavin and which functions in dehydrogenation reactions catalyzed by flavoproteins; abbreviated as FAD in its oxidized form and as $FADH_2$ in its reduced form.

flavin enzyme *See* flavoprotein.

flavin-linked dehydrogenase A dehydrogenase that contains a flavin nucleotide as a prosthetic group.

flavin mononucleotide The flavin nucleotide, riboflavin-5′-monophosphate, which is a coenzyme form of the vitamin riboflavin and which functions in dehydrogenation reactions catalyzed by flavoproteins; abbeviated as FMN in its oxidized form and as $FMNH_2$ in its reduced form.

flavin nucleotide A collective term for flavin mononucleotide and flavin adenine dinucleotide.

flavocoenzyme FLAVIN NUCLEOTIDE.

flavodoxin One of a group of flavoproteins that are similar to ferredoxins in their properties and that can replace ferredoxins in many reactions; flavodoxins differ from ferredoxins in that they contain FMN instead of iron. *See also* bioflavonoids.

flavone *See* bioflavonoid.

flavonoid See bioflavonoid.

flavonol *See* bioflavonoid.

flavoprotein A conjugated protein in which the nonprotein portion is a flavin nucleotide, such as the dehydrogenases that use either FMN or FAD as a coenzyme. FMN is linked noncovalently to the protein, while FAD may be linked either covalently or noncovalently. Some flavoproteins are more complex and may also contain either metal ions or heme in addition. *Abbr* FP.

flavor Whimsical name for the quality that distinguishes the electron from its neutrino, the up-quark from the down-quark, and so on. *See also* elementary particles.

Fletcher factor PREKALLIKREIN.

flexible active site An active site that can undergo conformational changes in the course of a reaction. *See also* induced-fit theory.

flexible polymer A polymer that can assume a number of different conformations of essentially identical energy but that is not a free-draining polymer.

flexion The degree of flexibility; used, for example, to describe changes in the conformation of phospholipid molecules in a bilayer that do not involve lateral diffusion, rotation, or flip-flop.

flickering cluster model A model for the structure of water according to which the water structure results primarily from cooperative-type hydrogen bonding between the water molecules, such that short-lived clusters of extensively hydrogen-bonded molecules form and break repeatedly.

flight hormones Epinephrine and norepinephrine.

flip *See* spin flip.

flip-flop 1. The transverse diffusion of a molecule through the lipid bilayer of a biological membrane; involves a rotation of the molecule through the plane of the lipid bilayer so that it moves from one monolayer to the other. 2. A model proposed to explain the anticooperative effects of certain oligomeric enzymes. It assumes that the subunits of the enzyme exist in two different conformational states and move back and forth between these two conformations. *Aka* half-of-the-sites reactivity. 3. Any other oscillatory mechanism, involving two molecular forms, such as the interconversion between the caldesmon–calmodulin and the caldesmon–F–actin complexes.

flip-flop chromatography A column chromatographic technique in which a number of polar and nonpolar solvents are used as eluents in order of alternating polarity beginning with the most extreme polar and nonpolar solvents. As a result, the tail ends of the polarity distribution of the sample are successively extracted, leaving behind material with a more restricted polarity range which increases the selectivity of subsequent eluants. The polar solvents are passed through the column in one direction and the nonpolar solvents are passed through the column in the opposite direction.

floating beta lipoprotein A lipoprotein that occurs in some conditions of hyperlipoproteinemia and that has an unusually low density due to its high content of triglycerides.

floating receptor model MOBILE RECEPTOR MODEL.

flocculation The precipitation of finely divided particles in the form of fleecy masses.

flocculation reaction The flocculation that is obtained in a precipitin reaction when soluble antigen–antibody complexes are formed in both the antigen excess zone and the antibody excess zone.

Florisil Trademark for a magnesium silicate adsorbent that is used in the column chromatography of lipids.

Flory–Huggins lattice theory A theory used in calculating the chemical potential of relatively concentrated polymer solutions.

flotation The movement of molecules in solution under the influence of a centrifugal field and toward the center of rotation. Such systems lead to the formation of inverted peaks

in the schlieren optical system, and the peaks rise from the bottom of the cell toward the meniscus as centrifugation proceeds.

flotation coefficient The analogue of the sedimentation coefficient for molecules that undergo flotation; a negative sedimentation coefficient. *Sym* s_f. *See also* standard flotation coefficient.

flow birefringence The birefringence that is caused by the orientation of asymmetric molecules in a solution that is subjected to flow and shear; used for determining rotational diffusion coefficients, and commonly measured by placing the solution between two concentric cylinders, of which the outer one is being rotated, and by passing polarized light through the solution.

flow cell A small container, such as a cuvette, that has an inlet and an outlet so that liquid can flow through it; permits the analysis of liquids as a function of time and/or of changing composition.

flowchart FLOWSHEET.

flow dichroism The dichroism that results from the orientation of asymmetric macromolecules when a non-Newtonian liquid, which contains the macromolecules, is subjected to shear.

flowering hormone A hormone that is synthesized in the leaves of plants and that is involved in the formation of flowers. So far, the hormone has been demonstrated physiologically but has not yet been isolated or chemically characterized.

flower model The viral RNA coding for the coat protein of MS-2 phage; so called because it has a double-stranded "stem" terminating in a set of double-stranded "petal" loops.

flow method *See* rapid flow technique; stopped flow technique.

flow potential STREAMING POTENTIAL.

flow quenching A rapid flow technique in which the enzyme and the substrate are mixed in the usual manner, but the mixture then flows into a second mixing chamber rather than into an observation cell. The enzymatic reaction is stopped rapidly in the second chamber by mixing a chemical quenching reagent with the enzyme and the substrate. A reactant or a product of the reaction is then determined by any convenient method.

flowsheet A chart, involving symbols and abbreviations, that is used to outline the steps, and the order in which they are to be performed, for the synthesis of a compound, the purification of an enzyme, the isolation of a protein, or some other multistep process. *Aka* flowchart; road map.

FLS Fibrils long spacing.

fluctuation analysis A study of voltage-gated channels that is based on the flutuations in current, produced by the nonperfect opening and closing of the channels; the fluctuations are produced by channels that do not open or close smoothly, or by channels that do not operate in perfect unison. *Aka* noise analysis.

fluctuation spectroscopy An analysis of the spectrum of fluctuations of a given system such as the fluctuations in reactant concentration of a chemical reaction or the fluctuations in electric current of an excitable membrane.

fluctuation test A statistical analysis for proving that selective variants, such as phage- or drug-resistant mutants, arise spontaneously and not as a result of exposure to the phage, drug, or other agent.

fluctuation theory 1. A theory, proposed by Linderstrom-Lang, according to which proteins fluctuate continuously between a large number of closely related conformational states. *Aka* motility model. 2. A theory of light scattering by solutions that is based on the fluctuations in the concentrations of solute and solvent molecules in small volume elements of the solution.

fluence *See* photon fluence; energy fluence.

fluid compartment The total amount of fluid in the body that either is located in a particular type of tissue or has a particular composition. The two major fluid compartments are the intracellular fluid and the extracellular fluid; the extracellular fluid consists of several sub-compartments such as interstitial fluid, cerebrospinal fluid, blood plasma, and lymph.

fluidity The reciprocal of viscosity.

fluidity gradient The increase in the range of motion of the fatty acyl chains of the phospholipids in biological membranes toward the methyl ends of the chain, that is, toward the center of the bilayer.

fluid mosaic model A model of biological membranes in which amphipathic lipids and globular proteins are arranged in an alternating mosaic pattern throughout the membrane. The phospholipids and glycolipids are arranged primarily in the form of a lipid bilayer. Integral proteins are "dissolved" in the bilayer. Some proteins are located at one or the other of the two surfaces of the lipid bilayer. Other proteins are embedded in the hydrophobic matrix of the bilayer while still others may extend throughout the membrane, from one side to the other. The lipids, as well as some of the proteins, possess a degree of fluidity that allows some lateral movement within the lipid matrix of the membrane. *Aka* lipid–globular protein mosaic model.

fluor 1. A liquid or a solid that is used in scintillation counters and that emits a flash of light when it is excited by radioactive or other radiation 2. FLUOROCHROME.

fluoresce To exhibit fluorescence.

fluorescein A yellow dye, chemically related to eosin, that is used as a fluorochrome in immunofluorescence.

fluorescence The emission of radiation by an excited molecule that occurs as a result of an electronic transition whereby the molecule returns from the excited state to the ground state. The emitted radiation is usually of a different and longer wavelength than that of the exciting radiation, and the time interval between excitation and emission is of the order of 10^{-9} to 10^{-7} s. *See also* phosphorescence.

fluorescence depolarization Fluorescence in which the exciting light is polarized and the emitted light is partially depolarized; the degree of depolarization of the emitted light is then measured.

fluorescence enhancement A method for studying the binding of fluorescent ligands to antibodies by determining the difference between the fluorescence of the free ligands and that of the antibody-bound ligands.

fluorescence microphotolysis A technique for studying the translational diffusion of membrane proteins and lipids, and the transport of solutes through membranes and within cells. The principle of the method involves equilibrating cells with a fluorescent solute, illuminating the cells with an attenuated laser beam, and measuring the fluorescence with a fluorescence microscope. The intensity of the laser beam is then suddenly increased so that intracellular solutes are photolysed and rendered nonfluorescent. As fresh, nonphotolysed chromophores move into the cells from the outside, the fluorescence inside the cells increases and is monitored. *Aka* fluorescence photobleaching recovery (FPR); fluorescence recovery after photobleaching (FRAP).

fluorescence microscope A microscope in which structures are illuminated with ultraviolet light, or short-wave visible light, and are made visible by fluorescence.

fluorescence microscopy A type of microscopy in which the specimen is fluorescent and emits light rather than transmitting or reflecting it; used for tissues stained with a fluorochrome and for immunofluorescence studies.

fluorescence photobleaching recovery FLUORESCENCE MICROPHOTOLYSIS.

fluorescence polarization Fluorescence in which the exciting light is not polarized and the emitted light is partially polarized; the degree of polarization of the emitted light is then measured.

fluorescence quenching A method for studying the binding of haptens to antibodies. The aromatic amino acids of the antibody are first excited and caused to fluoresce; subsequently the fluorescence is decreased when the antibody is allowed to combine with the hapten and an energy transfer takes place from the excited antibody to the hapten. The method can likewise be used to study binding reactions with other proteins.

fluorescence recovery after photobleaching FLUORESCENCE MICROPHOTOLYSIS.

fluorescent antibody An antibody that is covalently linked to a fluorescent dye, such as fluorescein or rhodamine, and that has retained its immunochemical activity.

fluorescent antibody technique A technique for locating either antigens or antibodies in a microscopic preparation of cells or tissues by treating the preparation with the corresponding fluorescent antibodies or fluorescent antigens. *See also* direct fluorescent antibody technique; indirect fluorescent antibody technique; anticomplement fluorescent antibody technique.

fluorescent antigen An antigen that is covalently linked to a fluorescent dye, such as fluorescein or rhodamine, and that has retained its immunochemical activity.

fluorescent screen A plate coated with a material, such as calcium tungstate or zinc sulfide, which fluoresces upon irradiation.

fluoridation The addition of fluoride to water supplies in an attempt to decrease dental caries; the final fluoride concentration is usually 1 mg/L.

fluorimeter Variant spelling of fluorometer.

fluorimetry Variant spelling of fluorometry.

fluorine An element that is essential to humans and animals. Symbol, F; atomic number, 9; atomic weight, 18.9984; oxidation state, -1; most abundant isotope, ^{19}F.

fluorochrome A substance that, when irradiated with light of a certain wavelength, emits light of a longer wavelength; a fluorescent compound, particularly one used to stain biological specimens.

1-fluoro-2,4-dinitrobenzene *See* Sanger reaction.

fluorography *See* solid scintillation fluorography.

fluoroimmunoassay An immunoassay employing antigens labeled with a fluorochrome. *Abbr* FIA.

fluorometer An instrument for the measurement of fluorescence that contains both a light source for supplying the excitation energy and a light detector for measuring the emission energy; filter fluorometers and spectrofluorometers are the two basic types.

fluorometry The measurement of fluorescence that may include a determination of one or more of the following: (a) the concentration

of a fluorescent compound; (b) the relative efficiencies of various exciting wavelengths to cause fluorescence; (c) the relative intensities of various wavelengths in the emitted fluorescent light; and (d) the probability that an absorbed photon will generate an emitted photon in fluorescence.

fluorophenylalanine An amino acid analogue of phenylalanine that can be incorporated into protein during protein synthesis.

fluorophore A potentially fluorescent group of atoms in a molecule.

fluorosis A condition caused by excessive intake of fluorine, usually derived from drinking water, and characterized by the occurrence of mottled teeth.

5-fluorouracil A pyrimidine analogue that is used in cancer chemotherapy; an antitumor agent that inhibits the enzyme thymidylate synthetase. *Abbr* FU.

flush ends *See* restriction enzyme.

flu virus INFLUENZA VIRUS.

flux 1. The metabolic rate with respect to a particular substrate in a given tissue; equal to AV/K_m where A is the substrate concentration in the tissue, V is the maximum velocity, and K_m is the Michaelis constant. 2. The rate of flow of either matter or radiation; equal to the number of particles (or the mass) or the number of photons that pass through a unit area per unit time. *See also* glycolytic flux.

flux ratio method A technique that is useful for the interpretation of complexities in enzyme mechanisms. It resembles a product inhibition technique but, rather than examining effects on initial rates, it examines the fate of individual product molecules participating in inhibitory reactions. Thus, for the reaction A + B $\rightleftharpoons$ P + Q, the ratio of two fluxes, one involving the conversion P $\rightarrow$ A and the other that of P $\rightarrow$ B can be determined and plotted as a function of the concentration of A or B, respectively. The resulting curves can be interpreted as supporting a random or an ordered mechanism.

F-mediated transduction SEXDUCTION.

fMet-tRNA N-Formylmethionyl tRNA.

FMN Flavin mononucleotide.

FMNH₂ Reduced flavin mononucleotide.

Fm protease A proteolytic enzyme, isolated from *Flavobacterium meningosepticum*, that cleaves peptide bonds in which the carbonyl group is donated by proline or methionine.

FNPA 4-Fluoro-3-nitrophenyl azide; a photoaffinity labeling compound that selectively binds to the active sites of protein molecules in antibodies and in acetylcholine binding sites on intact membranes.

foam The colloidal dispersion of a gas in a liquid.

foam cells Lipid-swollen cells found in atheromas. Such cells have been enlarged with droplets of cholesterol esters so that their cytoplasm appears vacuolated.

foam fractionation A chromatographic technique in which the stationary phase is the surface of bubbles and the mobile phase is the liquid flowing between the bubbles. *See also* emulsion fractionation.

Foerster's theory A theory for the dipole–dipole transfer of electronic excitation energy between a fluorescent energy donor and a suitable energy acceptor; the theory postulates that the rate of energy transfer depends on the inverse sixth power of the distance between the donor and the acceptor.

folacin 1. A generic descriptor for all folates and related compounds that exhibit qualitatively the biological activity of tetrahydropteroylglutamic acid. 2. FOLIC ACID.

folate A generic descriptor for the family of compounds that contain the pteroic acid nucleus.

folate coenzyme Tetrahydrofolic acid or one of its derivatives.

Folch method A method for the isolation of lipids from either tissues or fluids by extraction with chloroform/methanol/water mixtures.

foldback DNA DNA in which a single-stranded segment has folded back upon itself to form a hydrogen-bonded region. The intrastrand hydrogen bonding results from the occurrence of inverted repeats (palindromes). Simple inverted repeats result in a structure known as hairpin DNA; interrupted inverted repeats result in a structure known as stem-and-loop DNA. In the latter, the loop represents the segment between the inverted repeats and constitutes and a non-hydrogen-bonded region. When either hairpin or stem-and-loop structures form in double-stranded DNA, they extend outward from the double helix and, since they are formed by both strands, give rise to cross-shaped configurations known as cruciform DNA.

foldback elements Transposable elements in *Drosophila* that contain extensive amounts of foldback DNA.

folded chromosome Bacterial DNA that has been isolated by gentle techniques so that DNA breakage and protein denaturation have been avoided; a compact structure containing protein and supercoiled DNA.

folding *See* protein folding.

fold purification *See* purification (2).

folic acid Pteroylglutamic acid; a widely distributed vitamin of the vitamin B complex. The coenzyme forms of folic acid are derivatives of tetrahydrofolic acid and they function in the metabolism of one-carbon fragments. *Abbr* F.

folic acid coenzyme FOLATE COENZYME.

folic acid conjugate One of a group of folic acid derivatives that contain from two to seven glutamyl residues per molecule.

folic acid reductase The enzyme tetrahydrofolate dehydrogenase.

Folin–Ciocalteau reaction A colorimetric reaction for tyrosine that is used for the quantitative determination of proteins; based on the production of a blue color upon treatment of the sample with a complex phosphomolybdotungstic acid reagent.

folinic acid N^5-Formyltetrahydrofolic acid; a reduced and formylated derivative of folic acid that is more stable to air oxidation than is the parent compound.

Folin method LOWRY METHOD.

Folin reaction A colorimetric reaction for amino acids that is based on the production of a red color by treatment of an alkaline solution of the sample with 1,2-naphthoquinone-4-sulfonate.

Folin–Wu method One of a group of analytical procedures for the determination of glucose or other components in blood. In the case of glucose determination, the proteins are precipitated with tungstic acid and the protein-free filtrate is heated with an alkaline copper tartrate solution, followed by treatment with phosphomolybdic acid.

follicle-stimulating hormone The gonadotropic protein hormone, secreted by the anterior lobe of the pituitary gland, that stimulates the growth of ovarian follicles and the secretion of estradiol in the female and spermatogenesis in the male. *Abbr* FSH. *Aka* follitropin.

follicle-stimulating hormone releasing hormone The hypothalamic hormone that controls the secretion of the follicle-stimulating hormone. *Abbr* FSHRH; FRH. *Aka* follicle-stimulating hormone releasing factor (FSHRF; FRF).

follitropin FOLLICLE-STIMULATING HORMONE.

following strand LAGGING STRAND; *see* DNA replication.

food additive A substance that is added to, and not naturally present in, a food; includes those substances that are added in the preparation of a fortified food exclusive of sugar, salt, and vinegar.

food chain A sequence of organisms that feed one upon the other in succession and that provide for the transfer of food energy from the simpler to the more complex organisms.

food groups The four basic categories of nutrients that are recommended for daily inclusion in the diet: (a) milk group, (b) meat group, (c) vegetable and fruit group; and (d) bread and cereal group.

food web A system of interlocking food chains.

footprinting A technique for identifying protein-binding regions in DNA; based on the principle that such regions, when protein is bound to them, are protected against endonuclease action. The DNA is allowed to bind a specific protein and is then subjected to brief endonuclease action, so that each DNA molecule receives not more than one single-strand break (nick); these nicks are made only in the unprotected regions. After denaturation, the resulting single-stranded segments are separated according to their size by gel electrophoresis, yielding a series of bands. A control DNA sample, without the bound protein, is subjected to the same treatment and the two distributions of bands are compared. For every susceptible phosphodiester bond, a band will be found in both the sample and the control gels. But the sample gel will lack certain bands, corresponding to phosphodiester bonds in the protected region. The missing bands in the sample gel define the size of the DNA segment that had been protected by interaction with the specific protein. *See also* DNase protection; photofootprinting; dimethyl sulfate protection.

Forbe's disease GLYCOGEN STORAGE DISEASE TYPE III.

forbidden clone hypothesis The hypothesis of autoimmunity according to which the normal tolerance of an animal to self antigens is due to the death, in fetal life, of the clones that are responsible for the synthesis of the corresponding autoantibodies. If, as a result of a mutation, such normally forbidden clones reemerge during the adult life of the animal, they would then lead to the synthesis of auto-antibodies and produce autoimmunity.

forbidden transition A transition between energy states of either an atom or a molecule that is forbidden on the basis of the quantum mechanical selection rules; in practice this means that the transition occurs at a negligibly small rate.

forced dialysis A variant of equilibrium dialysis in which solvent, containing free ligand, is forced through the protein-containing compartment, and the effluent is then analyzed for the concentration of residual free ligand.

forced diffusion The diffusion that takes place when free diffusion is modified by the application of an external force such as an electrical or a centrifugal force.

force-feeding The feeding of experimental animals by forcing food into them; frequently performed by insertion of a tube through the nose into the stomach.

forescattering *See* forward scattering.

formal electrode potential The electrode potential of a solution that contains equal concentrations of both the oxidized and the reduced forms of the substance of interest, and that contains all other substances at specified concentrations. *Aka* formal potential.

formalin An aqueous 37% (w/v) solution of formaldehyde that is used as a fixative in microscopy and as a reagent for reacting with the amino groups of proteins and nucleic acids.

formality The concentration of a solution expressed in terms of the number of formula weights of solute in one liter of solution. *Sym* F.

formal solution A solution that contains one formula weight of solute per liter of solution.

formamide The amide of formic acid that reacts with the free amino group of adenine. In double-stranded DNA, this leads to disruption of A-T base pairs and denaturation of the DNA.

formation constant The equilibrium constant for the formation of a metal ion–ligand complex.

formation reaction A chemical reaction, frequently hypothetical, in which a compound is considered to be formed from its constituent elements; useful for calculating various thermodynamic quantities.

formazan A sugar derivative formed by the reaction of a carbohydrate phenylosazone with a diazo compound.

formed elements A collective term for the erythrocytes, leukocytes, and thrombocytes of blood. When applied to urine, the term also includes crystals, bacteria, ova, and parasites.

forme fruste An atypical, mild, or incomplete form of a pathological condition, such as an abnormality, a disease, or a syndrome.

formic fermentation MIXED ACID FERMENTATION.

formimino group The grouping —CH=NH.

formol titration The titration of an amino acid, peptide, or protein in the presence of formaldehyde. The formaldehyde reacts with, and lowers the pK of, the amino groups so that the region in the titration curve in which the amino groups are being titrated can be identified.

formose reaction An autocatalytic reaction in which formaldehyde, in the presence of alkaline catalysts, gives rise to various carbohydrates. The reaction has been considered as a model for prebiotic synthesis of carbohydrates.

formula weight The sum of the atomic weights, expressed in grams, in the formula of a compound; identical to the molecular weight for those substances that exist as true molecules.

formycin One of a group of pyrimidine antibiotics, produced by *Nocardia interforma*.

formylation The introduction of a formyl group into an organic compound.

formyl group The grouping —CH=O that is derived from formic acid.

N-formylkynurenine *See* trytophan dioxygenase.

N-formylmethionine A formylated form of methionine that, when bound to a specific transfer RNA molecule, serves as the initiating aminoacyl-tRNA in the translation of bacterial systems. *See also* N-acetylserine.

N-formylmethionyl-tRNA The initiating aminoacyl-tRNA molecule in the translation of bacterial systems. *Abbr* fMet-tRNA.

N^5-formyltetrahydrofolic acid FOLINIC ACID.

Forssman antigen A heterogenetic antigen located in the red blood cells and in the tissue cells of various organisms.

Forssman hapten The nonprotein portion of the Forssman antigen; a glycosphingolipid (pentahexosyl ceramide; a ceramide pentasaccharide) that contains 2 mol N-acetyl galactosamine (GalNac), 2 mol galactose (Gal), and 1 mol glucose (Glc) per mole of hapten. The hapten has the structure GalNac-GalNac-Gal-Gal-Glc-sphinganine.

fortified food Food to which nutrients have been added over and above those occurring naturally; may refer either to the increase in concentration of a naturally occurring nutrient, or to the addition of a different nutrient.

FORTRAN Acronym for *for*mula *tran*slation; a high-level computer language used primarily for mathematical computations.

forward mutation A mutation that changes a normal (wild-type) allele into an abnormal (mutant) allele; a mutation that leads to an altered phenotype which differs from the wild type.

forward scattering The scattering of radiation in the direction of the beam of radiation and away from the source of the radiation.

fossil Any remains of a living organism from a past geological period; frequently restricted to remains that have been petrified (converted to stone).

fossil fuel Fuel derived from fossilized organic material; coal, oil, and natural gas are examples.

Fouchet's test A test for bilirubin in urine that is based on the production of a green color by treatment of urine with ferric chloride and trichloroacetic acid.

Fould's rules A set of six general principles that describe the progression of tumors in animals: (1) Different tumors in the same animal progress independently. (2) Progression occurs independently in different characters of the same tumor. (3) Progression occurs in latent tumor cells and in tumor cells in which growth is arrested. (4) Progression is continuous or discontinuous and proceeds gradually or by abrupt changes. (5) Progression follows one of alternate paths of development. (6) Progression does not always reach an endpoint within the lifetime of the host. *See also* tumor progression.

founder cells One or more ancestral cells that have become specialized for development into a given type of differentiated tissue. Thus, there are founder cells for bone cells, liver cells, cartilage cells, and so on.

founder effect The principle that when a small group establishes itself as a separate and isolated entity, it carries only a fraction of the genetic variability of the parent population.

four-carbon plants *See* C_4 plants.

Fourier analysis The theory of representing functions of a variable as the sum of a series of sine and cosine terms of the type $a_j \sin(2\pi j/\lambda_j)$ and $a_j \cos(2\pi j/\lambda_j)$ where $j = 0, 1, \ldots$.

Fourier synthesis The process of computing the form of a function from the values of its coefficients in a Fourier series (an infinite series of sine and cosine terms); used in deriving the electron density distribution of a molecule from x-ray diffraction data.

Fourier transform infrared spectroscopy A powerful spectroscopic technique for collection of infrared absorption data that has three major advantages over ordinary infrared spectroscopy: (a) it permits the simultaneous passage of all infrared frequencies through a sample so that spectral data are obtained very rapidly; (b) it does not require slits or filters, resulting in greater intensities of radiation and hence greater sensitivity for the analysis of mixtures; and (c) it permits collection of more accurate and precise data since the instrument is continually calibrated by an internal laser. *Abbr* FTIR.

Fourier transform nuclear magnetic resonance A nuclear magnetic resonance technique in which all resonances are excited simultaneously by a short radio frequency pulse that encompasses all proton frequencies in the sample. The resulting time domain signal, containing the mixed frequencies of sample resonances, is Fourier transformed in a computer to reveal each resonance in the normal frequency domain. Advantages of the method include increased sensitivity, shortened resolution time, and double resonance capability. *Abbr* FTNMR.

FP Flavoprotein.

F-1-p Fructose-1-phosphate.

F-6-p   Fructose-6-phosphate.

F-pilus   (*pl* F-pili). A hollow tube that is formed by a male bacterium and that serves for the transfer of its DNA to a female bacterium during conjugation. *Aka* sex pilus.

F plasmid   FERTILITY FACTOR.

F′ plasmid   F′ FACTOR.

FPLC   Fast protein, peptide, and polynucleotide liquid chromatography.

FPR   Fluorescence photobleaching recovery.

fractal dimension   A measure of the degree of irregularity (roughness) of the surfaces of protein molecules. The fractal dimension (D) varies from 2 for a completely smooth surface to 3 for a completely space-filling surface, and is defined by $D = 2 - (d \log A_s/d \log R)$, where A_s is the molecular surface area and R is the radius of the probe.

fractile   The value of a statistical variable below which the indicated fraction of the measurements of the frequency distibution falls.

fraction   1. A preparation derived from a biological source and composed of one or more components. 2. A discrete portion of material obtained by fractionation.

fractional centrifugation   DIFFERENTIAL CENTRIFUGATION.

fractional distillation   The slow distillation of a liquid that permits the collection of fractions (portions) which distill at different temperatures.

fractional electrical transport   An electrochemical technique for determining whether an "active principle" for which an assay is available, is a strong acid, a strong base, an amphoteric compound, a nonelectrolyte, or a combination of these. The technique is based on subjecting dilute solutions of the "active principle" to electrolysis under high potentials and in cells that can be cut into a number of compartments in which the principle can then be assayed.

fractional precipitation   The stepwise separation of substances, particularly macromolecules, from a solution; based on precipitating the substances in the order of their increasing solubilities by changing the ionic strength, the pH, the dielectric constant, etc., of the solution.

fractional saturation   The fraction of occupied binding sites; the average number of ligands (v) bound per molecule of protein divided by the number of binding sites (n) per molecule of protein for the particular ligand; $\overline{Y}_s = v/n$, where $\overline{Y}_s$ (or $\overline{Y}$) is the fractional saturation.

fractional sterilization   The sterilization of a material by means of short, intermittent periods of heating that are separated by periods of incubation. Spores are allowed to germinate during the incubation period and are then killed by the subsequent heating period. The method is used for materials that cannot tolerate the long exposure to the high temperatures that are used in ordinary sterilization procedures.

fractional turnover rate   The reciprocal of the turnover time.

fraction-antibody binding   Fab FRAGMENT.

fractionated dose   A dose of radiation that is administered in the form of a series of short exposures.

fractionation   The separation of sample material into discrete portions.

fraction collector   An automatic device for the collection of consecutive portions of a liquid, as those obtained from a chromatographic column or from a density gradient. The collection may be based on time, volume, weight, or number of drops.

fraction-crystalline   Fc FRAGMENT.

fractogram   *See* sedimentation field flow fractionation.

fracture faces   The exposed sample surfaces obtained by the freeze fracture method. The face representing the hydrophobic interior of the cytoplasmic half of the bilayer is called the P face; the face representing the hydrophobic interior of the external half of the bilayer is called the E face.

fracture label   A modification of freeze fracture electron microscopy in which the cleavage surfaces produced are cytochemically labeled.

fragile site   A site at which a chromosome is particularly likely to become broken. Some fragile sites are those at which chromosomal translocations occur frequently.

fragile X syndrome   The most common form of inherited mental retardation in humans which occurs in about 1 out of 2000 newborn males; it is associated with a fragile site on the X chromosome.

fragmentation reaction   A chemical reaction in which there is a cleavage of one or more carbon-to-carbon bonds, leading to a breakage in the carbon skeleton of the molecule.

fragment ion   Any one of the positively charged fragments obtained in mass spectrometry when the original molecule is broken up.

fragment length mapping   A technique that aids in determining the base sequence of a nucleic acid molecule that is synthesized in vitro by a nuclease-free polymerase. The incubation of the polymerase with substrate for varying lengths of time results in the synthesis of polynucleotide fragments of different size, all of which contain the same 5′-terminal. The fragments are separated according to their size by means of polyacrylamide gel electrophoresis, and a specific oligonucleotide is then located in the nucleic acid by determining the minimum fragment length that is required to ensure the presence of the oligonucleotide.

fragment map RESTRICTION MAP.

fragment reaction An in vitro assay of the enzyme peptidyl transferase; involves measuring the ribosome-dependent transfer of the growing polypeptide chain to puromycin as evidenced by the release of peptidyl puromycin.

fragmin A calcium-sensitive protein that binds to monomeric actin (G-actin) and thereby suppresses actin filament (F-actin) formation.

frame of reading *See* reading frame.

frameshift mutation A mutation that leads to an alteration in the normal relation between nucleotide readout from a nucleic acid and amino acid sequence in the corresponding protein. A point mutation that results in a deletion or an insertion of a nucleotide constitutes a frameshift mutation; in this case, the codons will be unchanged up to the point of the mutation and will specify correct amino acids, but all subsequent codons will be altered and will specify incorrect amino acids.

frameshift suppression The suppression of a frameshift mutation.

framework model A molecular model in which the atoms are represented by their nuclei and by their bonds and in which the bond angles have their actual values.

framework regions Those parts of the variable regions of an immunoglobulin molecule that are not hypervariable regions; the variability of amino acid sequences in framework regions is much less than that in hypervariable regions.

Franck–Condon principle The principle that a change in the electronic configuration of a molecule as a result of the emission or the absorption of a photon requires about one-hundredth the length of time that is required for a vibration of the molecule; consequently, the atomic nuclei do not alter their positions appreciably during the emission or the absorption of a photon.

Frank–Evans iceberg *See* iceberg.

FRAP Fluorescence recovery after photobleaching.

fraudulent DNA A DNA molecule into which purine and/or pyrimidine analogues have been incorporated that do not occur naturally.

fraudulent nucleotide A nucleotide into which purine and/or pyrimidine analogues have been incorporated that do not occur naturally.

frayed end The terminal region of a double-stranded nucleic acid molecule in which the bases are not perfectly complementary.

free calcium The sum of ionized calcium and calcium complexes in serum. Calcium complexes represent the small amount of calcium that is bound to bicarbonate, phosphate, citrate, or other anions.

free diffusion The diffusion across a boundary in a solution or that in a gas phase as opposed to the diffusion across a membrane or some other porous medium.

free-draining polymer A polymer in solution that is coiled and flexible in such a fashion that the solvent held within the polymer is free to travel at its own velocity rather than at the velocity of the polymer; consequently, a free-draining polymer is not well approximated by a solid particle in solution. *Aka* free-draining coil.

free electrophoresis MOVING BOUNDARY ELECTROPHORESIS.

free energy That component of the total energy of a system that can do work under conditions of constant temperature and pressure; known as the Gibbs free energy (G) and expressed by the thermodynamic function $G = H - TS$, where H is the enthalpy, T is the absolute temperature, and S is the entropy. *See also* Helmholtz free energy.

free energy change The difference between the free energy of formation of the products and that of the reactants in a chemical reaction; denoted ΔG. The free energy change for a reaction at pH 7.0 is the biochemical free energy change and is denoted $\Delta G'$. *See also* standard free energy change.

free fatty acids Nonesterified fatty acids. *Abbr* FFA.

free insulin IMMUNOREACTIVE INSULIN.

freely jointed chain An artificial, mathematical model for a polymer in which a string of vectors of fixed length represents the bonds between atoms. The atoms are not included in the chain and the chain has no volume. The angle between each pair of bond vectors is free to assume all values with equal probability and all rotations about the bonds are equally likely. The direction of each bond vector is completely uncorrelated to every other bond vector so that the polymer is freely jointed. *Aka* random walk chain; random flight chain.

freely rotating chain An artificial, mathematical model for a polymer that is based on the freely jointed chain model but includes the constraint of fixed bond angles; rotations about these skeletal bonds are, however, completely unhindered.

free radical An atom or a group of atoms that has an odd number of electrons; an atom or a group of atoms containing an unpaired electron.

free rotation The sterically unhindered rotation of an atom or a group of atoms about a single bond.

freeze-blowing A modification of the freeze-clamp technique that is used for the preparation of brain specimens; entails the rapid removal and rapid freezing of brain tissue from conscious animals.

freeze-clamp technique A technique for determining the concentration and the compartmentation of metabolites under conditions that closely resemble those of the in vivo state; performed by abruptly stopping the metabolic reactions in a tissue specimen by pressing the tissue into a thin, frozen wafer between two aluminum blocks, previously cooled in liquid nitrogen, and then analyzing the tissue for the metabolites.

freeze-cleaving FREEZE-FRACTURING.

freeze-drying LYOPHILIZATION.

freeze-etching A modification of the freeze-fracturing technique in which the specimen is suspended in water prior to freezing, rather than in glyercol or in some other cryoprotective agent. The cleaved surfaces can then be etched by a brief heating of the specimen to $-100°C$ in a vacuum; this removes some of the ice bound to the surfaces in the specimen by sublimation.

freeze-fracturing A technique used to prepare specimens for electron microscopy. The specimen is frozen and then fractured with a knife edge to yield cleavage surfaces; the surfaces are then replicated by metal casting and are examined by conventional transmission electron microscopy after the original cellular material has been digested away. Analysis of the resultant micrographs provides three-dimensional information of ultrastructural details with a resolution of 0.1 to 0.2 nm.

freeze-stop technique FREEZE-CLAMP TECHNIQUE.

freeze-substitution The dehydrating of tissue specimens by freezing them in a cold organic solvent such as propane or isopentane.

freeze-thawing The disruption of cells by repeated freezing and thawing of a cell suspension.

freezing 1. A restriction in the rotation of a molecule that enhances its binding to another molecule, as in the freezing of a substrate that binds to an enzyme, or in the freezing of ribosomes that bind to mRNA. 2. The transition of a liquid to a solid that is produced by a lowering of the temperature.

freezing-point osmometer An osmometer, the operation of which is based on the lowering of the freezing point of a solvent by the addition of a solute.

French press An apparatus used for the disruption of cells and for the preparation of cell-free extracts. The cells, in suspension, are subjected to high pressures by means of a piston and the suspension is then forced through a small orifice. The sudden decrease in pressure leads to an explosion of the cells. Some cell breakage is also caused by the shearing forces.

Freon Trademark for a group of fluorinated alkanes used in refrigeration and aerosols. There is concern that escape of freon to the stratosphere leads to reactions that deplete the ozone layer which shields the earth from excessive ultraviolet irradiation.

frequency The number of times that an event recurs per unit time, such as the number of vibrations per second. For electromagnetic radiations, the frequency is equal to the speed of light divided by the wavelength of the radiation.

frequency distribution A systematic graphical arrangement of statistical data that shows both the classes into which a variable is divided and the frequencies of these classes.

frequency factor See Arrhenius equation.

frequency of recombination A measure of the distance between loci on a genetic map that is equal to the number of recombinants divided by the total number of progeny.

frequency shift See chemical shift.

Freund's adjuvant An adjuvant that is used as either a complete or an incomplete adjuvant; complete Freund's adjuvant consists of mineral oil, detergent, and dead mycobacteria; incomplete Freund's adjuvant consists of mineral oil and detergent.

FRF Follicle-stimulating hormone releasing factor; See follicle-stimulating hormone releasing hormone.

FRH Follicle-stimulating hormone releasing hormone.

frictional coefficient See rotational frictional coefficient; translational frictional coefficient.

frictional force The force that is exerted on a particle in solution as a result of friction; it is equal to the product of the velocity of the particle and its frictional coefficient.

frictional ratio The ratio of the experimentally determined translational frictional coefficient of a macromolecule to the translational frictional coefficient of a sphere that has the same molecular weight as the macromolecule. The frictional ratio is commonly divided into two factors which measure the contribution of hydration and the contribution of molecular asymmetry. See also Oncley equation.

Friedman test A test, analogous to the Aschheim–Zondek test, except that the voided urine is injected intravenously into rabbits.

Friend leukemia virus A mouse leukemia virus that belongs to the group of leukoviruses.

front See solvent front.

frontal analysis See frontal chromatography.

frontal chromatography A column chromatographic technique in which the sample is passed continuously into the column and only the component that emerges first from the column can be obtained in pure form.

fronting LEADING.

frontside attack The mechanism of a chemical displacement reaction that proceeds with retention of configuration.

frozen accident theory A theory of the evolution of the genetic code according to which the code evolved by chance until the current codon assignments had been developed; once this was achieved, the code was essentially prevented from any further evolution because the code provided the organism with selective advantages and any changes in the code would have been deleterious for the organism. *See also* lethal mutation model; specific interaction theory.

frozen replica method FREEZE-ETCHING.

FRS Ferredoxin-reducing substance.

Fru Fructose.

fructan A homopolysaccharide of fructose that occurs in plants.

fructose A six-carbon ketose that, together with glucose, makes up a molecule of sucrose. *Abbr* Fru.

fructose-1,6-bisphosphate A metabolite that is cleaved in glycolysis to two triose phosphates, glyceraldehyde-3-phosphate and dihydroxyacetone phosphate. *Abbr* FBP. *Aka* fructose-1,6-diphosphate (FDP).

fructose intolerance A genetically inherited metabolic defect in humans that is due to a deficiency of the enzyme fructose-1-phosphate aldolase.

fructose-6-phosphate A metabolite that is phosphorylated in glycolysis to fructose-1,6-bisphosphate.

fructosuria A genetically inherited metabolic defect in humans that is characterized by the presence of excessive amounts of fructose in the urine and that is due to a deficiency of the enzyme fructokinase.

fruit fly A fly of the genus *Drosophila*, an organism widely used for genetic research.

fruit-ripening hormone The gas ethylene that occurs widely in plant tissues and that accelerates the ripening of fruit; it is formed from methionine.

fruit sugar FRUCTOSE.

FSF Fibrin-stabilizing factor.

FSH Follicle-stimulating hormone.

FSHRF Follicle-stimulating hormone releasing factor. *See* follicle-stimulating hormone releasing hormone.

FSHRH Follicle-stimulating hormone releasing hormone.

FST-DNA First-step transfer DNA.

F′ strain A bacterial strain in which an augmented fertility factor has become incorporated into the bacterial chromosome. *See also* F′ factor.

F test A statistical test for comparing the variances of two sets of results by means of their *F* value (the ratio of the two variances); widely used to test for the identity of the means of two populations.

FTIR Fourier transform infrared spectroscopy.

FTNMR Fourier transform nuclear magnetic resonance.

FU 5-Fluorouracil.

Fuc Fucose.

fucolipid Any fucose-containing glycolipid.

fucose A deoxysugar that occurs in some bacterial cell walls; 6-deoxy-L-galactose. *Abbr* Fuc.

fucosidosis A genetically inherited metabolic defect in humans that is characterized by cerebral degeneration, muscle spasticity, and an accumulation of H-isoantigen; a lipid storage disease that is due to a deficiency of the enzyme alpha fucosidase.

full agonist A ligand that binds to a receptor and causes a maximum biological response. *See also* agonist; partial agonist.

Fuller's earth Any clay that has adequate decolorizing capacity and that contains aluminum magnesium silicate.

fulvic acid The complex mixture of acid-soluble and alkali-soluble substances that are extracted from the organic matter of soil. *See also* humic acid; humin.

fumagillin An antibiotic, produced by *Aspergillus fumigatus*, that has no activity against bacteria or fungi but does act against amoeba. It also inhibits the development of some bacteriophages and has some antiviral activity in tissue culture.

fumarase The enzyme that catalyzes the reversible hydration of fumaric acid to malic acid in the citric acid cycle.

fumarate pathway A catabolic pathway whereby either phenylalanine or tyrosine is converted to fumaric acid, which then feeds into the citric acid cycle.

fumaric acid The unsaturated, dicarboxylic acid that is formed from succinic acid in the citric acid cycle.

functional death The death of an organism that results from the inability of a gene, or a group of genes, to carry out their functions.

functional group A reactive atom or a group of atoms in a molecule that has specific properties; aldehyde, ketone, amino, hydroxyl, carboxyl, and sulfhydryl groups are some examples.

functional group isomer One of two or more isomers that have the same molecular composition but differ from each other in the type of functional group that they contain.

function of state   *See* state function.

fungal   Of, or pertaining to, fungi.

fungicide   An agent that kills fungi.

fungus (*pl* fungi)   A plant protist that is non-photosynthetic and that is devoid of chlorophyll; fungi generally contain a mycelium and are frequently coenocytic.

furan   A heterocyclic compound, the structure of which resembles the ring structure of the furanoses.

furanose   A monosaccharide having a five-membered ring structure.

furanoside   A glycoside of a furanose.

fused gene   A hybrid gene produced by linking a gene of interest (for example, a mammalian gene) to some other gene (for example, a plasmid gene) using methods of recombinant DNA technology. *Aka* hybrid gene. *See also* fusion gene.

fused protein   A hybrid protein molecule, consisting of two linked and different proteins, and produced from a fused gene. *Aka* hybrid protein.

fused ring   A ring that has two or more atoms in common with another ring.

fused rocket immunoelectrophoresis   *See* rocket electrophoresis.

fusel oil   A group of compounds formed as side products during alcoholic fermentation; the mixture consists mainly of amyl, isoamyl, isobutyl, and propyl alcohols.

fusidic acid   A steroid antibiotic, produced by *Fusidium coccineum*, that inhibits protein synthesis in both prokaryotes and eukaryotes by interfering with elongation factor G (translocase).

fusiform   Spindle-shaped; tapered at each end.

fusion   *See* plasmid fusion; replicon fusion; gene fusion; nuclear fusion.

fusion gene   A hybrid gene consisting of parts of two others genes. A fusion gene can be formed by deletion of a chromosomal segment between two genes or by crossing over. *See also* fused gene.

fusogenic agent   An agent, such as polyethylene glycol or Sendai virus, that induces cell fusion.

futile cycle   A substrate cycle in which the two opposing reactions occur at comparable rates in the same cell. Such a cycle accomplishes nothing except the waste of the free energy difference between the two reactions or, possibly, the generation of some heat. For example, the reaction glucose + ATP $\rightleftharpoons$ glucose-6-phosphate + ADP together with the reaction glucose-6-phosphate + H_2O $\rightleftharpoons$ glucose + P_i leads only to the net reaction of ATP $\rightleftharpoons$ ADP + P_i.

fuzzy coat   CELL COAT.

F value   1. A ratio of two variances; *See* F test. 2. The time required, when treating an aqueous suspension at 121 °C, to heat inactivate (kill) the entire population of viable bacterial cells or spores.

Fv fragment   The N-terminal portion of the Fab fragment of the immunoglobulins; it consists of the variable portions of one heavy and one light chain.

G

g 1. Gram. 2. Gravity; used to describe centrifugal forces. *See also* relative centrifugal force. 3. *g* value.

G 1. Guanine. 2. Guanosine. 3. Glycine. 4. Gibbs free energy. 5. Glucose. 6. Gauss. 7. G value.

G_0 *See* G protein.

G_1 *See* cell cycle.

G_2 *See* cell cycle.

G 17, G 34 *See* gastrin.

GA 1. Glyceric acid. 2. Glutamic acid.

GABA γ-Aminobutyric acid; 4-aminobutyric acid.

GABAergic neuron *See* inhibitory transmitter.

GABA shunt *See* γ-aminobutyrate bypass.

G-actin The monomeric, water-soluble, and globular form of actin.

gal Gallon.

Gal Galactose.

galactan One of a group of polymers of galactose that occur in plants, have high molecular weights, and are usually unbranched; agar and carrageenan are two examples.

galactin PROLACTIN.

galactocerebroside A galactose-containing cerebroside.

galactolipid A galactose-containing lipid.

galactosamine The amino sugar of galactose that occurs in glycolipids and in chondroitin sulfate.

galactose A six-carbon aldose that is widespread in animals and that is a component of cerebrosides and gangliosides. *Abbr* Gal.

galactosemia A genetically inherited metabolic defect in humans that is characterized by the inability of an infant to metabolize the galactose that is derived from the lactose in milk; due to a deficiency of the enzyme galactose-1-phosphate uridyl transferase. The condition leads to an accumulation of galactose, a loss of inorganic phosphate, liver failure, and mental retardation.

galactose operon *See* gal operon.

galactose tolerance test A liver function test that measures the ability of the liver to remove galactose from the blood stream and to convert it to glycogen; performed in a similar manner as a glucose tolerance test.

β-galactosidase *See* beta galactosidase.

galactoside permease LACTOSE PERMEASE.

galactosphingolipid GALACTOCEREBROSIDE.

Galactostat Trade name for a galactose oxidase reagent.

galactosuria The presence of excessive amounts of galactose in the urine; a consequence of galactosemia.

galactozymase The enzyme system responsible for the inducible fermentation of galactose in *E. coli*; consists of galactokinase, galactose-1-phosphate uridyl transferase, and UDP-glucose epimerase.

galaptins Low molecular weight, β-galactoside-specific, animal lectins believed to function in cell–cell recognition and adhesion by a bridging mechanism.

GalN Galactosamine.

GalNAc *N*-Acetylgalactosamine.

gal operon The operon in *E. coli* that functions in the metabolism of galactose and codes for the enzymes galactokinase, galactose transferase, and galactose epimerase.

gamblegram A diagrammatic representation of the composition of body fluids. It consists of two rectangles, placed side by side, one for cations and one for anions. The various ions are indicated within each rectangle by means of blocks, with the height of each block being proportional to the concentration of the ion in the given body fluid.

gamete A mature, haploid germ cell.

gametic number HAPLOID NUMBER.

gametocyte A cell that is destined to develop into a gamete; an oocyte or a spermatocyte.

gametogenesis The formation of gametes; oogenesis or spermatogenesis.

gametophyte The haploid individual in the life cycle of an organism exhibiting alternation of generations.

gamma 1. *n* A microgram. 2. *adj* Denoting the third carbon atom from the carbon atom that carries the principal functional group of the molecule. *Sym* γ.

gamma aminobutyrate bypass *See* γ-aminobutyrate bypass.

gamma aminobutyric acid *See* γ-aminobutyric acid.

gamma chain 1. The heavy chain of the IgG immunoglobulins. 2. One of the two types of polypeptide chains present in fetal hemoglobin.

gamma globulin 1. A protein that belongs to a specific fraction of serum proteins. 2. IMMUNOGLOBULIN.

gamma ray An electromagnetic radiation of short wavelengths that is frequently emitted by radioactive isotopes and that consists of photons that originate in the nucleus of an atom.

gamma ray spectrometer An instrument for measuring the distribution of the energies of gamma rays.

gamone A plant sex hormone produced by a gamete; gamones facilitate fertilization and act as chemotactic agents. *See also* pheromone.

ganglion (*pl* ganglia) A mass of nervous tissue consisting principally of numerous nerve cell bodies.

ganglio series glycosphingolipids *See* glycosphingolipids.

ganglioside A ceramide oligosaccharide that contains at least one residue of sialic acid in addition to the other sugars; a glycosphingolipid containing sialic acid; a sialosylglycosyl sphingolipid (sialoglycosphingolipid). Gangliosides are especially abundant in the gray matter of the brain and in the thymus gland.

gangliosidosis *See* generalized gangliosidosis; Tay–Sachs disease.

gap filament A very thin filament seen in electron micrographs in the gap between the A-band and the I-band of striated muscle; believed to be composed of accidental sarcomere constituents that appear as a result of muscle stretching.

gap-filling The extension and linking of Okazaki fragments in DNA replication.

gap junction A communicating junction between two cells such that small molecules can pass directly from the interior of one cell to the interior of the other cell. The interacting plasma membranes are separated by a gap of 2 to 4 nm, hence the term gap junction. A gap junction differs from a chemical synapse in that the communication between the two cells is direct rather than indirect. Gap junctions consist of clusters of transmembrane channels through which the molecules move. Each channel is called a connexon. Each connexon is formed by a hexameric protein spanning the cell membrane. An interconnecting aqueous channel is formed by apposition of the two hexameric structures (connexons) in the two cells. *See also* synapse; cell junction.

GAR Glycinamide ribonucleotide; an intermediate in the biosynthesis of purines.

gargoylism A genetically inherited metabolic defect in humans that is characterized by skeletal deformities and mental deficiency and that is due to a defect in mucopolysaccharide storage, resulting in the presence of excessive amounts of chondroitin sulfate in the urine. *See also* Hunter's syndrome; Hurler's syndrome.

Garrod's disease ALKAPTONURIA.

GAS General adaptation syndrome.

gas amplification 1. The process whereby the ions formed in an ionization chamber produce more ions from the gas molecules in the chamber and thereby increase the electrical current that is measured. The magnitude of the current produced depends on the applied potential; as the potential is increased, the current rises above a saturation current, passes through a proportional region, a limited proportional region, a Geiger–Mueller region, and ultimately reaches a level at which there is a continuous discharge in the chamber. 2. The ratio of the actual charge collected at the electrode of an ionization chamber to the charge produced in the chamber by the initial radiation.

gas chromatogram The tracing of a gas chromatographic separation; a plot of detector response versus either time or volume of carrier gas. Each component, or mixture of components, is represented by a peak in a differential chromatogram and by a step in an integral chromatogram.

gas chromatography 1. Column partition chromatography in which the stationary phase is a solid and the mobile phase is a gas. *Abbr* GSC. 2. Column partition chromatography in which the stationary phase is an inert carrier, coated with an essentially nonvolatile liquid, and the mobile phase is an inert gas which sweeps volatile compounds through the column. *Abbr* GLC.

gas chromatography–mass spectrometry The combination of gas chromatographic and mass spectrometric techniques in which components are separated by a gas chromatography column and are then fed into a mass spectrometer. *See also* mass fragmentography.

gas constant The physical constant that is derived from the ideal gas law, $PV = nRT$, where R is the gas constant [1.987 cal/(deg)(mol)], P is the pressure, V is the volume, T is the absolute temperature, and n is the number of moles.

gaseous exposure method WILZBACH METHOD.

gaseous group A group of compounds that have very low melting points and that are believed to have occurred in the original gas dust of the solar nebula. *See also* earthy group; icy group.

gas flow counter A radiation counter in which the ionization detector must be flushed continually with the counting gas, used for gas amplification, to eliminate air leakage into the detector.

gas holdup HOLDUP VOLUME.

gas ionization The formation of ion pairs from gases that are subjected to an ionizing radiation.

gas liquid chromatography *See* gas chromatography (2).

gasohol A fuel, chiefly for internal combustion engines, composed of a 9:1 ratio by volume of unleaded gasoline and ethyl alcohol.

gasometry The measurement of gas volumes and/or gas pressures.

gas solid chromatography *See* gas chromatography (1).

gas storage The storage of fruit at decreased oxygen levels and at increased carbon dioxide levels compared to those in air; used in an attempt to prolong the storage life of the fruit.

gastric Of, or pertaining to, the stomach.

gastric inhibitory peptide A polypeptide gastrointestinal hormone that is a potent inhibitor of gastric secretion and motility and that also functions in the regulation of insulin release. *Abbr* GIP.

gastric juice The digestive juice that consists of the secretion from the stomach and that contains hydrochloric acid and pepsinogen. *Aka* gastric fluid.

gastricsin A proteolytic enzyme present in gastric juice.

gastrin A peptide hormone of 17 amino acids that is produced by the gastric mucosa and that stimulates the secretion of gastric juice. Human gastrin is known as gastrin I, little gastrin, or G 17; a modified form, in which a sulfate group is esterfied to tyrosine 12, is known as gastrin II. Human gastrin also occurs in the form of a 34-amino acid peptide, believed to be a precursor of gastrin I; it is known as big gastrin, gastrin II, or G 34. A shortened form of gastrin I, consisting of 13 amino acids is known as minigastrin.

gastrinoma A gastrin-producing tumor which may occur in the pancreas, the duodenum, or the stomach.

gastrointestinal Of, or pertaining to, the stomach and the intestine.

gastrone A glycoprotein in gastric juice that, when injected into animals, inhibits gastric secretion.

gate 1. A channel in a biological membrane that can open and close in a transient fashion. 2. A cutoff level for pulses in scintillation counting. *See also* differential counting; integral counting.

gated channel A membrane channel whose permeability is regulated; a mediated transport system. Two major types involve the opening of a channel in response to (a) changes in the membrane potential (voltage-gated channels) or (b) extracellular binding of a ligand to a specific receptor (ligand-gated channels). *Aka* gated pore.

gating The opening and closing of a channel in a biological membrane; believed to be brought about by conformational changes of the integral membrane proteins, lining the channel, as a result of the binding of substrates or other ligands.

gating current The electric current set up in a gate as a result of the movement of ions through the gate.

gauche conformation A stable, staggered conformation at which there is a potential energy minimum and interatomic interactions are minimized.

Gaucher's disease A genetically inherited metabolic defect in humans that is characterized by an accumulation of cerebrosides in tissues and by an enlargement of the spleen and the liver; due to a deficiency of the enzyme glucocerebrosidase.

gauss A unit of magnetic field strength. *Sym* G.

Gauss error function NORMAL DISTRIBUTION.

Gaussian chain FREELY JOINTED CHAIN.

Gaussian distribution NORMAL DISTRIBUTION.

Gaussian error curve NORMAL DISTRIBUTION.

Gay-Lussac equation The simple equation for alcoholic fermentation, established in 1815 by Gay-Lussac: $C_6H_{12}O_6 \rightarrow 2CO_2 + 2C_2H_5OH$.

GC Gas chromatography.

G + C/A + T ratio *See* A + T/G + C ratio.

GC box A nucleotide sequence, present upstream from the CAAT box in a few eukaryotic promoters. It has the consensus sequence GGGCGG.

GC content The amount of guanine plus cytosine in a nucleic acid, expressed in mole percent. In Watson–Crick-type, double-stranded DNA, the GC content plus the AT content equals 100 mol% and the melting out temperature of the DNA increases with increasing GC content.

GC-MS Gas chromatography–mass spectrometry.

GC pair A guanine–cytosine base pair.

GC-rich nucleus A region in a double-helical nucleic acid structure that is rich in guanine and cytosine; GC-rich nuclei are denatured more slowly than regions that are rich in either adenine and thymine, or adenine and uracil.

GDH 1. Glucose dehydrogenase. 2. Glycerophosphate dehydrogenase. 3. Glutamate dehydrogenase.

GDP 1. Guanosine diphosphate. 2. Guanosine-5′-diphosphate.

GDPM Guanosine diphosphate mannose.

GEF Gel electrofocusing.

gegenion COUNTERION.

Geiger–Mueller counter A radiation counter, of the ionization chamber type, that is designed for operation in the Geiger–Mueller region. *Aka* Geiger counter.

Geiger–Mueller plateau The counting plateau

obtained with a Geiger–Mueller counter.

Geiger–Mueller region That portion of the characteristic curve of an ionization chamber in which, during gas amplification, maximum gas amplification is obtained, and the collected charged is independent of the size of the initial charge produced by the radiation.

gel A solid colloidal dispersion consisting of a network of particles and a solvent that is immobilized in this network.

Gelarose Trademark for a group of agarose gels that are used in gel filtration chromatography.

gelatin *See* parent gelatin.

gelation Gel formation; the transition from a sol to a gel.

gelation protein CROSS-LINKER.

gel blot hybridization *See* blotting.

gel chromatography GEL FILTRATION CHROMATOGRAPHY.

gel diffusion *See* single diffusion; double diffusion.

gel electrofocusing Isoelectric focusing in which a gel is used as the supporting medium for the pH gradient; gel electrofocusing is faster and requires smaller amounts of sample and reagents than isoelectric focusing in a density gradient. *Abbr* GEF.

gel electrophoresis Zone electrophoresis in which a gel is used as the supporting medium.

gel exclusion chromatography GEL FILTRATION CHROMATOGRAPHY.

gel filtration chromatography 1. A column chromatographic technique in which the stationary phase consists of gel particles of controlled size and porosity, as those prepared from polymeric carbohydrates. Molecules are fractionated on such a column on the basis of their size and shape and, hence, their rates of diffusion into the gel particles; smaller molecules of a given shape diffuse more rapidly into the gel and move more slowly through the column than larger ones. 2. Molecular sieving in which aqueous systems are used. *Abbr* GFC. *Aka* gel filtration.

gel immunofiltration *See* immuno-gel filtration.

gellan gum A high molecular weight, water-soluble, heteropolysaccharide, produced by fermentation of some species of *Pseudomonas*, that forms a gel similar to that formed by agar.

gel osmometer An osmometer, the operation of which is based on changes in the dimensions of a gel.

gel permeation chromatography 1. GEL FILTRATION CHROMATOGRAPHY. 2. Molecular sieving in which nonaqueous systems are used. *Abbr* GPC.

gelsolin A protein, isolated from macrophages, that binds to, and fragments, actin filaments; a capping protein.

gem- Combining form meaning geminal.

geminal Referring to two substituents on the same carbon atom. *Abbr* gem.

gene The unit of heredity that occupies a specific locus on the chromosome. A sequence of bases along a DNA molecule (in certain viruses, an RNA molecule). The gene may be a functional unit (cistron; structural gene), a mutational unit (muton), or a recombinational unit (recon).

gene activation The process of genetic induction in which an individual molecule brings about the transcription of one or more structural genes. *See also* operon hypothesis.

gene amplification The selective replication of specific genes disproportionately to the extent of their occurrence in the genome. During development, some genes become amplified in specific tissues permitting the synthesis of vast quantities of a specific gene product. As an example, the genes coding for rRNA increase in number about 4000-fold during oogenesis in the toad *Xenopus laevis*.

gene bank GENE LIBRARY.

gene cloning *See* recombinant DNA technology.

gene cluster MULTIGENE FAMILY.

gene conversion The asymmetrical segregation of genes during replication that leads to an apparent conversion of one gene into another.

gene dosage The number of times that a particular gene occurs in the nucleus of a cell.

gene duplication The process whereby one or more genes from one chromosome are integrated by crossing over into a second chromosome that already carries the same genes. This results in a duplication, or in higher states of multiplication, of the material present in the chromosome. Gene duplication may be partial or virtually complete, and is considered to be responsible for the similarities in amino acid sequences among groups of proteins, such as the immunoglobulins, the hemoglobins, and the haptoglobins.

gene expression 1. The multistep process, and the regulation of this process, by which the product of a gene is synthesized; the flow of genetic information. 2. GENE ACTIVATION.

gene family MULTIGENE FAMILY.

gene frequency A measure of the proportion of an allele in a population; equal to the number of loci at which a particular allele is found divided by the total number of loci at which it could occur.

gene fusion *See* fused gene; fusion gene.

gene hypothesis The hypothesis that, in the development of life, nucleic acids were formed first and proteins arose later; based in part on the attributes of life shown by nucleic acids in that they are able to replicate, code for protein, and undergo mutation.

gene insertion Any process for the introduction of a gene or genes from some source into a cell; such processes include recombinant DNA technology, cell fusion, transformation, conjugation, and transduction.

gene library A clone library which contains a large number of representative nucleotide sequences from all sections of the DNA of a given genome; a random collection of DNA fragments from a single organism, linked to vectors, and cloned in a suitable host. The DNA from the organism of interest is fragmented (enzymatically or mechanically), the fragments are linked to suitable vectors (plasmids or viruses), the modified vectors are introduced into host cells, and the latter are cloned. A gene library contains both transcribed DNA fragments (exons) as well as nontranscribed fragments (introns, spacer DNA). Retrieval of specific DNA sequences from a gene library frequently involves screening by means of a probe. *Aka* colony bank; DNA library. *See also* genomic library; probe.

gene linkage *See* linkage.

gene locus *See* locus.

gene machine Any one of a variety of manual, semiautomatic, or automatic instruments used for the synthesis of deoxyoligonucleotides. Recent advances, including solid-phase synthesis on silica-bound supports and stable deoxynucleoside phosphoramidites as synthons, permit the rapid synthesis of segments containing over 100 deoxynucleotides in good yield.

gene mapping 1. The assignment of a specific gene to a specific chromosome. 2. The determination of the sequence of specific genes, and their relative distances from each other, on a given chromosome.

gene pair The identical or nonidentical alleles of a specific gene at a given locus on homologous chromosomes in a diploid cell.

gene pool The sum total of all the genes in a population of sexually reproducing organisms.

gene product 1. The polypeptide chain translated from the mRNA molecule transcribed from the given gene. 2. The nontranslated RNA molecule (for example, rRNA) transcribed from the given gene.

gene 32 protein A phage T4 protein that was the first single strand binding protein to be discovered.

general acid–base catalysis The catalysis in solution in which the catalysts are various acidic and/or basic species that serve as proton donors and/or proton acceptors. *See also* specific acid–base catalysis.

general adaptation syndrome The sequence of reactions that are initiated by an increased secretion from the adrenal cortex in response to a stress and that allows an animal to adapt to the stress. *Aka* GAS.

generalized anaphylaxis A severe, reagin-mediated, systemic reaction to an allergen that is characterized by itching, edema, wheezing respiration, rapid and weak pulse, and falling blood pressure; may result rapidly in shock and death.

generalized gangliosidosis A genetically inherited metabolic defect in humans that is characterized by mental retardation, skeletal deformities, an enlargement of the liver, and an accumulation of gangliosides; a sphingolipidosis that is due to a deficiency of the enzyme beta galactosidase which removes galactosyl groups from both gangliosides and keratin.

generalized transduction Bacterial transduction in which any of the genes of the donor bacterium may become transduced.

generally labeled Designating a compound that is randomly labeled at various positions in the molecule; usually refers to tritiated compounds. *Sym* G.

general recombination Genetic recombination in which the exchange of genetic material takes place between two homologous DNA sequences, most commonly between two copies of the same chromosome. *Aka* homologous recombination; legitimate recombination. *See also* site-specific recombination.

generation cycle CELL CYCLE.

generation time The time required by a cell for the completion of one cycle of growth. *Sym* T_g. *See also* doubling time.

generation-time hypothesis The hypothesis that protein and DNA evolution is faster in species having unusually short generation times (many generations per unit time) than in species with unusually long generation times (few generations per unit time).

gene redundancy GENE REITERATION.

gene reiteration The occurrence of multiple copies of the same gene on a given chromosome.

gene repetition *See* gene redundancy; gene reiteration; dosage repetition; variant repetition.

generic Of, or pertaining to, a genus.

genesis The origin or the coming into existence of anything; the formation or the production of anything.

gene splicing *See* recombinant DNA technology.

gene substitution The replacement of one allele by another allele of the same gene.

gene synthesis *See* gene machine.

gene therapy The introduction of a functional gene or genes from some source into a living cell to correct for a genetic disease. The genetic material may be introduced into the recipient cell by any of the processes of gene insertion.

genetic Of, or pertaining to, genetics.

genetic anticode A tabular arrangement of tRNA anticodons that resembles that of mRNA codons in outlay. The array indicates that certain amino acid properties can be correlated with certain properties of the anticodons. This has been taken to support the notion that the genetic code evolved as a result of necessary chemical–physical relations between amino acids and nucleotides.

genetic block A metabolic block that results from a mutation.

genetic code The specification of a sequence of amino acids in a protein by a sequence of nucleotides in a nucleic acid; the set of codons that specify the amino acids and carry the information for protein synthesis.

genetic code dictionary The set of 64 codons, 61 of which code for amino acids commonly occurring in proteins, and 3 of which are termination codons.

genetic complementation *See* complementation.

genetic cross 1. The mating of two organisms that results in the formation of genetic recombinants. 2. The production of progeny containing genotype of two or more parents, as in the simultaneous infection of bacteria with several types of phages. 3. The progeny derived from two or more parents by mating or by other means.

genetic disease A hereditary disease that arises from an abnormality in the genetic makeup of an organism, as that caused by the presence of a mutant gene. *See also* inborn error of metabolism; genetotrophic disease.

genetic drift The random change of gene frequency due to chance fluctuations, particularly that occurring in small populations.

genetic engineering The experimental or industrial approaches involving modification of the genome of a living cell that do not involve normal sexual or asexual transmission of genetic material; the manipulation of the genetic complement of a cell with the aim of altering the functions or the products of the cell. Thus, a cell may be designed that is capable of new functions or that is capable of synthesizing either new chemicals or more of existing chemicals. The design of microorganisms capable of degrading wastes or toxic substances or the design of microorganisms

capable of synthesizing hormones or new antibiotics, are examples. The term is often used synonymously with recombinant DNA technology. *See also* biotechnology; gene therapy.

genetic equilibrium The condition in which a given gene frequency remains constant throughout successive generations.

genetic expression The phenotypic aspect of a gene; the active function of a gene.

genetic fine structure The location of mutable sites on a chromosome as determined by fine-structure genetic mapping.

genetic induction GENE ACTIVATION.

genetic information The hereditary information contained in a sequence of nucleotides in either chromosomal DNA or chromosomal RNA.

genetic linkage *See* linkage.

genetic locus *See* locus.

genetic map The linear arrangement of mutable sites, or of genes, on a chromosome that is deduced from genetic recombination experiments.

genetic marker A mutable site on a chromosome that, when mutated, leads to gross and visible changes in the organism. *See also* biochemical marker.

genetic material The chromosomal nucleic acid, predominantly DNA but at times RNA, that carries the information for the synthesis of proteins and for the synthesis of other nucleic acids.

genetic recombination *See* recombination.

genetic reversion *See* reversion (1).

genetics The science that deals with heredity.

genetic system The genetic material of a given species; includes the genetic components, their organization, and the methods of transmission of genetic information.

genetotrophic disease A disease in which there are genetic insufficiencies that may be prevented, or at least ameliorated, by an increased supply of one or more nutrients. Some genetic diseases appear to be genetotrophic in origin and supply of the nutrient partially restores the impaired activity of the defective enzyme. Mental retardation is also believed to be, in part, genetotrophic in origin.

genetotrophic principle The principle that the nutritional requirements of an organism are determined by its genetic makeup.

gene transfection *See* transfection.

gene walking CHROMOSOME WALKING.

genic Of, or pertaining to, genes.

genin The aglycone portion of a saponin.

genome 1. A complete single set of the genetic material of a cell or of an organism; the complete set of genes in a gamete. 2. The single

DNA molecule of bacteria, phages, and most animal and plant viruses.

genomic blotting *See* Southern blotting.

genomic library A clone library, established much as a gene library, but containing varying amounts of the total DNA segments from a given genome. *See also* gene library.

genopathy GENETIC DISEASE.

genophore The genetic material (nucleic acid) in prokaryotes and viruses that is the equivalent of the chromosome in eukaryotes but lacks associated histones.

genotype 1. The genetic makeup of an organism; the totality of the genes of an organism. 2. A group of organisms that have an identical genetic makeup. *See also* phenotype.

genotypic Of, or pertaining to, genotype.

genotypic adaptation The preferential growth of a genotypically varied organism.

genotypic variation A rare mutation that involves only a few organisms in a population and that leads to a new genotype.

gentamycin An aminoglycoside antibiotic, produced by *Micromonospora*, that is similar to streptomycin in its mechanism of action. *Var sp* gentamicin.

gentiobiose A disaccharide of D-glucose in which the glucose molecules are linked by means of a $\beta(1 \rightarrow 6)$ glycosidic bond.

genus (*pl* genera) A taxonomic group that includes one or more closely related species.

geobiochemistry BIOGEOCHEMISTRY.

geometrical isomer One of two isomers that differ from each other in the configuration of the groups attached to two carbon atoms that are linked by a double bond; the groups may be on the same side (cis; syn) with respect to the plane of the double bond, or they may be on opposite sides (trans; anti).

geometric mean A special kind of "average"; for a set of *n* positive numbers, the geometric mean is equal to the *n*th root of their product. Thus, the geometric mean of 4, 6, and 9 is equal to $(4 \times 6 \times 9)^{1/3}$.

geotaxis A taxis in which the stimulus is gravity.

geotropism A tropism in which the stimulus is gravity.

geriatrics The branch of medicine that deals with the diseases, debilities, and problems of old age and aging people. *See also* gerontology.

GERL Golgi-endoplasmic reticulum lysosomes; a hydrolase-rich region of the endoplasmic reticulum, associated with the Golgi apparatus. Originally considered to be a region from which lysosomes arise; now believed to be the last Golgi compartment in which the sorting of all proteins into separate pathways (membrane proteins, secretory pro-

teins, lysosomal enzymes) takes place. *Aka* trans-Golgi network (TGN); trans-tubular network.

germ cell A reproductive cell; a cell that can be fertilized when it is mature and that can reproduce the organism; an ovum or a spermatozoon, or any of their antecedents. *See also* somatic cell.

germ-free animal An animal reared in a bacteria-free environment.

germicidal agent An agent that kills microorganisms.

germicide GERMICIDAL AGENT.

germinal cell A cell that develops into a gamete upon meiosis; an oocyte or a spermatocyte.

germinal mutation A mutation in a germinal cell.

germination 1. The overall process, consisting of activation, initiation, and out-growth, whereby a spore is converted to a vegetative cell. 2. The second stage in the conversion of a spore to a vegetative cell that is characterized by the rehydration of the spore and by the loss of dipicolinic acid and glycopeptides. *Aka* initiation. 3. The beginning of growth of a seed or of a reproductive body after a period of dormancy.

germ layer *See* ectoderm, endoderm, mesoderm.

germ line The gametes and their antecedents.

germ-line theory A theory of the origin of the genes that code for the variable regions of antibody molecules and that allow for the great diversity of antibodies. According to this theory all cells, including lymphocytes, have the same set of genes as those in the germ cells from which the individual arose. All the genes for all the antibodies that an individual can make are, therefore, already present in the fertilized egg and are transmitted through the germ line.

germ plasm The genetic material in the germ cells; the sum total of the genes transmitted to the offspring through the germ cells.

gerontology The science that deals with the physical processes and phenomena of old age and aging people. *See also* geriatrics.

gerovital A buffered solution of procaine hydrochloride, better known as Novocain, a painkiller used by dentists. As the name indicates, the compound is being promoted as one that alleviates symptoms of aging and it has been designated by some as vitamin H_3 or vitamin GH_3. These claims have not been supported and gerovital is not recognized as a vitamin by United States and Canadian drug authorities.

gestagen PROGESTIN.

gestation Pregnancy.

gestin PROGESTIN.

GeV Giga (10^9) electron volts.

g factor *See* g value.

G factor 1. GUANOSINE TRIPHOSPHATASE. 2. TRANSLOCASE.

GFC Gel filtration chromatography.

GFR Glomerular filtration rate.

GGE Gradient gel electrophoresis.

GH Growth hormone.

GHIF *See* growth hormone regulatory hormone.

GHIH *See* growth hormone regulatory hormone.

ghost 1. An erythrocyte that has lost some or all of its cytoplasmic content; prepared by allowing the erythrocyte to swell in distilled water so that its permeability is increased and the cytoplasmic material leaks out. 2. A T-even phage that has lost some or all of its DNA; prepared by subjecting the phage to an osmotic shock and digesting the DNA of the ruptured phage with deoxyribonuclease. 3. A spheroplast that has lost essentially all of its intracellular material; prepared by growing gram-negative bacteria in the presence of penicillin or by digesting the cell wall of gram-negative bacteria with lysozyme.

ghost peak An unexpected peak that is present in a gas chromatogram and that is usually due to a contaminant.

GHRF *See* growth hormone regulatory hormone.

GHRH *See* growth hormone regulatory hormone.

GHRIF *See* growth hormone regulatory hormone.

GHRIH *See* growth hormone regulatory hormone.

G_i *See* G protein.

giant chromosome POLYTENE CHROMOSOME.

giant messenger-like RNA HETEROGENEOUS NUCLEAR RNA.

giant RNA HETEROGENEOUS NUCLEAR RNA.

gibberellic acid The parent compound of the gibberellins.

gibberellin One of a group of widely occurring plant hormones that stimulate the growth of leaves and shoots; chemically, they are diterpenoid acids, derived from a tetracyclic skeleton.

gibberellin antagonist Any compound that functions by counteracting the effect of gibberellin; this includes competitive inhibitors (antigibberellins), growth retardants (such as chlorocholine chloride), and other plant hormones (such as abscisic acid).

Gibbs–Donnan equilibrium The unequal distribution of diffusible ions that is established at equilibrium on the two sides of a membrane if one side contains a nondiffusible ion.

The unequal distribution of the diffusible ions becomes more pronounced with an increase in the concentration and/or the charge of the nondiffusible ion; it becomes less pronounced with increasing concentrations of the diffusible ions added initially to that side of the membrane that does not contain the nondiffusible ion. *Aka* Gibbs–Donnan effect; Gibbs–Donnan membrane equilibrium.

Gibbs–Duhem equation An equation that relates the chemical potential of different components in a system. At constant temperature and pressure the equation can be written as $dG = \sum_i \mu_i dn_i + \sum_i n_i d\mu_i$, where G is the free energy, and μ_i and n_i are the chemical potential and the number of moles of component i. Since $dG = \sum_i \mu_i dn_i$, the equation simplifies to $\sum_i n_i d\mu_i = 0$.

Gibbs free energy *See* free energy.

Gibbs–Helmholtz equation An equation that describes the variation in free energy changes as a function of temperature at constant pressure (P). It can be formulated as $(\partial G/\partial T)_p = -\Delta S$ or as $[\partial(\Delta G/T)/\partial T]_p = -\Delta H/T^2$ where ΔG is the change in free energy, T is the absolute temperature, ΔH is the enthalpy change, and ΔS is the entropy change.

Gibbs phase rule PHASE RULE.

GIF *See* growth hormone regulatory hormone.

giga- Combining form meaning one billion (10^9) and used with metric units of measurement. *Sym* G.

gigantism ACROMEGALY.

GIH *See* growth hormone regulatory hormone.

Gilbert–Maxam method See Maxam–Gilbert method.

Gilbert's disease CRIGLER–NAJJAR SYNDROME.

GIP Gastric inhibitory peptide.

Girard's reagent One of a group of reagents used to extract ketosteroids from urine; reagent T is trimethylammonium acetyl hydrazide chloride, and reagent P is pyridinium acetyl hydrazide chloride.

G_k *See* G protein.

GK Glycerokinase.

gland A cell or an organ that produces specific substances that are secreted outside the cell or the organ, either directly or through ducts, and that are used in other parts of the body or are eliminated from the body.

glandular Of, or pertaining to, glands.

glass electrode An electrode that has a thin glass membrane incorporated into its design and that is sensitive to the hydrogen ion concentration of solutions; it is used for the measurement of pH.

Glc Glucose.

GLC Gas liquid chromatography.

GlcA Gluconic acid.

GlcN Glucosamine.

GlcNAc *N*-Acetylglucosamine.

GlcUA Glucuronic acid.

GLDH Glutamate dehydrogenase.

GLI Glucagon-like immunoreactivity.

gliadin A seed protein of wheat.

glial cell A cell of neural tissue that plays an important role in controlling the chemical environment of neurons.

Gln 1. Glutamine. 2. Glutaminyl.

globin The polypeptide chain that is associated with an iron–porphyrin group in both hemoglobin and myoglobin.

globinometer An instrument for measuring the amount of oxyhemoglobin in blood.

globin zinc insulin The zinc salt of a mixture of globin and insulin that is less soluble than insulin alone. *See also* NPH insulin.

globoid leukodystrophy KRABBE'S DISEASE.

globoprotein One of a group of cell surface glycoproteins that react with antibodies to globosides.

globo-series glycosphingolipids *See* glycosphingolipids.

globoside A ceramide oligoglycoside that contains two or more simple sugars, amino sugars, and *N*-acetyl amino sugars.

globular Spherical.

globular actin G-ACTIN.

globular protein A protein in which the polypeptide chain (or chains) is coiled in three dimensions to form a more or less globular molecule. The polypeptide chain is held together through intrachain bonds such as hydrogen bonds, hydrophobic bonds, electrostatic bonds, and disulfide bonds, and varying lengths of the polypeptide chain may be in helical configuration. Globular proteins have diverse functions and occur in the form of enzymes, storage proteins, transport proteins, and so on. *See also* fibrous protein.

globulin A globular and simple protein that is either insoluble or sparingly soluble in water, is soluble in dilute salt solutions, and is precipitated by ammonium sulfate at 50% saturation.

glomerular filtrate The filtrate, free of cells and colloidal particles, that is produced from blood by the glomeruli of the kidney.

glomerular filtration rate The rate at which the glomerular filtrate is produced; a measure of the efficiency of the kidney that is generally expressed in terms of the clearance of a substance, such as inulin, which is metabolically inert, freely filterable, and neither absorbed nor secreted by the kidney. *Abbr* GFR.

glomerulus (*pl* glomeruli). *See* nephron.

glove box A sealed box of glass or plastic that has two or more gloves fitted into its sides; allows for the safe manipulation of the contents of the box without breaking the atmos-

pheric seal. *Aka* glove bag.

Glu 1. Glutamic acid. 2. Glutamyl.

GluA Glucuronic acid.

glucagon A polypeptide hormone that antagonizes the action of insulin and leads to an increase in the level of blood sugar by stimulating glycogenolysis; it is secreted by the islets of Langerhans in the pancreas.

glucagon-like immunoreactivity ENTEROGLUCAGON.

glucalogue A monosaccharide analogue of glucose, such as deoxyglucose, methylglucose, or galactose.

glucan A homopolysaccharide composed of glucose units.

glucocerebroside A glucose-containing cerebroside.

glucocorticoid A 21-carbon steroid hormone that is secreted by the adrenal cortex and that acts primarily on carbohydrate, lipid, and protein metabolism. Glucocorticoids include corticosterone, cortisone, and cortisol, and lead to protein catabolism, gluconeogenesis from the amino acids thus formed, lipid mobilization, and an increase in ketone bodies. In addition, glucocorticoids also have antiallergic and anti-inflammatory effects.

glucocorticoid receptors A group of cytosolic receptors that mediate the action of glucocorticoids.

glucocorticosteroid GLUCOCORTICOID.

glucogenesis The synthesis of glucose from precursors other than glycogen.

glucogenic Of, or pertaining to, glucogenesis.

glucokinase An enzyme that catalyzes the phosphorylation of glucose to glucose-6-phosphate. It has different molecular and kinetic properties than hexokinase and is found almost exclusively in the liver. *See also* hexokinase.

glucolipid A glucose-containing lipid.

glucomannan A heteropolysaccharide composed of glucose and mannose residues.

glucone The glucose moiety of a glucoside.

gluconeogenesis The synthesis of glucose from noncarbohydrate precursors such as amino acids, intermediates of glycolysis, or intermediates of the citric acid cycle.

gluconeogenic Of, or pertaining to, gluconeogenesis.

glucoplastic amino acid An amino acid whose degradation products can contribute to gluconeogenesis.

glucopyranose Glucose that has a six-membered ring structure resembling that of the compound pyran.

glucosamine An amino sugar of glucose that is a component of chitin and occurs in vertebrate tissues. *Abbr* GlcN.

glucosaminoglycan *See* proteoglycan.

glucose The six-carbon aldose that is the major sugar in the blood and a key intermediate in metabolism. *Abbr* Glc; G.

glucose–alanine cycle A cyclic set of reactions whereby glucose is converted to alanine in muscle and alanine is reconverted to glucose in the liver. The reactions are connected via the blood which transports alanine from muscle to liver and glucose from liver to muscle. The cycle serves two functions: (a) transport of amino groups from amino acids in muscle to the liver where they can be converted to urea; (b) supply of muscle with glucose derived from amino acids catabolized in the liver.

glucose effect CATABOLITE REPRESSION.

glucose electrode One of a number of membrane probes that permit the determination of glucose by means of enzymes incorporated into the electrode.

glucose-6-phosphatase The enzyme that catalyzes the hydrolysis of glucose-6-phosphate to glucose and inorganic phosphate and that is responsible for maintaining the glucose level in the blood. The enzyme is found only in the liver, the kidneys, and the intestine, and acts as a member of a multienzyme system which includes glucose-6-phosphotranslocase (T_1), an inorganic phosphate translocase (T_2), and a glucose translocase (T_3).

glucose-6-phosphate dehydrogenase The pyridine-linked dehydrogenase that catalyzes the first reaction of the hexose monophosphate shunt whereby glucose-6-phosphate is oxidized to 6-phosphoglucone-δ-lactone. *Abbr* G6PDH; G6PD.

glucose-6-phosphate dehydrogenase deficiency A genetically inherited metabolic defect in humans that is due to a deficiency of the enzyme glucose-6-phosphate dehydrogenase. Afflicted individuals show primaquine sensitivity, develop Favism upon eating broad beans, and appear to be protected from malaria transmitted by *Plasmodium falciparum*.

glucose-regulated protein One of a number of proteins in animal cells whose synthesis is regulated by the glucose concentration in the culture medium. *Abbr* GRP.

glucose repression CATABOLITE REPRESSION.

glucose tolerance factor A low molecular weight (500 daltons), water soluble, organic complex of chromium that is required by humans and animals for normal glucose tolerance. *Abbr* GTF.

glucose tolerance test A measure of the rate at which glucose is metabolized; used as a screening test for diabetes. A dose of glucose is administered to a fasting individual and the blood glucose concentration is then determined as a function of time. In normal individuals, the glucose concentration rises to a maximum within about 30 min and drops back to the initial level within about 2 h. In diabetic individuals, the glucose concentration rises to a higher value and does not drop back as rapidly as is the case for normal individuals.

β-glucosidase *See* amygdalin.

glucoside A glycoside of glucose.

glucosphingolipid GLUCOCEREBROSIDE.

Glucostat Trade name for a glucose oxidase reagent.

glucosuria The presence of excessive amounts of glucose in the urine.

glucosylation The introduction of a glucose residue into an organic compound.

glucosylceramide lipidosis GAUCHER'S DISEASE.

glucosyl group A glucose residue that is linked to another group or molecule by means of a glycosidic bond.

glucuronate pathway A metabolic pathway for the conversion of glucose to xylulose-5-phosphate that is operative in higher plants, mammals, crustaceans, and yeast; it is apparently not a major pathway for the oxidation of glucose. The pathway serves to provide vitamin C in plants and in those animals capable of synthesizing this vitamin and ties in with the pentose phosphate pathway. A specific block in the glucuronate pathways leads to the human disease idiopathic pentosuria. *Aka* glucuronic acid oxidation pathway; glucuronate–gulonate pathway; glucuronate–xylulose cycle.

glucuronic acid A sugar acid of glucose. *Abbr* GluA; GlcUA.

glucuronic acid oxidation pathway GLUCURONATE PATHWAY.

β-glucuronidase *See* laetrile.

glucuronide A compound formed by linking glucuronic acid to another compound by means of a glycosidic bond; many toxic compounds are detoxified by being converted to a glucuronide and are then excreted in this form.

glumitocin A peptide hormone, secreted by the posterior lobe of the pituitary gland, that is related in structure and function to oxytocin and occurs in some fish.

GluNH$_2$ 1. Glutamine. 2. Glutaminyl.

Glusulase Trademark for an enzyme preparation, obtained from the intestinal juice of snails (*Helix pomatia*), that contains a mixture of β-glucuronidase and sulfatase.

glutamate dehydrogenase An enzyme that catalyzes the addition of NH_3 to α-ketoglutarate to form glutamate in ammonia fixation and catalyzes the removal of NH_3

from glutamate to form α-ketoglutarate in amino acid deamination. *Abbr* GDH.

glutamic acid   An aliphatic, acidic, and polar alpha amino acid. *Abbr* Glu; E; GA.

glutamic semialdehyde   A derivative of glutamic acid in which only one of the two carboxyl groups has been converted to an aldehyde group.

glutamine   An aliphatic, polar alpha amino acid; the amide of glutamic acid. *Abbr* Gln; $GluNH_2$; Q.

glutamine antagonists   Analogues of glutamine that serve as competitive inhibitors in enzyme reactions involving glutamine.

γ-glutamyl cycle   A cyclic set of reactions, present in some tissues, that serves to transport amino acids across the cell membrane; uses reactions of glutathione metabolism and a membrane-localized enzyme, γ-glutamyl transpeptidase (γ-glutamyl transferase). *Aka* Meister cycle.

glutaredoxin   A heat-stable protein that acts in conjunction with glutathione in the formation of deoxyribonucleotides from ribonucleotides.

glutaric acid   A five-carbon dicarboxylic acid that is structurally similar to succinic acid and that is a competitive inhibitor of succinate dehydrogenase; it is an intermediate in lysine metabolism.

glutaric aciduria   One of two genetically inherited metabolic defects in humans that are characterized by a large excretion of glutaric acid. Type I involves progressive neurological degeneration and is due to a deficiency of the enzyme glutaryl-CoA dehydrogenase. Type II involves hypoglycemia and fatal neonatal acidosis; its enzyme deficiency is not known.

glutarimide antibiotics   A group of antibiotics that includes cycloheximide, streptimidone, and streptovitacin.

glutathione   A widely distributed tripeptide, γ-glutamyl-cysteinyl-glycine, that serves as a coenzyme for some enzymes and is also thought to function as an antioxidant in protecting the sulfhydryl groups of enzymes and other proteins. Reduced glutathione is synonymous with glutathione and is abbreviated GSH; oxidized glutathione is a dimer of two glutathione molecules, linked by means of a disulfide bond, and is abbreviated GSSG.

glutathione-S-transferase   *See* mercapturic acids.

glutelin   A simple, globular protein of plant origin that is insoluble in water, alcohol, or salt solutions, but is soluble in dilute solutions of acids or bases.

gluten   The principal protein of wheat; the properties of gluten permit the formation of leavened bread.

Glx   The sum of glutamic acid and glutamine when the amide content is either unknown or unspecified.

Gly   1. Glycine. 2. Glycyl.

glycan   POLYSACCHARIDE.

glycaric acid   An acid formed by the oxidation of both the terminal —CHO group and the terminal —CH_2OH group of an aldose to —COOH groups.

glycation   The nonenzymatic glycosylation of proteins; the reaction of glucose with the alpha and epsilon amino groups of proteins.

glycemia   The presence of glucose in the blood.

glyceraldehyde   A three-carbon aldose, a phosphorylated derivative of which is an intermediate in glycolysis; serves as a reference compound for the assignment of D and L configurations to amino acids, carbohydrates, and related compounds.

glycerate pathway   An anaplerotic reaction sequence whereby glyoxylate is converted to 3-phosphoglyceric acid.

glyceride   ACYLGLYCEROL.

glycerin   GLYCEROL.

glycerol   A three-carbon trihydroxy alcohol that occurs in many lipids; phosphorylated derivatives of glycerol are intermediates in glycolysis.

glycerol fermentation   The formation of small amounts of glycerol during alcoholic fermentation.

glycerolipid   A lipid derived from glycerol.

glycerol phosphate shuttle   A shuttle involving glycerol-3-phosphate, dihydroxyacetone phosphate, and glycerol phosphate dehydrogenase. The shuttle achieves the oxidation of cytoplasmic NADH to NAD^+ at the expense of the reduction of mitochondrial FAD to $FADH_2$. As a result, for every NADH formed in the cytoplasm, and then oxidized, 2 ATPs are formed inside the mitochondria as $FADH_2$ feeds into the electron transport system.

glycerol phosphatide   PHOSPHOGLYCERIDE.

glycerone   The ketone derived from glycerol; dihydroxyacetone.

glycerophospholipid   Any derivative of glycerophosphoric acid that contains at least one *O*-acyl, *O*-alkyl, or *O*-alkenyl group attached to the glycerol residue. The common glycerophospholipids are named as derivatives of phosphatidic acid (phosphatidyl choline, phosphatidyl serine, and phosphatidyl ethanolamine). *Aka* phosphatidate.

glycerophosphoric acid   PHOSPHOGLYCERIC ACID.

glycine   The simplest alpha amino acid (NH_2CH_2COOH); it can be classified as an aliphatic and polar amino acid. *Abbr* Gly; G. *Aka* aminoacetic acid; glycocoll.

glycine–allantoin cycle   A cyclic set of reactions that leads to the synthesis of urea and that occurs in the lung fish and in some urea-accumulating plants; involves glycine for purine biosynthesis, formation of allantoin from purine degradation, and conversion of allantoin to glycine.

glycine cleavage enzyme A multienzyme system that catalyzes the conversion of glycine and tetrahydrofolic acid to N^5,N^{10}-methylene tetrahydrofolic acid and CO_2. A deficiency of this enzyme system leads to nonketotic hyperglycinemia.

glycine–succinate cycle SHEMIN CYCLE.

glycinin The major protein component of soybeans.

glycitol ALDITOL.

glyco- Combining form meaning carbohydrate.

glycoaldehyde group GLYCOLALDEHYDE GROUP.

glycoalkaloids *See* saponin; steroid alkaloids.

glycocalyx CELL COAT.

glycocholic acid The compound formed by the conjugation of cholic acid and glycine; one of the bile salts.

glycocoll GLYCINE.

glycoconjugates Collective term for glycolipids and glycoproteins; refers to these lipids and proteins together with their oligosaccharide components.

glycocyamine GUANIDINOACETIC ACID.

glycogen A highly branched homopolysaccharide of D-glucose units that is the major form of storage carbohydrate in animals; the glucose units are linked by means of $\alpha(1 \rightarrow 4)$ and $\alpha (1 \rightarrow 6)$ glycosidic bonds.

glycogenesis The synthesis of glycogen; the anabolism of glycogen.

glycogen granule A cytoplasmic storage particle of glycogen that also contains proteins and enzymes that function in the synthesis and breakdown of glycogen.

glycogenic Of, or pertaining to, glycogenesis.

glycogenic amino acid An amino acid that can serve as a precursor of pyruvic acid, glucose, and glycogen in metabolism.

glycogenolysis The breakdown of glycogen; the catabolism of glycogen.

glycogenosis (*pl* glycogenoses) GLYCOGEN STORAGE DISEASE.

glycogen phosphorylase The enzyme that catalyzes the successive hydrolytic removal of glucose residues, in the form of glucose-1-phosphate, from the nonreducing end of glycogen; this reaction is the first step for the utilization of glycogen in glycolysis. *See also* phosphorylase.

glycogen storage disease One of a group of genetically inherited metabolic defects in humans that are characterized by an abnormal accumulation of liver and tissue glycogen and that are due to deficiencies of enzymes which

function in glycogen metabolism. The enzymatic deficiencies and the synonymous names of the eight known types of glycogen storage diseases are as follows: type I, glucose-6-phosphatase, von Gierke's disease; type II, α-1,4-glucosidase, Pompe's disease; type III, amylo-1,6-glucosidase, Cori's disease, limit dextrinosis; type IV, amylo $(1,4 \rightarrow 1,6)$ transglucosylase, Andersen's disease, amylopectinosis; type V, muscle glycogen phosphorylase, McArdle's disease; type VI, liver glycogen phosphorylase, Hers' disease; type VII, muscle phosphofructokinase; type VIII, liver phosphorylase kinase.

glycogen synthase The enzyme that catalyzes the synthesis of the straight chains of glycogen from UDP-glucose. Two forms of glycogen synthase exist, a dependent form, denoted D, and an independent form, denoted I. *See also* dependent form; independent form.

glycoglycerolipid A glycolipid containing one or more glycerol residues.

glycol A 1,2-diol such as ethylene glycol, $HOCH_2$—CH_2OH.

glycolaldehyde group The grouping CH_2OH—CO—.

glycolic acid cycle A cyclic set of reactions whereby glyoxylate (CHO—COO^-) is converted to glycolate (CH_2OH—COO^-) which is then reoxidized to glyoxylate.

glycolipid Any compound containing one or more monosaccharide residues linked via a glycosidic bond to a lipid component such as an acylglycerol, a sphingoid, or a ceramide.

glycoloyl group GLYCOLALDEHYDE GROUP.

glycolyl group GLYCOLALDEHYDE GROUP.

glycolysis 1. The anaerobic degradation of carbohydrates whereby a molecule of glucose is converted by a series of steps to two molecules of lactic acid. The overall reaction sequence yields a limited amount of ATP and is commonly divided into two stages. Stage I refers to the ATP-requiring reactions whereby glucose is converted to glyceraldehyde-3-phosphate; stage II refers to the ATP-yielding reactions whereby glyceraldehyde-3-phosphate is converted to lactic acid. *Aka* anaerobic glycolysis. 2. The sequence of reactions from glucose to pyruvic acid that is common to carbohydrate catabolism under both aerobic and anaerobic conditions. *Aka* aerobic glycolysis.

glycolytic Of, or pertaining to, glycolysis.

glycolytic flux The rate of glycolysis particularly in reference to the energy requirements of the cell or the organism. Thus, if an organism requires energy and can obtain it only from glucose, the resultant glycolytic flux would be high.

glycolytic pathway GLYCOLYSIS.

glycone The carbohydrate portion of a glycoside.

glyconeogenesis The synthesis of glycogen from noncarbohydrate precursors, such as fat or protein.

glyconic acid An acid formed by the oxidation of the terminal —CHO group of an aldose to a —COOH group.

glycopeptide A peptide that is linked covalently to a carbohydrate, as that which is polymerized to form the peptidoglycan of bacterial cell walls.

glycophorin A major sialoglycoprotein in the red blood cell membrane; it contains sialic acid, carbohydrate (60%), and protein (40%). Glycophorin is an integral membrane protein that spans the entire width of the membrane. The probable structure of glycophorin is that of a single polypeptide chain to which are attached short carbohydrate chains. The latter represent a large proportion of the total mass of the molecule and protrude on the extracellular side of the membrane. *Aka* PAS-1.

glycophosphoglyceride A compound composed of phosphatidic acid, the phosphate group of which is linked to a sugar.

glycophosphosphingolipid A phosphosphingolipid that is a derivative of a ceramide and that contains both sugars and phosphate esters.

glycoprotein A conjugated protein in which the nonprotein portion is a carbohydrate that is linked covalently to the protein. The attached carbohydrate is either a monosaccharide or a relatively short oligosaccharide. Frequently, the amount of carbohydrate per molecule is small (about 4%). Some glycoproteins contain only one or a few carbohydrate groups; others have numerous oligosaccharide side chains which may be linear or branched. Glycoproteins are generally either secreted into body fluids or are membrane proteins; that is, they have an extracellular location and function. Glycoproteins include enzymes, hormones, carriers, lectins, membrane proteins, antibodies, and so on. *See also* proteoglycan.

glycosaminoglycan The carbohydrate moiety of proteoglycans; consists of repeating disaccharide units in which either D-glucosamine or D-galactosamine is always present. Glycosaminoglycans generally contain uronic acid and sulfate groups, linked via ester or amide bonds. Six distinct classes are recognized: hyaluronic acid, chondroitin sulfate, dermatan sulfate, keratan sulfate, heparin, and heparan sulfate.

glycosidase CARBOHYDRASE.

glycoside A mixed acetal (or ketal) derived from the cyclic hemiacetal (or hemiketal) form of an aldose (or a ketose); a compound formed by replacing the hydrogen of the hydroxyl group of the anomeric carbon of the carbohydrate with an alkyl or an aryl radical.

glycosidic bond The bond between the anomeric carbon of a carbohydrate and some other group or molecule; the C–O bond in disaccharides and the C–N bond in nucleosides are two examples. *Aka* glycosidic link.

glycosome A proposed compartment that contains the enzymes of glycolysis; such an organelle, containing all of the glycolytic enzymes, has been observed in trypanosomes.

glycosphingolipid Any compound containing a sphingoid and one or more monosaccharide units; glycosylsphingoids and glycosylceramides are examples. Most polyglycosyl ceramides have one of the following core structures: (a) ganglio series, ceramide-glucose-galactose-N-acetylgalactosamine-galactose; (b) globo series, ceramide-glucose-galactose-galactose-*N*-acetylgalactosamine; (c) lacto series, ceramide-glucose-galactose-*N*-acetylgalactosamine-galactose. See also glycosyl ceramide.

glycosuria The presence of excessive amounts of glucose and/or other reducing sugars in the urine. When the particular carbohydrate has been identified, more specific terms (glucosuria, fructosuria, etc.) are used to describe the condition.

glycosylase *See* DNA glycosylase.

glycosylation The covalent attachment of carbohydrates to a molecule. Used particularly for the attachment of carbohydrates to secretory proteins (forming glycoproteins) which occurs in the lumen of the rough endoplasmic reticulum and which is belived to play a role in guiding secretory proteins to their particular cellular destinations. Glycosylated proteins are of two types. Some contain N-linked oligosaccharides (oligosaccharides linked to the amide group of asparagine); others contain O-linked oligosaccharides (oligosaccharides linked to the hydroxyl group of serine, threonine, or hydroxylysine). The oligosaccharide is transferred to the protein via a carrier, dolichol phosphate. *See also* undecaprenol; signal hypothesis.

glycosyl ceramide A carbohydrate-containing derivative of a ceramide. Monoglycosyl ceramides contain one carbohydrate group per molecule and include cerebrosides and sulfatides; oligoglycosyl ceramides contain 2 to 10 carbohydrate groups per molecule and include cytosides, globosides, and gangliosides; polyglycosyl ceramides contain more than 10 carbohydrate groups per molecule. Mono-, oligo-, and polyglycosyl ceramides are also known either as ceramide mono-, ceramide oligo-, and ceramide polyglycosides, or as ceramide mono-, ceramide oligo-, and ceramide polysaccharides. *See also* glycosphingolipid.

glycosyl glyceride A glycerolipid that is a glycoside of a diacylglycerol.

glycosyl group A sugar residue that is linked to another group or molecule by means of a glycosidic bond.

glycosyllipid GLYCOLIPID.

glycosyltransferase One of a group of enzymes that catalyze the transfer of a glycosyl group (a mono- or oligosaccharide) from a glycosylnucleotide to some acceptor (a carbohydrate, peptide, or lipid) having a suitable OH group. *See also* one enzyme-one linkage hypothesis.

glycuronic acid An acid formed by the oxidation of the terminal —CH_2OH group of an aldose to a —COOH group.

glycyrrhizinic acid A sweet compound in licorice roots that has mineralocorticoid activity. Its aglycone is a pentacyclic triterpene called glycyrrhetic acid (glycyrrhetin). *Aka* glycyrrhizic acid.

glyoxalase The enzyme that catalyzes the conversion of methyl glyoxal to lactic acid.

glyoxalate The ionized form of glyoxal (glyoxalic acid; glyoxylic acid); the compound O=CH—COO⁻. *Var sp* glyoxylate.

glyoxisome PEROXISOME.

glyoxylate Variant spelling of glyoxalate.

glyoxylate bypass A set of two enzymatic reactions whereby isocitric acid is converted to malic acid; the operation of these reactions results in a modified citric acid cycle, known as the glyoxylate cycle.

glyoxylate cycle A modified citric acid cycle in which the reaction sequence between isocitric acid and malic acid is altered and acetate is used both as a source of energy and as a source of intermediates. The cycle occurs in some plants and in some microorganisms, and requires the input of two molecules of acetyl coenzyme A; it leads to the synthesis of a molecule of succinic acid which is used for the synthesis of carbohydrates and other cell components. *Aka* glyoxylate shunt; dicarboxylic acid cycle.

glyoxylic acid reaction HOPKINS—COLE REACTION.

glyoxysome A cytoplasmic organelle of plants that contains the enzymes of the glyoxylate cycle.

gm Gram; the preferred abbreviation is g.

G-M counter Geiger—Mueller counter.

Gm group A group of allotypic antigenic sites on the gamma chain of human IgG immunoglobulins.

GMP 1. Guanosine monophosphate (guanylic acid). 2. Guanosine-5'-monophosphate (5'-guanylic acid).

G myeloma protein An abnormal IgG immunoglobulin that is produced by individuals suffering from multiple myeloma.

gnotobiosis The rearing of gnotobiotic animals.

gnotobiota The known microfauna and microflora of a gnotobiotic animal.

gnotobiotic animal 1. GERM-FREE ANIMAL. 2. A germ-free animal that has purposely been infected with one or more known bacterial species.

GnRF Gonadotropin releasing factor; *See* luteinizing hormone releasing hormone.

goiter An abnormal enlargement of the thyroid gland.

goitrogen A goitrogenic compound; an antithyroid compound; a compound that inhibits the formation of thyroxine and triiodothyronine.

goitrogenic Causing goiter.

goitrogenic glycoside A plant toxin that is a glycoside and that causes hyperthyroidism.

Goldberg–Hogness box TATA BOX.

Goldman equation An equation for calculating the potential of a membrane that is permeable to several ionic species; based on the ionic concentrations on both sides of the membrane and on the permeability coefficients of the ions.

Golgi apparatus A cytoplasmic organelle, composed of cisternae and vesicles, that functions in the collection and in the subsequent secretion of substances synthesized by the cell; an example of such substances are the proteins that are synthesized by the ribosomes attached to the endoplasmic reticulum. *Aka* Golgi body; Golgi complex; Golgi material.

Golgi-endoplasmic reticulum lysosomes *See* GERL.

Golgi stack A set of flattened, disk-shaped cisternae of the Golgi apparatus that form a structure resembling a stack of plates. *Aka* dictyosome.

gonad A sex gland; an ovary or a testis.

gonadal hormones SEX HORMONES.

gonadoliberin GONADOTROPIN RELEASING HORMONE.

gonadotrophic hormone Variant spelling of gonadotropic hormone.

gonadotropic hormone GONADOTROPIN.

gonadotropin A hormone that stimulates the gonads; gonadotropins are secreted by the anterior lobe of the pituitary gland, the placenta, and the endometrium. Gonadotropins include follicle-stimulating hormone (FSH), luteinizing hormone (LH), human menopausal gonadotropin (HMG), prolactin (PRL), and human chorionic gonadotropin (HCG). *Var sp* gonadotrophin.

gonadotropin releasing hormone LUTEINIZING

HORMONE RELEASING HORMONE.

gonane The parent hydrocarbon skeleton of the steroids; the tetracyclic ring system without the angular methyl groups and the aliphatic side chain.

gonosome A motile germ cell of animal origin; a spermatozoon.

goodness of fit The agreement between an observed set of values and a second set, derived wholly or partly on the basis of theoretical or hypothetical considerations.

Good's buffers A group of buffers, proposed by N. E. Good, that are especially suited for research involving biochemical systems in the physiological pH range. *See also* biological buffers.

gossypol An aromatic triterpene found in cotton seed (*Gossypum hirsutum*).

GOT Glutamate–oxaloacetate transaminase. *Aka* aspartate transaminase.

gougerotin A pyrimidine antibiotic, produced by *Streptomyces gougeroti*, that inhibits protein synthesis in both prokaryotic and eukaryotic systems.

gout A metabolic disease that is characterized by an increase in the concentration of uric acid in the serum and by its precipitation as sodium urate in various tissues of the body; a form of acute arthritis. One type of gout is a genetically inherited metabolic defect due to a deficiency of the enzyme hypoxanthine-guanine phosphoribosyl transferase.

Gouy interferometer An interferometer in which constructive and destructive interference of light that has passed through a horizontal slit results in a series of horizontal light and dark fringes that are progressively compressed in one direction.

GP Glycerophosphate. *Aka* glycerol-1-phosphate; α-glycerophosphate.

G-1-P Glucose-1-phosphate.

G-3-P Glyceraldehyde-3-phosphate.

G-6-P Glucose-6-phosphate.

GPC Gel permeation chromatography.

G6PD Glucose-6-phosphate dehydrogenase.

GPDH Glyceraldehyde phosphate dehydrogenase.

G6PDH Glucose-6-phosphate dehydrogenase.

G protein One of a number of guanosine triphosphate (GTP)-binding, regulatory proteins that serve as membrane bound transducers of chemically and physically coded information; they are intermediaries in transmembrane signaling pathways that consist of three proteins: receptor, G protein, and effector. The G protein becomes activated upon binding GTP. The latter is subsequently slowly hydrolyzed to GDP. When the hydrolysis is complete, the regulatory effect of the G protein is terminated and it is then available for reactivation by binding GTP. A G protein has been isolated from human red blood cells (designated G_k); it functions in the activation of K^+ channels in heart cells. Another G protein (called transducin) is a peripheral membrane protein, occurring in retinal rods, and plays a part in the biochemistry of vision. Other G proteins include G_s and G_i, which regulate adenyl cyclase, and G_0, a G protein of unknown function. *See also* amplifier enzyme.

GPT Glutamate-pyruvate transaminase.

gradient The change in the value of a property per unit distance in a specified direction.

gradient-coupled active transport SECONDARY ACTIVE TRANSPORT.

gradient curve The plot of refractive index gradient, which is proportional to the concentration gradient, versus distance; obtained with the schlieren optical system.

gradient elution A column chromatographic technique in which the composition of the eluent is changed continuously, usually with respect to ionic strength and/or pH.

gradient-flow method A method for studying the kinetics of enzymatic reactions by means of a flow system and a gradient, such as a linear substrate gradient.

gradient gel electrophoresis An electrophoretic technique in which the particles move across a concentration gradient by passing through a gel of progressively decreasing pore size. *Abbr* GGE.

gradient layer A layer, the composition of which forms a gradient; used in thin-layer chromatography and thin-layer electrophoresis.

gradient mixer A device for preparing density gradients.

gradient plate technique A technique for isolating antibiotic resistant mutants by plating the culture on a petri dish containing a solid medium with a gradient of antibiotic concentration. The petri dish is prepared by first pouring agar into it, with one edge of the plate being elevated. After hardening, the dish is leveled and a second quantity of agar, containing the antibiotic, is poured into it.

gradient sievorptive chromatography Any sievorptive chromatographic technique in which there is a gradient of small molecules across the chromatographic support.

graduated Divided into equal units by means of lines, such as the scale of a thermometer.

Graffi leukemia virus A mouse leukemia virus that belongs to the group of leukoviruses.

graft TRANSPLANT.

graft copolymer A synthetically produced copolymer in which homopolymer branches of one monomer unit (for example, vinyl polymers) are "grafted" onto a homopolymer chain of another monomer unit (for example, cellulose). Thus, the polymer has a structure such as

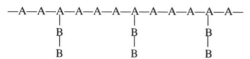

graft rejection The cell-mediated immune response, triggered by the antigens of the transplanted tissue, that leads to destruction of the transplant.

graft-versus-host reaction The disease in the recipient of a transplant that is caused by the transfer of immunocompetent cells together with the transplant from the donor; as a result, the immunocompetent cells of the transplant form antibodies to the tissue antigens of the recipient.

gram A metric unit of weight, originally defined as the mass of 1 cm^3 of water at 4 °C; now defined as one-thousandth of the SI base unit of mass, the kilogram. Originally designated gm, now designated g.

gram-atom GRAM-ATOMIC WEIGHT.

gram-atomic weight The atomic weight expressed in grams; the weight of an element in grams that is numerically equal to its atomic weight.

gram-equivalent weight The weight of a substance in grams that will either release or combine with 1 g of hydrogen or 8 g of oxygen.

gramicidin One of a group of linear polypeptide antibiotics, produced by *Bacillus brevis*. They carry a formyl group at the N-terminal and an ethanolamine group at the C-terminal. Gramicidins are active against gram-positive bacteria and function by increasing the permeability of the cell membrane to ions. *See also* tyrocidin.

gramicidin S A cyclic polypeptide antibiotic, produced by *Bacillus brevis*; an ionophorous antibiotic that also acts as an uncoupler of oxidative phosphorylation. Gramicidin S is actually misnamed since it belongs to the tyrocidins.

gram-mole MOLE.

gram-molecular weight MOLE.

gram-negative Designating a bacterium that does not retain the initial Gram stain but retains the counterstain. Gram-negative bacteria possess a relatively thin cell wall that is not readily digested by the enzyme lysozyme, and in which the peptidoglycan layer is covered with an outer membrane consisting of lipopolysaccharide, protein, and lipoprotein. The space between the outer and the inner (cell, cytoplasmic) membranes is called the periplasmic space.

gram-positive Designating a bacterium that retains the initial Gram stain and is not stained by the counterstain. Gram-positive bacteria generally possess a relatively thick and rigid cell wall that is readily digested by the enzyme lysozyme, and that consists of a layer of peptidoglycan.

Gram stain A set of two stains that are used to stain bacteria; the staining depends on the composition and the structure of the bacterial cell wall. *See also* gram-negative; gram-positive.

grand unified theory Any one of a number of theories that unify the Weinberg–Salam theory and quantum chromodynamics of particle physics.

granule A small grain or particle, such as a starch or a glycogen granule.

granulocyte POLYMORPHONUCLEAR LEUKOCYTE.

granulose 1. *n* A polysaccharide, composed of glucose, that resembles amylopectin and that occurs in species of *Clostridium*. 2. *adj* Having a granular appearance.

granum (*pl* grana) A stack of thylakoid disks in a chloroplast.

grape sugar GLUCOSE.

GRAS list Acronym for generally regarded as safe; a list of compounds, compiled by the Food and Drug Administration, that have been in wide use for a long time without any evidence of adverse effects to humans.

grating *See* diffraction grating.

gratuitous induction Enzyme induction in which the synthesis of an inducible enzyme is brought about in response to compounds other than the natural substrates of the enzyme.

Grave's disease A pathological condition caused by thyroid hyperfunction and characterized by protruding eyeballs and by goiter.

gravimetric analysis A method of chemical analysis that is based on separating the substance of interest from other components and weighing either the purified substance or one of its derivatives.

greasy spots The matching, nonpolar surfaces of subunits that aid in the association of the subunits to form an oligomeric protein.

greater membrane The components that are external to the unit membrane structure, particularly the carbohydrate-rich layer that is present on the exterior of many cells.

Greco-Latin square An extension of a Latin square in which pairs of elements (such as Greek and Roman letters) appear only once in each row and in each column.

Greek-key structure A protein conformation of antiparallel β-sheets that resembles a specific pattern found in Greek pottery. *Aka* Greek key topology.

greenhouse effect The environmental phenomenon in which an increase in the concentration of atmospheric gases (such as CO_2 and methane) could lead to a decrease of radiative heat loss and to a rise in the earth's temperature resulting, possibly, in the melting of the polar ice caps.

Greenstein hypothesis A hypothesis of cancer according to which tumors are characterized by a general convergence of enzyme patterns which leads to a biochemical uniformity of tumor tissues.

GRF 1. Growth hormone releasing factor. *See* growth hormone regulatory hormone. 2. Growth factor for guinea pigs.

GRH *See* growth hormone regulatory hormone.

grid 1. A two-dimensional network of uniformly spaced horizontal and vertical lines. 2. A screen for mounting specimens in electron microscopy.

GRIF *See* growth hormone regulatory hormone.

GRIH *See* growth hormone regulatory hormone.

grisein An iron-containing antibiotic, produced by *Streptomyces griseus*; a cyclic polypeptide that contains cytosine and that is active against gram-negative bacteria.

griseofulvin A polyketide antibiotic, produced by a number of *Penicillium* species, that is used as a fungicide.

groove *See* major groove; minor groove.

Gross leukemia virus A mouse leukemia virus that belongs to the group of leukoviruses.

Grotthus–Draper law FIRST LAW OF PHOTOCHEMISTRY.

ground state The normal, unexcited, lowest-energy level of a nucleus, an atom, or a molecule.

ground substance The gel-like and mucopolysaccharide-containing matrix of connective tissue.

group activation The transfer of a high-energy group (such as a pyrophosphate moiety) from one compound to another.

group frequency band An infrared absorption band that is characteristic of a particular chemical group in a molecule, such as the $C=O$, $C=C$, $C-H$, or $C=N$ group. *See also* skeletal band.

group specificity The selectivity of an enzyme that allows it to catalyze a reaction with a group of related substrates.

group transfer agent COENZYME.

group transfer potential The free energy change for the hydrolysis reaction in which a given group of atoms is removed from a compound.

group transfer reaction A reaction, other than an oxidation–reduction reaction, in which a functional group is transferred from one molecule to another.

group translocation The active transport of a solute across a membrane that is coupled to a chemical modification of the solute. The transport of monosaccharides by the bacterial phosphotransferase system is an example; here the monosaccharides are phosphorylated and transported as such.

growing point REPLICATING FORK.

growth curve A plot of either the number of cells or the cell mass of a growing culture as a function of time.

growth factor 1. Any factor, such as a mineral, a vitamin, or a hormone, that promotes the growth of an organism. 2. A specific substance that must be present in a growth medium to permit cell multiplication. 3. A substance that has hormone-like properties and acts as a mitogen, stimulating cell division and multiplication. Some growth factors act on a variety of cells, others act on a single cell type only. *See also* specific compounds.

growth hormone The protein hormone that is secreted by the anterior lobe of the pituitary gland, stimulates body growth, and affects many aspects of metabolism. *Abbr* GH.

growth hormone regulatory hormone One of two hypothalamic hormones (or factors) that, respectively, stimulate or inhibit the release of growth hormone from the pituitary gland. The growth hormone releasing hormone (or factor) is abbreviated variously as GRH (GRF), GHRH (GHRF), or SRH, SHRH (SRF) where S stands for somatotropin. The growth hormone release-inhibiting hormone (or factor) is abbreviated variously as GIH (GIF), GHIH (GHIF), GRIH (GRIF), GHRIH (GHRIF), SIH (SIF), or SRIH, SHRIH (SRIF).

growth hormone release-inhibiting hormone *See* growth hormone regulatory hormone.

growth hormone releasing hormone *See* growth hormone regulatory hormone.

growth medium *See* medium.

growth rate constant The relative increase in either the number of cells or the cell mass per unit time; specifically, $dN/dt = \alpha N$, where α is the growth rate constant, and dN is the increase in the number of cells or the cell mass in the time dt.

growth retardant A synthetic compound that inhibits the growth of plants.

growth substance AUXIN.

growth vitamin VITAMIN A.

GRP Glucose-regulated protein.

G_s *See* G protein.

GSC Gas–solid chromatography.

GSH Glutathione.

GSSG Oxidized gluthathione; *see* glutathione.

GTF Glucose tolerance factor.

GTP 1. Guanosine triphosphate. 2. Guanosine-5′-triphosphate.

GTPase Guanosine triphosphatase.

GTP-binding protein G PROTEIN.

Gua Guanine.

GU-AG rule The observation that the base sequence of an intron begins with GU and ends with AG; GU represents the 5′-end of the donor junction and AG represents the 3′-end of the acceptor junction.

guanidinium group GUANIDO GROUP.

guanidino acetic acid An intermediate in the biosynthesis of creatine that is formed by transfer of the guanido group from arginine to glycine.

guanidino group GUANIDO GROUP.

guanido group The basic grouping that is present in the amino acid arginine:

$$NH_2{-}C{-}NH{-}$$
$$\|$$
$$NH$$

guanine The purine, 2-amino-6-oxypurine, that occurs in both RNA and DNA. *Abbr* G; Gua.

guanine-nucleotide-binding protein G PROTEIN.

guanosine The ribonucleoside of guanine. Guanosine mono-, di-, and triphosphate are abbreviated, respectively, as GMP, GDP, and GTP. The abbreviations refer to the 5′-nucleoside phosphates unless otherwise indicated. *Abbr* Guo; G.

guanosine-3′, 5′-cyclic monophosphate A cyclic nucleotide, commonly called cyclic GMP, that is formed from GTP in a reaction catalyzed by the enzyme guanylate cyclase. Cyclic GMP is present intracellulary at very low concentrations and is believed to function as an antagonist of cyclic AMP in systems composed of opposing reactions that are controlled in both directions, such as muscle contraction–muscle relaxation, and glycogen synthesis–glycogen breakdown. *Abbr* cGMP, *Aka* cyclic guanylic acid.

guanosine polyphosphates MAGIC SPOTS.

guanosine triphosphatase The enzyme that catalyzes the hydrolysis of GTP, as in the process of peptide bond formation during protein synthesis. *Abbr* GTPase.

guanosine-5′-triphosphate A high-energy compound that is required for peptide bond formation in protein synthesis. *Abbr* GTP.

guanylate cyclase The enzyme that catalyzes the formation of guanosine-3′,5′-cyclic monophosphate from guanosine-5′-triphosphate.

guanylic acid The ribonucleotide of guanine.

guest *See* host–guest system.

gum An excretion of certain plants that usually contains polysaccharides composed of glucuronic acid, galactose, arabinose, and, at times, other sugars.

gum arabic The gum produced by trees of the genus *Acacia*

Guo Guanosine.

gutta A rubber-like polyterpene containing about 100 isoprene units; produced from the latex of *Palaquium gutta*. A mixture of gutta and resins is called gutta-percha. A mixture of gutta and triterpene alcohols is called chicle and is used in the preparation of chewing gum.

g value A factor used in electron paramagnetic resonance to relate the frequency of the absorbed radiation to the strength of the applied magnetic field. The magnitude of this factor is a measure of the extent to which an electron interacts with other electrons or nuclei.

G value A measure of the sensitivity of a compound to undergo a reaction subsequent to and as a result of exposure to an ionizing radiation; equal to the number of molecules sensitized by the radiation per 100 eV absorbed.

gyrase *See* DNA gyrase.

gyratory shaker A shaker that provides a rotational motion.

gyromagnetic coefficient The ratio of the magnetic moment of a nucleus to its angular momentum spin; used in nuclear magnetic resonance.

gyromagnetic ratio A constant that is characteristic of an individual atomic nucleus and that is related to the energy that must be absorbed by this nucleus before it can undergo a transition. *See also* nuclear magnetic resonance.

H

h 1. Hour. 2. Planck's constant. 3. Hecto.

H 1. Hydrogen. 2. Enthalpy. 3. Histidine. 4. Henry. 5. Magnetic field strength.

H1, H2A, H2B, H3, H4 *See* histones; nucleosome.

^{3}H Tritium; the radioactive isotope of hydrogen that has a half-life of 12.3 years and is a weak beta emitter.

HA Hydroxyapatite.

HAA Hepatitis-associated antigen; the Australia antigen.

habituation The phenomenon that an organism frequently shows a decreased response to repeated stimuli, such as a decreased transmission of nerve impulses through a synapse upon repeated electrical stimulations; may be related to the nature of memory.

hadron *See* elementary particles.

haem British spelling of heme.

haemoglobin British spelling of hemoglobin.

Hageman factor The plasma zymogen that is activated by surface contact or by the kallikrein system at the start of the intrinsic pathway of blood clotting.

HAI Hemagglutination inhibition.

hairpin bend BETA BEND.

hairpin DNA *See* foldback DNA.

hairpin loop The looped structure of hairpin DNA.

Hakamori methylation A method for the exhaustive methylation of carbohydrates, glycolipids, and glycoproteins by means of methylsulfinyl carbanions.

Haldane coefficient The change in the concentration of hemoglobin-bound hydrogen ions with changing concentration of hemoglobin-bound oxygen at a constant pH; the number of protons taken up per O_2 released.

Haldane effect The release of protons that accompanies the oxygenation of hemoglobin.

Haldane–Oparin hypothesis The hypothesis that the origin of life on earth was preceded by a very long period of abiogenic evolution. During this period simple organic compounds were formed first, primarily from gases in the atmosphere, and more complex compounds were then formed from them by a variety of reactions, occurring primarily in the seas. These reactions and compounds then gave rise to macromolecules which ultimately assembled to form structures that were the forerunners of living cells.

Haldane relation An expression that relates the equilibrium constants of an enzymatic reaction to the kinetic constants of the forward and the reverse reactions. For a simple, one-substrate enzymatic reaction the Haldane relation is $K_{eq} = V_f K_{mr}/V_r K_{mf}$, where K_{eq} is the equilibrium constant, V is the maximum velocity, and K_m is the Michaelis constant; the subscripts f and r refer to the forward and the reverse reaction, respectively. *Aka* Haldane equation.

half-band width The width of an absorption band at the point at which the absorption equals one-half of the maximum absorption.

half-cell HALF-REACTION.

half-life 1. The time required for one-half of either the mass or the number of atoms of a radioactive substance to undergo radioactive decay. *Aka* radioactive half-life. 2. The time required for one-half of the mass of a substance to be either metabolized or excreted by an organism. *Aka* biological half-life. 3. The time required for one-half of the mass of a reactant to undergo chemical reaction. For a first-order reaction, $t_{1/2} = 0.693/k$, where $t_{1/2}$ is the half-life and k is the rate constant.

half-mustard A monofunctional sulfur or nitrogen mustard.

half-of-the-sites reactivity The phenomenon, observed with a number of regulatory enzymes, in which the reaction at one site of the enzyme prevents the reaction at a second site so that, at any given time, only one-half of the potential sites of the enzyme participate in the reaction. *Aka* half-of-the-sites phenomenon; flip-flop.

half-period HALF-TIME.

half-reaction A reaction in which there is either a gain or a loss of electrons; a half-reaction in which there is a gain of electrons can take place only in the presence of a second half-reaction in which there is a corresponding loss of electrons, and vice versa.

half-reaction time *See* cot value.

half-site editing A procedure for in vitro mutagenesis that is designed to make precise plasmid constructions; based on hybridization of a mutagenic oligonucleotide primer to a single-stranded template DNA, followed by polymerization with DNA polymerase I (Klenow fragment).

half-thickness HALF-VALUE LAYER.

half-time The time required to achieve one-half of the maximum of a reaction.

half-time of exchange The time required for

one-half of the exchangeable atoms to be exchanged in a reaction that involves the exchange of atoms.

half-value dose The dose of radiation or toxic compound that produces deaths in 50% of the cells, or loss of infectivity in 50% of the virus particles, in a test group within a specified time; analogous to the median lethal dose for animals.

half-value layer The thickness of an absorbing material that reduces the intensity of a beam of incident radiation to one-half of its original intensity. *Abbr* HVL.

half-wave potential The polarographic potential at which one-half of the maximum current is obtained.

halide A binary compound formed from a halogen and either a metal or an organic radical; a fluoride, a chloride, a bromide, or an iodide.

hallucinogenic drug A substance that, if taken in appropriate doses, produces distortion of perception, vivid images, or hallucinations. *Aka* hallucinogen. *See also* narcotic drug.

halobacteria A family of aerobic, extreme halophiles which grow only in media containing at least 15–20% sodium chloride.

halogen An element of group VIIA in the periodic table that consists of the elements fluorine F, chlorine Cl, bromine Br, iodine I, and astatine At.

halogenation The introduction of a halogen into an organic compound.

halogen quenching The quenching of an ionization detector that occurs when a halogen gas is added to the mixture used for gas amplification. *See also* internal quenching.

halophile An organism that grows only in solutions of either moderate or high salt concentration. Extreme halophiles constitute one group of archaebacteria. They are found in salty habitats such as the Great Salt Lake and the Dead Sea. *See also* archaebacteria.

Hamilton syringe A finely tooled syringe for the delivery of volumes in the microliter range.

Hammett equation One of two equations that describe the effect of meta and para substituents of an aromatic compound on either the rate or the equilibrium constant of a reaction. Specifically, $pk^0 - pk = \rho\sigma$, where k and k^0 are either the rate constants or the equilibrium constants for the substituted and unsubstituted compound, respectively, and pk and pk^0 are the corresponding negative logarithms of these constants; ρ is a constant that characterizes the reaction with respect to its sensitivity to electron supply at the reaction site and that is independent of the substituent; and σ is a constant that characterizes a substituent with respect to its electron-withdrawing

power and that is independent of the nature of the reaction. It is called the substituent constant.

Hammond postulate An attempt to link reaction rate and product stability by considering the energy level and structure of the transition state of the reaction. The structure and energy of the transition state are taken to be close to those of the nearest stable species. It is postulated, therefore, that for an endergonic reaction, the transition state resembles a product while, for an exergonic reaction, the transition state resembles a reactant.

handedness The property of a helix of being either right-handed or left-handed; chirality.

handle method An early method for sequencing nucleic acids in which the polynucleotide strand was labeled at each end and then cleaved to yield fragments of varying sizes. *Aka* double-label method.

hands-on experience Experience gained by the actual manipulation and use of instruments and other items of equipment.

Hanes plot A graphical treatment of enzyme kinetics data in which [S]/v is plotted as a function of [S], where [S] is the substrate concentration, and v is the velocity of the reaction; a single reciprocal plot of the Michaelis–Menten equation. *Aka* Hanes–Wilkinson plot; Hanes–Woolf plot.

Hanes report Acronym for Health And Nutritional Examination Survey; an extensive dietary survey involving 24-hour recalls from 28,000 Americans, aged 1 to 74, from 1971 to 1974. The survey indicated that many Americans may be consuming suboptimal amounts of iron, calcium, vitamin A, and vitamin C.

hanging drop technique A technique for the microscopic examination of live microorganisms that avoids possible distortion of the organisms which may be caused by drying or fixing; performed by suspending the microorganisms in a drop of fluid on a concave microscope slide.

hanging mercury drop electrode An electrode used for fast-sweep polarography.

hanging strip electrophoresis An electrophoretic technique in which paper strips are used that are suspended from a central support in the form of an inverted V, with each end of the paper dipping into a buffer compartment.

H antigen 1. A protein antigen of bacterial flagella. 2. A precursor of the A and B antigens of the human ABO blood group system. 3. HISTOCOMPATIBILITY ANTIGEN.

H-2 antigens *See* major histocompatibility complex.

Hanus iodine number An iodine number determined by the use of a solution of iodine in glacial acetic acid, with iodine bromide serving as an accelerator of the reaction.

haploid number The fundamental number of chromosomes that comprise a single set; the number of chromosomes in a genome; the gametic chromosome number. *Sym* N.

haploid state The chromosome state in which each type of chromosome is represented only once; the state of a cell or an organism having only one set of chromosomes. *Aka* haploidy.

hapten A substance that can react selectively with antibodies of the appropriate specificity but that stimulates the production of these antibodies in an animal only when it is coupled to a carrier.

haptenic antigen HAPTEN.

haptenic group The hapten plus the amino acid through which it is covalently linked to a protein carrier; the haptenic group may constitute either all, or part of, an antigenic determinant.

hapten inhibition test A test that measures the extent to which a hapten inhibits an antigen–antibody reaction; performed by allowing the hapten to bind to and block the antigen binding sites of the appropriate antibody, and then adding antigen that is directed against the same antibody.

hapten protection The shielding of the antigen binding sites of an antibody by allowing a hapten to bind to them. The reaction may be used to specifically label the antibody by reacting the antibody–hapten complex first with an unlabeled reagent, and, after removal of the hapten, with the same, but labeled, reagent.

haptoglobin One of several plasma glycoproteins that bind hemoglobin and aid in its transport and conservation.

hard center An electrophilic center in a molecule which, when attacked by a nucleophile, leads to a transition state that involves the formation of a normal bond. *See also* soft center.

hard clot The final, insoluble blood clot that is formed by the cross-linking of fibrin molecules in the presence of calcium ions and the enzyme fibrin stabilizing factor.

hardening The hydrogenation of oils; the conversion of triacylglycerols (triglycerides), rich in unsaturated fatty acids, to those rich in saturated fatty acids.

Harden–Young ester FRUCTOSE-1,6-BISPHOSPHATE.

hard ligand A small atom, or a group of atoms, of low polarizability; oxygen and nitrogen atoms are examples. *See also* soft ligand.

hard soap A sodium salt of a long-chain fatty acid; hard soaps are less water soluble than soft soaps.

hardware The physical equipment used with computers; the electronic, magnetic, and mechanical items such as cabinets, tubes, transistors, and wires.

hard water Water that contains appreciable concentrations of calcium, magnesium, and iron ions; these ions form insoluble soaps that are ineffective as surface-active agents.

hard x rays High-frequency x rays that have short wavelengths and great penetrating power.

Harris–Ray test A test for vitamin C that is based on the titrimetric reduction of the dye 2,6-dichlorophenol indophenol.

Hartnup's disease A genetically inherited metabolic defect in humans that is associated with mental retardation and that is due to a deficiency of the enzyme tryptophan pyrrolase.

harvest The collection of bacterial cells at a particular stage of growth.

hashish 1. The dried resin from the glandular hairs of the female hemp plant (*Cannabis*); contains Δ^1-tetrahydrocannabinol (THC), which is psychoactive; a common narcotic. 2. MARIJUANA.

Hatch–Slack–Kortschak pathway An alternative pathway to the Calvin cycle of photosynthesis in which phosphoenolpyruvate is carboxylated to yield oxaloacetate as the first product; the cycle is operative in C_4 plants. *Abbr* HSK pathway; C_4 pathway.

HAT medium A growth medium for animal cells in tissue culture; contains hypoxanthine, aminopterin, and thymidine and is useful for the selection of hybrid cells.

H^+-ATPase One of a group of adenosine triphosphatases that do not require other cations for activation and that occur in both prokaryotic and eukaryotic organisms. One class of H^+-ATPases is designated as F_0F_1-ATPases (ATP synthases). This class includes the mitochondrial enzyme which functions in the reverse direction (synthesis of ATP) and is driven by the proton gradient resulting from electron transport. *See also* F_0F_1-ATPase.

hawkinsin *See* tyrosinosis.

Haworth projection A representation of carbohydrates in which the ring structures are drawn as regular hexagons or pentagons, in a plane that is perpendicular to the plane of the paper; the attached atoms or groups of atoms are indicated as being either above or below the plane of the ring. *Aka* Haworth formula.

Hayflick limit The number of times that animal cells appear to be capable of dividing in tissue culture before they reach the crisis stage; this amounts to 30–50 divisions for mouse and human cells. *See also* crisis (2).

hazard The likelihood of toxic injury to a living system by a harmful chemical under the circumstances of its intended use.

Hb Hemoglobin; also abbreviated as HHb.

Related compounds are abbreviated as follows: HbA, adult hemoglobin; HbCO, carbon monoxide hemoglobin; HbF, fetal hemoglobin; HbO_2, $HHbO_2$, oxyhemoglobin; HbS, sickle cell hemoglobin; HbC, hemoglobin in which lysine has replaced glutamic acid at position 6 in the beta chain; HbH, hemoglobin that is a tetramer of four identical beta chains; HbM, hemoglobin in which at least two of the heme groups have been oxidized to hemin groups (Fe^{3+}).

HB-Ag   Hepatitis B-antigen.

H band   H ZONE.

HB$_s$-Ag   Hepatitis B surface antigen.

HBV   Hepatitis B virus; *See* hepatitis.

HCG   Human chorionic gonadotropin.

H chain   1. One of the two types of polypeptide chains of lactate dehydrogenase isozymes; denoted H, since the tetramer of H chains is found predominantly in heart tissue. 2. HEAVY CHAIN.

HCI   Heme-controlled inhibitor.

H-2 complex   The major histocompatibility complex of mouse. *See* major histocompatibility complex.

HCR   1. Hemin-controlled repressor. 2. Host-cell reactivation.

HCS   Human chorionic somatomammotropin.

Hcy   Homocysteine.

HD$_{50}$ Median hemolytic dose.

HDL   High-density lipoprotein.

H-DNA   An unusual form of DNA that consists of a mixture of single-stranded and triple-stranded sequences.

HDP   1. Hexose diphosphate. 2. Helix-destabilizing protein; *See* single-strand binding protein.

head   1. The hexagonal, DNA-containing structure of a T-even phage. 2. The activated portion of a condensing unit. 3. The globular portion of the myosin molecule. 4. The 5′-phosphate end of an oligo- or a polynucleotide strand. 5. ROTOR. 6. A structural domain of prokaryotic 30S ribosomal subunits. 7. The polar, ionic, portion of a fatty acid or a phospholipid molecule. 8. The region of a sperm that contains a condensed haploid nucleus.

headful mechanism   A mechanism of packaging DNA in a phage head that involves cutting the DNA from long concatemers. The cuts are not made at specific points, determined by particular base sequences, but rather at points, determined by the amount of DNA that can fit into a head. The mechanism can account for the circular permutation of phage DNA.

head growth   *See* headward growth.

headpiece   *See* supermolecule.

head space analysis   A gas chromatographic technique for substances that are sufficiently volatile so that a determination of their concentration in the vapor phase can be used as a measure of their concentration in the liquid phase; involves an analysis of the vapor in equilibrium with the liquid phase (called the head space).

head-to-head condensation   The condensation of two molecules by way of their activated ends, as in the condensation of two molecules of acetaldehyde to form acetoin.

head-to-tail condensation   The condensation of two molecules by way of the active end of one molecule and the passive end of the other molecule. The condensation of isoprene units in the biosynthesis of cholesterol, and the polymerization of amino acids in the formation of peptides and proteins are two examples.

headward growth   The polymerization mechanism in which the passive tail of a monomer adds to the activated head of a chain, thereby making its own head the receptor for the next addition of monomer. *Aka* head polymerization. *See also* head-to-tail condensation.

heat capacity   The quantity of heat required to raise the temperature of a given amount of substance by 1 °C.

heat content   ENTHALPY.

heat labile   Descriptive of a molecule that loses its activity upon heating to moderate temperatures of about 50 °C.

heat labile citrovorum factor   10-Formyltetrahydrofolic acid. *Abbr* HLCF.

heat-sensitive enzyme   An enzyme that loses its stability and activity as the temperature is raised; due to progressive denaturation of the polypeptide chain as the temperature is increased.

heat shock proteins   Proteins that are synthesized by an organism following a heat shock (a shift-up in growth temperature). This includes proteins not synthesized at all at the lower temperature and proteins synthesized at increased rates at the higher temperature. Heat shock proteins protect the organism against thermal damage in some unknown way; their synthesis is under control of the HTP-regulon. *Abbr* hsp, HSP, HTP.

heat shock puffs   A set of chromosome puffs induced in *Drosophila* upon exposure to elevated temperatures.

heat shock response   The transcriptional and translational activity of an organism induced by a heat shock (a shift-up in the growth temperature). The response includes complete repression of the synthesis of some proteins, synthesis of some proteins not made at the lower temperature at all, and increased rates of synthesis of other proteins. The heat shock response is widespread among both prokaryotes and eukaryotes. *See also* HTP regulon.

heat stable Descriptive of a molecule that retains its activity upon heating to moderate temperatures of about 50 °C.

heavy Labeled with a heavy isotope.

heavy atom method ISOMORPHOUS REPLACEMENT.

heavy chain One of two polypeptide chains that are linked to two light chains to form the immunoglobulin molecule. The molecular weight of a light chain is about 25,000 and that of a heavy chain is about 50,000. The heavy chains of the IgA, IgD, IgE, IgG, and IgM immunoglobulins are denoted, respectively, as α, δ, ϵ, γ, and μ chains.

heavy-chain class switching *See* class switching.

heavy-chain disease A disorder in which free Fc fragments of immunoglobulin heavy chains are present in the serum.

heavy hydrogen DEUTERIUM.

heavy isotope An isotope that contains a larger number of neutrons in the nucleus than the more frequently occurring common isotope of that element.

heavy label A heavy isotope that is generally introduced into a molecule to facilitate its separation from identical molecules containing the more frequently occurring isotope.

heavy meromyosin The fragment of the myosin molecule that consists of the globular head and a portion of the tail of the molecule. *Abbr* H-meromyosin; HMM. *Aka* $(F_1 + F_2)$ fragment.

heavy metal contamination The presence of trace amounts of heavy metals, such as lead and zinc, in the water and/or in the chemicals that are used in the preparation of media and other solutions.

heavy ribosome 1. A ribosome labeled with a heavy isotope. 2. POLYSOME.

heavy strand 1. A polynucleotide chain labeled with a heavy isotope. 2. The naturally occurring polynucleotide chain of a duplex that has a greater density than the complementary chain.

heavy water Deuterium oxide; D_2O.

hecto- Combining form meaning one hundred and used with metric units of measurement. *Sym* h.

Hehner number The percentage of nonvolatile, water-insoluble fatty acids in a fat. *Aka* Hehner value.

Heidelberger curve PRECIPITIN CURVE.

height equivalent to a theoretical plate The length of a gas chromatographic column, some other chromatographic column, or a distillation column divided by its efficiency in terms of theoretical plates; the length of a column over which the separation effected is

equivalent to that of one theoretical plate. *Abbr* HETP; EHTP. *See also* theoretical plate.

Heinz body INCLUSION BODY (2).

Heisenberg uncertainty principle *See* uncertainty principle.

Heitler–London theory VALENCE BOND THEORY.

HeLa cells Cells that have been derived from a human carcinoma of the cervix from a patient named Henrietta Lack; the cells have been maintained in tissued culture since 1953.

helical Of, or pertaining to, a helix.

helical content The proportion of hydrogen-bonded base pairs in a nucleic acid, or the proportion of hydrogen-bonded amino acid residues in a protein.

helical cross The cross-like x-ray diffraction pattern that is obtained with fibrous helical material such as DNA.

helical virion A virus, such as tobacco mosaic virus, in which the capsid is a helix that forms a hollow cylinder.

helicase An enzyme that binds ahead of the replicating fork in the discontinuous replication of DNA and that catalyzes the energy-dependent unwinding of the duplex. The enzyme has ATPase activity and hydrolyzes 2 molecules of ATP per DNA base pair broken. The helicase activity in *E. coli* is called the Rep protein.

helicene A chiral polyaromatic hydrocarbon that has a helical shape.

helicity Helical structure.

heliotropism PHOTOTROPISM.

helix A coiled structure, or spiral, that is described by the thread of a bolt or the turns of a tubular spring; the curve that is traced on the surface of a cylinder by the rotation of a point that cuts the elements of the cylinder at a constant oblique angle. A helix is said to be left-handed or right-handed depending on whether it corresponds to the thread of a left-handed or a right-handed bolt. *See also* alpha helix; collagen helix; Watson–Crick model.

3_{10} helix A variant of the alpha-helix structure of proteins that is occasionally seen in short stretches of the polypeptide chain. In this structure, the separation between successive hydrogen bonds is shortened by one amino acid residue, and there are only three amino acid residues per turn of the helix.

helix-breaking amino acid An amino acid that, wherever it occurs in the polypeptide chain, interrupts the alpha-helical structure and creates a bend in the chain. This is always the case for proline and hydroxyproline since the imino group, after being tied up in a peptide bond, no longer carries a hydrogen atom and hence cannot serve as a hydrogen bond donor. Breaks in the helix can also be produced by amino acids with bulky side chains and by electrostatic repulsion of charged functional groups of the amino acids.

helix–coil transition The transition of a polymer from a helical configuration to that of a random coil, as in the denaturation of helical proteins and double-stranded DNA.

helix-destabilizing protein SINGLE-STRAND BINDING PROTEIN.

helix nucleation The slow step, involving formation of a short stretch of double helix, in the renaturation of double-stranded DNA or RNA from the single strands. This step is followed by rapid zippering as the double helix grows to maximize base-pairing interactions. *See also* nucleation.

helix winding number LINKING NUMBER.

Heller's test A qualitative test for protein that is based on the formation of a white precipitate at the interface between concentrated nitric acid and a test solution which is layered over it.

Helmholtz double layer IONIC DOUBLE LAYER.

Helmholtz free energy That component of the total energy of a system that can do work under conditions of constant temperature and volume; expressed by the thermodynamic function $A = E - TS$, where A is the Helmholtz free energy, E is the internal energy, T is the absolute temperature, and S is the entropy.

helper factors Protein regulatory factors isolated from helper T cells.

helper phage *See* helper virus.

helper T cells A group of T cells that help specific T or B lymphocytes respond to an antigen and that can activate some nonlymphocyte cells, such as macrophages. *Aka* helper T lymphocytes.

helper virus A virus that, by infecting a cell and supplying a missing product, allows the simultaneous infection of that cell by a defective virus.

hemadsorption The adsorption of a substance to red blood cells.

hemagglutination The agglutination of red blood cells.

hemagglutination inhibition The inhibition of hemagglutination; used for assaying hemagglutinating viruses by adding antiviral antibodies to a mixture of virus particles and red blood cells. *Abbr* HI.

hemagglutinin An agglutinin of red blood cells.

hematin FERRIHEME.

hematinometer HEMOGLOBINOMETER.

hematocrit 1. The relative volume of blood occupied by the erythrocytes and expressed in cubic centimeters per 100 cc of blood. 2. An apparatus for measuring the relative volume

of cells and plasma in blood, usually by means of centrifugation.

hematology The science that deals with blood and blood-forming organs.

hematolysis HEMOLYSIS.

hematopoiesis HEMOPOIESIS.

hematopoietic HEMOPOIETIC.

hematoside One of a class of monosialogangliosides; a compound having the structure: ceramide-glucose-galactose-*N*-acetylneuraminic acid.

hematuria The presence of either blood or red blood cells in the urine.

heme 1. Any tetrapyrrolic chelate of iron, generally an iron porphyrin complex, in which four coordination positions of iron are occupied. 2. A protoheme; an iron–porphyrin complex that has a protoporphyrin nucleus, specifically one containing protoporphyrin IX, which is the oxygen-binding portion of the hemoglobin molecule. Heme A (protoheme A) is the prosthetic group of cytochrome *a*; heme C (protoheme C) is the prosthetic group of cytochrome *c*.

heme-controlled inhibitor HEMIN-CONTROLLED REPRESSOR.

heme–heme interactions The cooperative interactions between the hemes of the subunits of hemoglobin with respect to the binding of oxygen.

heme iron An iron atom that is coordinately bound in a porphyrin.

hemel An aziridine mutagen.

heme pocket The hydrophobic crevice in a hemoglobin subunit in which the heme is located.

heme protein *See* hemoprotein.

hemerythrin A nonheme, iron-containing, respiratory pigment of sipunculid worms and other marine invertebrates.

heme synthetase FERROCHELATASE.

hemiacetal A compound formed by either an inter- or an intramolecular reaction between an aldehyde group and an alcohol group.

hemicellulose A heterogeneous group of branched polysaccharides that serve to cement plant fibers together. Each polysaccharide has a long linear backbone of sugar units of one type, linked β $(1 \rightarrow 4)$, and short side chains, composed of other sugars, that protrude from the backbone.

hemichrome FERRIHEMOCHROME.

hemichromogen FERRIHEMOCHROME.

hemidesmosome A cell junction that resembles spot desmosomes but, instead of joining adjacent epithelial cell membranes, serves to join the basal surface of epithelial cells to the underlying basal lamina. *See also* cell junction.

hemiglobin METHEMOGLOBIN.

hemiketal A compound formed by either an inter- or an intramolecular reaction between a ketone group and an alcohol group.

hemin FERRIHEME.

hemin-controlled repressor A protein kinase that functions in the translational control of globin synthesis in reticulocytes. The enzyme catalyzes the phosphorylation of initiation factor 2 (eIF-2) and thereby prevents eIF-2 from complexing with another protein (called eIF-2 stimulating protein or ESP). This prevents the initiation of translation since the latter requires formation of a ternary complex between the initiator-tRNA, eIF-2, and ESP. The hemin-controlled repressor is activated by a cAMP-dependent protein kinase which is activated by hemin. Thus, if there is excess hemin present, the hemin-controlled repressor is not activated, and globin synthesis can begin. *Abbr* HCR. *Aka* heme-controlled inhibitor.

hemizygous gene A gene that is present only once among the chromosomes of a cell, such as a sex-linked gene.

hemochromatosis A disease that is caused by excessive iron absorption from the intestine and by excessive iron deposition in various organs. The organs thus affected show pathological damage and the disease is frequently associated with glucosuria. *See also* bronzed diabetes.

hemochrome A low-spin compound of heme in which the fifth and sixth coordination positions of the iron are occupied by strong-field ligands.

hemochromogen FERROHEMOCHROME.

hemoconcentration An increase in the concentration of red blood cells in the blood; a decrease in the concentration of plasma.

hemocyanin A copper-containing respiratory pigment of mollusks and crustaceans.

hemocyte 1. A blood cell. 2. An ameboid blood cell of insects.

hemodialysis The dialysis of blood by means of a semipermeable membrane.

hemodialyzer An artificial kidney; an apparatus that circulates blood from and back into the body by passing it through a series of semipermeable membranes which remove waste products in a manner analogues to the operation of the kidney.

hemodilution A decrease in the concentration of red blood cells in the blood; an increase in the concentration of plasma.

hemoflavoprotein A heme-containing flavoprotein.

hemoglobin The oxygen-transporting protein of the blood that consists of four polypeptide chains, two alpha and two beta chains, each one surrounding a heme group. Hemoglobin occurs in several normal forms such as adult and fetal hemoglobin, and in various abnormal forms, such as sickle cell hemoglobin and hemoglobin C. *Abbr* Hb; HHb. *See also* Hb.

hemoglobinemia The presence of free hemoglobin in blood plasma.

hemoglobinometer An instrument for the visual or the photoelectric measurement of the hemoglobin content of blood.

hemoglobinopathy A genetically inherited metabolic defect in humans that is characterized by the presence of a structurally altered hemoglobin.

hemoglobin switching The changeover, during the development of an organism, from the synthesis of one type of hemoglobin to the synthesis of a different type.

hemoglobinuria The presence of either free hemoglobin or closely related pigments in the urine.

hemoglobin variant *See* variant (1).

hemolymph The circulatory fluid of various invertebrates that is functionally comparable to the blood and lymph of vertebrates.

hemolysate The suspension of lysed cells obtained upon hemolysis.

hemolysin 1. An antibody that, in the presence of complement, causes hemolysis. 2. A substance, such as the bacterial toxin streptolysin, that causes hemolysis.

hemolysis The lysis of red blood cells.

hemolytic Of, or pertaining to, hemolysis.

hemolytic anemia An anemia that is characterized by an excessive destruction of red blood cells; several hemolytic anemias are genetically inherited metabolic defects due to a deficiency of a glycolytic enzyme.

hemolytic antibody HEMOLYSIN (1).

hemolytic immune body An antibody to a surface antigen of erythrocytes; an amboceptor. *Abbr* HIB.

hemolytic plaque assay *See* plaque technique.

hemolytic system The mixture of red blood cells and antibodies against them that is used in the complement fixation test.

hemolyze To lyse red blood cells.

hemopexin A plasma glycoprotein that binds heme and aids in its disposal.

hemophilia A genetically inherited, sex-linked metabolic defect in humans that is characterized by prolonged clotting times of blood and that is caused by a deficiency of the antihemophilic factor. Hemophilia A is caused by a deficiency of antihemophilic factor A, and hemophilia B is caused by a deficiency of antihemophilic factor B.

hemopoiesis The formation of red blood cells.

hemopoietic Of, or pertaining to, hemopoiesis.

hemopoietine ERYTHROPOIETIN.

hemoprotein A conjugated protein that contains a heme as a prosthetic group.

hemorrhagic disease The pathological condition that is characterized by bleeding and by long clotting times, and that is caused by a deficiency of vitamin K.

hemorrhagic disease of the newborn A temporary hemorrhagic disease, characterized by vitamin K deficiency, that occurs in an infant and that persists only until a bacterial flora is established in the infant's intestine.

hemosiderin A water-insoluble protein that serves to store iron in the form of ferric hydroxide in the liver, the spleen, and the bone marrow.

hemosiderosis An increase in the storage forms of iron, such as hemosiderin; the condition may be relative or absolute. Relative hemosiderosis represents a redistribution of the body's iron with less in the red cell mass and more in the iron stores; absolute hemosiderosis (or iron overload) represents an increase in the total iron content of the body with the extra iron being laid down in storage compounds.

hempa An aziridine mutagen.

Henderson–Hasselbalch equation An equation that relates the pH of a solution to the dissociation constant of a weak acid, and to the concentrations of the proton donor form HA and the conjugated proton acceptor form A^- of the acid. Specifically, $pH = pK_a + \log([A^-]/[HA])$, where pK_a is the negative logarithm of the dissociation constant and brackets indicate molar concentrations.

Henle's loop *See* nephron.

Henri equation An equation, analogous to that obtained by integration of the Michaelis–Menten equation, that is applicable to an enzymatic reaction if the decrease in reaction rate as a function of time is due solely to the decrease in enzyme saturation as the concentration of the substrate decreases. The equation was reported by Victor Henri in 1902, prior to the derivation of the Michaelis–Menten equation.

Henri–Michaelis–Menten equation MICHAELIS–MENTEN EQUATION.

henry A unit of inductance, equal to the inductance in which an induced electromotive force of 1 V is produced when the inducing current is changed at the rate of 1 A/s. *Sym* H.

Henry's function The product of the radius of a macromolecule and its reciprocal ion–atmosphere radius.

Henry's law The law that the solubility of a gas in a liquid, at constant temperature, is proportional to the partial pressure of the gas.

hepadnavirus A virus belonging to a family that includes the hepatitis B virus and similar viruses; contains circular DNA which is only partially duplex. The name is derived from hepatotropic DNA virus.

heparan sulfate A glycosaminoglycan closely related to heparin and containing the same disaccharide repeating units; it differs from heparin in being smaller and less sulfated. Heparan sulfate is extracellularly distributed and is found in the lungs, arterial walls, and many cell surfaces.

heparin A glycosaminoglycan that contains two types of disaccharide repeating units; one consists of D-glucosamine and D-glucuronic acid, the other consists of D-glucosamine and L-iduronic acid. Heparin is highly sulfated and has anticoagulant activity; it combines with and activates antithrombin III which can then bind to and inhibit many clotting factors. Heparin occurs in the lungs, the liver, the skin, and the intestinal mucosa.

hepatectomy The surgical removal of the liver.

hepatic Of, or pertaining to, the liver.

hepatic porphyria A porphyria that is characterized by the formation of excessive amounts of heme precursors in the liver.

hepatitis An inflammation of the liver, usually caused by a viral infection but sometimes caused by a toxic compound. Two forms of viral hepatitis have been recognized. Type A (infectious or epidemic hepatitis) is transmitted through the mouth and the intestine and has an incubation period of 2–6 weeks. Type B (serum hepatitis or homologous serum jaundice) is transmitted via blood transfusions and has an incubation period of 2–6 months; it is characterized by the presence of Australia antigen in the serum. The hepatitis B virus (HBV) is now considered to be a member of the family of hepadnaviruses.

hepatitis-associated antigen AUSTRALIA ANTIGEN.

hepatitis B antigen AUSTRALIA ANTIGEN.

hepatitis B surface antigen AUSTRALIA ANTIGEN.

hepatocyte A liver cell.

hepatoflavin An impure preparation of riboflavin from liver.

hepatolenticular degeneration Wilson's disease.

hepatoma A carcinoma of liver cells.

HEPES *N*-2-Hydroxyethylpiperazine-*N'*-2-ethanesulfonic acid; used for the preparation of biological buffers in the pH range of 6.8 to 8.2. *See also* biological buffers.

HEPPS A biological buffer, also designated as EPPS. *See* EPPS.

HEP strain A high egg passage strain; a viral strain that has been passed repeatedly from one chick embryo to another (serial passage).

hepta- Combining form meaning seven.

heptamer An oligomer that consists of seven monomers.

heptose A monosaccharide that has seven carbon atoms.

herbicide A chemical that kills herbs, especially weeds.

hereditary code *See* genetic code.

hereditary disease *See* genetic disease.

hereditary material *See* genetic material.

heroin A semisynthetic narcotic drug, made by converting morphine to diacetylmorphine; a highly addictive narcotic.

herpesvirus An enveloped, icosahedral virus that contains double-stranded DNA and that infects humans and lower animals. Type I herpes simplex causes cold sores; type II herpes simplex is associated with genital lesions and is sexually transmitted.

herpes zoster The recurrent form of the disease that is produced by varicella virus in a host that was previously infected by the virus.

herring bodies Neurosecretory granules that contain neurohypophyseal hormones and neurophysins and that are located within axonal dilatations in the posterior lobe of the pituitary gland.

Hers' disease GLYCOGEN STORAGE DISEASE TYPE VI.

Hershey–Chase experiment An experiment that demonstrates that, during phage infection, the phage DNA is injected into, and the phage protein coat remains outside of, the bacterial cell. The experiment is performed by infecting *E. coli* cells with ^{32}P- or ^{35}S-labeled T-even phages, followed by removal of the empty phage heads and the phage tails by shearing in a blender.

Hershey circle A double-stranded, circular DNA molecule formed by hydrogen bonding between cohesive ends.

hertz A unit of frequency; one cycle per second. *Sym* Hz.

Hess's law *See* thermochemistry.

het A partially heterozygous phage.

hetaeron *See* zwitterion-pair chromatography.

HETE Hydroxyeicosatetranoic acid; a compound obtained by reduction of HPETE (hydroperoxyeicosatetranoic acid), which is derived from arachidonic acid.

hetero- Combining form meaning different or other.

heteroallelic Pertaining to alleles of a gene that have mutations at different sites.

heteroantibody An antibody to a heteroantigen.

heteroantigen An antigen that is immunogenic in a given animal species and that is either produced by synthetic reactions or is derived from plants, microorganisms, or other animal species.

heteroatom An atom in a ring structure that is not a carbon atom.

heteroauxesis ALLOMETRY.

heterocapsidic virus *See* segmented genome.

heterocatalytic Pertaining to the catalysis of the reaction of one substance by another, different substance, as distinct from autocatalytic.

heterochiral Descriptive of the relation between isometric chiral molecules that are only improperly congruent. *See also* homochiral.

heterochromatin The condensed chromatin that is not very active in RNA synthesis and that stains strongly during interphase.

heterochromosome A chromosome that consists primarily of heterochromatin.

heterocyclic Of, or pertaining to, an organic compound that has a ring structure which consists of carbon atoms and one or more noncarbon atoms.

heterocyclic atom HETEROATOM.

heterocysts Cells that occur in some blue-green algae and that appear to have a specialized function of fixing molecular nitrogen.

heterocytotropic antibody A cytotropic antibody that binds to and sensitizes target cells of a species that differs from the one in which the antibody was produced.

heterodimer A protein composed of two nonidentical polypeptide chains.

heterodisperse Consisting of macromolecules that differ greatly in their size.

heteroduplex 1. A double-stranded DNA in which the two strands have different hereditary origins; produced during genetic recombination by base pairing between single strands from different parental duplexes. 2. A double-stranded nucleic acid molecule formed in vitro by annealing two single strands of different origins that have some, but not complete, complementary base sequences. A double-stranded phage DNA molecule, formed by hybridizing one strand from a wild-type phage with one strand from a mutant of this phage, is an example. A hybrid, formed by annealing eukaryotic mRNA with corresponding single-stranded DNA is another example. *See also* homoduplex.

heteroenzyme One of two or more isodynamic enzymes derived from different sources, such as alcohol dehydrogenase from yeast and alcohol dehydrogenase from liver. *See also* isodynamic enzyme; isozyme.

heterofermentation 1. Fermentation in which more than one product is formed. 2. HETEROLACTIC FERMENTATION.

heterofermentative lactic acid bacteria Lactic acid bacteria that produce in fermentation less than 1.8 mol of lactic acid per mole of glucose; in addition to lactic acid, these organisms produce ethanol, acetate, glycerol, mannitol, and carbon dioxide. *See also* homofermentative lactic acid bacteria.

heterogeneic XENOGENEIC.

heterogeneity 1. The state of a preparation of macromolecules in which the macromolecules differ with respect to size, charge, structure, or other properties. 2. The state of a system in which there are two or more phases.

heterogeneity index A measure of antibody heterogeneity that is based on the assumption of an essentially random distribution of the binding affinity of antigen binding sites and hapten molecules. It can be expressed as $r/n = (K_0 c)^a/[1 + (K_0 c)^a]$, where a is the heterogeneity index, n is the antibody valence, r is the average number of hapten molecules bound per molecule of antibody, c is the concentration of free hapten molecules, and K_0 is the average intrinsic association constant (the association constant for that concentration of hapten at which half of the antibody sites are occupied by hapten molecules). The equation can be rewritten as $\log[r/(n - r)] = a \log c + a \log K_0$, and a can be determined from a plot of $\log[r/(n - r)]$ versus $\log c$. The heterogeneity index varies from 0 to 1; a value of 1 indicates homogeneous antibody. *See also* Hill equation; Hill plot.

heterogeneous Of, or pertaining to, heterogeneity.

heterogeneous catalysis Catalysis in a system that consists of two or more phases.

heterogeneous nuclear RNA 1. The total pool of extrachromosomal RNA fragments in the nucleus of a eukaryotic cell; consists of molecules of widely varying sizes and includes primary transcripts, partly processed transcripts, discarded intron RNA, ubiquitous RNA, and small nuclear RNA. 2. The pool of primary RNA transcripts and partly processed transcripts in the nucleus of a eukaryotic cell. *Abbr* hnRNA.

heterogenetic Widely distributed in many species.

heterogenetic antigen An antigen that is produced by several, phylogenetically unrelated species.

heterogenic process An unsymmetrical formation of a covalent bond in which both of the electrons of the bond are contributed by one of the bonding atoms.

heterogenote A merozygote in which corresponding alleles on its endogenote and exogenote are different.

heteroglycan HETEROPOLYSACCHARIDE.

heterograft A transplant from one individual to another individual of a different species.

heteroimmune phage One of two phages that

are sensitive to different immunity substances.

heteroimmune superinfection The superinfection of a bacterium with a heteroimmune phage, as in the infection of a lysogenic bacterium with a phage that is insensitive to the immunity substance of the prophage.

heteroimmunity 1. The immune reactions in which the antigens and the antibodies are derived from different animal species. 2. The immune reactions in which the antigens are derived from plants, microorganisms, or synthetic reactions.

heterokaryon A cell that contains two or more nuclei, derived from genetically different sources, as a result of cell fusion that is not accompanied by fusion of the nuclei; commonly occurs in fungi.

heterolactic fermentation *See* heterofermentative lactic acid bacteria.

heterologous 1. Pertaining to subcellular fractions, such as transfer RNA, ribosomes, and enzymes, that have been isolated from different species. 2. Pertaining to an antigen and an antibody (or an antiserum) when the antibody has been produced in response to the administration of a different antigen. 3. Pertaining to an antigen, an antibody, or an antiserum when the antigens and antibodies are derived from different species. 4. Pertaining to genetically dissimilar individuals of the same species; xenogeneic.

heterologous association An association of identical protein subunits in which the interacting surfaces of the subunits are not identical.

heterologous bonds The nonidentical interactions between subunits linked via heterologous associations.

heterologous desensitization The decreased sensitivity of cells to stimulation that is brought about by a variety of ligands and not just the one to which the cells had been previously exposed. *See also* desensitization (3).

heterologous graft HETEROGRAFT.

heterologous interference The viral interference induced by either an active or a suitably inactivated virus against a virus of a different taxonomic group.

heterologous mischarging The reaction whereby an aminoacyl-tRNA synthetase from one source (for example, yeast) aminoacylates the "wrong" tRNA from another source (for example, *E. coli*).

heterolysis The cleavage of a chemical bond between two atoms in which both of the two electrons constituting the bond move to one of the atoms; a reaction of the type $R:X \rightleftharpoons R^- + X^+$

heterolysosome An organelle produced by the

fusion of a lysosome, originating from the Golgi apparatus, with other vesicles or cytoplasmic bodies.

heterolytic process An unsymmetrical covalent bond breakage in which one of the two bonded atoms leaves with both of the electrons that constitute the bond.

heteromultimer HETEROPOLYMER.

heterophagic Other-digesting, as in the digestion by lysosomes of exogenous material taken up by the cell.

heterophagosome An endocytotic vacuole that has fused with other cellular vesicles, or with cytoplasmic bodies containing particulate matter, but that has not fused with a lysosome to form a digestive vacuole.

heterophile antibody An antibody that reacts with antigens of more than one species.

heterophile antigen HETEROGENETIC ANTIGEN.

heteroploid state The chromosome state in which the number of chromosomes differs from that of the characteristic diploid (or haploid) state. *Aka* heteroploidy.

heteropolar AMPHIPATHIC.

heteropolar bond POLAR BOND.

heteropolymer A polymer composed of two or more types of monomers.

heteropolysaccharide A polysaccharide composed of two or more types of monosaccharides.

heteropycnosis The variation in the degree of condensation of either different chromosomes or of parts of the same chromosome; used as the basis for classifying chromatin into euchromatin and heterochromatin.

heteroreactivation The reactivation of a poxvirus by another poxvirus of a different immunological subgroup. *See also* reactivation (2).

heteroside A glycoside, the "aglycone" portion of which is not a carbohydrate. *See also* holoside.

heterosomal aberration An interchromosomal aberration. *See also* chromosomal aberration.

heterospecific antibody An artificially produced antibody that has one antigen binding site that is specific for one antigen and a second antigen binding site that is specific for a different antigen.

heterothallic Descriptive of organisms, such as certain fungi and algae, that exist in the form of two mating types so that only gametes from strains of opposite mating type can fuse to form zygotes.

heterotherm POIKILOTHERM.

heterotopic transplant A transplant from one site to another on the same organism.

heterotroph A cell or an organism that requires a variety of carbon-containing compounds from animals and plants as its source of carbon, and that synthesizes all of its carbon-containing biomolecules from these compounds and from small inorganic molecules.

heterotropic interactions Cooperative interactions between binding sites involving ligands of different kinds. The interactions associated with the binding of substrate molecules, activators, and inhibitors to an allosteric enzyme are an example.

heterovalent resonance Resonance in which the various resonance structures do not have the same number of chemical bonds.

heterozygosity The state of having different alleles at one or more loci in the homologous chromosomes.

heterozygote A zygote that carries different alleles at one or more loci in the homologous chromosomes and that does not breed true; a cell or an organism that carries two different alleles of the same gene.

heterozygous Of, or pertaining to, a heterozygote.

HETP Height equivalent to a theoretical plate.

heuristic process A problem-solving process in which solutions are discovered by evaluating the progress made toward the final solution as by means of a controlled trial and error method, exploratory methods, the sequencing of investigations, etc. The process is valuable for stimulating further research, calculations, and the like, even though individual steps may remain unproven or be incapable of proof. *See also* algorithm; stochastic process.

hexa- Combining form meaning six.

hexagonal capsomer *See* capsomer.

hexamer 1. An oligomer that consists of six monomers. 2. HEXAGONAL CAPSOMER.

hexokinase The enzyme that catalyzes the conversion of glucose to glucose-6-phosphate in the first step of glycolysis; it is found in all tissues and exists in three isoenzyme forms (types I, II, and III). *See also* glucokinase.

hexon A capsomer composed of 6 protomers; a morphological subunit of some viruses.

hexon antigen An antigen of the hexon capsomer of adenoviruses.

hexosamine An amino sugar of a six-carbon monosaccharide.

hexosaminidase An enzyme, composed of two subunits, that functions in the metabolism of gangliosides. A deficiency of this enzyme leads to Tay–Sachs and Sandhoff diseases.

hexosan A polysaccharide of hexoses.

hexose A monosaccharide that has six carbon atoms.

hexose diphosphate pathway GLYCOLYSIS.

hexose monophosphate oxidative pathway The formation of ribose-5-phosphate from glucose-

6-phosphate in the initial reactions of the hexose monophosphate shunt.

hexose monophosphate pathway HEXOSE MONOPHOSPHATE SHUNT.

hexose monophosphate shunt The metabolic pathway that requires the input of six molecules of glucose-6-phosphate and leads to the complete oxidation of one molecule of glucose-6-phosphate to carbon dioxide, water, and phosphate. The pathway functions to generate reducing power in the form of NADPH, allows for the interconversion of monosaccharides, and is linked to the fixation of carbon dioxide in photosynthesis. *Abbr* HMS; HMP shunt.

hexose phosphoketolase pathway A catabolic pathway of glucose that is related to the hexose monophosphate shunt and that occurs in some bacteria.

hexuronic acid A sugar acid, formed by oxidation of the C-6 group of a hexose to a carboxyl group.

Hfr strain HIGH-FREQUENCY OF RECOMBINATION STRAIN.

HFT HIGH-FREQUENCY TRANSDUCTION.

HFT lysate High-frequency transduction lysate; a lysate prepared by induction of a prophage that possesses an unusually high transducing power.

Hg Mercury.

HGF Hyperglycemic-glycogenolytic factor.

HGG Human gamma globulin.

HGH Human growth hormone.

HGPRT Hypoxanthine-guanine phosphoribosyl transferase; *See* Lesch–Nyhan syndrome.

HHbO$_2$ OXYHEMOGLOBIN.

HI Hemagglutination inhibition.

5-HIAA 5-Hydroxyindoleacetic acid.

HIB Hemolytic immune body.

high-angle x-ray diffraction LARGE-ANGLE X-RAY DIFFRACTION.

high-copy number *See* copy number.

high-density lipoprotein A plasma lipoprotein that has a density of 1.063–1.210 g/mL and that is designated as HDL. An increase in the concentration of HDL is believed to be linked to a decrease in the incidence of atherosclerosis. HDL contain about 33% protein, 29% phospholipid, 30% cholesterol and cholesterol esters, and 8% triglycerides; they have moleculear weights of about 2 to 4×10^5 and are classified as the α-fraction on the basis of electrophoresis. *See also* lipoprotein.

high-efficiency liquid chromatography HIGH-PERFORMANCE LIQUID CHROMATOGRAPHY.

high egg passage strain *See* HEP strain.

high-energy bond A covalent bond, the hydrolysis of which under standard conditions yields a large amount of free energy. The term refers to the large negative free energy change associated with the hydrolysis reaction, and not to the bond energy. High-energy bonds are commonly denoted by a squiggle (~).

high-energy compound A compound that, upon hydrolysis under standard conditions, yields a large amount of free energy. High-energy compounds are frequently those in which a phosphate group is removed by hydrolysis and the free energy change of this reaction is of the order of 7 kcal/mol or more. *See also* high-energy bond.

high-energy ion scattering RUTHERFORD BACKSCATTERING.

high-energy phosphate donor A high-energy compound that can function as a phosphoryl group donor to a low-energy phosphate acceptor by way of the ADP-ATP phosphoryl group carrier system. *Aka* high-energy phosphate compound.

high-frequency of recombination strain A bacterial strain that has a high frequency of recombination due to the fact that the episomal fertility factor has become incorporated into the bacterial chromosome. *Abbr* Hfr strain.

high-frequency transduction Transduction in which the phages that are capable of transducing constitute a high proportion of the total phage population. *Abbr* HFT.

high-level language A programing language that allows a person to give instructions to a computer in an English-like text rather than by means of a numerical (binary) code of ones and zeros.

high-level promoter A promoter that can undergo a promoter-up mutation.

high-lipid lipoprotein LOW-DENSITY LIPOPROTEIN.

highly repetitive DNA *See* repetitive DNA.

high-mannose glycoproteins *See* glycosylation.

high-mannose oligosaccharides A group of N-linked oligosaccharides that contain up to 6 mannose residues, in addition to those in the core region. These oligosaccharides occur in ovalbumin, thyroglobulin, and in some cell-surface glycoproteins. They may have 2, 3, or 4 branches and are then designated as bi, tri-, or tetraantennary.

high-mobility group A group of abundant, nonhistone chromosomal proteins; so called, because they are relatively small and highly charged and, therefore, move rapidly in electrophoresis. *Abbr* HMG.

high molecular weight kininogen A plasma glycoprotein of high molecular weight (150,000 daltons) that functions in the initiation of the intrinsic pathway of blood coagulation. *Abbr* HM$_w$K.

high-mutability gene MUTATOR GENE.

high-performance liquid chromatography A column chromatographic technique that is rapid and provides high resolution; it can be used with the various modes of liquid chromatography such as gel filtration, adsorption, partition, and ion exchange. The apparatus consists of a column, a hydraulic system, a detector, and a recorder. The liquid is forced in series and under pressure through a column that is maintained at a constant temperature, through a detector, and then into a fraction collector. *Abbr* HPLC.

high polymer A polymer of very high molecular weight, such as a polymer occurring in plastics, rubber, fibers, or human tissues.

high-potential iron protein An iron–sulfur protein that has a high standard reduction potential (E_0') which may be of the order of +0.4 V as opposed to one of the order of −0.4 V for a low-potential iron protein. High-potential iron proteins contain ($Fe_3^{3+} — Fe^{2+}$) in their iron clusters as opposed to the ($Fe_2^{3+} — Fe_2^{2+}$) which occurs in the low-potential iron proteins. *Abbr* HiPiP. *Aka* high-potential iron–sulfur protein.

high-pressure liquid chromatography HIGH-PERFORMANCE LIQUID CHROMATOGRAPHY.

high-quality proteins Proteins from animal sources (eggs, milk, meat, and fish) that supply adequate amounts of all of the essential amino acids.

high-resistance-leak method A method for amplifying the ion current produced in an ionization chamber.

high-sensitivity liquid chromatography HIGH-PERFORMANCE LIQUID CHROMATOGRAPHY.

high-speed liquid chromatography HIGH-PERFORMANCE LIQUID CHROMATOGRAPHY.

high-speed sedimentation equilibrium MENISCUS DEPLETION SEDIMENTATION EQUILIBRIUM.

high spin The state of a complex in which there is a maximum of unpaired electrons; referred to as a state of essentially ionic bonding and ascribed to certain hemoproteins.

high-temperature protection regulon *See* HTP regulon.

high-temperature–short-time method The pasteurization of material by heating it at 71.7°C for 15 s. *Abbr* HTST method.

high-voltage electrophoresis An electrophoretic technique in which the applied electric field is greater than 20 V/cm and in which separations are achieved in short times; useful for the separation of low molecular weight compounds such as amino acids, peptides, and nucleotides.

high-yielding strain PRODUCTION STRAIN.

Hill-Bendall scheme Z SCHEME.

Hill coefficient The coefficient n_H in the Hill equation; it is also the slope in a Hill plot. *See*
also Hill equation; Hill plot.

Hill equation An equation that describes the binding of ligands to a protein; common forms of the equation are (i) $r/n = K'[S]^{n_H}/(1 + K'[S]^{n_H})$; (ii) $r/(n - r) = K'[S]^{n_H}$; (iii) $\log [r/(n - r)] = n_H \log [S] + \log K'$; (iv) $Y_s = [S]^{n_H}/(K + [S]^{n_H})$; (v) $\log[Y_s/(1 - Y_s)] = n_H \log[S] - \log K$, where r is the number of moles of ligand S bound per mole of total protein, n is the number of binding sites of a given type per molecule of protein, K' is the intrinsic association constant for this type of site, Y_s is the fraction of total binding sites occupied by ligand S ($Y_s = r/n$), K is the intrinsic dissociation constant ($K = 1/K'$), and n_H is the interaction factor for the sites which varies from 1, for noninteracting sites, to n, for highly cooperative binding.

Hill plot A graphical representation of binding data based on the Hill equation that is used for determining intrinsic association constants and for determining the number of binding sites of a given type per molecule of protein; consists of a plot of $\log[r/(n - r)]$ as a function of $\log[S]$ or a plot of $\log[Y_s/(1 - Y_s)]$ as a function of $\log[S]$. *See also* Hill equation.

Hill reaction The light reaction in photosynthesis that is carried out in the presence of an artificial electron acceptor; the reaction: $2H_2O + A = 2 AH_2 + O_2$, where A is the electron acceptor.

Hill reagent The artificial electron acceptor in the Hill reaction.

hinge point The position in an analytical ultracentrifuge cell at which the concentration, during sedimentation equilibrium, is equal to the initial concentration of the solution.

hinge region 1. That portion of the IgG immunoglobulin molecule that is adjacent to the two disulfide bonds that link the two heavy chains together; the hinge region is near the sites of action of papain and pepsin and is believed to be a flexible region that permits the molecule to open up in the form of a Y.2. One of two regions in the predominantly helical rod section of the myosin molecule that is less stable than the rest of the molecule, and that is more susceptible to proteolytic attack; a trypsin-sensitive region of myosin.

HiPiP High-potential iron protein.

hippuric acid *N*-Benzoylglycine; the form in which benzoic acid is detoxified and excreted in the urine.

Hirano bodies Paracrystalline inclusions found in the brains of humans exhibiting senile dementia and other neurodegenerative diseases; thought to be stacked sheets of membrane-bound ribosomes, derived from partially degraded rough endoplasmic reticulum of neurons (Nissl substance).

His 1. Histidine. 2. Histidyl.

H isoantigen A pentahexosyl ceramide that accumulates in individuals suffering from fucosidosis.

His operon The operon in *Salmonella typhimurium* that consists of the genes that code for the 10 enzymes involved in the biosynthesis of histidine.

histamine A pharmacologically active mediator of the allergic response that causes vasodilation, increased capillary permeability, and contraction of smooth muscle. Histamine is formed by decarboxylation of histidine and is widely distributed in mammalian tissues, particularly in mast cells from which it is released during the allergic response. *See also* H_1 receptors; H_2 receptors.

histidine A heterocyclic, basic, and polar alpha amino acid that contains the imidazole ring system. *Abbr* His; H.

histidinemia A genetically inherited metabolic defect in humans that is characterized by elevated blood and urine levels of histidine, and that is due to a deficiency of the enzyme histidase that functions in histidine catabolism.

histochemistry The science that deals with the chemical constitution and the chemical changes of tissues and cells by combining the techniques of biochemistry and histology.

histocompatibility Immunological tolerance to transplanted tissue. *See also* major histocompatibility complex.

histocompatibility antigen A tissue antigen of the donor of an allograft that induces transplantation immunity in the recipient of the transplant; one of a group of cell-surface glycoprotein antigens that are involved in the rejection of transplanted tissues between two genetically dissimilar individuals of the same species.

histocompatibility gene A gene responsible for the production of a histocompatibility antigen.

histogram A graphical representation of a frequency distribution by means of rectangles, the widths of which represent the class interval and the heights of which represent the frequencies of the different classes.

histohematin CYTOCHROME.

histoincompatibility Immunological intolerance to transplanted tissue that results from antigenic differences between the donor and the recipient of a transplant.

histological chemistry *See* histochemistry.

histology The branch of anatomy that deals with the structure and the properties of tissues, as examined by staining and microscopy.

histolysis The destruction of tissues.

histone A basic, globular, and simple protein that is characterized by its high content of arginine and lysine. Histones are found in association with nucleic acids in the nuclei of eukaryotic cells; they are classified into five major groups, on the basis of their arginine and lysine content: H1 (28% lys, 2% arg). H2A (11% lys, 9% arg), H2B (16% lys, 6% arg), H3 (10% lys, 13% arg), H4 (10% lys, 14% arg).

histoplasmin A crude and sterile filtrate, derived from mycelia of the fungus *Histoplasma capsulatum*, that is used for intradermal injections as a test for delayed-type skin hypersensitivity.

hit theory *See* target theory.

HIV *See* AIDS virus.

hive The inflammation of human skin produced during cutaneous anaphylaxis.

HL One of four chloroform-soluble fractions from liver (HL-1 through HL-4) that contain growth factors for *Lactobacillus helveticus* and *Lactobacillus lactis* and that may be related to folic acid and lipoic acid.

HLA antigens *See* major histocompatibility complex.

HLA complex Human-leukocyte associated complex; the major histocompatibility complex in humans.

HLB Hydrophilic–lipophilic balance.

HLCF Heat-labile citrovorum factor.

H locus A chromosomal site in humans that is responsible for synthesis of an enzyme, fucosyl transferase, that functions in the biosynthesis of ABO blood group antigens.

HMC 5-Hydroxymethylcytosine; a minor base.

H-meromyosin Heavy meromyosin.

HMG 1. β-Hydroxyl-β-methyl glutarate. 2. Human menopausal gonadotropin. 3. High-mobility group.

HMM Heavy meromyosin.

HMP 1. Hexose monophosphate pathway. 2. Hexose monophosphate.

HMP shunt Hexose monophosphate shunt.

HMS Hexose monophosphate shunt.

H mutant HOST-RANGE MUTANT.

HM_wK High molecular weight kininogen.

HMW-MAP High molecular weight microtubule-associated protein. A family of high molecular weight proteins (about 300,000 daltons) that appear to function in the assembly of microtubules.

HnRNA Heterogeneous nuclear RNA.

Hodge scheme A schematic summary of various enzymatic and nonenzymatic mechanisms of the browning reactions.

Hodgkin's disease A malignant disorder characterized by painless, progressive enlargement of lymphoid tissue.

Hofmeister series LYOTROPIC SERIES.

Hogness box TATA BOX.

holandric Designating a trait that appears only in males.

holandric gene A gene that is located only on the Y chromosome and that appears only in males.

holdback carrier A nonradioactive compound that is added to a sample to prevent either the coprecipitation or the adsorption of a soluble radioactive compound. *Aka* holdback agent.

holdup volume The gas chromatographic retention volume of a nonadsorbed component; the volume of mobile phase that must be eluted before a component, which is not retarded by the stationary phase, can be eluted.

hole 1. An energy level that is not occupied by a particle even though adjacent energy levels are filled. 2. An unoccupied position in a crystal, a metal, or a liquid where there should normally be either an electron or an atom.

hole zones Thin regions in collagen fibers, resulting from the presence of gaps between staggered collagen fibrils.

holism The doctrine that higher levels of organization cannot be understood or predicted from a knowledge of lower levels; the entirety is greater than the sum of the parts. *See also* reductionism.

Holliday model A model, proposed by R. Holliday, that accounts for crossing over between homologous chromosomes in terms of a series of breakage and reunion events, involving X-shaped (chi) structures. The crossed strand regions are called Holliday junctions and the chi structures are called Holliday intermediates.

hollow fiber technique A technique for dialyzing, desalting, concentrating, and fractionating solutions of macromolecules; entails the use of bundles of semipermeable, hollow-bore fibers that function as molecular sieves and that have pores of controlled dimensions.

holo- Combining form meaning whole or entire.

holocrine gland A gland that produces a secretion that consists of altered secretory cells of the gland itself.

holoenzyme 1. An entire conjugated enzyme that consists of a protein component, or apoenzyme, and of a nonprotein component, or cofactor, which may be either a coenzyme or an activator. 2. An entire oligomeric enzyme that consists of two or more subunits and that retains some activity even when one (or several) of the subunits is missing; DNA-dependent RNA polymerase and DNA-dependent DNA polymerase III are two examples.

hologynic Designating a trait that appears only in females.

holophytic nutrition A mode of nutrition, as that of green plants, that requires only inorganic compounds for growth and maintenance of the organism.

holoside A glycoside whose "aglycone" portion is another carbohydrate; an oligosaccharide. *See also* heteroside.

holothurin ANIMAL SAPONIN.

holozoic nutrition A mode of nutrition, as that of animals, that requires organic compounds for growth and maintenance of the organism.

homeomorph One of a group of possible and reasonable reaction mechanisms that obey the same rate equation.

homeoplastic graft HOMOGRAFT.

homeostasis 1. The constancy of the internal environment of an organism; the steady state with respect to functions, tissues, and fluids of the organism. 2. The processes involved in the regulation and maintenance of the internal environment of an organism.

homeostatic Of, or pertaining to, homeostasis.

homeothermic Variant spelling of homoiothermic.

homeotic genes Genes that establish the diverse pathways by which an embryonic segment develops a distinct adult phenotype. Mutations of homeotic genes result in the replacement of one development pattern by a different, but homologous one. *Var sp* homoeotic genes.

homeotic mutation A mutation that results in transformation of parts of an organism into structures that are appropriate for other positions; the sprouting of a wing from the head of a fruitfly is an example. *Var sp* homoeotic mutation.

homo- 1. Combining form meaning the same or alike. 2. Prefix meaning one additional —CH_2— group; the next higher homologue of the compound indicated.

homoallelic Pertaining to alleles of a gene that have mutations at identical sites.

homoamino acid An amino acid that has an additional —CH_2— group compared to another amino acid; an amino acid homologue.

homocellular transport The transport across the cell membrane that moves material into or out of the cell. *See also* intracellular transport; transcellular transport.

homochiral Descriptive of the relation between isometric chiral molecules that are properly congruent. The molecules in an enantiomerically pure material are homochirally related.

homochromatography A chromatographic technique in which a group of compounds are separated by development with a solvent that contains either the same or related compounds; applied, for example, to the paper chromatographic separation of a mixture of labeled nucleotides from a hydrolysate of *E. coli* 5S RNA by development with a solvent containing a mixture of unlabeled nucleotides from a hydrolysate of yeast RNA.

homocodonic amino acid One of four amino acids that have codons in which all three bases are identical: glycine (GGG), proline (CCC), lysine (AAA), and phenylalanine (UUU).

homocopolymer A copolymer in which individual chains are composed entirely of one type of monomer, as in poly dA:dT.

homocysteine A homologue of cysteine that contains one —CH$_2$— group more than cysteine. *Abbr* Hcy.

homocystinuria A genetically inherited metabolic defect in humans that is associated with mental retardation and that is characterized by the presence of excessive amounts of homocystine in the urine; due to a deficiency of the enzyme cystathionine synthase which functions in the metabolism of cysteine.

homocytotropic antibody A cytotropic antibody that binds to and sensitizes target cells in the same species in which it was produced. *See also* reagin.

homodimer A protein that is composed of two identical polypeptide chains.

homoduplex DNA A double-stranded DNA molecule, formed by complementary base pairing of single strands from the same DNA sample. The DNA formed by annealing ordinary *E. coli* DNA or that formed by annealing ^{15}N-labeled *E. coli* DNA are two examples. *See also* heteroduplex.

homoeologous chromosomes Chromosomes that are only partially homologous.

homoeotic genes HOMEOTIC GENES.

homoeotic mutation HOMEOTIC MUTATION.

homofermentation 1. Fermentation in which only one major product is formed. 2. HOMOLACTIC FERMENTATION.

homofermentative lactic acid bacteria Lactic acid bacteria that produce in fermentation 1.8 to 2.0 mol of lactic acid per mole of glucose. *See also* heterofermentative lactic acid bacteria.

homogenate The suspension prepared by homogenization of tissues, cells, or cellular components.

homogeneity 1. The state of a preparation of macromolecules of a single type (e.g., of one enzyme) in which the macromolecules are identical with respect to size, charge, structure, and all other properties. 2. The state of a system in which there is only one phase.

homogeneous Of, or pertaining to, homogeneity.

homogeneous catalysis Catalysis in a system that consists of one phase.

homogenic process A symmetrical formation of a covalent bond in which each of the two bonding atoms contributes one electron to the shared pair of electrons that constitutes the bond.

homogenization The disruption of tissues, cells, or cellular components and their reduction to particles of small size so that a relatively uniform suspension is obtained.

homogenizer A device, frequently a tube with a finely tooled pestle, that is used for homogenization.

homogenote A merozygote in which corresponding alleles on its endogenote and exogenote are identical.

homogentisic acid 2,5-Dihydroxyphenylacetic acid; an intermediate in the catabolism of phenylalanine and tyrosine.

homoglycan HOMOPOLYSACCHARIDE.

homograft ALLOGRAFT.

homoimmune phage One of two phages that are sensitive to the same immunity substance.

homoimmunity The resistance of a bacterium carrying a prophage (a lysogenic bacterium) to infection by a phage that is of the same kind as the prophage.

homoiothermic Descriptive of an organism that has a nearly constant body temperature irrespective of the temperature of its environment; it can regulate its body temperature by means of its metabolism. *Aka* warm-blooded.

homokaryon A fungus cell that contains two or more nuclei of only one genotype. *Var sp* homocaryon.

homolactic fermentation *See* homofermentative lactic acid bacteria.

homologous 1. Pertaining to subcellular fractions such as tRNA, ribosomes, and enzymes that have been isolated from the same species; isologous. 2. Pertaining to genetically dissimilar individuals of the same species; allogeneic. 3. Pertaining to proteins or nucleic acids from different species that have identical or similar functions, such as the hemoglobins of various vertebrate species. 4. Pertaining to two proteins that show sequence homology. 5. Pertaining to an antigen and an antibody (or an antiserum) when the antibody has been produced in response to the administration of that antigen. 6. Pertaining to an antigen and an antibody (or an antiserum) when the antigens and the antibodies are derived from the same species; isologous. 7. Pertaining to chemical compounds that are members of a homologous series. 8. Pertaining to DNA molecules that have identical or nearly identical base sequences.

homologous chromosomes Chromosomes that occur in pairs, with one each derived from the male and female parent, and that contain the same linear gene sequences so that each gene is present in duplicate.

homologous desensitization The decreased sensitivity of cells to stimulation that is brought about only by that ligand to which the cells had been previously exposed. *See also* desensitization (3).

homologous disease GRAFT-VERSUS-HOST REACTION.

homologous enzyme variants Enzyme variants that have similar molecular structures and catalytic properties. *See also* homologous (3).

homologous hapten A hapten that closely resembles the haptenic group of the immunogen.

homologous interference The inhibition of the multiplication of one virus by an active or a suitably inactivated different virus which belongs to the same taxonomic group.

homologous recombination Genetic exchange between DNA molecules that have identical or nearly identical base sequences. *Aka* general recombination.

homologous series A series of organic compounds in which each member differs from the preceding one by having one additional —CH_2— group.

homologous serum jaundice *See* hepatitis.

homologue 1. A member of a homologous series. 2. One of a pair of homologous chromosomes. *Var sp* homolog.

homolysis The cleavage of a chemical bond between two atoms in which one of the two electrons constituting the bond moves to one atom and the second electron moves to the other atom; a reaction of the type R:X = R· + X·.

homolytic process A symmetrical covalent bond breakage in which each of the two bonded atoms leaves with one of the electrons of the pair that constitutes the bond.

homomultimer HOMOPOLYMER.

homophilic bonding The attraction of like groups for each other, as that of nonpolar groups in a hydrophobic bond or that of polar groups in an electrostatic bond.

homopolar bond NONPOLAR BOND (1).

homopolymer A polymer composed of only one type of monomer.

homopolymer tail joining A method for joining two double-stranded DNA molecules by means of homopolymer sequences that are attached through the action of the enzyme terminal nucleotidyl transferase. Poly(dA) tails are put on one DNA molecule and poly(dT) tails are put on the other. The two types of tails are then allowed to anneal and DNA ligase is used to complete the linking of the two DNA molecules.

homopolynucleotide A polynucleotide composed of only one type of nucleotide.

hompolypeptide A polypeptide composed of only one type of amino acid.

homopolysaccharide A polysaccharide composed of only one type of monosaccharide.

homoreactant antibody An autoantibody, occurring in rabbits, that appears to be directed against antigenic determinants of the autologous IgG immunoglobulin that are hidden in the intact IgG molecule but are exposed when it is digested with the enzyme papain.

homoreactivation The reactivation of a poxvirus by another poxvirus of the same immunological subgroup. *See also* reactivation (2).

homoserine The homologue of serine that contains one —CH_2— group more than serine.

homosomal aberration An intrachromosomal aberration. *See also* chromosomal aberration.

homospecific antibody An antibody in which both of the antigen binding sites are specific for the same antigen. *See also* heterospecific antibody.

homosteroid A steroid-like molecule; a modified steroid in which a ring has been expanded by the introduction of one or more —CH_2— groups; a homologous steroid.

homothallic Descriptive of organisms, such as certain fungi and algae, that exist in the form of one mating type so that gametes from a single strain can fuse to form zygotes.

homotope A monomer that can take the place of another monomer in a definable segment of a polymer; thus serine and glycine which occupy position 9 of the A chain of ox and sheep insulin, respectively, are said to be homotopes.

homotropic interactions Cooperative interactions between binding sites involving ligands of one kind. The interactions associated with the binding of 4 oxygen molecules by a molecule of hemoglobin and those associated with the binding of several substrate molecules by an allosteric enzyme are two examples. In the latter case, the substrate functions both as a substrate and as an effector.

homozygosity The state of having identical alleles at one or more loci in the homologous chromosomes.

homozygote A zygote that carries identical alleles at one or more loci in the homologous chromosomes and that breeds true; a cell or an organism that carries two identical alleles of the same gene.

homozygous Of, or pertaining to, a homozygote.

Hoogsten base pairs A proposed set of base pairs in which the hydrogen bonds involve purine atom N_7 rather than N_1 as in Watson–Crick base pairs. Specifically, the proposed hydrogen bonds are as follows: AT: N_6 (adenine) – O_4 (thymine); N_7 (adenine) – N_3 (thymine); GC: O_6 (guanine) – N_4 (cytosine); N_7 (guanine) – N_3 (cytosine).

hook model A model for the attachment of the prophage to the bacterial DNA. According to this model, the prophage is joined through a point to the uninterrupted bacterial DNA, somewhat in the manner of a hook. *See also* insertion model.

hopanoids A group of pentacyclic, sterol-like molecules found in many bacteria and some plants. Like the sterols, they are synthesized from mevalonic acid, which is converted to squalene; the latter is then cyclized to yield the hopane nucleus. Bacterial hopanoids appear to play a role in maintaining the stability of membranes.

Hopkins–Cole reaction A colorimetric reaction for tryptophan and other compounds containing the indole ring; based on the formation of a violet color upon treatment of the sample with glyoxylic acid and sulfuric acid.

Horecker cycle HEXOSE MONOPHOSPHATE SHUNT.

horizontal evolution The simultaneous, parallel evolution of many sequences in the gene complement of a single species. The process whereby a population splits into two or more subgroups which then evolve into separate and distinct species. This contrasts with the more common evolution of a single gene in two or more diverging species.

horizontal strip electrophoresis An electrophoretic technique in which paper strips are used that are supported in a horizontal frame, with each end of the paper dipping into a buffer compartment.

horizontal transmission The transmission of viruses between individual hosts of the same generation. *Aka* horizontal infection.

hormonad One of a group of proposed multienzyme complexes that function in the biosynthesis of steroids such as estradiol, cortisol, and aldosterone.

hormone A regulatory substance that is synthesized by specialized cells of an organism, that is active at low concentrations, and that exerts its effect either on all of the cells of the organism or only on certain cells in specific organs. Hormones act by binding to macromolecular receptors that are located either on the cell membrane or inside responsive cells. Hormones have three major functions: (a) an integrative function that deals with the interrelations between different hormones and with the interrelations between hormones and the nervous system, the blood flow, the blood pressure, and other factors; (b) a mor-

phogenetic function that deals with the control of the type and rate of growth of various tissues; and (c) a regulatory function that deals with the maintenance of a constant internal environment with respect to the intra- and extracellular fluids. An animal hormone is a substance such as a polypeptide, a protein, or a steroid, that is secreted principally by an endocrine gland and that is transported by way of the circulation to target organs or target tissues; there the hormone exerts its effect either directly or indirectly and helps to regulate such overall physiological processes as metabolism, growth, and reproduction. A plant hormone is an organic compound that controls growth or some other function at a site removed from its place of production in the plant. Five major types of plant hormones have been identified: auxin, cytokinin, gibberellin, abscisic acid, and ethylene. All of the plant hormones are pleiotropic; that is, they have multiple effects. *See also* acceptor (1).

hormonogen PROHORMONE.

horror autotoxicus An early synonym for autoimmunity, so named because it was originally believed to be nonexistent; it is now known to be widespread.

host 1. An organism upon or in which a parasite lives. 2. The recipient of a transplant. 3. The cell used to propagate a vector and its insert.

host-cell reactivation The restoration of the activity of an ultraviolet-irradiated DNA phage by means of the excision–repair mechanism of the host, subsequent to the infection of the host cells by the phage. *Abbr* HCR.

host-controlled modification HOST-INDUCED MODIFICATION.

host-controlled restriction A host-induced modification that results in the formation of a restricted virus.

host–guest system A system in which a molecule having a well-defined and relatively rigid cavity (the host) binds ions or small molecules (the guest) to sites within this cavity. Examples of naturally occurring hosts are the cyclodextrins and examples of synthetic ones are the crown ethers. Host–guest systems have been used to study membrane aspects and to determine the helix-forming tendencies of amino acids.

host-induced modification A change in the properties of a virus that is brought about by the propagation of the virus in the host cells. In phages, the change amounts to a chemical alteration such as that produced by glucosylation or methylation of the phage DNA; in animal viruses, the change amounts to an incorporation of host cell membrane components into the viral envelope.

host range The spectrum of hosts that can be infected by a specific virus.

host-range mutant A mutant virus that can adsorb to, and infect, cells that are resistant to the wild-type virus.

hot Containing one or more radioactive isotopes.

hot spot A site on the DNA molecule at which mutational changes or recombinational events occur at a much higher frequency than is the case for the average site. The term is used particularly for sites in the rII region of T4 phage DNA.

housekeeping genes Genes that provide for general cell function and that are theoretically expressed in all cells and are continuously transcribed; the genes coding for the enzymes of glycolysis and the citric acid cycle are examples. *See also* luxury genes.

Houssay animal A hypophysectomized and depancreatized animal that is used in endocrinological studies.

HPETE Hydroperoxyeicosatetranoic acid; a compound derived from arachidonic acid by oxidation with lipoxygenase.

HPG Human pituitary gonadotropin.

HPIEC High-performance ion-exchange chromatography. *See* high-performance liquid chromatography.

HPL Human placental lactogen.

HPLC High-performance liquid chromatography; also used as an abbreviation for high-pressure liquid chromatography.

HPr A cytoplasmic, heat-stable protein that functions in the transfer of glucose across a bacterial membrane. *See also* enzyme I; enzyme II.

hr Hour.

H1 receptors Receptors, present in some smooth muscles, that are affected by histamine and that cause dilation of arterioles and constriction of venules and bronchioles.

H2 receptors Receptors, present in the gastric mucosa, that are affected by histamine and that cause an increase in the secretion of gastric HCl.

HSA Human serum albumin.

HSK-pathway Hatch–Slack–Kortschak pathway.

HSLC High-speed liquid chromatography.

HSP Heat shock proteins.

H substance A polysaccharide precursor in the biosynthesis of the ABO blood group antigens.

5-HT 5-Hydroxytryptamine.

HTLV-III *See* AIDS virus.

HTP High-temperature protection; also used as an abbreviation for heat shock proteins.

5-HTP 5-Hydroxytryptophan.

HTP regulon High-temperature protection regulon; the regulon responsible for production of heat shock proteins.

HTST method High-temperature–short-time method.

hu Dihydrouridine.

Hudson's rule An empirical rule for assigning anomeric configurations to carbohydrates. The rule states that the α-anomer of a given carbohydrate is that anomer which has a more positive specific rotation in the D-series and a more negative specific rotation in the L-series of the carbohydrates than the other anomer of the same carbohydrate.

Huebl number IODINE NUMBER.

Huebl's iodine solution A solution of iodine and mercuric chloride that is used in the determination of iodine numbers.

Huefner's quotient The ratio of the absorbance at one wavelength to that at another wavelength for a solution containing two hemoglobin derivatives (such as oxyhemoglobin and deoxyhemoglobin, or oxyhemoglobin and carbon monoxide hemoglobin). Such absorbance ratios can be used to determine the relative proportions of the two hemoglobin derivatives.

human chorionic gonadotropin A gonadotropic hormone, produced by the placenta, that has similar biological effects to luteinizing hormone. *Var sp* human chorionic gonadotrophin. *Abbr* HCG.

human lactogen PLACENTAL LACTOGEN.

human menopausal gonadotropin A glycoprotein of the anterior lobe of the pituitary gland that is formed in increasing amounts during menopause and that has follicle-stimulating hormone activity. *Var sp* human menopausal gonadotrophin. *Abbr* HMG.

humectant A hygroscopic substance that is added to other materials, such as food, to ensure a desired level of moisture content.

humic acid A complex mixture of acid-insoluble and alkali-soluble substances that are extracted from the organic matter of soil. *See also* fulvic acid; humin.

humin The heterogeneous fraction of soil organic matter that cannot be extracted with base. *See also* humic acid; fulvic acid.

humor 1. A chemical substance, such as acetylcholine, that is formed in the body and that acts locally. 2. A fluid or a semifluid substance of an animal or a plant, such as the vitreous humor of the eye.

humoral 1. Of, or pertaining to, a humor. 2. Involving only soluble factors as distinct from being cell-mediated.

humoral immunity Immunity that is due to circulating antibodies in the blood and antibodies secreted onto mucous surfaces in contrast to cellular immunity; it involves extracellular phases of infection and is associated with B lymphocytes. *Aka* humoral immune response; humoral antibody response.

humus The major insoluble portion of the organic substances in soil that is produced by the decomposition of animal and vegetable matter.

Hund's rule The principle that when two or more empty atomic orbitals of equal energy are available, one electron will enter each orbital until all of the orbitals are half-filled (that is, contain one electron). Only after that has occurred are the second electrons added to the orbitals.

Hunter's syndrome A mucopolysaccharidosis due to a deficiency of the enzyme iduronate sulfatase that is involved in the catabolism of dermatan sulfate and heparan sulfate.

H₂Urd Dihydrouridine.

Hurler's syndrome An autosomally linked form of gargoylism; a mucopolysaccharidosis due to a deficiency of the enzyme α-L-iduronidase and characterized by an accumulation of dermatan and heparan sulfates. It appears in infancy and involves steady mental and physical deterioration.

Huxley–Hanson model SLIDING FILAMENT MODEL.

Huxley–Simmons model ROWBOAT MODEL.

HVL Half-value layer.

hv region Hypervariable region.

hyaline VITREOUS.

hyaloplasm CYTOSOL.

hyaluronic acid A glycosaminoglycan of connective tissue that is composed of D-glucuronic acid and *N*-acetyl-D-glucosamine; it aids in blocking the spread of invading microorganisms and toxic substances and serves as a lubricant and shock absorbant. It occurs in connective tissue, synovial fluid, vitreous humor, and cartilage.

hyaluronidase The enzyme, present in snake venom and in bacteria, that catalyzes the hydrolysis of hyaluronic acid and thereby decreases the effectiveness of hyaluronic acid for blocking the spread of invading microorganisms and toxic substances in the tissues.

hybrid antibody An artificially produced antibody molecule that is composed of fragments, such as intact light and heavy chains, which are derived from two purified and different antibodies.

hybrid-arrested translation A screening procedure that is similar to that of hybrid-selected translation; it is based on the ability of the DNA to hybridize with, and remove, a translatable mRNA, thereby leading to the disappearance of the corresponding protein from the in vitro translation products.

hybridase RIBONUCLEASE H.

hybrid cell HYBRIDOMA.

hybrid duplex A double-stranded nucleic acid molecule that is produced by hybridization.

hybrid gene FUSED GENE.

hybrid hemoglobin An artificially produced hemoglobin molecule that contains either globin chains or subunits which are derived from different sources.

hybridization 1. A technique for assessing the extent of sequence homology between single strands of nucleic acids. The technique is based on allowing the polynucleotide strands to form double-helical segments through hydrogen bonding between complementary base pairs. The greater the extent of complementarity between the strands, the greater is the extent of formation of double-helical segments. The polynucleotide strands may be those of single-stranded nucleic acids or they may be derived from denatured double-stranded nucleic acids. The hybrids formed can be of the DNA/DNA, RNA/RNA, or DNA/RNA type. *See also* Southern blotting; northern blotting. 2. The reconstitution of an oligomeric protein, such as hemoglobin, or of a molecular aggregate, such as a ribosome, from separate and different monomers and/or component parts. The monomers and/or the component parts may be either chemically or mutationally altered ones, or they may be derived from different sources. 3. The fusion of two cells in tissue culture to form a hybrid cell, as in the production of a hybridoma. The hybrid cell contains nuclei and cytoplasm from the two different cells. Cell fusion is promoted by the adsorption of certain viruses (such as Sendai virus), by chemicals (such as polyethylene glycol), or by other means (such as electrofusion). 4. The rearrangement and combination of atomic orbitals in a compound that is due to the effects of either neighboring or bonded atoms.

hybridoma A hybrid cell, produced by the fusion of two cells, and useful for the production of monoclonal antibodies. Antibody-producing lymphocytes of limited life span in cell culture are fused with tumor lymphocytes that are "immortal" in cell culture. From the resulting heterogeneous mixture of hybrid cells, those hybrids are selected that have both the ability to produce a specific antibody, and the ability to multiply indefinitely in tissue culture. These hybridomas are then propagated and provide a permanent and stable source of a single monoclonal antibody. *See also* hybridization (3).

hybrid protein FUSED PROTEIN.

hybrid-selected translation A screening procedure for clones from a cDNA library that can be used if the mRNA, corresponding to a desired gene sequence, can be translated in vitro to produce an identifiable protein product. The method involves hybridization of mRNA and DNA under R-looping conditions, separation of the hybrids formed, and translation of the hybrids in vitro.

hydratase An enzyme that catalyzes the reversible hydration of a double bond.

hydrate A compound that contains one or more molecules of water in loose combination.

hydrated electron *See* solvated electron.

hydration 1. The process whereby water molecules surround, and bind to, solute ions and solute molecules in solution. 2. The formation of a hydrate.

hydration factor *See* Oncley equation.

hydration shell The layer of water molecules that are bound to an ion or a molecule in solution. *Aka* hydration sphere.

hydrazinolysis The cleavage of a polypeptide or a protein by treatment with hydrazine. The C-terminal amino acid is released as the free amino acid and can be identified; all the other amino acid residues are converted to acyl hydrazine derivatives.

hydri-, hydro- Proposed prefixes for the mixture of hydrogen isotopes that occurs in nature. *See also* proti-, proto-.

hydride ion The anion $H{:}^-$; a proton with an associated pair of electrons.

hydrion A hydrogen ion; a proton.

hydrocarbon An organic compound consisting only of carbon and hydrogen.

hydrocortisone CORTISOL.

hydrodynamic Of, or pertaining to, the motion of fluids and the force on, as well as the motion of, particles that are immersed in these fluids.

hydrodynamic method A physical method for studying molecules, particularly macromolecules, on the basis of their movement in solution; includes such methods as sedimentation, diffusion, electrophoresis, and viscosity.

hydrodynamic unit The unit that moves in solution and that consists of the solute particle together with the solvent that is bound tightly to it.

Hydrogemonas *See* Knallgas reaction.

hydrogen An element that is essential to all plants and animals. Symbol, H; atomic number, 1; atomic weight, 1.00797; oxidation states, -1, $+1$; most abundant isotope, ^1H; the stable isotope, ^2H; the radioactive isotope, ^3H, half-life, 12.26 years, radiation emitted, beta particles.

hydrogenase An enzyme that catalyzes the reduction of a substrate by molecular hydrogen.

hydrogenation The introduction of hydrogen into an organic compound.

hydrogen bond The attractive interaction that occurs between a covalently linked hydrogen atom and a neighboring atom or group of atoms. The hydrogen atom is linked covalently to an electronegative atom referred to as the donor, and is attracted to an electronegative atom or group of atoms referred to as the acceptor; the acceptor is frequently an oxygen or a nitrogen atom. The hydrogen bond is weaker and has a smaller bond energy than a covalent bond; it has a major electrostatic component and a minor covalent component. The hydrogen bond occurs both intra- and intermolecularly and it is known as a bifurcated bond if the hydrogen atom is attracted simultaneously to two acceptor atoms.

hydrogen carrier An electron carrier that undergoes oxidation–reduction reactions by either the loss or the gain of hydrogen atoms.

hydrogen electrode An electrode at which hydrogen gas is in equilibrium with hydrogen ions in solution. The hydrogen electrode serves as the primary standard for determinations of pH and electrode potentials.

hydrogen exchange *See* deuterium exchange.

hydrogen ion concentration *See* pH.

hydrogen ion equilibrium The dissociation and association reactions of hydrogen ions, particularly those pertaining to the functional groups in a protein.

hydrogen ion titration curve A titration curve, particularly that of a protein, in which the number of hydrogen ions that are either bound or dissociated per molecule is plotted as a function of pH.

hydrogen isotope exchange DEUTERIUM EXCHANGE.

hydrogenosome A subcellular organelle of some anaerobic parasitic protozoans that contains pyruvate synthase and hydrogenase and that produces molecular hydrogen by using protons as terminal electron acceptors.

hydrogen peroxide A toxic compound (H_2O_2) that is the substrate of the enzyme catalase.

hydrogen transport system ELECTRON TRANSPORT SYSTEM.

hydrolase An enzyme that catalyzes a hydrolysis reaction. *See also* enzyme classification.

hydrolysate The solution that contains the mixture of compounds obtained by hydrolysis. *Var sp* hydrolyzate.

hydrolysis (*pl* hydrolyses) The reaction of a substance with water in which the elements of water (H, OH) are separated: (a) the breakage of a molecule into two or more smaller fragments by the cleavage of one or more covalent bonds of acid derivatives; the elements of water are incorporated at each cleavage point such that one of the products combines with the hydrogen of the water while the other product combines with the hydroxyl group of the water; (b) the formation of the undissociated form of a weak electrolyte through the reaction of the ion of that electrolyte with either protons or hydroxyl ions.

hydrolysis constant The equilibrium constant for a hydrolysis reaction.

hydrolytic Of, or pertaining to, hydrolysis.

hydrolytic deamination Nonoxidative deamination.

hydrolytic enzyme HYDROLASE.

hydrolyze To carry out a hydrolysis reaction.

hydrometer An instrument for measuring the specific gravity of liquids.

hydron *See* proti-, proto-.

hydronation *See* proti-, proto-.

hydronium ion The hydrated proton H_3O^+.

hydroperoxide A compound that contains the grouping —OOH.

hydroperoxyl radical The free radical $HO_2\cdot$; an oxidizing agent that is readily reduced to hydrogen peroxide (H_2O_2).

hydrophilic 1. POLAR. 2. Descriptive of the tendency of a group of atoms or of a surface to become either wetted or solvated by water. *See also* lyophilic.

hydrophilic–lipophilic balance A characteristic parameter of a detergent, defined as the ratio of the mass (in atomic mass units) of the hydrophilic portion of the molecule to that of the hydrophobic portion. *Abbr* HLB.

hydrophobic 1. NONPOLAR. 2. Descriptive of the tendency of a group of atoms or of a surface to resist becoming either wetted or solvated by water. *See also* lyophobic.

hydrophobic bond HYDROPHOBIC INTERACTION.

hydrophobic chromatography A column chromatographic technique for the separation of proteins in which the chromatographic support consists of a nonpolar material, such as a hydrocarbon-coated agarose. Fractionation is achieved primarily on the basis of the interactions that take place between accessible hydrophobic pockets or regions on the proteins and the hydrocarbon chains of the support.

hydrophobic effect The phenomenological aspects of hydrophobic interactions, that is, the attraction of nonpolar molecules or nonpolar groups for each other. The basis for these interactions is due, largely, to the gain in entropy experienced by the surrounding water molecules as the interacting molecules or groups aggregate together.

hydrophobic interaction The attractive force between nonpolar molecules or nonpolar groups of molecules which leads to the association of these molecules or groups in an aqueous environment. Hydrophobic interactions occur due to the gain in entropy of the water which results when nonpolar groups associate and liberate the water molecules that were previously organized around the nonpolar groups. Hydrophobic interactions are unique in that their strength increases as the temperature is raised in the range of 0 to 60 °C.

hydrophobic interaction chromatography HYDROPHOBIC CHROMATOGRAPHY.

hydrophobicity constant A measure of the hydrophobic contribution of a substituent to a parent compound (S); defined as follows: $\pi = \log(P/P_0)$, where π is the hydrophobicity constant for substituent R; P_0 is the ratio of the solubility of the parent compound (H–S) in *n*-octanol to that in water; and P is the ratio of the solubility of the substituted compound (R–S) in *n*-octanol to that in water.

hydroponics The growth of plants in aqueous solutions that contain chemical nutrients.

hydrops fetalis *See* Thalassemia.

hydrosol A sol in which the dispersion medium is water.

hydrosphere The aqueous envelope of the earth; includes oceans, lakes, streams, underground water, and water vapor in the atmosphere.

hydroxide ion The anion OH^-. *Aka* hydroxyl ion.

hydroxy acid A carboxylic acid in which one or more hydrogen atoms of the alkyl group has been replaced by a hydroxyl group; lactic acid is an example.

hydroxyallysine A derivative of hydroxylysine in which the epsilon amino group has been converted to an aldehyde group; hydroxyallysine undergoes an aldol condensation with allysine during the crosslinking of collagen chains.

hydroxyapatite A calcium phosphate–calcium hydroxide complex, $Ca_{10}(PO_4)_6(OH)_2$, that is useful as an adsorbent for the purification of proteins and that also binds double-stranded DNA. The mineral of bone has an approximate composition corresponding to hydroxyapatite. *Abbr* HA.

β-hydroxybutyric acid A compound that can be formed from acetyl coenzyme A and that is one of the ketone bodies.

5-hydroxyindoleacetic acid A catabolite of serotonin. *Abbr* 5-HIAA.

β-hydroxyisovaleric aciduria A genetically inherited metabolic defect in humans that is due to a deficiency of the enzyme β-methylcrotonyl CoA carboxylase.

hydroxylamine A chemical mutagen that leads to the deamination of cytosine, which then forms a base pair with adenine.

hydroxylase MONOOXYGENASE.

hydroxylation The introduction of a hydroxyl

group into an organic compound.

hydroxyl group The radical —OH. *See also* alcoholic hydroxyl group; phenolic hydroxyl group.

hydroxyl value The number of milligrams of potassium hydroxide that are chemically equivalent to the hydroxyl content of 1g of fat.

hydroxylysine An alpha amino acid that is derived from lysine and occurs in collagen and gelatin. *Abbr*, Hyl; Hylys.

5-hydroxymethylcytosine A pyrimidine that is found in the DNA of T-even phages where it takes the place of cytosine and is hydrogen-bonded to guanine.

β-hydroxy-β-methylglutarate A compound that, in the form of β-hydroxy-β-methylglutaryl CoA (HMG-CoA), is an important intermediate in the biosynthesis of cholesterol and ketone bodies. *Abbr* HMG.

hydroxymethyl group The one-carbon fragment —CH_2OH.

hydroxyproline An alpha imino acid that is derived from proline and occurs in collagen and gelatin. *Abbr* Hyp; Hypro.

hydroxyprolinemia A genetically inherited metabolic defect in humans that is characterized by high plasma and urine concentrations of hydroxyproline and that is due to a deficiency of the enzyme hydroxyproline oxidase.

5-hydroxytryptamine SEROTONIN.

5-hydroxytryptophan An intermediate in the biosynthesis of serotonin. *Abbr* 5-HTP.

hydroxyurea A derivative of urea that inhibits ribonucleotide reductase and DNA replication; it does not inhibit DNA repair.

Hyflow supercel Trademark for a preparation of diatomaceous earth.

hygrometer An instrument for measuring the moisture content of a gas, particularly that of the atmosphere.

hygroscopic Descriptive of a substance that readily absorbs moisture from the atmosphere.

Hyl 1. Hydroxylysine; also abbreviated Hylys. 2. Hydroxylysyl.

Hyp 1. Hydroxyproline; also abbreviated Hypro. 2. Hydroxyprolyl. 3. Hypoxanthine.

hyper- Prefix meaning excessive.

hyperacidity Excessive acidity, particularly that due to increased concentrations of gastric hydrochloric acid which is associated with "heartburn," "indigestion," and peptic ulcers.

hyperalaninemia One of a number of genetically inherited metabolic defects in humans that are due to blocks in the metabolism of pyruvate, such as those resulting from a deficiency of pyruvate carboxylase and pyruvate dehydrogenase.

hyperammonemia A genetically inherited metabolic defect in humans that is due to a deficiency of the enzyme ornithine carbamoyl transferase, in the case of hyperammonemia I, and due to a deficiency of the enzyme carbamoyl phosphate synthase, in the case of hyperammonemia II.

hyperargininemia A genetically inherited metabolic defect in humans that is characterized by elevated levels of arginine in the blood and in the urine, and that is due to a deficiency of the urea cycle enzyme arginase. *Aka* argininemia.

hyperbilirubinemia CRIGLER–NAJJAR SYNDROME.

hyperbolic inhibition Inhibition that yields a curve that is convex upward when either slopes or intercepts from a primary plot are plotted as a function of inhibitor concentration.

hyperbolic kinetics *See* Michaelis–Menten kinetics.

hypercalcemia The presence of excessive amounts of calcium in the blood.

hypercalcemic factor PARATHORMONE.

hypercapnia The presence of excessive amounts of carbon dioxide in the blood.

hyperchloremia The presence of excessive amounts of chloride in the blood.

hyperchloremic metabolic acidosis Metabolic acidosis in which there is an increase in the concentrations of both protons and chloride ions.

hyperchlorhydria The presence of excessive amounts of hydrochloric acid in the gastric juice.

hypercholesterolemia *See* familial hypercholesterolemia; cholesterolemia.

hyperchromic effect The increase in the ultraviolet absorbance of a solution containing either DNA or RNA that occurs upon either the denaturation or the degradation of the nucleic acid. The effect is due to changes in the electronic interactions between the bases and is generally measured at 260 nm.

hyperchromicity The increase in absorbance that is due to the hyperchromic effect and that is measured at a specified wavelength.

hyperchromism 1. The increase in absorbance that is due to the hyperchromic effect and that is measured between two wavelengths, or between two forms, or between two states, of the absorbing substance. 2. HYPERCHROMICITY.

hyperfine splitting The breakup of a spectral peak into a series of peaks. In electron paramagnetic resonance this is due to the interaction of electrons with other electrons or nuclei; in nuclear magnetic resonance this is due to spin–spin coupling.

hyperglycemia The presence of excessive amounts of glucose in the blood.

hyperglycemic factor GLUCAGON.

hyperglycemic–glycogenolytic factor GLUCAGON.

hyperglycinemia A genetically inherited metabolic defect in humans that is associated with mental retardation and that is due to a deficiency of the enzyme propionate carboxylase. *See also* ketotic hyperglycinemia; nonketotic hyperglycinemia.

hyperimmunity Immunity that is characterized by the presence of large amounts of antibodies in the blood, and that may be produced by repeated injections of antigens into an animal organism.

hyperinsulinism The presence of excessive amounts of insulin in the body, resulting in hypoglycemia.

hyperkalemia HYPERPOTASSEMIA.

hyperlipemia LIPEMIA.

hyperlipoproteinemia The presence of excessive amounts of lipoproteins in the blood.

hyperlysinemia A genetically inherited metabolic defect in humans that is associated with mental retardation and that is due to a deficiency of the enzyme lysine-ketoglutarate reductase.

hypermagnesemia The presence of excessive amounts of magnesium in the blood.

hypermethioninemia A genetically inherited metabolic defect in humans that is due to a deficiency of the enzyme methionine adenosyl transferase which catalyzes the conversion of methionine to S-adenosyl methionine.

hypermodified nucleosides A group of more than 20 complex nucleosides that occur in transfer RNA. These nucleosides are more complex than simple methylated derivatives of the common nucleosides and do not include compounds such as dihydrouridine, pseudouridine, and ribothymidine but include compounds such as queuosine and wyosine.

hypermorph A mutant gene that has a similar, but stronger, effect than its wild-type gene.

hypernatremia The presence of excessive amounts of sodium in the blood.

hyperon A nuclear particle that has a mass greater than that of a nucleon.

hyperoxaluria A genetically inherited metabolic defect in humans that is characterized by a high urinary excretion of oxalate and deposition in the tissues of insoluble calcium oxalate. Two types are known, one due to a deficiency of α-ketoglutarate-glyoxylate carboligase and one due to a deficiency of D-glycerate dehydrogenase.

hyperphenylalaninemia PHENYLKETONURIA.

hyperphosphatemia The presence of excessive amounts of phosphate in the blood.

hyperplasia The increase in the size of a tissue or an organ, excluding tumor formation, that results from an increase in the number of the component cells.

hyperploid state The aneuploid state in which the chromosome number is greater than that of the characteristic euploid number. *Aka* hyperploidy.

hyperpolarization A change in membrane potential that is the opposite of depolarization. An increase in membrane potential; the membrane potential becomes more negative than it is in the normal resting state.

hyperpotassemia The presence of excessive amounts of potassium in the blood.

hyperprolinemia A genetically inherited metabolic defect in humans that is due to a deficiency of the enzyme proline oxidase, in the case of hyperprolinemia I, and due to a deficiency of the enzyme pyrroline-5-carboxylate reductase, in the case of hyperprolinemia II.

hypersensitive response ALLERGIC RESPONSE.

hypersensitive sites Specific regions in eukaryotic DNA that are cleaved by deoxyribonuclease at higher rates than the bulk regions of the DNA; these sites frequently correlate with transcriptional activity of particular genes and may constitute regulatory regions in the DNA.

hypersensitivity 1. The altered immunological state that is produced in humans and animals by their previous exposure to an antigen and that is characterized by the occurrence of different and pathological reactions upon their subsequent exposure to either the same antigen or a structurally related substance. *Aka* allergy. 2. The above average response of a human or an animal to a drug.

hypertensin ANGIOTENSIN.

hypertensinogen ANGIOTENSINOGEN.

hypertension High blood pressure.

hypertensive agent An agent that brings about an increase in blood pressure.

hyperthermia A higher than normal temperature.

hyperthyroidism Excessive activity of the thyroid gland that leads to an increase in oxygen consumption and to an increase in the overall metabolic rate.

hypertonic contraction An alteration in the water and electrolyte balance in the body in which there is a decrease in the volume of the extracellular fluid but no equivalent loss of electrolytes, so that the osmotic pressure of the extracellular fluid increases.

hypertonic expansion An alteration in the water and electrolyte balance in the body in which there is an increase in the electrolyte concentration and a less than equivalent increase in the volume of the extracellular fluid, so that the osmotic pressure of the extracellular fluid increases.

hypertonic solution A solution that has a higher osmotic pressure than another solution.

hypertrophy The increase in the size of a tissue or an organ, excluding tumor formation, that is due to either an increase in the volume of the component cells, or an increase in the functional activity of the tissue or the organ.

hyperuricemia The presence of excessive amounts of uric acid in the blood.

hypervalinemia A genetically inherited metabolic defect in humans that is due to a deficiency of the enzyme valine transaminase.

hypervariable codon A codon that has undergone an excessive number of mutations when it is considered in the context of the biochemical evolution of proteins.

hypervariable regions A number of small regions, present in the variable regions of both the light and the heavy immunoglobulin chains, in which the bulk of the variability of amino acid sequences is found. The remainder of the variable regions are relatively constant in amino acid sequence and are called framework regions.

hyperventilation Excessive aeration of the blood in the lungs.

hypervitaminosis A pathological condition resulting from an overdose of a vitamin, such as that produced by an overdose of either vitamin A or vitamin D.

hypha (*pl* hyphae) One of a number of filamentous and branched tubes of a fungus that contain cytoplasm and that form a network known as mycelium.

hypo- Prefix meaning less than.

hypo Sodium thiosulfate, a commonly used photographic fixer; a reagent for the removal of unexposed and unreduced silver halide grains from a photographic film.

hypobaria A lower than normal atmospheric pressure.

hypobetalipoproteinemia A genetically inherited metabolic defect in humans that is characterized by low concentrations of plasma low-density lipoproteins (LDL).

hypocalcemia A deficiency of calcium in the blood.

hypocalcemic factor CALCITONIN.

hypocapnia A deficiency of carbon dioxide in the blood.

hypochloremia A deficiency of chloride in the blood.

hypochlorhydria A deficiency of hydrochloric acid in the gastric juice.

hypocholesterolemia A deficiency of cholesterol in the blood.

hypochromic anemia An anemia that is characterized by a decrease in the hemoglobin concentration of the red blood cells.

hypochromic effect The decrease in the ultraviolet absorbance of a solution containing either DNA or RNA compared to the absorbance of the same solution but containing either the denatured or the degraded nucleic acid. The effect is due to changes in the electronic interactions between the bases and is generally measured at 260 nm.

hypochromicity The decrease in absorbance that is due to the hypochromic effect and that is measured at a specified wavelength.

hypochromism 1. The decrease in absorbance that is due to the hypochromic effect and that is measured between two wavelengths, between two forms, or between two states, of the absorbing substance. 2. HYPOCHROMICITY.

hypofibrinogenemia A pathological condition characterized by low levels of fibrinogen in the blood.

hypogammaglobulinemia A genetically inherited metabolic defect in humans that is characterized by an almost complete absence of specific immunoglobulins. *See also* agammaglobulinemia.

hypoglycemia A deficiency of glucose in the blood.

hypoglycemic factor INSULIN.

hypoglycin A derivative of propionic acid that is the toxic principle of the ackee fruit; it inhibits the enzyme isovaleryl CoA dehydrogenase and hence gives rise to the same symptoms present in the genetic disease isovaleric acidemia.

hypohaptoglobinemia A genetically inherited metabolic defect in humans that is characterized by decreased levels of haptoglobins in the blood.

hypokalemia HYPOPOTASSEMIA.

hypomagnesemia A genetically inherited metabolic defect in humans that is characterized by a deficiency of magnesium in the blood and that is due to a failure of intestinal absorption of magnesium ions.

hypomorph A mutant gene that has a similar, but weaker, effect than its wild-type gene.

hyponatremia A deficiency of sodium in the blood.

hypophase technique KLEINSCHMIDT TECHNIQUE.

hypophosphatasia A genetically inherited metabolic defect in humans that is characterized by deficient bone formation and that is due to a deficiency of the enzyme alkaline phosphatase.

hypophosphatemia A deficiency of phosphate in the blood.

hypophyseal Of, or pertaining to, the pituitary gland.

hypophysectomy The surgical removal of the pituitary gland.

hypophysin An extract made of the posterior lobe of the pituitary gland.

hypophysis PITUITARY GLAND.

hypoplasia 1. The arrested development of a tissue or an organ. 2. The decrease in the size of a tissue or an organ that results from a decrease in the number of the component cells.

hypoploid state The aneuploid state in which the chromosome number is less than that of the characteristic euploid number. *Aka* hypoploidy.

hypopotassemia A deficiency of potassium in the blood.

hyposensitivity The below average response of a human or an animal to a drug.

hypostatic gene A gene, the expression of which is suppressed by another gene in epistasis.

hypotension Low blood pressure.

hypotensive agent An agent that brings about a lowering of the blood pressure.

hypothalamic hormone A hormone that is produced by the hypothalamus and that either stimulates or inhibits the release of another hormone. Hypothalamic stimulatory hormones stimulate the release of corticotropic hormone, luteinizing hormone, follicle-stimulating hormone, thyrotropic hormone, and growth hormone; hypothalamic inhibitory hormones inhibit the release of prolactin and melanocyte-stimulating hormone. *Aka* hypothalamic factor; hypothalamic regulatory hormone. *See also* chalone.

hypothalamus A basal part of the brain.

hypothermia A lower than normal temperature.

hypothesis (*pl* hypotheses) A statement describing a relation that may be either true or untrue; the statement appears to be confirmed by experimental facts but is not as certain as a theory. A hypothesis is a tentative explanation or description of observed phenomena that remains to be critically tested by further experiments.

hypothyroidism Deficient activity of the thyroid gland that leads to a decrease in the oxygen consumption and to a decrease in the overall metabolic rate. *Aka* myxedema.

hypotonic contraction An alteration in the water and electrolyte balance in the body in which there is a decrease in the volume of the extracellular fluid and a more than equivalent decrease of electrolytes, so that the osmotic pressure of the extracellular fluid decreases.

hypotonic expansion An alteration in the water and electrolyte balance in the body in which there is an increase in the volume of the extracellular fluid but no equivalent increase in electrolytes, so that the osmotic pressure of the extracellular fluid decreases.

hypotonic solution A solution that has a lower osmotic pressure than another solution.

hypotrophy 1. Subnormal growth. 2. The degeneration and loss of function of a tissue or an organ.

hypoventilation Deficient aeration of the blood in the lungs.

hypovitaminosis The deficiency disease that results from the inadequate dietary intake of one or more vitamins.

hypoxanthine The purine 6-hydroxypurine that occurs in transfer RNA and that is formed in catabolism by the deamination of adenine. *Sym* I.

hypoxia A condition in which the oxygen concentration of the blood and the tissues is below normal either as a result of environmental deficiency or as a result of impaired respiration and circulation.

hypro Hydroxyproline.

hypro-protein Hydroxyproline in its protein form as distinct from free hydroxyproline.

hypsochromic group A group of atoms that, when attached to a compound, shifts the absorption of light by the compound to shorter wavelengths.

hypsochromic shift A shift in the absorption spectrum of a compound toward shorter wavelengths.

hysteresis The phenomenon in which changes of a property in one direction can be reversed back to the initial value of the property, but in so doing take on intermediate values that differ from those taken on in the forward direction. A plot of such a system results in two curves, coming together at both ends and forming a hysteresis loop in between.

hysteretic Of, or pertaining to, hysteresis.

hysteretic enzyme An enzyme that appears to undergo very slow changes in either kinetic or molecular properties after the addition or the removal of a ligand; believed to be due to conformational changes of the enzyme as a result of the change in ligand concentration.

Hz Hertz.

H zone The central, less dense portion of the A band of the myofibrils of striated muscle. *Aka* H band.

I

I 1. Inosine. 2. Hypoxanthine. 3. Isoleucine. 4. Iodine. 5. Ionic strength. 6. Luminous intensity. 7. Electric current.

IAA 1. Indoleacetic acid. 2. Monoiodoacetamide.

Ia antigens I-region-associated antigens.

i-assay The construction of a thermal denaturation profile by measurements of the system after it has been rapidly returned to standard conditions; the assay measures the reversibility of the transition. *See also* d-assay.

iatrogenic disease A disease caused by treatment or by diagnostic procedures; a disease brought about by medical personnel, medical procedures, or exposure to the environment of a health care facility.

I band The transverse light band that is seen in electron microscope preparations of the myofibrils from striated muscle and that is formed by the thin filaments.

ic 1. *adj* Intracutaneous. 2. *adv* Intracutaneously.

ICD Isocitrate dehydrogenase; an enzyme of the citric acid cycle.

iceberg A cluster of water molecules that have become stabilized around a nonpolar group; the process is accompanied by a loss of entropy, and the water molecules possess greater crystallinity than that of the molecules in ordinary water.

I-cell disease A genetically inherited metabolic defect in humans that results in early death and that is due to a deficiency of the enzyme *N*-acetylglucosamine-1-phosphotransferase and other lysosomal enzymes; a mucolipidosis. It is called I-cell disease because of the formation of unusual inclusion bodies in fibroblasts. *Aka* mucolipidosis II.

ICH Interstitial cell hormone.

ichthyotocin ISOTOCIN.

icosadeltahedron The solid obtained by subdividing the surfaces of an icosahedron into smaller equilateral triangles.

icosahedral Of, or pertaining to, an icosahedron.

icosahedral virion A virus, such as poliovirus or adenovirus, in which the capsid has the shape of an icosahedron.

icosahedron A symmetrical polyhedron having 12 vertices and 20 faces, with each face being an equilateral triangle; frequently descriptive of the structure of viruses.

ICP Inductively coupled plasma.

ICSH Interstitial cell stimulating hormone.

icterus index A liver function test that measures the approximate level of bilirubin in either serum or plasma; based on a comparison of the color of either diluted serum or diluted plasma with a standard solution of potassium dichromate.

ictotest A semiquantitative test for bilirubin in the urine; based on the production of a blue-purple color by diazotization of bilirubin.

icy group A group of compounds that have neither very high nor very low melting points and that are believed to have occurred in the original gas dust of the solar nebula. *See also* earthy group; gaseous group.

i.d Inside diameter.

ID$_{50}$ Median infectious dose.

ideal gas A gas, the behavior of which is accurately described by the various gas laws.

ideal solution A dilute solution in which all of the solutes follow Raoult's law; a solution such that the chemical potential μ_i of each component is given by $\mu_i = \mu_i^0 + RT\ln X_i$, where μ_i^0 is the standard chemical potential (the potential of the substance in its pure state), R is the gas constant, T is the absolute temperature, and X_i is the mole fraction of component i.

identical twins MONOZYGOTIC TWINS.

identifier sequence One of a group of nucleotide sequences, about 80 base pairs long, that are transcribed in the brain. They are located in the introns of genes that are expressed in neural tissues and are believed to control the tissue-specific expression of genes. *Abbr* ID sequence.

idiogram A diagrammatic representation of the chromosome complement of an individual according to the size and/or a numbering system of the chromosomes.

idiopathic 1. Denoting a disease of unknown cause. 2. Peculiar to an individual.

idiopathic hypertension ESSENTIAL HYPERTENSION.

idiopathic pentosuria PENTOSURIA.

idiotype An antigenic determinant, or a collection of antigenic determinants, peculiar to an individual immunoglobulin molecule. Idiotypes represent regions in or near the antigen-binding site of an antibody that are themselves capable of acting as antigens, thereby stimulating the production of anti-bodies. It is believed that the production of a particular antibody is kept in check by the corresponding anti-idiotype antibodies (idiotype suppression) and that the normal immune system involves an interlocking network of antibodies directed at one another's idiotypes. *See also* allotype; isotype.

idiotype suppression Suppression of the synthesis of idiotype antibodies that is brought about by anti-idiotype antibodies. The process involves activation of suppressor T cells by the anti-idiotype antibodies. *See also* idiotype.

idiotypic antibody ANTI-IDIOTYPE ANTIBODY.

idiotypic marker An antigenic determinant in the antigen-binding site of an idiotypic antibody.

idling reaction The reaction that takes place on a ribosome when an uncharged tRNA molecule becomes bound at the A-site. In addition to the temporary stoppage of polypeptide chain growth, there is an idling reaction in which ATP serves as a pyrophosphate donor leading to the conversion of GDP to $_{pp}G_{pp}$ and $_{ppp}G_{pp}$.

IDP 1. Inosine diphosphate. 2. Inosine-5′-diphosphate.

ID sequence Identifier sequence.

IE Immunoelectrophoresis.

IEF Isoelectric focusing.

IEMA Immunoenzymometric assay.

IEP Isoelectric point.

IF 1. Initiation factor. 2. Intrinsic factor. 3. Isoelectric focusing. 4. Intermediate filament.

IFMA Immunofluorometric assay.

IFN Interferon.

I form Independent form.

Ig Immunoglobulin.

IgA A human immunoglobulin that constitutes about 15% of the total serum immunoglobulins, has a variable sedimentation coefficient of 7–11S, exists as a monomer or as a dimer (MW 180,000–500,000), contains about 8% carbohydrate, does not activate complement, and does not cross the placenta. IgA is the major immunoglobulin in secretions. *Aka* γA.

IgD A human immunoglobulin that constitutes less than 1% of the total serum immunoglobulins, has a sedimentation coefficient of 6.8–7.9S and a molecular weight of 180,000, contains about 13% carbohydrate, does not activate complement, and does not cross the placenta. The physiological function of IgD is unknown. *Aka* γD.

IgE A human immunoglobulin that constitutes less than 1% of the total serum immunoglobulins, has a sedimentation coefficient of 8.2S and a molecular weight of 190,000, contains about 11% carbohydrate, does not activate complement, and does not cross the placenta.

IgE binds to mast cells and basophils, and is associated with allergic reactions. *Aka* γE.

IGF Insulin-like growth factor.

IgG The most predominant human immunoglobulin which constitutes about 80% of the total serum immunoglobulins, has a sedimentation coefficient of 6.5–7.0S and a molecular weight of 150,000, contains about 3% carbohydrate, activates complement, and crosses the placenta. IgG aids in phagocytosis, and is the major immunoglobulin produced during the secondary immune response. *Aka* γG.

IgM A human immunoglobulin that constitutes about 5% of the total serum immunoglobulins, has a sedimentation coefficient of 19S, exists as a pentamer (MW 950,000), contains about 12% carbohydrate, is very efficient in activating complement but does not cross the placenta. IgM is the first type of antibody produced in response to an antigen (primary response) and binds antigens efficiently because of its multiple binding sites; it is found primarily in blood plasma. *Aka* γM.

IHF Integration host factor.

IL-2 Interleukin-2.

ILA Insulin-like activity.

Ile 1. Isoleucine. 2. Isoleucyl.

ileum The third and lowest portion of the small intestine.

Ilkovic equation A basic equation of polarography.

Illegitimate recombination The process whereby a transposon is inserted into recipient DNA; so called because it requires no homology between donor and recipient DNA.

illicit transport The entry of a substance into a cell by means of a transport system that is designed for another substance; the transport of the derivative of a compound, where the compound itself cannot be transported, is an example.

im 1. *adj* Intramuscular. 2. *adv* Intramuscularly.

imbalance theory A theory of cancer according to which all organisms must evolve systems of growth regulation; a tumor arises when there is a failure in this regulation, and the imbalance in favor of growth passes a certain threshold.

imbibition The uptake of fluid by a substance, as the uptake of water by a colloidal system.

ImD$_{50}$ Median immunizing dose.

imidazole group The heterocyclic ring system of the amino acid histidine. *Aka* imidazolium group.

imine An organic compound that contains the grouping

$$-N=C-CH-$$

imine–enamine tautomerism The tautomerism that is due to a shift of a hydrogen atom so that one of the isomers is an imine and the other is an enamine.

imino acid An acid derived from an imine; proline and hydroxyproline are alpha imino acids in which the nitrogen of the imino group and the carboxyl group are attached to the same carbon atom.

imino group The grouping —NH—.

2-iminothiolane A cross-linking reagent that forms a covalent bridge between two epsilon amino groups of lysine residues in a protein.

immediate early RNA PREEARLY RNA.

immediate-type hypersensitivity An allergic response that occurs soon, generally within a few minutes, after the administration of an antigen to an animal organism; the response is mediated by circulating antibodies. *See also* anaphylaxis; type 2 reaction; type 3 reaction.

immersion oil An oil that has a refractive index of about 1.52 and that is used with an oil immersion objective in microscopy.

immiscible Incapable of being mixed.

immobile phase STATIONARY PHASE (2).

immobilized enzyme An enzyme that is physically confined while it carries out its catalytic function. This may occur naturally, as in the case of particulate enzymes, or it may be produced artificially by chemical or by physical methods. In the chemical methods, the enzyme is linked covalently to a support. These methods include attachment of the enzyme to a water-insoluble support, incorporation of the enzyme into a growing polymer chain, or cross-linking of the enzyme with a multifunctional low molecular weight reagent. In the physical methods, the enzyme is not linked covalently to a support. These methods include adsorption of the enzyme to a water-insoluble matrix, entrapment of the enzyme within either a water-insoluble gel or a microcapsule, or containment of the enzyme within special devices equipped with semipermeable membranes.

immune Of, or pertaining to, an organism that has been immunized.

immune adherence The attachment of a complex, composed of particulate antigens, antibodies, and complement, to the surfaces of nonsensitized particles such as erythrocytes, platelets, yeast, or starch granules.

immune adsorbent *See* immunoadsorbent.

immune antibody ACQUIRED ANTIBODY.

immune clearance IMMUNE ELIMINATION.

immune competent cell *See* immunocompetent cell.

immune complex ANTIGEN–ANTIBODY COMPLEX.

immune conglutination A conglutination reaction caused by an immunoconglutinin.

immune conglutinin *See* immunoconglutinin.

immune cytolysis The lysis of cells by antibodies in the presence of complement.

immune deficiency diseases A group of diseases linked to deficiencies in the immune system. Thus, for example, a deficiency of either purine nucleoside phosphorylase or adenosine deaminase leads to a decrease in the number of lymphocytes. *See also* AIDS.

immune elimination The stage in an immune response during which the antigen is rapidly removed from the blood as a result of its combination with the antibody.

immune globulin *See* immunoglobulin.

immune hemolysis Hemolysis that results from complement fixation.

immune lysis IMMUNE CYTOLYSIS.

immune opsonin *See* opsonin.

immune reaction The reaction between a specific antigen and an antibody.

immune response The formation of antibodies in an animal organism in response to an immunization and the reactions of these antibodies with the antigens used in the immunization; may involve humoral immunity or cell-mediated immunity.

immune response gene A gene that controls the ability of lymphocytes to produce an immune response upon stimulation by specific antigens. *Abbr* Ir gene.

immune serum (*pl* immune sera). ANTISERUM.

immune surveillance theory *See* immunological surveillance theory.

immunity 1. The resistance of an individual or an animal to a specific disease, infecting agent, or toxic antigen. 2. The capacity of lysogenic bacteria to withstand infection by phage particles that are of the same kind as the prophage of the bacteria.

immunity substance A cytoplasmic factor that is formed under the control of a phage gene and that confers immunity on a lysogenic bacterium against infection by a phage of the same type as its prophage. The immunity substance also functions as a repressor of the vegetative replication of the prophage in that bacterium.

immunization 1. The administration of an antigen to an animal organism to stimulate the production of antibodies by that organism. 2. The administration of antigens, antibodies, or lymphocytes to an animal organism to produce the corresponding active, passive, or adoptive immunity.

immuno- Combining form meaning immunology.

immunoadsorbent An insoluble material that is used for the purification of antibodies by adsorbing them from a serum; a gel for trapping antibodies, or an inert solid, to which either antigens or haptens have been covalently linked, are two examples.

immunoassay An assay that utilizes antigen–antibody reactions for the determination of biochemical substances.

immunoblast A blast cell that is a forerunner of an immunocyte.

immunochemistry The science that deals with the chemical aspects of immunology and combines the techniques of biochemistry and immunology.

immunochromatography IMMUNO-GEL FILTRATION.

immunocompetent cell A cell that has the capacity to recognize antigens and/or to synthesize antibodies.

immunoconglutinin An antibody that is specific for antigenic determinants which are exposed in fixed complement, but which are unavailable for reaction in free complement. *See also* conglutinin.

immunocore electrophoresis A separation technique that is based on the combined use of electrophoresis and immunodiffusion. Antigens or antibodies are first separated by disc gel electrophoresis, and the gel column is then extruded. A core of the gel column is removed and replaced with either an antiserum or a solution containing antigens. Following this, the gel is incubated to allow for the formation of precipitin bands of antigen–antibody complexes.

immunocyte 1. IMMUNOCOMPETENT CELL. 2. An immunocompetent lymphocyte.

immunocyte adherence A technique for detecting cells that carry antibodies on their surfaces either because they produce the antibodies or because the antibodies have become bound to the cells. The cells are reacted with the corresponding antigens or with other cells that are coated with soluble antigens. An antibody-bearing cell binds the antigens and forms a rosette-type structure, and the number of these rosettes is then determined microscopically. *Aka* rosette technique.

immunodeficiency *See* immune deficiency diseases.

immunodiffusion A method for carrying out the precipitin reaction in a gel that is based on the diffusion of antibody and/or antigen molecules through the gel. *See also* double diffusion; single diffusion.

immunodominant 1. Descriptive of that part of an antigenic determinant that binds most strongly to the antibody. 2. Descriptive of that part of the antigenic determinant that elicits the greatest immune response.

immunoelectroadsorption A method for measuring antibody concentrations in serum. A layer of the appropriate antigens is ad-

sorbed onto a glass slide with the aid of an electrical current, and the antibodies in the given serum are then adsorbed onto the antigens. The thickness of the antibody layer is determined and it provides a measure of the antibody concentration in the serum.

immunoelectrofocusing One of several separation techniques that are based on the combined use of gel electrofocusing and either immunodiffusion or immunoelectrophoresis; performed by incorporating either all or fractions of the gel from a gel electrofocusing experiment into a gel to be used for either immunodiffusion or immunoelectrophoresis.

immunoelectronmicroscopy The use of electron microscopy in conjunction with immunochemical methods, as in the staining of electron microscope specimens with ferritin-labeled antibodies.

immunoelectrophoresis A technique for identifying antigens in complex mixtures by first separating the antigens in one dimension by means of gel electrophoresis, and then allowing them to react with antibodies by means of two-dimensional double diffusion through the gel; a pattern of precipitin arcs is thereby produced. *Abbr* IE.

immunoenzymometric assay A variation of an enzyme immunoassay that is based on the use of enzyme-labeled antibodies. *Abbr* IEMA. *See also* ELISA.

immunoferritin FERRITIN-LABELED ANTIBODY.

immunofiltration The purification of an immunological solution by passing it through an immunoadsorbent.

immunofluorescence *See* direct fluorescent antibody technique; indirect fluorescent antibody technique; anticomplement fluorescent antibody technique.

immunofluorometric assay A variation of a fluoroimmunoassay that is based on the use of antibodies labeled with a fluorochrome. *Abbr* IFMA.

immuno-gel filtration A separation technique that is based on the combined use of immunodiffusion and thin-layer gel filtration.

immunogen 1. ANTIGEN. 2. A substance capable of producing an immune response that leads to the synthesis of antibodies.

immunogenetics The branch of immunology that deals with the interrelations of immunological reactions and the genetic makeup of an organism.

immunogenic Capable of producing an immune response.

immunogenicity ANTIGENICITY.

immunoglobulin 1. A protein of animal origin that has a known antibody activity. 2. A protein that is closely related to an antibody by its chemical structure and by its antigenic specificity. 3. A protein of immunological signi-ficance, such as myeloma or Bence–Jones pro-tein. *See also* IgA, IgD, etc.

immunoglobulin chains *See* heavy chain; light chain.

immunoglobulin fold The very similar three-dimensional structure of the various domains of an immunoglobulin molecule; a "sandwich-type" structure that consists of two antiparal-lel pleated sheets.

immunoglobulin genes Genes that code for the light and heavy chains of the immunoglobulin molecules.

immunoisoelectric focusing IMMUNOELECTRO-FOCUSING.

immunoliposome An artificial organelle, pro-duced by chemically coupling a monoclonal antibody to a liposome.

immunologic IMMUNOLOGICAL.

immunological Of, or pertaining to, immuno-logy.

immunological competence *See* competence (2).

immunological enhancement *See* enhancement.

immunological equivalence The amounts of antigen and antibody that precipitate each other in the equivalence zone of the precipitin curve.

immunological inhibition The competitive in-hibition of antibody formation that is pro-duced either by the administration of anti-bodies against the stimulating antigen, or by the administration of unconjugated hapten together with active hapten–protein conju-gates.

immunologically competent cell *See* immuno-competent cell.

immunological memory The enhanced capacity of an animal to respond to a second dose of antigen that is characteristic of the secondary immune response.

immunological paralysis 1. IMMUNOLOGICAL TOLERANCE. 2. Immunological tolerance pro-duced by large doses of polysaccharide anti-gens.

immunological rejection The destruction of foreign cells and tissues, which are either in-oculated or transplanted into a recipient from a donor, by a specific immune reaction.

immunological suppression *See* immunosup-pression.

immunological surveillance theory A theory according to which cell-mediated immunity represents a defense mechanism that evolved in order to recognize and destroy newly emerged cancer cells and cells containing pathogens. As a result, the frequency of naturally arising tumors is less than that ex-pected from the spontaneous mutation rate of DNA.

immunological tolerance The decrease in, or the loss of, the ability of an animal to produce an immune response upon the administration of a particular antigen. Tolerance is induced by prior exposure of the animal to the same antigen and does not affect the ability of the animal to respond to other antigens.

immunological unresponsiveness 1. IMMUNOLO-GICAL TOLERANCE. 2. Immunological tolerance produced by large doses of protein antigens (overloading).

immunological zoo The set of reagents re-quired in the complement fixation test.

immunology The science that deals with resist-ance to disease.

immunoprecipitation The precipitation of either antigens or antibodies as a result of a precipitin reaction.

immunoprecipitation test A test that permits the identification of a bacterial colony that secretes a specific protein; involves growing the colonies on an agar surface that contains antibodies to the protein. If the protein is secreted, an antigen–antibody reaction will take place and a precipitin precipitate will be formed around the colony producing the pro-tein.

immunoradiometric assay A variation of a radioimmunoassay that is based on the use of labeled antibodies for the assay of unlabeled antigens. *Abbr* IRMA; IRA.

immunoreactive insulin That fraction of serum insulin that is not protein-bound and that is readily neutralized by anti-insulin serum; the bulk of this insulin is called little insulin and the rest is called big insulin. *Abbr* IRI.

immunosedimentation An analytical technique that involves ultracentrifugation, using acrylamide-containing sucrose gradients, fol-lowed by immunodiffusion in agarose gel. The gradients obtained after ultracentrifugation are photopolymerized in the tubes and the gel thus formed is then removed from the tube, cut into slabs, and incorporated into an agar-ose gel for immunodiffusion.

immunoselection A technique for isolating cell-line variants that lack certain antigens; in-volves treating the cells with a specific anti-serum and complement. This treatment leads to the death of all cells except those few that lack the corresponding antigens. These living, variant cells can then be isolated.

immunosorbent An immunological sorbent. *See also* absorption (4); adsorption (2).

immunosuppressant *See* immunosuppressive agent.

immunosuppression The prevention of an im-mune response by physical, chemical, or biological means.

immunosuppressive agent A physical, chemical, or biological agent that prevents the production of an immune response by an antigen.

immunotherapy The treatment of a disease by immunization.

IMP 1. Inosine monophosphate (inosinic acid). 2. Inosine-5′-monophosphate (5′-inosinic acid).

impedance The resistance to the flow of an alternating electric current; equal to the ratio of the complex potential difference applied across a circuit element to the complex current flowing through the element. *Sym* Z.

imperfect excision The excision of a genetic element from DNA, such as a prophage or an insertion sequence, in which the segment excised is either larger or smaller than the actual genetic element.

impermeable Not permeable.

impervious Impenetrable; not permeable.

imphilyte An immobilized amphipathic ampholyte.

implantation The artificial introduction of material, such as a tissue transplant or an encapsulated enzyme, into an organism.

import 1. Transport of material into a cell. 2. PROTEIN IMPORT.

inactivated vaccine A vaccine that contains microorganisms or virus particles that have been inactivated; they are no longer capable of causing disease.

inactivation The destruction of biological activity.

inactivation cross section A measure of the sensitivity of a target to inactivation by irradiation that is equal to the product of the quantum yield and the absorption cross section.

inactivation probability The ratio of the target molecular weight, calculated from radiation inactivation experiments, to the actual molecular weight.

inanition The exhaustion that results from either the lack of food or the inability to assimilate it.

inapparent infection A transient viral infection that does not produce overt disease symptoms. *See also* latent infection (2).

inborn error of metabolism A genetically inherited metabolic defect that results in the synthesis of a modified enzyme or other protein, or in the complete lack of synthesis of an enzyme or other protein. *See also* genetic disease.

inbred strain A strain of experimental animals produced by sequential brother–sister matings over many generations so that all the individuals are genetically identical.

inbreeding The crossing of plants or animals that are closely related genetically.

incapsidate ENCAPSIDATE.

inchworm theory The theory that, during the enzymatic binding of aminoacyl-tRNA to the A-site of the ribosome, a kink is created in the mRNA. The kink may be formed by rotation of the tRNA (which is hydrogen-bonded to the mRNA) about its long axis. Rotation of the tRNA results in the amino acid being brought into close proximity of the peptide in the P-site so that a peptide bond can now be formed. When the kink in the mRNA is straightened out, as a result of GTP and translocase activity, the peptidyl-tRNA is translocated from the A- to the P-site. This movement of the mRNA along the ribosome has been compared to that of an inchworm moving across a surface.

incident Falling on or striking.

incision A break in a single strand of a nucleic acid; a nick.

incision enzyme An enzyme that catalyzes the cleavage of a polynucleotide chain; an endonuclease.

inclusion A discrete mass within the cell of either foreign or metabolically passive material.

inclusion body 1. A mass of virus particles within the cell of an animal that is infected with the virus. 2. An aggregate formed in vivo by a number of abnormal hemoglobins; these aggregates deform the red blood cell and shorten its life span. *Aka* Heinz body.

incompatibility *See* histoincompatibility.

incompatible plasmids Two plasmids that cannot coexist in the same cell as is the case in superinfection immunity.

incomplete antibody An antibody that does not give the serologic reactions of precipitation and agglutination, and that can only be demonstrated either indirectly by means of special techniques or by means of its in vivo biological effect.

incomplete antigen HAPTEN.

incomplete lectin A lectin that requires the presence of some other substance for full activity.

incomplete oxidation The oxidation of organic compounds such that partially oxidized organic compounds are the final products; the term may refer either to a group of reactions or to a single reaction.

incomplete protein A protein that is deficient in one or more of the amino acids commonly found in proteins.

incubation The maintaining of organisms, reaction mixtures, and the like in an incubator or in some other constant temperature environment.

incubation mixture A reaction mixture that is maintained at a specified temperature.

incubation period 1. The time interval between the invasion of an organism by a virus or by a bacterium and the appearance of overt disease symptoms. 2. The length of time that a reaction mixture is maintained at a constant temperature.

incubator A constant temperature chamber used principally to provide a controlled environment for growth of cells and organisms.

independent assortment The random distribution of genes, located on nonhomologous chromosomes, during gametogenesis.

independent binding The binding of ligands to a macromolecule such that the binding of one ligand to one binding site has no effect on the binding of subsequent ligands to other sites on the same molecule. The ligands and the binding sites may be of one type each or they may be of different types. Independent binding is noncooperative binding.

independent form The dephosphorylated form of the enzyme glycogen synthase that is active in the absence of glucose-6-phosphate.

independent variable A quantity that can assume any arbitrarily chosen value independent of the values of other related variables.

indeterminacy principle UNCERTAINTY PRINCIPLE.

indeterminate error An error in measurement which is due to the fact that all physical measurements require a degree of estimation in their evaluation; such errors can be decreased in magnitude but cannot be eliminated entirely.

index fossil A fossil that is of widespread occurrence in one or in a few contiguous geological layers and that can be used to correlate the ages of geological deposits at various locations.

index of discrimination The ratio of the activities of a hormone that is assayed by two different methods.

index of hydrogen deficiency A measure of the degree of unsaturation in a molecule; equal to the number of pairs of hydrogen atoms that must be removed from the formula of a saturated hydrocarbon, which has the same number of carbon atoms as a given compound, to yield the molecular formula of the given compound.

index of precision 1. The standard deviation of the responses to a hormone divided by the slope of the dose–response curve. 2. The reciprocal of the slope of the dose–response curve of a hormone assay.

index of refraction *See* refractive index.

indicator A weak acid or a weak base in which the proton donor and the proton acceptor species have different colors; used for indicating end points of acid–base titrations, since the color of the indicator is determined by the relative concentrations of the two species, which, in turn, are determined by the pH of the solution. *Aka* acid–base indicator.

indicator enzyme MARKER ENZYME.

indicator strain SENSITIVE STRAIN.

indicator virus A virus that has been heated so that its neuraminidase activity is destroyed but its hemagglutinating capacity is retained.

indirect-acting bilirubin The water-insoluble and unconjugated form of bilirubin that does not give a color reaction with diazotized sulfanilic acid unless alcohol is added first to solubilize the bilirubin. *See also* direct-acting bilirubin.

indirect calorimetry A method for determining the basal metabolic rate of an animal from measurements of the amount of nitrogen excreted in the urine, the volume of oxygen inhaled, the volume of carbon dioxide exhaled, and the respiratory quotient. *See also* direct calorimetry.

indirect complement fixation test A complement fixation test for determining antibodies that fail to fix guinea pig complement; entails the addition of an antiserum of known ability to fix guinea pig complement to an antigen–antibody–guinea pig complement system, followed by the addition of a hemolytic system.

indirect Coombs' test A Coombs' test in which the red blood cells are coated with antibody in vitro. *See also* direct Coombs' test.

indirect coupling COTRANSPORT.

indirect effect The change brought about in a molecule as a result of its interaction with molecules, radicals, atoms, or electrons that are produced by a direct interaction of atoms or molecules with radiation. *See also* direct effect.

indirect fluorescent antibody technique A fluorescent antibody technique in which the antigen of interest is first reacted with its immunoglobulin and then with a fluorescent antibody against the immunoglobulin. This technique is more sensitive than the direct one, since it multiplies the number of fluorochromes per antigen that is being stained. The technique can also be applied to the staining of antibodies. *Aka* sandwich technique; antiglobulin method. *See also* direct fluorescent antibody technique.

indirect induction CROSS-INDUCTION.

indirect mutagenesis *See* mutagenesis.

indirect photoreactivation The recovery of cells from the damage caused by their irradiation with ultraviolet light that is brought about by exposure of the damaged cells to light in the wavelength range of 310 to 370 nm; may be due to an inhibition of growth which allows more time for the repair of the damaged DNA.

indispensable amino acid ESSENTIAL AMINO ACID.

indispensable enzyme ESSENTIAL ENZYME.

indispensable fatty acid ESSENTIAL FATTY ACID.

indispensable gene ESSENTIAL GENE.

indole The aromatic and heterocyclic ring system of the amino acid tryptophan. *See also* antipromoter.

indoleacetic acid An auxin that also has vasoconstrictor activity. *Abbr* IAA.

indole alkaloids *See* alkaloids.

induced dipole moment A dipole moment produced in a substance by the application of an external electric or magnetic field.

induced enzyme *See* inducible enzyme.

induced-fit theory A modification of the lock and key theory for the binding of a substrate to an enzyme. According to this theory, the active site is not preformed but, rather, is formed as a result of an interaction between the substrate and the enzyme. The substrate first induces a conformational change in the enzyme such that the catalytic and the binding groups of the enzyme achieve the required active site orientations which then permits the substrate to become bound to the enzyme.

induced hypersensitivity The hypersensitivity produced in an animal by contact with an antigen.

induced mutation A mutation produced by the intentional exposure of an organism to a mutagen, in contrast to a spontaneous mutation.

induced radioactivity Radioactivity that is artificially produced by bombardment of nuclei with high-velocity particles.

induced tumor A tumor that arises subsequent to, and as a result of, the exposure of an organism to a carcinogen.

inducer 1. The substance that brings about the synthesis of an inducible enzyme in the process of enzyme induction and that is generally either a substrate of the enzyme or a compound which is structurally similar to the substrate. 2. The substance used to induce an allergic state in an animal.

inducer T cells HELPER T CELLS.

inducible enzyme An enzyme that is normally either absent from a cell or present in very small amounts, but that is synthesized in appreciable amounts in response to an inducer in the process of enzyme induction. *See also* constitutive enzyme.

inducible system The regulatory system consisting of the components that function in enzyme induction. *See also* enzyme induction.

inductance The property of a conductor by which an electromotive force is induced in it by variations in an inducing electric current.

induction 1. ENZYME INDUCTION. 2. The stimulation of a lysogenic bacterium that causes it to shift to a lytic cycle, to produce infective phage particles, and ultimately to lyse in a burst; the derepression of a prophage. 3. The reasoning from particulars to generals. 4. MORPHOGENIC INDUCTION. 5. INDUCTIVE EFFECT.

induction effect DIPOLE-INDUCED DIPOLE INTERACTION.

induction period 1. LAG PERIOD. 2. *See* presteady-state kinetics.

induction profile The pattern of enzyme induction that is produced by a given inducer.

induction ratio The ratio of the concentration of the induced enzyme to that of the basal enzyme. *Abbr* IR.

inductive effect The partial charge induced in an atom, or in a group of atoms, of a molecule as a consequence of the electron-withdrawing and electron-donating properties of neighboring atoms and groups of atoms.

inductively coupled plasma An analytical technique for the analysis of multiple elements that has greater sensitivity than atomic absorption and that is useful for the analysis of trace elements in biological samples. The method involves the use of an argon plasma, at temperatures of about 9000 K, for emission spectrometry and mass spectrometry. A liquid sample is converted to an aerosol that is carried into the very hot plasma region where the various elements are excited to emission temperatures. Either the emitted light intensities or the masses of the various species present are then determined.

inductive phase The time interval between the administration of an antigen to an animal and the appearance of antibodies in the serum.

inductor 1. A substance that increases the rate of a chemical reaction and that is used up during the reaction. 2. A substance that acts like an organizer in affecting the development of embryonic and other undifferentiated tissues.

inelastic collision A collision in which there is a loss of kinetic energy; the sum of the kinetic energies of the colliding particles after the collision is less than that before the collision; some kinetic energy is lost in the form of excitation or ionization.

inert Chemically and/or physiologically inactive.

infantile myxedema CRETINISM.

infantile paralysis POLIOMYELITIS.

infarct An area of coagulation necrosis in a tissue due to local ischemia resulting from obstruction of circulation to the area, most commonly by a thrombus or an embolus.

infarction 1. An infarct. 2. Formation of an infarct.

infection 1. The introduction of an infective agent, such as a virus or a bacterium, into a host cell or a host organism. 2. The condition produced by the introduction of an infective agent into a cell or an organism.

infectious center *See* infective center.

infectious disease Any disease that can be transmitted from one person, animal, or plant to another.

infectious hepatitis *See* hepatitis.

infectious mononucleosis An acute infectious disease, primarily of lymphoid tissue, that is caused by the Epstein–Barr virus. It is characterized by enlarged lymph nodes, an enlarged spleen, and the presence of abnormal leukocytes in the blood.

infectious nucleic acid A purified viral nucleic acid that can infect a host cell and lead to the production of infective viral particles.

infectious RNA VIROID.

infectious titer The number of infective units in a viral sample. *Aka* infective titer.

infectious transfer The rapid spread of extrachromosomal episomes from donor to recipient cells in a population of bacterial cells.

infective center The phage particle or the phage-infected bacterium that forms a single plaque in the plaque assay.

infective unit A virus particle in a viral sample that leads to the infection of a host cell. *Aka* infectious unit.

infectivity The capacity of bacteria and viruses for interacting with, and altering, host cells.

infinite dilution A solute concentration of zero to which the values of physical parameters are commonly extrapolated.

infinite thickness SATURATION THICKNESS.

infinite thinness A layer of solid radioactive material that is so thin that self-absorption is negligible and that the sample can be counted as if it were infinitely thin.

influent ELUENT.

influenza virus The flu virus that causes respiratory infections and that belongs to the group of myxoviruses. *See also* antigenic drift.

influx Inward flow, as that into a cell.

informational molecule A molecule that carries genetic information in the form of specific sequences of building blocks. The term is generally restricted to DNA, messenger RNA, viral RNA, and to the anticodon segment of transfer RNA.

information theory The branch of science that deals with the measurement, processing, and transmission of information; an extension of thermodynamics and probability theory that attempts to resolve information into a series of binary (yes/no) decisions.

informofers Globular protein particles with which HnRNA is complexed.

informosome A cytoplasmic messenger RNA–nonribosomal protein complex that occurs in eukaryotic cells. It is believed to function in the protection of mRNA against nuclease activity and in the transport of mRNA from the nucleus to the cytoplasm.

infra- Prefix meaning under or below.

infrared dichroism The dichroism of polarized infrared light that is used in the study of polypeptides.

infrared spectrum That part of the electromagnetic spectrum that covers the wavelength range of about 7.5×10^{-5} to 4.2×10^{-2} cm and that includes photons that are emitted or absorbed during vibrational and rotational transitions. *Abbr* IR spectrum.

infusion An extract made by soaking a substance, such as meat or a plant, in water.

ingest To take material, such as food, into either a cell or a body.

INH Isonicotinic acid hydrazide. *See* isoniazid.

inheritance of acquired characteristics The hereditary transmission of structural changes in organisms that is postulated by the Lamarckian theory.

inherited immunity NATURAL IMMUNITY.

inhibin A postulated protein hormone, of testicular origin, believed to function in the feedback regulation of gonadotropic secretion, particularly that of follicle-stimulating hormone (FSH); inhibin is believed to be a feedback inhibitor of FSH secretion. The existence of inhibin is not yet universally accepted.

inhibition A decrease in the extent and/or the rate of an activity, or the complete abolition of an activity.

inhibition analysis The study of secondary agents that can bring about the reversal of the inhibition of an enzyme; based on measurements of the precursors and products of the inhibited enzymatic reaction, and on measurements of substances that increase the enzyme concentration or influence the rates at which metabolites or inhibitors are destroyed.

inhibition coefficient The lowest concentration of a bacteristatic agent that inhibits bacterial growth under defined conditions.

inhibition constant INHIBITOR CONSTANT.

inhibition index A measure of the potency of an antimetabolite; equal to the ratio of the concentration of antimetabolite that is required to inhibit the effect of an essential metabolite, to the concentration of the metabolite.

inhibition ratio The ratio of the amount of an antivitamin that is required to inhibit the effect of a given amount of a vitamin.

inhibition zone The antigen excess zone in a precipitin curve.

inhibitor An agent that produces inhibition.

inhibitor constant The equilibrium dissociation constant of the reaction $EI \rightleftharpoons E + I$, where E is the enzyme, and I is the inhibitor. *Sym* K_i.

inhibitor source material A plasma glyceride that inactivates factor VIII of blood clotting.

inhibitor stop technique A method for measuring transport across the mitochondrial membrane. Involves addition of a transportable metabolite to the mitochondrial suspension, followed by addition of a specific inhibitor at timed intervals. The inhibitor must stop the transport immediately and completely upon addition to the mitochondria. The mitochondria are then separated from the medium and assayed for the transportable metabolite.

inhibitory autacoid *See* autacoid.

inhibitory medium A medium that contains one or more substances designed to inhibit the growth of one or more types of organisms.

inhibitory transmitter A substance that is released by one neuron and that serves to dampen, or inhibit completely, the firing of another, target, neuron The compound γ-aminobutyric acid (GABA), which is secreted by special neurons referred to as GABAergic neurons, is an example. *Aka* inhibitory neurotransmitter.

initial heat The heat produced by a muscle when it is stimulated by either an electric shock or a nerve impulse.

initial velocity The reaction velocity at the early stages of an enzymatic reaction; measured before the substrate concentration has decreased significantly (usually meaning that less than 5% of the initial substrate has been utilized) and while the concentrations of the products are low, so that the reverse reaction can be neglected. The initial velocity is given by the tangent, at the origin, to the curve that is obtained by plotting reaction velocity as a function of time. *Sym* v. *Aka* initial rate; initial steady-state rate; instantaneous velocity.

initiation 1. The process of chain initiation during protein synthesis, specifically the formation of the ribosome–mRNA–initiator tRNA complex. 2. The first stage in a two-stage or multistage mechanism of carcinogenesis during which a normal cell is converted to a precancerous cell by the action of a carcinogen. Initiation involves a brief and irreversible interaction between a carcinogen and the genetic material of the target tissue. This results in a molecular lesion, or mutation, that may transform some cells to an abnormal state but does not generate a clinically observable tumor unless acted upon by a promoter. Initiation and promotion are regulated independently at different times by different agents. 3. The first stage of transcription. 4. The second stage in the germination of a spore. *See also* germination. 5. The first step in a chain reaction.

initiation codon The codon AUG that codes for the binding of the initiator tRNA for the initiation of protein synthesis; in prokaryotic systems, the codon binds *N*-formylmethionyl-tRNA and in eukaryotic systems the codon binds methionyl-tRNA. *Aka* initiator codon; start codon.

initiation complex The initial complex formed during protein synthesis which, in bacterial systems, consists of *N*-formylmethionyl-tRNA, mRNA, and a 30S ribosomal subunit. *Aka* initiator complex.

initiation factor 1. One of several protein factors that function in the initiation of protein synthesis. In prokaryotes there are 3 such factors, in eukaryotes there are at least 9. Prokaryotic and eukaryotic initiation factors are designated as IF and eIF, respectively. 2. The sigma subunit of RNA polymerase.

initiation point The site on the DNA molecule at which DNA replication begins.

initiator 1. A structural gene that forms part of the replicon and that forms a product, believed to be a protein, that interacts with the replicator and initiates the replication of the DNA that is attached to the replicator. 2. The product that is formed by the initiator structural gene and that interacts with the replicator to initiate DNA replication. 3. PRIMER. 4. A carcinogenic agent that brings about the first stage in a two-stage or multistage mechanism of carcinogenesis; generally a substance that interacts with cellular DNA. *See also* replicon.

initiator locus *See* replicon.

initiator transfer RNA The transfer RNA molecule that is reponsible for the initiation of protein synthesis; in both prokaryotic and eukaryotic systems this is methionine-tRNA. In prokaryotes, it is bound to the initiator codon as *N*-formylmethionyl-tRNA, in eukaryotes it is bound as methionyl-tRNA.

innate immunity NATURAL IMMUNITY.

inner compartment MATRIX.

inner filter effect The decrease in the intensity of light passing through a sample that is due to the absorption of light by the sample. The effect constitutes a source of error in emission spectroscopy, as in fluorescence, since under these conditions only some of the molecules of the sample will be excited by the incident beam.

inner membrane 1. The internal mitochondrial

membrane which is the site of the electron transport system. 2. The membrane forming the thylakoid disks in chloroplasts. 3. The cell membrane (cytoplasmic membrane) of bacteria; it houses the systems for active transport, oxidative phosphorylation, and the biosynthesis of certain macromolecules. 4. The inner membrane of the nuclear envelope that is connected via nuclear pores to the outer membrane.

inner-membrane particle F_0F_1-ATPase.

inner-membrane sphere The elementary particle, or supermolecule, that is observed when inner mitochondrial membranes are examined with the electron microscope.

inner orbital An orbital that functions in the bonding of a low-spin complex.

innervation The distribution, or supply, of nerves to a particular tissue or organ.

inner volume The volume of solvent that is trapped within the gel particles of the bed in gel filtration chromatography.

Ino Inosine.

inoculation 1. The introduction of an inoculum into a culture or a culture medium. 2. The introduction of a substance into a cell or an organism, particularly the introduction of immunological substances for the purpose of producing immunity.

inoculum (*pl* inocula) A mass or a suspension of either cells or viruses that is used to initiate the growth of a new culture or to infect another culture.

inorganic 1. Pertaining to compounds other than those of carbon. 2. Pertaining to substances other than those derived from plant, animal, or microbial sources.

inorganic phosphate An anion, or a mixture of anions, derived from orthophosphoric acid (H_3PO_4). *Sym* P_i;P.

inorganic pyrophosphate An anion, or a mixture of anions, derived from pyrophosphoric acid ($H_4P_2O_7$). *Sym* PP_i;PP.

inosine The ribonucleoside of hypoxanthine. Inosine mono-, di-, and triphosphate are abbreviated, respectively, as IMP, IDP, and ITP. The abbreviations refer to the 5′-nucleoside phosphates unless otherwise indicated. *Abbr* Ino;I.

inosinic acid The ribonucleotide of hypoxanthine.

inositol An optically and biologically active, cyclic sugar alcohol; frequently classified with the B vitamins, since it is a growth factor for some organisms. *Aka i*-inositol; *meso*-inositol; *myo*-inositol; inosite.

inositolphospholipid Any phospholipid derived from inositol or related compounds.

inotropic effect An effect on the contractility of muscular tissue, especially that of the heart.

insect hormones A group of substances that direct the life cycle of insects; the post-embryonic development of insects is controlled by the activation hormone, the molting hormone, and the juvenile hormone.

insecticide An agent that kills insects.

insert PASSENGER.

insertase An enzyme that catalyzes the attachment of the proper base to an apurinic or apyrimidinic site (AP site).

insertion 1. A mutation in either DNA or RNA in which one or more extra nucleotides are inserted into a polynucleotide chain. 2. INTEGRATION (c). 3. GENE INSERTION.

insertional inactivation The inactivation of a gene that results from the insertion of a segment of foreign DNA into the coding sequence of the gene. The inactivation can be detected by a plating test and can serve as a means for isolating a plasmid that contains foreign DNA. *See also* recombinant DNA technology.

insertion element INSERTION SEQUENCE.

insertion model A model, proposed by Campbell, for the attachment of the prophage to the bacterial DNA. According to this model, the prophage first becomes attached in cyclic form to the bacterial DNA. The bacterial DNA is then interrupted by opening the ring at the point of contact, and the prophage is inserted linearly into the bacterial DNA which is attached to it at both ends. *See also* hook model.

insertion sequence 1. A small, transposable element in bacteria (a small transposon) that can insert into several sites in a genome. Insertion sequences function in the transposition of segments which they flank. They usually contain genes coding for proteins that function in their transposition but contain no other genes and have no other known effect than to function in transposition. The termini of each insertion sequence consist of inverted repeats. *Aka* IS element. 2. The C-terminal hydrophobic tail of a free protein that allows it to become inserted in, and translocated across, a biological membrane.

insertion vector *See* lambda cloning vector.

insertosome A small segment of DNA (800–1400 bp) that can insert itself randomly into the chromosome of *E. coli* to cause polar mutations analogous to those produced by phage Mu-1.

inside-out particles Subcellular particles, produced by disruption of mitochondria, that consist chiefly of fragments of the inner mitochondrial membrane which have become released to form vesicles.

inside-out protein A protein (such as bacteriorhodopsin) that has an organization of amino acid residues that is opposite that of typical soluble, globular proteins; a protein in which polar amino acids tend to be concentrated in the interior of the molecule while nonpolar amino acids predominate on the exterior side of the molecule.

inside–outside transition   FLIP-FLOP (1).

in situ   In the normal, natural location or position.

in situ hybridization   A technique for localizing specific DNA segments within intact chromosomes; involves incubating chromosomes, devoid of RNA and protein, with tritium-labeled nucleic acid and visualizing the hybridized segments by radioautography.

in situ hybridization assay   COLONY HYBRIDIZATION.

insoluble enzyme   An enzyme that is linked covalently to a water-insoluble support, such as agarose or polyacrylamide, without destruction of its activity. Applications of such enzymes occur in batch-type reactions, adsorption chromatography, and gel filtration. *See also* immobilized enzyme.

insoluble fibrin   HARD CLOT.

inspire   To inhale.

instability factor   One of a set of arbitrary values assigned to substituents, interactions, and the like, that affect the stability of monosaccharide conformations. The values obtained by summing these factors permit an estimation of the relative stabilities of different conformations of a given compound.

instructive theory   A theory of antibody formation according to which the information for antibody synthesis results from an instructive effect by the antigen, rather than from a genetic determination; the antigen instructs the biosynthetic machinery to synthesize specific antibodies which, in the absence of the antigen, would either not be formed at all or be formed only very rarely as a result of a chance event. *See also* selective theory.

insulin   A protein hormone that lowers the level of blood sugar and stimulates the utilization of glucose by affecting the rate of transport of glucose across the cell membrane. Insulin is secreted by the islets of Langerhans in the pancreas and also has anabolic effects in protein and lipid metabolism.

insulinase   A protease that hydrolyzes both of the separated A and B chains of insulin but does not attack the intact insulin molecule.

insulin-dependent diabetes   *See* diabetes.

insulin-independent diabetes   *See* diabetes.

insulin-like activity   A group of substances in serum that have some of the biological properties of insulin, but do not react with insulin antibodies. *Abbr* ILA.

insulin-like growth factor   *See* somatomedin.

insulinoma   A benign, insulin-secreting tumor of the beta cells of the pancreas.

insulin resistance index   The sum of the values for the blood glucose concentration at 60, 90, and 120 min as determined in an insulin tolerance test. *Abbr* IRI.

insulin shock   A condition characterized by anxiety, delirium, and convulsions, that is occasionally produced upon the administration of insulin to an animal; it results from a lowering of the blood sugar by insulin to a level below that which is required for normal functioning of the brain.

insulin stimulation test   A test for assessing the integrity of the hypothalamus–pituitary–adrenal system; based on the fact that, in normal individuals, an intravenous injection of insulin that is sufficient to lower blood glucose, will also result in a rapid and marked rise of plasma ACTH and cortisol.

insulin tolerance test   A test for evaluating insulin resistance and certain endocrine disorders. The test is performed by placing an individual on a carbohydrate diet, followed by an injection of insulin and a determination of the blood sugar level as a function of time.

integral counting   The counting, in a scintillation counter, of pulses that are above a certain level of intensity; the pulses are selected by means of a discriminator that rejects pulses of lower intensity.

integral discrimination   The selection of pulses that takes place in integral counting.

integral dose   The total, cumulative dose of radiation received by an individual.

integral proteins   Membrane proteins that are integrated into the structure of the cell membrane. They may extend partially or completely through the phospholipid bilayer of the membrane. They usually cannot be removed from the membrane without the use of drastic conditions which disrupt the membrane. *Aka* intrinsic proteins.

integrase   A phage-specific enzyme that catalyzes the site-specific exchange occurring when a prophage is inserted into, or excised from, a bacterial chromosome; in the excision process, an accessory host protein (excisionase) is also needed.

integrated circuit   CHIP.

integrated state   The state of an episome in which it is incorporated into, and replicates with, the chromosome. *See also* integration.

integrating circuit   An electronic circuit for measuring the total number of ionizations and the resulting electrical currents that are produced in an ionization chamber in a given time interval. *See also* differentiating circuit.

integration   The incorporation of one DNA segment into another as in (a) the incorporation of donor DNA into recipient DNA in genetic recombination; (b) the incorporation of episomal DNA into chromosomal DNA; and (c) the insertion of prophage DNA into the host bacterial DNA.

integration efficiency The frequency of incorporation of foreign DNA into recipient bacterial DNA; used particularly for foreign DNA incorporation during transformation.

integration host factor EXCISIONASE.

integrator gene *See* Britten–Davidson model.

integrin One of a group of receptors that bind the RGD sequence of adhesive proteins. Integrins are heterodimeric proteins with two membrane-spanning subunits. *See also* RGD.

intensive property A property of a system, such as density or concentration, that is intrinsic to the system and that does not depend on the amount of substance involved. *See also* extensive property.

inter- Prefix meaning between.

interacting flows *See* interaction of flows.

interacting sites *See* cooperative binding.

interaction of flows The interdependence of the diffusion of different solutes in a solution that contains two or more solutes; due to the effect of the concentration gradient of one solute on the diffusion of another.

interaction of heme groups *See* heme–heme interaction.

interactive Descriptive of a computer in which a two-way interaction goes on between the user and the computer.

interactive graphics The interaction of an individual with a computer, the output of which is shown in graphical form; the interaction occurs while the computer is in operation, and with the computer being coupled to oscilloscopes or other display devices.

interallelic complementation INTRAGENIC COMPLEMENTATION.

inter-alpha-globulin THYROXINE-BINDING GLOBULIN.

interband A region between two adjacent bands in a polytene chromosome.

intercalary deletion A deletion in which genetic material is lost from some part of the chromosome other than its end; a deletion occurring at the end of a chromosome is called a terminal deletion.

intercalation The process whereby a flat molecule, such as an acridine dye, becomes inserted between two adjacent, stacked bases in a double-helical nucleic acid. Intercalation results in a frameshift mutation, since the subsequent replication of the nucleic acid leads to either a deletion or an insertion of a nucleotide.

intercept replot *See* secondary plot.

intercistronic region That section of DNA, in a polycistronic transcription unit, that lies between the termination point of one gene and the initiation point of the next gene.

interconversion 1. The change of one metabolite into another. 2. The change of one enzyme form into another. *See* covalently modified enzyme.

interconvertible enzyme An enzyme capable of undergoing covalent modification by another enzyme; glycogen phosphorylase, which can be converted enzymatically from the a to the b form, and vice versa, is an example.

interesterification The formation of a new ester by the reaction of an ester with an acid, an alcohol, or another ester; these three reactions are known as acidolysis, alcoholysis, and ester interchange.

interface 1. The boundary between two phases. 2. COMPUTER INTERFACE.

interface centrifugation A centrifugal technique in which there are two immiscible phases in the centrifuge tube, and transfer of solute particles occurs from one phase to the other.

interfacial tension The surface tension at the interface between two liquid phases.

interfacial test RING TEST.

interference 1. The mutual effect upon meeting of two sets of waves, such as light waves, that results in neutralization (destructive interference) at some points, and in reinforcement (constructive interference) at other points. 2. The effect of a crossover at one locus on a chromosome on the probability of a crossover at another locus; the interference is said to be positive if the probability of the second crossover is decreased and it is said to be negative if that probability is increased. 3. VIRAL INTERFERENCE.

interference filter A filter that allows the passage of a narrow range of wavelengths, about 10 to 20 nm wide. It is made by depositing semitransparent silver films on both sides of a dielectric so that constructive and destructive interference take place when light passes through the films.

interference fringe One of either the light or the dark bands that are produced by the interference or the diffraction of light.

interference microscope A microscope that utilizes interference effects and that permits the observation of transparent objects and the measurement of refractive indices.

interference optics INTERFEROMETRIC OPTICAL SYSTEM.

interfering virus A virus that interferes with the multiplication of another virus.

interferogram The photographic record of an interference pattern.

interferometer An instrument for the precise

determination of wavelengths or distances; based on the separation of light in the instrument into two parts that travel unequal paths and that, when reunited, interact with each other to produce an interference pattern.

interferometric optical system An optical system that focuses ultraviolet light passing through a solution in such a fashion that a photograph of interference fringes is obtained. A boundary in the solution appears as a break in the interference fringes and measurements are made on the photographic plate by counting the interference fringes. An optical system of this type, incorporating a Rayleigh interferometer, is used in the analytical ultracentrifuge. *See also* Gouy interferometer.

interferons A family of proteins, occurring in a large number of vertebrates, which have a variety of cell regulatory functions. They affect cell motility and immunological processes; they also interfere with the replication of various viruses (hence the name). Interferons are species-specific and can generally not be detected unless induced. They are induced in response to a variety of agents including viruses, microorganisms, and endotoxins. Upon induction, interferons circulate to neighboring cells which they stimulate to make antiviral proteins that prevent the translation of viral mRNA. Human interferons are classified into three, antigenically distinct, types: α- (or leukocyte), β- (or fibroblast), γ- (or immune) interferons. The names indicate the types of cells that synthesize the particular interferon.

intergenic complementation Complementation that is produced by two mutant chromosomes that carry a mutation in different genes. The unmutated gene of one chromosome makes up for, or complements, the mutated gene of the other chromosome so that the necessary gene products of both genes are formed, and the wild-type allele of each gene is expressed.

intergenic suppression The restoration of a genetic function, which was lost by mutation, through a second mutation in a gene other than the one that sustained the primary mutation; often refers to the ability of a tRNA molecule to recognize a termination codon or two different sense codons.

intergenic suppressor mutation *See* suppressor mutation.

interkinesis The interphase between the first and the second division in meiosis.

interleukin-2 A lymphokine produced naturally by some helper T cells in response to infection. It causes a rapid amplification of the number of T cells at the site of infection and may be used in tissue culture to produce

antigen-specific T cell lines. It has also been used to reduce the mass of certain tumors and to prevent metastasis. *Var sp* interleuken-2. *Abbr* IL-2. *See also* cytokines; lymphokines.

intermediary metabolism 1. The enzyme-catalyzed reactions in cells whereby nutrients are transformed and energy is extracted from them for the growth and maintenance of the cells. 2. The sum total of all the chemical transformations of the nutrients in an animal subsequent to the absorption of the nutrients into the blood.

intermediate A compound that participates in a reaction and that occurs between the starting materials and the final products of the reaction; in metabolism, an intermediate occurs between the nutrients on the one hand, and the cellular components and waste products on the other hand.

intermediate filaments Intracellular fibers, having a diameter of about 8–12 nm, which is between that of microfilaments and microtubules. Intermediate filaments are heterogeneous in their protein composition and are an important component of the cytoskeleton. In general, a given class of filaments is characteristic of a specific cell type. For example, keratin filaments are found in epithelial cells, neurofilaments in neurons, and vimentin filaments in fibroblasts. *Abbr* IF.

intermedin MELANOCYTE-STIMULATING HORMONE.

internal conversion 1. A mode of radioactive decay in which gamma rays that emanate from a nucleus collide with an orbital electron, transfer all their energy to it, and then eject the electron from the atom. 2. A mode of vibrational deexcitation in which the energy of an excited electronic state of a molecule is dissipated by conversion to vibrational energy of a lower electronic state that has the same multiplicity (i.e., singlet to singlet or triplet to triplet).

internal conversion electron CONVERSION ELECTRON.

internal energy The energy within a system. In chemistry, this usually refers only to those types of energy that can be modified by chemical processes; these include translational, vibrational, and rotational energies of molecules, energy involved in chemical bonding, and energy involved in noncovalent interactions between molecules.

internal gas counter A radiation counter in which a radioactive gaseous sample is counted by being incorporated into the detector-filling gas mixture.

internal indicator An indicator that is added to the titration vessel in which a liquid is being titrated.

internalization *See* coated pit; receptosome.

internal monooxygenase A monooxygenase in which the cosubstrate that incorporates the second oxygen atom is itself a product of the reaction.

internal protein A protein that is complexed with DNA and occurs in the head of T-even phages.

internal quenching The quenching of an ionization detector that occurs when a specific gas, such as butane or chlorine, is added to the mixture used for gas amplification. *See also* organic quenching.

internal radiation The radiation emitted by radioactive substances that are deposited in the tissues.

internal resolution site A region, in some transposable elements, at which a site-specific exchange of genetic elements occurs.

internal respiration *See* respiration (3).

internal-sample scintillation counter A scintillation counter in which the sample and the fluor are in intimate contact, as in a liquid scintillation counter.

internal standard A standard that is added to, and treated with, the sample.

internal standardization A method for determining the counting efficiency of radioactive samples that is based on counting the sample both by itself and together with a known amount of added isotope.

internal volume INNER VOLUME.

International Union of Biochemistry The organization that standardizes biochemical nomenclature, symbols, etc. *Abbr* IUB.

International Union of Pure and Applied Chemistry The organization that standardizes chemical nomenclature, symbols, atomic weights, etc. *Abbr* IUPAC.

international unit An arbitrarily defined measure for the activity of a natural substance such as an enzyme, a hormone, or a vitamin; for enzymes, an international unit is identical to an enzyme unit. *Sym* IU.

interphase The period between two successive mitotic divisions; composed of successive G_1, S, and G_2 phases of the cell cycle.

interphase cycle CELL CYCLE.

interpolation The estimation of the value of a function between two known values without using the equation of the function itself.

interrupted gene A gene containing intervening sequences (introns).

interrupted mating experiment An experiment designed to study the transfer of genetic information during bacterial conjugation; performed by withdrawing samples at various times from a culture of mating bacteria and separating the mating bacteria by intense agitation in a blender. The bacterial fractions thus obtained contain recipient bacteria which have received varying amounts of the chromosome from the donor bacteria.

interrupted trough technique An immunoelectrophoretic technique for comparing the antigens in two mixtures.

intersome A collective term for the intermediate-size particles that either are naturally occurring precursors of ribosomes or are produced from ribosomes by the stepwise removal of proteins.

interstice A small space between two structures such as cells, organs, or tissues.

interstitial Of, or pertaining to, interstices.

interstitial cell hormone LUTEINIZING HORMONE.

interstitial cell-stimulating hormone LUTEINIZING HORMONE.

interstitial volume The total volume of the mobile phase within the length of the column in column chromatography.

intersystem crossing A nonradiative transition of a molecule from one energy state to another that has a different multiplicity; the transition from an excited singlet state to a triplet state by a decrease in the energy of the molecule as a result of collisions is an example.

intervening sequence INTRON.

intervent A solute that intervenes in, and reduces, the interactions between macromolecules.

intervent dilution chromatography A chromatographic technique for the separation of strongly interacting macromolecules that is based on forcing the macromolecules repeatedly across a boundary in an ion-exchange gel. On one side of the boundary the intervent concentration is high, so that the macromolecules are relatively independent of each other and macromolecular complexes dissociate; on the other side of the boundary the intervent concentration is low, so that the macromolecules are in an aggregated form and can be separated by adsorptive and ion-exchange processes.

interwound helix A supercoiled duplex in which the axis is twisted around itself.

intestinal juice The digestive juice that consists of the secretion of the intestinal mucosa and that is discharged into the small intestine; contains enterokinase, lipases, peptidases, carbohydrases, nucleases, and phosphatases. *Aka* intestinal fluid.

intima The inner lining of a blood vessel.

in toto Totally; altogether.

intoxication The abnormal state of an animal produced by relatively large amounts of a chemical agent such as a poison, a drug, or a vitamin.

intra- Prefix meaning within.

intracellular accumulation period The period, during viral infection, that extends from the end of the eclipse to the first appearance of extracellular virus particles.

intracellular digestion Digestion within the cell in which the lysosomes play an important role; the processes of pinocytosis and phagocytosis.

intracellular titer The total titer of phage particles minus the extracellular titer.

intracellular transport The transport across membranes of subcellular organelles, such as the transport across mitochondrial or chloroplast membranes. *See also* homocellular transport; transcellular transport.

intracellular virus CELL-ASSOCIATED VIRUS.

intracistronic complementation INTRAGENIC COMPLEMENTATION.

intracistronic suppression INTRAGENIC SUPPRESSION.

intracristael space The intermembrane space between the inner and the outer mitochondrial membranes.

intradermal injection An injection into the skin.

intragenic complementation Complementation that is produced by two mutant chromosomes that carry a mutation in the same gene but at different sites. Such complementation may arise when the product of the genes is a multimeric protein and when the individual gene products, or monomers, are nonfunctional but combine to produce an aggregate, or multimer, that is functional and nearly normal in its properties.

intragenic recombination The genetic recombination between mutons of the same cistron.

intragenic suppression The restoration of a genetic function, which was lost by mutation, through a second mutation in the same gene that sustained the primary mutation but at a different site in that gene.

intragenic suppressor mutation *See* suppressor mutation.

intramuscular injection An injection into the muscle. *Abbr* im.

intraperitoneal injection An injection into the peritoneal cavity. *Abbr* ip.

intrapleural injection An injection into the chest fluid.

intrathecal injection An injection into the spinal cord.

intravenous injection An injection into a vein. *Abbr* iv.

intravital staining The staining of living cells without killing them.

intrinsic activity A measure of the potency of an agonist that binds to a receptor; equal to the maximum response the agonist is capable of evoking and expressed as a fraction of the response evoked by a full agonist.

intrinsic association constant The association constant that describes the binding of a ligand to a particular site on a protein molecule, provided that all sites of this type present on the same molecule are identical and noninteracting. *Aka* intrinsic binding constant.

intrinsic blood coagulation INTRINSIC PATHWAY.

intrinsic Cotton effect A Cotton effect that is caused by the protein itself and not by a small molecule that is bound to the protein.

intrinsic dissociation constant The reciprocal of an intrinsic association constant.

intrinsic factor A glycoprotein in the gastric juice that combines with free vitamin B_{12} and aids in its absorption from the intestine. *Aka* intrinsic factor of Castle.

intrinsic fluorescence Fluorescence that is caused by the aromatic amino acids of the protein itself and not by a small molecule that is bound to the protein. *See also* autofluorescence.

intrinsic heterogeneity The heterogeneity of antibodies that results from intrinsic factors, such as the different genes that are responsible for the synthesis of antibodies.

intrinsic pathway The series of reactions in blood clotting that involve factors normally present in the circulation and that are initiated by surface contact in the blood capillaries or by kallikrein. The pathway proceeds in the form of a cascade mechanism, with each stage involving protease activation of a zymogen to its enzymatically active form. Major steps of the pathway are as follows (Roman numerals indicate factor numbers; the subscript a indicates the active form of the factor):

$$
\begin{array}{c}
\text{XII} \\
\text{Kallikrein/surface} \rightarrow\downarrow \quad \text{XI} \\
\text{factor} \quad \text{XII}_a \rightarrow\downarrow \quad \text{IX} \\
\text{XI}_a \rightarrow\downarrow \quad \text{X} \\
\text{IX}_a \rightarrow\downarrow \quad \text{II} \\
\text{VII} \nearrow \text{VII}_a \quad \text{IV} \nearrow \quad \text{X}_a \rightarrow\downarrow \quad \text{I} \\
\text{VIII} \nearrow \text{VIII}_a \quad \text{V} \rightarrow \text{V}_a \quad \text{II}_a \rightarrow\downarrow \\
\text{XIII} \rightarrow \text{XIII}_a \rightarrow \text{I}_a \\
\downarrow \\
\text{Fibrin clot}
\end{array}
$$

The sequence from factor X_a on is common to both the intrinsic and the extrinsic pathways. The common names of the factors are listed under factor. *Aka* intrinsic system.

intrinsic proteins INTEGRAL PROTEINS.

intrinsic system INTRINSIC PATHWAY.

intrinsic viscosity The limiting value, at infinite dilution, of the reduced viscosity; it is equal to the product of the partial specific volume of the solute and its viscosity increment. *Sym* $[\eta]$.

intron An intervening sequence in the DNA of eukaryotic genes. Such sequences are transcribed into RNA but are then excised and are not translated. The term is also used for the excised RNA sequence. Introns and exons (translated sequences) make up split genes. *Aka* intervening sequence (IVS).

intron intrusion The breakup of a functional gene through the insertion of an intron into it. Intron intrusion, exon shuffling, and junctional sliding have been proposed to help explain the evolutionary diversification of genes.

intron-mediated recombination *See* exon shuffling.

intussusception A mode of growth in which new material is incorporated within an existing matrix. The deposition of material in a plant cell wall or in a cell membrane are two examples.

inulin A homopolysaccharide of D-fructose that occurs in some plants and that is used for measurements of renal clearance.

in utero Within the uterus.

in vacuo In a vacuum.

invagination 1. The process of infolding and pocket formation. 2. The structure formed by infolding, as that produced by an infolded membrane.

invariant residues Amino acid residues that show sequence homology; amino acids that are identical in their location and sequence in a protein of a given type, isolated from various species. The term can likewise be applied to nucleotides in a nucleic acid.

invariants Identical amino acid sequences in a protein of a given type, isolated from various species.

invasin HYALURONIDASE.

invasiveness The spreading of bacteria, viruses, or cancer cells from the site of infection, or the site of the original tumor, to other sites and other tissues in the same organism.

inverse isotope dilution analysis A technique for determining the amount of a labeled compound of known specific activity by the addition of a known amount of the same, but unlabeled, compound. The mixture of labeled and unlabeled compounds is then isolated and its specific activity is measured.

inverse square law The law stating that a property, such as radiation intensity or Coulomb's force, varies inversely as the square of the distance from a given point.

inverse substrate A substrate for an enzyme in which the arrangement of the site-specific groups is reversed, compared to that of the normal substrate.

inversion 1. The hydrolysis of sucrose to an equimolar mixture of glucose and fructose; so called because the optical rotation changes sign as the hydrolysis proceeds. 2. A chromosomal aberration in which a block of genes is rotated by 180° so that the sequence of genes in that block is inverted.

inversion of configuration The change from one enantiomeric configuration about an asymmetric carbon atom to the other in the course of a chemical reaction; requires the breaking of covalent bonds in a compound and the remaking of these bonds in the reverse sense.

inversion symmetry The symmetry of a body that exists when identical structures are produced by an inversion about an inversion center.

invertase The enzyme sucrase that catalyzes the hydrolysis of sucrose to glucose and fructose. *See also* inversion (1).

invertebrate 1. *n* An animal that lacks a backbone. 2. *adj* Of, or pertaining to, an animal that lacks a backbone.

inverted repeat PALINDROME.

inverted repetition PALINDROME.

inverted terminal repeats Identical, or closely related, nucleotide sequences that have opposite orientations (run in opposite directions); they occur at the ends of some transposons.

invert sugar The equimolar mixture of glucose and fructose that is formed by the hydrolysis of sucrose. *See also* inversion (1).

Inv group A group of allotypic antigenic sites in the constant region of the kappa chains of human immunoglobulins.

in vitro Outside a living organism; pertaining to conditions of or to experiments with a perfused organ, a tissue slice, cells in tissue culture, a homogenate, a crude extract, or a subcellular fraction. *See also* in vivo.

in vitro complementation The in vitro demonstration of intragenic complementation by the mixing of nonfunctional monomeric proteins and the formation of a functional multimeric protein.

in vitro marker An induced mutation of mammalian cells that permits their phenotypic detection in tissue culture.

in vitro mutagenesis The introduction of mutations into DNA by treatment of the DNA with chemical or physical mutagens. The mutated DNA is then assayed for biological activity using either in vitro (cell-free) or in vivo (plasmid) systems.

in vitro packaging The process of encapsidating naked DNA, by means of phage lambda packaging proteins and preheads, to produce infectious particles.

in vitro protein synthesis CELL-FREE AMINO ACID INCORPORATING SYSTEM.

in vivo Within a living organism; pertaining to conditions of or to experiments with a whole animal, an intact plant, an intact organ, or a population of microbial cells. *See also* in vitro.

in vivo marker A naturally occurring mutation of mammalian cells that permits their phenotypic detection in tissue culture.

iodide pump The active transport mechanism that concentrates iodide in the thyroid gland. *aka* iodide trapping mechanism.

iodine An element that is essential to humans and several classes of animals and plants. Symbol, I; atomic number, 53; atomic weight, 126.9044; oxidation states, -1, $+1$, $+5$, $+7$; most abundant isotope, ^{127}I, a radioactive isotope, ^{131}I, half-life, 8.1 days, radiation emitted, beta particles and gamma rays.

iodine number A measure of the extent of the unsaturation in a fat that is equal to the number of grams of iodine taken up by 100 g of fat. The greater the iodine number, the greater the extent of unsaturation in the fat. *Aka* iodine value.

iodophor A complex formed between iodine and a surface active compound. Iodophors have both antimicrobial activity, due to the iodine, and detergent properties, due to the surface active compound; they are widely used as disinfectants. *Var sp* iodophore.

iodopsin The visual pigment, consisting of cone opsin plus retinal, that occurs in mammals and other vertebrates and that has an absorption maximum at 562 nm.

ion An atom, a group of atoms, or a molecule that carries an electrical charge. *See also* ionization.

ion antagonism The phenomenon in which one or more ions lead to an activation of a reaction or a set of reactions, while one or more similar ions lead to an inhibition of the same reaction or set of reactions.

ion atmosphere The region surrounding a charged atom, a charged group of atoms, or a charged molecule in which there is a statistical preference for ions of opposite charge.

ion carrier *See* ionophore.

ion chamber IONIZATION CHAMBER.

ion channel *See* ionophore.

ion chromatography Originally defined as a liquid chromatographic technique involving the separation of ions by means of a low-capacity ion-exchange column and an analysis of the effluent by conductivity measurements. The term is now used to describe the separation of ions by means of high-performance liquid chromatography (HPLC); a variety of columns can be used and a variety of detection methods (conductivity, fluorescence, refractive index, etc.) can be employed.

ion cloud ION ATMOSPHERE.

ion cluster An aggregate of ion pairs that is formed in the immediate vicinity of a primary ionizing event, and that is produced by the secondary electrons which are generated by the incident ionizing particle or by the incident photon. *See also* ion pair (1).

ion–dipole interaction The attractive or repulsive electrical force between an ion and a dipole; the energy of such interactions is proportional to r^{-4}, where r is the distance between the ion and the dipole.

ion electrode *See* ion-selective electrode.

ion etching The exposure of a biological specimen to a beam of inert ions, such as those of argon, prior to an examination of the specimen with the scanning electron microscope.

ion-exchange chromatography A chromatographic technique in which molecules are separated on the basis of their charge. The stationary phase is an ion-exchange resin, and the mobile phase is an aqueous solution. Ion-exchange chromatography is usually performed in columns, and the charged molecules are retarded in their movement through the column depending on the sign and the magnitude of their charge. *Aka* ion exchange.

ion-exchange resin A high molecular weight, insoluble, branched, ionized polymer that is used as a support in ion-exchange chromatography; may be natural or synthetic. *Aka* ion exchanger.

ion exclusion A process whereby strong electrolytes are separated from weak electrolytes or from nonelectrolytes by passage of the mixture through an ion-exchange resin; the strong electrolytes are excluded from the ionized resin and elute more readily than the other substances. *See also* ion retardation.

ion-filtration chromatography A chromatographic technique that is based on the combined use of gel filtration and ion-exchange chromatography; performed by eluting macromolecules in the sieving range from a crosslinked ion-exchange gel under precisely controlled conditions of pH and ionic strength. Performed, for example, by using cross-linked DEAE-cellulose as the support in column chromatography.

ionic Of, or pertaining to, ions.

ionic bond The attractive force, described by Coulomb's law, between a cation and an anion or between a cationic and an anionic group of atoms; the cationic and anionic groups may be on the same molecule or on different molecules.

ionic contact distance The distance between two oppositely charged and interacting ions at which the attractive force due to the opposite charges is just balanced by the repulsive force due to the orbital electrons of the two ions.

ionic detergent *See* anionic detergent; cationic

detergent; detergent.

ionic double layer The region encompassing the charges on an atom, a group of atoms, or a molecule together with the layer of counterions that surrounds these charges.

ionic migration ISOTACHOPHORESIS.

ionic mobility analysis ISOTACHOPHORESIS.

ionic orbital An orbital that functions in the bonding of a high-spin complex.

ionic sieving Gel filtration chromatography of ions.

ionic strength A measure of the ionic concentration of a solution that is equal to $\frac{1}{2}\Sigma c_i Z_i^2$, where c_i is the concentration of the ith ion and Z_i is its charge. *Sym* I; $\Gamma/2$; μ.

ion–induced dipole interaction The attractive electrical force between an ion and a dipole which is induced by the ion; the energy of such interactions is proportional to r^{-4}, where r is the distance between the ion and the induced dipole.

ion–ion interaction The attractive or repulsive electrical force between two ions that is described by Coulomb's law. *See also* electrostatic interactions.

ionizable Capable of undergoing ionization.

ionization 1. The breakup of a molecule into two or more ions when it is dissolved in water. 2. The formation of a charged group in a molecule by either the association or the dissociation of a group, most commonly a proton. 3. The formation of an ion pair by either an ionizing particle or a strong electrostatic field.

ionization chamber A chamber for the production and collection of ions and for the measurement of the electric currents produced by these ions. The ions are commonly produced by bombardment of a gas with an ionizing radiation. Simple ionization chambers are used especially for measurements of x or gamma rays; ionization chambers with gas amplification are used for measurements of alpha and beta rays, and are incorporated in proportional and Geiger–Mueller counters.

ionization constant The dissociation constant for a reaction in which ions are produced, as in the dissociation of an acid to form a proton and an anion.

ionization detector An instrument for the detection of ionizing events, such as those occurring in an ionization chamber.

ionization energy The energy required to remove an orbital electron from an atom; expressed in terms of the energy required to remove a mole of electrons (kcal/mol or kJ/mol). *Aka* ionization potential.

ionization interference The interference that occurs in atomic absorption spectrophotometry when the atoms, in addition to being dissociated from the molecule, are also excited by the flame; as a result, the excited atoms emit radiation that is of the same wavelength as that which is being absorbed.

ionization potential IONIZATION ENERGY.

ionization track The trace of ion pairs that are produced by either an ionizing particle or a photon as it passes through matter, commonly water vapor.

ionized In the form of an ion or ions.

ionized calcium The physiologically active form of calcium in the serum that consists of unbound, ionized calcium.

ionizing energy *See* ionization energy.

ionizing event Any process that produces an ion or a group of ions as a result of interaction of matter with radiation; the formation of ions in an ionization chamber or in an irradiated cell are examples.

ionizing particle A charged particle that has sufficient energy to dislodge an orbital electron from an atom and to produce an ion pair.

ionizing radiation The electromagnetic or corpuscular radiation that produces ion pairs in the matter through which it passes; types of ionizing radiation include x rays, protons, neutrons, alpha particles, and high-speed electrons.

ionogenic Capable of forming ions; an amino group, for example, is ionogenic.

ionography ZONE ELECTROPHORESIS.

ionomer A cross-linked copolymer of an unsaturated hydrocarbon, such as ethylene, and a vinyl monomer with an acid group, such as methacrylic acid. These polymers contain covalent bonds between elements of the chain and ionic bonds between the chains; they contain both positively and negatively charged groups.

ionone ring The cyclic structural component of the carotenoids.

ionophore A lipid-soluble compound that increases the permeability of a membrane to metal ions by interacting with the metal ions and mediating their transport across the membrane. Many ionophores are antibiotics and two major types are ion carriers and ion channels. An ion carrier, such as the antibiotic valinomycin, has a hydrophilic cavity that binds that ion, forming a complex. This shields the ion from the hydrophobic region of the lipid bilayer in the membrane, makes the ion effectively lipid soluble, and facilitates its transport across the membrane. An ion channel, such as the antibiotic gramicidin, does not complex the metal ion but rather provides a tube that spans the membrane and through which the cation can pass.

ionophoresis 1. The electrophoresis of small ions. 2. ELECTROPHORESIS.

ionophorous antibiotic *See* ionophore.

ionotropy The ionization of a tautomeric compound in which a charged atom or a charged radical separates from an unsaturated molecule, thereby forming an oppositely charged fragment. The process is referred to as anionotropy or cationotropy depending on whether the separated atom or radical has a negative or a positive charge. *Aka* ion shifting.

ion pair 1. The positive ion formed by the action of ionizing radiation together with the orbital electron that was ejected from the uncharged atom when the ion was formed. 2. The two ions consisting of either an anion and a cation that has remained in the secondary hydration shell of the anion, or a cation and an anion that has remained in the secondary hydration shell of the cation.

ion-pair chromatography A form of ion-exchange chromatography in which the charged species of interest (such as a protonated amine) forms an ion pair with an appropriate counterion (such as Cl^- or ClO_4^- ions) that is partitioned between the stationary and the mobile phases; the ion pair is then eluted. *See also* zwitterion pair chromatography.

ion pairing SPECIFIC ADSORPTION.

ion product of water The product of the hydrogen and hydroxyl ion concentrations, in moles per liter, in pure water at 25°C; approximately equal to 10^{-14}. *Sym* K_w.

ion pump *See* pump.

ion retardation A process whereby nonelectrolytes are separated from electrolytes by passage of the mixture through a specially prepared ion-exchange resin; the resin absorbs electrolytes preferentially so that the nonelectrolytes elute more readily. *See also* ion exclusion.

ion-selective electrode An electrode that responds in a reproducible manner to changes in the activity of a specific ion and that is almost insensitive to changes in the activities of other ions in the solution. The selectivity of the electrode results from the incorporation of an ion exchanger, a specially formulated glass, or a specific crystal into the design of the electrode. *Aka* ion-specific electrode.

iontophoresis An electrophoretic technique in which drugs, in ionic form, move through the skin (or some other structure) under the influence of an electric field applied across the skin.

ion-translocating antibiotic *See* ionophore.

ion-transporting antibiotic *See* ionophore.

ip 1. *adj* Intraperitoneal. 2. *adv* Intraperitoneally.

IP Isoelectric point.

I protein A protein that regulates the ATPase active site of F_0F_1-ATPase.

IPTG Isopropylthiogalactoside.

IR 1. Infrared. 2. Induction ratio. 3. Inverted repeat.

IRA Immunoradiometric assay.

I region The section of DNA that contains the immune response genes (Ir genes).

I region-associated antigens A group of cell-surface glycoprotein antigens that are coded for by genes located in the DNA region that also contains the immune response genes (I region). *Abbr* Ia antigens.

Ir gene Immune response gene.

IRI 1. Immunoreactive insulin. 2. Insulin resistance index.

iridescent virus IRIDOVIRUS.

iridovirus A naked, icosahedral virus that contains double-stranded DNA and that is parasitic in a wide range of insects. Iridoviruses, when present in masses, show the property of iridescence (displaying colors like the rainbow).

IRMA Immunoradiometric assay.

iron An element that is essential to all animals and plants. Symbol, Fe; atomic number, 26; atomic weight, 55.847; oxidation states, +2, +3; most abundant isotope, ^{56}Fe; a radioactive isotope, ^{59}Fe, half-life, 45 days, radiation emitted, beta particles and gamma rays.

iron-binding globulin TRANSFERRIN.

ironophore SIDEROPHORE.

iron overload *See* hemosiderosis.

iron porphyrin HEME.

iron protein *See* nitrogenase.

iron–sulfur cluster An aggregate of iron and sulfur atoms in an iron–sulfur protein that mediates a one-electron transfer reaction. *Aka* iron–sulfur center.

iron–sulfur protein A protein in which iron is bound by means of sulfur-containing ligands of the protein. Three known types of iron–sulfur proteins are those in which (a) a single iron atom is linked to four cysteine sulfur atoms as in rubredoxin; (b) two iron atoms are linked to four cysteine sulfur atoms and to two bridging, acid-labile sulfur atoms as in plant-type ferredoxin; and (c) four iron atoms are linked to four cysteine sulfur atoms and to four bridging, acid-labile sulfur atoms as in bacterial-type ferredoxin and in high-potential iron–sulfur proteins. *Sym* FeS.

irradiation The exposure to radiation.

irradiation chimera *See* radiation chimera.

irreversible antagonist A ligand that binds to a receptor with high affinity, elicits no biological response, and subsequently becomes covalently linked to the receptor.

irreversible inhibitor   An inhibitor that binds to an enzyme (usually covalently) in an irreversible reaction so that the inhibition cannot be reversed by attempts to remove the inhibitor from the enzyme by such processes as dialysis or ultrafiltration. It may, however, be possible to restore the enzymatic activity by removing the inhibitor through a chemical reaction. *See also* active site-directed irreversible inhibitor; reactivation; reversible inhibitor.

irreversible process   A process in equilibrium thermodynamics in which a system goes from an initial equilibrium state to a final equilibrium state through stages that are not equilibrium states. In such a process, the net entropy change for the system plus its surroundings is greater than zero.

irreversible reaction   A chemical reaction that (a) proceeds to completion and cannot be reversed, (b) proceeds essentially to completion because of an equilibrium constant that is much larger than 1.0, or (c) is made to proceed overwhelmingly in one direction because of other factors.

irreversible shock   A shock that has progressed so far that it cannot be reversed even by the proper therapy.

irreversible thermodynamics   A branch of thermodynamics that deals with changes between nonequilibrium states of open systems.

irritability   EXCITABILITY.

ischemia   A localized anemia; a deficiency of blood in a tissue or an organ due to a functional constriction or an obstruction of a blood vessel. May be produced, for example, by the narrowing of arteries as a result of a spasm or disease.

IS element   INSERTION SEQUENCE (1).

Ising model   A model for interacting sites on a protein, enzyme, or other polymer, in which the sites are arranged in a linear array to form a one-dimensional lattice. *Aka* Ising chain.

islets of Langerhans   Clusters of hormone-secreting cells in the pancreas some of which, the alpha cells, secrete glucagon while others, the beta cells, secrete insulin.

iso-   1. Combining form meaning equal or like. 2. Prefix in the Cleland nomenclature of enzyme kinetics denoting a mechanism in which a stable enzyme form must isomerize to a different form before further reaction can occur. 3. Combining form meaning genetically different individuals of the same species. 4. Combining form meaning genetically identical individuals.

isoacceptor transfer RNA   One of two or more transfer RNA molecules that differ in their anticodon structure but are specific for the same amino acid. *Abbr* iso-tRNA. *Aka* isoac-cepting transfer RNA.

isoagglutination   An agglutination reaction of isoagglutinins.

isoagglutinin   An antibody that is formed in one individual and that causes the agglutination of cells derived from another individual of the same species.

isoagglutinogen   A cell surface antigen that can induce the formation of isoagglutinins (homologous antibodies) in another individual of the same species.

isoallele   An allele that is very similar to the normal allele and that can be distinguished from it only by means of special tests.

isoalloxazine   The heterocyclic ring structure that occurs in riboflavin and in the flavin nucleotides.

isoantibody   An antibody that is formed in one individual in response to antigens derived from another individual of the same species.

isoantigen   An antigen that is derived from one individual and that is immunogenic in other individuals of the same species.

isoascorbic acid   An isomer of ascorbic acid that has only slight vitamin C activity; erythorbic acid.

isobacteriochlorin   A specific porphyrin structure that occurs in siroheme.

isobar   1. A graph or an equation that describes changes of temperature or volume at constant pressure. 2. One of two or more nuclides that have the same mass but have a different number of protons, and hence possess different atomic numbers.

isobaric ions   Ions, as those produced in a mass sepctrometer, that have the same mass.

isobaric process   A process in which the pressure is maintained constant.

isobologram analysis   A graphical method for determining the extent and type of interaction between several drugs. Inspection of the plots allows deductions regarding the occurrence of antagonism, synergism, or lack of interaction. In the case of three drugs, one plots the concentration of each drug along the x, y, and z axes, respectively, and examines the shape of the surface thus formed.

isocapsidic virus   *See* segmented genome.

isocaudamers   One of a group of restriction endonucleases that recognize different nucleotide sequences but produce identical cohesive (sticky) ends. The term means "same tail."

isochore   A graph or an equation that describes changes in temperature or pressure at constant volume.

isochoric process   A process in which the volume is maintained constant.

isochromosome   A chromosome having two

identical arms that contain homologous loci but in reverse sequence.

isocitric acid An isomer of citric acid that is formed from *cis*-aconitic acid in the reactions of the citric acid cycle.

isocoding mutation A point mutation in DNA that leads to an altered, but synonym codon in mRNA. As a result, the same amino acid is incorporated into protein in response to the mutated codon as was incorporated in response to the original codon. *See also* silent mutation.

isocratic elution The chromatographic elution with one solvent of constant composition as opposed to a gradient elution.

isodense Of equal density.

isodensity equilibrium centrifugation DENSITY GRADIENT SEDIMENTATION EQUILIBRIUM.

isodesmic structure A structure of indefinite composition or description; a protein self-associating system in which aggregates of varying composition form, or a crystal structure in which no distinct groups are formed within the structure and no bond is stronger than any other, are some examples of isodesmic structures.

isodiametric Having equal diameters, as a cell that has essentially the same length and width.

isodynamic enzyme One of two or more different enzymes or enzyme forms that catalyze the same reaction but differ in their properties. *See also* heteroenzyme; isozyme; multiple forms of an enzyme; pseudoisoenzyme.

isodynamic law The generalization, enunicated by Rubner in 1878 and disproven since, that foodstuffs can replace each other in proportion to their caloric value; thus one food, containing a certain number of calories, is equivalent to any other food that contains the same number of calories, regardless of the composition of the foods.

isoelectric analysis ISOELECTRIC FOCUSING.

isoelectric condensation ISOELECTRIC FOCUSING.

isoelectric equilibrium electrophoresis ISOELECTRIC FOCUSING.

isoelectric focusing An electrophoretic technique for fractionating amphoteric molecules, particularly proteins, that is based on their distribution in a pH gradient under the influence of an electric field that is applied across the gradient. The molecules distribute themselves in the gradient according to their isoelectric pH values. Positively charged proteins are repelled by the anode and negatively charged proteins are repelled by the cathode; consequently, a given protein moves in the pH gradient and bands at a point where the pH of the gradient equals the isoelectric pH of the protein. The pH gradient is produced in a chromatographic column by the electrolysis of

amphoteric compounds and is stabilized by either a density gradient or a gel. *Abbr* IEF.

isoelectric fractionation ISOELECTRIC FOCUSING.

isoelectric point The pH at which a molecule has a net zero charge; the pH at which the molecule has an equal number of positive and negative charges, which includes those due to any ions bound by the molecule. The isoelectric pH is operationally defined as the pH at which the molecule does not move in an electric field. *Sym* pI. *Aka* isoelectric pH. *See also* isoelectrophoretic point; isoionic point.

isoelectric precipitation A fractional precipitation of proteins; a mixture of proteins is adjusted to the isoelectric pH of one of them so that all or most of the protein is precipitated.

isoelectric protein A protein at its isoelectric pH; a protein that has a net charge of zero.

isoelectric separation ISOELECTRIC FOCUSING.

isoelectric spectrum The distribution pattern of the proteins that are separated by isoelectric focusing.

isoelectronic Describing two atoms, two groups of atoms, two ions, or two molecules that have the same number and arrangement of valence electrons.

isoelectrophoretic point The pH at which the electrophoretic mobility of a protein is zero; this pH may coincide with the theoretical isoelectric pH of the protein, depending on the surface structure of the protein, the ionic strength, and the nature of the ionic double layer around the protein. *Aka* isoelectrophoretic pH. *See also* isoelectric point; isoionic point.

isoenzyme ISOZYME.

iso fatty acid A fatty acid that is branched at the penultimate carbon atom at the hydrocarbon end of the molecule.

isofocusing ISOELECTRIC FOCUSING.

isoforms Two or more isomeric forms of a protein or of a protein subunit; two or more molecular forms of functionally related proteins that differ only slightly in their structure. Examples of isoforms include the two forms of the α-catalytic subunit of the Na^+, K^+-ATPase and the two forms of ribonuclease (U_2A and U_2B) from the fungus *Ustilago*. In the latter case, the two enzymes differ in only one peptide bond involving either the α-(U_2A) or the β-(U_2B) carboxyl group of aspartic acid.

isofunctional enzymes Two or more different enzymes that catalyze the conversion of the same substrate to the same product. Such enzymes may be involved at the first step of a branched metabolic pathway and may be subject to feedback inhibition; the end product of one branch inhibits one of the enzymes while the end product of the second branch inhibits the other enzyme.

isogeneic SYNGENEIC.

isogenic Referring to genetically identical individuals.

isogenous Having the same origin.

isograft SYNGRAFT.

isohemagglutinin A hemagglutinin that has the capacity to react with isoantigens on the surface of red blood cells.

isohormone One of two or more different forms of the same hormone.

isohydric shift A set of two reactions, occurring in the red blood cell, whereby the oxygenation and deoxygenation of hemoglobin during respiration are linked to the reversible ionization of carbonic acid. As a result, the intracellular pH of the red blood cell remains essentially constant despite the fact that protons participate in both reactions. *Aka* isohydric transport; isohydric carriage.

isoimmunity The immunity acquired through reactions of antigens and antibodies that are derived from different individuals of the same species.

isoionic dilution The dilution of a polyelectrolyte solution with a salt solution that has the same ionic strength as the polyelectrolyte solution, so that the total concentration of mobile counterions remains constant.

isoionic point The pH at which the number of positive and negative charges of a protein that arise exclusively from proton exchange are equal to each other. The isoionic pH is operationally defined as either (a) the pH of a solution of isoionic protein in water, or (b) the pH of a protein solution that does not change with increasing concentrations of the protein. *Aka* isoionic pH. *See also* isoelectric point; isoelectrophoretic point.

isoionic protein A protein that has an equal number of protonated basic groups and deprotonated acidic groups; operationally defined as a protein from which all bound ions have been removed by electrodialysis or by mixed-bed ion-exchange chromatography. Such a protein contains no ions other than those arising from the dissociation of the solvent.

isokinetic gradient A density gradient in which the sedimenation coefficient, for all particles of a given density, is constant throughout the gradient. An isokinetic gradient can be approximated by a properly chosen exponential density gradient such that both the density and the viscosity of the gradient change linearly with concentration.

isokinetic sedimentation Sedimentation in an isokinetic gradient.

isolable Capable of being isolated. *aka* isolatable.

isolate To separate and purify a substance.

isolated rat uterus assay A bioassay for neurohypophyseal hormones that is based on measurements of the contractility of rat uterus.

isolated system A thermodynamic system that exchanges neither matter nor energy with its surroundings.

isolectins Closely related forms of an individual lectin; a group of closely related proteins the synthesis of which is under direct genetic control. Isolectins appear to be different aggregates of distinct subunits much as is the case for isoenzymes.

isoleucine An aliphatic, branched, nonpolar alpha amino acid that contains six carbon atoms. *Abbr* Ile; I.

isologous 1. HOMOLOGOUS (1,6). 2. SYNGENEIC.

isologous association An association of identical protein subunits in which the interacting surfaces of the subunits are identical.

isologous bonds The identical interactions between subunits linked via isologous associations.

isologous cell lines Cell lines derived from identical twins or from highly inbred animals.

isomer One of two or more compunds that have the same molecular composition but have different molecular structures and hence possess different properties.

isomerase An enzyme that catalyzes the interconversion of one isomer into another. *See also* enzyme classification.

isomeric Of, or pertaining to, isomers.

isomeric transition The transition of one nuclear isomer to another that is accompanied only by the emission of gamma rays from the nucleus, and during which there is no change in either the atomic mass or the atomic number of the nuclide *Abbr* IT.

isomerism The phenomenon in which two or more compounds have the same molecular composition but have different molecular structures and hence possess different properties.

isomerization The interconversion of one isomer into another; the formation of an isomer.

isomer number The total number of possible stereoisomers that have the same structural formula.

isometric Having the same dimensions.

isometric contraction The exertion of force by a muscle without a shortening in the length of the muscle; may be achieved by stimulating a muscle while maintaining it mechanically at constant length.

isometric virus A virus, the structure of which can be described by a symmetrical polyhedron of the isometric crystal system; this system is characterized by three identical axes at right angles to each other and includes the cube and the regular octagon.

isomorphic Morphologically alike; having the same shape and structure. *Aka* isomorphous.

isomorphous replacement A method for solving the phase problem in an x-ray diffraction study of a protein by using a heavy atom derivative of the protein. The introduction of the heavy atom or atoms into the protein must be isomorphous, so that there is no change in the conformation of the protein nor any change in the size or in the symmetry of the unit cell. A comparison of the diffraction patterns of the original protein and of its heavy metal derivative permits the localization of the heavy atom in the unit cell and hence permits a determination of phase angles. The determination of all the phase angles requires the use of several heavy atom derivatives.

isoniazid Isonicotinic acid hydrazide; a synthetic antibiotic that is active against *Mycobacterium tuberculosis*. It also has antivitamin activity against both nicotinic acid and vitamin B_6. *Abbr* INH.

isonicotinic acid hydrazide ISONIAZID.

isoosmolar solution ISOTONIC SOLUTION.

isoosmotic ISOSMOTIC.

isopeptide bond A covalent amide bond formed between the epsilon amino group of lysine and the side chain carboxyl group of aspartic or glutamic acid. The bond has been found in polymerized fibrin and in native wool; it is not hydrolyzed by human proteases but is hydrolyzed by bacterial proteases in the intestine.

isophane insulin NPH INSULIN.

isophile antibody An antibody that reacts only with antigens that are derived from the same species.

isopiestic technique A technique for measuring the binding of solute and/or solvent molecules to a macromolecule in a solution in which only one of the solvent components, generally water, is volatile. A number of solutions of known initial composition, and all containing the same volatile solvent component, are allowed to equilibrate with each other with respect to the activity of the volatile solvent component. Equilibration may be carried out in a vacuum desiccator and the equilibrium compositions of the solutions are determined by weighing; any change in composition from an initial to an equilibrated solution is due to the transfer of the volatile solvent component.

isoplith ISOSTICH.

isopolar substitution The replacement of one amino acid in a protein by another amino acid of similar polarity; the replacement of a polar (nonpolar) amino acid by another polar (nonpolar) amino acid. Many mutations lead to isopolar substitutions as a result of the degeneracy of the genetic code.

isoprene The five-carbon compound, 2-methyl-1,3-butadiene, that occurs in the structure of several biochemically important compounds, including coenzyme Q, vitamin A, and vitamin K.

isoprene rule The principle that the hydrocarbon skeletal structure of open-chain and cyclic terpenes can be considered to have been constructed by head-to-tail joining of isoprene units.

isoprenoid 1. *n* A compound containing two or more isoprene units or derivatives of isoprene units. 2. *adj* Of, or pertaining to, isoprene.

isopropylthiogalactoside A gratuitous inducer of the enzyme beta galactosidase. *Abbr* IPTG.

isoproteins Multiple forms of a specific protein. Isoproteins may be coded by a single genetic locus (allelic isoproteins), such as abnormal hemoglobins and major histocompatibility complex antigens, or they may be coded by multiple genetic loci (nonallelic isoproteins), such as the lactate dehydrogenase, hexokinase, actin, and globin families of proteins.

isoprotic point ISOIONIC POINT.

isopycnic Having the same density, specifically buoyant density.

isopycnic gradient centrifugation DENSITY GRADIENT SEDIMENTATION EQUILIBRIUM.

isoquinoline alkaloids *See* alkaloids.

isorheic Of, or pertaining to, any period during gas chromatography in which there is a constant flow rate of carrier gas.

isorotation rule The rule, proposed by Hudson in 1902, which states that the optical rotation of a carbohydrate can be approximated by a sum of two factors; one factor represents a contribution due to the anomeric carbon, and the other factor represents a contribution due to the rest of the molecule. The rule further states that these two contributions are approximately independent of each other and have similar values for similar molecules.

isosbestic point The wavelength at which the absorptivity of two or more compounds is identical; the wavelength at which the absorption spectra of two or more compounds intersect when the absorbance is measured for solutions of equimolar concentrations.

isoschizomer One of a group of restriction endonucleases, isolated from different organisms, that recognize the same DNA base sequence for cleavage but do not necessarily cleave the DNA at the same position within that sequence.

isosemantic substitution The incorporation of an amino acid into a polypeptide chain that results from a mutation in which the normal codon of an amino acid has been mutated to a synonym codon; consequently, the amino acid that is incorporated into the polypeptide chain is identical to the normal one, but is incorporated in response to one of its synonym codons.

isosmotic Having the same osmotic pressure.

isosmotic solution ISOTONIC SOLUTION.

isostere 1. A compound that has apparently different chemical characteristics from those of a naturally occurring essential metabolite, but that can substitute for the metabolite because it fits a specific binding site for that metabolite; the binding site may be the active site of an enzyme, the receptor site of a hormone, or a similar site. 2. One of two or more compounds that are similar in their physical properties and that have the same number and arrangement of valence electrons; compounds such as CO_2 and N_2O or N_2 and CO are examples.

isosteric Of, or pertaining to, isosteres.

isostich A cluster of either purine or pyrimidine oligonucleotides, as one of the clusters obtained by the hydrolysis of apurinic or apyrimidinic acid. *Var sp* isostiche.

isotachophoresis An electrophoretic technique in which the sample is introduced into a capillary tube between a leading electrolyte and a terminator electrolyte that have mobilities that are, respectively, greater and smaller than those of any of the sample ions; a displacement analysis in which the separation and the sharpness of the fronts between two substances are a function of the properties of the substances, and not of their concentration in the original sample. *Abbr* ITP.

isotactic polymer A polymer in which the monomers have been polymerized in a stereospecific manner so that all the R groups of the monomers are on one side of the plane which contains the main chain.

isotherm A graph or an equation that describes changes of volume or pressure at constant temperature. *See also* adsorption isotherm.

isothermal process A process in which the temperature is maintained constant.

isotocin A peptide hormone of nine amino acids that is secreted by the posterior lobe of the pituitary gland of bony fishes and that is related to oxytocin in its structure and in its function.

isotone One of two or more nuclides that have the same number of neutrons but have a different number of protons and hence possess different atomic numbers.

isotonic contraction 1. An alteration in the water and electrolyte balance in the body in which there is an increase in the volume of the extracellular fluid and an equivalent increase in electrolyte, so that the osmotic pressure of

the extracellular fluid remains unchanged. 2. The exertion of force by a muscle that is accompanied by a shortening in the length of the muscle.

isotonic expansion An alteration in the water and electrolyte balance in the body in which there is a decrease in the volume of the extracellular fluid and an equivalent decrease in electrolyte, so that the osmotic pressure of the extracellular fluid remains unchanged.

isotonic solution A solution that has the same osmotic pressure as another solution.

isotope One of two or more nuclides of the same element that have the same number of protons and the same atomic number, but that have a different number of neutrons and different atomic masses.

isotope derivative method A quantitative analytical method that may be used as an enzyme assay; based on adding an isotopically labeled compound to a sample, reacting it with the sample, and determining the labeled derivative that is formed.

isotope dilution analysis The quantitative determination of a substance by means of isotopes. *See also* direct isotope dilution analysis; double isotope dilution analysis; inverse isotope dilution analysis; modified inverse isotope dilution analysis.

isotope effect The effect of an isotope on the rate and/or the mechanism of a reaction. *See also* primary isotope effect; secondary isotope effect; kinetic isotope effect.

isotope exchange The replacement of one isotope in a compound by another. *See also* deuterium exchange.

isotope incorporation The introduction of an isotope into a compound by synthesizing the compound in the presence of one or more labeled precursors.

isotopically enriched Descriptive of a substance in which the relative amounts of one or more isotopes has been increased.

isotopic competition A method for demonstrating that compound X is a precursor of metabolite Y. A labeled form of X is added to the in vivo, or the in vitro, system in the presence of a general, unlabeled carbon source, and the extent of label in Y is then determined. The appearance of label in Y indicates that X is a direct precursor of Y.

isotopic dilution analysis *See* isotope dilution analysis.

isotopic enrichment *See* isotopically enriched.

isotopic label The isotope that serves to label a compound.

isotopic tracer TRACER (1).

isotopic trapping A method for demonstrating that compound X is an intermediate in a metabolic pathway. A labeled precursor of the pathway is added to the in vivo or the in vitro system, followed by the addition of large amounts of unlabeled X. The appearance of label in X indicates that it is an intermediate of the pathway at a position subsequent to that of the labeled precursor.

iso-tRNA   Isoacceptor transfer RNA.

isotropic   Of, or pertaining to, isotropy.

isotropic band   I BAND.

isotropy   The constancy in the physical properties of a substance, regardless of the direction in which these properties are measured. *Aka* isotropism.

isotype   One of a group of identical antigenic determinants of a given serum protein or immunoglobulin that occur in all individuals of the same species. *See also* idiotype; allotype.

isovalent resonance   Resonance in which the various resonance structures have the same number of bonds.

isovaleric acidemia   A genetically inherited metabolic defect in humans that is characterized by elevated levels of α-ketoisovaleric acid in blood and urine, and that is due to a deficiency of the enzyme isovaleryl CoA de-

hydrogenase. *Aka* isovaleric aciduria.

isozyme   1. One of two or more isodynamic enzymes derived from a single homogeneous biological source; isozymes may occur within a single species, a single organism, or a single cell. *Aka* pseudoisoenzyme. 2. One of two or more multiple forms of an enzyme that arise from genetically determined differences in primary structure; excludes enzymes derived by modification of the same primary structure. *Aka* isoenzyme. *See also* heteroenzyme; isodynamic enzyme; multiple forms of an enzyme; pseudoisoenzyme.

IT   Isomeric transition.

iterative   Marked by, or involving, repetition or reiteration.

ITP   1. Inosine triphosphate. 2. Inosine-5'-triphosphate. 3. Isotachophoresis.

IU   International unit.

IUB   International Union of Biochemistry.

IUPAC   International Union of Pure and Applied Chemistry.

iv   1. *adj* Intravenous. 2. *adv* Intravenously.

IVS   Intervening sequence; intron.

J

J 1. Joule. 2. Coupling constant.

Jablonski diagram The energy diagram of a molecule in which discrete molecular energy states are indicated by a series of horizontal and parallel lines.

Jacob and Monod hypothesis *See* operon hypothesis.

J chain A third type of immunoglobulin chain (in addition to light and heavy chains) found in molecules of the IgA and IgM type; a polypeptide (MW 15,000) that serves to join (hence the term J chain) monomeric units in the multimeric structures of IgA and IgM.

Jerne plaque technique *See* plaque technique.

jet lag The desynchronization of biological rhythms that occurs when an individual crosses a number of time zones rapidly (as in a jet), particularly when traveling west to east. Recovery requires several days and complete reestablishment of the normal circadian rhythms usually requires about a week.

J genes A group of genes that code for parts of the light and heavy immunoglobulin chains in humans and mice. The term derives from the fact that these genes help join one of the genes for the variable region to one of the genes for the constant region; hence, J genes play an important role in the mechanism leading to antibody diversity.

JH Juvenile hormone.

Jimpy mice Mice having the Quaking mutation.

Jimpy mutation An X-linked recessive mutation in mice that produces a myelin-deficient animal.

Johnston–Ogston effect The changes in the values of the sedimentation coefficients and of the apparent concentrations that are obtained for two or more components when they are present in a mixture as compared to the values obtained when each component is present alone. The effect is due to the dependence of the sedimentation coefficient on concentration, and it leads to a decrease in the observed sedimentation coefficients for both the slow and the fast components in a mixture. The apparent concentration, which is proportional to the area under the ultracentrifuge peak, will be increased for the slow component and will be decreased for the fast component.

joinase LIGASE.

joining enzyme 1. LIGASE. 2 DNA LIGASE.

joining genes J GENES.

joint transduction LINKED TRANSDUCTION.

joint transformation LINKED TRANSFORMATION.

Jones–Mote reaction A delayed-type hypersensitivity reaction of the skin that is of low intensity and that is produced only by several daily injections of antigen.

joule The standard unit of work, energy and heat that is equal to 10^7 erg. The proportionality between joules and calories is as follows: 1 calorie = 4.184 joules. *Sym* J.

jumping gene TRANSPOSABLE ELEMENT.

jump method *See* temperature jump method; pH jump method.

junction *See* cell junction.

junctional complex Any specialized region of intercellular adhesion that can be discerned with the electron microscope; a desmosome is an example.

junctional sliding The change in the location of the junction between an intron and an exon that is brought about by an increase or a decrease in the length of the exon, and that is observed for different members of a given gene family. Junctional sliding, exon shuffling, and intron intrusion have been proposed to help explain the evolutionary diversification of genes.

junction potential The potential that arises either across the junction between two half cells or across the boundary between two solutions of different concentrations; it is due to differences in diffusion rates of the ions on the two sides.

junk DNA 1. Unusual DNA, containing branches, which can arise during the in vitro replication of DNA by DNA polymerase I (Kornberg enzyme). 2. SELFISH DNA.

justification hypothesis A hypothesis that attributes mental retardation in phenylketonuria (PKU) to an inability of the heterozygous mother to deliver an appropriate amount of tyrosine to the PKU fetus who, in turn, is unable to correct for this deficiency because of its genetic constitution.

juvenile hormone One of a group of insect hormones, composed of modified linear isoprenoid units, that promote larval development; produced by neuroendocrine structures known as corpora allata and hence also referred to as allatum hormone. *Abbr* JH.

juvenile-onset diabetes *See* diabetes.

K

k 1. Kilo. 2. Boltzmann constant. 3. Rate constant.

K 1. Equilibrium constant. 2. Degree Kelvin. 3. Potassium. 4. Lysine. 5. Abbreviation for kilobyte (1000 bytes) which, actually, contains 1024 bytes.

K_0 Average intrinsic association constant.

$K_{0.5}$ The substrate concentration at which one-half of the maximum velocity for an allosteric enzyme is obtained. *See also* K_m.

K' Apparent equilibrium constant; an equilibrium constant that is based on molar concentrations, as opposed to the thermodynamic equilibrium constant K that is based on activities.

K_a Dissociation constant of an acid.

KAF Conglutinogen activating factor; an enzyme, present in serum, that modifies the bound C3b component of complement in conglutination.

kairomone *See* allomone.

kallidin *See* kinin.

kallidin I LYSYL-BRADYKININ.

kallidin II BRADYKININ.

kallikrein One of a group of plasma serine proteases that catalyze the formation of kinins from kininogens. Kallikrein can also initiate the intrinsic pathway of blood clotting.

kanamycin An aminoglycoside antibiotic, produced by *Streptomyces kanamyceticus* It inhibits protein synthesis by binding to the 30S ribosomal subunit and causes misreading of the genetic code.

K antigens Polysaccharide antigens of bacterial capsules.

kaolin A clay that consists principally of hydrated aluminum silicate.

kappa RECIPROCAL ION-ATMOSPHERE RADIUS.

kappa chain One of the two types of light chains of the immunoglobulins.

kappa particles *See* killer paramecia.

karyogram A photographic representation of the chromosome complement of an individual according to the size and/or a numbering system of the chromosomes.

karyokinesis The division of the cell nucleus in either mitosis or meiosis.

karyolymph NUCLEOPLASM.

karyon NUCLEUS (1).

karyoplasm NUCLEOPLASM.

karyosome 1. A nucleolus or a nucleolus-like body. 2. A subnuclear body, found in *Dro-*

sophila and in some plant species, that contains DNA.

karyotheca NUCLEAR ENVELOPE.

karyotype 1. The sum of the characteristics (size, number, shape, etc.) of the chromosomes of an individual or a species. 2. The sum of the characteristics of the somatic metaphase chromosomes of an individual or a species, frequently described by photomicrographs arranged as in an idiogram.

kasugamycin An antibiotic, produced by *Streptomyces kasugaensis*, that inhibits the binding of tRNA^{fmet} to ribosomes in prokaryotes.

kat Katal.

μkat Microkatal.

katabolism Variant spelling of catabolism.

katal The amount of enzymatic activity that converts one mole of substrate per second; the katal (kat) is related to an enzyme unit (U) by the relation 1 kat = 6×10^7 U. In some cases "conversion of 1 mol of substrate" is equivalent to the number of reaction cycles which equals the number of carbon atoms in 0.012 kg of the nuclide ^{12}C.

kat F unit A unit of catalase activity no longer in use and equal to the unimolecular rate constant divided by the number of grams of enzyme under specified conditions.

katharometer A thermal conductivity detector that is similar to a thermistor and that is used for determining the composition of gas mixtures in gas chromatography and in studies of basal metabolism.

K_b Dissociation constant of a base.

kb Kilobase; a unit of length in nucleic acids equal to 1000 nucleotides.

KB cells An established cell line of cells derived from a human carcinoma in 1954 and maintained in tissue culture since then.

kbp Kilobase pairs; a unit of length in nucleic acids equal to 1000 base pairs.

kcal Kilocalorie; 1000 (small) calories; a large calorie.

K-capture 1. The capture of a K-shell electron by an atomic nucleus. 2. ELECTRON CAPTURE.

k_{cat} The catalytic rate constant of an enzyme reaction; the turnover number, which is equal to $V_{max}/[E]$ where V_{max} is the maximum velocity and [E] is the molar concentration of active sites on the enzyme.

K cells Killer cells.

kd, kD, kDa Kilodalton; 1000 daltons.

Keilin–Hartree particles A particulate submitochondrial preparation from heart muscle that can carry out electron transport, but cannot carry out oxidative phosphorylation.

Kelvin temperature scale ABSOLUTE TEMPERATURE SCALE.

kemptide A synthetic heptapeptide substrate (leu-arg-arg-ala-ser-leu-gly) that contains an exposed serine residue which can be phosphorylated enzymatically. *Aka* phosphate acceptor peptide.

Kendall's compound A designation for some steroids; Kendall's compounds B, E, and F refer, respectively, to corticosterone, cortisone, and cortisol.

K_{eq} Equilibrium constant.

keratan sulfate A glycosaminoglycan in which the repeating disaccharide contains D-galactose and *N*-acetyl-D-glucosamine and variable amounts of sulfate groups. It occurs in cornea, cartilage, and intervertebral disks.

keratin One of a group of diverse fibrous scleroproteins occurring in hair, wool, nails, and other epidermal structures.

keratin filaments Intermediate filaments, composed of various keratins and characteristic of epithelial cells.

keratinization The formation of keratin-rich horny sections and skin appendages.

keratohyalin One of several proteins that constitute the matrix in which keratin fibers are embedded; the resultant fiber–matrix complex is found in the epidermis, hair, nails, and related structures.

kernicterus A pathological condition in newborn humans that is characterized by discoloration and degeneration of brain tissue and by elevated serum levels of unconjugated bilirubin.

Kerr effect ELECTRIC BIREFRINGENCE.

ketal A compound derived from a ketone and two alcohol molecules by splitting out a molecule of water.

ketimine An organic compound that has the general formula

$$\begin{array}{c} R' \\ | \\ R\!-\!C\!=\!NH \end{array}$$

keto- 1. Combining form meaning ketose. 2. Combining form meaning ketone.

ketoacid An organic acid that also carries a ketone group.

ketoacidosis An acidosis that is brought about by the accumulation of ketone bodies; a pathological state that is characterized by both ketosis and acidosis.

β-ketoadipate pathway A branched pathway in bacteria that leads to succinyl-coenzyme A, by way of β-ketoadipate, from either *cis,cis*-muconate or from β-carboxy-*cis,cis*-muconate.

keto-enol tautomerism The tautomerism that is due to a shift of a hydrogen atom so that one of the isomers is a ketone and the other is an enol.

ketofuranose A ketose in furanose form.

ketogenesis The formation of ketone bodies.

ketogenic amino acid An amino acid that can serve as a precursor of acetyl coenzyme A and ketone bodies in metabolism.

α-ketoglutarate dehydrogenase A multienzyme system that is similar in its structure and in its properties to the pyruvate dehydrogenase system and that catalyzes the oxidation of α-ketoglutaric acid to succinic acid in the citric acid cycle.

α-ketoglutarate pathway The catabolic pathway whereby arginine, proline, histidine, glutamine, and glutamic acid enter the citric acid cycle by way of α-ketoglutaric acid.

α-ketoglutaric acid A dicarboxylic acid that is an intermediate in the citric acid where it is formed from isocitric acid.

ketol moiety GLYCOLALDEHYDE GROUP.

ketone An organic compound that contains a ketone group.

ketone body One of the three compounds, acetoacetic acid, acetone, and β-hydroxybutyric acid, that arise from acetyl coenzyme A and that may accumulate in excessive amounts as a result of starvation, diabetes mellitus, or other defects in carbohydrate metabolism.

ketone group The carbonyl group attached to two carbons; the grouping

$$\begin{array}{c} O \\ \| \\ -\!C\!- \end{array}$$

ketonemia The presence of excessive amounts of ketone bodies in the blood.

ketonuria The presence of excessive amounts of ketone bodies in the urine.

ketopyranose A ketose in pyranose form.

ketose A monosaccharide, or its derivative, that has a ketone group.

ketosis The presence of excessive amounts of ketone bodies in the body.

ketosteroid One of a group of compounds that are known as either 17-ketosteroids or neutral-17-oxosteroids and that represent degradation products of steroids; they are excreted in the urine and provide an index of androgen production in the body. Major ketosteroids are androsterone, dehydroisoandrosterone, and etiocholanolone.

ketostix test A rapid, semiquantitative test for the determination of ketone bodies in urine or serum by means of impregnated paper strips.

keto sugar A carbohydrate that carries a ketone group.

ketotic Of, or pertaining to, ketosis.

ketotic hyperglycinemia A syndrome that is characterized by severe ketoacidosis and hyperammonemia in addition to hyperglycinemia and hyperglycinuria; it is associated with a deficiency of an enzyme, other than that of the glycine cleavage system, such as propionyl-CoA carboxylase or methylmalonyl-CoA mutase.

keV Kiloelectronvolt; 1000 eV.

key enzyme An enzyme that is unique to a metabolic pathway which has several enzymes in common with other pathways.

kg Kilogram; 1000 g.

K_h Hydrolysis constant.

K_i Inhibitor constant.

kidney stone *See* calculus.

kieselguhr A fine-grain diatomaceous earth.

Kiliani–Fischer synthesis The reaction whereby a one-carbon fragment is added to a carbonyl group by means of cyanide, as in the extension of a monosaccharide from a pentose to a hexose.

killed vaccine A vaccine that consists of originally infectious bacteria or viruses that have been rendered noninfectious; a vaccine of either killed bacteria or of inactivated viruses.

killer cells Lymphocyte-like cells, which are not B or T lymphocytes, that kill a variety of tumor cells and virus-infected cells but only after previous immunization. *See also* natural killer cells; ADCC.

killer paramecia Strains of *Paramecium aurelia* that are capable of killing other (sensitive) strains of *Paramecia*. The killer strains possess kappa particles in their cytoplasm; these are released and ingested by the sensitive strains, leading to their death. Kappa particles are symbiotic bacteria and those with killing activity carry defective DNA phages.

killer particle A double-stranded RNA plasmid that occurs in killer strains of the yeast *Saccharomyces cerevisiae*. The killer particle contains genes for synthesis of a killer toxin and is the only known plasmid that does not contain DNA.

killer plasmid KILLER PARTICLE.

killer strain A strain of cells that kills sensitive cells of the same species; such strains have been isolated from both yeast and *Paramecium*. The former secrete a toxic substance (killer toxin) and the latter secrete a toxic particle (kappa particle).

killer T cells CYTOTOXIC T CELLS.

killer toxin The toxic protein encoded by the double-stranded RNA of the killer particle in yeast (*Saccharomyces cerevisiae*).

killing efficiency INACTIVATION PROBABILITY.

killing titer The titer of a phage suspension that is determined from the number of bacterial cells before infection and the number of cells surviving after infection.

kilo- Combining form meaning one thousand and used with metric units of measurements. *Sym* k.

kilobase A unit of length equal to either 1000 bases in a single-stranded nucleic acid (kb) or 1000 base pairs in a double-stranded nucleic acid (kbp).

kilobase pair *See* kbp.

kilobyte *See* K (5).

kilocalorie A large calorie; 1000 small calories. *Abbr* kcal; Cal.

kinase An enzyme that catalyzes the transfer of a phosphoryl group from ATP, and occasionally from other nucleoside triphosphates, to another compound.

kinematic viscosity The dynamic viscosity of a liquid divided by the density of the liquid. The unit of kinematic viscosity is the stoke.

kinesis The movement of an organism in response to a stimulus such that the latter controls the rate (not the direction) of movement. *See also* taxis.

kinetic Of, or pertaining to, kinetics.

kinetic advantage The phenomenon that hexoses, combined in a disaccharide, and amino acids, combined in a dipeptide, are taken up faster by the intestine than the free monosaccharides or the free amino acids.

kinetic analysis The study of a chemical reaction by making kinetic measurements. The velocity of the reaction is measured under various conditions for the purpose of deducing possible reaction mechanisms.

kinetic assay CONTINUOUS ASSAY.

kinetic coefficient A rate constant that depends on the concentration of either a reactant or a product.

kinetic complexity DNA COMPLEXITY.

kinetic constant RATE CONSTANT.

kinetic energy correction A correction term that is applied to viscosity measurements made in a capillary viscometer and that takes into account the kinetic energy that is acquired by the liquid during its flow through the capillary.

kinetic equation RATE EQUATION.

kinetic isolation The state of a chemical reaction in which it is not affected by changes in another reaction that ties in with it; may be produced, for example, by the preceding reaction either discharging a product at zero concentration or accepting a substrate that is already present at a saturation level.

kinetic isotope effect The effect of an isotope on the rate of a reaction that is due to the differences in mass of the isotopes in question.

kinetic law RATE EQUATION.

kinetic pK A pK that does not represent a real ionization step but rather is composed of a ratio of rate constants; a kinetic pK occurs in enzymology when there is a change of a rate-determining step with pH.

kinetic proofreading A proofreading mechanism that permits a discrimination between correct and incorrect enzyme substrates (or intermediates) on the basis of kinetic considerations. The distinctive characteristic of kinetic proofreading is that there is no hydrolytic site on the enzyme at which hydrolysis of undesired intermediates would take place. Instead, it is postulated that an incorrect intermediate diffuses away from the enzyme and into solution where it is hydrolyzed nonenzymatically at a faster rate than it reacts to give products on the surface of the enzyme. *See also* proofreading; proofreading function.

kinetics 1. The science that deals with the rate behavior of physical and chemical systems. 2. The rate behavior of a physical or a chemical system.

kinetin 6-Furfurylaminopurine; a cytokinin.

kinetochore 1. CENTROMERE. 2. One of two faces of the centromere that point toward the spindle poles and to which the spindle fibers attach.

kinetogene A plasmagene that is attached to a kinetosome.

kinetoplast 1. An organelle, located within the mitochondria of trypanosomes, that consists of linear and circular DNA and that is believed to code for some mitochondrial components; it is located near the kinetosome. 2. The specialized mitochondrion of trypanosomes that is located near the kinetosome.

kinetosome The self-duplicating cytoplasmic organelle to which a cilium or a flagellum is attached. Kinetosomes are responsible for the synthesis of cilia and flagella.

King–Altman procedure An algorithmic method for deriving complicated steady-state rate equations for enzyme reactions. The method involves construction of a reaction scheme (master pattern) in which all relevant enzyme forms (free and complexed) are connected by appropriate arrows, and the latter are annotated with the corresponding unitary rate constants. The method provides a shortcut for obtaining expressions for the concentrations of various enzyme species and these expressions are then substituted into an appropriate equation for the net steady state velocity.

kinin One of a group of vasoactive peptide hormones that are formed from kininogens by the action of enzymes known as kallikreins.

Kinins are potent vasodilators, leading to an increase in the diameter of blood vessels, hypotension, and increased capillary permeability. Bradykinin is a nonapeptide and kallidin (lysyl bradykinin) is a decapeptide.

kininase A peptidase acting on kinin; two are known (I and II).

kininogen One of a group of plasma globulins that are precursors of the vasoactive kinins. They are secretory proteins, synthesized primarily in the liver, and have a number of other functions in addition to their role as kinin precursors. *See also* cystatins.

kininogenase KALLIKREIN.

K_{int} Intrinsic dissociation constant.

K_{ion} Ionization constant.

Kirkwood–Shumaker interactions The attractive charge-fluctuation interactions between ionized macromolecules that result from fluctuations in their dipole moments caused by the movement of protons over the surfaces of the macromolecules. Since at any instant only a random fraction of the ionizable groups on the macromolecule are fully ionized, the protons move in random fashion from one group to another; for example, from such groups as $—NH_3^+$ to such groups as $—CO_2^-$ in the case of a protein. The fluctuations in the dipole moment of one molecule then induce fluctuations in the dipole moment of a neighboring molecule which results in an attractive interaction between the two molecules.

Kirschner value The number of milligrams of butyric acid in the fraction of volatile water-soluble fatty acids obtained from 5 g of fat.

Kjeldahl method A procedure for determining protein nitrogen by digesting the sample with concentrated sulfuric acid in the presence of a catalyst. The protein nitrogen is converted to ammonia that is distilled over into standard acid, and the excess acid is then titrated. The Kjeldahl method is used especially for the standardization of other protein determinations. *Aka* Kjeldahl digestion.

Kjeldahl nitrogen The protein nitrogen determined by means of the Kjeldahl method.

Kleinschmidt technique A technique for preparing monmolecular films of DNA for electron microscopy in which the DNA is stabilized by complexation with a basic protein, such as cytochrome c.

Klenow fragment A fragment of DNA polymerase I (Kornberg enzyme) which contains the DNA polymerizing activity and the $3' \rightarrow 5'$ exonuclease activity but lacks the $5' \rightarrow 3'$ exonuclease activity of the intact enzyme. It is prepared by the proteolytic digestion of DNA polymerase I and has been used for DNA sequencing. *Aka* Klenow polymerase; Klenow enzyme.

Klotz plot A plot of binding data in which the average binding number (v) is plotted as a function of the logarithm of the free ligand (S) concentration; (a plot of v versus log[S]). *Aka* Bjerrum formation function.

K_m Michaelis constant.

Knallgas bacteria Bacteria of the genus *Hydrogemonas* which utilize the Knallgas reaction.

Knallgas reaction The reaction between hydrogen and oxygen, forming water, that is utilized as an energy-yielding reaction by bacteria of the genus *Hydrogemonas*.

KNF model Koshland, Nemethy, and Filmer model; *See* sequential model.

knife and fork model KORNBERG MECHANISM.

knife breaker An apparatus for breaking glass into sharp pieces which are then used as knife edges to cut tissue specimens for electron microscopy.

Knoop's hypothesis The hypothesis, proposed by Knoop in 1905, that fatty acids are oxidized in metabolism by means of successive removals of two-carbon fragments in the form of acetic acid. *See also* beta oxidation.

Knop's solution A solution that contains the major inorganic constituents that are required for the growth of plant cells: calcium nitrate, potassium chloride, magnesium sulfate, and potassium dihydrogen phosphate. The solution will support plant growth if it is supplemented with a carbon source, trace elements, vitamins, etc.

Koagulations Vitamin VITAMIN K.

Koch phenomenon A delayed-type hypersensitivity reaction in which skin inflammation to injected tubercle bacilli is more intense in a previously infected individual than in a noninfected one.

Koettstorfer number SAPONIFICATION NUMBER.

Kok effect The phenomenon in which the quantum efficiency above the compensation point differs from that below it. *See also* compensation point.

Kornberg enzyme The DNA-dependent DNA polymerase, first isolated by Kornberg in 1958 from *E. coli*, that is believed to function primarily in the repair–synthesis of DNA and secondarily in DNA replication. *Aka* DNA pol I.

Kornberg mechanism A mechanism for the in vitro replication of DNA that requires the action of three enzymes, endonuclease, DNA polymerase, and DNA ligase, and allows for the essentially simultaneous replication of both strands of an antiparallel duplex DNA.

Koshland, Nemethy, and Filmer model SEQUENTIAL MODEL.

Kostoff genetic tumor A plant tumor that develops spontaneously in certain interspecific

hybrids of the genus *Nicotiana*.

Kovats retention index system A system for characterizing retention volumes in gas chromatography in which the retention volume of a compound is compared with the retention volumes of a series of saturated aliphatic hydrocarbons, chromatographed on the same column as that used for the compound.

Krabbe's disease A genetically inherited metabolic defect in humans that is associated with mental retardation and that is characterized by an accumulation of galactocerebrosides and by the almost complete absence of myelin; due to a deficiency to the enzyme galactosyl ceramide β-galactosidase. *Aka* Krabbe's leukodystrophy.

Krafft temperature The temperature at which the solubility of a surfactant is sufficient to achieve the critical micelle concentration. Since unassociated surfactant molecules have limited solubility while micelles are highly soluble, the solubility of a surfactant increases sharply above the Krafft temperature.

Kranz anatomy The arrangement of mesophyll and bundle sheath cells in C_4 plants.

Krebs cycle CITRIC ACID CYCLE.

Krebs fluid A manometer fluid containing NaBr, Triton X-100, and Evan's blue; it has a density of 1.033 g/mL.

Krebs–Henseleit cycle UREA CYCLE.

Krebs–Kornberg cycle GLYOXYLATE CYCLE.

Krebs–Ringer solution A modified Ringer's solution that contains magnesium sulfate and phosphate buffer in addition to the other components of Ringer's solution.

K region A bond in an aromatic hydrocarbon that is active in addition reactions; the presence of such bonds has been correlated with the carcinogenic activity of some hydrocarbons. *See also* L region.

kringle A triple-looped, disulfide-cross-linked, domain that may occur once or several times in a given protein. Kringles have been found in many plasma proteinases and they are believed to play a role in binding involving membranes, proteins, or phospholipids and in the regulation of the proteolytic activity of enzymes.

Kronig–Kramer transformation An equation that permits the interconversion of circular dichroism data and optical rotatory dispersion data.

K_s Substrate constant.

K system A system composed of a regulatory enzyme and an effector, such that the effector alters the substrate concentration at which one-half of the maximum velocity of the reaction is obtained but does not alter the value of the maximum velocity.

Kuhn statistical length The length of a polymer

that is equal to twice its persistence length.

Kunitz inhibitor A soybean trypsin inhibitor.

Kupffer cells Liver macrophages.

kurtosis The extent to which a frequency distribution is "peaked"; the extent to which the curve shows a steeper or shallower ascent in the neighborhood of the peak relative to a normal distribution with the same parameters.

K_w Ion product of water.

kwashiorkor A disease caused by nutritional deficiency, primarily protein deficiency, that occurs in children between the ages of one and three. The disease generally occurs in children of underdeveloped areas of the tropical belt when the children are weaned from the breast and are placed on a low-protein diet.

Kwok's disease CHINESE RESTAURANT SYNDROME.

kymograph An apparatus that consists of a smoked drum and that is used for recording physiological responses.

kynureninase An enzyme that functions in tryptophan metabolism and requires pyridoxal phosphate as a coenzyme.

kynurenine An intermediate in the catabolism of tryptophan that is excreted in the urine in cases of vitamin B_6 deficiency. *See also* tryptophan dioxygenase.

L

l 1. Levorotatory. 2. Liter.

L 1. L-Configuration. 2. Leucine. 3. Linking number. 4. Designating a ribosomal protein from the large ribosomal subunit. 5. Designating the three-dimensional structure of tRNA. 6. Liter.

LAA *See* LATS-P.

label 1. A radioactive or a stable isotope that is introduced into a molecule. 2. A group of atoms or a molecule that is linked covalently to another molecule for purposes of identification.

labeled compound A compound containing a label.

label triangulation method A method for determining the quaternary structure of protein complexes. The subunits of the complex are separated, labeled with heavy atoms, and allowed to aggregate and reform the complex. The distances between the subunits are then measured by some technique such as x-ray diffraction, neutron diffraction, or spectroscopy.

labile Unstable; readily undergoing change.

labile factor PROACCELERIN.

labile methyl group A methyl group that can be transferred from one compound to another by transmethylation, such as the methyl group attached to either a nitrogen or a sulfur atom of certain organic compounds.

labile phosphate group A phosphate group that is readily liberated from a compound by hydrolysis at 100°C in 1 N HCl for 7 to 10 min. The terminal phosphate group of nucleoside triphosphates and the phosphate groups of pyrophosphate fall into this category; the phosphate group of ordinary esters generally requires longer times for hydrolysis. *Aka* 7-min phosphate; 10-min phosphate.

labile proteins Proteins that are lost rapidly upon starvation and whose loss does not lead to any measurable impairment of function.

labile sulfur Sulfur, such as that in some iron–sulfur proteins, that is readily released from the protein in the form of H_2S when the pH is lowered to about 1.0.

lac operon The operon in *E. coli*, isolated in pure form in 1969, that consists of three genes that code for the enzymes that function in the hydrolysis and transport of β-galactosides, particularly lactose. The genes code for beta galactosidase, beta galactoside permease, and beta galactoside transacetylase.

lactalbumin The fraction of whey proteins that consists principally of beta lactoglobulin and alpha lactalbumin.

lactam The keto form of a cyclic amide that contains the grouping

$$—NH—\overset{|}{C}=O$$

lactam–lactim tautomerism The tautomerism that is due to a shift of a hydrogen atom so that one of the isomers is a lactam and the other isomer is a lactim.

lactate acidosis A metabolic acidosis that is characterized by elevated levels of lactic acid along with a decrease of blood pH; may result from overproduction of lactate, underutilization of lactate, or both. Congenital lactate acidosis is a hereditary disease in which there is decreased gluconeogenesis.

lactate dehydrogenase The pyridine-linked dehydrogenase that catalyzes the oxidation of lactic acid to pyruvic acid and that occurs in the form of five isozymes. *Abbr* LDH.

lactation 1. The formation of milk by the mammary gland. 2. The period following childbirth during which milk is formed by the mammary gland.

lacteal 1. Of, or pertaining to, milk. 2. An intestinal lymph vessel that transports chyle.

lactic acid The hydroxy acid that is formed from pyruvic acid when glycolysis proceeds under anaerobic conditions.

lactic acid bacteria Bacteria of different genera that are characterized by the production of lactic acid as their main metabolic product. *See also* homofermentative lactic acid bacteria; heterofermentative lactic acid bacteria.

lactic acid fermentation *See* homofermentative lactic acid bacteria; heterofermentative lactic acid bacteria.

lactic acidosis *See* lactate acidosis.

lactim The enol form of a cyclic amide that contains the grouping

$$—N=\overset{|}{C}—OH$$

Lactobacillus bulgaricus factor PANTETHEINE.

Lactobacillus casei factor FOLIC ACID.

lactochrome LACTOFLAVIN.

lactoferrin *See* siderophilin.

lactoflavin An impure preparation of riboflavin from milk.

lactogen *See* placental lactogen.

lactogenic Stimulating lactation; having prolactin activity.

lactogenic hormone PROLACTIN.

lactoglobulin *See* beta lactoglobulin.

lactonase An enzyme that catalyzes the hydrolysis of a lactone.

lactone An intramolecular ester that is formed by elimination of a molecule of water between a hydroxyl group and a carboxyl group.

lactone rule A rule, proposed by Hudson in 1910, which states that the specific rotation of the γ- or δ-lactone of a D-aldonic acid is positive when the hydroxyl group on carbon atom 4 or 5 of the acid is to the right in the Fischer formula, and that the specific rotation is negative when the hydroxyl group is to the left.

lacto-ovo vegetarian An individual who is not a strict vegetarian (vegan) and who will eat plant and dairy products as well as eggs.

lactopoiesis LACTATION.

lactose A disaccharide of galactose and glucose that is present in milk. *Aka* milk sugar.

lactose intolerance In some cases, a genetically inherited metabolic defect in humans due to a deficiency of the enzyme lactase; in other cases, a condition that develops in adults. *See also* milk intolerance.

lactose operon *See* lac operon.

lactose permease A beta galactoside permease that controls the rate of entry of lactose from the medium into the *E. coli* cell. A transmembrane protein that is the carrier for lactose and that functions as a proton symport system; one proton is cotransported across the membrane for every molecule of lactose transported into the cell. Lactose permease represents an active transport system.

lacto-series glycosphingolipids *See* glycosphingolipids.

lactose synthetase The enzyme system composed of two proteins, denoted A and B, that catalyzes the synthesis of lactose from glucose and UDP-galactose in the mammary gland.

lactose tolerance test A test used for evaluating intestinal lactase activity by a measurement of the level of blood glucose as a function of time following the administration of a dose of lactose to an individual.

lactosuria The presence of excessive amounts of lactose in the urine.

lactotropin PROLACTIN.

lacto vegetarian An individual who is not a strict vegetarian (vegan) and who will eat both plant and dairy products.

lacuna 1. A small space or cavity in a tissue. 2. A clean area in a bacterial lawn as that produced by addition of a growth inhibitor.

ladder sequencing Any one of the nucleic acid sequencing techniques (such as the Maxam–Gilbert and Sanger–Coulson methods) that entails "reading" the nucleotide sequences off the bands of an electrophoresis gel which has the appearance of a ladder.

laetrile The compound 1-mandelonitrile-β-glucuronic acid. It has been claimed, but not supported experimentally, that laetrile has therapeutic value in the treatment of cancer due to the cyanide, present in laetrile, which acts specifically to destroy the cancer cells. Normal animal cells contain the enzyme beta glucuronidase, which hydrolyzes laetrile to glucuronic acid and mandelonitrile, but do not contain the enzymes required to release cyanide from mandelonitrile. *See also* vitamin B$_{17}$; amygdalin.

lagging strand *See* DNA replication.

lag period 1. The delay in cell-free protein synthesis that occurs when synthetic polyribonucleotides are used as messenger RNA and that is thought to be caused by the absence of an initiating codon in the polyribonucleotide. 2. INDUCTIVE PHASE.

lag phase The phase of growth of a bacterial culture that precedes the exponential phase and during which there is only little or no growth.

lake To lyse erythrocytes by suspending them in a hypotonic medium.

Laki–Lorand factor FACTOR XIII.

Lamarckian theory A theory of evolution, proposed by Lamarck in 1809 and no longer accepted, according to which the environment leads to structural changes in organisms, especially through new or increased uses of organs and through disuse and atrophy of other organs; these acquired characteristics are then transmitted to the offspring. *Aka* Lamarckism.

lambda 1. A temperate phage that infects *E. coli* and that contains linear, double-stranded DNA. 2. A microliter. *Sym* λ.

lambda chain One of the two types of light chains of the immunoglobulins.

lambda cloning vector A lambda phage that is genetically engineered so that it can accept foreign DNA and serve as a vector in recombinant DNA experiments. This involves cleavage of the phage DNA with restriction endonucleases and insertion of the foreign DNA. Vectors that have a single site at which the phage DNA is cleaved, and the foreign DNA inserted, are called insertion vectors; vectors that have two sites, which span a segment of DNA that can be excised and replaced with foreign DNA, are called substitution or replacement vectors.

lambda pipet A pipet for the transfer of volumes in the microliter range.

Lambert's law The law which forms part of the Beer–Lambert law. It states that the intensity of monochromatic light passing through an absorbing medium decreases exponentially with increasing thickness of the absorbing medium.

lamella (*pl* lamellae). 1. LAMINA. 2. The thylakoid membrane.

lamellar LAMINAR.

lamellar bone Bone that consists of ordered, parallel bundles of collagen fibers and that is the normal type of bone found in adults. There are two types of lamellar bone: (a) cortical bone, which has a dense structure and is found in the shafts of long bones; (b) trabecular (or cancellous) bone, which is more porous and is found at the end of long bones.

lamellipodia RUFFLED EDGES.

lamin One of a group of proteins found in the polymeric network (lamina) that underlies the nuclear membrane.

lamina (*pl* laminae) A thin layer; a thin plate.

laminar Arranged in the form of layers or plates.

laminar flow The undisturbed flow of a liquid along a tube; flow without obstacles. *See also* turbulent flow.

laminar-flow burner PREMIX BURNER.

laminin A large, complex, noncollagenous glycoprotein, synthesized by a variety of cell types, that is a major component of basement membranes; believed to be a critical adhesive protein of hepatocytes.

Lamm equation CONTINUITY EQUATION.

Lamm scale displacement method A method for measuring diffusion coefficients from the distortion of a transparent scale that is placed in front of a diffusion cell and that is photographed with light that has passed through the cell. The displacements of the scale lines are proportional to the refractive index gradient in the cell.

lampbrush chromosomes Very large chromosomes found in amphibians; they are neither polytene nor unusually condensed chromosomes but have a fuzzy structure resulting from the presence of hundreds of paired loops extending laterally from the main axis of each chromosomes.

Lande G factor G VALUE.

Landsteiner's rule The rule that a blood group antigen and its antibody do not coexist in one individual.

Langmuir adsorption isotherm An equation that describes the adsorption of a gas onto a solid and that has the same mathematical form as the Michaelis–Menten equation.

Langmuir trough A surface tension balance.

language *See* computer language.

lanolin The wax of sheep wool; a complex mixture of fatty acids, alcohols, fats, and waxy substances, including esters of steroids (such

as cholesterol and lanosterol), and long-chain aliphatic alcohols.

lanosterol The first biosynthetic sterol formed from squalene; the immediate precursor of cholesterol and the parent steroid in animals.

LAP Leucine aminopeptidase.

Laplacian distribution NORMAL DISTRIBUTION.

lard factor VITAMIN A.

large-angle x-ray diffraction A method of x-ray diffraction in which the scattering of the x rays is measured at large angles; used for the analysis of small molecular spacings, as those between individual atoms.

large calorie A kilocalorie; 1000 small calories. *Abbr* kcal; Cal.

lariat form A "tailed circle"; *See* sigma structure.

laser Acronym for light amplification by stimulated emission of radiation; a device capable of producing intense beams of monochromatic light.

laser dye A dye that is useful for production of laser beams at a specific wavelength (lasing wavelength) or range of wavelengths.

laser microprobe Part of a microscopic technique in which a laser beam is used to vaporize a very small tissue area; the vapor is then analyzed spectrographically.

lasing *See* laser dye.

lasing droplet A droplet that emits laser radiation when irradiated with a laser beam. The emitted radiation highlights the liquid-air interface and shows the changes in droplet size, shape, and orientation.

late enzyme A virus-specific enzyme that is transcribed from a late gene.

late gene A viral gene that is transcribed late after the infection of a host cell by the virus. *Aka* late function gene. *See also* late protein.

late mRNA A virus-specific messenger RNA that is synthesized by RNA polymerase late after the infection of the host cell by the virus.

latence The mean time, generally measured in days, between the exposure of animals to a carcinogen and the appearance of tumors. In humans, latence ranges from 10 to 20 years.

latency The phenomenon of an inactive particulate enzyme that either becomes active after it is exposed, while still attached to an insoluble matrix, or becomes active after it is detached from the insoluble matrix. *See also* crypticity.

latent enzyme An enzyme, especially a particulate enzyme, that exhibits activity only when the conditions are changed. *See also* latency.

latent image The invisible image that is produced on a photographic film when it is exposed to radiation and that is rendered visible by subsequent photographic development of the film.

latent infection 1. A bacterial infection that does not produce overt disease symptoms and in which the bacteria cannot be detected by currently available techniques. 2. A persistent viral infection that does not product overt disease symptoms. *See also* inapparent infection.

latent iron-binding capacity UNSATURATED IRON-BINDING CAPACITY.

latent period 1. The time interval between the infection of a bacterium by a phage and the first appearance of extracellular phage progeny. 2. The time interval between the injection of a sensitizing antiserum into an animal and the time at which a second injection will elicit an anaphylactic response. 3. LATENCE.

latent virus A tumor virus that has pathogenic activity and that has infected a host, but that does not lead to manifestations of the disease for a period of time.

late protein A virus-specific protein that is transcribed from a late gene; late proteins are typically structural proteins of the virus as distinct from early proteins which are viral enzymes.

lateral Of, pertaining to, or directed to, the side; away from the center line toward the left or right side.

lateral diffusion Diffusion that proceeds sideways; used specifically for the movement of proteins and lipids within the plane of the lipid bilayer of biological membranes.

lateral phase separation The phenomenon that, at certain temperatures, a system may separate itself into two phases, a "solid" and a "liquid" phase that coexist in one plane. The term is applied particularly to the phospholipid bilayer of biological membranes where it is believed that, at certain temperatures, the phospholipid molecules separate, within the plane of the membrane, into two phases, one rich in lipids of higher melting points ("solid" phase) and one rich in lipids of lower melting points ("liquid" phase).

lathyrism A disease caused by ingestion of poisons such as β-aminopropionitrile, which is found in *Lathyrus odoratus* peas. These poisons inhibit the enzyme lysyl oxidase which initiates the reactions whereby collagen molecules are cross-linked. Lathyretic individuals have skeletal deformities and a high excretion of hydroxyproline due to the increased turnover of collagen.

Latin square An array of elements that are distributed over an equal number of columns and rows, such that any element occurs only once in each column and once in each row; used for the design of experiments in which the effects of combinations of variables are to be studied. A 4 × 4 Latin square consists of four columns, four rows, and four elements,

spread over a total of 16 positions.

LATS Long-acting thyroid stimulator.

LATS-P LATS-protector; an immunoglobulin that competes with LATS for binding to LAA. LAA is a LATS-absorbing activity which, upon binding LATS, inactivates the latter. LAA, LATS, and LATS-P are all found in the sera of patients with Grave's disease.

lattice A three-dimensional network of elements that are arranged in geometric patterns, such as the atoms in a crystal lattice or the antigens and antibodies in an antigen-antibody lattice.

lattice microfilaments *See* microfilaments.

lattice theory The theory according to which the reaction between antigens and antibodies leads to the formation of an insoluble antigen–antibody network. Each antibody has at least two binding sites for antigens, while each antigen has many binding sites for antibodies. Upon mixing antigens and antibodies an insoluble network, or lattice, is formed in which each antibody is bound to at least two antigens and each antigen is bound to several antibodies.

Laue pattern An x-ray diffraction pattern that is produced by nonmonochromatic x rays. *Aka* Laue photograph.

lauric acid A saturated fatty acid that contains 12 carbon atoms.

LAV *See* AIDS virus.

Lavoisier and Laplace law *See* thermochemistry.

law A statement that describes a general truth or a general relation; a principle.

lawn The layer of bacterial cells growing on a solid medium.

law of constant heat summation *See* thermochemistry.

law of mass action The law that, at a given temperature, the rate of a chemical reaction is directly proportional to the active masses of the reactants. The rate of the reaction is proportional to the product of the molar concentrations of the reactants (more precisely, the activities), with each concentration raised to a power equal to the number of reactant molecules of the corresponding type which participate in the reaction.

law of parsimony OCCAM'S RAZOR.

law of pH monotonicity The principle that, in convection-free electrophoresis, the pH increases monotonically from the anode to the cathode. The principle is of interest for isoelectric focusing.

laws of thermochemistry *See* thermochemistry.

layered metabolic pathway TIERED METABOLIC PATHWAY.

layer line One of the parallel lines of spots that

are obtained by the rotating crystal method of x-ray diffraction.

LBF *Lactobacillus bulgaricus* factor.

LBM Lean body mass.

LC Liquid chromatography.

LC$_{50}$ The concentration of toxic compound that is lethal to 50% of the organisms to be tested under the test conditions in a specified time. *Aka* lethal concentration. *See also* median lethal dose.

LCAT lecithin–cholesterol acyltransferase.

L cell A cell belonging to a strain of normal mouse fibroblast cells that have been maintained in tissue culture for many years. *See also* L-form.

L-chain Light chain.

L-configuration The relative configuration of a molecule that is based upon its sterochemical relation to L-glyceraldehyde.

LD Lethal dose.

LD$_{50}$ Median lethal dose.

LDCF Lymphocyte-derived chemotactic factor.

LDH Lactate dehydrogenase.

LDL Low-density lipoprotein.

LDL receptor A large membrane glycoprotein that binds low-density lipoproteins (LDL) and leads to their internalization; this decreases the chance of arterial wall invasion and atherosclerotic plaque formation by the LDL particles.

lead A highly toxic and cumulative element in humans and animals; it inhibits many steps in heme biosynthesis as well as other enzymes. Symbol, Pb; atomic number, 82; atomic weight, 207.2; oxidation states, +2, +4; most abundant isotope, ^{208}Pb; a radioactive isotope, ^{209}Pb, half-life, 3.3 h, radiation emitted, beta particles.

leader A segment of nucleotides at the 5′-end of mRNA, preceding the AUG initiation codon at which translation begins. The leader is an untranslated segment of mRNA that varies in length but always contains part or all of a unique nucleotide sequence (Shine–Dalgarno sequence) which base pairs with 16S rRNA and thereby places the initiation codon in the proper orientation relative to the ribosome for the initiation of translation. In addition to the ribosomal RNA binding sequence, a leader may also contain regulatory signals, such as an attenuator region. *Aka* leader sequence.

leader peptide A peptide, coded for by a sequence of nucleotides in the leader segment of mRNA, and believed to function in the regulation of attenuation. *Aka* leader polypeptide.

leader protein SIGNAL SEQUENCE.

leader sequence LEADER.

leader sequence peptide SIGNAL SEQUENCE.

leading The chromatographic and electrophoretic phenomenon in which a peak appears lopsided and a band, or a spot, appears ill-defined; due to the fact that in the support the front edge of the region that contains the component is diffuse while the back edge is sharp. *See also* tailing; trailing.

leading reactant The first substrate that is bound to an enzyme in an ordered mechanism.

leading strand *See* DNA replication.

leading substrate The substrate, in a multisubstrate enzyme system, that is the first to bind to the enzyme.

leaflet Each of the two monolayers of a lipid bilayer.

leakage The loss of material across a cell membrane.

leak current The transfer of charge that occurs when an electrolyte moves through a tube and that is due to back conduction by ion diffusion and to electroosmosis.

leaky gene HYPOMORPH.

leaky mutant A mutant that has a leaky gene.

leaky patch model A model of the conformational changes produced in a cell membrane by the action of complement. According to this model, the phospholipid bilayer of the cell membrane is temporarily disrupted either by direct enzymatic action of complement or through the production of a lytic substance by complement. The disrupted membrane represents a leaky patch that allows the passage of water and ions across the membrane. *See also* doughnut model.

leaky protein A protein, formed by a mutant gene, that has a fraction of the activity of the normal protein.

lean body mass A measure of the composition of the body that is equal to the weight of an individual after removal of excess fat (adipose tissue). *Abbr* LBM.

LEAP Linked enzyme assay procedure.

least squares *See* method of least squares.

leaving group A group of atoms that is displaced from a carbon atom by the attack of a nucleophile in a nucleophilic substitution reaction. *See also* S$_N$1 mechanism; S$_N$2 mechanism.

Le Chatelier's principle The principle that a system in chemical equilibrium reacts to any change in its conditions by establishing a new equilibrium position in such a manner that the effect of the change is minimized.

lecithin PHOSPHATIDYL CHOLINE.

lecithin–cholesterol acyltransferase The enzyme that converts cholesterol in high-density lipoproteins to cholesteryl esters; it transfers an acyl group from phosphatidylcholine to cholesterol. *Abbr* LCAT.

lectins A large group of proteins or glycoproteins that have two or more binding sites and that bind to specific carbohydrate-containing receptor sites on animal cells. They were originally isolated from plants and shown to agglutinate red blood cells (hence, they were called phytohemagglutinins) but have since been isolated from all types of organisms and shown to bind to many kinds of animal cells. Lectins are proteins of nonimmune origin that agglutinate cells and/or precipitate complex carbohydrates; they are especially abundant in plants, particularly in the seeds of legumes. Some lectins are integrated into cellular membranes, others occur in soluble form; they have antibody-like activity in that they cause red blood cell agglutination. Some lectins are mitogenic and stimulate lymphocyte transformations; some agglutinate malignant cells preferentially; some are highly toxic and some seem to be involved in cell–cell recognition. Because of the specificity of the lectin/carbohydrate-containing receptor interaction, lectins are useful for locating and isolating such receptors and for replacing antisera in immunological studies.

LEED Low-energy electron diffraction.

left splicing junction *See* splicing junctions.

leghemoglobin A red pigment in the root nodules of leguminous plants; a heme protein that resembles hemoglobin structurally and functionally. Leghemoglobin combines reversibly with oxygen and is essential for symbiotic nitrogen fixation in the root nodules where it serves to transport oxygen to the bacteroids.

legitimate recombination GENERAL RECOMBINATION.

Leidig cells Variant spelling of Leydig cells.

leiotonin system A regulatory mechanism of smooth muscle contraction that involves a number of proteins (called leiotonins) and that does not appear to involve phosphorylation of myosin light chains.

LEIS Low-energy ion scattering.

lente insulin A slightly soluble form of insulin that is produced by crystallizing insulin from an acetate buffer in the presence of zinc ions. *See also* NPH insulin.

LEP strain Low egg passage strain; a viral strain that has been passed only a few times from one chick embryo to another (serial passage).

lepton *See* elementary particles.

Lesch–Nyhan syndrome A genetically inherited metabolic defect in humans that is due to a deficiency of the purine salvage enzyme, hypoxanthine-guanine phosphoribosyl transferase; characterized by an overproduction of uric acid and severe neurological disorders.

lesion 1. A pathological change in a tissue. 2. A genetic defect; a mutation. *See also* biochemical lesion.

-less mutant Combining form indicating an auxotrophic mutant, as in "thymineless mutant."

LET Linear energy transfer.

lethal Fatal; causing death.

lethal concentration *See* LC_{50}.

lethal dose The dose of a toxic compound, virus, etc., that kills 100% of the animals in a test group within a specified time. *Abbr* LD.

lethal gene A gene, the expression of which leads to the death of the organism that carries the gene. *Aka* lethal factor.

lethal hit The hit by a photon or by an ionizing particle that kills a cell or inactivates a virus.

lethal mutation A mutation that leads to the premature death of the organism that possesses the mutation.

lethal mutation model A model for the evolution of the genetic code according to which the code evolved so as to minimize the possibilities of deleterious mutations. *See also* frozen accident theory.

lethal synthesis The process whereby an enzyme catalyzes a metabolic reaction with a compound, other than its normal substrate, and leads to formation of a product that is lethal to the organism. Used specifically for the biological conversion of fluoroacetate to fluorocitrate. The former, by itself, has no toxic effects on cells but, once it is converted metabolically to fluorocitrate, becomes a potent inhibitor of the enzyme aconitase in the citric acid cycle. As a result, fluoroacetate is among the most deadly simple compounds known.

LETS protein Large external transformation-sensitive protein; fibronectin.

letter A nucleotide in a codon.

Leu 1. Leucine. 2. Leucyl.

leucine An aliphatic, branched, nonpolar alpha amino acid that contains six carbon atoms. *Abbr* Leu; L.

leucine aminopeptidase An aminopeptidase that acts on most amino acids and that catalyzes the sequential hydrolysis of a polypeptide chain from the N-terminal. *Abbr* LAP.

leuco- Combining from meaning white or colorless.

leucocidin One of a group of extracellular proteins (such as Streptolysin-S), produced by pathogenic species of *Staphylococcus* and *Streptococcus*, that kill leukocytes of certain species and exhibit toxic or lytic activity agains other cells, such as erythrocytes.

leucocyte Variant spelling of leukocyte.

leuco dye The colorless form of a dye.

leuco methylene blue The colorless, reduced form of methylene blue.

leucomycin A macrolide antibiotic.

leucoplast Variant spelling of leukoplast.

leucovirus Variant spelling of leukovirus.

leucovorin FOLINIC ACID.

Leu-enkephalin *See* enkephalin.

leukemia One of a group generally fatal, cancerous diseases of the blood. The disease affects the white blood cells, which are usually overproduced, and the blood-forming organs.

leukocyte A white blood cell that protects the organism against infection by eliminating invading bacteria through phagocytosis.

leukocyte inhibitory factor A lymphokine that inhibits the migration of polymorphonuclear leukocytes. *Abbr* LIF.

leukocytosis An abnormal increase in the number of circulating white blood cells.

leukopenia An abnormal decrease in the number of circulating white blood cells.

leukoplast A plastid that does not contain pigments.

leukopoiesis The formation of white blood cells.

leukosis An abnormal proliferation of one or more of the leukocyte-forming tissues.

leukotrienes A group of biologically active molecules, formed by leukocytes, macrophages, and other cells and tissues in response to immunological and nonimmunological stimuli. They produce contraction of bronchial smooth muscles, stimulation of vascular permeability, attraction and activation of leukocytes, and are involved in asthma and allergy. Chemically, leukotrienes are derived from arachidonic acid, or other unsaturated fatty acids, and are noncyclic as distinct from the prostaglandins. They are classified as follows: LTA (epoxide form); LTB (hydroxide form); LTC (linked to glutathione, glutamyl-cysteinyl-glycine), LTD (linked to cysteinyl-glycine); LTE (linked to cysteine); the last three categories represent peptolipids.

leukovirus An enveloped, complex, animal virus that contains single-stranded RNA. Leukoviruses mature by budding from cytoplasmic membranes and some leukoviruses are oncogenic.

lev Levorotatory.

levan A homopolysaccharide of fructose.

level of significance A value, generally set in terms of percentages, that indicates the level at which a statistical test may yield an erroneous result. Thus, a 5% level of significance indicates that there is a 5% chance that the particular value is incorrect.

levodopa L-Dopa.

levorotatory Having the property of rotating the plane of plane-polarized light to the left, or counterclockwise, as one looks toward the light source. *Abbr* lev; l.

levulose D-Fructose; a levorotatory monosaccharide.

Lewis acid An atom, an ion, or a molecule that acts as an electron pair acceptor.

Lewis acid–base catalysis The catalysis in solution in which the catalysts are Lewis acids and/or Lewis bases.

Lewis base An atom, an ion, or a molecule that acts as an electron pair donor.

Lewis factor One of two antigens, designated Le^a and Le^b, that are closely related to those of the ABO blood group system. The two factors constitute the Lewis blood group system. *Aka* Lewis antigen.

LEXA LexA repressor.

LexA repressor *See* SOS repair.

Leydig cells Cells in the testes that are the major source of androgens in the male and that store cholesterol in the form of cholesterol esters.

L1 family *See* long interspersed repeated sequences.

L-form One of number of bacterial variants of different genera that lack a rigid mucopeptide cell wall and that are capable of growth and multiplication. L-forms are formed spontaneously or as a result of a variety of treatments such as temperature shock, osmotic shock, or the presence of antibiotics which inhibit cell wall synthesis; in L-forms, the cell wall is either defective or totally absent.

LFT Low-frequency transduction.

LFT lysate Low-frequency transduction lysate; a lysate prepared by induction of a prophage that possesses normal, low transducing power.

LH Luteinizing hormone.

LHR Liquid holding recovery.

LHRF Luteinizing hormone releasing factor; *See* luteinizing hormone releasing hormone.

LHRH Luteinizing hormone releasing hormone.

LIBC Latent iron-binding capacity.

liberin A releasing hormone; also used as a suffix for the names of the hypothalamic releasing hormones.

library *See* clone library; gene library; genomic library; cDNA library.

librational motion A rotational oscillation about an equilibrium position.

Liebermann–Burchard reaction A colorimetric reaction for cholesterol and related sterols that is based on the successive production of a red, blue, and blue-green color upon treatment of the sample with acetic anhydride and concentrated sulfuric acid.

LIF Leukocyte inhibitory factor.

life The sum of the properties that distinguish animals, plants, and microorganisms from nonliving matter, such as metabolism, reproduction, growth, excitability, movement, function, and complexity; the state of existence of a functioning cell, a group of cells, or an organism.

lifeboat response The escape of a prophage from a doomed host cell through formation of an infectious viral particle (induction).

life cycle 1. The sequence of the developmental stages of an organism from its formation to its death, or from any specified stage to the recurrence of that stage. 2. CELL CYCLE.

lifetime 1. For a chemical species: the reciprocal of the sum of all the rate constants for the reactions that the species can undergo $(1/(k_1 + k_2 + \cdots))$. 2. For a process: the reciprocal of the first-order rate constant of the process.

ligand 1. An atom, a group of atoms, or a molecule that binds to a metal ion. 2. An atom, a group of atoms, or a molecule that binds to a macromolecule.

ligand chromatography A column chromatographic method in which metal ion complexes, as those formed with amino acids, are chromatographed on an ion-exchange resin.

ligand-exchange chromatography A column chromatographic technique in which one ligand in the mobile phase replaces another ligand bound to the column much as one ion replaces another ion in ion-exchange chromatography. The method has been used to resolve racemic mixtures.

ligand field theory An extension of the crystal field theory to include covalent character in the bonding. The theory describes the way in which the electrons of a metal ion in a metal ion–ligand complex reduce the repulsion of ligand electrons through angular polarization.

ligand-gated channel See gated channel.

ligandin See mercapturic acids.

ligand-induced endocytosis A mechanism for the selective uptake by cells of specific ligands which bind to receptors on the cell surface and are then internalized via coated pits. See also coated pits.

ligand–receptor internalization See coated pits; receptosome.

ligase An enzyme that catalyzes the joining together of two different molecules or of two ends of the same molecule in a reaction that is coupled to the hydrolysis of a pyrophosphate bond (high-energy bond) in ATP or some other nucleoside triphosphate. See also DNA ligase; RNA ligase; enzyme classification.

ligate To bind a ligand.

ligation Formation of a $3',5'$-phosphodiester bond that links two adjacent nucleotides in the same nucleic acid strain in either DNA or RNA. See also splicing.

ligature 1. The act of tying off a vessel or a duct, typically a blood vessel. 2. A thread, cord, or band, used for tying off a vessel or a duct.

light 1. n A form of electromagnetic radiation that has both wave and particle aspects. 2. adj Unlabeled, as opposed to being labeled with a heavy isotope. 3. adj Labeled with a light isotope, as opposed to containing either the natural or a heavy isotope.

light chain One of two polypeptide chains that are linked to two heavy chains to form the immunoglobulin molecule. The molecular weight of a light chain is about 25,000 and that of a heavy chain is about 50,000. The light chains of type K and type L immunoglobulins are known respectively, as κ and λ chains. Abbr L-chain.

light chopper See chopper.

light compensation point The light intensity, at a given CO_2 concentration, at which the rate of carbon dioxide fixation by a plant (photosynthesis) is equal to the rate of photorespiration. See also CO_2 compensation point.

lightening hormone An octapeptide neurohormone, produced by the eye stalk glands of crustaceans. The hormone is released in response to visual stimuli and acts on pigment granules in the skin, allowing the organism to adjust its color to match that of the surroundings.

light-harvesting Chl a/b protein The complex, composed of chlorophylls a and b, carotenoid, and one or more polypeptides, that functions in conjunction with photosystems I and II in photosynthesis of green plants.

light-harvesting molecules ANTENNA MOLECULES.

light isotope An isotope that contains a smaller number of neutrons in the nucleus than the more frequently occurring isotope.

light label A light isotope that is generally introduced into a molecule to facilitate its separation from identical molecules containing the more frequently occurring isotope.

light meromyosin The terminal tail fragment of the myosin molecule. Abbr L-meromyosin; LMM. Aka F_3 fragment.

light microscope An ordinary microscope that consists of an optical system for use with visible light.

light path The thickness of sample through which light passes for optical measurements.

light reaction A photosynthetic reaction or reaction sequence that depends directly on light energy and that serves to convert light energy into chemical energy.

light respiration PHOTORESPIRATION.

light scattering The dispersion of light rays by matter in directions other than that of the incident beam; commonly refers to the light scattered by solutions of macromolecules and to the use of this scattering for determining molecular weights of the solute macromolecules. *See also* Mie scattering; Raman effect; Rayleigh scattering.

light strand 1. A polynucleotide chain that is either not labeled with a heavy isotope or that is labeled with a light isotope. 2. The naturally occurring polynucleotide chain of a duplex that has a lower density than the complementary chain.

lignification The formation of lignin.

lignin The complex phenylpropanoid polymer that strengthens the cellulose framework of wood fibers and vascular plants.

lignoceric acid A saturated fatty acid that contains 24 carbon atoms.

limit dextrin The branched core of amylopectin or glycogen that remains after digestion of the carbohydrate with either alpha or beta amylase.

limit dextrinosis GLYCOGEN STORAGE DISEASE TYPE III.

limited chromosome A chromosome that does not occur in nuclei of somatic cells but only in nuclei of germ cells.

limited proportional region That portion of the characteristic curve of an ionization chamber that is above the proportional region and similar to it except for the fact that the amplification that can be achieved has a limiting value.

limiting amino acid That essential amino acid that is present in a given protein in the smallest amount; the ratio of the amount of the limiting amino acid in the protein to the amount required by the organism provides an estimate of the nutritional value of the protein. *See also* chemical score.

limiting current The maximum current obtainable in a polarographic system which, in simple cases, may be equal to the diffusion current.

limiting reagent That reagent (reactant) in a chemical reaction whose amount determines the amount of product formed; the reactant that is used up completely in the reaction.

limiting velocity MAXIMUM VELOCITY.

limiting viscosity number INTRINSIC VISCOSITY.

lincomycin An antibiotic, produced by *Streptomyces lincolnensis*, that inhibits protein synthesis in prokaryotes by inhibiting peptidyl transferase.

Linderstrøm–Lang column A density gradient column, consisting of varying mixtures of two miscible organic liquids that are immiscible with water, (bromobenzene and kerosene, for example). The column is calibrated with drops of known aqueous solutions and the density of an unknown aqueous solution is then determined by measuring the position, in the column, of a drop of the unknown solution.

line 1. Abbreviation for long interspersed repeated sequence. 2. A pure-breeding group of homozygous individuals that have a distinctive phenotype.

linear absorption coefficient The fractional decrease in the intensity of a beam of radiation per unit thickness of the absorber.

linear accelerator An instrument for imparting high kinetic energy to subatomic particles that are made to move in a long and straight path.

linear chain OPEN CHAIN.

linear correlation A relation between two variables so that, as one increases, the other either increases or decreases, and a plot of one variable against the other yields a straight line. *See also* regression line.

linear density gradient A density gradient in which the density increases in such a fashion that a plot of density versus distance in the tube yields a straight line.

linear dichroism The dichroism that occurs when linearly polarized light is absorbed by partially or completely oriented molecules.

linear electric field effect The change in the electron paramagnetic resonance properties of a sample by the application of an electric field across the sample.

linear energy transfer The energy dissipation of a radiation as it passes through a tissue or other matter; generally expressed either in terms of kiloelectronvolts per micron, or in terms of megaelectronvolts per centimeter divided by the density of the substance in grams per cubic centimeter. *Abbr* LET.

linear growth The growth of a culture such that a plot of the number of cells (or the cell mass) as a function of time yields a straight line; may be brought about, for example, by regulating the supply of a critical nutrient by dropwise addition or diffusion.

linear inhibition Inhibition that yields a straight line when either slopes or intercepts from a primary plot are plotted as a function of inhibitor concentration.

linear metabolic pathway A metabolic pathway of the type $A \rightarrow B \rightarrow C$.

linear polymer A polymer composed of open, unbranched chains.

linear polyphosphate One of a group of compounds that contain a polymetaphosphate grouping, H_2PO_3—$(HPO_3)_n$—H_2PO_4, and that have been used to drive the polymerization of amino acids in studies on the origin of life. *Aka* polymetaphosphate.

linear regression *See* regression line.

linear-sweep polarography Oscillographic polarography in which the entire potential scan is applied to the dropping mercury electrode either once as a single sweep, or several times as a multisweep, during the life of a single mercury drop.

linear transformation A mathematical transformation of an equation into one for a straight line. The conversion of the Michaelis–Menten equation into its Lineweaver–Burk formulation is an example.

linear velocity The straight line distance moved per unit time.

line emission The emission of light of either one or several specific wavelengths that is produced in flame photometry by a given ion.

Line 1 family See long interspersed repeated sequences.

line of best fit See method of least squares.

line of stability The line drawn through the band that represents the stable nuclides in a plot of the number of protons versus the number of neutrons in the nuclides.

line spectrum A spectrum in which either the absorption or the emission of radiation is limited to only a few wavelengths.

line splitting See spin–spin splitting.

Lineweaver–Burk plot The double reciprocal plot of the Michaelis–Menten equation; a plot of $1/v$ versus $1/[S]$, where v is the velocity of the reaction and $[S]$ is the substrate concentration.

linkage 1. Any association of genes in inheritance that exceeds that to be expected from the independent assortment and that is due to their being located on the same chromosome; linkage is assessed by the tendency of two markers to remain together during recombination. 2. COUPLED REACTIONS.

linkage group A group of linked genes; all the genes located on the same chromosome.

linkage map A scale representation of a chromosome that shows the relative positions of all its known genes.

linked assay COUPLED ASSAY.

linked enzyme assay procedure A method for the detection of proteinases and proteinase inhibitors; involves coupling a reporter enzyme to a substrate protein which, in turn, is immobilized on a particulate support. Upon incubation with a proteinase, peptides containing an active enzyme label are released from the support. The labeled peptides are collected by centrifugation or filtration and are used, in a second incubation, to transfer molecules of the reporter enzyme substrate, thereby providing great amplification of the initial peptide bond hydrolysis by the proteinase. Abbr LEAP.

linked gene A gene showing linkage; a gene on the same chromosome as another gene.

linked reactions COUPLED REACTIONS.

linked transduction A bacterial transduction in which there is a simultaneous transfer of two or more genes that lie close together on a chromosome.

linked transformation A bacterial transformation in which there is a simultaneous transfer to the bacterium of two or more genes that lie close together on a chromosome.

linker DNA 1. A section of DNA that connects adjacent nucleosomes in a chromosome and to which is bound a molecule of histone H1. 2. A short, synthetic, double-stranded DNA segment that contains a site cleavable by a restriction endonuclease. Linkers are useful in binding to double-stranded DNA segments and in inserting the latter into double-stranded DNA molecules. If the DNA segment contains several restriction endonuclease sites, it is known as a polylinker.

linking number The total number of times that the two strands of a double helix of a closed, circular DNA molecule cross each other. The linking number is a topological property that can change only if one or both strands are nicked and then rejoined; it is an integer and is positive for right-handed helical regions and negative for left-handed helical regions. It is defined as follows: $L = W + T$ and $\Delta L = \Delta W + \Delta T$, where L is the linking number, W is the writhing number, and T is the twisting number. Aka linkage number; winding number.

link protein See proteoglycan aggregates.

linoleic acid An unsaturated fatty acid that contains 18 carbon atoms and two double bonds. See also promoter.

linolenic acid An unsaturated fatty acid that contains 18 carbon atoms and three double bonds.

lipamino acid Variant spelling of lipoamino acid.

lipase An enzyme that catalyzes the hydrolysis of fats to glycerol and fatty acids.

lipectomy The surgical removal of adipose tissue.

lipemia The presence of excessive amounts of lipid in the blood.

lipid One of a heterogeneous group of compounds that are synthesized by living cells and that are sparingly soluble in water but are soluble in nonpolar solvents; they can be extracted from tissues by nonpolar solvents, and they have as a major part of their structure long hydrocarbon chains that may be branched or unbranched, straight or cyclic, saturated or unsaturated. Various classifications of lipids are in use, including (a) simple lipids (neutral fats and waxes), complex lipids (phospholipids, sphingolipids, glycolipids, etc.), derived lipids (steroids, vitamins, carotenoids, etc.); and (b) neutral lipids (neutral fats, waxes, carotenoids, etc.), amphipathic lipids (glycerolipids, sphingolipids, etc.), redox lipids (quinones, etc.).

lipid bilayer A layer of amphipathic lipid molecules that is two molecules thick and that is believed to form most or all of the central portion of biological membranes. In a lipid bilayer that is surrounded by a polar environment, the nonpolar parts of the lipid molecules are directed inward and the polar parts are on the outside.

lipide LIPID.

lipidemia LIPEMIA.

lipid–globular protein mosaic model FLUID–MOSAIC MODEL.

lipid imbibition theory A theory of atherosclerosis according to which the formation of atheromatous plaques is caused by the uptake of lipids, such as cholesterol, from the blood stream by the walls of the arteries.

lipid intermediate The compound, undecaprenyl phosphate, that functions in peptidoglycan synthesis.

lipid mobilization See mobilization.

lipidosis (*pl* lipidoses). 1. One of a number of genetically inherited or acquired diseases that are characterized by the deposition of one or more types of lipids in specific tissues or organs. 2. One of a number of genetically inherited metabolic defects in humans that are characterized by lipid accumulation in specific tissues or organs and that result from a defect in the metabolism of glycosphingolipids. *See also* Fabry's disease; Gaucher's disease; Krabbe's disease; Niemann–Pick disease; Tay–Sachs disease.

lipid peroxidation The nonenzymatic oxidation of fatty acids, especially unsaturated ones, to hydroperoxides (R—O—O—H) by strong oxidizing agents such as hydrogen peroxide, the superoxide anion radical ($O_2^-\cdot$), or the hydroxy radical ($\cdot OH$).

lipid-soluble vitamin FAT-SOLUBLE VITAMIN.

lipid solvent A nonpolar solvent, such as chloroform, acetone, or methanol, that will extract lipids from tissues.

lipid storage disease LIPIDOSIS.

lipid vesicle LIPOSOME.

lipin Obsolete term for lipid.

lipo- Combining form meaning lipid.

lipoamide The dipeptide-like structure, formed by linking a molecule of lipoic acid through its carboxyl group to the ε-amino group of a lysine residue, that is part of the enzyme for which lipoic acid serves as a coenzyme. *See also* lipoylprotein; lysyl-lipoamide.

lipoamino acid 1. A compound formed by joining a fatty acid or a long-chain alcohol to an amino acid by means of either an ester or an amide bond. 2. An amino acid ester of phosphatidylglycerol.

lipocaic A substance, secreted by the pancreas, that prevents the fatty infiltration of liver and stimulates the oxidation of fatty acids.

lipochondrion (*pl* lipochondria). 1. DICTYOSOME. 2. A temporary storage form of lipids in which they are absorbed by epithelial cells in the Golgi apparatus.

lipochrome A naturally occurring, fat-soluble pigment.

lipocortin A protein (MW 37,000) that inhibits phospholipase A_2, apparently by sequestering the phospholipid substrate. The physiological significance of lipocortin is unclear.

lipofuscin AGE PIGMENT.

lipogenesis The biosynthesis of fatty acids from acetyl coenzyme A.

lipoglycan A polysaccharide that contains covalently linked lipids. Lipoglycans differ from lipopolysaccharides in their composition and cellular location; they appear to be integral components of the cytoplasmic membranes of various bacterial species.

lipoic acid A compound that is generally classified with the B vitamins, since it is a growth factor for some microorganisms; it functions as a coenzyme in the multienzyme systems that catalyze the oxidative decarboxylation of pyruvic acid to α-ketoglutaric acid and of α-ketoglutaric acid to succinic acid.

lipoid 1. LIPID. 2. Resembling a fat or an oil.

lipolysis The hydrolysis of lipids.

lipolytic Of, or pertaining to, lipolysis.

lipolytic hydrolysis LIPOLYSIS.

lipoma A benign tumor of adipose tissue.

lipopeptide 1. A compound formed by joining a fatty acid or a long-chain alcohol to a peptide by means of either an ester or an amide bond. 2. A peptide ester of phosphatidylglycerol.

lipophilic NONPOLAR.

lipophilic Sephadex A Sephadex preparation that can be used with organic solvents.

lipophilic stain A stain for lipids.

lipophilin A proteolipid that is the major membrane protein of brain myelin.

lipopolysaccharide A water-soluble, lipid–polysaccharide complex, that is an important component of the outer membrane of gram-negative bacteria. A lipopolysaccharide molecule consist of a heteropolysaccharide chain linked covalently to a glycolipid. The glycolipid (lipid A) consists of a disaccharide, substituted with long-chain fatty acids. The heteropolysaccharide consists of a core polysaccharide, which is similar or identical in closely-related bacteria, and an O-specific chain (O-antigen), which determines the identity of cell surface antigens. The overall structure of a lipopolysaccharide can thus be represented by [glycolipid–heteropolysaccharide] or [lipid A–core polysaccharide–O-specific chain]. *Abbr* LPS.

lipoprotein One of a group of conjugated, water-soluble proteins in which the nonprotein portion is a lipid; the lipid is usually a glyceride, a phospholipid, cholesterol, or a combination of these. The lipid component is tightly bound (prosthetic group). Lipoproteins occur in blood plasma, cell cytoplasm, cell membranes, cell organelles, and egg yolk. Blood plasma lipoproteins function in the transport and distribution of lipids and are classified into 5 major groups on the basis of their density: chylomicrons, very low-density lipoproteins (VLDL), low-density lipoproteins (LDL), high-density lipoproteins (HDL), and very high-density lipoproteins (VHDL). *See also* specific lipoprotein groups; proteolipid.

lipoprotein lipase An extracellular enzyme that is most active within the capillaries of adipose tissue, cardiac muscle, and skeletal muscle; it catalyzes the hydrolysis of the 1 or 3 ester bond of di- and triacyl glycerols (di- and triglycerides) present in chylomicrons and very low-density lipoproteins. *Abbr* LPL. *Aka* clearance factor; clearing factor.

lipoprotein tissue factor THROMBOPLASTIN.

liposarcoma A malignant tumor of adipose tissue.

liposome An artifically prepared, cell-like structure in which one or more bimolecular layers of phospholipid enclose one or more aqueous compartments; a membrane-bound vesicle, frequently formed by dispersion of phospholipid in aqueous salt solutions. *See also* vesicle.

lipotrophic hormone LIPOTROPIC HORMONE.

lipotrophin LIPOTROPIC HORMONE.

lipotropic Descriptive of a compound that can contribute methyl groups for the synthesis of choline and that can prevent or alleviate a fatty liver condition that results from a dietary deficiency of choline.

lipotropic agent A compound, such as choline or methionine, that aids in the transport of fat and thereby prevents or alleviates the condition of fatty infiltration of the liver.

lipotropic hormone A polypeptide hormone, secreted by the anterior lobe of the pituitary gland, that stimulates the mobilization of lipids, especially fatty acids, from lipid deposits. *Var sp* lipotrophic hormone. *Abbr* LPH.

lipotropin LIPOTROPIC HORMONE.

lipovitellenin 1. A degraded form of a low-density lipoprotein component that is present in hens' egg yolk. 2. LOW-DENSITY FRACTION.

lipovitellin 1. A high-density lipoprotein in hens' egg yolk. Two such proteins, denoted α and β, have been isolated; they are similar in their composition except for their content of protein-bound phosphorus. 2. A high-density yolk lipoprotein from any species.

lipoxins A group of biologically active, leukocyte-derived, arachidonic acid metabolites.

lipoxygenase A key enzyme in the biosynthesis of leukotrienes; it catalyzes the first reaction in the lipoxygenase pathway whereby arachidonic acid is oxidized to a hydroperoxide which is then converted to leukotrienes.

lipoyl dehydrogenase DIAPHORASE.

lipoyllysine LIPOAMIDE.

lipoylprotein A conjugated protein in which lipoic acid is covalently bound to the protein by way of an amide link between its carboxyl group and the ϵ-amino group of a lysine residue in the protein. *See also* lipoamide.

lipuria The presence of lipid in the urine.

liquefying amylase ALPHA AMYLASE.

liquid chromatography A collective term for liquid–liquid, liquid–solid, paper, thin-layer, ion-exchange, and molecular sieve chromatography. *Abbr* LC.

liquid crystal A phase that has a mobility like that of a liquid and a high degree of order like that in a crystal. Liquid crystals exhibit aspects of both the liquid and the solid states but also possess properties that are not found in either state. They are classified as either lyotropic or thermotropic depending on the principal way by which the order of the parent solid state is destroyed: in lyotropic ones, by means of solvent action; in thermotropic ones, by means of heat. Lyotropic liquid crystals are frequently two-component systems, composed of amphipathic molecules and water. These can have a variety of structures involving, for example, lamellar, cubic, hexagonal, cylindrical, or micellar packing of the molecules. Thermotropic liquid crystals are further classified as either nematic (having a thread-like pattern) or smectic (having greasy or soapy properties); there are 3 known types of nematic structures (ordinary, cholesteric, and blue phase) and 9 known types of smectic structures (5 structured and 4 unstructured smectics). Nematic liquid crystals typically consist of molecules that are parallel, can rotate about their axes, and can move both up and down and from side to side. Smectic liquid crystals typically consist of molecules arranged in strata or layers which can slide over one another; the molecules in each layer can move from side to side, as well as forward and backward, but cannot move up or down from one layer to the next. In structured smectics, the molecules in each layer form a regular two-dimensional lattice; in unstructured smectics, the molecules in each layer are positioned randomly. *Aka* mesophase; mesomorphic phase.

liquid holding recovery The phenomenon that bacterial cells, when allowed to stand in buffer following their irradiation with ultraviolet light, show an increased viability as compared to cells that are plated out immediately after irradiation. *Abbr* LHR.

liquid hybridization Nucleic acid hybridization that is carried out in solution and that requires separation of the hybrid double-stranded molecules from any unreacted, single-stranded molecules left. *See also* filter hybridization.

liquid junction potential *See* junction potential.

liquid–liquid chromatography Partition chromatography in which the mobile phase is a liquid and the stationary phase is an inert support, coated with a liquid. *Abbr* LLC.

liquid medium A solution of nutrients.

liquid protein diet A diet based on protein hydrolysates.

liquid scintillation The emission of light flashes by a solution containing a fluorescent chemical when the chemical is struck by either an ionizing particle or a photon; used as a method for measuring the radioactivity of a sample dissolved in the solution. When used in this fashion, the light flashes are transformed into electrical pulses by means of a photomultiplier tube and are then counted.

liquid scintillation counter A radiation counter in which incident ionizing particles or incident photons are counted by the scintillations that they induce in a liquid fluor; the sample and the fluor are either dissolved in a common solvent or one is suspended in a solution of the other. *Abbr* LSC.

liquid–solid chromatography Adsorption chromatography in which the mobile phase is a liquid and the stationary phase is a solid. *Abbr* LSC.

liquid surfactant membrane A water-immiscible phase that contains emulsion-size droplets and that consists of surfactants, a hydrocarbon solvent, and other compounds; the droplets may be used to hold a reagent or to encapsulate an enzyme.

liter A metric unit of volume equal to 1 dm^3. *Sym* L, l.

lithiasis The formation of calculi, particularly biliary and urinary ones.

lithocholic acid A bile acid that has one hydroxyl group.

lithogenic Leading to the formation of calculi;

stone-producing.

lithosphere The solid, mineral part of the earth.

lithotroph A cell or an organism that uses inorganic compounds as electron donors for its energy-yielding, oxidation–reduction reactions.

little gastrin *See* gastrin.

little insulin That fraction of free serum insulin that is indistinguishable from pancreatic insulin.

little t *See* t antigen.

live Alive, not dead; viable.

liver The principal metabolic organ of animals that is capable of carrying out all the major metabolic reactions and that has a variety of functions in both anabolism and catabolism. It is a large gland, located in the abdominal cavity.

liver filtrate factor PANTOTHENIC ACID.

liver function test A quantitative determination of either a metabolite or an enzyme that is used in evaluating the functional capacity of the liver. Liver function tests include measurements of serum bilirubin concentration, alkaline phosphatase activity, and galactose tolerance.

liver profile The composite results obtained from a battery of liver function tests.

live-timing A method of timing, used in scintillation counters, in which the timing device is turned off during the interval that is required for the electronic processing of a pulse. *See also* clock-timing.

live vaccine A vaccine that consists of infectious bacteria or of viruses, the virulence of which has been attenuated.

living Possessing the properties of life; alive, not dead.

LLC Liquid–liquid chromatography.

LLD factor VITAMIN B_{12}.

LLF Laki–Lorand factor.

L-L factor Laki–Lorand factor.

L-meromyosin Light meromyosin.

LMM Light meromyosin.

ln Natural logarithm.

LNPF Lymph node permeability factor.

LnRNA Low molecular weight nuclear RNA; see heterogeneous nuclear RNA.

loading 1. The in vitro process whereby cellular structures, such as erythrocyte ghosts or mitochondria, are made to take up or accumulate specific substances. 2. The process of applying a sample to a chromatographic or electrophoretic support.

load test A tolerance test in which an individual is given a dose of a specific metabolite and the urinary concentration of this metabolite, or of a related compound, is determined as a function of time. A phenylalanine load test is used as a diagnostic tool for phenylketonuria, and a tryptophan load test is used as a diagnostic tool for schizophrenia. *See also* glucose tolerance test; galactose tolerance test.

Lobry De Bruyn–Alberta van Eckenstein transformation The interconversion of monosaccharides that occurs in alkaline solution as a result of the formation of enediol intermediates.

local hormone AUTACOID.

localized bond A chemical bond involving only two atoms.

localized infection A viral or a bacterial infection in which the infective agents remain primarily at the site of entry into the host.

localized orbital A molecular orbital that is spread only over two bonding atoms.

locant The portion of a chemical name that designates the position of an atom or a group in a molecule; β in β-naphthylamine, *m* in *m*-xylene, and 2 in 2-butanol, are examples.

lock and key theory A theory of the mechanism of an enzymatic reaction according to which the substrate binds to the enzyme to form an enzyme–substrate complex. The binding site on the enzyme is preformed and is called the active site; it is structurally complementary to the substrate, so that the substrate fits onto the enzyme much as a key fits into a lock. *See also* flexible active site; induced fit theory.

Locke's solution A solution that is similar in composition to Ringer's solution; it contains (in w/v) 0.9% sodium chloride, 0.024% calcium chloride, 0.042% potassium chloride, 0.02% sodium bicarbonate, and 0.1% glucose.

locoweed One of a number of plants that contain selenium or other poisons and that cause alkali disease when eaten by animals.

locus (*pl* loci) A place or a position, particularly that occupied by a gene on a chromosome.

log Logarithm; also denoted $\log_{10}$.

logarithm The exponent that indicates the power to which a fixed number has to be raised to produce a given number; the fixed number is known as the base. In ordinary, or Briggsian, logarithms the base is 10; in natural, or Naperian, logarithms the base is the constant *e*. Ordinary logarithms are denoted log or $\log_{10}$; natural logarithms are denoted ln or $\log_e$.

logarithmic growth EXPONENTIAL GROWTH.

logarithmic paper Graph paper in which both scales have been distorted to allow the plotting of the logarithm of one variable versus the logarithm of a second variable.

logarithmic phase EXPONENTIAL PHASE.

$\log_e$ Natural logarithm.

logit A logarithmic quantity, defined as logit $(x) = \log [x/(1 - x)]$ when $0 < x < 1$; used, for example, in animal studies where *x* is the probability of obtaining a particular response to a given dosage.

log–normal distribution A positively skewed distribution of a random variable that, when subjected to a logarithmic transformation, tends to take on the shape characteristic of a normal distribution.

log phase Logarithmic phase.

Lohmann reaction The reversible reaction, catalyzed by the enzyme creatine kinase, in which ATP and creatine are formed from ADP and phosphocreatine.

London dispersion forces DISPERSION FORCES.

long-acting thyroid stimulator A thyroid-stimulating substance that exerts its effect over a long period of time; it is an immunoglobulin (IgG) which stimulates thyroid plasma membrane adenyl cyclase and is present in many individuals suffering from hyperthyroidism. *Abbr* LATS.

Long cat A pancreatectomized and adrenalectomized cat that is used in endocrinological studies.

long-chain base A term used to describe sphinganine, its homologues, stereoisomers, and hydroxy or unsaturated derivatives.

long-chain fatty acid thiokinase A fatty acid thiokinase that catalyzes the activation of fatty acids having more than 12 carbon atoms to fatty acyl coenzyme A.

long interspersed repeated sequences A family of long (6–7 kb) repetitive sequences that occur in the form of 20,000–50,000 copies per mammalian genome. They constitute the Line 1 (L1) family and appear to consist largely of retroposons. *Abbr* LINE.

longitudinal Lengthwise; parallel to the long axis of a body or a structure.

longitudinal relaxation SPIN–LATTICE RELAXATION.

long patch pathway A mechanism for DNA repair in both prokaryotes and eukaryotes in which the repaired DNA segments are relatively long.

long period interspersion A DNA structure that is characterized by relatively long (over 300 bp) segments of moderately repetitive DNA alternating with relatively long (over 1000 bp) segments of nonrepetitive DNA.

long-range hydration The hydration by water molecules that are located in the secondary hydration shell.

long-range interactions The attractive and repulsive forces between atoms and molecules that do not decrease rapidly with distance. *See also* strong interactions.

long spacing fibrils *See* fibrils long spacing.

long spacing segments *See* segments long spacing.

Longsworth scanning method An optical method for obtaining a photograph that depicts the refractive index gradient of a boundary; produced by photographing the boundary while a knife edge moves upward in front of the camera lens and while the photographic plate is driven horizontally at a speed that is proportional to that of the knife edge.

long terminal repeat One of two long sections of double-stranded DNA (250–1200 bp) that occur when a double-stranded DNA molecule is synthesized by reverse transcriptase from the RNA strand of a retrovirus. The two long terminal repeats (LTR) have an identical sequence, composed in part of sections unique to the 3'- and 5'-ends of the viral RNA. One LTR occurs at each end of the linear duplex DNA intermediate that is formed. LTRs are believed to provide functions fundamental to the expression of most eukaryotic genes, such as promotion, initiation, and polyadenylation of transcripts. *Abbr* LTR.

loop *See* arm; omega loop.

looped rolling circle model A variation of the rolling circle model for the replication of duplex circular DNA in which the progeny is a single-stranded circular molecule.

loose coupling The state of cellular respiration in which the mitochondria are characterized by having a low acceptor control ratio.

low-angle x-ray diffraction SMALL-ANGLE X-RAY DIFFRACTION.

low background counter A specially shielded radiation counter in which the level of background counts has been greatly reduced.

low-copy number *See* copy number.

low-density fraction 1. A low-density fraction of hens' egg yolk that contains about 89% lipid and two lipoproteins which differ primarily in their size. 2. LIPOVITELLENIN.

low-density lipoprotein A plasma lipoprotein that has a density of 1.006–1.063 g/mL. An increase in the concentration of low-density lipoproteins (LDL) is believed to be linked to an increase in the incidence of atherosclerosis. LDL contain about 22% protein, 22% phospholipid, 46% cholesterol and cholesterol esters, and 10% triacylglycerols (triglycerides). LDL have molecular weights of about 3×10^6, a flotation coefficient of 0–12S, and are classified as the beta fraction on the basis of electrophoresis. *Abbr* LDL. *See also* lipoprotein.

low egg passage strain *See* LEP strain.

low-energy compound A compound that, upon hydrolysis under standard conditions, yields a small amount of free energy; the standard free energy change for the hydrolysis reaction is less than 7 kcal/mol.

low-energy electron diffraction A technique for the study of surface structures in which a beam of low-energy electrons is used to probe the surface. The electrons striking the surface penetrate only the outermost layer of atoms and are diffracted by the atoms. The diffraction pattern provides information on the arrangement and the spacing of the surface atoms. *Abbr* LEED.

low-energy ion scattering A technique for the study of surfaces in which a low-energy beam of noble gas ions is directed at the surface. Measurements of the energy loss and scattering angles provide information as to the positions and identities of atoms in the first few layers of the surface. *Abbr* LEIS.

low-energy phosphate acceptor A low-energy compound that can function as an acceptor for the phosphoryl group transferred from high-energy phosphate donors by way of the ADP–ATP phosphoryl group carrier system. *Aka* low-energy phosphate compound.

low-frequency transduction Transduction in which the phages that are capable of transducing constitute a small proportion of the total phage population. *Abbr* LFT.

low gate The cutoff level in integral discrimination.

low-level promoter A promoter than can undergo a promoter-down mutation.

low-lipid lipoprotein HIGH-DENSITY LIPOPROTEIN.

low-order antibody INCOMPLETE ANTIBODY.

low-potential iron protein *See* high-potential iron protein.

low-quality proteins Proteins from plant sources that have a low content (limiting amount) of one or more of the essential amino acids.

Lowry method A modification of the Folin–Ciocalteau reaction that is used as a colorimetric reaction for the quantitative determination of proteins.

low-speed sedimentation equilibrium SEDIMENTATION EQUILIBRIUM.

low spin The state of a complex in which there is a maximum of paired electrons; referred to as a state of essentially covalent bonding and ascribed to certain hemoproteins.

low-temperature heat method The pasteurization of material by heating it at 61.6 °C for 30 min. *Abbr* LTH method.

LPC Lysophosphatidyl choline; *See* lysophosphoglyceride.

LPE Lysophosphatidyl ethanolamine; *See*

lysophosphoglyceride.

LPH Lipotropic hormone.

LPL Lipoprotein lipase.

LPS Lipopolysaccharide.

L region Two reactive para positions in an aromatic hydrocarbon; the presence of such a region has been correlated with the lack of carcinogenic activity of some hydrocarbons. *See also* K region.

LRF Luteinizing hormone releasing factor; *See* luteinizing hormone releasing hormone.

LRH Luteinizing hormone releasing hormone.

L-rRNA RNA of the large ribosomal subunit.

LS antigen An antigen that can be dissociated from a poxvirus and that consists of a heat-labile and a heat-stable component.

LSC 1. Liquid–solid chromatography. 2. Liquid scintillation counter. 3. Liquid scintillation counting.

LSD Lysergic acid diethylamide.

L-shaped structure *See* L-type structure.

L/S ratio Lecithin/sphingomyelin ratio; a quantity that is measured in the amniotic fluid and that provides information regarding the maturity of the fetal lungs. When the L/S ratio is less than 2.0, there is great risk that the baby will develop respiratory distress syndrome.

l-strand WATSON STRAND.

LT 1. Lymphotoxin 2. Leukotriene.

LTA *See* leukotriene.

LTB *See* leukotriene.

LTC *See* leukotriene.

LTD *See* leukotriene.

LTE *See* leukotriene.

LTH 1. Lactogenic hormone. 2. Luteotropic hormone.

LTH method Low-temperature heat method.

LTR Long terminal repeat.

L-type structure The three-dimensional structure of tRNA, formed by folding of the cloverleaf structure to yield two helical, double-stranded branches, at right angles to each other, in the shape of an L. The structure is maintained by means of tertiary base pairs (tertiary hydrogen bonds).

Lubrol A nonionic detergent.

luciferase The enzyme that functions in the bioluminescence reactions of the firefly; it catalyzes the oxidation of luciferin with the concomitant release of visible light.

luciferin The substrate of the enzyme luciferase. *See also* luciferase.

Lucite PLEXIGLASS.

Lugol's solution A solution containing (in w/v) 5% iodine and 10% potassium iodide.

lumen 1. A passageway in a small tube or duct. 2. A unit of luminous flux; equal to the quantity of visible light falling on 1 cm^2 at a distance of 1 cm from a light source of 1 cd.

lumichrome A blue fluorescent compound formed by photolysis of riboflavin in acidic solution.

lumiflavin A yellow-green fluorescent compound formed by photolysis of riboflavin in basic solution.

luminescence The emission of light that results from chemical reactions within, or a flow of energy into, an emitter, rather than from an increase in temperature. *See also* fluorescence; phosphorescence; luciferase.

luminescent Of, or pertaining to, luminescence.

lumirhodopsin A structurally altered form of rhodopsin that is produced after the exposure of rhodopsin to light and prior to its conversion to metarhodopsin.

lumisome A subcellular organelle, consisting of a membrane-bound vesicle, that is the site of bioluminescence in some organisms.

lumisterol A compound produced from ergosterol by irradiation with ultraviolet light.

lung surfactants A group of substances, manufactured and secreted by lung tissue, that are essential for normal functioning of the lungs. They have detergent action and serve to maintain the proper surface tension of the alveoli. The major component is an unusual phosphatidylcholine (dipalmitoylphosphatidylcholine).

Luria–Delbrueck fluctuation test *See* fluctuation test.

Luria–Latarjet experiment An experiment for determining the sensitivity of intracellular vegetative phage to irradiation. The phage-infected bacteria are irradiated at different stages of the phage multiplication cycle with varying doses of ultraviolet light or other radiation. The cells are then analyzed for the fraction of surviving infective phage particles.

luteal Of, or pertaining to, the corpus luteum.

luteinizing hormone A gonadotropic protein hormone, secreted by the anterior lobe of the pituitary gland, that stimulates the final ripening and rupture of the ovarian follicles and the secretion of progesterone by the corpus luteum; it also stimulates the production of testosterone in the male. *Abbr* LH. *Aka* lutropin.

luteinizing hormone releasing hormone The hypothalamic hormone that controls the secretion of luteinizing hormone. *Abbr* LRH; LHRH. *Aka* luteinizing hormone releasing factor (LRF; LHRF).

luteohormone PROGESTERONE.

luteotrophic hormone Variant spelling of luteotropic hormone.

luteotrophin Variant spelling of luteotropin.

luteotropic hormone PROLACTIN.

luteotropin PROLACTIN.

lutropin LUTEINIZING HORMONE.

LUV Large unilamellar vesicle; *See* vesicle.

luxury genes Genes that provide for specialized cell functions. The products of such genes are generally synthesized in large amounts only in specialized cell types, such as the synthesis of hemoglobin in erythrocytes and the synthesis of immunoglobulins in plasma cells. *See also* housekeeping genes.

LV Lipovitellin.

LVP Lysine vasopressin.

lyase An enzyme that catalyzes the cleavage of a molecule or the removal of a group; involves electron rearrangements (elimination reactions) but not hydrolysis or oxidation–reduction. Lyases cleave C—C, C—O, C—N, and similar bonds and create double bonds or rings in one of the products (or add a group to a double bond in the reverse reaction). *See also* enzyme classification.

lyate ion A solvent molecule minus a proton; in the case of water, the lyate ion is the hydroxyl ion.

lycopene A red pigment in tomatoes that is the parent compound of the carotenoids.

lymph The fluid, derived from the interstitial fluid, that bathes the tissues, circulates through the lymphatic vessels, and is ultimately discharged into the blood.

lymphatic 1. *n* A small lymph vessel. 2. *adj* Of, or pertaining to, lymph.

lymph node permeability factor A factor that occurs in extracts of lymph node cells and which, when injected into the skin, increases vascular permeability. *Abbr* LNPF.

lymphoblast A cell derived from a T lymphocyte upon antigenic stimulation.

lymphocyte A cell that occurs in lymphatic tissue, spleen, and blood and that is characterized by having a large round nucleus; a lymph cell. Lymphocytes are immunologically competent cells that recognize antigens and play a central role in the immune response. They are classified according to their origin and function into B cells and T cells. *See also* B cell; T cell.

lymphocyte-derived chemotactic factors Various lymphokines that stimulate leukocyte chemotaxis. *Abbr* LDCF.

lymphocyte transformation The formation of immunoblasts from lymphocytes, a process that is accompanied by rapid synthesis of DNA and RNA.

lymphocytosis An increase in the number of lymphocytes in the blood or in the tissue fluids.

lymphocytotoxin LYMPHOTOXIN.

lymphoid 1. Resembling lymph or lymphatic tissue. 2. LYMPHATIC.

lymphoid cell LYMPHOCYTE.

lymphokines A heterogeneous group of substances that are secreted by some helper T cells after they are primed by contact with an antigen. Lymphokines are not antibodies but are mediators of cellular immunity. They activate various white blood cells, including other lymphocytes. Examples of lymphokines are interleukin 2, some interferons, and migration inhibition factor (MIF). All lymphokines studied so far are proteins.

lympholysis The lysis of lymphocytes.

lympholytic agent An immunosuppressive agent that causes the destruction of lymphocytes.

lymphoma A tumor of lymphatic tissue.

lymphon The entire immune system of an individual, including lymphocytes, complement, and so on.

lymphopoiesis The formation of lymphocytes.

lymphosarcoma A malignant lymphoma composed of abnormal and immature lymphocytes.

lymphotoxin A lymphokine that has cytotoxic effects on various target cells.

lyochrom An impure preparation of riboflavin.

lyoenzyme SOLUBLE ENZYME.

lyolysis Solvolysis in which the solvent either donates or accepts protons.

lyonium ion A solvent molecule plus a proton; in the case of water, the lyonium ion is the hydronium ion.

lyophilic Descriptive of the tendency of a group of atoms or of a surface to become either wetted or solvated by the solvent. *See also* hydrophilic.

lyophilization The removal of water under vacuum from a frozen sample; a relatively gentle process for the removal of water in which the water sublimes from the solid to the gaseous state. *See also* cryosublimation.

lyophobic Descriptive of the tendency of a group of atoms or of a surface to resist becoming either wetted or solvated by the solvent. *See also* hydrophobic.

lyophobic bond LYOPHOBIC INTERACTION.

lyophobic interaction The association of nonpolar groups with each other in nonaqueous, but polar, solvents. These interactions are weaker than hydrophobic interactions, and their strength decreases with an increase in temperature.

lyotropic liquid crystal *See* liquid crystal.

lyotropic series An arrangement of cations and anions in a series according to their effect on the solubility of proteins; the ions are ordered according to decreasing salting-out efficiency and decreasing extent of hydration. *See also* salting out.

Lys 1. Lysine. 2. Lysyl.

lysate 1. The suspension of ruptured cells obtained upon lysis. 2. The suspension of phage particles released from ruptured host cells during the lytic cycle.

Lysenkoism A pseudoscientific doctrine, proposed by T.D. Lysenko and stressed in the Soviet Union between 1932 and 1965. Lysenko rejected the gene concept and believed in the inheritance of acquired characteristics (Lamarckism).

lysergic acid diethylamide The most potent psychotomimetic substance known; it induces a schizophrenic-like state in humans. A hallucinogenic drug that is believed to produce chromosomal aberrations. *Abbr* LSD.

lysin An antibody that can lead to cell lysis.

lysine An aliphatic, basic, and polar alpha amino acid; contains six carbon atoms and two amino groups. *Abbr* Lys; K.

lysine intolerance A genetically inherited metabolic defect in humans that is due to a deficiency of the enzyme L-lysine: NAD-oxidoreductase.

lysine-rich histone An older term for histone H1.

lysine vasopressin A vasopressin, occurring in hogs, in which the eighth amino acid residue has been replaced by a lysine residue. *Abbr* LVP.

lysine vasotocin A vasotocin, occurring in hogs, that contains a lysine residue at position 8.

lysis The rupture and dissolution of cells.

lysis from within The lysis of bacterial cells that occurs as a result of the intracellular multiplication of phage particles.

lysis from without The lysis of bacterial cells that occurs without intracellular phage multiplication and that is due to the large number of holes produced in the cell wall by lytic enzymes of the phage; the enzymes either are released into the medium or are contained within the adsorbed phage particles.

lysis inhibition The delay in the lysis of bacterial cells that occurs when a cell culture is inoculated with a phage and is then heavily reinoculated with the same phage a few minutes later.

lyso- *See* lysoglycerophospholipid.

lysochrome A substance that dissolves in lipids and thereby colors them.

lysogen A bacterium that contains a complete set of genes from a temperate phage; a lysogenic bacterium.

lysogenic Of, or pertaining to, lysogeny.

lysogenic bacterium 1. A bacterium that has survived the infection by a temperate phage and that has incorporated the prophage into the bacterial DNA; a lysogenized bacterium. 2. A bacterium that can be infected by a temperate phage.

lysogenic conversion The phage-mediated, phenotypic changes of a bacterium, as those relating to growth and morphology, that may accompany the infection of the bacterium by a temperate phage.

lysogenic cycle The sequence of reactions whereby a bacterial cell becomes infected with a temperate phage, incorporates the prophage into the bacterial DNA, and then divides. A variation of this set of reactions occurs when the phage DNA becomes a plasmid rather than a segment of the host chromosome.

lysogenic immunity IMMUNITY (2).

lysogenic response The incorporation of phage DNA into the bacterial DNA that follows the infection of the bacterium by a temperate phage.

lysogenic virus A virus that can become a prophage; a temperate phage.

lysogenization The production of lysogenic bacteria by infection of a sensitive bacterial strain with a temperate phage.

lysogenized bacterium LYSOGENIC BACTERIUM (1).

lysogeny The phenomenon of bacterial infection by temperate phages.

lysoglycerophospholipid A glycerophospholipid of phosphatidic acid in which one of the two acyl groups (fatty acid groups), attached to the carbon atoms of glycerol, has been removed by hydrolysis; these reactions are catalyzed by phospholipases and the compounds are hemolytic. *Aka* lysophosphoglyceride.

lysolecithin A lysophosphoglyceride of lecithin that causes lysis of erythrocytes.

lysophosphatide LYSOPHOSPHOGLYCERIDE.

lysophosphoglyceride LYSOGLYCEROPHOSPHOLIPID.

lysophospholipase *See* phospholipase.

lysophospholipid LYSOPHOSPHOGLYCERIDE.

lysosomal disease A genetically inherited metabolic defect in humans that is due to a deficiency of an enzyme located in the lysosomes; examples of lysosomal diseases are Fabry's, Gaucher's, Niemann–Pick's, Pompe's, and Wolman's diseases. *Aka* lysosomal storage disease. *See also* mucolipidosis; mucopolysaccharidosis.

lysosomal storage disease *See* lysosomal disease.

lysosome A membrane-enclosed cytoplasmic structure in eukaryotic cells that is rich in hydrolytic enzymes and functions in intracellular digestion. Lysosomes contain some 60 acid hydrolases, including glycosidases, nucleases, proteases, lipases, sulfatases, and phosphatases. Primary lysosomes are newly formed organelles that have not yet encountered substrates for digestion; secondary lysosomes are membranous sacs of diverse morphology formed by repeated fusion of primary lysosomes with a variety of other structures; they are frequently given special names such as digestive vacuole, multivesicular body, and autophagic vacuole.

lysosomotropic agent An agent that is rapidly taken up by cells and is selectively concentrated within the lysosomes; the uptake of chloroquine and NH_4Cl by adipocytes is an example.

lysozyme The enzyme that catalyzes the hydrolysis of polysaccharides that occur in the glycopeptide layer of bacterial cell walls; a bacteriolytic enzyme that is present in such biological fluids as egg white and saliva.

lysyl bradykinin *See* kinin.

lysyl–lipoamide The complex formed by the covalent linkage (via an amide bond) of lipoic acid (in either its oxidized, disulfide, or its reduced, sulfhydryl form) to an epsilon amino group of a lysine residue in a polypeptide chain.

lytic Of, or pertaining to, lysis.

lyticase ZYMOLYASE.

lytic cycle 1. The sequence of reactions whereby a virulent phage infects a bacterial cell, multiplies inside the cell, and ultimately leads to the lysis of the cell. 2. The sequence of reactions whereby a prophage is induced in an infected bacterial cell, multiplies inside the cell, and ultimately leads to the lysis of the cell.

lytic enzyme An enzyme that catalyzes a hydrolysis reaction; a hydrolytic enzyme.

lytic response The intracellular multiplication of a virulent virus that leads to the lysis of the infected cell.

lytic virus A virus that can cause cell lysis; a virulent virus.

M

m 1. Meta. 2. Milli. 3. Meter. 4. Prefix or subscript used to designate the modified form of an interconvertible enzyme. 5. Molal concentration.

M 1. Molar concentration. 2. Molecular weight. 3. Metal ion. 4. Methionine. 5. Mitosis. 6. Thioinosine. 7. The M phase of the cell cycle.

M13 A filamentous phage that contains circular, single-stranded DNA and infects *E. coli* by adsorbing to the F-pilus on the surface of the cell.

MAbs Monoclonal antibodies.

macroamylase An abnormal, high molecular weight (about 200,000 daltons) form of α-amylase that occurs in some individuals; probably consists of α-amylase bound to either IgA or some other (normal or abnormal) high molecular weight plasma protein.

macroevolution Evolution that extends over long time periods, involves large steps, and leads to marked changes in the genetic makeup of an organism.

macroglobulin A globulin, such as the IgM immunoglobulin, that has a molecular weight above 400,000.

macroglobulinemia The presence of excessive amounts of macroglobulins, specifically IgM immunoglobulins, in the blood.

macroion A charged macromolecule.

macrolide antibiotic One of a number of antibiotics, such as erythromycin and oleandomycin, that are similar in their structure, action, and antimicrobial spectrum, and that are characterized by having a large lactone ring that contains anywhere from 14 to 20 carbon atoms; they are produced by various strains of *Streptomyces* and inhibit protein synthesis.

macromineral An element that is required in the diet in relatively large amounts; Ca, P, Mg, Na, K, and Cl are examples.

macromolecule A high molecular weight molecule; a polymer.

macromutation A mutation in which a large segment of a chromosome is altered, as distinct from a point mutation.

macronucleus One of two types of reproductive nuclei found in many ciliate protozoa; they are large, contain nucleoli, are usually polyploid, and contain active DNA that undergoes transcription. A cell may contain one or more macronuclei. *See also* micronucleus.

macronutrient An essential nutrient that is needed by an organism in appreciable amounts; carbohydrates, lipids, and proteins are examples.

macrophage A cell, derived from the reticuloendothelial system, that functions in phagocytosis.

macrophage activation factor MIGRATION ENHANCEMENT FACTOR.

macrophage inhibition factor MIGRATION INHIBITION FACTOR.

macropinocytosis Pinocytosis that involves the formation of large vesicles, having diameters in excess of 300 nm.

macroscopic Descriptive of a size that is visible to the unaided eye.

macroscopic binding constant The binding constant for the binding of the first, second, or subsequent ligand to a macromolecule; K_1, K_2, etc. *See also* microscopic binding constant.

MAF Migration activation factor. *See* migration enhancement factor.

magainin One of two closely related peptides (23 amino acids each), isolated from frog skin, that kill a variety of bacteria, fungi, and protozoa; believed to be part of a defense system in vertebrates that is different from the immune system. The magainins appear to damage the cell membrane and work faster than antibiotics.

magic amino acids The set of 20 amino acids that are required by all living species for the biosynthesis of proteins. While more than 20 amino acids occur in proteins, these additional amino acids (such as hydroxyproline and phosphoserine) represent modifications of the "magic 20."

magic number 1. The number of different amino acids that occur in proteins. 2. The number of either the neutrons or the protons which, when they occur in an atomic nucleus, contribute to the stability of the nucleus.

magic spot One of two unusual nucleotides, originally discovered as spots on a radioautogram and designated I and II, that occur in microorganisms and that serve in a regulatory capacity; they are the effectors of the stringent response in bacteria and are believed to function as alarmones. The nucleotides are guanosine-5′-diphosphate, 3′-diphosphate ($_{pp}G_{pp}$) and guanosine-5′-triphosphate, 3′-diphosphate ($_{ppp}G_{pp}$) *Abbr* MS.

magnesium An element that is essential to all plants and animals. Symbol, Mg; atomic number, 12; atomic weight, 24.312; oxidation state, +2; most abundant isotope, ^{24}Mg; a radioactive isotope, ^{28}Mg, half-life, 21.3h, radiation emitted, beta particles and gamma rays.

magnetic affinity chromatography A variation of affinity chromatography in which the irreversibly bound molecule (or ligand) binds to its complementary molecule in solution and the resulting complex is removed from suspension by the application of a magnetic field. The method is particularly useful when dealing with marginally soluble materials.

magnetic circular dichroism Circular dichroism that is measured in the presence of a magnetic field. Magnetic circular dichroism depends on the nature of, and the couplings between, the ground and the excited states of the chromophore; it reflects the intrinsic geometry of the chromophore and is not sensitive to the surroundings of the chromophore. *Abbr* MCD. *See also* circular dichroism.

magnetic dipole A substance that has two magnetic poles; the separation between the poles is measured by the magnetic dipole moment.

magnetic dipole moment A measure of the tendency of a substance to become oriented in a magnetic field; equal to the product of the strength of the magnetic pole and the length of the magnet. The magnetic dipole moment depends on the presence or absence of paired electrons, since a spinning electron behaves as a magnet. Substances with paired electrons have no net magnetic dipole moment and are diamagnetic; substances with unpaired electrons have a permanent magnetic dipole moment and are paramagnetic.

magnetic field The space surrounding magnetic poles in which a mechanical force will be exerted on a magnet introduced into it.

magnetic moment MAGNETIC DIPOLE MOMENT.

magnetic resonance *See* electron paramagnetic resonance; nuclear magnetic resonance.

magnetic stirrer A plastic- or glass-coated magnet that is used as a stirring bar in conjunction with an appropriate electric motor.

magnetic susceptibility A measure of the tendency of a substance to become oriented in a magnetic field; the proportionality constant between the total magnetic moment per unit volume in the direction of an applied magnetic field and the magnetic field strength.

magnetic transition moment The magnetic moment that is induced in a molecule during a transition, such as that associated with the absorption of a photon, as a result of a rotation of charge.

magnetoliposome An artificial structure prepared by binding ferromagnetic particles to antibody-bearing liposomes.

magnetosome A small, subcellular, electron-dense particle that consists of magnetite (Fe_3O_4). A chain of these particles occurs in some aquatic bacteria and functions as a compass, orienting the bacterium in the same direction as the lines of force of the earth's magnetic field.

magnetosphere A region in space in which planetary magnetic fields govern local physical processes.

magnetotaxis A taxis in which the stimulus is a magnetic field. *See also* magnetosome.

magnetotactic Of, or pertaining to, magnetotaxis.

magnification The apparent enlargement of an object when viewed through a microscope; expressed as the number of times that the diameter of the object appears to have been enlarged.

magnitude *See* order of magnitude.

Maillard reaction One of a group of nonenzymatic reactions in which aldehydes, ketones, or reducing sugars react with amino acids, peptides, or proteins; these occur as part of the browning reactions of food.

main band DNA The bulk DNA, as opposed to the smaller fractions referred to as satellite bands. *See also* satellite DNA.

main diffusion coefficient The diffusion coefficient that a component has when no other diffusing components are present; used in the treatment of diffusion data in a system showing interaction of flows. *See also* cross-term diffusion coefficient.

mainframe computer A large computer, storing enormous amounts of information.

maize factor ZEATIN.

major gene A gene that has a marked phenotypic effect as opposed to a modifying gene.

major groove The deep and wide groove in Watson–Crick type DNA that is approximately 22 Å across.

major histocompatibility complex 1. A multigene locus that is responsible for the production of the histocompatibility antigens. The locus has been assigned different symbols in different organisms, such as human (HLA), mouse (H-2), chicken (B), dog (DLA), guinea pig (GPLA), and rat (Rt-1). The corresponding major histocompatibility antigens (MHC antigens) have been designated as

HLA antigens, H-2 antigens, and so on. *Abbr* MHC. 2. MAJOR IMMUNOGENE COMPLEX.

major immunogene complex A genetic region that includes the major histocompatibility complex but extends beyond it to include other functions of the immune system such as transplantation rejection, killing of virus-infected cells, and synthesis of complement. *Abbr* MIC.

major–minor code An early version of the genetic code according to which the central nucleotide of a codon is the major factor in positioning the correct amino acid on the template, while the two adjoining nucleotides are minor factors in selecting the amino acid.

MAK Methylated albumin–kieselguhr.

malaria An acute, and sometimes chronic, disease of human and other vertebrates that is caused by protozoans of the genus *Plasmodium* which are transmitted by mosquitoes of the genus *Anopheles*.

malate–aspartate shuttle A shuttle, the components of which are malic acid, α-ketoglutaric acid, glutamic acid, aspartic acid, and the enzymes malate dehydrogenase and glutamic–aspartic transaminase. The shuttle achieves the oxidation of cytoplasmic NADH at the expense of the reduction of mitochondrial NAD^+.

malathion An insecticide and nerve poison that forms a stable covalent intermediate with a serine residue in the active site of acetylcholinesterase.

male hormone TESTOSTERONE.

maleic acid The cis isomer of fumaric acid that is not an intermediate in the citric acid cycle.

male sex hormone An androgen that affects the development of secondary sex characteristics in the male. The principal male sex hormones are testosterone and dihydrotestosterone which are produced by the testes; other compounds that have male sex hormone activity, such as adrenosterone, are produced by the adrenal gland.

malic acid A dicarboxylic acid that is formed from fumaric acid in the reactions of the citric acid cycle.

malic enzyme The enzyme that catalyzes the anaplerotic reaction whereby pyruvic acid is carboxylated to malic acid.

malignant Descriptive of a tumor that metastasizes and endangers the life of the organism.

malnutrition A condition that is caused by inadequate quantity, quality, digestion, absorption, or utilization of ingested nutrients.

malo-lactic fermentation A type of fermentation, carried out by lactic acid bacteria, in which L-malic acid is converted to lactic acid and CO_2.

Maloney leukemia virus A retrovirus that produces lymphocytic leukemia in mice; it can be transmitted from mother to offspring via the milk.

malonic acid A competitive inhibitor of the enzyme succinate dehydrogenase; a three-carbon, dicarboxylic acid that closely resembles succinic acid.

maltose A disaccharide that is composed of two glucose residues linked by means of an α(1 → 4) glycosidic bond and that constitutes the repeating unit of starch.

malt sugar MALTOSE.

mammal A vertebrate of the class Mammalia that is characterized by possession of hair and mammary glands.

mammalian Of, or pertaining to, mammals.

mammalian expression vector A vector capable of producing large quantities of eukaryotic proteins; a shuttle vector that is usually first grown in *E. coli* and then in animals.

mammary tumor agent MOUSE MAMMARY TUMOR VIRUS.

mammotropin PROLACTIN.

Man Mannose.

mandelate pathway A degradative pathway of mandelic acid, catechol, and related aromatic compounds that occurs in *Pseudomonas fluorescens*.

manganese An element that is essential to all animals and plants. Symbol, Mn; atomic number, 25; atomic weight, 54.9380; oxidation states, +2, +3, +4, +7; most abundant isotope, ^{55}Mn; a radioactive isotope, ^{54}Mn, half-life, 313 days, radiation emitted, gamma rays.

mannan A homopolysaccharide of mannose that occurs in bacteria, molds, and plants.

Mannich reaction The condensation of an aldehyde and an amine with a nucleophilic carbon atom; believed to constitute part of the biosynthetic pathway of alkaloids.

mannitol A sugar alcohol derived from mannose.

mannose A six-carbon aldose. *Abbr* Man.

mannosidosis A genetically inherited metabolic defect in humans that is due to a deficiency of the enzyme α-mannosidase.

manometer An instrument for measuring the pressure of a liquid or a gas.

M antigen M PROTEIN (1–3).

mantissa The fractional part of a logarithm.

Mantoux test An intradermal tuberculin test for delayed-type hypersensitivity in humans.

MAO Monoamine oxidase.

MAOI Monoamine oxidase inhibitor.

map 1. *n* GENETIC MAP. 2. *n* CYTOGENETIC MAP. 3. *v* To establish the structure or structural details for a portion of a macromolecule, as in the mapping of an active site. 4. *v* To establish

MAP   288   Mason's theory

the location of either a mutable site on a genetic map or a gene on a cytogenetic map. *See also* peptide map; nucleotide map; restriction map.

MAP Microtubule-associated proteins.

map distance The distance between any two markers on a genetic map. *See also* map unit.

maple syrup urine disease A genetically inherited metabolic defect in humans that is associated with mental retardation and that is characterized by the presence of urinary ketoacids derived from valine, leucine, and isoleucine; due to a deficiency of the enzyme, ketoacid decarboxylase.

map unit A measure of distances along a linkage map that is equal to a recombination frequency of 1%. *Abbr* mu.

Marasmus A disease of infant starvation that is similar to kwashiorkor but involves, additionally, dietary deficiencies of carbohydrates and lipids; caloric starvation.

marigranules Synthetic organized particles produced from a mixture of glycine, acidic, basic, and aromatic amino acids using a modified sea medium in experiments that simulate chemical evolution in the primordial soup. *See also* marisomes.

marihuana Variant spelling of marijuana.

marijuana 1. The dried and chopped tips of the shoots of the female hemp plant (*Cannabis*); contains Δ'-tetrahydrocannabinol (THC) which is psychoactive; a common narcotic. *Var sp* marihuana. 2. HASHISH.

Mariotte flask A device used in column chromatography for maintaining a constant pressure head of eluent; consists of a liquid reservoir containing a tube that is open to the air.

marisomes Synthetic organized particles, with elastin-like properties, that are formed in experiments that simulate chemical evolution in the primordial soup. They are prepared by heating amino acids at 105 °C for several weeks under nitrogen in a seawater medium. Marisomes may represent a protocell-type, macromolecular complex, and are believed to be precursors of marigranules.

marker 1. A mutable site on a chromosome that is useful for cell identification and for genetic studies; the site of a gene of known function and known location on the chromosome. *See also* biochemical marker; genetic marker. 2. A group or a molecule that is linked chemically to another molecule for purposes of identification. 3. A reference substance that is used in a physical technique such as chromatography, electrophoresis, or density gradient centrifugation.

marker enzyme An enzyme, the intracellular location of which is known, so that an assay of the enzyme can be used as an aid in following the isolation and purification of subcellular fractions.

marker rescue The incorporation of a genetic marker from a mutated virus into the DNA of an active progeny virus during cross-reactivation.

Marker synthesis A partial synthetic procedure for the synthesis of steroid hormones developed by R.E. Marker; involves the conversion of diosgenin (the aglycone of the steroid saponin, dioscin) to progesterone.

Markov chain A Markov process in which either the time parameters or the values of the process are discontinuous.

Markovnikov's rule The rule, proposed by Markovnikov in 1905, that in the addition reaction of HX to an alkene, the acid hydrogen becomes attached to the carbon with fewer alkyl substituents. Conversely, the X group always bonds to the carbon with more alkyl substituents. The rule is applicable, for example, to some reduction reactions of double bonds catalyzed by pyridine-linked dehydrogenases. *Var sp* Markownikoff's rule.

Markov process A stochastic process such that the conditional probability distribution for the state at any future instant, given the present state, is unaffected by an additional knowledge of the past history of the system.

Maroteaux–Lamy syndrome A genetically inherited metabolic defect in humans that resembles Hurler's syndrome but does not involve loss of normal intelligence. A mucopolysaccharidosis due to a deficiency of the enzyme N-acetylgalactosamine sulfatase.

maser Acronym for microwave amplification by stimulated emission of radiation; a device capable of producing intense beams of microwave radiation.

masked mRNA Messenger RNA that is stored in large quantities in eukaryotic cells, particularly in unfertilized eggs (sea urchin eggs, for example). The mRNA is present in an inactive form; it is protected against digestion by nucleases and prevented from binding to ribosomes, apparently by being associated with proteins. Shortly after fertilization, the mRNA is activated and then translated.

masked residue An amino acid residue in a protein that is not accessible to, and cannot undergo a reaction with, specific reagents. The lack of activity may be due to the occurrence of the residue in the internal portion of the protein or to stereochemical, electrostatic, and other properties of its immediate environment.

masked virus A tumor virus that lacks pathogenic activity.

Mason's theory A theory that describes the first step in chemical carcinogenesis in terms of the electronic interactions between the carcinogen and either a protein or a nucleic acid molecule of the affected cell or organism.

mass absorption coefficient A measure of the absorption efficiency of gamma radiation, defined as the ratio of the linear absorption coefficient to the density of the absorber.

mass action *See* law of mass action.

mass-action ratio The concentration term [products]/[reactants] as it is used, for example, in calculating free energy changes of reactions. For the reaction $aA + b\,B \rightleftharpoons cC + dD$, the mass-action ratio is given by

$$\frac{[C]^c[D]^d}{[A]^a[B]^b}$$

where [ ] indicate actual, not equilibrium, concentrations.

mass balance equation CONSERVATION EQUATION.

mass chromatogram 1. A paper chromatogram of preparative-type separations in which the sample is streaked as a band on a large sheet of filter paper. 2. A reconstructed ion-current profile obtained by means of a computer from mass spectrometry data.

mass fragmentogram The photographic record obtained from mass fragmentography.

mass fragmentography The combined use of gas chromatography and mass spectrometry in which the effluent from a gas chromatographic column is fed into a mass spectrometer which serves as a detector for the recording of from one to three fragments at preselected mass-to-charge ratios; the mass spectrometer is preset to these mass-to-charge ratios and serves as a sensitive detector for small quantities of specific ions. *Aka* single-ion monitoring.

mass number The sum of the number of protons and neutrons in the nucleus of an atom. *Sym* A.

mass spectrogram The photographic record of a mass spectrum as obtained with a mass spectrograph.

mass spectrograph A mass spectrometer in which a photographic record of the mass spectrum is obtained.

mass spectrometer An instrument for the separation of charged particles according to their mass-to-charge ratios. Molecules and ions are fragmented by bombardment with electrons, and the ions thus formed are focused by means of electrostatic and magnetic fields and ultimately strike a photographic plate or some other detector. Both the mass-to-charge ratios and the relative amounts of the charged fragments can be determined.

mass spectrum A plot of the number of fragments as a function of their mass-to-charge ratio as measured with a mass spectrometer or a mass spectrograph.

mass stopping power LINEAR ENERGY TRANSFER.

mass unit *See* atomic mass unit.

mast cell A basophilic cell of connective tissue that contains heparin and histamine and that functions in immediate-type hypersensitivity.

master gland The pituitary gland.

master pattern *See* King–Altman procedure.

master plate The mounted piece of sterile velvet that is used in replica plating and that is covered with the original bacterial culture.

master–slave model A genetic model according to which each cistron is present in the form of many copies, joined end to end. In this linear group of genes, there is one copy (probably at the end) that is the "master" and all the other genes are the "slaves." The latter are corrected once per life cycle of the organism so that they conform to the master copy.

master strand SENSE STRAND.

mate killer A *Paramecium aurelia* cell that carries mu particles and that kills or injures sensitive paramecia with which it conjugates; the mu particles protect the organism against other mate killers. *See also* kappa particle.

mathematical model An equation that describes the behavior of an actual physical system and that is derived on the basis of theoretical considerations and pertinent numerical parameters.

mating pool *See* Visconte–Delbrueck hypothesis.

matrix (*pl* matrices). 1. A gel-like substance that fills the space between the cristae of mitochondria and that is the site of many of the enzymes of the citric acid cycle. 2. A rectangular array having *m* rows, each containing *n* numbers (called elements).

matrix interference The interference that occurs in atomic absorption spectrophotometry when light is absorbed either by the organic solvent in which the sample is dissolved, or by the solids that are formed from this solvent by its evaporation in the flame.

matrix method A mathematical method for treating a steady-state enzyme system in which the time derivatives of the differential equations describing this state are set equal to zero, and the resultant algebraic equations are then solved by matrix inversion.

matrix proteins Protein components of the outer membrane of gram-negative bacteria; they are characterized by their tight, but noncovalent, association with peptidoglycan, and can be released from the latter by extraction with sodium dodecyl sulfate.

maturase A protein that is encoded by an intron–exon combination and that helps catalyze the excision of the intron from its own primary RNA transcript. In this mechanism (known as splicing homeostasis), the maturase destroys its own mRNA and thereby limits its own level of activity. Maturase is believed not to be an enzyme but rather a factor that modifies the specificity of a splicing enzyme.

maturation 1. The development of a red blood cell from its formation to its final form as an erythrocyte. 2. The assembly of the different components of a virus that results in a complete and infectious virion. 3. The development of a spore. 4. SPLICING (2).

maturation-defective mutant A phage mutant that can synthesize DNA but not the viral structural proteins.

maturation factor 1. A substance in liver that aids the maturation of red blood cells; vitamin B_{12} or a combination of vitamin B_{12} and intrinsic factor. 2. MATURATION PROTEIN.

maturation protein A phage-specific protein of small RNA phages, such as MS2, f2, and R17, that is required for the production of complete and infectious phage particles. *Aka* maturation factor; A protein.

mature virion A fully assembled, infectious virus particle.

maturity-onset diabetes *See* diabetes.

max Maximum.

Maxam–Gilbert method A chemical method for sequencing DNA. The method entails labeling single-stranded DNA, derived from double-stranded DNA, with radioactive phosphate (^{32}P) at the 5′-end. The DNA preparation is then divided into several parts and each part is subjected to a different chemical cleavage procedure. These procedures break the DNA at specific points and the resulting fragments are separated by polyacrylamide gel electrophoresis (on the basis of their size) and detected by autoradiography. The method can also be adapted to sequencing RNA.

maximal medium A rich medium that contains all the necessary metabolites for the growth of cells and that frequently consists of a protein hydrolysate, a yeast extract, and inorganic salts.

maximum height–area method A method for calculating translational diffusion coefficients from the height of the peak of the gradient curve and from the area under the peak.

maximum permissible body burden The greatest amount of total, cumulative exposure of an individual to radioactive radiation that is permitted by federal safety standards.

maximum permissible concentration The greatest concentration of radioactive isotopes in air and water that is permitted by federal safety standards. *Abbr* MPC.

maximum permissible dose The greatest amount of radioactive radiation that an individual may receive over a given period of time according to federal safety standards. *Abbr* MPD.

maximum stationary phase STATIONARY PHASE (1).

maximum velocity The greatest velocity of an enzymatic reaction that is attainable with a fixed amount of enzyme under defined conditions; the velocity that is obtained when the enzyme is saturated with substrates and cosubstrates. *Sym* V_{max}; V.

Maxwell distribution A plot of the "fraction" of molecules of a gas or a liquid as a function of their kinetic energy. The "fraction" is equal to $(1/N)(dN/dE)$ where N is the number of molecules having a kinetic energy E.

Maxwell effect FLOW BIREFRINGENCE.

Maxwell's demon A whimsical creature, proposed by James Clerk Maxwell to illustrate a violation of the second law of thermodynamics. The demon is visualized as operating a gate between two chambers containing a gas in thermal equilibrium such that only fast molecules approaching from one direction, and only slow molecules approaching from the other, are let through. This would result in an increase in the average kinetic energy of the molecules in one chamber (a temperature increase) and a decrease of energy and temperature in the other chamber. Thus, a temperature difference would be produced spontaneously in a system in thermal equilibrium which is in violation of the second law. Leo Szilard pointed out later that Maxwell's argument was invalid since it did not consider the information processing activities of the demon. These consist of detecting the molecular velocities and deciding to open or close the gate. Hence, the total entropy change, that of the gas plus that of the demon and his measuring instruments, must be positive in accordance with the second law of thermodynamics.

Mb Myoglobin; related compounds are abbreviated as MbO_2 (oxymyoglobin) and MbCO (carbon monoxide myoglobin).

MBSA Methylated bovine serum albumin.

mc Millicurie.

MC Microtubule organizing center.

5-MC 5-Methylcytosine.

McArdle–Schmid–Pearson disease GLYCOGEN STORAGE DISEASE TYPE V.

McArdle's disease GLYCOGEN STORAGE DISEASE TYPE V.

MCD Magnetic circular dichroism.

M chain One of two types of polypeptide chains of lactate dehydrogenase isozymes; denoted M, since the tetramer of M chains is found predominantly in muscle tissue.

mCi Millicurie.

McLeod gauge A laboratory pressure gauge for measuring gas pressures of vacuum systems to as low as 10^{-6} mm Hg.

M component PARAPROTEIN.

MCP Methyl-accepting chemotaxis protein.

MDGC Multidimensional gas chromatography.

MDH Malate dehydrogenase.

mDNA Mitochondrial DNA.

MDR Morphology-dependent resonance.

Me Methyl group.

mean The value obtained by summing the values of a set of measurements and dividing the sum by the number of individual measurements in the set; arithmetic average.

mean activity coefficient *See* mean ionic activity coefficient.

mean deviation The arithmetic mean (average) of the absolute deviations from the mean.

mean free path The average distance traveled by a molecule, an ionizing particle, or some other particle between collisions in a gas at equilibrium.

mean generation time DOUBLING TIME.

mean ionic activity coefficient The average activity coefficient for the anion and cation of an electrolyte. It may be calculated from the Debye–Hueckel limiting law: $\log \gamma_\pm = -0.509|z_+z_-|\sqrt{I}$, where $\gamma_\pm$ is the mean ionic activity coefficient, $|z_+z_-|$ represents the product for the absolute values of the charges of the cation and anion, respectively, and I is the ionic strength.

mean life AVERAGE LIFE.

mean range The distance between the source of a radiation, particularly a source of alpha particles, and the point at which the intensity of the beam is reduced to one-half.

mean residue rotation The specific rotation of a polymer that is calculated on the basis of the concentration of the monomers, or residues, of the polymer rather than on the basis of the concentration of the intact polymer molecules. Specifically, $[m] = M_0[\alpha]/100$, where $[m]$ is the mean residue rotation, M_0 is the mean residue weight, and $[\alpha]$ is the specific rotation of the polymer. If the mean residue rotation is corrected to that in a medium having a refractive index of one, it is given by the expression $[m] = [3/(n^2 + 2)] \times M_0[\alpha]/100$, where n is the refractive index of the medium.

mean residue weight The average molecular weight of a monomer, or residue, in a polymer.

mean square VARIANCE.

mean square deviation The squared difference between an estimated and a theoretically correct value. *Aka* mean square error.

mean square displacement The average of the sum of the squared displacements of a molecule in a given time; the distances that a molecule moves (displacements) are squared, the squared values are added, and the sum is divided by the number of displacements.

mean square end-to-end distance ROOT MEAN SQUARE END-TO-END-DISTANCE.

mean square error MEAN SQUARE DEVIATION.

measuring pipet A pipet consisting of a tube of uniform diameter that is drawn out to a tip and is graduated uniformly along its length.

mechanism 1. A step-by-step description of a chemical or a physical reaction or reaction sequence; for a chemical reaction this includes the electron shifts and the bond-making and bond-breaking aspects of the reaction. 2. MECHANISTIC PHILOSOPHY.

mechanistic philosophy The doctrine that life and its phenomena are explicable entirely in terms of the laws and processes of chemistry and physics. *Aka* mechanism. *See also* vitalism.

mechanistic process A deterministic process in which each step is a necessary and direct result of a preceding step. *See also* stochastic process.

mechanistic theory 1. A theory according to which the evolution of the genetic code is based on a necessary physical–chemical relation between an amino acid and its codons; consequently, the code could have evolved in one, or at most in a few, possible ways. *See also* selective theory. 2. A theory according to which the formation of atheromatous plaques in atherosclerosis is due to the precipitation and/or the coagulation of one or more of the components of the blood.

mechanochemical coupling hypothesis CONFORMATIONAL COUPLING HYPOTHESIS.

median The value of a set of measurements around which the measurements are equally distributed; one-half of the measurements are numerically greater and one-half are numerically smaller than this value.

median effective dose The dose of a drug that produces therapeutic effects in 50% of the animals in a test group within a specified time. *Sym* ED_{50}. *See also* EC_{50}.

median hemolytic dose The dose of complement that produces hemolysis in 50% of a standardized suspension of sensitized erythrocytes. *Sym* HD_{50}.

median immunizing dose The dose of a vaccine or an antigen that produces immunity in 50% of the animals in a test group within a specified time.

median infectious dose The dose of bacteria or viruses that produces demonstrable infection in 50% of the animals in a test group within a specified time. *Sym* ID_{50}. *Aka* median infective dose.

median lethal dose The dose of bacteria,

viruses, or a toxic compound that produces deaths in 50% of the animals in a test group within a specified time. *Sym* LD_{50}. *See also* half-value dose.

median paralysis dose The dose of virus that produces paralysis in 50% of the animals in a test group within a specified time. *Sym* PD_{50}.

median tissue culture dose The dose of virus that produces tissue culture degeneration in 50% of the test units within a specified time. *Sym* TC_{50}; TCD_{50}. *Aka* tissue culture infectious dose.

median toxic dose The dose of a toxic agent that produces toxic effects in 50% of the animals in a test group within a specified time. *Sym* TD_{50}.

mediated transport The movement of a solute across a biological membrane that requires the participation of one or more transport agents. The transport may be passive (along a concentration gradient) or active (against a concentration gradient).

medicine 1. The science and art of diagnosing, treating, and preventing disease. 2. A substance used in either the treatment or the prevention of a disease.

medium (*pl* media) A liquid or a solid preparation of nutrients that is used for the maintenance and growth of microorganisms and cells and for the cultivation of tissues and organs.

medium-chain fatty acid thiokinase A fatty acid thiokinase that catalyzes the activation of fatty acids having 4 to 12 carbon atoms to fatty acyl coenzyme A.

medulla *See* adrenal medulla.

MEF Migration enhancement factor.

mega- Combining form meaning one million (10^6) and used with metric units of measurement. *Sym* M.

megadose A very large dose such as the gram quantities of vitamin C recommended by Linus Pauling for the prevention of colds.

megamitochondrion An enlarged mouse liver mitochondrion that is produced artificially by exposure of the animal to the drug cuprizone (biscyclohexanone oxaldihydrazone) or the antibiotic chloramphenicol.

Mehler reaction A photosynthetic reaction whereby hydrogen peroxide is formed from water and molecular oxygen.

meiosis The nuclear division of the germ cells of sexually reproducing organisms in which the chromosome number is halved; it occurs during gametogenesis in animals and during sporogenesis in plants.

meiospore A spore produced by meiosis.

meiotic Of, or pertaining to, meiosis.

meiotic drive An irregularity in the segregation of the chromosomes during meiosis that leads to alterations in the allele frequencies of a population.

meiotic effect The phenomenon in which the mutation rate during meiosis differs from that during mitosis.

Meister cycle γ-GLUTAMYL CYCLE.

melanic 1. Of, or pertaining to, melanism. 2. Having a dark pigmentation.

melanin pigment One of a group of dark coloring substances that are responsible for the pigmentation of the skin and that are formed in melanocytes by the oxidation of phenylalanine, tyrosine, and other aromatic compounds.

melanism The abnormal coloration of the skin or other tissues that is caused by the accumulation of melanin pigments.

melanocyte A cell that synthesizes melanin pigments in its cytoplasm.

melanocyte-stimulating hormone One of two peptide hormones, denoted α and β, that are produced by the posterior lobe of the pituitary gland and that have a darkening effect by causing the dispersion of melanin pigments in the melanocytes. *Abbr* MSH.

melanocyte-stimulating hormone regulatory hormone One of two hypothalamic hormones (or factors) that, respectively, stimulates or inhibits the release of melanocyte-stimulating hormone from the pituitary gland. The melanocyte-stimulating hormone releasing hormone (or factor) is abbreviated variously as MRH (MRF), or MSHRH (MSHRF). The melanocyte-stimulating hormone release-inhibiting hormone (or factor) is abbreviated variously as MIH (MIF), MRIH (MRIF), MSHIH (MSHIF), or MSHRIH (MSHRIF).

melanocyte-stimulating hormone release-inhibiting hormone *See* melanocyte-stimulating hormone regulatory hormone.

melanocyte-stimulating hormone releasing hormone *See* melanocyte-stimulating hormone regulatory hormone.

melanogen One of a group of compounds frequently observed in the urine of patients with malignant melanoma; believed to be either precursors of melanin or metabolic byproducts of melanin biosynthesis.

melanoma A malignant tumor derived from melanocytes.

melanosome A tyrosinase-containing intracellular organelle of melanocytes.

melanotrophin Variant spelling of melanotropin.

melanotropin MELANOCYTE-STIMULATING HORMONE.

melatonin A tryptophan-related hormone that is formed in the pineal gland and that reverses the darkening effect of the melanocyte-stimulating hormone by causing aggregation of the melanin granules in the melanocytes.

melitose RAFFINOSE.

melittin A linear polypeptide of 26 amino acids that is the major component of bee venom; a toxic and hemolytic peptide that is synthesized as an inactive precursor, promelittin.

melituria GLYCOSURIA.

melphalan A mutagenic, alkylating agent. *Var sp* melphalin.

melting 1. The transition of double-helical nucleic acid segments to random coil conformations that is produced by an increase in the temperature of the solution containing the nucleic acid and that is due to the breaking of the hydrogen bonds of the paired bases. *Aka* thermal denaturation; melting out. 2. The transition of a solid to a liquid that is produced by an increase in temperature.

melting curve THERMAL DENATURATION PROFILE.

melting out profile THERMAL DENATURATION PROFILE.

melting out temperature The temperature of a thermal denaturation profile at which one-half of the maximum change in absorbance (or other property) is obtained; the temperature at which one-half of the helical structure is lost. *Sym* T_m; $T_{1/2}$. *Aka* melting temperature.

melting point 1. MELTING OUT TEMPERATURE. 2. The temperature at which a solid changes to a liquid as a result of the application of heat; at this temperature the solid and the liquid are in equilibrium.

melting profile THERMAL DENATURATION PROFILE.

melting protein SINGLE-STRAND BINDING PROTEIN.

membrane carrier *See* carrier (3).

membrane electrode An electrode, such as the glass electrode, that has a membrane incorporated into its design.

membrane equilibrium GIBBS–DONNAN EQUILIBRIUM.

membrane filter A thin filter, made of nitrocellulose or other cellulose esters, that is used for the collection of microorganisms and of protein and nucleic acid precipitates.

membrane fluidity *See* fluid mosaic model.

membrane hydrolysis The effective hydrolysis of a protein, present on one side of a membrane, that is brought about by establishment of a pH gradient across the membrane (as in the Gibbs–Donnan equilibrium); involves conversion of a salt form of the protein (P) to a protonated or hydroxylated form. For example: $NaP + H_2O \rightleftharpoons HP + NaOH$; $PCl + H_2O \rightleftharpoons POH + HCl$.

membrane mimetic chemistry Chemical research of processes that take place in simple media and that mimic aspects of biological membranes; includes studies of aqueous micelles, reversed micelles, monolayers, black membranes, and the like.

membrane osmometer An osmometer that has a semipermeable membrane incorporated into its design and that is used for measurements of osmotic pressure and for determinations of number average molecular weights of macromolecules.

membrane potential The electrical potential across a membrane, particularly a biological membrane, that arises from the charges in the membrane and from the charges present on either side of the membrane. *See also* diffusion potential; dipolar potential; distribution potential; equilibrium potential.

membrane structure *See* Benson model; Davson–Danielli model; fluid mosaic model; supermolecule.

membrane transport The movement of materials across a biological membrane.

membrane trigger hypothesis An alternative to the signal hypothesis. According to this hypothesis, some proteins, which are synthesized on soluble polysomes, can assume two conformations, one that is more stable in aqueous solutions and another that is induced (triggered) by contact with the hydrophobic environment of the membrane. The soluble protein precursor is thought to undergo a conformational change as it inserts itself into the bilayer of the membrane.

membranochromic pigment One of a group of pigments, such as phenols and quinones, that impregnate the cell wall of plants.

membranolysis Lysis of the cell membrane.

membron A functioning, regulatable, translating complex of a polysome and a specific surface area of a biological membrane.

membron theory of cancer The theory according to which tumor cells differ from normal cells in the stability of their membrons.

memory 1. A device for storing information. 2. IMMUNOLOGICAL MEMORY. 3. The storage capacity of a computer.

memory cell A cell that is responsible for immunological memory. A lymphocyte in peripheral lymphoid tissue that is not actively engaged in making an immunological response to a given antigen but that is readily induced to do so, and to become an effector cell, by a later encounter with the same antigen.

memory response SECONDARY IMMUNE RESPONSE.

menadione Vitamin K_3.

menaquinone Vitamin K_2.

Mendelian Of, or pertaining to, Mendel or to his laws of heredity.

Mendelian character A character that follows Mendel's laws of inheritance.

Mendel's laws The laws of inheritance proposed by Mendel in 1866 and known as the law of segregation and the law of independent assortment.

mengo virus A virus that belongs to a subgroup of picornaviruses and that can cause fatal encephalitis in mice.

meniscus The flat or crescent-shaped interface between a liquid in a tube and air.

meniscus depletion sedimentation equilibrium A variation of the sedimentation equilibrium method in which the ultracentrifuge is operated at sufficiently high speeds so that all the macromolecules are sedimented out of the region near the meniscus. The method obviates a separate run to determine the initial concentration of the solution, is especially useful for monodisperse systems, and is generally used in conjunction with an interferometric optical system.

menu A list of computer commands that most ready-made programs will display at request.

meq Milliequivalent; also denoted mEq.

meractinomycin ACTINOMYCIN D.

mercaptan THIOL.

2-mercaptoethanol A sulfhydryl group containing compound that is used to protect the sulfhydryl groups of enzymes and other proteins against oxidation.

β-Mercaptoethylamine The sufhydryl group-carrying component of coenzyme A.

mercapto group The sulfhydryl (—SH) group.

6-mercaptopurine A purine analogue used in cancer chemotherapy; an antitumor agent. *Abbr* MP.

mercapturic acids A large group of substances formed by the detoxification of xenobiotics. The reactions involve conjugation of the xenobiotic with glutathione (catalyzed by glutathione-S-transferases, also called ligandins), followed by removal of the glutamyl and glycyl residues of glutathione and conversion of the residual compound (containing the cysteine residue of glutathione) to a sulfur-containing acid, called mercapturic acid.

mercerization The treatment of cellulose with 20% sodium hydroxide which produces a cellulose that has an increased capacity for dyes and a greater tensile strength.

mercurial An organic compound, such as a drug, that contains mercury.

meridian The direction on the film, in x-ray diffraction, that is parallel to the fiber axis and that passes through the x-ray beam when the film is considered to be wrapped cylindrically

about the fiber axis; a line that is perpendicular to the layer lines.

meridional 1. Of, or pertaining to, a meridian. 2. At right angles to the equator.

meridional reflection An x-ray diffraction spot that lies on the meridian.

merocrine gland A gland that produces a secretion without significantly damaging its secreting cells.

meromyosin One of two fragments produced from myosin by treating it with either trypsin or chymotrypsin. *See also* heavy meromyosin; light meromyosin.

merozygote A partially diploid zygote that contains one complete and one partial genome; produced as a result of a partial genetic exchange, as that which may occur during bacterial transformation, transduction, or conjugation. The genetic complement of the merozygote consists of the endogenote and the exogenote.

Merrifield method SOLID PHASE SYNTHESIS.

mersalyl A drug that inhibits the exchange of phosphate and hydroxyl ions across the inner mitochondrial membrane.

MES 2-(N-Morpholino)ethanesulfonic acid; used for the preparation of biological buffers in the pH range of 5.5 to 6.7. *See also* biological buffers.

mescaline A hallucinogenic drug that occurs naturally in a cactus and that has the structure of a phenylethylamine; a narcotic.

Meselson–Stahl experiment An experiment that provides support for the semiconservative mode of DNA replication. The experiment consists of labeling the DNA in a growing culture of *E. coli* with a heavy isotope, transferring the culture to an unlabeled medium for further growth, isolating the DNA at various stages of the growth curve, and analyzing the DNA by means of density gradient sedimentation equilibrium.

mesh size A standard screen for designating the particle size of ion-exchange resins, gels, and other chromatographic supports; a larger mesh size indicates a smaller particle diameter.

meso carbon A carbon atom to which are attached two identical and two different substituents. The two identical substituents bear a mirror image relation to each other and react differently with a given enzyme. *Aka* prochiral carbon.

meso compound An optical isomer that possesses asymmetric elements, such as asymmetric carbons, but has overall molecular symmetry and is, therefore, optically inactive.

mesoderm The middle of the three germ layers of an embryo from which connective tissues, muscles, blood and lymph tissues, and urinogenital organs develop.

mesomorphic Of, or pertaining to, a liquid crystal.

meson A subatomic particle that has a mass greater than that of a lepton but smaller than that of a nucleon.

mesophase LIQUID CRYSTAL.

mesophile An organism that grows at moderate temperatures in the range of 20 to 45 °C, and that has an optimum growth temperature in the range of 30 to 37 °C.

mesophilic Of, or pertaining to, mesophiles.

mesophyll cells The outer layer of cells in the leaves of C_4 plants; site of the preliminary fixation of CO_2 that precedes the reactions of the Calvin cycle which take place in the bundle sheath cells.

mesosome An infolding of the bacterial cell membrane. Mesosomes are believed to represent centers of cell respiration (electron transport system) and thus to function much as the inner mitochondrial membrane does in higher organisms. *Aka* chondrioid.

mesotocin A peptide hormone that is related to oxytocin in its structure and in its function; it is secreted by the posterior lobe of the pituitary gland and occurs in reptiles and amphibia.

mesotrophic lake A lake that has intermediate properties between those of an oligotrophic lake and those of a eutrophic lake.

Mesozoic era The geologic time period that extends from about 63 to 225 million years ago and that is characterized by the development of the reptiles.

message 1. A messenger RNA molecule. 2. The segment of a polycistronic messenger RNA molecule that codes for one polypeptide chain.

messenger ribonucleoprotein particle A messenger RNA molecule that is complexed with protein but is not associated with ribosomes. Such complexes have been isolated from the cytoplasm of eukaryotic cells. It is believed that at least some of these complexes are specific ones that function in the translation of mRNA in the absence of ribosomes. *Abbr* mRNP.

messenger ribonucleoproteins A complex of heat-shock proteins and small cytoplasmic RNA (scRNA) molecules. *Abbr* mRNP.

messenger RNA A single-stranded RNA molecule that is synthesized during transcription, is complementary to one of the strands of double-stranded DNA, and serves to transmit the genetic information contained in DNA to the ribosomes for protein synthesis. *Abbr* mRNA. *Aka* messenger.

messenger RNA hypothesis The hypothesis, proposed by Jacob and Monod, that an RNA molecule serves as the template for the synthesis of proteins; this RNA molecule, the messenger RNA, is transcribed from DNA, has a base sequence that is complementary to that of one of the strands of duplex DNA, and carries the genetic information from the DNA to the ribosomes where the proteins are synthesized.

Met 1. Methionine. 2. Methionyl.

meta- Prefix indicating two substituents on alternate carbon atoms of the ring in an aromatic compound. *Sym* m.

metabolic Of, or pertaining to, metabolism.

metabolic acidosis A primary acidosis that results from changes in the concentrations of acids and bases other than carbon dioxide and carbonic acid.

metabolic alkalosis A primary alkalosis that results from changes in the concentrations of acids and bases other than carbon dioxide and carbonic acid.

metabolic antagonist ANTIMETABOLITE.

metabolic balance study *See* balance study.

metabolic block A block in a biochemical reaction, generally due to a mutation, that results in the lack of synthesis of an enzyme or in the synthesis of a defective enzyme.

metabolic bypass METABOLIC SHUNT.

metabolic depression A general reduction of the synthesis of biosynthetic enzymes that may occur when cells are grown in a rich medium in which the growth rate is very rapid. The term is meant to differentiate this type of depression from one that is specific for a given metabolic pathway.

metabolic disease A pathological abnormality of metabolism such as acidosis, alkalosis, or an inborn error of metabolism.

metabolic pathway A sequence of consecutive enzymatic reactions that brings about the synthesis, breakdown, or transformation of a metabolite from a key intermediate to some terminal compound. A metabolic pathway may be linear, cyclic, branched, tiered, directly reversible, or indirectly reversible. *See also* linear metabolic pathway; cyclic metabolic pathway; etc.

metabolic poison A substance that inhibits a metabolic reaction.

metabolic pool *See* pool.

metabolic quotient A measure of the rate of uptake or discharge of a metabolite by a tissue or an organism. The uptake of oxygen, denoted Q_{O_2}, and the evolution of carbon dioxide, denoted Q_{CO_2}, are frequently expressed in terms of microliters taken up or evolved per hour per milligram dry weight of tissue. *Sym* Q.

metabolic shunt A pathway in metabolism that uses some reactions of a major metabolic pathway and bypasses others. The glyoxylate bypass and the γ-aminobutyrate bypass are two examples.

metabolic transformation BIOTRANSFORMATION.

metabolic turnover The rate at which cellular components are replaced by degradation and synthesis under steady-state conditions; turnover time.

metabolism 1. The sum total of all the chemical and physical changes that occur in a living system, which may be a cell, a tissue, an organ, or an organism. The reactions of metabolism are almost all enzyme-catalyzed and include transformation of nutrients, excretion of waste products, energy transformations, synthetic and degradative processes, and all the other functions of a living organism. Metabolism is broadly divided into anabolism, which encompasses the synthetic reactions, and catabolism, which encompasses the degradative reactions. 2. The sum total of all the chemical and physical changes in a living system with respect to one class of compounds, as in "amino acid metabolism."

metabolite Any reactant, intermediate, or product in the reactions of metabolism. Metabolites that are involved in those processes that are basically similar in all organisms and that are necessary for maintenance and survival are known as primary metabolites. This includes metabolites involved in the processes of growth (biosynthesis), energy production and transformation, and turnover of cellular components. Other metabolites, such as pigments, alkaloids, antibiotics, terpenes, and so on, that occur only in certain organisms and that serve no apparent biological function in the life of the organisms that produce them, are known as secondary metabolites.

metabolon 1. Proposed term to describe an organized chemical entity capable of growth and believed to have been formed in the course of chemical evolution prior to the development of protocells. 2. Proposed term to describe a supramolecular complex of sequential metabolic enzymes and cellular structural elements that exists within individual organelles and cell compartments. Ribosomes on the endoplasmic reticulum with bound mRNA and tRNAs, glycolytic enzymes bound to actin, citric acid cycle enzymes bound to the inner mitochondrial membrane, and the DNA replication complex are some examples.

metachromatic dye A dye that stains a tissue with two or more different colors depending on the extent to which the dye molecules are stacked on the chromotrope.

metachromatic granule VOLUTIN GRANULE.

metachromatic leukodystrophy A genetically inherited metabolic defect in humans that is associated with mental retardation and that is characterized by an accumulation of sulfatides; a sphingolipidosis that is due to a deficiency of the enzyme cerebroside sulfatase (sufatide sulfatase) which hydrolyzes sulfatides to inorganic sulfate and galactosylceramide.

metagon An RNA particle of the protozoan *Paramecium aurelia* which acts like a messenger RNA molecule in that organism, but replicates like an RNA virus when ingested by the protozoan *Didnium*.

metakentrin LUTEINIZING HORMONE.

metal-activated enzyme An enzyme that retains one or more metal ions in equilibrium with binding groups on its surface. Such metal ions are frequently lost during purification of the enzyme and must then be added back to restore the catalytic activity. *See also* metalloenzyme.

metal bridge complex *See* bridge complex.

metal chelate *See* chelate.

metalloenzyme A conjugated enzyme that contains one or more metal ions as prosthetic groups. Metalloenzymes generally fail to show activity enhancement upon addition of the free metal ions. *See also* metal-activated enzyme.

metalloflavoprotein A complex flavoprotein that contains a metal ion in addition to either FMN or FAD.

metalloporphyrin A complex composed of a metal ion that is chelated by a porphyrin.

metalloprotein A conjugated protein that contains one or more metal ions as prosthetic groups.

metallothionein A small, cytoplasmic, cysteine-rich (about 30 mol%) metal-binding protein that occurs in a wide variety of eukaryotic species including vertebrates, invertebrates, plants, and microorganisms. Metallothionein binds, and can be induced by, various metal ions such as Zn, Cd, Cu, and Hg. It is believed to be involved in the storage and regulation of Zn and Cu and in the detoxification of heavy metal ions. *Abbr* MT.

metal shadowing SHADOWCASTING.

metamorphosis A transformation in the form of an animal, such as that from larval to adult form in insects, or that from tadpole to adult form in amphibians.

metaphase The second stage in mitosis during which the chromosomes arrange themselves in an equatorial region.

metaphosphate The anionic radical PO_3^- of metaphosphoric acid, HPO_3.

metaprotein A denatured protein, formed by the action of acids or bases, that is soluble in weak acids and bases but is insoluble in neutral solutions.

metarhodopsin One of several structurally

altered forms of rhodopsin that are produced after the exposure of rhodopsin to light and prior to its dissociation into opsin and retinal₁.

metastable Describing an unstable condition or substance that changes readily either to one that is more stable, or to one that is less stable.

metastable nuclide An excited nuclear isomer that emits a gamma ray upon its return to the ground state.

metastable state 1. The excited state of an atom or a molecule that is characterized by a delayed emission of the excitation energy as the atom or the molecule returns to the ground state. 2. The excited state of a nuclear isomer that is characterized by the emission of a gamma ray as the nucleus returns to the ground state. 3. Any state of an ion or a molecule that is of short duration.

metastasis (*pl* metastases) 1. The detachment of cells from a tumor and their transport, by way of blood and the lymphatic system, to distant sites in the organism where they grow to form additional tumors. 2. The tumor formed at a site distant from that of the original tumor, and produced by detachment and transport of cells from the original tumor.

metastasize To invade by means of metastasis.

metastatic tumor A tumor that is undergoing metastasis; a metastasizing tumor.

metathesis A chemical reaction in which there is a double exchange of elements or groups, as in the reaction $AB + CD = AD + BC$.

Met-enkephalin *See* enkephalin.

methanogens A group of anaerobic archaebacteria that generate methane by reduction of CO_2; they are widely distributed but not commonly encountered since they are killed by oxygen and do not exist in the open. They can be isolated from the ocean bottom and from hot springs. *See also* archaebacteria.

methanolysis An alcoholysis reaction in which the alcohol is methanol; frequently used for the formation of fatty acid methyl esters which are then analyzed by gas chromatography.

MetHb Methemoglobin.

methemoglobin A hemoglobin molecule in which the iron has been oxidized to the trivalent state. *Abbr* MetHb.

methemoglobinemia A genetically inherited metabolic defect in humans that is characterized by high concentrations of methemoglobin in the blood and that is due to a deficiency of the enzyme NADH-methemoglobin reductase.

methenyl group The grouping $-CH=$.

methionine A sulfur-containing, nonpolar alpha amino acid. *Abbr* Met; M.

methionyl transfer RNA A transfer RNA molecule that exists in two forms, designated $tRNA^{met}$ and $tRNA^{fmet}$. The former is ordinary methionyl-tRNA, and the latter allows for the formylation of methionine after it has become attached to the tRNA; N-formylmethionyl-tRNA^{fmet} serves as the initiator aminoacyl-tRNA in bacterial protein synthesis.

method of continuous variation A method for studying the interaction between two components in solution, as in the formation of a duplex from two single strands. The method requires the plotting of a property that is characteristic of one component versus the mole fraction of that component in a two-component system. In the absence of interaction between the components, a straight line is obtained which connects the points corresponding to a mole fraction of one for each of the components. In the presence of interaction between the components, various deviations from a straight line are obtained.

method of least squares A method for fitting a straight line to a set of experimental points such that the sum of the squares of the deviations of the experimental points from the line has a minimum value; the line is referred to as a line of best fit.

method of optimal proportions A method for determining the equivalence zone of a precipitin reaction by measuring the proportion of antigen to antibody at which precipitation occurs most rapidly. The zone of optimal proportions measured in this fashion is generally near the equivalence zone of the precipitin curve. *See also* Dean and Webb method; Ramon method.

method of ultimate precision A method for measuring the absorbance of a solution in a single beam photometer with maximum sensitivity and with minimum error; measurements are made by adjusting the instrument to 0 and 100% transmission with two solutions of known concentration in the compound of interest.

methotrexate AMETHOPTERIN.

methoxatin *See* quinoprotein.

methyl-accepting chemotaxis protein One of a group of bacterial transmembrane proteins that are responsible for transmitting chemotactic signals across the cell membrane; they become methylated during the chemotactic response. *Abbr* MCP.

methylase An enzyme that catalyzes a methylation reaction; a methyltransferase.

methylated albumin–kieselguhr An adsorbent used for the chromatographic fractionation of nucleic acids; prepared by converting the carboxyl groups of the glutamic and aspartic acid residues of serum albumin to methyl esters, and then precipitating the methylated albumin onto kieselguhr particles. *Abbr* MAK.

methylated cap   CAP (1).

methylated xanthines   *N*-Methyl derivatives of xanthine; collective term for the purine alkaloids caffeine, theobromine, and theophylline.

methylating agent   *S*-ADENOSYLMETHIONINE.

methylation   The introduction of a methyl group ($-CH_3$) into an organic compound.

methylation analysis   EXHAUSTIVE METHYLATION.

methylcholanthrene   A carcinogenic hydrocarbon.

5-methylcytosine   A derivative of cytosine that is found in the DNA of certain higher plants and in tRNA.

O-methyl derivative   A carbohydrate derivative in which one or more hydroxyl groups have been methylated.

methylene blue   A dye used as an oxidation–reduction indicator; the oxidized form is blue, and the reduced form is colorless.

methylene blue technique   THUNBERG TECHNIQUE.

methylene group   The grouping $-CH_2-$.

methylferase   Methyltransferase; a methylase.

methyl green   A basic dye used in cytochemistry for the staining of DNA.

methyl group   The radical $-CH_3$. *Sym* Me.

ε-N-methyllysine   A rare amino acid that is present in the protein actin.

methylmalonic aciduria   METHYLMALONYL ACIDEMIA.

methylmalonyl acidemia   A genetically inherited metabolic defect in humans that is characterized by massive ketosis and that is due to a deficiency of the enzyme methylmalonyl-CoA carboxymutase. *Aka* methylmalonic aciduria.

methylneogenesis   The biosynthesis of methyl groups from one-carbon fragments.

methylol amino acid   An amino acid derivative, formed by reaction with formaldehyde. Examples include formation of a monomethylol derivative ($R-NH-CH_2OH$) by reaction with the imidazole group of histidine, and formation of a dimethylol derivative [$R-N(CH_2OH)_2$] by reaction with the alpha amino group of amino acids.

methylotroph   An organism that can utilize as its sole carbon source either one-carbon compounds (such as methane and methanol) or carbon compounds that contain no carbon-to-carbon bonds (such as dimethyl ether).

methylpherase   Variant spelling of methylferase.

methyl-poor transfer RNA   A transfer RNA molecule that contains less than the usual amount of methylated bases.

N⁵-methyl tetrahydrofolic acid   SERUM FOLATE.

methyl-trap hypothesis   A hypothesis that attempts to explain the interrelations between folic acid and vitamin B_{12} by assuming that a vitamin B_{12} deficiency results in the transformation of all of an individual's tetrahydrofolate to 5-methyltetrahydrofolate which is trapped as such. This is based on the fact that the enzyme 5-methyltetrahydrofolate methyltransferase requires a methyl-vitamin B_{12} coenzyme and that it catalyzes the major, and perhaps the only, metabolic reaction that can utilize 5-methyltetrahydrofolate.

MetMb   Metmyoglobin.

metmyoglobin   A myoglobin molecule in which the iron has been oxidized to the trivalent state. *Abbr* MetMb.

metopon   A semisynthetic drug, made by converting morphine to methyldihydromorphinone.

metric combining forms   *See* exa; peta; tera; giga; mega; kilo; hecto; deca; deci; centi; milli; micro; nano; pico; femto; atto.

met-tRNA   Methionyl-transfer RNA.

metyrapone test   A test for assessing the integrity of the hypothalamus–pituitary–adrenal system; based on the fact that, in normal individuals, oral administration of metyrapone will inhibit several enzyme systems including one involved in cortisol biosynthesis. The resultant decrease of cortisol stimulates the secretion of adrenocorticotropic hormone (ACTH).

MeV   Mega-electronvolt; 10^6 eV.

mevalonic acid   An intermediate in the biosynthesis of cholesterol and a precursor of squalene and other isoprenoid compounds.

mevinolinic acid   A fungal metabolite that is a potent inhibitor of the enzyme HMG-CoA reductase.

Meyerhof oxidation quotient   A measure of the Pasteur effect that is equal to the difference between the rates of anaerobic and aerobic glycolysis, divided by the rate of oxygen uptake.

MF   1. Mitogenic factor. 2. Maize factor. 3. Microfilaments.

MFO   Mixed-function oxidase.

mg   Milligram.

Mg   Magnesium.

MgATP   Magnesium ATP; magnesium chelated to adenosine-5′-triphosphate.

mg percent   Referring to a solution, the concentration of which is expressed in terms of the number of milligrams of solute per 100 mL of solution; widely used in clinical chemistry. *Sym* mg%.

MHC   1. Myosin heavy chain; *see* myosin. 2. Major histocompatibility complex.

MHC antigens   Cell surface glycoprotein antigens encoded by genes of the major histocompatibility complex. They determine the specificity of antigen recognition by lymphocytes. Class I MHC antigens have a wide tissue distribution and serve to distinguish body cells from invading cells. Class II MHC antigens are expressed only by some cell types; they are also known as Ia antigens or I region-associated antigens. They serve to distinguish immune system cells from other cells.

MHC associative recognition The property of most cytotoxic T cells of recognizing foreign antigens on the surface of cells only if the antigens are associated with self major histocompatibility complex glycoproteins, expressed on the same cell surface.

MHC molecules The antigens of the major histocompatibility complex; the MHC antigens.

MHD 1. Minimum hemagglutinating dose. 2. Minimum hemolytic dose.

mho The reciprocal of ohm; a measure of electrical conductance.

MIC 1. Major immunogene complex. 2. Minimum inhibitory concentration; the lowest concentration (generally in terms of micrograms per milliliter) at which an antibiotic is effective against a particular microorganism.

micellar Of, or pertaining to, micelles.

micelle An organized, spherical colloidal structure that consists of a large number of oriented surface-active molecules; micelles are generally charged and are typically formed by soaps and by phospholipids. In an aqueous micelle, formed in a polar solvent such as water, the nonpolar parts of the molecules are clustered in the interior of the spherical structure while the polar parts are distributed over the surface of the structure. In a reversed micelle, formed in a nonpolar solvent, the arrangement of the polar and nonpolar parts of the molecules is reversed.

micellisation The formation of micelles.

mic gene A gene capable of encoding an RNA complementary to all or part of a specific mRNA. Thus transcription of a mic(lpp) gene will result in the production of RNA complementary to the lpp mRNA.

Michaelis complex A structure in which the substrate is bound to an enzyme but the substrate is not in its transition state.

Michaelis constant A kinetic constant for a given substrate of an enzymatic reaction; it is numerically equal to that substrate concentration that yields one-half of the maximum velocity of the reaction at saturating concentrations of all cosubstrates. The Michaelis constant K_m is composed of the rate constants for the individual steps in the reaction; for the sequence

$$E + S \underset{k_{-1}}{\overset{k_{+1}}{\rightleftharpoons}} ES \xrightarrow{k_{+2}} E + P$$

it is given by

$$K_m = (k_{-1} + k_{+2})/k_{+1};$$

for the sequence

$$E + S \underset{k_{-1}}{\overset{k_{+1}}{\rightleftharpoons}} ES \underset{k_{-2}}{\overset{k_{+2}}{\rightleftharpoons}} EP \xrightarrow{k_{+3}} E + P$$

it is given by

$$K_m = \frac{(k_{-1}k_{-2} + k_{-1}k_{+3} + k_{+2}k_{+3})}{[k_{+1}(k_{+2} + k_{-2} + k_{+3})]}$$

where the k's are the rate constants, E is the enzyme, S is the substrate, and P is the product. *Sym* K_m.

Michaelis–Menten–Briggs–Haldane equation MICHAELIS–MENTEN EQUATION.

Michaelis–Menten equation The rate equation $v = V\,[S]/(K_m + [S])$, where v is the initial velocity of the reaction, V is the maximum velocity, $[S]$ is the substrate concentration, and K_m is the Michaelis constant. The equation is actually derived by means of the Briggs–Haldane treatment of enzyme kinetics.

Michaelis–Menten kinetics The kinetics of an enzymatic reaction that can be described by the Michaelis–Menten equation; such a reaction yields a typical hyperbolic curve when the velocity of the reaction is plotted as a function of the substrate concentration.

Michaelis–Menten treatment The treatment of enzyme kinetics that is based on the assumptions that (a) a rapid equilibrium is established between the enzyme, the substrate, and the enzyme–substrate complex, and (b) the velocity of the reaction is an initial velocity, proportional to the concentration of enzyme–substrate complex, so that the reverse reaction from products to enzyme–substrate complex can be neglected. The resulting rate equation has the form $v = V[S]/(K_s + [S])$, where v is the initial velocity of the reaction, V is the maximum velocity, $[S]$ is the substrate concentration, and K_s is the substrate constant. This equation has no specific name. The term Michaelis–Menten equation is used for the rate equation derived by means of the Briggs–Haldane treatment of enzyme kinetics.

Michaelis pH function An expression for the concentration of either the undissociated or the dissociated form of a substance which can undergo one or more ionizations by the loss of protons; the concentration of the substance is expressed in terms of the hydrogen ion concentration and the appropriate ionization constants.

mic RNA Messenger RNA inhibiting complementary RNA; a small synthetic RNA molecule that is complementary to the initiating region of a specific mRNA. Mic RNA molecules hybridize to the Shine–Dalgarno sequence.

micro- 1. Combining form meaning one-millionth (10^{-6}) and used with metric units of measurements. *Sym* μ 2. Combining form meaning microscopic or minute.

microaerophilic Descriptive of bacteria that grow best at partial pressures of oxygen that are considerably lower than those in air, but do not quite correspond to those of an anaerobic environment. Thus, the term describes bacteria that are incapable of growth either under aerobic conditions or in the complete absence of gaseous oxygen.

microbe A microorganism.

microbeam irradiation The irradiation of subcellular components with a very narrow beam of ionizing radiation or ultraviolet light.

microbial genetics The genetics of microorganisms.

microbial mining The use of microorganisms to leach metals from ores. The method has been used for the extraction of copper and uranium by means of *Thiobacillus ferrooxidans*. This organism carries out oxidations, yielding a mixture of H_2SO_4 and $Fe_2(SO_4)_3$ which then oxidizes and dissolves the minerals of interest.

microbiological assay The assay of a biochemical compound, such as a vitamin or an amino acid, that is based on measuring the growth of microorganisms for which the compound is an essential growth factor.

microbiology The study of microorganisms.

microbioscope A microscope that permits the observation of living tissues.

microbody A cytoplasmic organelle, occurring in eukaryotic cells, that consists of a number of functionally related enzymes contained within a membranous envelope. *See also* peroxisome; glyoxysome.

microcinematography The study of cells by means of motion pictures taken through a phase contrast microscope. In time-lapse microcinematography, cells are photographed at selected time intervals and the film is then projected at a faster speed to provide a better understanding of the time relations of cellular processes.

micrococcal nuclease An endonuclease, isolated from *Staphylococcus aureus*, that catalyzes the hydrolysis of phosphodiester bonds in DNA. The enzyme does not attack DNA that is in contact with protein and hence is useful for cleaving the DNA of eukaryotic chromatin at points located between nucleosomes.

microcomparator An optical instrument for making measurements on photographic plates as those used with schlieren and interferometric optical systems. The instrument resembles a toolmaker's microscope that is equipped with a screen upon which is reflected an enlarged image from a photographic plate, and measurements are made on the screen by means of cross hairs and micrometers.

microcrystalline wax A wax derived from petroleum and consisting of large numbers of small crystals.

microcurie One-millionth of a curie. *Sym* μCi; μc.

microdrop technique A technique for studying antibody formation in individual lymphocytes by suspending a single lymphocyte in a microdrop and determining whether a bacterium or a bacteriophage becomes immobilized in the drop.

microencapsulated enzyme An enzyme that is immobilized in a microcapsule, equipped with a semipermeable membrane, such that substrates and products, but not the enzyme, can diffuse across the membrane.

microencapsulation Any one of a number of techniques in which use is made of small capsules containing specific substances. Microencapsulation has been used for the delivery of drugs, pesticides, and food additives to biological systems, for boosting the output of monoclonal antibodies by hybridomas, and for studies of immobilized enzymes.

microenvironment A very small environment, as that surrounding a molecule or a functional group of a molecule.

microevolution Evolution that extends over short time periods, involves small steps, and leads to small changes in the genetic makeup of an organism.

microfilaments Thin, intracellular fibers having a diameter of about 5–7 nm and consisting largely of actin. Microfilaments are important components of the cytoskeleton. They are identical to the actin filaments of muscle and are contractile. They appear to exist in two forms, lattice microfilaments (a loose network of short, interconnected filaments) and sheath microfilaments (bundles of fibers).

microfiltration The filtration of a solution through a filter or a membrane that will retain suspended material of variable particle size.

microfossil A fossil of microorganisms.

β_2-microglobulin The smallest known plasma protein (MW 11,800). It is present in plasma in very small amounts and its amino acid sequence shows great sequence homology to the immunoglobulins; it constitutes a subunit of the HLA complex.

micrograph *See* electron micrograph.

microheterogeneity The state of a given preparation of macromolecules of one kind, especially proteins, in which the macromolecules exhibit slight differences with respect to their charge, state of aggregation, extent of denaturation, or other properties.

microincineration The combustion of the organic material of a thin tissue slice or of a cell suspension that is placed on a glass slide. The remaining ash provides information about the quality, quantity, and distribution of inorganic compounds in the sample.

microiontophoresis *See* iontophoresis.

micromanipulator An instrument for positioning and handling needles, electrodes, pipets, and the like for experimentation with microscopic specimens, including single cells.

micrometer 1. A device for measuring minute distances that is used in conjunction with optical instruments. 2. MICRON. *Sym* μm; μ.

micromethod A method of chemical analysis that requires very small amounts of sample and reagents.

micromineral A trace element that is required in the diet in relatively small amounts; Fe, Zn, Cu, I, F, Cr, Se, and Mo are examples.

micron A unit of length equal to 10^{-6} m and useful for describing cellular dimensions; a micrometer. *Sym* μm;μ.

micronucleus One of two types of reproductive nuclei found in many ciliate protozoa; they are small, do not contain nucleoli, are usually diploid, and contain inactive DNA that does not undergo transcription. A cell may contain one or more micronuclei. *See also* macronucleus.

micronutrient An essential nutrient that is needed by an organism in minute amounts; vitamins and minerals are examples.

microorganism An organism that is too small to be seen with the naked eye.

microperoxisome *See* peroxisome.

micropinocytosis Pinocytosis that involves the formation of small vesicles that can be visualized only by electron microscopy.

microprocessor The central arithmetic and logic unit of a computer, together with its associated circuitry, scaled down so that it fits on a single silicon chip (sometimes several) holding tens of thousands of transistors, resistors, and similar circuit elements. A microprocessor holds all the elements for manipulating the data and for performing the arithmetic calculations of the computer.

microscope An instrument for magnifying and visualizing objects that are too small to be seen with the naked eye.

microscope electrophoresis PARTICLE ELECTROPHORESIS.

microscopic binding constant A binding constant that applies to a single binding site; an intrinsic association constant. *See also* macroscopic binding constant.

microscopic reversibility *See* principle of microscopic reversibility.

microsequencing The sequencing of a protein or a nucleic acid that requires only very small amounts of sample.

microsomes A heterogeneous subcellular fraction obtained by disruption of cells by homogenization. Microsomes are closed lipoprotein vesicles of variable size and shape, formed by resealing of the torn endoplasmic reticulum and other membranes. Microsomes derived from the rough endoplasmic reticulum are called rough microsomes and are studded with ribosomes. Microsomes lacking attached ribosomes are called smooth microsomes; these are derived in part from the smooth endoplasmic reticulum and in part from fragments of the cell membrane, the Golgi apparatus, and mitochondria.

microspectrophotometer A cytophotometer that consists of a microscope and a spectrophotometer.

microsphere A sperical cell-like structure that is formed spontaneously from proteinoids under suitable conditions and that is believed by some to have been a forerunner of primitive cells.

microspikes Hairlike, contractile extensions of cells that are believed to act as sensory devices by which cells explore their environment; they are thin (about 0.1 μm diameter), long (5–50 μm), and contain actin filaments oriented with the same polarity.

microsurgery MICRURGY.

microtome An instrument for cutting thin sections of tissues, about 1 to 10 μm in thickness, for staining and microscopic examination.

microtomy The methodology connected with the use of a microtome.

microtrabecular network A lattice of very thin, interlacing filaments which interconnect the three major components of the cytoskeleton—microtubules, microfilaments, and intermediate filaments.

microtubule-associated proteins High molecular weight proteins (MW 200,000–300,000) that are associated with, and enhance the polymerization of, microtubules. *Abbr* MAP. *See also* tau proteins.

microtubule organizing center The structure in the cell from which microtubules regrow after they have been depolymerized experimentally; the nucleating center that gives rise to microtubular arrays. Centrioles and kinetosomes have this function in some organisms. *Abbr* MTOC.

microtubules Long, cylindrical tubes that are composed of bundles of small filaments, called protofilaments. The latter are formed by end-to-end association of tubulin molecules. Microtubules have an inside diameter of about 15 nm, an outside diameter of about 25 nm, and are an important component of the cytoskeleton in general and of cilia and flagella in particular. *Abbr* MT.

microvilli Small, finger-like projections of the cell membrane, frequently packed together like the bristles of a brush (brush border membrane).

micrurgy Microsurgery performed on a specimen that is viewed through a microscope; frequently entails operations on single cells with the aid of a micromanipulator.

middle mesophase A mesophase consisting of cylindrical micelles, arranged in a hexagonal array.

middle-repetitive DNA *See* repetitive DNA.

midpoint potential The electrode potential, measured at 25 °C and 1 atm, at the midpoint of an oxidation–reduction titration curve; the electrode potential at which the oxidant and the reductant are present at equal concentrations.

Mie scattering The scattering of light by spherical particles that are neither very large nor very small in comparison to the wavelength of the incident light.

MIF 1. Melanocyte-stimulating hormone release-inhibiting factor; *see* melanocyte-stimulating hormone regulatory hormone. 2. Migration inhibition factor.

migration 1. The movement of a molecule in either electrophoresis or chromatography. 2. An intramolecular rearrangement of atoms, groups of atoms, or bonds.

migration enhancement factor A lymphokine that enhances the migration of macrophages. *Abbr* MEF. *Aka* macrophage activation factor (MAF). *See also* lymphokine.

migration inhibition factor A lymphokine that inhibits the migration of macrophages from a region where T cells have been activated; it also activates nearby macrophages to become more efficient at phagocytosis. *Abbr* MIF. *Aka* macrophage inhibition factor.

MIH See melanocyte-stimulating hormone regulatory hormone.

milieu intérieur The internal environment that consists of the extracellular fluid that surrounds the tissue cells of multicellular organisms.

milk agent MOUSE MAMMARY TUMOR VIRUS.

milk intolerance A syndrome of abdominal cramps, pain, and diarrhea that may occur in adults upon drinking milk; especially widespread among orientals and blacks. The syndrome is due to progressive loss of the enzyme lactase, which hydrolyzes lactose to glucose and galactose, from the brush border membrane of the small intestine. The unhydrolyzed and unabsorbed lactose passes into the large intestine where it is fermented by intestinal bacteria.

milk letdown response A physiological response that enables a nursing infant to obtain milk from the breast; initiated by suckling at the breast which leads to release of oxytocin. The latter results in constriction of specialized cells that encircle the mammary gland and cause expulsion of the milk.

milk sugar LACTOSE.

Miller experiment An experiment in which organic compounds are synthesized under conditions believed to simulate those that have existed during the early stages of chemical evolution on the earth. The experiment demonstrates the formation of amino acids and other compounds from a mixture of reducing gases that have been subjected to an electric discharge.

Miller index (*pl* Miller indices) One of three numbers that designate a plane in which the atoms of a crystal lie.

Miller spread A method for mounting chromosomes for electron microscopy, developed by O.L. Miller; involves rupturing the nuclei and centrifuging the chromosomes onto a membrane-coated grid.

Miller tree The electron microscopic pattern of a gene, coding for rRNA, that is undergoing transcription by RNA polymerase.

milli- Combining form meaning one-thousandth (10^{-3}) and used with metric units of measurements. *Sym* m

millicurie One-thousandth of a curie. *Sym* mCi; mc.

milliequivalent One-thousandth of a gram equivalent weight. *Sym* meq; mEq.

milligram percent *See* mg percent.

millimicron A unit of length equal to 10^{-9}m; a nanometer. *Sym* nm; mμm; mμ.

Millipore filter Trademark for a group of synthetic filters having pores of specified diameter.

Millon reaction A colorimetric reaction for tyrosine and other phenolic compounds that is based on the production of a red color upon treatment of the sample with a solution of mercurous and mercuric nitrates in concentrated nitric acid.

min 1. Minute. 2. Minimum.

mineralization The conversion of organic matter to inorganic matter, and the infiltration of organic matter by inorganic matter.

mineral nutrients *See* macrominerals; microminerals.

mineralocorticoid A 21-carbon steroid hormone, such as deoxycorticosterone or aldosterone, that is secreted by the adrenal cortex and that acts primarily on water and electrolyte metabolism by stimulating the retention of sodium and the excretion of potassium.

mineralocorticoid receptors A group of cytosolic receptors that mediate the action of mineralocorticoids.

mineralocorticosteroid MINERALOCORTICOID.

mineral oil 1. Paraffin oil. 2. Any oil derived from nonliving sources such as coal, petroleum, or shale.

minicell A spherical, anucleate bacterial body that results from an abnormal fission near the polar extremity of a parent cell. Such cells do not contain a bacterial chromosome and they lack a number of DNA-associated enzymes, such as DNA-dependent RNA polymerase, but they may contain plasmid DNA.

minichromosomes 1. Beaded structures of some viral DNAs that resemble the nucleosome structure of eukaryotic chromosomes. 2. Miniature derivatives of chromosomes, constructed by joining restriction endonuclease-generated fragments of the chromosomes.

minigastrin *See* gastrin.

minigene One of a number of chromosome segments that code for the variable regions of the heavy and light chains of the immunoglobulins.

minimal deviation hepatoma A hepatoma that is cancerous but resembles normal liver cells so closely in its enzyme content and in its histological and other properties that it can be compared with normal cells.

minimal medium A synthetic medium that contains only those compounds essential for the growth of the wild-type organism and that is incapable of supporting the growth of auxotrophs; contains inorganic compounds but no organic compounds other than a carbon source such as a sugar. *Abbr* MM.

minimal stable length The minimum size of the segment of nucleotide pairs that must be formed during renaturation, with the base-pairing in perfect register, to lead to the formation of duplex DNA or RNA. *See also* snapback.

minimum hemagglutinating dose The smallest dose of a virus that produces complete agglutination of a standard volume of red blood cells within a specified time. *Abbr* MHD.

minimum hemolytic dose The smallest dose of complement that produces complete lysis of a standard volume of sensitized red blood cells within a specified time. *Abbr* MHD.

minimum lethal dose The smallest dose of bacteria, a virus, or a toxic compound that produces deaths in 100% of the animals in a test group within a specified time. *Abbr* MLD.

minimum molecular weight The molecular weight of a substance that is determined by an assay for some structural element on the assumption that there must be at least one such structural element per molecule of the substance. The structural element may be a metal ion, a functional group, a ligand, a monomer, etc.

minimum protein (nitrogen) requirement The minimum amount of protein (or nitrogen) required daily by an adult to compensate for the amount lost by excretion; approximately 40 g protein/day.

minisome The smallest ribonucleoprotein particle that is either a naturally occurring precursor of ribosomes, or is produced from ribosomes by the stepwise removal of proteins.

minor base One of a group of purines and pyrimidines that generally occur only in small amounts in most nucleic acids but that are found in relatively large amounts in transfer RNA; many of the minor bases are methylated derivatives of the commonly occurring purines and pyrimidines. *See also* hypermodified nucleosides.

minor groove The shallow and narrow groove in Watson–Crick type DNA that is approximately 12 Å across.

minority codon MODULATING CODON.

minus end *See* actin filament.

minus strand *See* plus strand.

minute A mutant of *Drosophila* that is characterized by prolonged cell proliferation resulting in small adult organisms that have a number of abnormal features.

minute phage One of a group of small phages that contain single-stranded DNA and that are either sperical phages, such as ØX-174 and S-13, or filamentous phages, such as M-13 and fd.

7-minute phosphate *See* labile phosphate.

10-minute phosphate *See* labile phosphate.

minute plaque mutant A plaque-type mutant that produces very small plaques.

mirror image ENANTIOMER.

misacylation MISCHARGING.

mischarging The covalent linking of an amino acid to a transfer RNA molecule that is specific for a different amino acid; acylation of a tRNA with a wrong (noncognate) amino acid; misacylation.

miscible Capable of being mixed.

miscoding MISTRANSLATION.

miscopying The occurrence of an error during transcription.

misincorporation The incorporation of either a wrong monomer or an analogue into a polymer; used specifically with regard to the synthesis of DNA.

mismatching MISPAIRING.

mismatch repair The repair of mispaired regions in DNA by excision repair; *see* cut and patch repair.

mispairing The occurrence of a base in one strand of a double-stranded nucleic acid molecule that is not complementary to the base in the corresponding position in the second strand, resulting in incomplete base-pairing.

misreading The occurrence of an error during translation.

misreplication The occurrence of an error during replication.

missense codon A codon that has been altered from its normal sense form in which it codes for one amino acid to a codon that codes for a different amino acid.

missense mutation A mutation in which a normal sense codon is altered so that it becomes a missense codon. As a result, a different amino acid will be incorporated in the corresponding position of the polypeptide chain.

missense suppression The suppression of a missense mutation.

mistranslation The incorporation of an amino acid into a polypeptide chain in response to a codon for a different amino acid.

MIT Monoiodotyrosine.

Mitchell hypothesis CHEMIOSMOTIC COUPLING HYPOTHESIS.

mit gene Mitochondrial gene; a gene of mitochondrial DNA.

MIT genes Yeast mitochondrial genes that code for proteins that function in the electron transport system and oxidative phosphorylation.

mitochondrial ATPase See F_0F_1-ATPase.

mitochondrial DNA A circular, histone-free, double-stranded DNA molecule that is located in the mitochondria, usually in the form of 5–10 copies per mitochondrion. It codes for components of mitochondrial protein synthesis and for proteins that function in the electron transport system and oxidative phosphorylation. The genetic code of mitochondrial DNA differs slightly from the "universal" code. *Abbr* mtDNA.

mitochondrial Eve The postulated female ancestor of modern humans (*Homo sapiens sapiens*), believed to have lived in Africa some 200,000 years ago. The ancestor is inferred by reconstructing phylogenetic trees, based on an analysis of mitochondrial DNA, cut into fragments by means of restriction enzymes. Since mitochondria pass from generation to generation only through the female line (mitochondria are not present in that part of the male sperm, the head, from which the DNA is injected into the female egg), the phylogenies inferred from mitochondrial DNA data essentially trace maternal inheritance; ultimately, a single female is reached at the root of the phylogenetic tree, hence the term mitochondrial Eve.

mitochondrial shuttle See shuttle.

mitochondrion (*pl* mitochondria) A subcellular organelle in aerobic eukaryotic cells that is the site of cellular respiration and that carries out the reactions of the citric acid cycle, electron transport, and oxidative phosphorylation. Mitochondria have a high degree of biochemical autonomy; they contain DNA and ribosomes, carry out protein synthesis, and are capable of self-replication.

mitogen An agent that causes cells to divide and multiply; a stimulant of mitosis.

mitogenic factor A lymphokine that stimulates lymphocyte division.

mitomycins A group of antibiotics, produced by *Streptomyces casepitosus*, that are toxic for bacteria and mammalian cells and that possess antitumor activity. Mitomycin C is the principal member of the group. It binds to guanine residues in DNA and cross-links the complementary strands, thereby preventing DNA replication.

mitoplast A mitochondrion from which the outer membrane has been removed; the intact inner membrane plus the matrix.

mitoribosome A ribosome of mitochondrial origin.

mitosis (*pl* mitoses) The division of the nucleus of eukaryotic cells which occurs in four stages designated prophase, metaphase, anaphase, and telophase. *Abbr* M.

mitotic Of, or pertaining to, mitosis.

mitotic crossover MITOTIC RECOMBINATION.

mitotic cycle 1. MITOSIS. 2. CELL CYCLE.

mitotic index The fraction of cells undergoing mitosis in a given sample.

mitotic poison A compound that prevents mitosis.

mitotic recombination The crossing over between homologous chromosomes during mitosis which leads to segregation of heterozygous alleles.

mixed acid fermentation The fermentation of glucose that is characteristic of *E. coli* and related bacteria and that yields formic, acetic, lactic, and succinic acids, as well as a number of other products.

mixed amino acid fermentation The fermentation of amino acids, as that occurring in putrefaction.

mixed anhydride See acid anhydride.

mixed bed demineralizer A mixture of cation and anion exchange resins that is used for the removal of ions in the preparation of deionized water. *Aka* mixed bed ion exchanger; mixed bed resin.

mixed complex A metal ion complex that contains two or more different ligands.

mixed function oxidase MONOOXYGENASE.

mixed indicator strain A mixture of two related bacterial strains, such as a wild-type and a mutant, that is used in determining the relative amounts of two corresponding viral genotypes in a mixed population of virions.

mixed inhibition The inhibition of an enzyme that cannot be fully described in terms of one of the basic types of inhibition. In the Cleland nomenclature of enzyme kinetics most mixed-type inhibitions are considered to be varieties of noncompetitive inhibition.

mixed lactic fermentation HETEROLACTIC FERMENTATION.

mixed micelle A micelle composed of more than one type of compound. Many detergents solubilize biological membranes by forming mixed micelles consisting of detergent–lipid or detergent–lipid–protein complexes.

mixed-order reaction A chemical reaction, the observed rate of which cannot be fully described by a simple first-, second-, or third-order rate equation.

mixed surface film A monolayer composed of two or more different components.

mixed triglyceride A triacylglycerol that contains two or three different fatty acids.

mixed-type inhibitor An enzyme inhibitor that alters both the maximum velocity of the reaction and the Michaelis constant of the enzyme.

mixed vaccine A vaccine that contains antigens derived from different infectious agents.

mixotroph A cell or an organism that uses simultaneously autotrophic and heterotrophic metabolic processes; it uses organic compounds as carbon sources and inorganic compounds as energy sources.

mixotropic series A series of solvents arranged in the order of their relative polarity, and hence their miscibility with water.

ml Milliliter.

mL Milliliter.

MLC Myosin light chain; *See* myosin.

MLD 1. Minimum lethal dose. 2. Median lethal dose.

M line The dark line that bisects the H zone of the myofibrils of striated muscle.

mlRNA Messenger-like RNA; a synonym of heterogeneous nuclear RNA.

MLV Multilamellar vesicle; *See* vesicle.

mM Millimolar concentration.

MM Minimal medium.

M macroglobulin An abnormal immunoglobulin of the IgM type that is produced by individuals suffering from Waldenstroem's macroglobulinemia.

MMTV Mouse mammary tumor virus.

Mn Manganese.

M_n Number-average molecular weight.

MN blood group system A human blood group system that consists of two antigens, M and N, which are present in glycophorins A and B, respectively; the two glycophorins differ in the

sequence of the first 5 amino acid residues.

Mo Molybdenum.

MO Molecular orbital.

mobile carrier *See* ionophore.

mobile genetic elements TRANSPOSABLE ELEMENTS.

mobile ion carrier *See* ionophore.

mobile phase The liquid or gas phase that is the bulk moving phase in chromatography.

mobile receptor model A proposed model to explain the functioning of receptors. According to this model, a receptor can exist in many states: free, complexed with its biological effector enzyme, or bound to one of a number of different ligands. Moreover, a given ligand–receptor complex can interact with several different effector enzymes, resulting in various distinct ternary complexes. *See also* two-state model.

mobility *See* electrophoretic mobility.

mobility shift analysis An early method for sequencing oligodeoxyribonucleotides by means of the shifts in electrophoretic mobility; involves two-dimensional electrophoresis homochromatography of partial degradation products.

mobility spectrum The distribution pattern of compounds that are separated by electrophoresis.

mobilizable plasmid A plasmid that can prepare its DNA for transfer to a recipient cell.

mobilization 1. The release of lipids, particularly fatty acids, from adipose tissue and their conversion to lipids that are transported by the blood. 2. The preparation of DNA for transfer that occurs when a plasmid is transferred from a donor to a recipient cell.

mobilizing lipase A lipase that functions in the mobilization of fatty acids from adipose tissue.

modal class The category in a statistical distribution that contains a larger number of observations or measurements than any other category.

mode The value of the variable that has the maximum frequency in a statistical distribution.

model A three-dimensional representation of a molecule.

model system A system that is studied and that is considered to either simulate or be representative of other systems in which the same or similar reactions take place.

MODEM Acronym for modulator–demodulator; a device that allows a computer to communicate with another computer via telephone lines.

moderately repetitive DNA *See* repetitive DNA.

moderate virus An animal virus that resembles

a temperate phage in its properties and that establishes a stable complex with the host cell.

moderator   A substance that alters the rate of an enzymatic reaction; an activator or an inhibitor.

moderator protein   CALMODULIN.

modification   *See* processing.

modification allele   *See* modification gene.

modification and restriction   *See* restriction–modification system.

modification enzyme   An enzyme that catalyzes the introduction of minor bases into RNA or DNA and that functions by modifying a normal base subsequent to its insertion into the polynucleotide strand.

modification gene   A gene the product of which is a modification enzyme.

modification methylase   A modification enzyme that catalyzes the methylation of DNA. *See also* restriction–modification system.

modified air storage   GAS STORAGE.

modified base   MINOR BASE.

modified inverse isotope dilution analysis   A variation of the inverse isotope dilution analysis in which the amount of labeled material is determined by means of a second radioactive substance.

modifier   1. MODIFYING GENE. 2. EFFECTOR.

modifying gene   A gene that effects the expression of another, nonallelic gene.

modulating codon   A codon that codes for a rare transfer RNA molecule and that does not lead to the insertion of an amino acid into the growing polypeptide chain during translation. Instead, such a codon acts as a regulatory agent, leading either to release of the ribosome and interruption of translation, or to a slowing down of the rate of translation. *Aka* modulator codon.

modulation   1. The regulation of the frequency with which a specific gene is transcribed. 2. The decrease in the rate of translation of a messenger RNA brought about by a modulating codon. 3. The control of a regulatory enzyme by means of an effector.

modulator   EFFECTOR.

modulator protein   CALMODULIN.

modulator transfer RNA   A rare transfer RNA molecule that is coded for by a modulating codon.

modulus of precision   A measure of the closeness with which observations are clustered; as it increases, the width of the peak of the normal error curve becomes smaller. The modulus of precision (*h*) is defined as $h = \pm 1/\sigma\sqrt{2}$, where σ is the standard deviation. *Abbr* MOP.

MoFd   Molybdoferredoxin; *see* nitrogenase.

Mo-Fe protein   Molybdoiron protein; *see* nitrogenase.

Moffit plot   A plot based on the Moffit–Yang

equation in which $[m'](\lambda^2 - \lambda_0^2)$ is plotted as a function of $(\lambda^2 - \lambda_0^2)^{-1}$ so that a straight line is obtained for a chosen value of λ_0.

Moffit–Yang equation   An equation that describes the variation of optical rotation with wavelength; specifically, $[m'] = a_0\lambda_0^2/(\lambda^2 - \lambda_0^2) + b_0\lambda_0^4/(\lambda^2 - \lambda_0^2)^2$, where $[m']$ is the reduced mean residue rotation, λ is the wavelength, and a_0, b_0, and λ_0 are constants. *Aka* Moffit equation.

Mohr pipet   A measuring pipet in which the calibration marks are contained between two marks on the stem of the pipet and do not extend to the tip of the pipet.

MOI   Multiplicity of infection.

moiety   1. One of two approximately equal parts. 2. One of two parts.

mol   Symbol for mole.

molality   The concentration of a solution expressed in terms of the number of moles of solute per 1000 g of solvent.

molal solution   A solution that contains one mole of solute dissolved in 1000 g of solvent.

molar absorbancy index   MOLAR ABSORPTIVITY.

molar absorptivity   The absorbance of a one molar solution when the light path through the solution is 1 cm; frequently denoted by the symbol ϵ. The molar absorptivity is related to the absorption cross section *s* by the equation $s = 3.8 \times 10^{-21}\epsilon$.

molar activity   A measure of enzymatic activity that is equal to the number of katals per mole of enzyme. *See also* molecular activity.

molar ellipticity   A measure of circular dichroism that is equal to 3300 times the difference between the molar extinction coefficients for the left and right circularly polarized light beams.

molar extinction coefficient   MOLAR ABSORPTIVITY.

molar growth yield   The dry weight of bacteria, in grams, that is obtained per mole of substrate utilized by the bacteria during their growth.

molarity   The concentration of a solution expressed in terms of the number of moles of solute per liter of solution.

molar mass   MOLE.

molar rotation   The optical rotation of a solute that is calculated on the basis of its molar concentration and that is corrected to that in a medium having a refractive index of one. Specifically, $[m'] = [3/(n^2 + 2)] \times M[\alpha]/100$, where $[m']$ is the molar rotation, M is the molecular weight of the solute, $[\alpha]$ is the specific rotation of the solute, and n is the refractive index of the medium. Also used in its uncorrected form as $[m'] = M[\alpha]/100$.

molar solution   A solution that contains one mole of solute in one liter of solution.

mold   1. A fungus characterized by having long

mycelia. 2. TEMPLATE.

mole 1. The molecular weight expressed in grams; the weight of a compound in grams that is numerically equal to its molecular weight. 2. The amount of substance that contains as many elementary entities as there are carbon atoms in 0.012 kg of the nuclide ^{12}C (Avogadro's number). *Sym* mol.

molecular 1. Of, or pertaining to, molecules. 2. Indicating the molecularity of a reaction when used with the prefixes mono-, bi-, or ter-.

molecular activity A measure of enzyme activity equal to the number of moles of substrate (or the number of equivalents of the group concerned) that are transformed into products per minute per mole of enzyme at optimal substrate concentration. *See also* molar activity; catalytic center activity.

molecular biology 1. The science that deals with the study of biological processes at the molecular level, particularly with respect to the physical–chemical properties and changes of cellular components and the relationship of these properties and changes to biological phenomena. Nerve impulse conduction, vision, membrane transport, and molecular genetics are some of the topics of molecular biology. 2. The science that deals with molecular genetics; the replication and transcription of both DNA and RNA, and the translation of RNA.

molecular clock EVOLUTIONARY CLOCK.

molecular cloning *See* cloning (1).

molecular disease A disease that can be traced to a change in a single type of molecule; sickle cell anemia, which is caused by an abnormal hemoglobin, is an example.

molecular evolution CHEMICAL EVOLUTION.

molecular-exclusion chromatography GEL FILTRATION CHROMATOGRAPHY.

molecular fossil CHEMICAL FOSSIL.

molecular genetics The study of genetics at the molecular level; the replication and transcription of both DNA and RNA, and the translation of RNA.

molecular hybrid See hybridization (1, 2).

molecular ion A molecule that has lost one electron, as that produced in a mass spectrometer.

molecularity The number of reactant molecules that participate in a chemical reaction.

molecular mass MOLECULAR WEIGHT.

molecular mimicry The formation of eclipsed antigens by a parasite.

molecular orbital A composite orbital in a molecule that is derived from the overlapping, hybridized or unhybridized, atomic orbitals of the component atoms. *Abbr* MO.

molecular orbital theory The theory of chemical bonding that is developed by considering the bonding atomic nuclei to occupy their equilibrium positions in the molecule, and then feeding orbital electrons into the resultant force field. When covalent bonds are formed, the atomic orbitals of individual atoms combine and overlap to form molecular orbitals which are a property of the whole molecule and not of single atoms. *Abbr* MO theory.

molecular photosensitization *See* photosensitization.

molecular radioautography High-resolution radioautography, as that applied to isolated DNA strands.

molecular rotation The rotation of an entire molecule about an axis.

molecular sieve 1. A substance used for fractionating molecules according to their size by means of either gel filtration or gel permeation chromatography. 2. ZEOLITE.

molecular sieve chromatography 1. GEL FILTRATION CHROMATOGRAPHY. 2. GEL PERMEATION CHROMATOGRAPHY.

molecular sieve coefficient The partition coefficient of a solute in gel filtration chromatography; equal to the ratio of the equilibrium concentration of the solute within the gel to its concentration in the mobile phase.

molecular surface The continuous surface of a molecule, equal to the sum of the contact surface and the reentrant surface.

molecular taxonomy CHEMOTAXONOMY.

molecular vibration The stretching or bending of the bonds between atoms that results in a displacement of the atomic nuclei but does not affect their equilibrium positions.

molecular weight 1. The sum of the atomic weights of all the atoms in a molecule. 2. The sum of the atomic weights of all the atoms in a molecular aggregate such as an oligomeric protein, a ribosome, or a virus. *Abbr* M; M_r; MW; mol wt. *Aka* particle weight.

molecular weight average *See* average molecular weight.

molecule The smallest unit of a compound; a structural unit of matter that retains all its properties, has an independent existence, and is composed of like or unlike atoms.

molecule microscope A microscope for the study of surface materials by molecular beam techniques. The microscope reveals spatial variations in the evaporation of neutral molecules from the surface of a sample that is exposed to reduced pressure. The evaporating molecules may be part of the sample, may have been applied previously to the surface, or may be passed through the sample during the microscopic observation.

mole fraction A measure of concentration expressed in terms of the number of moles of a substance divided by the total number of moles of all the substances in a solution or in a mixture.

mole percent A measure of concentration expressed in terms of the number of moles of a substance per 100 mol of all related substances.

Molisch test A test for carbohydrates that is based on the production of a purple color upon treatment of the sample with concentrated sulfuric acid and α-naphthol.

Moloney leukemia virus A mouse leukemia virus that belongs to the leukovirus group.

molting hormone ECDYSONE.

mol wt Molecular weight.

molybdenum An element that is essential to humans and several classes of animals and plants. *Symbol*, Mo; atomic number, 42; atomic weight, 95.94; oxidation state, +6; most abundant isotope, ^{98}Mo; a radioactive isotope, ^{99}Mo, half-life, 66.7 h, radiation emitted, beta particles and gamma rays.

molybdoferredoxin *See* nitrogenase.

molybdoiron protein *See* nitrogenase.

molybdoprotein A conjugated protein containing molybdenum as a prosthetic group.

molybdopterin A molybdenum-containing pterin that occurs, together with a heme, in the two identical subunits of rat liver sulfite oxidase. *Abbr* MPT.

monellin A basic and carbohydrate-free protein, consisting of two polypeptide chains of 44 and 50 amino acids, respectively. The protein has very high specificity for sweet taste receptors so that it is approximately 100,000 times sweeter than sugar on a molar basis.

monestrous Having one estrous cycle per year.

mongolism A congenital abnormality characterized by imbecility and due to the presence of one of the autosomes in the triploid rather than in the diploid state.

monitor A detector for determining a physical or a chemical variable either periodically or continuously; frequently refers either to a radiation detector for measuring the amount of ionizating radiation or of radioactive contamination, or to a spectrophotometer for measuring the absorbance of visible or ultraviolet light.

mono- Prefix meaning one.

monoacylglycerol An acylglycerol formed by the esterification of a glycerol molecule with one fatty acid molecule. *Aka* monoglyceride.

monoamine oxidase A flavoprotein enzyme that catalyzes the oxidative deamination of monoamines such as epinephrine and norepinephrine. *Abbr* MAO.

monobactam A monocyclic, bacterially produced, antibiotic that is similar to the β-lactam antibiotics but has a simpler structure; the four-membered ring, containing an amide group, is not fused to another ring.

monobasic Descriptive of a compound that contains one hydrogen atom that is replaceable by a metal (such as K_2HPO_4) or an acid that can furnish one hydrogen ion (such as HCl).

monochromatic radiation Electromagnetic radiation of a single wavelength; electromagnetic radiation in which all of the photons have the same energy.

monochromator An instrument for the isolation of narrow bandwidths of radiation by means of filters, prisms, or diffraction gratings.

monocistronic messenger RNA An mRNA molecule that carries the information for the synthesis of only one polypeptide chain.

monoclonal antibodies Immunoglobulins derived from a single clone of cells. Monoclonal antibodies are chemically and structurally identical; they have a single amino acid sequence and are specific for a single antigenic determinant. They can be obtained by propagation of a hybridoma in culture. *Abbr* MAbs. *See also* hybridoma.

monocyclic cascade The simplest cascade system involving one interconvertible enzyme and two converter enzymes. Each converter enzyme requires allosteric activation; one converter enzyme catalyzes the modification of the interconvertible enzyme and the other converter enzyme catalyzes its demodification.

monocyte A large amoeboid, phagocytic leukocyte derived from the bone marrow and containing one large nucleus.

monodentate Designating a ligand that is chelated to a metal ion through one donor atom.

monodisperse Consisting of macromolecules that are all alike in size.

Monod, Wyman, and Changeux model *See* concerted model.

monoenergetic radiation Radiation that consists of either photons or particles and in which all of the photons or all of the particles have the same energy.

monoenoic fatty acid A fatty acid that has one double bond.

monoesterase An enzyme that catalyzes the hydrolysis of an ester which is formed by the esterfication of one of the hydroxyl groups of phosphoric acid.

monogenic Involving a single gene.

monoglyceride MONOACYLGLYCEROL.

monolayer 1. A monomolecular layer formed either at a surface or at an interface. 2. A single layer of cells that are growing on a surface.

monomer 1. The repeating unit in a polymer. 2. The basic unit in a molecular aggregate, regardless of the number of polypeptide chains or the number of subunits of which it is composed; thus the 70S ribosome is a monomer while the 30S and ⁻0S ribosomes are subunits and the 100S ribosome is a dimer. 3. The individual polypeptide chain in an oligomeric protein. 4. A protein that is composed of a single polypeptide chain. 5. STRUCTURAL UNIT. *See also* protomer; subunit.

monomolecular layer A layer, one molecule thick.

monomolecular reaction A chemical reaction in which one molecule of a single reactant is converted into products.

monomorphism The occurrence of only one form or one shape. *See also* doctrine of monomorphism.

mononuclear complex A metal ion–ligand complex that is formed from a single metal ion. *See also* polynuclear complex.

mononucleosis An increase in the number of mononuclear leukocytes in the blood. *See also* infectious mononucleosis.

mononucleotide 1. A single nucleotide. 2. A compound that is structually related to a nucleotide, such as flavin mononucleotide or nicotinamide mononucleotide, and contains one phosphate group.

mononucleotide binding domain ROSSMAN FOLD.

monooxygenase An enzyme that catalyzes a reaction with molecular oxygen in which only one of the oxygen atoms is introduced into a compound. *Aka* hydroxylase.

monoploid state The chromosome state in which the number of chromosomes is the basic one in a polyploid series; the haploid state. *Aka* monoploidy.

monoprotic acid An acid that has one dissociable proton.

monosaccharide A polyhydroxy alcohol containing either an aldehyde or a ketone group; a simple sugar.

monose MONOSACCHARIDE.

monosome 1. The complex that consists of a single ribosome attached to a strand of messenger RNA. 2. A ribosome that has dissociated from a polysome and that contains no tRNA, mRNA, or peptidyl-tRNA. 3. A chromosome that lacks a homologue.

monospecific antiserum A purified antiserum that reacts with only one type of antigen or one type of antigenic determinant.

monovalent 1. Having a valence of one. 2. Descriptive of an allosteric enzyme that responds to only one effector.

monozygotic twins Twins that are genetically identical and that are derived from one fertilized egg; they are formed by a division of the embryo into two halves at some stage of its development.

montanic acid A saturated fatty acid that contains 28 carbon atoms.

Monte Carlo method The solution of a mathe-matical problem by sampling methods; involves constructing an artificial stochastic model of the problem and then performing sampling experiments on it. The calculation of the time course of a reaction from the probability that the reaction will occur during a given time interval is an example.

MOP Modulus of precision.

MOPS 2-(*N*-Morpholino)propanesulfonic acid; used for the preparation of biological buffers in the pH range of 6.5 to 7.9. *See also* biological buffers.

MOPSO 3-(*N*-Morpholino)-2-hydroxypropane-sulfonic acid; used for the preparation of biological buffers in the pH range of 6.2 to 7.6. *See also* biological buffers.

Morawitz theory An early formulation of the blood clotting mechanism in terms of a two-stage process, consisting of the activation of prothrombin to thrombin and the conversion of fibrinogen to fibrin.

Morgan unit A measure of the distance between genes on a chromosome that is equal to a crossover value of 100%.

Morner's test 1. A test for tyrosine that is based on the production of a green color upon treatment of the sample with sulfuric acid and formaldehyde. 2. A test for cysteine that is based on the production of a purple color upon treatment of the sample with sodium nitroprusside.

morphine An opium alkaloid that is formed from codeine as the poppy ripens; a narcotic drug which acts as an anesthetic without decreasing consciousness and is the most powerful analgesic known.

morphine rule The principle that the narcotic properties of morphine do not require the entire structure of the morphine molecule but only an aromatic ring attached to a quaternary carbon atom and the presence of a tertiary amine, two carbons farther away.

morphinomimetic Mimicking the action of morphine.

morphogen A substance that triggers growth and differentiation of cells and tissues by virtue of its concentration; all-trans retinoic acid, a derivative of vitamin A, appears to be such a substance.

morphogene A gene that functions, either directly or indirectly, in growth control and morphogenesis. Genes for hormones, inducers and mitogens are examples.

morphogenesis The developmental processes that lead to the mature size, form, and structure of organelles, cells, tissues, organs, or whole organisms.

morphogenetic Of, or pertaining to, morphogenesis. *Aka* morphogenic.

morphogenetic gene A gene that plays a role in morphogenesis through some function other than that of specifying the synthesis of a structural protein.

morphogenic induction The determination of the differentiation of one cell mass of an embryo by its interaction with another cell mass of the same embryo.

morphology The science that deals with the structures and forms of organisms.

morphology-dependent resonance Narrow spectral resonances in the light scattered from transparent or weakly absorbing microparticles (droplets and fibers); produced when the electromagnetic modes of the microparticles are excited by some incident radiation. *Abbr* MDR.

morphopoiesis MORPHOGENESIS.

morphopoietic gene MORPHOGENETIC GENE.

mortichemistry THANATOCHEMISTRY.

mosaic An individual composed of two or more genetically different types of cells derived from the same zygote. *See also* chimera.

mosaic theory The theory of hypertension that describes the regulation of blood pressure in terms of the interactions between eight variables placed diagrammatically at the corners of an octagon. The eight variables are chemical factors, neural factors, elasticity, cardiac output, viscosity, vascular caliber, volume, and reactivity.

Mossbauer effect The inelastic collision between a nucleus and a gamma ray in which the gamma ray is absorbed by and excites the nucleus; the nucleus remains in the excited state for a brief period of time (10^{-6} to 10^{-10} s) and subsequently returns to its ground state with the emission of the gamma ray.

Mossbauer spectrometer An instrument for detecting small changes in the interaction between a nucleus and its environment that are produced by variations in temperature, pressure, or chemical state.

MO theory Molecular orbital theory.

motilin A candidate hormone in the small intestine that, when injected into animals, leads to a marked increase in gastric and intestinal motility.

motility model FLUCTUATION THEORY (2).

motor end plate NEUROMUSCULAR JUNCTION.

motor neuron A neuron that conveys impulses to a muscle, resulting in muscle contraction.

mottled enamel A pitted and corroded form of tooth enamel that can be produced by the drinking of water that contains excessive amounts of fluoride ions.

mottled plaque A plaque produced by the joint growth of two related phages, such as two mutants, in the same infectious center.

mouse antialopecia factor INOSITOL.

mouse L cells *See* L cells.

mouse leukemia virus An oncogenic virus that contains single-stranded RNA and causes leukemia in mice; the Friend, Graffi, Gross, Moloney, and Rauscher leukemia viruses, which belong to the leukovirus group, are examples.

mouse mammary tumor virus An oncogenic virus that is transmitted through the milk and that causes mammary cancer in mice; it contains single-stranded RNA and belongs to the group of leukoviruses. *Abbr* MMTV. *Aka* murine mammary tumor virus. *See also* A-, B-, and C-type particles.

mouse satellite DNA A satellite DNA that has been isolated from a variety of mouse tissues and that constitutes about 10% of the total mouse DNA; it consists of highly repetitive DNA that contains about one million copies of a segment some 400 nucleotide pairs in length.

moving boundary analysis ISOTACHOPHORESIS.

moving boundary centrifugation Centrifugation in which an initially uniform solution is centrifuged so that boundaries are formed in the solution and move across the centrifuge cell; generally performed in an analytical-type ultracentrifuge.

moving boundary electrophoresis Electrophoresis, performed in a Tiselius apparatus, in which an initially uniform solution is partially separated so that boundaries are formed that move toward or away from an electrode.

moving zone centrifugation DENSITY GRADIENT CENTRIFUGATION.

mp Melting point; also abbreviated m.pt.

6-MP 6-Mercaptopurine.

MPD Maximum permissible dose.

M-protein 1. A galactoside carrier protein in the permease system of *E. coli*. *See also* lactose permease. 2. A major cell surface antigen of *Brucella*. 3. A cell wall antigen in virulent strains of *Streptococcus*. 4. A structural protein present in the M line of the myofibrils of striated muscle.

MPT Molybdopterin.

M$_r$ Relative molecular mass; molecular weight.

MR 1. Multiplicity reactivation. 2. Metabolic rate.

MRF *See* melanocyte-stimulating hormone regulatory hormone.

MRH *See* melanocyte-stimulating hormone regulatory hormone.

MRIF *See* melanocyte-stimulating hormone regulatory hormone.

MRIH *See* melanocyte-stimulating hormone regulatory hormone.

mRNA Messenger RNA.

mRNA coding triplet CODON.

mRNA strand ANTICODING STRAND.

mRNP 1. Messenger ribonucleoprotein particle. 2. Messenger ribonucleoproteins.

MS 1. Magic spot. 2. Mass spectrometry. 3. Multiple sclerosis.

MS-2 A small, tailless, icosahedral phage that contains single-stranded RNA and infects *E. coli.*

MSD mice Myelin-synthesis-deficient mice; mice that have the Quaking mutation.

msDNA Multicopy single-stranded DNA; an unusual satellite DNA, originally detected in myxobacteria, that occurs in multiple copies per genome.

MSG Monosodium glutamate. *See also* Chinese restaurant syndrome.

MSH Melanocyte-stimulating hormone.

MSHIF *See* melanocyte-stimulating hormone regulatory hormone.

MSHIH *See* melanocyte-stimulating hormone regulatory hormone.

MSHRF *See* melanocyte-stimulating hormone regulatory hormone.

MSHRH *See* melanocyte-stimulating hormone regulatory hormone.

MSHRIF *See* melanocyte-stimulating hormone regulatory hormone.

MSHRIH *See* melanocyte-stimulating hormone regulatory hormone.

MS/MS Tandem mass spectrometry.

MT 1. Metallothionein. 2. Microtubules.

MTA Mammary tumor agent.

mtDNA Mitochondrial DNA.

mt mRNA Mitochondrial messenger RNA.

MTOC Microtubule organizing center.

mt rRNA Mitochondrial ribosomal RNA.

mt tRNA Mitochondrial transfer RNA.

mu 1. Map unit. 2 Mate killer.

mu chain The heavy chain of the IgM immunoglobulins. *Sym* μ.

mucilage A complex, colloidal, carbohydrate material that is derived from plants; it can form gels and has adhesive properties.

mucin A mucoprotein secreted by mucous glands and mucous cells.

mucin clot The clot, composed of hyaluronic acid and small amounts of protein, that is formed upon acidification of some biological fluids such as the vitreous humor of the eye and the synovial fluid.

muco Combining from meaning amino sugar.

mucoid MUCOPROTEIN.

mucolipidosis One of a group of genetically inherited metabolic storage diseases in humans in which both mucopolysaccharides and lipids accumulate in tissues. The diseases are characterized by a normal urinary concentration of mucopolysaccharides and by symptoms of sphingolipid and/or glycolipid storage abnormalities. I-cell disease is an example.

mucopeptide A peptide that is covalently linked to an amino sugar.

mucopolysaccharide GLYCOSAMINOGLYCAN.

mucopolysaccharide storage disease MUCOPOLY-SACCHARIDOSIS.

mucopolysaccharidosis One of a number of genetically inherited metabolic storage diseases in humans that are characterized by excessive accumulation and excretion of the oligosaccharides of proteoglycans; the defects are due to deficiencies of specific lysosomal enzymes. Hunter's syndrome, Hurler's syndrome, and Sanfillipo's syndrome are some examples.

mucoprotein PROTEOGLYCAN.

mucosa MUCOUS MEMBRANE.

mucosal block The permeability barrier of the intestinal mucosa to the absorption of iron from the intestine.

mucous Of, or pertaining to, mucus.

mucous gland A gland that secretes mucus.

mucous membrane An epithelial membrane, the surface of which is bathed by mucus.

mucus The viscous secretion of mucous glands that consists largely of mucin and water and that serves to bathe mucous membranes.

mull A two-phase mixture made by grinding a solid sample and dispersing it in a suitable organic solvent; used for the analysis of some samples by infrared spectroscopy.

multi- Combining form meaning many.

multichannel analyzer A scintillation spectrometer that can record pulses in a number of different channels.

multicomponent survival curve MULTITARGET SURVIVAL CURVE.

multicomponent virus COVIRUS.

multicyclic cascade A system, produced by the coupling of two or more monocyclic cascades. The extrinsic and intrinsic pathways of blood clotting and the regulation of glycogen phosphorolysis are some examples.

multidimensional gas chromatography A gas chromatographic technique that involves the use of two or more separate columns for a given separation. The columns may be connected serially but are most commonly used by means of column switching in which an inadequately resolved fraction from one column is shunted to a second column for more complete resolution. The technique is particularly useful for the analysis of complex mixtures. *Abbr* MDGC.

multienzyme complex 1. MULTIENZYME SYSTEM. 2. METABOLON (2).

multienzyme system The structural and functional entity that is formed by the association of several different enzymes which catalyze a sequence of closely related reactions; the aggregate may contain one or more molecules of a given enzyme. *Aka* enzyme complex; multienzyme complex.

multifactorial Of, or pertaining to, a polygene; polygenic.

multifunctional protein A protein that has two or more different catalytic and/or binding functions on a single polypeptide chain.

multigene family A group of genes, derived by duplication and variation from a common ancestral gene, that are located on the same or on different chromosomes. A multigene family exhibits four properties: multiplicity, close linkage, sequence homology, and related phenotypic functions. The genes coding for histones, immunoglobulins, and hemoglobins are examples of multigene families.

multiheaded protein MULTIFUNCTIONAL PROTEIN.

multihit survival curve 1. A survival curve that describes a radiation phenomenon in which two or more photons must be absorbed by one target before the viability of the active unit is lost. 2. MULTITARGET SURVIVAL CURVE.

multilamellar vesicle *See* vesicle.

multimer OLIGOMER.

multiparticle virus COVIRUS.

multiphasic zone electrophoresis DISC GEL ELECTROPHORESIS.

multiple alleles A group of three or more alternative alleles, any one of which may occur at the same locus on a chromosome.

multiple binding MULTIPLE EQUILIBRIA.

multiple codon recognition The binding of a given molecule of tRNA to more than one codon, as postulated by the Wobble hypothesis.

multiple component virus COVIRUS.

multiple development A chromatographic technique, used particularly with paper or thin-layer chromatography, in which the sample is developed repeatedly with either the same or different solvents.

multiple displacement mechanism NONSEQUENTIAL MECHANISM.

multiple equilibria The interactions that occur between the macromolecule that has several binding sites and the ligands that bind to these sites.

multiple-event curve MULTIHIT SURVIVAL CURVE.

multiple factor hypothesis The hypothesis that quantitative traits, such as size and weight, result from the cumulative effect of a group of genes. *See also* polygene.

multiple forms of an enzyme A collective term for all the proteins that possess the same enzyme activity and that occur naturally in a single species; includes genetically independent proteins, heteropolymers, genetic (allelic) variants, proteins conjugated with other groups, proteins derived from one polypeptide chain, polymers of a single subunit, and forms differing in conformation.

multiple gene POLYGENE.

multiple-hit survival curve *See* multihit survival curve.

multiple inhibition analysis A kinetic analysis of the interactions of two or more inhibitors of an enzymatic reaction. The analysis indicates whether the inhibitors are mutually exclusive or whether they can bind simultaneously to the enzyme and, if so, whether the binding of one inhibitor to the enzyme facilitates or hinders the binding of another.

multiple myeloma A malignant disease of antibody-producing plasma cells in which single, specific cells have undergone neoplastic transformation. These cells proliferate to produce excessive amounts of specific proteins, including the Bence–Jones protein, myeloma globulins, and fragments of the various classes of normal immunoglobulins.

multiple sclerosis A human demyelination disease in which the myelin sheath of nerves undergoes destruction. The cause of this is uncertain but a genetic predisposition for the disease occurs. In the absence of myelin, axons fail to conduct nerve impulses. The disease is characterized by partial paralysis, changes in speech, and inability to walk. *Abbr* MS. *Aka* demyelination disease.

multiplet A multiple peak, as that obtained in nuclear magnetic resonance.

multiplication cycle The sequence of steps from the infection of a cell by a virus to the formation of new virus particles and their release from the cell.

multiplicity *See* enzyme multiplicity; transfer RNA multiplicity; spin multiplicity.

multiplicity of infection 1. The number of virus particles that have either adsorbed to, or infected, cells in a culture, divided by the total number of cells in the culture. 2. The number of virus particles added to a culture divided by the total number of cells in the culture. *Abbr* MOI.

multiplicity reactivation The restoration of the activity of a virus that carries a lethal mutation by the simultaneous infection of a host cell with this and one or more other mutant viruses. The process involves a genetic exchange whereby a viable genome is produced from the undamaged sections of the mutant, and from otherwise nonviable, genomes. *Abbr* MR. *See also* cross-reactivation.

multistep induction theory A theory of carcinogenesis according to which cancer is induced as a result of a number of steps or stages. It is believed that, at the very least, there are two such steps, initiation and promotion. Initiation involves the acquisition of mutations, and promotion involves the expression of these mutations. *Aka* multistage induction theory.

multisubstrate enzyme system An enzyme system involving more than one substrate. The mechanism of a multisubstrate enzyme system may be random or ordered and may be sequential or nonsequential.

multitarget survival curve A survival curve that describes a radiation phenomenon in which two or more photons must be absorbed by two or more targets before the viability of the active unit is lost.

multivalent POLYVALENT.

multivalent allosteric enzyme An allosteric enzyme, the activity of which can be altered by more than one effector.

multivalent allosteric inhibition The inhibition of an allosteric enzyme by two or more negative effectors; the inhibition of a multivalent allosteric enzyme.

multivalent feedback inhibition CONCERTED FEEDBACK INHIBITION.

multivalent vaccine *See* polyvalent vaccine.

multivesicular body A secondary lysosome that contains a large number of vesicles.

muon *See* elementary particles.

mu particle *See* mate killer.

muramic acid A compound derived from glucosamine and lactic acid, the acetylated form of which is a major building block of the bacterial cell wall.

muramidase LYSOZYME.

Murayama hypothesis The hypothesis that the replacement of glutamic acid by valine at position 6 in the beta chains of sickle cell hemoglobin permits the formation of intermolecular hydrophobic bonds which lead to a head-to-tail stacking of hemoglobin molecules; the filaments thus formed distort the red blood cell and convert it to a sickle cell.

Murchison meteorite A large meteorite that fell near Murchison, Australia, on September 28, 1969. Analysis of the meteorite provided support for the notion that amino acids can be synthesized in outer space by mechanisms similar to those postulated to have been involved in their synthesis on the primitive earth.

murein PEPTIDOGLYCAN.

murexide test A test for purines that is based on the production of a red color upon treatment of the sample with concentrated nitric acid and then with ammonium hydroxide.

murine Of, or pertaining to, mice and rats.

murine leukemia Leukemia in mice that is produced by a mouse leukemia virus.

murine leukemia virus MOUSE LEUKEMIA VIRUS.

murine mammary tumor virus MOUSE MAMMARY TUMOR VIRUS.

muropeptide The repeating unit in peptidoglycan that consists of *N*-acetylglucosamine, *N*-acetylmuramic acid, and a tetrapeptide side chain.

muscarine A toxin from the mushroom *Amanita muscaria. Var sp* muscarin.

muscarinic receptor A synaptic acetylcholine receptor to which muscarin binds, thereby mimicking the action of acetylcholine. Such receptors are found at smooth muscle end plates and in the brain. *See also* decamethonium.

muscarinic synapse A synapse containing muscarinic receptors.

muscle A contractile organ of the body. *See also* red muscle; white muscle.

muscle contraction *See* contraction; sliding filament model; rowboat model.

muscle fiber A long, multinucleated cell of striated muscle.

muscle hemoglobin MYOGLOBIN.

muscle phosphorylase *See* phosphorylase a; phosphorylase b.

muscle sugar INOSITOL.

muscular dystrophy One of a group of genetically inherited metabolic defects in humans that are characterized by painless progressive degeneration and atrophy of muscle without involvement of the nervous system. Major types of muscular dystrophy involve degeneration of the musculature of girdles and limbs.

mushroom sugar TREHALOSE.

mustard 1. SULFUR MUSTARD. 2. NITROGEN MUSTARD.

mustard oil glycoside A plant toxin that is a sulfur-containing glycoside.

mutability The capacity to undergo mutation.

mutability spectrum MUTATIONAL SPECTRUM.

mutable Capable of undergoing mutation.

mutable gene An unstable gene that is characterized by a high rate of spontaneous mutation.

mutable site Any site along the chromosome at which a mutation can occur.

mutagen A physical or chemical agent that is capable of inducing mutations; a mutagen raises the frequency of mutation above that due to spontaneous mutations.

mutagenesis The production of mutations. Mutagenesis is said to be direct if the mutagenic agent leads to mutations only at the site of the DNA damage. Mutagenesis is said to be indirect if mutations occur at other sites as well because of relaxed fidelity of replication or other causes. Many environmental chemical carcinogens are believed to lead to indirect mutagenesis.

mutagenic Capable of inducing mutations.

mutagenic agent MUTAGEN.

mutagenic repair SOS REPAIR.

mutagenize To expose nucleic acids, viruses, cells, or organisms to a mutagen.

mutant 1. A cell, a virus, or an organism that

carries a gene that has undergone mutation 2. A gene that has undergone mutation.

mutant allele *See* mutant gene.

mutant gene A gene that has undergone a mutation; the modified nucleotide sequence of a wild-type gene.

mutant protein A protein formed from a mutant gene.

mutarotation The change in optical rotation with time that occurs when an optical isomer, such as a carbohydrate, is dissolved in water and is converted to an equilibrium mixture of several different optical isomers.

mutase An enzyme that catalyzes the intra-molecular transfer of a group, specifically a phosphate group.

mutation 1. The process whereby a gene undergoes a structural change leading to a sudden and stable change in the genotype of a cell, a virus, or an organism; any heritable change in the genome of a cell, a virus, or an organism other than that due to the incorporation of genetic material from other sources; any change in the base sequence of DNA. 2. MUTANT GENE. 3. The cell, the virus, or the organism that carries a mutant gene.

mutational load The genetic inadequacy of a population as a result of the mutational accumulation of deleterious genes.

mutational spectrum The genetic map of the point mutations that either arise spontaneously or are produced by exposure to a mutagen.

mutation distance The smallest number of mutational changes that is required to change one nucleotide sequence in DNA into a different one.

mutation frequency 1. The proportion of mutants of a given type in a population of growing cells or organisms. 2. MUTATION RATE.

mutation frequency decline The phenomenon that ultraviolet-induced reversions from auxotroph to prototroph decrease as a function of time if they are determined under conditions whereby postirradiation protein synthesis is inhibited. *See also* mutation stabilization.

mutation index An estimate of the mutation frequency; it is equal to the proportion of mutants in a population of cells that have been grown from an inoculum that contained such a small number of organisms that it is unlikely that any mutants were among them.

mutation pressure The continued production of specific mutants.

mutation rate The total number of mutations, or the number of mutations of a specified kind, that are produced in a population of cells or organisms per cell division, or per replication, over a given period of time.

mutation stabilization The phenomenon that ultraviolet-induced reversions from auxotroph to prototroph achieve a constant value as a function of time if they are determined under conditions allowing for postirradiation protein synthesis. *See also* mutation frequency decline.

mutation theory *See* somatic mutation theory.

mutator gene A gene that increases the mutation rate of other genes, as in a system where the mutator gene produces a DNA polymerase that makes errors during replication.

mutator mutant A mutant carrying a mutator gene.

mutator strain A strain carrying a mutator gene.

mutein A mutant protein such as the cross-reacting material.

muton The unit of genetic mutation; the smallest section of a chromosome, which may be as small as a single nucleotide, a change in which can result in a mutation.

$\bar{M}_v$ Viscosity-average molecular weight.

$\bar{M}_w$ Weight-average molecular weight.

MW Molecular weight.

MWC model Monod, Wyman, and Changeux model; *see* concerted model.

myasthenia gravis A disease, characterized by profound muscular weakness, that is due to the formation of antibodies to the receptors for acetylcholine. Binding of the antibodies to the receptors decreases the level of active receptors in the tissues and decreases the efficiency of neuromuscular transmissions.

mycelium (*pl* mycelia) The vegetative structure of a fungus that consists of a multinucleate mass of cytoplasm, enclosed within a branched network of filamentous tubes known as hyphae.

myco- Combining form meaning fungus.

mycobacteria A genus of aerobic, gram-positive bacteria that occur in soil, water, and the tissues of various animals; includes the causative agents of tuberculosis and leprosy.

mycobactin One of a group of siderophores of the hydroxamic acid type that are found in mycobacteria.

mycolic acid One of a group of complex long-chain hydroxy fatty acids that contain from 60 to 90 carbon atoms, have varying degrees of branching and unsaturation, and contain cyclopropane rings.

mycology The branch of botany that deals with fungi.

mycoplasma A genus of primitive bacteria that are the simplest known, independently living organisms, and that differ from other bacteria in not having a cell wall.

mycorrhiza 1. A stable symbiotic association between a fungus and the root of a plant. 2. The root-fungus structure in the symbiotic relationship of a fungus and the root of a plant.

mycoside A lipid composed of a long-chain, highly branched, and hydroxylated hydrocarbon that is terminated by a phenol to which a trisaccharide is linked by means of a glycosidic bond.

mycosis A disease caused by a fungus.

mycosterol A sterol of fungi.

mycotoxins Toxic substances produced by fungi that are harmful to other organisms, especially vertebrates, including humans. Those toxic for humans and animals include aflatoxins and ergot alkaloids. The term is also used frequently to include toxins produced by other microorganisms such as the endotoxins and exotoxins of bacteria.

myelination The formation and the deposition of the myelin sheath around an axon.

myelin basic protein A major myelin protein of the mammalian central nervous system. Upon injection into guinea pigs, rabbits, or rats, it induces allergic autoimmune encephalomyelitis, an inflammation of the brain and spinal column. The protein is rich in basic amino acids. *Aka* encephalitogenic protein; myelin protein A1. *See also* Cop 1.

myelin protein A1 MYELIN BASIC PROTEIN.

myelin sheath The lipid-rich, insulating covering of an axon that is formed by wrapping the plasma membrane of a Schwann cell around the axon. Myelin contains lipids, especially shingolipids, proteins, polysaccharides, salts, and water.

myeloma A tumor of cells that are derived from the hematopoietic tissue of bone marrow; a tumor in which a single antibody-secreting plasma cell, programmed to make one type of antibody, has multiplied to yield a large clone of cells, each of which produces only that antibody. As a result, the tumor secretes large amounts of a single species of antibody.

myeloma proteins Pathological immunoglobulin proteins, so called because they are formed in large amounts in patients with multiple myeloma. Myeloma proteins include the Bence–Jones protein, myeloma globulins (which have typical immunoglobulin structures with light and heavy chains), and fragments of the various classes of normal immunoglobulins.

myeloperoxidase A lysosomal enzyme of phagocytic leukocytes that aids in the destruction of alien objects by forming hypochlorite from hydrogen peroxide and chloride ions.

myelosome An organelle formed by the fusion of myelinated nerve cell axons during homogenization; contains plasma membrane vesicles encased in a myelin sheath that is open at both ends.

myo- Combining form meaning muscle.

myocardial Of, or pertaining to, the heart muscle.

myocyte 1. A muscle cell. 2. A contractile cell, especially one in sponges.

myofibril A small, contractile, threadlike structure of striated muscle; the myofibrils are arranged in parallel bundles within the cytoplasm of a muscle fiber.

myofibrillar ghost FIBRIL GHOST.

myofilament A minute, contractile, threadlike structure of striated muscle; the myofilaments are arranged in parallel bundles within a myofibril and consist of thin and thick filaments.

myogen An aqueous extract of the sarcoplasm of striated muscle that consists largely of glycolytic enzymes.

myogenesis The formation of muscle tissue.

myoglobin The oxygen-storing protein of muscle that consists of a single polypeptide chain surrounding a heme group, and that is closely related to the monomeric unit of hemoglobin. *Abbr* Mb.

myograph An instrument for recording the forces of muscular contractions.

myohematin CYTOCHROME.

myoinositol INOSITOL.

myokinase ADENYLATE KINASE.

myoneme A bundle of intracellular protein filaments found in some ciliates and believed to account for contractility.

myoneural Of, or pertaining to, muscles and nerves.

myoneural junction NEUROMUSCULAR JUNCTION.

myosin The most abundant protein of the myofilaments of striated muscle and the protein that forms the thick filaments. Myosin is an asymmetric molecule composed of two heavy chains (MHC) and two pairs of light chains (MLC). Each heavy chain consists of a long, alpha-helical section attached to a globular head. In the intact myosin molecule, the two heavy-chain alpha-helical sections are intertwined, forming a long rod-like, fibrous tail, from which the two heads project. Each head is a complex of the globular portion of a heavy chain and two light chains (one molecule of each type). Both the ATPase activity and the actin-binding capacity of myosin reside in the globular head portion of the molecule.

myosin B ACTOMYOSIN.

myosin filament A thick filament of striated muscle from which cross-bridges protrude that link the thick filament to the thin filaments; a myofilament.

myosin subfragment SF_1 FRAGMENT.

myotropic activity The anabolic effect of androgens on nitrogen metabolism that leads to nitrogen retention of the body and to a limited increase in muscle strength and development.

myria- 1. Combining form meaning ten thousand and used with metric units of measurements. *Sym* my. 2. Combining form meaning a great number.

myristic acid A saturated fatty acid that contains 14 carbon atoms and that occurs in animal fat.

myxedema 1. HYPOTHYROIDISM. 2. The dry, waxy type of skin swelling found in individuals that have marked hypothyroidism; due to infiltration of the skin with proteoglycans.

myxovirus A large, enveloped animal virus that contains single-stranded RNA. Myxoviruses are divided into two subgroups: the orthomyxoviruses, which include the influenza virus, and the paramyxoviruses, which include the mumps virus.

$\overline{M}_z$ Z-average molecular weight.

MZE Multiphasic zone electrophoresis.

N

n 1. Refractive index. 2. Nano. 3. Neutron.

N 1. Nitrogen. 2. Normal concentration. 3. Nucleoside. 4. Asparagine. 5. Haploid number. 6. Neutron number. 7. Newton. 8. Avogadro's number. 9. Nominally labeled.

Na Sodium.

NA Noradrenaline.

NAcneu *N*-Acetylneuraminic acid.

NAD⁺ Nicotinamide adenine dinucleotide. *Aka* DPN⁺.

NADH Reduced nicotinamide adenine dinucleotide. *Aka* DPNH.

NADH–coenzyme Q reductase complex COMPLEX I.

NADH–dehydrogenase complex COMPLEX I.

NADH : ubiquinone oxidoreductase COMPLEX I.

Na⁺,K⁺-ATPase The adenosine triphosphatase that is located in the cell membrane and that functions in the active transport of sodium and potassium ions. The enzyme is vectorial in its action and requires the presence of both sodium and potassium ions. The Na⁺,K⁺-ATPase is a primary active transport mechanism.

naked virion A virion that consists of a nucleocapsid that is not surrounded by a membrane.

nalidixic acid A synthetic antibiotic that inhibits prokaryotic DNA replication; it inhibits the DNA gyrase of *E. coli.*

NAN *N*-Acetylneuraminic acid.

NANA *N*-Acetylneuraminic acid.

nano- Combining form meaning one-billionth (10⁻⁹) and used with metric units of measurements. *Sym* n.

Naperian logarithm NATURAL LOGARITHM.

naphthol blue black AMIDO BLACK 10B.

narcosis A state of stupor produced by a narcotic drug; loss of consciousness.

narcotic 1. Of, or pertaining to, narcosis. 2. NARCOTIC DRUG.

narcotic drug A substance that acts predominantly on the central nervous system and that produces, depending on the dose, sedation, stimulation, sleep, loss of consciousness, or relief of pain. According to the World Health Organization (1964), narcotic drugs are classified as follows: (a) alkaloids (LSD, mescalin, opium, etc.); (b) barbiturates and other sleeping drugs; (c) alcohol; (d) cocaine; (e) hashish and marijuana; (f) hallucinogens; (g) stimu-

lants or antidepressants (such as amphetamines).

nascent Being formed; in the process of being synthesized, particularly in reference to the synthesis of macromolecules.

nascent polypeptide chain 1. A polypeptide chain that is in the process of being formed and that is attached to a transfer RNA molecule which, in turn, is bound to a ribosome. 2. A complete, newly synthesized, polypeptide chain prior to any post-translational modification.

nascent RNA 1. An RNA molecule that is in the process of being formed. 2. A complete, newly synthesized, RNA molecule prior to any post-transcriptional modification.

Natelson microgasometer An instrument for the manometric measurement of the oxygen, carbon monoxide, and carbon dioxide content of blood.

National Formulary A pharmacopeia published in the United States. *Abbr* N.F.

native 1. Descriptive of a protein or a nucleic acid molecule in its natural, in vivo state as opposed to its denatured state. 2. Descriptive of a protein or a nucleic acid molecule that has been isolated by mild procedures so that it is undenatured, or only slightly denatured, and is taken to represent the in vivo state of the molecule.

native conformation 1. The in vivo conformation of a macromolecule. 2. The normal conformation of a macromolecule that has been isolated by mild procedures, shows no apparent structural alterations, and is investigated under suitable conditions of pH, temperature, and ionic strength.

native immunity NATURAL IMMUNITY.

native plasma Plasma obtained without the addition of an anticoagulant.

natriuretic hormone A postulated hormone that increases the excretion of sodium ions in the urine by inhibiting reabsorption of the sodium ions by the kidney. The hormone is also believed to act on sodium ion transport of other tissues. Specifically, the hormone is thought to change the ion concentration of smooth muscles in small blood vessels, making them contract more readily and thereby raising the blood pressure. *Aka* third factor.

natural abundance The relative proportion of

an isotope in nature, based on the sum of the concentrations of all the other isotopes of that particular element.

natural antibody An antibody that is present in the blood and that is capable of reacting with specific antigens even though the organism had no known exposure to those antigens.

natural auxin INDOLEACETIC ACID.

natural chain elongation The elongation of a chain by tailward growth.

natural immunity The immunity that is characteristic of an organism and that is a result of the genetic makeup of the organism. Currently an outmoded concept since it is now believed that all immunity ultimately requires exposure to an antigen and, hence, is acquired immunity.

natural immunization An immunization brought about by natural exposure, as by inhalation, ingestion, skin contact, or infection.

natural killer cells Small lymphocyte-like cells that are present in normal individuals of various mammalian species. These cells spontaneously kill a variety of tumor cells and virus-infected cells and their activity is enhanced by interferon. They are known as natural killer cells because their cytolytic action does not require prior immunization. *Abbr* NK cells. *See also* killer cells.

natural logarithm A logarithm to the base *e*. *Abbr* $\log_e$; ln. *See also* logarithm.

natural pH gradient The pH gradient that is used in isoelectric focusing and that is produced during the experiment by the electrolysis of carrier ampholytes. The carrier ampholytes band at positions where the isoelectric pH of the ampholyte equals the pH of the solution; a pH gradient is thereby established which, once formed, is stable for prolonged times.

natural product 1. A secondary metabolite that has no known function. 2. Any organic compound produced by a living organism.

natural selection The principle, proposed by Darwin in 1859, that natural processes favor those members of the species that are better adapted to their environment and tend to eliminate those that are unfitted to their environment. Thus, the "fittest" organisms survive and through successive generations changes become established that lead to the production of new types and new species. *Aka* Darwinian selection.

nature's antacid SECRETIN.

n_D Refractive index, measured with the light of a sodium D line.

NDP 1. Nucleoside diphosphate. 2. Nucleoside-5'-diphosphate.

nearest-neighbor base frequency analysis A method for assessing base sequences in nuc-

leic acids by a comparison of the frequencies with which any pair of adjacent bases occurs in these nucleic acids. The method is used especially for DNA and, in that case, requires the synthesis of DNA by DNA polymerase in the presence of a DNA template and the four 5'-deoxyribonucleoside triphosphates, one of which is labeled with ^{32}P in its alpha phosphorus. The synthesized product is digested to the 3'-deoxyribonucleoside monophosphates which results in a shift of the ^{32}P label from the incorporated nucleotide to its nearest neighbor in the synthesized polynucleotide strand. The experiment is performed four times, using a different, labeled deoxyribonucleoside triphosphate at each time. The extent of label in the isolated 3'-deoxyribonucleoside monophosphates can then be used to calculate the frequencies of pairs of adjacent bases in the newly synthesized DNA.

nearest-neighbor cooperative model A model for the cooperative binding of multiple ligands to a macromolecule that is based on two constants; a nucleation constant K that represents the binding constant to a site whose nearest neighbors are unoccupied; and a growth constant K' that represents a binding constant to a site adjacent to an occupied one. The ratio $q = K'/K$ is a measure of the degree of cooperative binding. When $q > 1$, the binding shows positive cooperativity; when $q < 1$ it shows negative cooperativity (anticooperativity); and when $q = 0$ the binding is that of the nearest-neighbor exclusion model.

nearest-neighbor exclusion model EXCLUDED SITE BINDING.

nebulin A protein in striated muscle that may help to keep the myosin thick filaments centered within the sarcomere during force generation.

nebulizer ATOMIZER.

necrosis The sum of the morphological changes in a group of cells, a tissue, or an organ that is indicative of cell death; the pathological death of cells that is due to the progressive degradative action of enzymes.

NEFA Nonesterified fatty acids.

negative catalysis Catalysis that leads to a decrease in the rate of a chemical reaction. This may occur, for example, through the tight binding of an intermediate to an enzyme; as a result, the intermediate is prevented from reacting with a "wrong" compound by having the energy of activation for that reaction increased due to its binding to the enzyme.

negative complementation The inhibition of the activity of a subunit of a wild-type oligomeric protein by a mutant allele-type subunit.

negative contrast staining NEGATIVE STAINING.

negative control The prevention of a biological activity by the presence of a specific molecule; the prevention by a repressor of either inducible enzyme synthesis or initiation of mRNA synthesis are two examples. *See also* negative regulation.

negative cooperativity Cooperative binding in which the binding of one ligand to one site on the molecule decreases the affinity for the binding of subsequent ligands to other sites on the same molecule. *Aka* anticooperativity.

negative effector *See* effector.

negative electron ELECTRON (1).

negative feedback A feedback mechanism, as in many biological systems, in which a large output of the system leads to a decrease in the subsequent output while a small output of the system leads to an increase in the subsequent output.

negative gene control NEGATIVE REGULATION.

negative hydration Hydration in which the water molecules in the primary hydration shell of an ion have greater mobility than those in pure water.

negative labeling The process whereby a specific residue in a protein is either masked or protected so that it will not undergo a reaction with a given reagent, and all the remaining, available, and unprotected residues are then labeled by reaction with the reagent.

negative phase The stage that follows the administration of a second dose of antigen to an animal during which there is a temporary decrease in the concentration of free antibodies in the circulation due to the fact that the added antigens combine with preexisting circulating antibodies.

negative polarity Descriptive of a single-stranded RNA or single-stranded DNA molecule that has bases in a sequence that is complementary to the sequence in the corresponding mRNA molecule.

negative regulation The regulation of gene expression in which transcription is turned off by the presence of an inhibitor (repressor) on the DNA; an anti-inhibitor (inducer) is required to remove this inhibition and permit mRNA synthesis to take place.

negative staining A staining technique, used in electron microscopy, in which the material to be examined is mixed with an electron-dense substance, such as phosphotungstic acid, and appears transparent against an opaque background.

negative strand virus A virus containing single-stranded RNA that has negative polarity. The RNA strand (called a negative strand) has a base sequence that is complementary to that found in the viral mRNA. Therefore, the infecting RNA cannot code directly for pro-teins. Instead, it must first be copied by an RNA-dependent RNA polymerase to yield a translatable mRNA molecule that does code for proteins. *See also* plus strand; virus.

negative supercoil NEGATIVE SUPERHELIX.

negative superhelix *See* superhelix.

negatron A negatively charged electron.

negentropy The negative, or lack, of entropy. Since a positive information change can be considered to be equivalent to a negative entropy change, information can be considered to be the negative of entropy or negentropy. The information content of a molecule can, therefore, be considered to represent negentropy.

neighbor-exclusion principle EXCLUDED SITE BINDING.

neighboring group effect The effect of a group in a molecule on a nucleophilic displacement reaction in which the molecule participates; the effect is due to the group functioning as an internal nucleophile for an intramolecular displacement reaction.

neighbor restoration The reactivation of cells, damaged by irradiation with ultraviolet light, that is brought about when the cells are incubated in a liquid medium at high concentrations of cells from either the same or other strains.

NEM *N*-Ethylmaleimide.

nematic liquid crystal *See* liquid crystal.

nematosome A cytoplasmic inclusion occurring in certain neurons.

Nembutal Trademark for sodium pentobarbital; used as an anesthetic for mammals.

neoantigen An antigen that has acquired new antigenic specificity either by some modification of the original molecule and/or by the coupling of a hapten.

neobiogenesis The repeated formation of life from nonliving, inorganic matter.

neocarcinostatin *See* antitumor proteins.

neolysosome An acid phosphatase-containing vesicle that develops into a mature lysosome; derived from tubules associated with the Golgi apparatus.

neomorph A mutant gene that produces a qualitatively new effect which is not produced by the wild-type gene.

neomycin An aminoglycoside antibiotic, produced by *Streptomyces fradiae*, that inhibits protein synthesis by binding to the 30S ribosomal subunit; it also causes misreading of the genetic code.

neonatal 1. Of, or pertaining to, a neonate, 2. Of, or pertaining to, the period immediately following birth.

neonate A newborn.

neontology The science that deals with life of the current geologic era; the study of extant (living) species. *See also* paleontology.

neoplasia The pathological condition characterized by tumor formation and tumor growth.

neoplasm A new and abnormal growth; a proliferation of cells that is not subject to the usual limitations of growth; a tumor. *See also* benign neoplasm; malignant neoplasm.

neoplastic Of, or pertaining to, a neoplasm.

neosome A collective term to describe both intersomes and minisomes.

neotenin ALLATUM HORMONE.

nephelometry The quantitative determination of a substance in suspension that is based on measurements of the light scattered by the suspended particles at right angles to the incident beam. *See also* turbidimetry.

nephrectomy The surgical removal of a kidney.

nephron The structural and functional unit of the kidney; consists of the glomerulus, Bowman's capsule, Henle's loop, and the proximal and distal tubules.

Nernst equation An expression that relates the actual electrode potential E of a given redox couple to its standard electrode potential E_0 and to the concentrations of the oxidant and the reductant. For a half-reaction, the expression is

$$E = E_0 + (0.06/n)\log([Ox]/[Red])$$

where n is the number of electrons participating in the half-reaction, $[Ox]$ is the concentration of oxidant, and $[Red]$ is the concentration of reductant.

nerve An elongated structure of nervous tissue that consists of nerve fibers enclosed within a sheath, and that serves to connect the nervous system with other organs and tissues of the body.

nerve-end particle SYNAPTOSOME.

nerve fiber The process of a neuron.

nerve gas A mixture of compounds, including diisopropylfluorophosphate, that react with the serine hydroxyl group of the enzyme acetylcholinesterase and thereby inhibit the transmission of nerve impulses.

nerve growth factor A polypeptide that has hormone-like properties; a mitogen that produces hypertrophy and hyperplasia of nerve cells, growth of nerve cell processes, and an increase in the metabolism of various nerve cells. The nerve growth factor occurs as a monomer (MW 13,000) and as a dimer (MW 26,000); it has protease activity and is similar to insulin in its structure. *Abbr* NGF.

nerve impulse An electrical stimulus that passes along a nerve and that leads to excitation of the nerve along the way.

nerve impulse conduction The passage of a nerve impulse along a nerve cell.

nerve impulse transmission The passage of a nerve impulse from one nerve cell to another.

nervonic acid An unsaturated fatty acid that has 24 carbon atoms and one double bond; Δ^{15}-tetracosenoic acid; a constituent of cerebrosides.

nervous system The nervous tissue that, in vertebrates, consists of the central and the peripheral nervous systems.

Nesslerization The treatment of a sample with Nessler's reagent.

Nessler's reagent A solution of mercuric iodide and potassium iodide in potassium hydroxide that is used for the colorimetric determination of nitrogen in biological materials.

net charge The charge of a macromolecule obtained by summing the number of positively and negatively charged functional groups. Thus, a protein that, at a given pH, has 25 positively charged amino groups ($-NH_3^+$) and 13 negatively charged carboxyl groups ($-COO^-$), is said to have a net charge of +12.

net protein utilization A measure of the nutritional value of a protein that is based on how well the protein is digested and how well it is utilized once the amino acids have been absorbed into the system. It is defined as $(N_{retention}/N_{intake}) \times 100 = [(N_{intake} - N_{output})/N_{intake}] \times 100$. *Abbr* NPU. *See also* biological value.

network *See* computer network.

Neu Neuraminic acid.

NeuAc *N*-Acetylneuraminic acid.

Neuberg ester FRUCTOSE-6-PHOSPHATE.

Neuberg fermentation The alcoholic fermentation of yeast, either in the absence of added compounds (Neuberg's first form of fermentation) or in the presence of bisulfite (Neuberg's second form of fermentation).

NeuNAc *N*-Acetylneuraminic acid.

neural Of, or pertaining to, nerves.

neuraminic acid A compound, derived from mannosamine and pyruvic acid, the acetylated form of which is a major building block of animal cell coats; the compound 5-amino-3,5-dideoxy-D-glycero-D-galacto-nonulosonic acid. *Sym* Neu.

neuraminidase The enzyme that catalyzes the cleavage of *N*-acetylneuraminic acid from mucopolysaccharides; the enzyme is present on the surface of certain viruses and destroys the receptor activity of many cells for these viruses.

neuraminosyl group SIALOSYL GROUP.

neurilemmal cell SCHWANN CELL.

neurite A general term for a process of an embryonic nerve cell. These processes are difficult to identify as being either axons or dendrites and hence are referred to simply as neurites.

neuroaminoyl group SIALOYL GROUP.

neurochemistry The science that deals with the biochemistry of nervous tissue.

neuroendocrine Of, or pertaining to, both the nervous and the endocrine systems.

neurofibril A small, threadlike structure in a neuron.

neurofilaments *See* intermediate filaments.

neurohormone A hormone that is released into the circulation at nerve endings and that acts upon cells located at some distance from its point of release. Examples include the hypothalamic regulatory hormones and the hormones of the neurohypophysis. *See also* neurotransmitter.

neurohypophyseal Of, or pertaining to, the posterior lobe of the pituitary gland.

neurohypophysis The posterior lobe of the pituitary gland that produces vasopressin, oxytocin, and melanocyte-stimulating hormone.

neuroleukin A neurotrophic factor that promotes the survival of some neurons in tissue culture; it is also a lymphokine, produced by lectin-stimulated T cells.

neurological mutant A mutant that leads either to pronounced malformations of the central nervous system or to pronounced abnormalities in locomotion.

neurology The science that deals with the structure and function of the nervous system.

neuromuscular Of, or pertaining to, both nerves and muscles.

neuromuscular junction The junction between the axon of a motor neuron and a skeletal muscle fiber.

neuron A nerve cell; the structural and functional unit of the nervous system that consists of a cell body and its processes, the axon and the dendrites.

neuronal uptake A mechanism for terminating the biological action of catecholamines; involves an active transport whereby the catecholamines are moved across the membrane of sympathetic nerves and are then either stored or metabolized.

neuropeptide A peptide that is active in the nervous system. Some neuropeptides function as neurotransmitters. *See also* opioid.

neurophysin One of a group of carrier proteins to which the hormones oxytocin and vasopressin become bound noncovalently. These hormone–neurophysin complexes are stored in the neurohypophysis and are discharged upon stimulation.

neuroplasm The cytoplasm of a neuron.

neurosecretion 1. The secretion of chemical substances, such as neurohormones and neurohumors, by nerve cells. 2. The chemical substances secreted by nerve cells.

neurosecretory granule A particle, derived from the posterior lobe of the pituitary gland,

that contains oxytocin, vasopressin, and neurophysin.

neurosecretory spheres Small spherical structures, $0.1–0.2 \ \mu m$ in diameter, that are synthesized in the axoplasm of specialized neurons.

neurosecretosome SECRETOSOME.

Neurospora crassa The red bread mold; a fungus used for biochemical and genetic studies.

neurotensin A peptide neurotransmitter, consisting of 13 amino acids, that is present in the gastrointestinal tract; a vasodilator and an inhibitor of gastric secretion and intestinal motility.

neurotoxin A toxin that acts specifically on nervous tissues.

neurotransmitter A small molecule that is liberated at nerve endings and that diffuses to neighboring cells where it triggers a specific response; a chemical messenger between a neuron and a target cell. Acetylcholine, which functions in the transmission of nerve impulses, is an example. Other neurotransmitters include epinephrine, norepinephrine, dopamine, and γ-aminobutyrate. *See also* neurohormone.

neurotropic virus A virus, the target organ of which is the nervous system.

neurotypy The molecular heterogeneity of neurofilament proteins.

neutral 1. Being neither acidic nor basic; a neutral solution has a pH of 7.0. 2. Being neither positively nor negatively charged; a neutral atom has an equal number of protons and orbital electrons.

neutral amino acid An amino acid that has one amino group and one carboxyl group.

neutral fat An ester formed from a molecule of glycerol and one to three molecules of fatty acids; a mono-, di-, or triglyceride.

neutral glycolipids Glycolipids whose polar head groups consist only of neutral sugars.

neutralization 1. The reaction between an acid and a base, forming water. 2. The inactivation of a soluble antigen, such as a toxin, or a particulate antigen, such as a virus, by reaction with the appropriate antibodies.

neutral lipid A lipid, such as a glyceride or a steroid, that is devoid of pronounced polar groups. *Abbr* NL.

neutral mutation 1. A mutation that produces no change in the adaptive value of an organism. *See* neutral theory of molecular evolution. 2. A mutation that has no measurable phenotypic effect.

neutral sugars Simple sugars as opposed to sugar acids, amino sugars, and similar derivatives.

neutral theory of molecular evolution A theory of evolution according to which most evolutionary changes at the molecular level (nucleotide substitutions in functional genes) are caused not by positive Darwinian selection but rather by random drift of mutant genes that are selectively neutral (equivalent) or nearly neutral. A number of mutations of a given gene are considered to be neutral if they are all equally effective (or essentially so) for the adaptive value, or fitness, of the organism as regards survival and reproduction. Whichever mutation is retained is considered to be due to a random fixation of a particular gene and not due to a selection process. *Aka* neutral theory; neutral gene theory.

neutrino A subatomic particle that has zero charge and essentially zero mass; it accounts for that part of the energy of beta decay that is not associated with the emitted beta particle. *See also* elementary particles.

neutron A neutral, subatomic particle of the nucleus that has a mass of 1.009 amu; it is equivalent to a combined proton and electron. *Sym* n.

neutron-activated phosphorus–Bakelite plaque A radioactive disk that contains ^{32}P atoms and that is used as a source of beta particles.

neutron activation analysis *See* activation analysis.

neutron capture The capture of a neutron by an atomic nucleus that frequently occurs during the production of artificial radioactive isotopes.

neutron contrast matching technique A technique for investigating macromolecular structures; based on measuring neutron scattering by particulate matter in solutions containing varying concentrations of light and heavy water.

neutron number The number of neutrons in the nucleus of an atom. *Sym* N.

neutron scattering A technique for investigating macromolecular structures that permits a determination of the radius of gyration; based on measurements of the scattering of a neutron beam by the atomic nuclei of a deuterated protein. Neutrons are scattered by the nuclei of atoms while x rays are scattered by the extranuclear electrons. *Aka* neutron diffraction.

Newcastle disease virus A virus that infects the respiratory tract of birds and that belongs to the group of paramyxoviruses.

Newman projection The representation of a molecule in which the arrangement of the atoms is such as would be seen by an observer viewing the molecule from one end along the carbon-to-carbon bond closest to the observer. When viewed in this fashion, some atoms appear to be fully or partially hidden by other atoms and are called eclipsed, while other

atoms are clearly visible and are called staggered.

newton The SI unit of force; $kg\ m\ s^{-2}$. *Sym* N.

Newtonian fluid A fluid,the viscosity of which is independent of the rate of shear.

nexin A protein (MW 150,000) that forms the "links" between doublets of microtubules (A and B subfibers) in the axoneme structure of cilia and flagella. *See also* protease nexin.

N.F. National Formulary; used to denote a chemical that meets the specifications set out in the National Formulary.

n-fold helix A helix with *n* residues per turn of the helix.

n-fold symmetry Having an *n*-fold axis of rotational symmetry. *See also* axis of rotational symmetry.

NGF Nerve growth factor.

NHI Nonheme iron.

niacin 1. NICOTINIC ACID. 2. A generic descriptor for pyridine-3-carboxylic acid an its derivatives that exhibit qualitatively the biological activity of nicotinic acid.

niacinamide NICOTINAMIDE.

nick A break in a single strand of a nucleic acid, particularly a break in a single strand of a double-stranded nucleic acid. *See also* cut.

nickase An endonuclease that introduces single-strand breaks into double-stranded DNA.

nick-closing enzyme *See* topoisomerase.

nicked-circle A double-stranded circular DNA molecule that contains one or more single-strand breaks in one or both strands.

nicked DNA A DNA molecule having one or more breaks in either one or both of its strands.

nickel An element that is essential to humans and animals. Symbol, Ni; atomic number, 28; atomic weight, 58.70; oxidation states, +1, +2, +3, +4; most abundant isotope, ^{58}Ni; a radioactive isotope, ^{65}Ni, half-life, 2.6 h, radiation emitted, beta particles.

nicking-closing enzyme *See* topoisomerase.

nick translation An in vitro procedure in which the polymerase activity and the $5' \rightarrow 3'$ exonuclease activity of DNA polymerase I function simultaneously and at comparable rates. As a result, a given nick in the DNA moves along the DNA molecule in the direction of synthesis ($5' \rightarrow 3'$). The method is useful for producing radioactive DNA of high specific activity. This is done by carrying out the reaction in the presence of radioactively labeled deoxyribonucleoside triphosphates.

Nicol prism One of two prisms used in a polarimeter. *See also* analyzer; polarizer.

nicotiana alkaloids A group of pyridine alkaloids occurring mainly in the tobacco plant

(*Nicotiana*); includes nicotine, nicotyrine, anabasine, and others. *Aka* tobacco alkaloids.

nicotinamide The amide of nicotinic acid.

nicotinamide adenine dinucleotide A coenzyme form of the vitamin nicotinic acid; a coenzyme for pyridine-linked dehydrogenases. *Abbr* NAD^+; DPN^+.

nicotinamide adenine dinucleotide phosphate A coenzyme form of the vitamin nicotinic acid; a coenzyme for pyridine-linked dehydrogenases. *Abbr* $NADP^+$; TPN^+.

nicotinamide mononucleotide A precursor of nicotinamide adenine dinucleotide in which nicotinamide is linked to ribose-5-phosphate. *Abbr* NMN^+.

nicotine A pyridine alkaloid that belongs to the group of nicotiana alkaloids and that is considered to be carcinogenic.

nicotinic acid A B vitamin, the deficiency of which causes the disease pellagra and the coenzyme forms of which are NAD^+ and $NADP^+$. Nicotinic acid is unique among the B vitamins in that it can be synthesized in animal tissues (from tryptophan).

nicotinic acid amide NICOTINAMIDE.

nicotinic receptor A synaptic acetylcholine receptor to which nicotine binds, thereby mimicking the action of acetylcholine. Such receptors are found at skeletal muscle end plates and at autonomic ganglia. *See also* α-bungarotoxin.

nicotinic synapse A synapse containing nicotinic receptors.

Niemann–Pick disease A genetically inherited metabolic defect in humans that is associated with mental retardation and that is characterized by an accumulation of sphingomyelin in the tissues; it is due to a deficiency of the enzyme sphingomyelinase.

nif genes Genes involved in nitrogen fixation.

nigericin An ionophore that transports potassium and hydrogen ions.

night blindness An early manifestation of vitamin A deficiency in which the retinal rods have an elevated visual threshold and do not respond normally to faint light.

night vision The capacity to see in dim light that is due to the rods in the retina.

NIH National Institutes of Health; an agency of the U.S. Public Health Service.

NIH shift The reaction, discovered at the National Institutes of Health, in which a proton is shifted from the para position in phenylalanine to the meta position in tyrosine during the hydroxylation of phenylalanine to tyrosine.

nine plus two arrangement A characteristic arrangement of microtubules in eukaryotic cilia and flagella; 9 pairs of tubules are arranged in a circle around 2 central tubules.

ninhydrin reaction The reaction of ninhydrin (triketohydrindene hydrate) with the free alpha amino groups of amino acids, peptides, or proteins; the reaction yields colored compounds that are useful for the chromatographic detection and for the quantitative determination of amino acids and peptides.

nisin A peptide antibiotic, produced by *Streptococcus lactis*, that consists of 34 amino acids.

Nissl substance The rough endoplasmic reticulum of neurons.

nitratase NITRATE REDUCTASE.

nitrate A salt of nitric acid.

nitrate ammonification See nitrate respiration.

nitrate assimilation The reduction of nitrate to nitrite and, hence, to ammonia which is assimilated (ammonia fixation); occurs in green plants and the reducing power is provided by respiration.

nitrate dissimilation *See* dissimilatory reduction; nitrate respiration.

nitrate reductase The enzyme that catalyzes the reduction of nitrate to nitrite in nitrate assimilation by plants and fungi; it is a molybdenum-containing flavoprotein.

nitrate reduction The reduction of nitrate to nitrite, ammonia, or molecular nitrogen, that is carried out in nature by bacteria and fungi. *See also* nitrate respiration.

nitrate respiration The reduction of nitrate either to ammonia, which is not assimilated (ammonification of nitrate), or to NO, N_2, and similar gases that are excreted (denitrification). In all of these reactions, which occur in various bacterial species, nitrate serves as the terminal electron acceptor, instead of oxygen, for cellular respiration.

Nitrazine paper Trademark for a dye-impregnated paper that is used for estimation of pH values.

nitrification The oxidation of ammonia to nitrite or nitrate that is carried out in nature by nitrifying bacteria.

nitrifying bacteria Bacteria that oxidize ammonia to nitrite and that oxidize nitrite to nitrate.

nitrile A compound containing the grouping $-C \equiv N$.

nitrite A salt of nitrous acid.

nitrite reductase The enzyme that catalyzes the reduction of nitrite to ammonia in nitrate assimilation by plants.

nitrocellulose filter *See* membrane filtration.

nitrofuran One of a group of synthetic antibiotics that are active against many gram-positive and gram-negative bacteria; they are derivatives of 5-nitrofuran.

nitrogen An element that is essential to all plants and animals. Symbol, N; atomic number, 7, atomic weight, 14.00067; oxidation states, -1, $+1$, $+2$, $+3$, $+4$, $+5$; most abundant isotope, ^{14}N; a stable isotope, ^{15}N.

nitrogenase The nitrogen-fixing enzyme system that catalyzes the reduction of atmospheric nitrogen to ammonia and that is a complex of two proteins both of which are required for activity. One of these proteins is a tetramer (MW about 200,000–270,000); it contains iron, molybdenum, and labile sulfur, and is known as Mo–Fe protein, component I, molybdoferredoxin, FeMo protein, molybdoiron protein, or azofermo. The other protein is a dimer (MW about 60,000); it contains iron and labile sulfur and is known as Fe protein, component II, iron protein, azoferredoxin, or azofer.

nitrogen balance The difference between the nitrogen intake and the nitrogen excretion of an animal. The nitrogen balance is denoted as zero, positive, or negative depending on whether the intake is equal to, greater than, or smaller than the excretion.

nitrogen catabolite repression The repression by ammonia of the synthesis of various enzymes that function in nitrogen metabolism.

nitrogen cavitation *See* cavitation (2).

nitrogen cycle The cyclic set of reactions involving plants, animals, and bacteria, whereby (a) atmospheric nitrogen is converted to inorganic compounds and these are then converted to complex organic compounds; and (b) the organic compounds are broken down to inorganic compounds which ultimately yield atmospheric nitrogen.

nitrogen equilibrium The state of an animal in which the nitrogen balance is equal to zero.

nitrogen fixation The conversion of atmospheric nitrogen to ammonia by means of either a biological or a synthetic reaction.

nitrogen mustard Di(2-chloroethyl)methylamine; a chemical mutagen and alkylating agent. *See also* alkylating agent.

nitrogenous Nitrogen-containing.

nitrogenous base 1. A purine or a pyrimidine. *Aka* nitrogen base. 2. A nitrogen-containing basic compound.

nitrogen rule The principle of mass spectrometry which states that a compound, containing an odd number of nitrogen atoms, has an odd-numbered molecular ion. The logic behind this rule derives from the fact that nitrogen is trivalent, thus requiring an odd number of hydrogen atoms in the molecule.

o-nitrophenol See o-nitrophenyl galactoside.

p-nitrophenol See p-nitrophenyl phosphate.

o-nitrophenyl galactoside A synthetic compound used for the assay of beta galactosidase. The enzyme hydrolyzes *o*-nitrophenyl galactoside to galactose and *o*-nitrophenol.

The latter is intensely colored and is determined spectrophotometrically. *Abbr* ONPG.

p-nitrophenyl phosphate A synthetic substrate for assaying both acid and alkaline phosphatase activity; these enzymes catalyze the removal of a phosphate group from *p*-nitrophenyl phosphate and the *p*-nitrophenol that is formed is then determined spectrophotometrically. *Abbr* PNPP.

nitroprusside reaction A colorimetric reaction for cysteine and free sulfhydryl groups in a protein that is based on the production of a red color upon treatment of the sample with sodium nitroprusside and ammonia.

nitrosamine One of a class of mutagenic and carcinogenic alkylating agents that have the structure shown below. They are formed by reaction between amines and nitrogen oxides. There is concern that nitrates, used in food preservation, may become converted to nitrosamines in the process of food preparation and/or as a result of metabolic reactions.

$$\begin{array}{c} R \\ \backslash \\ N{-}N{=}O \\ / \\ R \end{array}$$

nitrosation A chemical reaction that results in the introduction of a nitroso group ($-\ddot{N}{=}O$) into a compound.

nitroso group The grouping $-\ddot{N}{=}O$.

nitrous acid A chemical mutagen that leads to the deamination of purines and pyrimidines and that converts adenine to hypoxanthine and cytosine to uracil. Since hypoxanthine base pairs with cytosine, the effect of nitrous acid treatment is that an original adenine is read as guanine and an original cytosine is read as uracil.

nitrous acid mutant A mutant produced by treatment of a nucleic acid with nitrous acid; such mutants have been produced, for example, by treating the RNA from tobacco mosaic virus with nitrous acid, infecting tobacco leaves with the mutated RNA, and isolating the mutant viral particles formed in the infected leaves. *See also* nitrous acid.

nkat Nanokatal.

NK cells Natural killer cells.

NL Neutral lipid.

Nle Norleucine.

N-linked oligosaccharides *See* glycosylation.

nm Nanometer; also indicated as mμm (millimicron).

NMN$^+$ Nicotinamide mononucleotide.

NMP 1. Nucleoside monophosphate. 2. Nucleoside-5′-monophosphate.

NMR Nuclear magnetic resonance.

NMR desert A nuclear magnetic resonance technique in which the differential relaxation rate of a nucleus, observed before and after a specific deuterium substitution for a proton, is converted into an internuclear distance through the use of rotational correlation times obtainable from a C-13 measurement. Desert is an acronym for deuterium substitution effect on relaxation time.

nodal compound The compound that is common both to a linear metabolic pathway and to its branch.

node 1. NODAL COMPOUND. 2. The crossover region in the figure-of-eight structure of superhelical DNA. The node is called positive when the left strand in the upper part of the 8 is closest to the viewer; the node is called negative when that same strand is in back of the other strand. Thus,

NOD mouse Nonobese diabetic mouse; a strain of mice that develop diabetes mellitus and that are characterized by destruction of pancreatic beta cells and by lymphocyte infiltration of pancreatic islets. The NOD mouse appears to have a unique major histocompatibility complex.

nodoc ANTICODON.

nodule bacteria *See* Rhizobium.

NOE Nuclear Overhauser effect.

NOESY Nuclear Overhauser effect spectroscopy; *See* nuclear Overhauser effect.

noise The background interference in an instrument that may be caused, for example, by electronic, optical, or chemical disturbances.

noise analysis FLUCTUATION ANALYSIS.

nojirimycin An antibiotic, synthesized by several strains of *Streptomyces*, that is a potent inhibitor of α-glucosidases; a glucose analogue (5-amino-5-deoxy-D-glucose) in which an NH— group substitutes for the oxygen atom in the pyranose ring.

Nomarski differential interference microscope A microscope that provides a three-dimensional view of an object and that permits the observation of transparent structures in the living cell. Like the phase contrast microscope, it employs the differences in the phase of the light wave as it passes through different parts of the specimen.

nomenclature A system of names, designations, and symbols that are used in a given discipline.

nominally labeled Designating a compound in which some, and usually a significant amount, of the label is at a given position or positions in the molecule, but for which no further information is available as to the extent of label,

if any, at other positions in the molecule. *Sym* N.

nomogram An alignment chart for the rapid determination of a variable from the given values of two or more variables. A typical nomogram consists of a minimum of three scales such that when known values of two scales are connected by a straight line, the line intersects the third scale at the sought value. The hemoglobin concentration in blood, for example, can be determined from a scale of the specific gravity of plasma and a scale of the specific gravity of whole blood.

nona- Combining form meaning nine.

nonagglutinating antibody INCOMPLETE ANTIBODY.

non-AIS-suppressible insulin INSULIN-LIKE ACTIVITY.

nonamer An oligomer that consists of nine monomers.

nonbasic chromosomal proteins Nonhistone proteins associated with the DNA in chromosomes; acidic or neutral proteins such as specific enzymes or regulatory proteins. *Aka* nonhistone chromosomal proteins.

nonbonding interaction NONCOVALENT INTERACTION.

nonbonding orbital A molecular orbital that contains electrons that take little part in the actual bonding of the atomic nuclei. Such orbitals usually have energies intermediate between those of bonding orbitals (sigma, pi) and those of antibonding orbitals (sigma star, pi star).

noncoding strand ANTICODING STRAND.

noncollisional energy transfer The energy transfer from an excited molecule to another molecule that occurs when the two molecules remain farther apart than the contact distance that they attain during molecular collisions.

noncompetitive inhibition The inhibition of the activity of an enzyme that is characterized by an increase in the slope of a double reciprocal plot (1/velocity versus 1/substrate concentration) and by a decrease in the maximum velocity compared to those of the uninhibited reaction.

noncompetitive inhibitor An inhibitor that produces noncompetitive inhibition and that in general is structurally unrelated to the substrate. A noncompetitive inhibitor can bind to either the enzyme or the enzyme–substrate complex. *See also* degree of inhibition.

noncoupled pump A pump for the transport of one solute across a membrane that also drives the transport of a second solute across the same membrane in the opposite direction and in such a fashion that the transport of the second solute is physically independent of the pump.

noncovalent bond A bond between atoms and/ or molecules that does not involve shared pairs of electrons and that is due to other types of interactions. Examples of such bonds are electrostatic bonds, hydrogen bonds, and hydrophobic bonds.

noncovalent interaction An interaction between atoms and/or molecules that does not involve the formation of chemical bonds and that is based on the formation of noncovalent bonds.

noncovalent structure Collective term for the secondary, tertiary, and quaternary structure of a protein.

noncyclic electron flow The light-induced, photosynthetic electron flow in which the electrons flow from water, or some other electron donor, to $NADP^+$, or some other electron acceptor; in chloroplasts, it is the electron flow from water through photosystems II and I to $NADP^+$.

noncyclic photophosphorylation The synthesis of ATP that is coupled to the noncyclic electron flow of photosynthesis.

nonelectrolyte A substance that does not dissociate into ions in water; solutions of nonelectrolytes do not conduct an electric current.

nonenergized conformation CONDENSED CONFORMATION.

nonequilibrium thermodynamics IRREVERSIBLE THERMODYNAMICS.

nonessential amino acids Amino acids that an organism can synthesize from various intermediates in metabolism and that, therefore, do not have to be obtained through the diet.

nonessential enzyme An enzyme that is not required for either the growth or the survival of a cell or an organism.

nonessential gene A gene, the product of which is a nonessential enzyme.

nonexclusive binding The binding to an allosteric enzyme that takes place when both the relaxed and the constrained forms of the enzyme are present in significant amounts. *See also* concerted model.

nonexclusive binding coefficient The ratio of the intrinsic dissociation constants for the substrate in the concerted model of allosteric enzymes; specifically, the ratio K_R/K_T, where K_R and K_T are the intrinsic dissociation constants for the substrate binding site on a protomer in the R state and T state, respectively.

nonheme iron Iron that occurs in biological systems but that is not in the form of a heme group. *Abbr* NHI.

nonheme-iron chromophore A pair of iron atoms, as in plant-type ferredoxin, that are located close enough to each other in a mole-cule so that they can engage in antiferromagnetic coupling; such atoms possess a characteristic absorption spectrum and a distinct electron paramagnetic resonance spectrum.

nonheme-iron protein A conjugated protein that contains iron but not in the form of a heme group. *See also* iron–sulfur protein.

nonhistone chromosomal proteins NONBASIC CHROMOSOMAL PROTEINS.

noninducible enzyme CONSTITUTIVE ENZYME.

nonionic detergent A surface-active agent that has polar and nonpolar groups but carries no charges. *Aka* nonionic surface-active agent.

nonionized acids COMBINED ACIDITY.

nonketotic hyperglycinemia A genetically inherited metabolic defect in human that is characterized by mental retardation and a high urinary excretion of glycine; a severe neonatal disease that is due to a deficiency of the glycine cleavage enzyme.

nonmediated transport Transport across a biological membrane that does not involve a transport agent; simple diffusion.

non-Mendelian inheritance CYTOPLASMIC INHERITANCE.

non-Newtonian fluid A fluid, the viscosity of which depends on the rate of shear; solute molecules of a non-Newtonian fluid, especially asymmetric molecules, tend to become oriented as the rate of shear is increased.

nonnucleic acid base A base that rarely, if ever, occurs in a nucleic acid under normal circumstances.

nonose A monosaccharide that has nine carbon atoms.

nonoverlapping code A genetic code in which each nucleotide is used in only one codon for the synthesis of a given nucleic acid molecule.

nonpalindromic helix A helix that has no end-to-end symmetry so that both ends are different either in composition and/or by virtue of the helix having a sense of direction. The rotation of such a helix by 180° about an axis perpendicular to the longitudinal axis of the helix produces a structure that is not identical to that before rotation.

nonparametric statistics Statistical calculations that are not based on any prior assumptions with respect to the variable and the probability distribution of the data.

nonpermissible substitution RADICAL SUBSTITUTION.

nonpermissive cell 1. A cell in which a conditional lethal mutant cannot grow. 2. A cell that does not support the lytic infection by a virus.

nonpermissive conditions RESTRICTIVE CONDITIONS.

non-plasma-specific enzyme An enzyme that is present in blood plasma but has no known specific function in the plasma.

nonpolar Lacking polarity; lacking a permanent dipole moment.

nonpolar amino acid An amino acid that has a nonpolar side chain.

nonpolar bond 1. A covalent bond in which the electron pair or pairs of the bond are held with equal strength by the two bonded atoms. 2. HYDROPHOBIC BOND.

nonpolar solvent A solvent that is devoid of significant concentrations of charged groups and/or dipoles.

nonprecipitating antibody COPRECIPITATING ANTIBODY.

nonproductive complex An enzyme–substrate complex in which the substrate is bound to the enzyme in such a fashion that catalysis is impossible and that products cannot be formed.

nonprotein amino acid An amino acid that rarely, if ever, occurs in a protein under normal circumstances. Many such amino acids occur in nature in the form of precursors of normal (protein) amino acids, as intermediates in catabolic pathways, and as D-enantiomers of normal amino acids.

nonprotein nitrogen The nitrogen in serum that is not present in the form of serum proteins. *Abbr* NPN.

nonprotein respiratory quotient The respiratory quotient that is calculated on the basis of a volume of oxygen equal to the total volume utilized minus that utilized for protein catabolism, and on the basis of a volume of carbon dioxide equal to the total volume produced minus that produced by protein catabolism.

nonreciprocal recombination GENE CONVERSION.

nonreducing end That end of an oligo- or polysaccharide that does not contain a hemiacetal or hemiketal grouping and, hence, will not reduce certain inorganic ions in solution.

nonreducing sugar A sugar that does not contain an aldehyde or potential aldehyde group, and hence, will not reduce certain inorganic ions in solution.

nonrepetitive DNA UNIQUE DNA.

nonsaponifiable lipid A lipid that cannot be hydrolyzed with alkali to yield soap as one of the products; steroids and terpenes are two major nonsaponifiable lipids.

nonsecretor An individual who does not secrete water-soluble forms of the blood group substances in body fluids.

nonselective medium A medium that allows the growth of all genotypes present.

nonsense codon A codon that does not code for an amino acid; now called a termination (stop) codon.

nonsense mutation A mutation in which a normal codon that specifies an amino acid is changed to one of the three termination codons or vice versa.

nonsense suppression The suppression of a nonsense mutation.

nonsense suppressor A gene that codes for a tRNA molecule which is altered in its anticodon such that it is able to recognize a nonsense (termination, stop) codon. The chain termination function of a stop codon is thereby suppressed and the polypeptide chain is extended beyond its normal end ("read through").

nonsequential mechanism PING-PONG MECHANISM.

nonspecific immunity The immunity that is produced by nonimmunological mechanisms such as lysozyme action, phagocytosis, or interferon action.

nonsuppressible insulin-like activity INSULIN-LIKE ACTIVITY.

nonsymbiotic nitrogen fixation The conversion of atmospheric nitrogen to ammonia by organisms, such as photosynthetic bacteria or blue-green algae, without the participation of plants.

nontransmissible plasmid A plasmid that cannot be transferred from one cell to another.

nonviable Descriptive of a cell or an organism that is dead and incapable of reproduction.

nopaline *See* opine.

nor Prefix used in steroid nomenclature to indicate elimination of an angular methyl group or elimination of a methylene group from a side chain. *See also* norsteroid.

NOR Nucleolus organizer region; *See* nucleolus organizer.

noradrenaline NOREPINEPHRINE.

n orbital Nonbonding orbital.

norepinephrine A catecholamine hormone that is secreted by the adrenal medulla and that has a biological activity that is similar to that of epinephrine but less pronounced. *Var sp* norepinephrin.

n orientation The sense of insertion of a DNA fragment into a vector such that the genetic maps of both the fragment and the vector have the same orientation. *See also* u orientation.

Norit Trademark for a purified charcoal made from birch; used for the decolorization of solutions and for the adsorption of compounds in adsorption chromatography. *Var sp* Norite.

Norit eluate factor FOLIC ACID.

norleucine An amino acid analogue that can be incorporated into protein during protein synthesis. A straight-chain isomer of leucine.

normal amino acids STANDARD AMINO ACIDS.

normal configuration The configuration of steroids in which substituents at positions 5 and

10 are cis with respect to the plane of rings A and B.

normal distribution A continuous frequency distribution characterized by a bell-shaped curve and described by the equation $Y = (1/\sigma\sqrt{2\pi})e^{-(x-m)^2/2\sigma^2}$ where m is the mean, σ is the standard deviation, e is the base of natural logarithms, π is a constant equal to 3.1416..., and Y is the height of the ordinate for a given value of X on the abscissa. Different values of m shift the curve along the abscissa without changing its shape. Different values of X change the shape of the curve without changing the position of the center.

normal electrode potential STANDARD ELECTRODE POTENTIAL.

normal enzyme An enzyme, the substrates of which are metabolites normally occurring within the organism, as distinct from a drug-metabolizing enzyme, the substrates of which are compounds foreign to the organism.

normal error curve NORMAL DISTRIBUTION.

normal frequency distribution NORMAL DISTRIBUTION.

normality The concentration of a solution expressed in terms of the number of gram-equivalent weights of solute in one liter of solution. *Sym* N.

normalized substrate concentration REDUCED SUBSTRATE CONCENTRATION.

normalizing The adjustment of data to an arbitrary standard; the normalizing of a spectrum, for example, is done by multiplying the observed absorbance values at all measured wavelengths by a factor that is equal to the ratio of the desired absorbance to the observed absorbance at one particular wavelength.

normal-phase chromatography *See* partition chromatography.

normal saline PHYSIOLOGICAL SALINE.

normal solution A solution that contains one gram-equivalent weight of solute per liter of solution.

normal temperature and pressure STANDARD TEMPERATURE AND PRESSURE.

normal value The amount of a chemical constituent in, or the value of a physical property of, a body fluid or an excretion that is found in 95% of a population of clinically normal and apparently healthy individuals.

norsteroid A steroid-like molecule; a modified steroid in which a ring has been contracted. *See also* nor.

northern blotting A variation of the Southern blotting technique in which RNA fragments are separated electrophoretically, transferred to a special paper which binds them covalently, and are then located by hybridization with probes of radioactive RNA or single-stranded DNA. *Aka* northern transfer; northern hybridization. *See also* blotting.

norvaline A straight-chain isomer of valine.

notatin Glucose oxidase; a flavoprotein enzyme that catalyzes the oxidation of glucose to the delta lactone and that can be isolated in a highly active form from the mold *Penicillium notatum*.

nothing dehydrogenase effect An abnormality in the electrophoretic determination of lactate dehydrogenase isozymes in which blank preparations, from which substrate has been omitted, exhibit faint replicas of the normal isozyme pattern. The effect is believed to be due to the presence of alcohol dehydrogenase in the enzyme preparation.

novobiocin An antibiotic, produced by *Streptomyces niveus*, that inhibits DNA replication mainly in gram-positive bacteria.

np Nucleotide pair.

NP antigen A nucleoprotein core antigen of poxviruses.

NPH insulin A neutralized zinc salt of protamine insulin developed by Hagedorn. The salt is insoluble and, when injected into an animal, provides a slowly adsorbed insulin depot so that fewer injections of insulin are required in clinical treatments of diabetes.

n–pi star transition The excitation of an electron from an *n* orbital to a pi star orbital.

NPN Nonprotein nitrogen.

NPU Net protein utilization.

nRNA Nuclear RNA.

NSF National Science Foundation.

NSILA Nonsuppressible insulin-like activity.

N-terminal The end of a peptide or of a polypeptide chain that carries the amino acid that has a free alpha amino group; in representing amino acid sequences, the N-terminal is conventionally placed on the left side. *Aka* N-terminus.

NTP 1. Nucleoside triphosphate. 2. Nucleoside-5′-triphosphate. 3. Normal temperature and pressure.

nu body NUCLEOSOME.

Nuc Nucleoside.

nuclear Of, or pertaining to, the nucleus of either an atom or a cell.

nuclear body NUCLEOID (1).

nuclear column A column of cell nuclei that are immobilized with small pieces of membrane filters and through which a solvent is passed.

nuclear cycle CELL CYCLE.

nuclear division KARYOKINESIS.

nuclear duplication MITOSIS.

nuclear emulsion A photographic emulsion that has been specially sensitized for the detection of alpha or beta particles; it is generally thicker and more concentrated in silver halide than ordinary photographic emulsions.

nuclear envelope The envelope that surrounds the nucleus of eukaryotic cells and that consists of two membranes, an inner and an outer one; both membranes are lipid bilayers and they are separated by a gap of 20–40 nm, known as the perinuclear space.

nuclear equivalent NUCLEOID.

nuclear fission A reaction in which a heavier atomic nucleus is broken up into two or more lighter and more stable nuclei with the release of large amounts of energy.

nuclear fusion A reaction between two atomic nuclei in which two lighter nuclei combine to form a heavier and more stable nucleus with the release of large amounts of energy.

nuclear isomer One of two or more nuclides that have the same atomic number and the same atomic mass but that have nuclei that are at different energy levels.

nuclear magnetic resonance A method for studying the interaction of an atomic nucleus, having an odd mass number and an odd number of protons, with the environment of the nucleus. A nucleus of this kind has a spin and a magnetic moment and may exist in one of several allowed energy levels. When placed in an applied magnetic field of suitable magnitude, the nucleus will undergo a transition from one energy level to another, accompanied by the absorption of electromagnetic radiation. The relative magnitudes of the applied magnetic field and of the absorbed radiation are interpreted in terms of the interaction of the nucleus with its environment. The method is used particularly for protons and, when so used, is also referred to as proton magnetic resonance. *Abbr* NMR.

nuclear membrane *See* nuclear envelope.

nuclear Overhauser effect The decrease in area of a given line in a nuclear magnetic resonance spectrum that is due to the transfer of energy from one proton to another; involves the change of a population of nuclei in a given energy level by saturating a nearby nucleus. The magnitude of the effect (that is, the decrease in the area of the line) increases the closer the two protons are to each other. The effect can, therefore, be used to measure intramolecular distances. *Abbr* NOE. *Aka* nuclear Overhauser enhancement.

nuclear reaction 1. A reaction taking place in the nucleus of a cell. 2. A physical reaction in which there are changes in the nuclei of reacting atoms as distinct from a chemical reaction in which there are changes in the orbital electrons.

nuclear reactor A device for the controlled use of nuclear reactions for the production of radioactive isotopes and energy. A fission reactor permits a controlled chain reaction to produce energy; a breeder reactor is designed to produce more fissionable material than it consumes; and a tokamak fusion reactor employs a doughnut-shaped magnetic field to hold the plasma in which fusion occurs.

nuclear resonance scattering MOSSBAUER EFFECT.

nuclear zone NUCLEOID.

nuclease An enzyme that catalyzes the hydrolysis of phosphodiester bonds in nucleotides and nucleic acids.

nucleated Possessing a nucleus.

nucleation 1. The formation of regions of three-dimensional structure in separate portions of a protein molecule prior to attainment of the complete tertiary structure of the molecule. 2. The formation of a crystal or an aggregate by the condensation of matter on minute particles that serve as nuclei. 3. The first step in the polymerization of actin and tubulin. *See also* helix nucleation. 4. The formation of bone around collagen.

nucleic acid A polynucleotide of high molecular weight that is synthesized by living cells. Nucleic acids occur as either DNA or RNA and may be either single-stranded or double-str..nded. DNA functions in the transfer of genetic information, and RNA functions in the biosynthesis of proteins.

nucleic acid bases *See* purine; pyrimidine.

nuclein The nucleoprotein discovered by Miescher in 1868.

nucleo- Combining form meaning nucleic acid.

nucleocapsid The protein coat of a virus together with the nucleic acid which it encloses. In some viruses, the nucleocapsid is contained within a lipoprotein membrane (enveloped); in others, it exists in naked form (nonenveloped).

nucleocidin A purine antibiotic, produced by *Streptomyces calvus*, that inhibits protein synthesis.

nucleo-cytoplasmic ratio The ratio of the volume of the nucleus to the volume of the cytoplasm for a given cell.

nucleodisome A structural fragment consisting of two nucleosomes connected via linker DNA.

nucleohistone A conjugated protein consisting of histone and nucleic acid.

nucleoid 1. A DNA mass, not bounded by a membrane, that occurs in a prokaryotic cell, a chloroplast, a mitochondrion, or a virus, and that is analogous to the nucleus of a eukaryotic cell. 2. The RNA core of an RNA virus that is surrounded by the protein capsid.

nucleolar organizer NUCLEOLUS ORGANIZER.

nucleolin The major nucleolar protein (MW 100,000) of exponentially growing eukaryotic cells; it is found in association with intranucleolar chromatin and preribosomal particles and is barely detectable in resting cells. Nucleolin is believed to play a direct role in pre-rRNA transcription and ribosome assembly. *Aka* C23.

nucleolus (*pl* nucleoli) An RNA-rich region in the cell nucleus that contains a variety of enzymes, the primary products of the genes present in the nucleolus organizer, and proteins associated with these genes. The nucleolus is the site where ribosomes are partially synthesized.

nucleolus organizer A portion of the chromosome that is associated with the nucleolus and that contains the genes for the synthesis of ribosomal RNA.

nucleon A constituent particle occurring within the atomic nucleus; a proton or a neutron.

nucleonics The application of nuclear phenomena to other fields.

nucleon number MASS NUMBER.

nucleophile An atom or a group of atoms that is electron pair donating.

nucleophilic Of, or pertaining to, either a nucleophile or a reaction in which a nucleophile participates.

nucleophilic catalysis Catalysis in which the catalyst donates a pair of electrons to a reactant.

nucleophilic displacement A chemical reaction in which a nucleophilic group attacks and displaces a susceptible group in a compound and then binds covalently to the compound at that site. *See also* S_N1 mechanism; S_N2 mechanism.

nucleophilic substitution NUCLEOPHILIC DISPLACEMENT.

nucleoplasm The protoplasm of the cell nucleus.

nucleoplasmin An acidic protein that has been isolated from the nuclei of many eukaryotic cells and that appears to function in the assembly of nucleosomes; its mechanism of action is unknown.

nucleoprotein A conjugated protein in which the nonprotein portion is a nucleic acid, and the protein portion is frequently either a histone or a protamine.

nucleosidase An enzyme that catalyzes the hydrolysis of a nucleoside to the pentose and the base.

nucleoside A glycoside composed of D-ribose or 2-deoxy-D-ribose and either a purine or a pyrimidine. *Abbr* Nuc; N.

nucleoside antibiotic One of a large number of antibiotics, many produced by species of *Streptomyces*, that contain a nucleoside as part of their structure.

nucleoside cyclic monophosphate A nucleotide in which the phosphoric acid residue is esterified to two hydroxyl groups on the sugar.

nucleoside diphosphate A high-energy derivative of a nucleoside in which a pyrophosphate group is esterified to a hydroxyl group of the sugar. *Abbr* NDP.

nucleoside diphosphate kinase An enzyme that catalyzes the transfer of a phosphate group from a nucleoside-5'-triphosphate to a nucleoside-5'-diphosphate.

nucleoside diphosphate sugar One of a group of compounds that consist of a sugar linked to a nucleoside diphosphate and that serve as glycosyl group donors in the biosynthesis of starch, glycogen, and other oligo- and polysaccharides. *Abbr* NuDP-sugar.

nucleoside monophosphate NUCLEOTIDE.

nucleoside monophosphate kinase An enzyme that catalyzes the transfer of a phosphate group from a nucleoside-5'-triphosphate to a nucleoside-5'-monophosphate.

nucleoside triphosphate A high-energy derivative of a nucleoside in which three phosphate groups are linked in succession to one hydroxyl group of the sugar. *Abbr* NTP.

nucleosome A repeating, bead-like structure of eukaryotic chromosomes; a macromolecular complex consisting of two molecules each of histones H2A, H2B, H3, and H4 around which is wrapped a DNA segment of 140 bp. Any two nucleosomes are connected via a segment of linker DNA to which a molecule of histone H1 is attached. *Aka* nucleosome core; platysome; core particle. *See also* chromatosome.

nucleosome phasing The sequence-specific placing of nucleosomes along the chromosome that leaves certain sections of DNA in association with histones and other sections as free linker DNA.

nucleosome spacing The distance between adjacent nucleosomes along the chromosome.

nucleotidase An enzyme that catalyzes the hydrolysis of a nucleotide to the nucleoside and orthophosphate.

nucleotide 1. The building block of the nucleic acids that consists of a nucleoside plus a phosphoric acid residue which is esterified to one of the hydroxyl groups of the sugar. 2. Any phosphorylated nucleoside, whether or not it is a genuine building block of nucleic acids. 3. One of a number of compounds, structurally somewhat related to a nucleotide. *See also* mono- and dinucleotides.

nucleotide anhydride NUCLEOTIDE COENZYME.

nucleotide-binding domain ROSSMAN FOLD.

nucleotide coenzyme One of a group of compounds, derived from uracil, cytosine, or thymine, that function as either coenzymes or substrates in carbohydrate and lipid metabolism; UDP-glucose, CDP-ethanolamine, and

TDP-ribose are some examples.

nucleotide exchange reaction The reaction, catalyzed by the enzyme nucleoside diphosphate kinase, in which a nucleoside diphosphate and a nucleoside triphosphate are converted to the opposite pair of nucleoside di- and triphosphates by the transfer of a phosphate group.

nucleotide map A fingerprint of nucleotides.

nucleotide pair Two nucleotides, in a double-stranded nucleic acid structure, that are linked by means of complementary base pairing (H bonds).

nucleotide sugar NUCLEOSIDE DIPHOSPHATE SUGAR.

nucleus (*pl* nuclei) 1. The structure in eukaryotic cells that contains the chromosomes. 2. The central core of an atom that consists of protons and neutrons. 3. The ring structure of an organic compound.

nuclide An atom that is characterized by the composition of its nucleus, having a specified atomic number and mass number; an atom that has a specific number of protons, a specific number of neutrons, and a specific energy content in its nucleus.

nuclidic mass ATOMIC MASS.

nude mice A strain of laboratory mice that are congenitally hairless and that lack a thymus (athymic); they are unable to make T cells and are virtually devoid of them. Nude mice must be raised in an essentially germ-free environment.

NuDP-sugar Nucleoside diphosphate sugar.

null allele An allele that does not produce a functional product.

null cells Lymphocytes that lack the cell surface markers characteristic of B and T lymphocytes.

null hypothesis A hypothesis stating that there is no difference between two values, such as between the means of two populations. The hypothesis is advanced to evaluate data; it is subsequently tested statistically and either accepted or rejected.

number-average molecular weight An average molecular weight that is weighted toward those molecules present in largest number; specifically, $\bar{M}_n = \Sigma n_i M_i / \Sigma n_i$, where n_i is the number of moles of component i, and M_i is the molecular weight of component i. *Sym* $\bar{M}_n$. *See also* average molecular weight.

number of degrees of freedom *See* degrees of freedom.

Nutrasweet *See* Aspartame.

nutrient A substance that promotes the growth, maintenance, function, and reproduction of a cell or an organism.

nutrient agar Nutrient broth gelled with agar; a solid medium for the growth of bacteria.

nutrient broth A liquid medium that contains a meat extract and that is used in bacteriology.

nutrient medium MEDIUM.

nutrilite A compound that fits the definition of a vitamin with the exception that it is required by a microorganism and not by an animal; a vitamin of microorganisms.

nutriment Nourishment; food.

nutrition The supplying to, and the intake and utilization by, an organism of all the necessary elements required by it for normal growth, maintenance, function, and reproduction.

nutritional calorie A large calorie; 1000 (small) calories. *Sym* kcal; Cal.

nutritional mutant AUXOTROPH.

nyctalopia NIGHT BLINDNESS.

Nylander's reagent A reagent that contains bismuth subnitrate, potassium–sodium tartrate, and potassium hydroxide, and that is used for the detection of reducing sugars; the sugars yield a black precipitate of metallic bismuth when treated with the reagent.

nylon Polyhexamethylene adipamide; a synthetic polymer.

O

o 1. Ortho. 2. Prefix or subscript used to designate the original (unmodified) form of an interconvertible enzyme.

O 1. Oxygen. 2. Orotic acid. 3. Orotidine.

OA Ovalbumin.

OAA Oxaloacetic acid.

OAc Acetyl group.

O antigens Polysaccharide antigens of the cell surface of gram-negative bacteria. *See also* K antigens; H antigens; lipopolysaccharide; rough strain; smooth strain.

obesity The state of being overweight, frequently due to overnutrition; usually interpreted as being 10% or more above the average standard weight for the individual.

objective The microscope lens closest to the specimen.

oblate ellipsoid of revolution An ellipsoid of revolution formed by rotation of an ellipse about its minor axis.

obligate Limited to living under a specific set of conditions.

obligate aerobe An organism or a cell that can grow only in the presence of molecular oxygen.

obligate anaerobe 1. An organism or a cell that can grow only in the absence of molecular oxygen. 2. An organism or a cell that lacks the ability to synthesize an oxygen-linked respiratory chain and must, therefore, utilize fermentation. *See also* strict anaerobe.

obligatory reactant LEADING REACTANT.

oc Abbreviation for ochre mutation.

Occam's razor A guiding principle for the selection of a hypothesis or a theory from among several alternative ones. The principle states that when several competing and possible alternative hypotheses or theories are presented, the simplest one is to be preferred and the more complex ones are to be discarded. *Aka* principle of simplicity.

occlude To produce an occlusion.

occlusion 1. The blocking of a passageway, such as the blocking of an artery. 2. The trapping of a substance by adsorption and adhesion, such as the trapping of soluble substances in a precipitate.

occult blood Small amounts of blood, as those in urine and feces, that cannot be detected visually.

occult virus A virus that has infected a host but cannot be detected.

occupation theory A theory according to which the action of drugs is due to their interaction with specific receptor sites in the organism that is being treated.

ochre codon The codon UAA; one of the three termination codons.

ochre mutation A mutation in which a codon is mutated to the ochre codon, thereby causing the premature termination of the synthesis of a polypeptide chain.

ochre suppression The suppression of an ochre codon.

ochronosis Abnormal pigmentation of cartilage and other connective tissue; may occur in alkaptonuria.

octa- Combining form meaning eight.

octamer An oligomer that consists of eight monomers.

octopine *See* opine.

octose A monosaccharide that has eight carbon atoms.

ocular The eyepiece of an optical instrument, such as that of a microscope.

ocytocin Variant spelling of oxytocin.

o.d. Outside diameter.

OD Optical density.

odd base MINOR BASE.

odd electron UNPAIRED ELECTRON.

odd-numbered fatty acid A fatty acid molecule that has an odd number of carbon atoms; refers to the total number of carbon atoms, those in the hydrocarbon portion of the molecule plus that in the carboxyl group. *Aka* odd-carbon fatty acid.

ODP 1. Orotidine diphosphate. 2. Orotidine-5′ -diphosphate.

oestradiol Variant spelling of estradiol.

oestrogen Variant spelling of estrogen.

-OEt Ethoxy group.

-ogen Suffix meaning inactive precursor of an enzyme.

Ogston hypothesis POLYAFFINITY THEORY.

ohm A unit of electrical resistance; equal to the resistance of a conductor that carries a current of 1 A when a potential difference of 1 V is applied across its terminals.

Ohm's law The law stating that the strength of a direct electric current is proportional to the potential difference and inversely proportional to the resistance.

oil A fat that is liquid at room temperature.

oiled coils Fibrillar segments in the elastin molecule that may be stretched from 2.0 to 2.5 times their length in the relaxed molecule and that are believed to account for the elasticity of the molecule.

oil-immersion objective An objective lens used to increase the resolution attainable with the light microscope; based on filling the space between the coverslide of the specimen and the objective with an oil that has the same refractive index as the coverslide.

Okazaki fragments A group of short DNA fragments that are produced during the initial stages of the discontinuous replication of DNA; they represent precursor fragments of the lagging strand that are subsequently joined by DNA ligase to form longer fragments. Okazaki fragments have been demonstrated, in the case of *E. coli*, by exposing the cells to tritiated thymine for short periods during their growth. *Aka* precursor fragments.

old age disease ALZHEIMER'S DISEASE.

old cells 1. Bacterial cells in the stationary phase of growth. 2. Cells that have been stored for a prolonged time.

old tuberculin Tuberculin prepared by concentrating and filtering a culture of the tubercle bacillus, *Mycobacterium tuberculosis. Abbr* OT.

old yellow enzyme A flavoprotein from yeast that catalyzes the oxidation of NADPH and that was isolated in 1932 by Warburg and Christian.

oleandomycin An antibiotic, produced by *Streptomyces antibioticus*, that is closely related to erythromycin.

olefin An unsaturated aliphatic hydrocarbon; an alkene.

oleic acid An unsaturated fatty acid that contains 18 carbon atoms and one double bond. *See also* promoter.

oleophilic HYDROPHOBIC.

oleophobic HYDROPHILIC.

oleosome A spherosome that is rich in lipids but devoid of acid phosphatase and other hydrolytic enzymes.

Olestra Trademark for a synthetic fat substitute not yet approved by the FDA.

oligodynamic effect The growth inhibitory effect of some metals (such as silver and copper) in their elementary form on bacteria; tested for by placing a metal disk on a freshly inoculated petri plate and measuring the zones of nongrowth surrounding each disk after incubation.

oligogene A gene that, by itself, produces a significant phenotypic effect as opposed to a polygene which, by itself, produces only a small effect.

oligomer A protein molecule that consists of two or more polypeptide chains, referred to as either monomers or protomers, linked together covalently or noncovalently. *See also* monomer; promoter.

oligomycin A polyene antibiotic produced by *Streptomyces diastatochromogenes*. It is an inhibitor of oxidative phosphorylation where it blocks ATP synthesis by inhibiting the F_0F_1-ATPase.

oligomycin-sensitivity-conferring factor The F_0 component of the F_0F_1-ATPase. *Abbr* OSCF.

oligonucleotide A linear nucleic acid fragment that consists of from 2 to 10 nucleotides joined by means of phosphodiester bonds; oligoribonucleotides consist of ribonucleotides, and oligodeoxyribonucleotides consists of deoxyribonucleotides.

oligopeptide A linear peptide that consists of from 2 to 10 amino acids joined by means of peptide bonds.

oligosaccharide A linear or branched carbohydrate that consists of from 2 to 10 monosaccharide units joined by means of glycosidic bonds.

oligosaccharin One of a group of plant regulatory molecules. They differ from plant hormones, which have multiple effects, in that each one regulates only a single plant function such as defense against disease, growth, reproduction, or differentiation. Oligosaccharins are fragments of the cell wall, released from the latter by enzymatic action.

oligotrophic lake A deep and clear-water lake, having a depth of 15 m or more, that has a plant population at various depths, and that has a low rate of nutrient supply in relation to its volume of water. In such a lake, both the biomass and the productivity are low. The bottom layers of the lake are saturated with dissolved oxygen throughout the year. *See also* eutrophic lake; mesotrophic lake.

O-linked oligosaccharides *See* glycosylation.

O-locus The locus of the operator.

-oma Suffix meaning tumor.

-OMe Methoxy group.

omega angle The torsion angle that denotes the rotation about the C^1—N peptide bond in proteins. *Sym* ω.

omega fraction CHYLOMICRONS.

omega loop A polypeptide chain conformation in globular proteins that is shaped like the Greek letter omega. A loop is a continuous segment of a polypeptide chain, folded back upon itself, and defined in terms of the (a) segment length, (b) absence of regular secondary structure, and (c) distance between segment termini.

omega oxidation An oxidative pathway of fatty acids in which the terminal methyl group of the fatty acid (usually one containing 8–12 carbon atoms) is oxidized first to a hydroxyl group and then to a carboxyl group, leading to the formation of a dicarboxylic acid; the pathway occurs in both animal and microbial cells.

omega protein The first topoisomerase to be discovered; a type I topoisomerase that occurs in *E. coli* and that relaxes only negatively supercoiled DNA. *Abbr* ω-protein.

ommochrome One of a group of pigments that are derived from tryptophan and that occur in the eyes of insects.

OMP 1. Orotidine monophosphate (orotidylic acid). 2. Orotidine-5′-monophosphate (5′-orotidylic acid).

Oncley equation An equation expressing the frictional ratio of a macromolecule as a product of two factors, one of which is a measure of the hydration, and the other is a measure of the asymmetry of the molecule. Specifically, $f/f_0 = (f/f_h)\,(f_h/f_0)$, where f, f_0 and f_h are, respectively, the frictional coefficients for the macromolecule, an anhydrous sphere, and a hydrated sphere; f_h/f_0 is the hydration factor, and f/f_h is the shape, or asymmetry, factor.

onco- Combining form meaning tumor.

oncogene A gene that has the potential to cause cancer; a gene that can bring about malignant transformation of cells. Such genes occur in oncogenic RNA viruses (oncornaviruses) and in normal cells. Cellular oncogenes are known as protooncogenes and contain introns while viral oncogenes do not. Cellular protooncogenes are converted to oncogenes by activation as a result of mutation or recombination with a viral genome. It is believed that the cellular genes were the progenitors of the viral oncogenes.

oncogenesis The origin and growth of a tumor.

oncogene theory A theory of cancer according to which normal cells contain latent retrovirus genes (protovirus; virogene) that become activated by carcinogens and radiation to yield viral oncogenes. These oncogenes then become determinants of cancer through the synthesis of specific enzymes and/or the synthesis of complete oncogenic virus particles. The information for the production of oncogenic viruses and malignant transformations of cells is, therefore, vertically transmitted through the germ line and is present in the DNA of all the cells of all the animals prone to cancer. *Aka* virogene theory; virogene–oncogene theory; protovirus theory. *See also* provirus hypothesis.

oncogenic Capable of inducing tumors.

oncogenic virus A DNA- or RNA-containing virus that can transform infected cells so that they proliferate in an uncontrolled fashion and may form a tumor. *See also* oncornavirus.

oncology The study of tumors.

oncolytic Capable of destroying tumor cells.

oncornavirus Acronym for oncogenic RNA virus. Oncornaviruses were originally characterized by (a) their content of a high molecular weight RNA genome (60S–70S RNA, 10^7 daltons.), (b) their banding at a particular density level in density gradient centrifugation, and (c) their content of RNA-dependent DNA polymerase (reverse transcriptase). The oncornaviruses were originally divided into three classes, denoted A, B, and C. Type A viruses constitute a small group of protein-encapsulated viruses that have not been shown to be oncogenic; they occur either in the cytoplasm, believed to be immature forms of type B viruses, or in body fluids, believed to be immature forms of type C viruses. Type B viruses have a somewhat eccentric nucleoid and glycoprotein surface spikes; they are associated primarily with the formation of carcinomas. Type C viruses have a roughly spherical nucleoid surrounded by an electron-translucent lipid layer; they infect a large number of animal species and cause leukemias, lymphomas, and sarcomas. The term oncornavirus is now also used as a synonym for retrovirus.

oncotic pressure The effective colloid osmotic pressure; it is equal to the difference between the osmotic pressure of the plasma proteins and that of the tissue fluid proteins.

one-carbon fragment A group of atoms or a compound that contains one carbon atom; examples are the methyl group ($-CH_3$), hydroxymethyl group ($-CH_2OH$), formyl group ($-CHO$), and formaldehyde ($HCHO$).

one-enzyme–one-linkage hypothesis The hypothesis that at least one specific glycosyltransferase is required for the synthesis of each specific type of glycosidic linkage found in nature.

one-gene–one-enzyme hypothesis The hypothesis that there is a large group of genes among the genes of an organism in which each gene codes for a specific enzyme or other protein. Since it is now known that many enzymes, as well as other proteins, consist of several polypeptide chains coded for by different genes, this hypothesis has been replaced by the one-gene–one-polypeptide chain hypothesis.

one-gene–one-polypeptide chain hypothesis The hypothesis that there is a large group of genes among the genes of an organism in which each gene codes for a specific polypeptide chain; the polypeptide chain may be part of a protein, or constitute an entire functional protein. In some cases, however, more than one gene may be involved in coding for a polypeptide chain. This is true for the immunoglobulins where separate genetic elements in the same chromosome code for different segments of the polypeptide chain. Such cases are described by the term two-genes–one-polypeptide chain.

one-hit theory A theory according to which the damage of one site on the erythrocyte membrane, resulting from a reaction with complement, is sufficient to bring about cell lysis.

one-phase chromatography FIELD FLOW FRACTIONATION.

one-sided test A test for which the region of rejection is wholly located at one end of the distribution of the test statistic.

one-sigma level The confidence interval corresponding to the standard deviation; a confidence interval such that there is a 68.27% chance that a measurement will fall within it.

one-step conditions A set of conditions that are used for infecting cells with viruses when it is desired to produce infected cells that contain only a single virus particle per cell. For bacterial cultures this may be achieved by incubating the cells briefly with the phage to allow attachment of the phage to the bacteria, and then diluting the phage–cell suspension drastically prior to additional incubation.

one-step growth experiment An experiment in which virus growth is carried out under one-step conditions.

one-step multiplication curve A curve that describes the production of progeny virus under one-step conditions.

one-tail test ONE-SIDED TEST.

ONPG *O*-Nitrophenyl galactoside.

Onsager's equations PHENOMENOLOGICAL EQUATIONS.

ontogenetic Of, or pertaining to, ontogeny.

ontogeny The development of an individual. *Aka* ontogenesis.

ontogeny recapitulates phylogeny RECAPITULATION THEORY.

oocyte A cell that develops into a mature ovum upon meiosis.

oocyte assay of mRNA An assay for the specificity of mRNA translation that is based on the injection of small amounts of exogenous RNA into *Xenopus* oocytes.

oogenesis The formation of a mature egg.

oolemma The cell membrane of an ovum.

ooplasm The cytoplasm of an ovum.

oosperm A fertilized ovum; a zygote.

opal codon The codon UGA, one of the three termination codons. *Aka* umber codon.

opal suppression The suppression of an opal codon.

Oparin's hypothesis The hypothesis that simple organic compounds were formed spontaneously during an early stage of the earth as a result of physical and chemical processes in the primitive atmosphere. These compounds are believed to have dissolved in the primitive oceans and to have led, by a large number of small spontaneous reaction steps, to the formation of macromolecules which ultimately gave rise to the first living cell. *See also* chemical evolution; biochemical evolution; biological evolution.

open chain A chain of atoms in which the two ends are not linked together covalently.

open circle *See* nicked circle.

open-circuit system A system for measurements of indirect calorimetry in which both the oxygen consumption and the carbon dioxide production are determined.

open culture CONTINUOUS CULTURE.

open gene A gene that is engaged in transcription.

open hemoprotein A hemoprotein in which the fifth and/or the sixth coordination positions of the heme are unoccupied.

opening transformation The transformation of a micellar membrane from one having small spaces between the micelles ("closed") to one having large spaces between them ("open").

open-promoter complex The conformation in transcription that follows the initial binding of RNA polymerase to the promoter and in which the RNA polymerase has undergone a conformational change and the two DNA strands have become locally unwound.

open reading frame A segment in mRNA that contains codons that can be translated into an amino acid sequence and that does not contain a termination codon.

open system A thermodynamic system that can exchange both matter and energy with its surroundings.

operating potential The potential at which a Geiger–Mueller plateau is obtained and at which a Geiger–Mueller counter is normally operated.

operational definition A definition that is based upon properties relevant to one or more specific experimental procedures or conditions, regardless of the possibility that different and more fundamental properties may apply to that which is defined.

operator A gene that is adjacent to a structural gene or to a group of contiguous structural genes and that controls the transcription of this gene or group of genes; the operator interacts with a specific repressor protein, thereby controlling the functioning of the adjacent gene or genes. *Aka* operator gene.

operator-constitutive mutant A mutant resulting from a constitutive mutation in which the operator gene has been mutated so that the repressor cannot combine with it; as a result, a previously inducible enzyme becomes a constitutive one.

operator gene OPERATOR.

operon A functional unit of transcription; the combination of operator and the adjacent structural gene or genes that are controlled by it. *Aka* transcription unit.

operon hypothesis The model, proposed by Jacob and Monod, according to which enzyme induction and enzyme repression result from the control of one or more structural genes by an adjacent operator gene. The operator is blocked ("turned off") during enzyme repression and is unblocked ("turned on") during enzyme induction. The blocking occurs through the binding to the operator of either a repressor or an aporepressor–corepressor complex. An inducer binds to the repressor and thus prevents the blocking of the operator; a corepressor binds to the repressor and leads to an even more effective blocking of the operator than that produced by the repressor alone. The synthesis of the repressor is controlled by a regulator gene which need not be adjacent to the operon, the term used to describe the combination of operator and structural gene or genes. Inducers are generally substrates of the enzyme or compounds similar to the substrates, while corepressors are generally products of the enzymatic reaction or compounds similar to the products.

operon network A system of interacting operons and their associated regulator genes such that the product of a structural gene from one operon acts as a repressor or an inducer for another operon.

opiate A narcotic that resembles opium in its action and that may be a natural, a semisynthetic, or a synthetic compound. Generally refers to the alkaloids derived from the juices of the opium poppy, as well as to derivatives of morphine and related compounds.

opiate receptors *See* opioid receptors, which is the preferred term.

opine One of a number of simple derivatives of amino acids, ketoacids, and sugars that are synthesized specifically by cells of crown gall tumors and that are coded for by Ti plasmid genes; examples are octopine and nopaline. *See also* crown gall tumor.

opioid Any compound that acts directly on narcotic receptors. A broad term that includes both the opiates and the opioid peptides.

opioid peptides A group of naturally occurring polypeptides that influence nerve transmissions in some parts of the brain. They are called opioid peptides since they bind to specific receptors that bind opiates, and thus they mimic some of the pharmacological properties of the opiates (morphine and related drugs) such as pain killing and alteration of mood perception. But, unlike the opiates, the opioid peptides are rapidly degraded after being released and do not accumulate in large enough amounts to induce the tolerance seen in morphine addicts. There are three major groups of opioid peptides: endorphins, enkephalins, and dynorphins. These are derived, respectively, from three precursors: proopiomelanocortin, proenkephalin A, and prodynorphin. *Aka* endogenous opiates.

opioid receptors Specific cell surface sites in the brain to which opioid peptides become bound. The term opioid receptor is preferred to opiate receptor since the opioid peptides are, in fact, the physiological ligands for these sites. It is a coincidence that opiates mimic the opioid peptides in their interaction with these receptors.

opium The dried exudate derived from the seeds of the oriental poppy *Papaver somniferum* that contains the narcotic drugs morphine and codeine.

opportunistic microorganism A microorganism that utilizes the opportunity, offered by generalized or localized defects in the antimicrobial defense mechanisms of a host, to inflict damage on that host.

opportunistic pathogen A microorganism that occurs normally in the body and does not cause infection in healthy individuals but may become pathogenic under certain conditions. This may take place, for example, when the individual's antimicrobial defense mechanisms are impaired as a result of one of a number of unrelated diseases.

opposing resonance The principle that a molecule is destabilized if two atoms compete for unshared electrons of a third atom. As an example, successive phosphorus atoms in ATP are considered to compete for the unshared electrons of the oxygen atom, sandwiched between them. This contributes to the instability of ATP and to its being a high-energy compound.

opposing rolling circle model A variation of the rolling circle model for the replication of duplex circular DNA. According to this model, replication begins with two nicks, not far from each other, and with one in each of the strands of the circular duplex. As a result, both strands grow in opposite directions and the molecule replicates in the form of two rolling circles that move in opposite directions.

opposing unidirectional reactions SUBSTRATE CYCLE.

opsin A protein, occurring in both the rods and the cones of the retina, that combines with either $retinal_1$ or $retinal_2$ to form the major visual pigments of vertebrates.

opsonic effect The enhancement of the phagocytosis of antigen–antibody complexes that is produced by antibodies.

opsonic index The ratio of the phagocytic index of an immune serum to that of a normal serum.

opsonin A substance that enhances phagocytosis by modifying the particles to be engulfed so that they are more readily taken up by the phagocytic cells. Some of these substances are antibodies, called immune opsonins, against surface antigens of the bacteria or the particles that are engulfed; others are nonantibody, heat-labile substances, related to components of complement.

opsonization The enhancement of phagocytosis by the action of opsonins.

optical activity 1. The capacity of a substance to interact with radiation in such processes as optical rotation and circular dichroism. 2. OPTICAL ROTATION.

optical antipode ENANTIOMER.

optical density ABSORBANCE.

optical isomer One of two or more isomers that differ from each other in their symmetry as a result of the presence of either asymmetric carbon atoms or overall molecular asymmetry; many, but not all, optical isomers exhibit optical rotation.

optically active Possessing optical activity.

optical path LIGHT PATH.

optical phonon *See* phonon.

optical quenching The quenching that occurs in liquid scintillation counting as a result of changes in the light-transmitting properties of the sample, such as those produced by the partial freezing of the sample solution, or by the fogging of the outside of the sample vial.

optical rotation The rotation of the plane of plane-polarized light by a substance when such light is passed through a solution containing the substance. Optical rotation is shown by a substance that can exist in the form of mirror images as a result of the presence of either asymmetric carbon atoms or overall molecular asymmetry. Because of this asymmetry, the substance has different refractive indices for left and right circularly polarized light and hence shows optical rotation.

optical rotatory dispersion The variation of optical rotation as a function of the wavelength of the light that is used for measuring it; optical rotatory dispersion is useful for studying the secondary structure of macromolecules. *Abbr* ORD.

optical system *See* absorption optical system; interferometric optical system; schlieren optical system.

optimal proportions *See* method of optimal proportions.

optimum pH The pH at which an enzyme exhibits maximal activity under specified conditions.

optimum temperature The temperature at which an enzyme exhibits maximal activity under specified conditions.

oral contraceptive One of a number of synthetic combinations of steroids designed to inhibit ovulation. Most preparations (known as "the pill") contain a mixture of progestin and estrogen.

oral hypoglycemic drug *See* Orinase.

oral insulin *See* Orinase.

orbital The probability distribution, or the wave function, for an electron that is in a particular energy level in either an atom or a molecule.

orbital electron capture *See* electron capture.

orbital steering The concept that the active sites of enzymes are so constructed that they align the orbitals of the substrate and the catalytic groups on the enzyme in an optimal fashion for entering the transition state.

orbivirus A virus belonging to a subgroup of arboviruses according to some authors. The virion contains 10 or more double-stranded, linear RNA molecules.

orchiectomy The surgical removal of a testis. *Aka* orchidectomy.

orcinol reaction A colorimetric reaction for carbohydrates that is based on the production of a green color upon treatment of the sample with orcinol and with ferric chloride dissolved in concentrated hydrochloric acid. The reaction is used particularly for pentoses and for the determination of RNA.

Ord Orotidine.

ORD Optical rotatory dispersion.

ordered mechanism The mechanism of an enzymatic reaction in which two or more substrates are added to the enzyme in an orderly fashion; thus, in the case of a reaction with two substrates, the formation of a ternary complex, composed of the enzyme and the two substrates, is preceded by the formation of a complex between the enzyme and one of the substrates.

order of a reaction *See* reaction order.

order of magnitude A range of magnitudes from some value to ten times that value; a factor of 10.

order parameter A quantity that contains information about molecular orientations in a membrane and that is useful primarily for testing theories or models of molecular conformation and motion in membranes. The order parameter can be obtained from NMR and ESR studies.

order with respect to concentration The order

of a chemical reaction that is determined on the basis of a number of experiments in which the initial reactant concentration is varied; the order of the reaction is then determined from a plot of the logarithm of the initial velocity of the reaction versus the logarithm of the initial reactant concentration.

order with respect to time The order of a chemical reaction that is determined on the basis of a single experiment in which a single initial reactant concentration is used; the order of the reaction is then determined from the decrease in the reactant concentration as a function of time.

ordinate The vertical axis, or y-axis, in a plane rectangular coordinate system.

O region CORE SEQUENCE.

organ A differentiated part of an organism that has a specific structure and performs specific functions.

organ culture The in vitro maintenance of an organ or of parts of an organ so that the structure and/or the function are retained.

organelle A specialized structure in the cell that has definite functions.

organic 1. Pertaining to the compounds of carbon. 2. Pertaining to an organ. 3. Pertaining to a living organism.

organicism HOLISM.

organic quenching The quenching of an ionization detector that occurs when organic compounds are added to the mixture used for gas amplification. *See also* internal quenching.

organism A living plant, animal, or protist.

organizer A portion of an embryo that, through a group of substances produced by it, affects the development (determination and differentiation) of another part of the embryo. *See also* inductor.

organogenesis The development of an organ or organs.

organotroph A cell or an organism that utilizes organic compounds as electron donors in its energy-yielding oxidation–reduction reactions.

organ-specific enzyme TISSUE-SPECIFIC ENZYME.

oriC A nucleotide sequence (245 bp) that constitutes the replication origin (ori site) in the *E. coli* chromosome.

orientation effect 1. The contribution to the catalytic activity of an enzyme that is due to the stereochemical arrangement of the reactants on the surface of the enzyme in a manner that favors their undergoing a reaction. *See also* proximity effect. 2. DIPOLE–DIPOLE INTERACTION.

origin 1. The point of sample application on an electrophoretic or a chromatographic support. 2. The point of intersection of the vertical and horizontal axes in a plane rectangular coordinate system. 3. REPLICATION ORIGIN.

original antigenic sin *See* doctrine of original antigenic sin.

origin of life The processes whereby biomolecules, subcellular structures, and ultimately, living cells have come into existence. *See also* biochemical evolution; biological evolution; chemical evolution.

origin of replication *See* replication origin.

Orinase Trademark for a synthetic sulfonyl urea derivative, tolbutamide, that has a hypoglycemic effect and is used in the control of diabetes. Orinase is administered orally and stimulates the secretion of insulin from the beta cells of the pancreas.

ori site REPLICATION ORIGIN.

Orn Ornithine.

ornithine A nonprotein, alpha amino acid that is an intermediate in the urea cycle. *Abbr* Orn.

ornithine cycle UREA CYCLE.

Oro 1. Orotic acid. 2. Orotate.

orosomucoid A conjugated plasma protein that contains a number of carbohydrates.

orotic acid A nonnucleic acid pyrimidine that is a precursor in the biosynthesis of the pyrimidine nucleotides of nucleic acids. *Abbr* O.

orotic aciduria A genetically inherited metabolic defect in humans that is characterized by the excretion of orotic acid and that is due to a deficiency of a protein that contains two enzymatic activities, orotate phosphoribosyl transferase and orotidylate decarboxylase.

orotidine The ribonucleoside or orotic acid. Orotidine mono-, di-, and triphosphate are abbreviated, repectively, as OMP, ODP, and OTP. The abbreviations refer to the 5′-nucleoside phosphates unless otherwise indicated. *Abbr* Ord; O.

orotidylic acid The nucleotide of orotic acid.

orphan drug According to the United States Department of Health and Human Services, an orphan drug is a drug that is used to prevent or treat a rare disease or condition, usually one that affects fewer than 200,000 persons in the United States. A drug that is not widely researched or available due to limited commercial interest or that has a limited market because of the rarity of the disease, may also qualify for "orphan" status.

orphan virus A virus, such as the echo virus, that occurs in healthy individuals and is not associated with any human disease.

orphon A dispersed single gene or pseudogene, derived from a multigene family (gene cluster). An unclustered tRNA gene and pseudogenes for histones and hemoglobins are some examples.

ortet The original single ancestor of a clone.

ortho- Prefix indicating two substituents of adjacent carbon atoms of the ring in an aromatic compound. *Sym* o.

orthochromatic dye A dye that stains cells or tissues with a single color as opposed to a metachromatic dye.

orthodox conformation A high-energy conformation of mitochondria that occurs in mitochondrial preparations containing little or no ADP and that is characterized by a mitochondrial matrix which is squeezed together tightly and which stains heavily. *See also* condensed conformation.

orthologous Descriptive of genes, or their protein products, that occupy the same genetic locus in different species. Only orthologous genes must be compared in constructing phylogenetic trees. *See also* paralogous.

orthomolecular medicine The maintaining of good health and the treatment of disease that is achieved by varying the concentrations in the human body of substances that are normally present in, and are required by, the body. The increased dietary intake of ascorbic acid for the control of respiratory infections is an example.

orthomyxovirus *See* myxovirus.

orthophosphate INORGANIC PHOSPHATE.

orthophospate cleavage The hydrolytic removal of an orthophosphate group from a nucleoside di- or triphosphate.

oryzamin VITAMIN B_1.

osazone *See* phenylosazone.

OSCF Oligomycin-sensitivity-conferring factor.

oscillating reactions Reactions that vary in their amplitude and frequency. Those that vary in a regular manner are called periodic; those with ever-changing and unpredictable frequencies and amplitudes are called chaotic. Oscillating reactions play an important role in biological clock functions, in inter- and intracellular signal transmissions, and in cellular differentiation.

oscillographic polarography A polarographic technique in which the polarographic wave is depicted on an oscilloscope.

osm 1. Osmole. 2. Osmolal.

osM Osmolar.

osmiophilic Descriptive of a specimen, prepared for electron microscopy, that has a tendency to take up the electron-dense stain of osmium tetroxide.

osmium tetroxide A compound used for fixing and staining specimens for electron microscopy.

osmol An osmole.

osmolality The concentration of a solution expressed in terms of the number of osmoles of solute per 1000 grams of solvent.

osmolal solution A solution that contains one osmole of solute dissolved in 1000 grams of solvent.

osmolarity The concentration of a solution expressed in terms of the number of osmoles in one liter of solution.

osmolar solution A solution that contains one osmole of solute in one liter of solution.

osmole A measure of the osmotically effective amount of solute; equal to the mole of the solute divided by the number of ions formed per molecule of the solute, on the assumption that electrolytes dissociate completely into ions in solution. Thus, one mole of sodium chloride is equivalent to two osmoles and one mole of phosphoric acid is equivalent to four osmoles. *Abbr* osmol; osm.

osmolute An osmotically active solute or solute particle.

osmometer An instrument for measuring osmotic concentration, osmotic pressure, and number average molecular weight.

osmophile An organism that grows preferentially in solutions that have high osmotic pressures.

osmophilic Of, or pertaining to, an osmophile.

osmoreceptor A receptor in the central nervous system that responds to changes in the osmotic pressure of the blood.

osmoregulation Regulation brought about by changes in osmolarity. Thus, for example, expression of the genes that code for the two major porin proteins in *E. coli* is regulated in opposite directions by osmolarity; as the latter is increased, the synthesis of one porin is increased and that of the other is depressed.

osmoregulator An organism that maintains a constant internal osmotic concentration irrespective of variations in the osmotic concentration of its external environment.

osmosis The movement of water or another solvent across a semipermeable membrane from a region of low solute concentration to one of higher solute concentration.

osmotic Of, or pertaining to, osmosis.

osmotic barrier PERMEABILITY BARRIER.

osmotic coefficient A factor that relates the observed osmotic pressure of a solution to that of an ideal solution.

osmotic concentration 1. OSMOLARITY. 2. OSMOLALITY.

osmotic potential The osmotic pressure that a solution is capable of producing if it were separated from the pure solute by a semipermeable membrane.

osmotic pressure The pressure that causes water or another solvent to move in osmosis from a solution having a low solute concentration to one having a high solute concentration; it is equal to the hydrostatic pressure that has to be applied to the more concentrated solution to prevent the movement of water (solvent) into it.

osmotic shock The sudden and drastic dilution of a suspension of cells, or their exposure to a hypotonic medium, that results in the movement of water into the cells and leads to cell rupture.

osmotic work The work performed by cells or cellular organelles in transporting substances across biological membranes, with the result that some substances accumulate inside the cell or inside the organelle, while others are eliminated. Work performed by moving a substance against a concentration gradient; concentration work.

ossification The formation of bone.

osteocalcin A small protein of 49 amino acids that is found in bone and that appears to function in the regulation of calcium in bone and teeth.

osteocyte A bone-forming cell.

osteomalacia The softening and bending of bones and the deficient bone mineralization that occurs in an adult due to a deficiency of vitamin D. *Aka* adult rickets.

osteoporosis A pathological condition characterized by progressive loss of both the organic matrix and the mineral content of bone; bones become brittle and the overall mass of the skeleton is reduced without any change in the ratio of bone mineral to bone matrix. *See also* rickets.

Ostwald–Folin pipet A blow-out pipet that is similar to a volumetric pipet but has the bulb closer to the tip of the pipet; used for the delivery of viscous fluids such as blood or serum.

Ostwald viscometer A simple capillary viscometer constructed in the shape of a U-tube with a small upper bulb in one arm, a larger lower bulb in the second arm, and a capillary connecting the two bulbs. The instrument is especially useful for measuring relative viscosities.

OT 1. Oxytocin. 2. Old tuberculin.

OTP 1. Orotidine triphosphate. 2. Orotidine-5′-triphosphate.

ouabain A cardiac glycoside that is a specific inhibitor of the sodium and potassium ion transport across the cell membrane.

Ouchterlony method A method for double immunodiffusion in which antigen and antibody solutions are placed in wells that are cut in agar and are spaced in one of a number of geometric arrangements. Both the antigens and the antibodies diffuse through the agar and interact to form precipitation arcs.

Oudin technique A method for single immunodiffusion in which the antibodies are mixed with agar in a narrow vertical tube and a solution of antigens is layered above the agar. The antigens diffuse through the gel and interact with the antibodies to form precipitation zones.

outbreeding The crossing of genetically unrelated plants or animals.

outer coat CELL COAT.

outer membrane 1. The layer of lipopolysaccharide, protein, and lipoprotein that is external to the peptidoglycan layer in the cell envelope of gram-negative bacteria. It provides a protective environment for certain hydrolytic enzymes and acts as a diffusion barrier against various compounds. 2. SHEATH (2). 3. The external membrane of the nuclear envelope that is continuous with the endoplasmic reticulum and that is often studded with ribosomes, engaged in protein synthesis. 4. The external membrane of mitochondria. 5. The external membrane of chloroplasts.

outer orbital An orbital that functions in the bonding of a high-spin complex.

outer sphere SECONDARY HYDRATION SHELL.

outer sphere activated complex An oxidation–reduction transition state that is formed by the transfer of electrons without a change in the inner coordination shells of the participating ions.

outer sphere complexing ligand A ligand that is within the secondary hydration shell of a metal ion and that participates in ion-pairing.

outer volume The volume of solvent that is contained in the spaces among the gel particles of the bed in gel filtration chromatography.

outgrowth The third stage in the conversion of a spore to a vegetative cell during which the spore core grows and the wall of the vegetative cell develops.

outlier An experimental value that clearly falls outside the range of values of a given population and should, therefore, be rejected.

ovalbumin The principal protein of egg white. *Abbr* OA.

ovariectomy The surgical removal of an ovary.

Overhauser effect *See* nuclear Overhauser effect.

overlap method The classical method for determining the amino acid sequence of a protein or the base sequence of a nucleic acid; involves partial digestion of the polymer into smaller fragments and elucidation of the amino acid or base sequence of each fragment. By using different ways to hydrolyze the original polymer and/or sections of it, fragments with varying degrees of overlap are produced and this helps in deducing the sequence of the original molecule.

overlapping code A genetic code in which one or more nucleotides of one codon also serve as nucleotides for an adjacent codon.

overlapping genes Genes that have some nucleotide sequences in common. The overlapping sequences may involve control genes or structural genes. In phage øX174, a number of different proteins are synthesized from overlapping genes by virtue of having different reading frames for the various genes.

overlap zones Thick regions in collagen fibers resulting from the overlap of staggered collagen fibrils.

overloading 1. The application of excessive amounts of material to an electrophoretic or a chromatographic support. 2. The addition of large amounts of a labeled or unlabeled intermediate to a system to study metabolic pathways. 3. The administration of large doses of protein antigens to an animal to produce immunological unresponsiveness.

overnutrition Malnutrition that results from an excessive intake of one or more nutrients.

overproducer A genetically engineered organism that produces large amounts of a given protein.

overspeeding technique The initial spinning of an analytical ultracentrifuge rotor at high speeds to decrease the time required to establish sedimentation equilibrium.

overwinding Positive supercoiling. *See* superhelix.

oviduct The duct that serves for the passage of eggs from the ovary to the uterus.

ovine Of, or pertaining to, sheep.

ovoflavin An impure preparation of riboflavin from egg white.

ovogenesis OOGENESIS.

ovomucoid A mucoprotein from egg white.

ovotransferrin *See* siderophilin.

ovulation inhibitor *See* oral contraceptive.

ovum (*pl* ova) The female reproductive cell; an egg cell; the female gamete.

oxaloacetate pathway A metabolic pathway whereby aspartic acid and asparagine are catabolized through conversion to oxaloacetic acid, which then feeds into the citric acid cycle.

oxaloacetic acid A dicarboxylic acid that initiates the reactions of the citric acid cycle by condensing with acetyl coenzyme A to form citric acid and coenzyme A. *Abbr* OAA.

oxalosis A pathological condition characterized by the deposition of calcium oxalate in the kidneys and other tissues.

oxaluria *See* primary oxaluria.

oxamycin CYCLOSERINE.

oxane ring *See* thromboxanes.

oxidant The electron acceptor species of a given half-reaction; the species that undergoes reduction in an oxidation–reduction reaction.

oxidase An enzyme that catalyzes the oxidation of a substrate, with molecular oxygen serving as the electron acceptor.

oxidation The change in an atom, a group of atoms, or a molecule that involves one or more of the following: (a) gain of oxygen; (b) loss of hydrogen; (c) loss of electrons.

oxidation number A measure of the oxidized or the reduced state of an atom. For monoatomic ions, the oxidation number is equal to the charge of the ion. For complex ions and for molecules containing covalent bonds, the oxidation number is equal to the number of electrons gained or lost when (a) the electrons of a covalent bond are assigned completely to the more electronegative atom, and (b) the sum of the oxidation numbers of all the atoms is set equal to either zero in the case of a neutral molecule, or to the charge of the ion in the case of a complex ion. For elements in the free state, the oxidation number is taken as equal to zero.

oxidation potential The electrode potential that measures the tendency of an oxidation–reduction reaction to occur through a loss of electrons. *Sym* E. *See also* electrode potential; standard electrode potential.

oxidation quotient *See* Meyerhof oxidation quotient.

oxidation–reduction enzyme OXIDOREDUCTASE.

oxidation–reduction potential The electrode potential expressed either as a reduction potential or as an oxidation potential.

oxidation–reduction reaction A reaction composed of two half-reactions, one of which is a reduction half-reaction and one of which is an oxidation half-reaction.

oxidation state OXIDATION NUMBER.

oxidative assimilation The metabolism of microorganisms in which a large fraction of the substrates that are being oxidized are converted to cellular components to permit growth; this contrasts with the metabolism in a fully grown animal or human where the bulk of the substrates are oxidized to carbon dioxide and water which are then excreted.

oxidative deamination A deamination reaction with a concomitant oxidation, as in the conversion of an alpha amino acid to an alpha ketoacid.

oxidative pathway of hexose metabolism HEXOSE MONOPHOSPHATE SHUNT.

oxidative phosphorylation The synthesis of ATP from ADP that is coupled to the operation of the mitochondrial electron transport system. *See also* conformational coupling hypothesis; chemical coupling hypothesis; chemiosmotic coupling hypothesis.

oxidizing agent OXIDANT.

oxidizing atmosphere An atmosphere that is rich in gases that are readily reduced; the present-day atmosphere of the earth, which is rich in oxygen, is an example.

oxidizing power The capacity of a substance to function as an oxidizing agent.

oxidoreductase An enzyme that catalyzes an oxidation–reduction reaction. *See also* enzyme classification.

oxidoreduction reaction An oxidation–reduction reaction.

oximeter An instrument for determining the degree of oxygenation of blood by a direct measurement of a translucent part of the organism.

oxirane EPOXIDE.

oxoacid KETOACID.

oxonium ion The ion $R—O^+—H$; ROH_2^+.

5-oxoproline PYROGLUTAMIC ACID.

oxoprolinuria A genetically inherited metabolic defect in humans that is characterized by the formation of large amounts of pyroglutamic acid from γ-glutamylcysteine; due to a deficiency of the enzyme that catalyzes the synthesis of glutathione from γ-glutamylcysteine. *Aka* pyroglutamic aciduria; 5-oxoprolinuria.

oxosteroid KETOSTEROID.

oxyanion A negatively charged carbonyl oxygen atom and, by extension, the transient tetrahedral intermediate of which it is a part:

$$—\overset{|}{\underset{|}{C}}—O^-$$

oxyanion hole A pocket in the active site of chymotrypsin into which the carbonyl oxygen of the susceptible peptide bond fits. *Aka* oxyanion pocket; oxyanion binding site.

oxybiontic Capable of using atmospheric (molecular) oxygen for growth. *Aka* oxybiotic. *See also* aerobic (2,3).

oxybiotin A synthetic analogue of biotin in which the sulfur has been replaced by oxygen and which can generally substitute for biotin.

oxycellulose Oxidized cellulose that is insoluble in water and common organic solvents; a cation exchanger.

oxygen An element that is essential to all plants and animals. Symbol, O; atomic number, 8; atomic weight, 15.9994; oxidation number, -2; most abundant isotope, ^{16}O; a stable isotope, ^{18}O.

oxygenase An enzyme that catalyzes an oxidation reaction by molecular oxygen, in which both of the oxygen atoms are inserted into a compound. *Aka* dioxygenase. *See also* monooxygenase.

oxygenation 1. The saturation of a liquid with oxygen, as in the oxygenation of blood. 2. The introduction of oxygen into a compound, as in the oxygenation of hemoglobin.

oxygen capacity of blood OXYGEN COMBINING POWER OF BLOOD.

oxygen carrier A pigment, such as hemoglobin, myoglobin, or hemerythrin, that combines with and serves to transport oxygen.

oxygen combining power of blood The volume of oxygen that is required for combining with all of the hemoglobin in 100 mL of blood.

oxygen content of blood The total amount of oxygen, both physically dissolved and combined with hemoglobin, that is present in 100 mL of blood and that is determined at partial pressures of oxygen corresponding to those of the blood when it flows through either the arteries or the veins.

oxygen cycle The group of reactions whereby oxygen is generated from water by photosynthetic organisms and is reconverted to water by heterotrophic organisms.

oxygen debt The extra amount of oxygen that is required by a mammal after a period of strenuous exercise. The oxygen is used to oxidize a portion of the lactic acid that accumulated during the exercise, and the energy thus obtained drives the conversion of the remainder of the lactic acid to glycogen.

oxygen dissociation curve OXYGEN SATURATION CURVE.

oxygen electrode One of a number of membrane probes that permit the determination of dissolved oxygen. *See also* Clark electrode.

oxygen saturation The percentage of oxygen actually bound to either hemoglobin or myoglobin at a given partial pressure of oxygen, compared to the maximum amount that can be bound at high partial pressures of oxygen.

oxygen saturation curve A plot of oxygen saturation of either hemoglobin or myoglobin as a function of the partial pressure of oxygen.

oxygen transferase DIOXYGENASE.

oxygen transport The carrying of oxygen by the blood from the lungs to the tissues.

oxyhemoglobin An oxygenated hemoglobin molecule. *Abbr* HbO_2; $HHbO_2$.

oxymyoglobin An oxygenated myoglobin molecule. *Abbr* MbO_2.

oxysome A macromolecular aggregate that functions as a unit in oxidative phosphorylation. It is referred to as either a lumped or a distributed oxysome depending on whether the components for oxidative phosphorylation are all present in one aggregated particle or are distributed over a number of different aggregated particles. *See also* complexes I–IV.

oxytetracycline *See* tetracyclines.

oxytocic hormone OXYTOCIN.

oxytocin A cyclic peptide hormone that consists of nine amino acids and that causes the contraction of smooth muscle; it is secreted by the posterior lobe of the pituitary gland. *Abbr* OT.

Oz group A group of allotypic antigenic sites in the constant region of the lambda chain of human immunoglobulins.

ozone The triatomic allotropic form of oxygen that is present in the stratosphere and that is generated by a silent electric discharge in oxygen or air.

ozonolysis The oxidation of a compound by means of ozone; used for locating the double bonds in a fatty acid molecule, since ozonolysis leads to a cleavage of the molecule at those points.

P

p 1. Para position. 2. Pico. 3. Proton. 4. Phospho-, as in p-creatine. 5. Phosphate, as in glucose-1-p.

P 1. Inorganic phosphate. 2. Phosphate in an abbreviation, as in ATP. 3. Phosphorus. 4. Product. 5. Probability. 6. Proline. 7. Poise. 8. Pressure. 9. Peta.

~P A phosphate group in a high-energy compound that, when removed by hydrolysis, results in a reaction that has a large, negative free energy change.

P_{II} P protein.

P_{450} See cytochrome P_{450}.

P_{680} The pigment of photosystem II of chloroplasts that functions in conjunction with the photolysis of water.

P_{700} The pigment of photosystem I of chloroplasts that is a special chlorophyll molecule, the properties of which are influenced by its environment.

^{32}P A radioactive isotope of phosphorus that is a strong beta emitter and has a half-life of 14.3 days.

Pa Pascal.

PA 1. Proton affinity. 2. Phosphatidic acid.

PAABS Pan-American Association of Biochemical Societies.

PABA p-Aminobenzoic acid; also abbreviated PAB.

PAbs Polyclonal antibodies.

pacemaker enzyme An enzyme that catalyzes a reaction that is essentially irreversible in the chemical sense; such enzymes frequently catalyze either the initial, the final, or a branch point reaction of a metabolic pathway. Pacemaker enzyme reactions are rate-limiting reactions. The committed step of a biosynthetic pathway is frequently such a reaction.

packaging The process of folding DNA so that the long molecule can fit into smaller spaces. Formation of nucleosomes and chromosomes represents packaging of DNA into eukaryotic cells. Likewise, folding and insertion of DNA into phage heads represents packaging of viral DNA. *See also* packing ratio.

packed cell volume HEMATOCRIT (1).

packing 1. *v* The process of setting up the stationary phase in column chromatography. 2. *n* The solid bed material in column chromatography; the stationary phase; the chromatographic support.

packing ratio The ratio of the length of the DNA molecule to the length of the subcellular structure containing it. Thus, in a human cell, the total length of the DNA is about 1 m and the total length of the 46 chromosomes is about 50×10^{-6} m, resulting in a packing ratio of $1/(50 \times 10^{-6})$ or 20,000/1.

PAF Platelet-activating factor.

PAGE Polyacrylamide gel electrophoresis.

Paget's disease A disease of unknown cause that is characterized by changes in bone remodeling (formation and resorption).

PAH p-Aminohippuric acid.

pair annihilation The production of a photon from the energy dissipated when the electron collides with a positron.

paired Designating bases in two nucleic acid strands, or in one strand folded back upon itself, that are linked by hydrogen bonding according to the base pairing rules.

paired electrons The two electrons, of opposite spin, that are normally present in an atomic orbital.

paired-ion chromatography ION-PAIR CHROMATOGRAPHY.

paired sera An acute serum and a convalescent serum.

paired source method A method for determining the resolving time of a radiation counter by counting two radiation sources, first separately and then together.

pair production A unique reaction of energy to mass transition in which a high-energy photon is converted to a positron and an electron in the strong magnetic field of an atomic nucleus.

PAL Pyridoxal phosphate.

Palade granule RIBOSOME.

paleobiochemistry A branch of biochemistry that deals with the study of the organic constituents of fossils.

paleontology The science that deals with life of earlier geologic eras and that is based on the study of plant and animal fossils; the study of extinct species. *See also* neontology.

paleoprotein One of a group of proteins isolated from fossils, particularly the exoskeleton of mollusks.

Paleozoic era The geologic time period that extends from about 225 to 600 million years ago and that is characterized by the development of land animals and plants.

palindrome A segment of DNA that has a

symmetrical structure; the sequence of bases in one strand (for example, in the 5′ to 3′ sense) is identical to that in the second strand, read in the same 5′ to 3′ sense. An example is

$$5'\ldots GAATTC\ldots 3'$$
$$3'\ldots CTTAAG\ldots 5'$$

palindromic helix   A helix that has end-to-end symmetry so that both ends are indistinguishable. A palindromic helix can be rotated by 180° about an axis perpendicular to the longitudinal axis of the helix and thereby produce a structure that is identical to that before rotation.

palmitic acid   A saturated fatty acid that contains 16 carbon atoms and is present in animal fat.

palmitoleic acid   An unsaturated fatty acid that contains 16 carbon atoms and one double bond.

PALP   Pyridoxal phosphate.

PAM   Percentage of accepted point mutations; the number of amino acid differences per 100 amino acid residues of a given protein. A unit that is used in comparing the sequence homology of proteins.

pancreas   An endocrine gland that secrets the hormones, insulin, and glucagon, and that is located behind the stomach.

pancreatectomy   The surgical removal of the pancreas.

pancreatic   Of, or pertaining to, the pancreas.

pancreatic cholera   A clinical syndrome, characterized by diarrhea, hypokalemia, and hypochlorhydria, that is due to the decreased secretion of gastric acid; appears to be brought about by the action of high concentrations of vasoactive peptide on the intestine, the pancreas, and the stomach.

pancreatic deoxyribonuclease   DEOXYRIBONUCLEASE I.

pancreatic diabetes   Diabetes that is caused by either a lesion or the removal of the pancreas.

pancreatic juice   The digestive juice that consists of the secretion of the pancreas and that is discharged into the small intestine; contains proteolytic enzymes, secreted as zymogens, nucleases, carbohydrases, and lipases. *Aka* pancreatic fluid.

pancreatic polypeptide   A polypeptide candidate hormone, discovered in the chicken and subsequently isolated from the F cells of the pancreas from a variety of species; it contains 36 amino acids and has marked gastrointestinal effects but its physiological role is unknown.

pancreatic ribonuclease   *See* ribonuclease A.

pancreatin   An acetone powder preparation of the pancreas that contains active pancreatic enzymes.

pancreatitis   An inflammation of the pancreas that may be acute or chronic.

pancreatropic   Exerting an effect on the pancreas.

pancreozymin   CHOLECYSTOKININ.

pandemic   Of, or pertaining to, an epidemic disease of unusually widespread occurrence.

Pandred's syndrome   A hereditary disease, associated with nerve deafness and characterized by a lower than normal incorporation of iodine into thyroglobulin.

pangamic acid   *See* vitamin B_{15}.

panspermy hypothesis   The hypothesis that life originated elsewhere in the universe before the solar system was formed and that it was later transported in some fashion through interstellar space and deposited on the earth in the form of heat-resistant spores or in some other form. *Aka* panspermia hypothesis.

pantetheine   An intermediate in the biosynthesis of coenzyme A in mammalian liver and in some microorganisms, and a growth factor for a species of lactic acid bacteria, *Lactobacillus bulgaricus*; consists of pantothenic acid linked to β-mercaptoethylamine.

P antigen   One of a number of glycosphingolipid blood group antigens of human erythrocytes.

pantothenic acid   One of the B vitamins, the coenzyme form of which is coenzyme A.

pantropic virus   A virus that affects more than one tissue or more than one organ.

PAP   3′-Phospoadenosine-5′-phosphate;   the compound formed from PAPS in the presence of a suitable acceptor; PAPS + acceptor ⇌ PAP + product.

papain   A proteolytic enzyme that is derived from papaya and that has a broad specificity.

paper chemistry   Chemical reactions that are combined in a stoichiometrically correct way but that have no basis in fact and do not portray the actual mechanisms involved.

paper chromatogram   The developed strip or sheet of filter paper that is obtained in paper chromatography.

paper chromatography   Partition chromatography in which the stationary phase is a moistened strip or sheet of filter paper and the mobile phase is a solvent that either ascends or descends along the paper. *Abbr* PC.

paper electrophoresis   Electrophoresis in which a strip or sheet of filter paper is used as a supporting medium; the filter paper is moistened with buffer and dips into two buffer compartments that contain the electrodes.

papergram   PAPER CHROMATOGRAM.

papilloma   A benign tumor of the skin or of a mucous surface; a wart.

papilloma virus   A small, naked, icosahedral

virus that contains double-stranded DNA and that produces papillomas in animals; belongs to the group of papovaviruses.

papovavirus A small, naked, icosahedral virus that contains double-stranded DNA; most papovaviruses produce either benign or malignant tumors.

PAPS 3'-Phosphoadenosine-5'-phosphosulfate. *See also* PAP.

para- Prefix indicating two substituents on opposite carbon atoms of the ring in an aromatic compound. *Sym* p.

parabiosis The natural or artificial joining of two organisms so that they are linked both anatomically and physiologically.

parabolic inhibition Inhibition that yields a curve that is concave upward when either slopes or intercepts from a primary plot are plotted as a function of inhibitor concentration.

paracasein *See* chymosin.

paracrine hormone A substance, such as a prostaglandin, a neurotransmitter, or a polypeptide growth factor, that acts on cells close to those that released it.

paracrystalline Descriptive of a solid that has a somewhat lesser order than the regular three-dimensional arrangement of the atoms in a true crystal. *See also* tactoid.

paradigm An example, model, or pattern; a scientific hypothesis that serves as an example; a known event that serves as an illustration for a more general phenomenon.

paraffin method A method of preparing a tissue specimen for microscopic examination in which the tissue is embedded in paraffin and then cut with a microtome to produce thin paraffin sections.

paraffin oil A mixture of saturated aliphatic hydrocarbons; a liquid petroleum derivative.

paraffin section *See* paraffin method.

paraffin wax A macrocrystalline wax obtained from petroleum.

parahematin FERRIHEMOCHROME.

parahemophilia A genetically inherited metabolic defect in humans that is caused by a deficiency of accelerator globulin, a factor in the blood clotting system.

parallel chains Two polypeptide chains that run in the same direction; both progress from the C-terminal to the N-terminal, or vice versa.

parallel spin The spin of two particles in the same direction.

parallel strands Two polynucleotide strands that run in the same direction; both progress from the 3'-terminal to the 5'-terminal, or vice versa.

paralogous Descriptive of genes, or their pro-

tein products, that occupy different genetic loci in different species. *See also* orthologous.

paralysis *See* immunological paralysis.

paralysis time COINCIDENCE TIME.

paramagnetic Descriptive of a substance that has unpaired electrons and that has a permanent magnetic dipole moment as a result of the magnetic properties of its spinning electrons; when such a substance is placed in a magnetic field, it tends to become oriented in line with the applied field.

Paramecium A genus of freshwater protozoans possessing kappa particles.

parameter 1. An independent variable through functions of which other variables may be expressed. 2. A quantity (such as a mean or a standard deviation) that is calculated for an entire population. *See also* statistic. 3. A property, associated with a molecule or some other particle, that can be expressed quantitatively and that has specific values; molecular weight, sedimentation coefficient, and axial ratio are some examples.

parametric statistics Statistical calculations that are based on prior assumptions with respect to the variable and the probability distributions of the data, and that are valid only if these assumptions hold.

paramylon A linear homopolysaccharide that occurs in *Euglena* and that is composed of D-glucose units linked by means of $\beta(1 \rightarrow 3)$ glycosidic bonds.

paramyosin A fibrillar element of some smooth muscles, such as the catch muscles of mollusks, whose main component is tropomyosin A.

paramyxovirus *See* myxovirus.

paranemic coiling The coiling of two threads in opposite directions so that they can be separated without uncoiling the threads. *See also* plectonemic coiling.

paranuclein Obsolete designation for a phosphorus-containing protein that is not a nucleoprotein.

paraprotein An abnormal, monoclonal immunoglobulin. Paraproteins are formed by proliferating concentrations of immunoglobulin producing cells and occur in the serum of individuals suffering from multiple myeloma or from Waldenstroem's macroglobulinemia. *Aka* M component.

paraproteinemia The presence of paraproteins in the blood.

parapyruvate A dimer of pyruvate that accumulates when pyruvate is stored.

parasite An organism that lives in or upon another organism from which it derives some or all of its nutrients.

parathion An insecticide and nerve poison that

forms a stable covalent intermediate with a serine residue in the active site of acetylcholinesterase.

parathormone A polypeptide hormone, secreted by the parathyroid glands, that stimulates the release of calcium from bone and leads to an increase in the level of calcium in the blood.

parathyrin PARATHORMONE.

parathyroidectomy The surgical removal of a parathyroid gland.

parathyroid gland One of two endocrine glands that secrete parathormone and that are adjacent to, or embedded within, the thyroid gland.

parathyroid hormone PARATHORMONE.

paratope ANTIGEN-BINDING SITE.

parenchymal cell A cell of the functional tissue of a gland or an organ.

parent 1. A cell in relation to the cells formed from it by cell division. 2. A molecule of DNA or a chromosome in relation to the molecules or the chromosomes formed from it by replication. 3. A virus in relation to its progeny. 4. A radioactive nuclide in relation to the nuclides formed from it by radioactive decay.

parenteral Referring to the introduction of a substance into an animal organism by ways other than that of the digestive tract, as in the case of an intradermal injection.

parent gelatin The product of the thermal helix-coil transition of collagen in solution; ordinary gelatin is derived from parent gelatin by heat denaturation or by degradation with acid or alkali.

parenthosome A septal pore structure in many basidiomycete fungi that consists of the pore and two hemispherical bulbs, one at each side of the pore. The structure is believed to be involved in the transport of structural components of the hypha.

parietal cell A hydrochloric acid-secreting cell of the gastric glands.

Park and Johnson method A method for determining reducing sugars based on the reduction of ferricyanide to ferrocyanide in alkaline solution.

Parkinson's disease A disorder of amino acid metabolism in humans that is not of genetic origin; it generally develops late in life and is characterized by a loss of dopamine, norepinephrine, and serotonin. The underlying pathology is impairment of the synthesis of these compounds. Treatment involves administration of L-dopa (levodopa) which is decarboxylated to dopamine, a reaction catalyzed by the enzyme dopa decarboxylase. *Aka* Parkinsonism.

Park nucleotide The compound UDP-*N*-acetylmuramyl pentapeptide, an intermediate in the biosynthesis of peptidoglycan.

paromomycin A broad-spectrum aminoglycoside antibiotic produced by *Streptomyces rimosus*.

parsimony principle OCCAM'S RAZOR.

parthenogenesis The development of an organism from an unfertilized egg; the ability to reproduce without fertilization; the activation of an egg in the absence of a sperm.

partial agonist A ligand that binds to a receptor and causes a less than maximum biological response even when it occupies all of the available receptor sites. *See also* agonist; full agonist.

partial digest PARTIAL HYDROLYSATE.

partial hydrolysate The solution that contains the mixture of substances obtained by partial hydrolysis. *Var sp* partial hydrolyzate. *Aka* partial digest.

partial hydrolysis Incomplete hydrolysis, as the hydrolysis of a polymer by an enzyme that is specific for only some of the bonds between the monomers.

partial inhibition The inhibition of the activity of an enzyme in which a saturating concentration of inhibitor cannot cause the reaction velocity to become zero.

partial molor quantity The rate of change of an extensive property of a substance with the number of moles of the substance.

partial pressure The pressure exerted by a gas when it is part of a mixture of gases; it is equal to the pressure that the gas would exert if it alone occupied the entire volume occupied by the mixture at the same temperature.

partial reactions A group of 4 reactions that are catalyzed by intact mitochondria and that are greatly affected by 2,4-dinitrophenol and oligomycin. The reactions are ATP hydroloysis (ATPase activity), phosphate–ATP exchange, phosphate–water exchange, and ADP–ATP exchange. These reactions are believed to represent steps in the mechanism by which ATP is synthesized in oxidative phosphorylation.

partial specific quantity The rate of change of an extensive property of a substance with the number of grams of the substance.

partial specific volume The rate of change of solution volume with the number of grams of solute; the volume increase of a very large volume of solution upon the addition of 1 g of solute. *Sym* $\bar{v}$.

particle diffusion The diffusion of ions through the granules of an ion-exchange resin which proceeds at such a rate that the diffusion controls the rate of ion-exchange taking place.

particle electrophoresis The electrophoretic movement of large colloidal particles in an electrophoresis cell that is mounted under a microscope; used for the direct measurement of the electrophoretic mobility of such particles.

particle immunoassay An immunoassay in which particles (erythrocytes, polystyrene spheres, gold sols, latex particles, dyes, etc.) are used as labels for either antigens or antibodies. *Abbr* PIA.

particle scattering factor A factor that allows for the angular dependence of light scattered by particles that have at least one dimension that is greater than one-twentieth of the wavelength of the incident light; the factor depends on the radius of gyration of the particles.

particle weight MOLECULAR WEIGHT (2).

particulate antigen An antigen that is part of an insoluble structure of a microbial or other cell, a virus, or some other particle.

particulate enzyme An enzyme that is bound to an insoluble structure of a cell, a cellular organelle, or some other particle.

particulate fraction A fraction consisting of cellular and intracellular insoluble structures but devoid of the intracellular fluids and their dissolved solutes.

partition chromatography Chromatography in which the distribution of compounds between a mobile and a stationary liquid phase is based on the solubilities of the compounds in the two phases; the stationary liquid phase is held in place by a porous solid such as a sheet of filter paper or a column of starch. Partition chromatography refers particularly to those cases in which the stationary phase is a more polar liquid than the mobile phase. Also referred to as normal-phase chromatography to distinguish it from reversed-phase chromatography. *Abbr* PC.

partition coefficient The ratio of the concentrations of a substance in two immiscible phases at equilibrium; the phases may be two liquids, or a liquid and a gas. In chromatography, the partition coefficient refers to the concentration ratio of the stationary to the mobile phase.

partitioner phase STATIONARY PHASE (2).

partition function A thermodynamic function that describes the distribution of molecules over all possible energy levels.

partition isotherm PARTITION COEFFICIENT.

partition law The law that a solute, added to a system composed of two immiscible phases, will distribute itself between the two phases according to its solubility in each phase; the phases may be two liquids, or a liquid and a gas.

parturition Childbirth; giving birth to young.

parvalbumin One of a group of muscle proteins that are found in lower vertebrates where they have a calcium-binding function much like troponin in higher organisms.

parvovirus A small, naked, icosahedral animal virus that contains single-stranded DNA. Some parvoviruses (subgenus A) are capable of autonomous replication; others (subgenus B) are capable of replication only in cells that are infected by adenoviruses; these are called adenovirus-associated viruses.

PAS 1. *p*-Aminosalicylic acid. 2. PA/S procedure.

PAS-1 GLYCOPHORIN.

pascal The SI unit of pressure (kg $m^{-1}s^{-2}$); 1 mm Hg = 133.3224 pascals. *Sym* Pa.

PASCAL A high-level computer language named after the seventeenth century French mathematician Blaise Pascal and designed to facilitate structured programming.

PA/S procedure Periodic acid/Schiff procedure. *Abbr* PAS.

PAS reagent SCHIFF'S REAGENT.

passage SUBCULTURE.

passenger The DNA segment that is spliced into a vector and that subsequently replicates along with the vector in a host cell in the cloning stage of recombinant DNA technology. *Aka* passenger DNA.

passenger virus A virus, the presence of which is not associated with the disease, if any, that is occurring in the tissue from which the virus is isolated.

passive agglutination The agglutination reaction of soluble antigens attached covalently, or noncovalently, to cell surfaces.

passive anaphylaxis The anaphylactic reaction produced in an animal by injecting it first with antibodies from another animal and then with the corresponding antigens.

passive diffusion *See* passive transport.

passive hemagglutination The passive agglutination of red blood cells.

passive hemolysis The lysis of red blood cells in the presence of complement that is brought about by antibodies to antigens that have been artificially attached to the surface of the cells.

passive immunity The immunity acquired by an animal organism as a result of the injection of antibodies into it.

passive mediated transport *See* mediated transport.

passive spread The spread of an electrical signal as a function of distance that occurs in an axon in the absence of any amplification through the opening of voltage-gated channels.

passive transfer 1. The transfer of immunity by

the injection of serum, antibodies, or lymphocytes from an immune individual to a normal one. 2. The transfer of hypersensitivity by the injection of serum, antibodies, or lymphocytes from an allergic individual to a normal one.

passive transport The movement of a solute across a biological membrane that is produced by diffusion, is directed downward in a concentration gradient, does not require carriers, and does not require the expenditure of energy. *Aka* simple diffusion.

Pasteur effect The inhibition of glycolysis and the decrease of lactic acid accumulation that is produced by increasing concentrations of oxygen. The effect is due to the inhibition of phosphofructokinase by the ATP formed as a result of oxidative phosphorylation. *See also* Meyerhof oxidation quotient; Crabtree effect.

pasteurization The brief heating of a food, such as milk or wine, that is designed to kill pathogenic microorganisms without actually sterilizing the food.

Pasteur–Liebig controversy A controversy that raged during the second half of the nineteenth century between Louis Pasteur and Justus Liebig. Pasteur believed that fermentation and similar processes were carried out by the metabolic activities of living cells; Liebig believed that such processes were due to chemical substances and that the reactions resulted from self-perpetuating instabilities in the solution that were initiated by exposure of the solution to air.

Pasteur pipet A drawn-out ungraduated pipet with a constriction in the wider part of the pipet for insertion of a cotton plug.

patch and cut repair A repair mechanism of DNA in which repair replication begins after the first incision by a nuclease, and the damaged segment is fully excised only after the repair replication is complete. The final step requires the action of DNA ligase to join the newly synthesized segment to the existing strand. *See also* cut and patch repair.

patching The clustering of membrane proteins on the surface of a cell that is brought about by their being cross-linked when either antibodies or lectins are added.

patch recording A method for measuring the current passing through a few (or a single) voltage-gated channels by stretching a small patch of membrane across the tip of a pipet.

patch test A test for delayed-type hypersensitivity to the cutaneous application of tuberculin.

pathogen A virus, a microorganism, or some other substance that can produce a specific disease.

pathogene VIROID.

pathogenesis The origin and development of a disease.

pathogenic Disease-producing.

pathogenicity The disease-producing capacity of a microorganism, a virus, or other substance.

pathogenic RNA VIROID.

pathological Of, or pertaining to, pathology.

pathological biochemistry The biochemistry of pathological tissues and fluids.

pathology The science that deals with the origin, nature, and course of diseases.

pathway *See* metabolic pathway.

pattern method A viral assay based on measuring the extent of hemagglutination by dilutions of the viral sample.

Patterson function A mathematical function used in the construction of a Patterson map.

Patterson map A graphical projection used in determining the positions of heavy atoms in x-ray diffraction patterns that have been obtained by the isomorphous replacement method.

paucidisperse Consisting of macromolecules that fall into a small number of classes with respect to their size.

Pauli exclusion principle The principle that no two electrons in an atom can be in the same detailed state described by the same four quantum numbers; as a result, a maximum of two electrons can occupy a single atomic orbital, and they must be of opposite spin.

Pauly reaction A colorimetric reaction for histidine and other imidazole compounds that is based on the production of a red color on treatment of the sample with an alkaline solution of diazotized sulfanilic acid.

Pauly's reagent A sulfanilamide-containing reagent that is used for the detection of tyrosine and other phenolic compounds in chromatograms.

pause The reduction in the rate of RNA chain elongation that occurs during the transcription of DNA by RNA polymerase at specific regions of the DNA; these regions are known as pausing sites.

PBB Polybrominated biphenyls; *See* polychlorinated biphenyls.

PBG Porphobilinogen.

PBI Protein-bound iodine.

P blood group *See* P antigen.

PC 1. Phosphatidylcholine. 2. Phosphocreatine. 3. Paper chromatography. 4. Partition chromatography. 5. Plastocyanin.

PCA 1. Passive cutaneous anaphylaxis. 2. Perchloric acid.

PCB Polychlorinated biphenyls.

PCMB *p*-Chloromercuribenzoic acid.

PCNA Cyclin.

P_{CO₂} 1. Carbon dioxide tension. 2. Partial pressure of carbon dioxide.

PCP Angel dust.

PCr Phosphocreatine.

pD A term that is equivalent to pH for a system that contains deuterons.

PD$_{50}$ Median paralysis dose.

PDC Pyruvate dehydrogenase complex.

^{32}P decay *See* radiophosphorus decay.

PDGF Platelet-derived growth factor.

PDH Pyruvate dehydrogenase. *See* pyruvate dehydrogenase complex.

PE 1. Phosphatidylethanolamine. 2. Polyethylene.

pectate lyase The enzyme that catalyzes the degradation of pectins whereby the glycosidic bonds between the monosaccharide residues are cleaved and water molecules are eliminated. *Aka* eliminase; transeliminase.

pectic acid Pectin in which the galacturonic acid residues are present as free acids rather than as methyl esters. *Aka* pectinic acid.

pectic substance Pectin, pectic acid, and related compounds.

pectins A group of heterogeneous, branched, and highly hydrated polysaccharides that contain large amounts of galacturonic acid, some of it in the form of methyl esters. Pectins are abundant in fruits and are used as gelling agents.

PEG Polyethylene glycol.

PEI-cellulose Polyethyleneimine-cellulose; an anion exchanger.

P element One of a group of transposable elements in the fruit fly, *Drosophila melanogaster*.

pellagra The disease caused by a deficiency of the B vitamin nicotinic acid; it is characterized by dermatitis, stomatitis, impaired digestion, and diarrhea.

pellagra-preventative factor NICOTINIC ACID.

pellet The material collected by sedimentation of a solution in a centrifuge tube.

pellicular Descriptive of a column chromatographic packing, used in high-performance liquid chromatography, that consists of small, spherical, glass particles which are coated with a thin layer of a chromatographic support such as a gel, an adsorbent, or an ion-exchange resin.

peloscope A device for obtaining samples of soil microorganisms; consists of several bundles of short (1–2 cm) rectangular capillaries that are filled with sterile water and are inserted vertically into the soil. After suitable periods, the peloscope is removed and examined microscopically.

Pendred's syndrome A genetically inherited metabolic defect in humans that is characterized by lower than normal incorporation of iodine into thyroglobulin. The specific biological defect is unknown.

penems A class of β-lactam antibiotics that act on the bacterial cell wall and are capable of killing resting cells. This is in contrast to penicillin which kills only growing cells by inhibiting cell wall synthesis.

penicillin One of a group of antibiotics, produced by the mold *Penicillium notatum*, that function by inhibiting the synthesis of the bacterial cell wall; penicillin G, or benzylpenicillin, is the most widely used of the group. *See also* β-lactam antibiotics; penems.

penicillinase The enzyme that catalyzes the hydrolysis of the amide linkage of the β-lactam structure in β-lactam antibiotics. The enzyme is produced by many gram-positive and gram-negative bacteria and is sometimes a constitutive and sometimes an inducible enzyme. *Aka* β-lactamase.

penicillin enrichment A method for concentrating bacterial auxotrophs. Wild-type or mutagenized cells are grown on a minimal medium and are subjected to penicillin-induced lysis. This destroys the growing cells but not the nongrowing auxotrophs. The latter can then be supplied with an enriched medium to allow their growth subsequent to the destruction of the penicillin by means of penicillinase. *Aka* penicillin method; penicillin selection technique.

Penning ionization electron spectroscopy A technique for investigating the environment of atoms in the few outermost layers of a solid by using metastable argon atoms that are allowed to collide with the solid surface. The argon atoms capture electrons from molecules of the solid to fill their 3*p* orbitals and, subsequently, eject 4*s* electrons.

penta- Combining form meaning five.

pentagonal capsomer *See* capsomer.

pentamer 1. An oligomer that consists of five monomers. 2. PENTAGONAL CAPSOMER.

pentitol A 5-carbon sugar alcohol such as ribitol.

penton A capsomer composed of five protomers; a morphological subunit of some viruses.

penton antigen An antigen of the penton capsomer of adenoviruses.

pentosan A polysaccharide of pentoses.

pentose A monosaccharide that contains five carbon atoms.

pentose cycle HEXOSE MONOPHOSPHATE SHUNT.

pentose nucleic acid RIBONUCLEIC ACID.

pentose oxidation cycle HEXOSE MONOPHOSPHATE SHUNT.

pentose phosphate carboxylase RIBULOSE-1,5-BISPHOSPHATE CARBOXYLASE.

pentose phosphate pathway HEXOSE MONOPHOSPHATE SHUNT.

pentose phosphoketolase pathway A pathway that is related to the hexose monophosphate shunt and that occurs in some bacteria.

pentosuria A genetically inherited metabolic defect in humans that is characterized by the presence of excessive amounts of L-xylulose in the urine; due to a deficiency of the enzyme L-xylulose dehydrogenase which leads to a block in the glucuronic acid oxidation pathway. *Aka* idiopathic pentosuria.

penultimate carbon The carbon atom preceding the last one in a chain.

PEP Phosphoenolpyruvic acid.

peplomer One of a group of proteins that project from the surface of the envelope of enveloped viruses (such as herpesvirus and arbovirus) forming a "fringe" over the lipoprotein coat surrounding the virion.

pepsin A proteolytic enzyme in the stomach that is unique because of its very low optimum pH. The major sites of pepsin action are peptide bonds in which the carbonyl group is contributed by either aromatic or acidic amino acids.

pepsin inhibitor A polypeptide fragment that inhibits pepsin and that has a molecular weight of about 3000; it is removed, together with other peptides, from pepsinogen in the course of its activation to pepsin.

pepsinogen The inactive precursor of pepsin.

peptic 1. Of, or pertaining to, pepsin. 2. Of, or pertaining to, the stomach and gastric digestion.

peptic peptides The peptides obtained by digestion of a protein with the enzyme pepsin.

peptidase 1. An enzyme that catalyzes the hydrolysis of peptide bonds in both peptides and proteins; a protease. 2. EXOPEPTIDASE.

peptide A linear compound that consists of two or more amino acids that are linked by means of peptide bonds.

peptide antibiotic An antibiotic, such as gramicidin or actinomycin D, that consists largely or entirely of a peptide.

peptide bond A covalent bond formed by splitting out a molecule of water between the carboxyl group of one amino acid and the amino group of a second amino acid. The grouping (a). *Aka* peptide link.

$$\overset{\displaystyle O}{\overset{\displaystyle \|}{—C—NH—}}$$
(a)

Aka peptide link.

peptide hormone One of a group of hormones that are peptides; oxytocin, vasopressin, and the hypothalamic hormones are some examples.

peptide map A fingerprint of peptides. *See* fingerprint.

peptide synthetase PEPTIDYL TRANSFERASE.

peptidoglycan The rigid framework of bacterial

cell walls that consists of a cross-linked network of mucopeptides; the mucopeptides are pentapeptides which are linked to the disaccharide N-acetylglucosamine-N-acetylmuramic acid.

peptidoglycolipid A compound composed of a peptide, a carbohydrate, and a lipid.

peptidolipid LIPOPEPTIDE.

peptidyl-puromycin A peptide attached to puromycin; formed between a growing polypeptide chain and puromycin when protein synthesis is inhibited by puromycin.

peptidyl site The site on the ribosome at which the peptidyl-tRNA is bound at a time when the next aminoacyl-tRNA becomes bound to the aminoacyl site. *Abbr* P site.

peptidyl transferase The enzymatic activity in protein synthesis that catalyzes peptide bond formation between the growing polypeptide chain and the next amino acid to be added. In prokaryotes, the active site of this enzyme consists of portions of several (presumably at least 10) different proteins of the 50S ribosomal subunit and is known as the peptidyl transferase center. No single protein having peptidyl transferase activity has yet been isolated.

peptidyl transferase center *See* peptidyl transferase.

peptidyl-tRNA A transfer RNA molecule with an attached peptide, as the tRNA with the attached growing polypeptide chain in protein synthesis.

peptidyl-tRNA site PEPTIDYL SITE.

peptolide *See* depsipeptide antibiotics.

peptolipid A lipid linked to an amino acid or a peptide; leukotrienes LTC, LTD, and LTE are examples.

peptone A partially hydrolyzed protein that is not precipitated by ammonium sulfate; a secondary protein derivative used as a component of microbiological culture media.

peptonization The enzymatic conversion of a protein into a peptone.

percentage average deviation The ratio of the average deviation to the mean, multiplied by 100; $100A/M$, where A is the average deviation, and M is the mean.

percentage error PERCENTAGE AVERAGE DEVIATION.

percentage law The law stating that the percentage of virus particles that are neutralized by a given antiserum is constant and does not depend on the virus titer, provided that the formation of antibody–virus complexes is reversible and that the antibody is present in excess.

percentile The value of a statistical variable below which the indicated percentage of the measurements of the frequency distribution fall; thus, the 10th percentile is that value below which 10% of the measurements fall.

percent ionization The degree of ionization multiplied by 100; percent hydrolysis and percent dissociation are defined likewise.

percent saturation The ratio of the salt concentration of a solution to that of a saturated solution of the same salt, multiplied by 100.

percent solution A measure of concentration that, unless otherwise indicated, refers to the number of grams of solute in 100 mL of solution (w/v). Other expressions for percent concentration are volume/volume (v/v) and weight/weight (w/w).

percent transmittance The ratio of the intensity of the transmitted light to that of the incident light, multiplied by 100; $100I/I_0$, where I is the intensity of the transmitted light, and I_0 is the intensity of the incident light. *Aka* percent transmission.

percutaneous Through the skin.

performance index A rating of ultracentrifuge rotors that expresses their relative effectiveness in accomplishing the complete sedimentation of a given material under idealized conditions.

perfusate The liquid leaving a perfused organ.

perfused organ An organ that either has been or is being subjected to perfusion.

perfusion The passage of blood, plasma, or other fluids through the blood vessels of an isolated organ or a tissue; used for metabolic studies and for keeping organs alive during organ transplantation.

perhydrocyclopentanophenanthrene The system of four fused rings that is the parent structure of the steroids.

periclinal Parallel to the surface or the circumference; surrounding.

pericyclic reaction A chemical reaction that is characterized by a concerted regrouping of bonding orbitals in the molecule and that proceeds by way of a cyclic transition state.

perikaryon The cytoplasmic cell body that surrounds the nucleus of a neuron.

perimysium The sheath, containing fat deposits and connective tissue, that encloses a mammalian muscle.

perinuclear space *See* nuclear envelope.

periodate oxidation The oxidative cleavage by periodate of the bond between two adjacent carbon atoms carrying any combination of hydroxyl, aldehyde, ketone, or primary amine groups. The periodate ion is reduced to the iodate ion ($IO_4^- \rightarrow IO_3^-$) and the reaction mixture is analyzed by titration. The reaction is used for elucidating the structure of an unknown carbohydrate.

periodic acid/Schiff procedure A staining procedure for polysaccharides in which period-ate oxidation of the polysaccharide is followed by staining with Schiff's reagent for aldehydes. *Abbr* PA/S procedure.

periodicity 1. An occurrence at regular intervals in either time or space; used to describe the spacing of diffraction spots in x-ray diffraction patterns of biopolymers. 2. The number of base pairs per complete turn of the DNA double helix.

periodic oscillations *See* oscillating reactions.

periodic polymer REPEATING POLYMER.

periodic table The arrangement of the chemical elements as a function of their atomic number; elements with similar properties are placed one under the other to form groups of elements.

peripheral At or near an outer surface or a boundary.

peripheral nervous system That part of the nervous system of vertebrates that consists of the nerves and the ganglia but excludes the components of the central nervous system. *Abbr* PNS.

peripheral proteins Membrane proteins that are loosely attached to the surface of the cell membrane on either the extracellular or the cytoplasmic side of the membrane. They are easily extracted by mild procedures which do not disrupt the membrane. *Aka* extrinsic proteins.

peripherals Items of equipment (usually hardware) that are external to the computer itself; disk drives and printers are examples.

periplasmic component *See* periplasmic permeases.

periplasmic enzyme A bacterial enzyme that exists either in free or in bound form in a region between the cell wall and the cell membrane.

periplasmic permeases Permeases that function in the periplasmic space and that are inactivated during osmotic shock because of the loss of an essential component that binds the transported solute and which is referred to as the periplasmic component. *Aka* shock-sensitive permeases.

periplasmic space The space between the inner (cell, cytoplasmic) membrane and the outer (lipopolysaccharide) membrane in gram-negative bacteria.

peristalsis The progressive wave-like movements occurring in the intestine and in other hollow, muscular structures that serve to mix the contents present in the structure and to move it forward.

peristaltic Of, or pertaining to, peristalsis.

peristaltic pump A pump that moves contents along a flexible tube by intermittently pressing on the tube from the outside; this may be achieved by the rotation of a cylinder with spaced protuberances over the tube.

peritrichous Descriptive of a bacterium that has flagella all over its surface.

permanent cell strain ESTABLISHED CELL LINE.

permanent dipole moment A dipole moment that is due to the structure of the substance and not due to the influence of an external electric or magnetic field.

permeability The property of a membrane that is measured by the qualitative and quantitative aspects of the passage of ions and molecules across it.

permeability barrier The limited ability, or the complete inability, of a substance to cross a biological membrane.

permeability coefficient The diffusion coefficient of a substance moving across a membrane, divided by the thickness of the membrane. It is equal to the number of moles of substance passing through a unit cross-sectional area of the membrane in unit time when there is a unit concentration difference across the membrane. *See also* Fick's first law.

permeability factor VITAMIN P.

permeable Descriptive of a membrane that permits the passage of both solutes and solvent across it.

permeant 1. *n* A substance that permeates. 2. *adj* Capable of permeating.

permeaphore A compound, located in the cell membrane, that aids in the transport of solutes across the membrane; a channel protein is an example.

permease 1. TRANSPORT AGENT. 2. The transport agent for galactosides in *E. coli.*

permeate To pass into or through a substance, such as a gel or a membrane.

permeation chromatography GEL FILTRATION CHROMATOGRAPHY.

PER method Protein efficiency ratio method.

permethylation An exhaustive methylation. *See also* Hakamori methylation.

permissible dose *See* maximum permissible dose.

permissible substitution CONSERVATIVE SUBSTITUTION.

permissive cell 1. A cell in which a conditional lethal mutant can grow. 2. A cell that supports the lytic infection by a virus.

permissive conditions Conditions (such as temperature or type of host) under which a conditional mutant can either grow or express its normal (nonmutant) phenotype.

permittivity An electrical unit that is identical to the dielectric constant when the electric field is static and of moderate intensity.

permselective membrane A membrane that is impermeable to water and that is selectively permeable to positive ions only or to negative ions only; such a membrane may be prepared

synthetically by incorporating a polyelectrolyte into a suitable matrix.

permutation Any ordered subset of a collection of *n* distinct objects. The number of permutations, each containing *r* objects, that can be formed from a collection of *n* distinct objects is given by

$$n\,(n-1)\cdots(n-r+1) = \frac{n!}{(n-r)!}$$

and is variously denoted as $_nP_r$, P_r^n, $P(n,r)$, or $(n)_r$.

permutation of the map CIRCULAR PERMUTATION.

permutite A synthetically produced, alkali metal– or alkaline earth–aluminum silicate that is used as an ion-exchange resin for water softening. *See also* zeolite.

pernicious anemia A fatal disease caused by the inadequate absorption of vitamin B_{12} from the intestine and characterized by changes in the hematopoietic system and by degeneration of the central nervous system.

per os By way of the mouth, as in the giving of food or medicine.

peroxidase An enzyme that catalyzes the oxidation of a substrate by using hydrogen peroxide as the electron acceptor.

peroxidasome A peroxisome that has an abnormally high peroxidase activity.

peroxidation *See* lipid peroxidation.

peroxide number A measure of the peroxide content of a fat as determined by the amount of iodine that is liberated from KI by the fat; equal to the number of milliliters of 0.002 N sodium thiosulfate required to titrate the iodine liberated per gram of sample.

peroxisome A membrane-enclosed, cytoplasmic organelle that contains a variety of enzymes that use or produce hydrogen peroxide (H_2O_2). These enzymes include catalase, urate oxidase, and D-amino acid oxidase. *Aka* peroxidosome; microbody; glyoxisome.

Perrin equation One of two complex equations that relate the axial ratio of either an oblate or a prolate ellipsoid of revolution to its frictional ratio.

persistence length A parameter related to the conformational rigidity of a worm-like coil; it is equal to the length of the projection of the end-to-end distance of the coil onto an axis tangential to one end of the coil.

persistent fraction 1. The fraction of stable virus–antibody complexes that is formed during the neutralization of a virus by antibodies and that contains nonneutralized virus particles. 2. The fraction of interferon-resistant virus particles.

persistent induction Enzyme induction in which enzyme synthesis does not drop off rapidly when the inducer is removed.

persistent virus A virus that is transmitted from plant to plant via insect vectors.

personal computer A small computer, inexpensive enough to be owned by an individual.

perspective formula A two-dimensional representation of a molecule in which bonds projecting forward with respect to the plane of the page are indicated by wedges, and bonds projecting backward are indicated by dotted lines.

perspiration 1. The excretion of fluid by the sweat glands. Perspiration is referred to as sensible or insensible depending on whether the loss of water through the skin is accompanied or unaccompanied by visible sweat formation. 2. The fluid excreted by the sweat glands.

perturbation *See* solvent perturbation.

pervaporation The evaporation of a solvent through a membrane; used for concentrating solutions.

PEST hypothesis The hypothesis that proteins that are rapidly degraded within eukaryotic cells contain regions rich in proline (P), glutamic acid (E), serine (S), and threonine (T).

pesticide A chemical, such as an insecticide or a herbicide, that kills forms of animal or plant life.

PET Positron emission tomography.

peta- Combining form meaning 10^{15} and used with metric units of measurements. *Sym* P.

PET genes A group of nuclear genes whose expression is required for the morphogenesis of mitochondria that have a competent respiration system. These genes may code for products that have a direct function in mitochondrial respiration and oxidative phosphorylation, or they may affect the oxidative metabolism of mitochondria indirectly.

petite mutant A mutant of the yeast *Saccharomyces cerevisiae* that grows in small colonies because of mutations affecting mitochondria. A segregational petite carries a mutated nuclear gene that results in mitochondrial defects; a cytoplasmic or vegetative petite carries a mutated mitochondrial gene. In cases where all of the mitochondrial-encoded gene products are missing, the yeast forms promitochondria. *See also* promitochondrion.

pet mutant An organism that carries a mutation in a PET gene; the segregational petite mutants of yeast are an example.

petri dish A flat, covered, glass container used for growing bacteria on a nutrient gel.

petri plate A petri dish containing agar mixed with cells and viruses for use in the plaque assay of viruses.

Petroff–Hausser counting chamber A special hollowed-out microscope slide that holds a known volume of liquid in a ruled grid and that is used for the direct counting of bacteria under a microscope.

PETT Acronym for positron emission transaxial tomography; a noninvasive technique that can be used to detect biochemical and physiological abnormalities in living tissue. A compound labeled with a positron emitting nuclide, such as ^{11}C, is administered to a subject by inhalation or injection. Inside the body, a positron travels only a short distance before it encounters an electron, its antimatter twin. The resulting annihilation produces two gamma–ray photons that shoot off in nearly opposite directions. All of the gamma-ray photons resulting from annihilation events in a given volume are picked up by a rotating assembly of detectors. A computer is used to reconstruct the spatial distribution of these annihilation events and hence to produce cross-sectional images of the radionuclide's distribution in the organ or the tissue being examined. *See also* tomography.

P face *See* fracture faces.

PFC Plaque-forming cell.

PFG, PFGE Pulsed field gel electrophoresis.

PFK Phosphofructokinase.

PFU Plaque-forming unit.

PG 1. Prostaglandin. 2. Phosphatidyl glycerol. 3. Phosphoglycerate.

PGA 1. Prostaglandin A; *See* prostaglandins. 2. Pteroylglutamic acid. 3. Phosphoglyceric acid.

PGAH$_4$ Tetrahydropteroylglutamic acid.

PGB Prostaglandin B; *See* prostaglandins.

PGC Prostaglandin C; *See* prostaglandins.

PGD Prostaglandin D; *See* prostaglandins.

PGE Prostaglandin E; *See* prostaglandins.

PGF Prostaglandin F; *See* prostaglandins.

PGG Prostaglandin G; *See* prostaglandins.

PGH Prostaglandin H; *See* prostaglandins.

PGI$_2$ Prostacyclin.

pH A measure of the hydrogen ion concentration in solution. The pH was originally defined as the negative logarithm, to the base 10, of the hydrogen ion concentration expressed in terms of equivalents (or grams, or moles) per liter; commonly, $pH = -log[H^+]$. The pH is now defined operationally by reference to standard solutions of assigned pH values. The difference in pH between an unknown and a standard solution is directly proportional to the difference in electromotive force between a cell containing the unknown solution and a cell containing the standard solution when these are measured with the same electrodes under identical conditions of temperature and pressure (a hydrogen or a glass electrode, and a reference electrode). *See also* pH scale; pH unit.

phage BACTERIOPHAGE.

phage conversion *See* conversion; lysogenic conversion.

phage cross The production of recombinant phage progeny, carrying genes of two or more parental phage types, that is brought about when a single bacterium is infected with two or more phages that differ in one or more of their genes.

phage induction *See* induction (2).

phage lambda *See* lambda.

phage lysate The suspension of newly synthesized phage particles obtained after phage-infected bacteria have undergone lysis.

phage lysozyme The enzyme, present in many phages, that has a specificity similar to that of lysozyme and disrupts the cell wall. It functions in the injection of phage nucleic acid into, and the release of progeny phage from, the bacterial cell.

phage M13 *See* M13.

phage MS-2 *See* MS-2.

phage Qβ *See* Q-beta.

phage T *See* T-even phage; T-odd phage.

phage φX174 *See* Phi X174.

phagocyte A cell that engulfs bacteria and other foreign particles by phagocytosis.

phagocytic Of, or pertaining to, phagocytosis or phagocytes.

phagocytic cell PHAGOCYTE.

phagocytic index 1. The average number of bacteria or particles that are taken up by phagocytosis per phagocytic cell; frequently measured in vitro. 2. The rate at which inert particles are removed from the blood and are taken up by the phagocytic cells; a measure of the activity of the reticuloendothelial system that is usually determined by means of an injection of carbon particles into an animal.

phagocytin A bactericidal, heat-stable protein that occurs in leukocytes and functions in phagocytosis.

phagocytosis The engulfment and destruction of foreign cells and foreign particulate matter by a cell.

phagolysosome An endocytotic vacuole that contains lysosomal hydrolytic enzymes and that is formed by the fusion of a phagocyte with one or more primary or secondary lysosomes. *Aka* phagolyosome.

phagosome An endosome that contains particulate material taken up by phagocytosis and destined for hydrolytic digestion.

phallotoxin One of a group of cyclic heptapeptides that are toxic components of the poisonous mushroom *Amanita phalloides*. *See also* amatoxin.

phantom A mass of material that approximates a tissue in its physical properties and that is used in determining the dose of radiation to be applied to the tissue.

pharmaceutical 1. *n* A drug. 2. *adj* Of, or pertaining to, pharmacy.

pharmaceutical chemistry The branch of chemistry that deals with the preparation, composition, and testing of drugs.

pharmacodynamics The branch of pharmacology that deals with the reactions between drugs and living structures, specifically the action and the fate of drugs in animal organisms.

pharmacogenetics The area of molecular genetics that deals with the genetic mechanisms that underlie individual differences in the response to drugs.

pharmacognosy A branch of pharmacology that deals with the identification of drugs.

pharmacokinetics The area of pharmacology that deals with the quantitative distribution of drugs in the body.

pharmacology The science that deals with the origin, the composition, and the identification of drugs, and with the effects of drugs on living systems.

pharmacopeia An official compilation of the names, the composition, and the medicinal doses of drugs, and of tests and procedures relating to these drugs. *Var sp* pharmacopoeia.

pharmacy The branch of pharmacology that deals with the origin, the composition, the preparation, and the dispensing of drugs.

phase A solid, liquid, or gaseous homogeneous substance that exists as a distinct and mechanically separable fraction in a heterogeneous system.

phase contrast microscope A microscope that converts differences in refractive index into visible variations of light intensity and permits the observation of transparent structures in the living cell; based on changes in the phase of the light wave as it passes through different parts of the specimen. *Aka* phase microscope.

phase partition A technique for the isolation and purification of subcellular fractions in which the material is allowed to partition itself between two or more immiscible or partially miscible phases. *See also* cross partition.

phase plate A plate that serves as a schlieren diaphragm in the schlieren optical system; it has a coating over half of its area so that incident light will be retarded by half a wavelength when it strikes this area.

phase problem A problem in the interpretation of x-ray diffraction patterns which is due to the fact that reflections from different sets of atomic planes can be evaluated with respect to their intensities but not with respect to their phase angles.

phase rule A mathematical generalization of

systems in equilibrium; expressed as $P + F = C + 2$, where P is the number of independent phases, F is the number of degrees of freedom, and C is the number of independently variable components. The expression can be used to assess the purity of a protein preparation on the basis of the solubility behavior of the protein.

phase shift mutation FRAMESHIFT MUTATION.

phase test A test for chlorophyll that is based on the change in color produced by treating chlorophyll with cold alcoholic potassium hydroxide.

phase transfer The transfer of an inorganic ion from one phase to another. The substance that brings about such transfer is called a phase transfer catalyst. Many types of organic reactions are subject to phase transfer catalysis.

phase variation A pheonomenon shown by some strains of *Salmonella typhimurium*. These organisms have two genes (called H1 and H2) that are responsible for the synthesis of two distinct types of flagellin which then lead to the formation of two distinct types of flagella. Upon prolonged growth of bacteria making one type of flagella, the bacteria making the other type of flagella will in time arise. This change in the type of flagellin (flagella) produced is known as phase variation and is a manifestation of the expression of alternate structural genes.

phasing *See* nucleosome phasing.

Phe 1. Phenylalanine. 2. Phenylalanyl.

pH electrode An electrode that is sensitive to the hydrogen ion concentration of solutions.

phene A phenotypic character that is controlled bv genes.

phenocopy 1. A nonhereditary change, resembling the change caused by a mutation, that occurs in the phenotype, but not in the genotype, of an organism. The change is brought about by nutritional or environmental factors and results in an organism that resembles another organism; the change leads to an effect that is characacteristic of that produced by a specific gene of the other organism. 2. The organism produced by a nonhereditary change in the phenotype of a parent organism.

phenol 1. Hydroxybenzene. 2. An aryl hydroxide; an aromatic alcohol.

phenol coefficient A measure of the sterilizing capacity of a compound; equal to the ratio, under standard conditions, of the minimal sterilizing concentration of the compound to the minimal sterilizing concentration of phenol.

phenolic group *See* phenolic hydroxyl group.

phenolic hydroxyl group A hydroxyl group attached to a benzene ring.

phenol reagent The reagent used in the Folin–Ciocalteau reaction.

phenolsulfonphthalein test A clinical test in which the renal blood flow and the secretory capacity of the renal tubules are assessed by means of the dye phenolsulfonphthalein.

phenome A collective term for all of the components of a living cell other than the genome.

phenomenase Any enzyme that is named after the phenomenon with which it is associated; disulfide bond rearrangease and ATP translocase are two examples. Such names are generally not approved by the Enzyme Commission of the IUB.

phenomenological coefficients A set of coefficients in the phenomenological equations that are functions of temperature, pressure, and composition.

phenomenological equations A set of equations of irreversible thermodynamics that represent the fluxes that take place within a system as a linear combination of driving forces and that describe the possible couplings which may give rise to new effects.

phenomic lag PHENOTYPIC LAG.

phenon A taxonomic grouping of organisms of similar phenotype; a group of organisms classified by methods of numerical taxonomy.

phenotype 1. The physical appearance and the observable properties of an organism that are produced by the interaction of the genotype with the environment. 2. A group of organisms that have the same physical appearance and the same observable properties. *See also* genotype.

phenotypic Of, or pertaining to, phenotype.

phenotypic adaptation The preferential growth of a phenotypically varied organism.

phenotypic curing The restoration of biological activity, lost through mutation, that is produced by changes in the environmental conditions; the changes result in a temporary alteration of transcription and/or translation that leads to the production of a functional protein even though the protein is specified by a mutant gene.

phenotypic lag The time between the exposure of an organism to a mutagen and the phenotypic expression of a mutation by that organism or by its progeny.

phenotypic mixing The production of a virus in which the phenotype differs from the genotype; may be achieved by infecting a bacterial cell with two related phages, such as two mutants of the same phage, so that components of one phage are incorporated into the structure of the second phage during phage assembly in the host cell. *See also* transcapsidation.

phenotypic suppression The suppression of a mutant phenotype by nongenetic factors at the level of mRNA transcription or ribosome translation. Thus, the presence of 5-fluorouracil or streptomycin may cause misreading in transcription and translation, respectively, and hence lead to suppression of certain mutations.

phenotypic variation A change that is within the range of the potential changes of a phenotype and that is shown by essentially all of the organisms in a population.

phenylaceturic acid A compound formed by the conjugation of phenylacetic acid with glycine; the form in which phenylacetic acid is detoxified and excreted in the urine.

phenylalanine An aromatic, nonpolar alpha amino acid. *Abbr* Phe; F.

phenylalanine hydroxylase The enzyme that catalyzes the synthesis of tyrosine from phenylalanine. *See also* phenylketonuria.

phenylalanine load test *See* load test.

phenylalanine tolerance index The composite of the concentrations of phenylalanine in serum after 1, 3, and 4h following the administration of a dose of phenylalanine to an individual; used in characterizing individuals who are suffering from, or are carriers for, phenylketonuria.

p-phenylenediamine A substrate used for assaying ceruloplasmin activity. *Abbr* PPD.

phenylhydrazone The compound formed by the reaction of a monosaccharide with equimolar amounts of phenylhdrazine.

phenylisothiocyanate *See* Edman degradation.

phenylketonuria A genetically inherited metabolic defect in humans that is characaterized by mental retardation, if the defect is not corrected for in childhood, and that is due to a deficiency of the enzyme phenylalanine hydroxylase. *Abbr* PKU.

phenylosazone The compound formed by the reaction of a monosaccharide with excess phenylhydrazine; phenylosazones are useful for the identification of unknown monosaccharides.

phenylpyruvic oligophrenia PHENYLKETONURIA.

phenylthiocarbamyl amino acid An intermediate in the formation of a phenylthiohydantoin amino acid. *Abbr* PTC-amino acid.

phenylthiohydantoin amino acid An amino acid derivative formed by the reaction of phenylisothiocyanate with the free alpha amino groups of amino acids, peptides, or proteins. *Abbr* PTH-amino acid. *See also* Edman degradation.

pH 5 enzyme AMINOACYL-tRNA SYNTHETASE.

pheophytin Chlorophyll without the chelated magnesium atom.

pheoporphyrin PHORBIN.

pherogram ELECTROPHEROGRAM.

pheromone A chemical that is produced and discharged by an organism and that elicits a physiological response in another organism of the same species; the sex attractant of insects is an example. *See also* allelochemicals; gamone.

pH 5 fraction A subcellular fraction obtained by centrifuging a suspension of broken cells at $100,000 \times g$ and then precipitating the tRNA and the aminoacyl-tRNA synthetases by adjusting the pH of the solution to 5. *See also* S-100 fraction.

pH gradient electrophoresis ISOELECTRIC FOCUSING.

PhIA Phosphoroimmunoassay.

phi angle The torsion angle that denotes the rotation about the $N—C^\alpha$ bond of the peptide backbone in proteins. *Sym* ϕ.

Philpot–Svensson optics A schlieren optical system that incorporates a cylindrical lens in its design.

phi X174 A phage that infects *E. coli* and that contains single-stranded, circular DNA. *Sym* ϕX174.

pH jump method A relaxation technique in which pH is the variable that disturbs the equilibrium of a system. *See also* relaxation technique.

phloem A group of specialized plant cells that have lost their nuclei and much of their cytoplasm. They are aligned to form tubes and function in the transport of organic compounds in the plant. *See also* xylem.

phloretin A derivative of trihydroxyacetophenone that acts as a competitive inhibitor of glucose transport across the erythrocyte membrane. *Aka* dihydronaringenin.

phloridzin A toxic glycoside that blocks the reabsorption of glucose by the kidney tubules.

pH meter An instrument for measuring pH values of solutions, commonly by means of a glass electrode and a reference electrode or by means of a combination electrode.

phonon A packet of energy associated with oscillating atoms; a quantized wave within a polymer. Oscillations of side chains at right angles to a polymer or the compression and stretching of a polymer along its axis generate waves called phonons; the former are called optical phonons and the latter are called acoustical phonons. An analysis of phonons is of use in the study of biological membranes.

phorbin The porphyrin derivative that consists of five rings and that constitutes the parent structure of the chlorophylls.

phorbol esters Esters of the alcohol phorbol (4,9,12-β,13,20-pentahydroxy-1,6-tigliadien-3-on) which occurs in croton oil (tiglium oil). Phorbol esters, such as TPA (12-*O*-tetradecanoyl phorbol-13-acetate) are potent tumor promoters. They also evoke pleiotypic responses in cultured cells, including stimulation of macromolecular synthesis and cell proliferation, stimulation of prostaglandin synthesis, loss of surface-associated fibronectin, alteration in cell morphology and permeability, and various other effects.

pho regulon   A regulon that responds to a deficiency of inorganic phosphate. The *E. coli* pho regulon consists of at least 24 genes scattered throughout the chromosome.

phosphagen   A high-energy phosphate compound, such as phosphocreatine or phosphoarginine, that serves as a storage form of free energy in muscle.

phosphatase   An enzyme that catalyzes the hydrolysis of the esters of orthophosphoric acid.

phosphate   1. An anionic radical of phosphoric acid, specifically one of orthophosphoric acid; inorganic phosphate. *Sym* P_i; P. 2. A salt of phosphoric acid.

phosphate acceptor peptide   KEMPTIDE.

phosphate bond energy   The free energy change of a reaction in which a phosphorylated compound is hydrolyzed to yield either inorganic phosphate or inorganic pyrophosphate as one of the products.

phosphate group   The molecule (a) or the radical (b) that, under physiological conditions, is dissociated to give the corresponding anionic forms. *See also* phosphoryl group.

$$
\begin{array}{cc}
\text{OH} & \text{OH} \\
| & | \\
\text{HO—P=O} & \text{—O—P=O} \\
| & | \\
\text{OH} & \text{OH} \\
(a) & (b)
\end{array}
$$

phosphate-group transfer   A reaction in which a phosphate group is transferred from one compound to another, as in the reactions involving high- and low-energy phosphate compounds; in actuality, a phosphoryl group, rather than a phosphate group, is transferred. *Aka* phosphoryl-group transfer.

phosphatemia   HYPERPHOSPHATEMIA.

phosphate potential   The concentration of ATP in a system divided by the product of the concentrations of ADP and inorganic phosphate; the term [ATP]/[ADP][P_i]. A measure of the energy state of cells. *See also* energy charge.

phosphate regulon   The pho regulon.

phosphate transfer potential   PHOSPHORYL TRANSFER POTENTIAL.

phosphatidal choline   A choline-containing plasmalogen.

phosphatidal ethanolamine   An ethanolamine-containing plasmalogen.

phosphatidal group   The parent structure of the plasmalogens; consists of a molecule of glycerol in which phosphoric acid is esterified to the first carbon, a fatty acid is esterified to the central carbon, and an α,β-unsaturated ether is linked to the third carbon of the glycerol.

phosphatidal serine   A serine-containing plasmalogen.

phosphatidase   PHOSPHOLIPASE.

phosphatidate   GLYCEROPHOSPHOLIPID.

phosphatide   1. GLYCEROPHOSPHOLIPID. 2. PHOSPHOLIPID.

phosphatidic acid   The parent compound of many phosphoglycerides; consists of a molecule of glycerol in which phosphoric acid is esterified to the first carbon and fatty acids are esterified to the remaining two carbons of the glycerol.

phosphatidylcholine   A major phosphoglyceride in higher plants and animals; consists of choline that is esterified to the phosphoric acid residue of phosphatidic acid. *Abbr* PC. *Aka* lecithin.

phosphatidylethanolamine   A major phosphoglyceride in higher plants and animals; consists of ethanolamine that is esterified to the phosphoric acid residue of phosphatidic acid. *Abbr* PE. *Aka* cephalin.

phosphatidylglycerol   A condensation product of phosphatidic acid and glycerol. *Abbr* PG.

phosphatidyl group   The group derived from phosphatidic acid by removal of a hydrogen from the phosphate group.

phosphatidylinositol   A phosphoglyceride present in biological membranes where it generally turns over rapidly upon external stimulation of the cells; consists of inositol that is esterified to the phosphoric acid residue of phosphatidic acid. *Abbr* PI.

phosphatidylinositol cycle   A cyclic set of reactions that characterizes somes hormone-receptor systems; involves degradation of phosphatidylinositol and its rapid resynthesis. The cycle may be coupled to calcium mobilization. The rapid resynthesis of phosphatidylinositol is known as the phosphatidylinositol response (PI response). *Abbr* PI cycle.

phosphatidylinositol response   *See* phosphatidylinositol cycle.

phosphatidylserine   A phosphoglyceride consisting of serine that is esterified to the phosphoric acid residue of phosphatidic acid. *Abbr* PS. *Aka* cephalin.

phosphatidyl sugar   GLYCOPHOSPHOGLYCERIDE.

phosphaturia   The presence of excessive amounts of phosphate in the urine.

phosphine oxide An organic compound containing the group

$$-\overset{|}{\underset{|}{P}}=O$$

phosphite-triester method A nonaqueous, solid-phase method for the synthesis of oligonucleotides; involves anchoring the growing chain at its 3'-end to a solid support, using a dimethoxytrityl group to protect the 5'-end, and then reacting the 5'-end with a 3'-phosphoramidite derivative of the next nucleotide to be polymerized.

3'-phosphoadenosine-5'-phosphosulfate *See* active sulfate (1).

phosphoarginine A phosphagen that is present in many invertebrates; consists of arginine in which a phosphate group is attached via a P—N bond to the guanido group of the amino acid.

phosphocozymase NICOTINAMIDE ADENINE DINUCLEOTIDE PHOSPHATE.

phosphocreatine A phosphagen occurring in the muscle of many vertebrates; consists of creatine in which a phosphate group is attached via a P—N bond to the guanido group of creatine. *Abbr* PC.

phosphodiester A compound consisting of two alcohols that are esterified to a molecule of phosphoric acid; a phosphoric acid molecule that is esterified twice.

phosphodiesterase An enzyme that catalyzes the hydrolysis of a doubly esterified phosphoric acid molecule, as that occurring in oligo- and polynucleotides; phosphodiesterases may be of either the endo- or the exonuclease type.

phosphodiester bond A linkage between two molecules by means of phosphoric acid to which each of the molecules is esterified once.

3',5'-phosphodiester bond The bond by which nucleotides are linked in both DNA and RNA; formed by esterification of the phosphoric acid residue, which is already esterified to the 5'-position of the sugar of one nucleotide, to the 3'-position of the sugar of an adjacent nucleotide. *Aka* 3',5'-phosphodiester link.

phosphoenolpyruvate carboxylase The enzyme that catalyzes the anaplerotic reaction whereby phosphoenolpyruvic acid is carboxylated to oxaloacetic acid.

phosphoenolpyruvic acid A high-energy compound, the dephosphorylation of which to pyruvic acid leads to the synthesis of ATP from ADP in the second stage of glycolysis. *Abbr* PEP.

phosphofructokinase The enzyme that cataly-zes the formation of fructose-1,6-bisphosphate from fructose-6-phosphate and ATP. A key regulatory enzyme in glycolysis; an allosteric enzyme whose inhibition by ATP forms the basis of the Pasteur effect. *Abbr* PFK.

phosphogluconate oxidative pathway HEXOSE MONOPHOSPHATE SHUNT.

phosphogluconate pathway HEXOSE MONOPHOSPHATE SHUNT.

phosphoglyceric acid A phosphate ester of glyceric acid, various forms of which are intermediates in glycolysis. *Abbr* PGA. *See also* 1,3-bisphosphoglycerate; 2,3-bis-phosphoglycerate.

phosphoglyceride GLYCEROPHOSPHOLIPID.

phosphoguanidine A high-energy compound, such as phosphocreatine or phosphoarginine, that contains a phosphorylated guanido group.

phosphoinositide INOSITOLPHOSPHOLIPID.

phosphoketolase The key enzyme of the phosphoketolase pathway; it catalyzes the cleavage of xylulose-5-phosphate to glyceraldehyde-3-phosphate and acetylphosphate.

phosphoketolase pathway A pathway of hexose and pentose degradation that occurs in various microorganisms, especially *Lactobacillus*. The pathway forms part of the reactions in heterolactic fermentation and involves phosphoketolase as the key enzyme.

phosphokinase KINASE.

phospholipase An enzyme that catalyzes the hydrolysis of fatty acids or other groups from phosphoglycerides. Phospholipase A_1 catalyzes the hydrolysis of the fatty acid from position 1; phospholipase A_2 catalyzes the hydrolysis of the fatty acid from position 2; phospholipase C catalyzes the cleavage of the bond between the phosphate group and glycerol; and phospholipase D catalyzes the hydrolysis of phosphoglycerides to yield phosphatidic acid. Phospholipases L_1 and L_2 remove the acyl group from the product formed by phospholipase A_1 and A_2, respectively. Phospholipases L_1 and L_2 are also known as lysophospholipases. *Aka* phosphatidase.

phospholipid A lipid that contains one or more phosphate groups (in mono- or diester form), particularly a lipid derived from glycerol, sphingosine, or inositol. Phospholipids are polar lipids that are of great importance for the structure and functioning of biological membranes. *Abbr* PL.

phospholipid effect The stimulation of ^{32}P incorporation into phospholipids that is brought about by acetylcholine or carbamylcholine; the effect has been observed in both the pancreas and the brain cortex.

phospholipoprotein A conjugated protein that

contains phospholipid and that is soluble in aqueous solutions.

phosphomonoesterase An enzyme that catalyzes the hydrolysis of a once-esterified phosphoric acid (phosphomonoester). The enzyme is called specific if it catalyzes the hydrolysis of a small number of phosphomonoesters, and it is called nonspecific if it catalyzes the hydrolysis of a large number of phosphomonoesters at similar rates.

phosphonolipid A lipid that contains a carbon-to-phosphorus bond; a lipid containing the phosphoryl group attached to a carbon atom

$$HO-\overset{\displaystyle O}{\overset{\displaystyle \|}{\underset{\displaystyle |}{\underset{\displaystyle OH}{P}}}}-C-$$

phosphonomycin An antibiotic that is a structural analogue of phosphoenol pyruvate and that inhibits one of the enzymes involved in peptidoglycan synthesis.

phosphopantetheine A derivative of 4'-phosphopantothenic acid that is the prosthetic group of the acyl carrier protein; the phosphate ester of N-(pantothenyl)-β-mercaptoethylamine. Intermediates in fatty acid synthesis are linked via a thioester bond to the sulfhydryl group of β-mercaptoethylamine. *Aka* 4'-phosphopantetheine. *See also* acyl carrier protein.

phosphoprotein A conjugated protein in which the nonprotein portion is a residue of phosphoric acid.

phosphoprotein phosphatase A converter enzyme that catalyzes the interconversion of the two allosteric forms of phosphorylase (a and b); it catalyzes the dephosphorylation of phosphorylase a to phosphorylase b. *Abbr* PP-1.

phosphor FLUOR.

phosphoramidate A salt or an ester of phosphoramidic acid.

phosphoramidic acid An amide of phosphoric acid; a compound having the structure

$$HO-\overset{\displaystyle O}{\overset{\displaystyle \|}{\underset{\displaystyle |}{\underset{\displaystyle OH}{P}}}}-NH_2$$

phosphoramidite chemistry A widely used chemical approach for the synthesis of phosphodiester bonds between deoxynucleoside residues; uses phosphite ester derivatives of nucleosides as starting material. A phosphite amide is known as a phosphoramidite. *See also* phosphite–triester method.

phosphorescence The emission of radiation by an excited molecule in which the excited molecule first undergoes an electronic transition to a long-lived excited state and then slowly returns from that state to the ground state, dissipating the excitation energy by the emission of radiation at the same time. The emitted radiation is of a different wavelength than that of the exciting radiation, and the time interval between excitation and emission is usually several seconds or longer. *See also* fluorescence.

α-5-phosphoribosyl-1-pyrophosphate A compound that is formed by the transfer of a pyrophosphate group of ATP to ribose-5-phosphate and that serves as a key intermediate in the biosynthesis of both purine and pyrimidine nucleotides. *Abbr* PRPP.

phosphorimetry The measurement of phosphorescence.

phosphoroclastic reaction A cleavage reaction by means of inorganic phosphate, as the cleavage of pyruvate by inorganic phosphate to acetyl phosphate, carbon dioxide, and hydrogen.

phosphoroimmunoassay An immunoassay employing antigens labeled with a phosphorescent compound. *Abbr* PhIA.

phosphorolysis The cleavage of a covalent bond of an acid derivative by reaction with phosphoric acid H_3PO_4, so that one of the products combines with the H of the phosphoric acid and the other product combines with the H_2PO_4 group of the phosphoric acid.

phosphorus An element that is essential to all plants and animals. *Sym*, P; atomic number, 15; atomic weight, 30.9738; oxidation states, $-3,+3,+5$; most abundant isotope,^{31}P; a radioactive isotope,^{32}P, half-life, 14.3 days, radiation emitted, beta particles.

phosphorylase An enzyme that catalyzes a phosphorolysis reaction; glycogen and starch phosphorylases are key regulatory enzymes in the catabolism of glycogen and starch. Glycogen phosphorylase exists in two allosteric forms, a and b, which are interconverted by the converter enzymes synthase-phosphorylase kinase (SPK) and phosphoprotein phosphatase (PP-1). Glycogen phosphorylase is a dimer; the a form is phosphorylated and is the more active form; the b form is dephosphorylated and is less active.

phosphorylase kinase SYNTHASE-PHOSPHORYLASE KINASE.

phosphorylase phosphatase PHOSPHOPROTEIN PHOSPHATASE.

phosphorylation The introduction of a phosphate group into a compound through the formation of an ester bond between the compound and phosphoric acid; more precisely referred to as the introduction of a phosphoryl group.

phosphorylation potential PHOSPHATE POTENTIAL

phosphoryl group The radical (a) which, under

physiological conditions, is dissociated to give the corresponding anionic forms:

$$OH$$
$$|$$
$$—P=O$$
$$|$$
$$OH$$
(a)

phosphoryl-group carrier   The ADP–ATP system that serves as an intermediate for phosphoryl-group transfer between high- and low-energy phosphate compounds.

phosphoryl-group transfer   A reaction in which a phosphoryl group is transferred from one compound to another, as in the reactions involving high- and low-energy phosphate compounds. *Aka* phosphate-group transfer.

phosphoryl transfer potential   The group transfer potential for the phosphoryl group.

phosphosphingolipid   *See* spingophospholipid.

phosphotransferase system   A bacterial transport system that moves monosaccharides across the cell membrane and phosphorylates them by using phosphoenolpyruvate as the phosphate donor. Since sugar phosphates do not penetrate the cell membrane, the monosaccharides are thus trapped inside the bacterial cell. *Abbr* PTS.

phosvitin   A phosphoglycoprotein in hen's egg yolk; a phosphorus-rich glycoprotein that contains about 10% phosphate and that is derived from vitellogenin. *Aka* phosphovitin.

phot   A unit of illumination that is equal to 1 lm/cm^2 of surface.

photo-   Combining form meaning light.

photoactivated cross-linking   A method for determining the regions in RNA that interact, and make contact, with a given protein. The protein is allowed to bind to the RNA and the complex is then irradiated with ultraviolet light which causes covalent bonds (cross-links) to form between bases in the RNA and amino acids in the protein. An endonuclease is then added to degrade most of the RNA outside the region protected by the cross-linked protein and oligonucleotides linked to the protein are subsequently identified. The method has been used, for example, to study the binding of aminoacyl-tRNA synthetase to its cognate tRNA.

photoaffinity labeling   Affinity labeling in which a chemical labeling reagent R—P is used such that R can bind specifically and reversibly to an active site, and P is a chemical group that is unreactive in the dark. Upon photolysis, R—P is converted to a highly reactive intermediate R—P*, which can then form a covalent bond with a group at the active site before R—P* dissociates from the site. In true photoaffinity labeling, the rate of formation of this covalent bond is much greater than the rate at which R—P* dissociates from the site; the reverse is the case for pseudophotoaffinity labeling. The latter is essentially identical to ordinary affinity labeling, with the exception that the labeling agent R—P* is produced by photolysis.

photoautotroph   A phototrophic autotroph.

photobleaching   FLUORESCENCE MICROPHOTOLYSIS.

photocell   A device, the electrical properties of which, such as voltage or resistance, are altered in response to changes in the intensity of light that impinges upon it.

photochemical   Of, or pertaining to, photochemistry.

photochemical action spectrum   A plot of photochemical efficiency, such as photosynthetic efficiency, as a function of the wavelength of the incident light.

photochemical effect   The initiation of a chemical reaction by the absorption of light.

photochemical reaction   A chemical reaction initiated by the absorption of light.

photochemical reaction center   *See* reaction center.

photochemical sensitizer   PHOTOSENSITIZER.

photochemistry   The area of chemistry that deals with the interaction of radiant energy and chemical processes. *See also* first law of photochemistry; second law of photochemistry.

photochrome   A molecule, or a group in a molecule, that is sensitive to light and shows the phenomenon of photochromism.

photochromism   The reversible change of color by a compound upon excitation by either ultraviolet or infrared radiation.

photocoupler   A photoreceptor that, upon stimulation by light, initiates a reaction that is driven by energy derived from the absorbed light energy. *See also* photosensor.

photodisintegration   A nuclear reaction in which an atomic nucleus absorbs a high-energy photon and ejects a neutron, a proton, or an alpha particle.

photodynamic action   The oxidation of biologically important molecules in the presence of molecular oxygen, a photodynamic substance, and visible light. *See also* photodynamic substance; photosensitization.

photodynamic dye   A pigment that can serve as a photodynamic substance.

photodynamic inactivation   The inactivation of a molecule, a virus, or a cell by photodynamic action.

photodynamic substance   A substance, frequently a pigment, that sensitizes a biologically important molecule toward oxidation and achieves this by absorbing light energy and transferring this energy by means of various mechanisms to the target molecule. *See also* photosensitization.

photoelectric cell PHOTOCELL.

photoelectric effect The ejection of an orbital electron from an atom as a result of the impingement on the atom of a photon of sufficient energy; all of the energy of the photon is used to eject the electron and to impart kinetic energy to it.

photoelectron The electron ejected from an atom in a photoelectric effect.

photofootprinting A technique for studying the sites of interaction between protein and DNA; based on the principle that ultraviolet light interacts with DNA only if the DNA has been distorted slightly from its double-helical configuration. If protein contacts with DNA cause the necessary distortion, then the sites of photoreaction correspond to those of protein contact with the DNA. The method can also be used for RNA and for protein–nucleic acid interactions in whole cells. *See also* footprinting.

photographic rotation technique A technique for determining the symmetry of a structure from its photograph; used for the determination of the symmetry of a virus from its electron micrograph. The micrograph is printed n times and the printing paper is rotated by $360°/n$ between successive exposures. A structure that has an n-fold radial symmetry will yield a photograph in which the details have been reinforced, while this is not the case for a structure that has an $(n - 1)$- or $(n + 1)$-fold radial symmetry.

photoheterotroph A phototrophic heterotroph.

photoinduction The development of fruiting bodies or spores in fungi that is brought about by exposure to light.

photoinhibition 1. The inhibition of photosynthesis by light. 2. The inhibition of growth and/or sporulation that is brought about in some microorganisms by exposure to light.

photoisomerization An isomerization brought about by the absorption of light.

photolithotroph An organism or a cell that utilizes primarily light as its source of energy, inorganic compounds as electron donors, and carbon dioxide as its source of carbon atoms.

photolyase PHOTOREACTIVATING ENZYME.

photolysis The fragmentation of a molecule into smaller parts by irradiation with light; the dissociation of water in photosynthesis is an example.

photolytic Of, or pertaining to, photolysis.

photometer 1. An instrument for the measurement of light scattered at different angles by means of a photomultiplier tube that can be

rotated around the sample cell. 2. An instrument for the direct measurement of light intensities; two basic types are the filter photometer and the spectrophotometer.

photometry The measurement of light intensity by means of a photometer; applicable to ultraviolet, visible, and infrared radiations.

photomicrograph A photograph taken through a light microscope.

photomicrography The methodology for obtaining photomicrographs.

photomorphogenesis The development of new tissues as a result of photoinduction.

photomultiplier tube An electronic tube that amplifies the beam of electrons released by the incident radiation and that is used in high-quality spectrophotometers.

photon A corpuscular unit of light; a quantum of light energy that is equal to hv, where h is Planck's constant $(6.625 \times 10^{-27}$ erg-s) and v is the frequency of the light in cycles per second. A photon has no charge and is believed to have a mass of exactly zero. *See also* elementary particles.

photoneutron A neutron ejected from an atomic nucleus by photodisintegration.

photon fluence The number of photons that cross a unit area; the photon fluence rate refers to the number of photons that cross a unit area per unit time.

photonuclear reaction PHOTODISINTEGRATION.

photoorganotroph An organism or a cell that utilizes primarily light as its source of energy, organic compounds as electron donors, and organic compounds as well as carbon dioxide as its source of carbon atoms.

photooxidation An oxidation reaction that is caused by light. The cleavage of water in association with photosystem II in photosynthesis is an example.

photoperiodism The periodicity in the response of an organism that results from changes in either light intensity or the length of days.

photophobia Lack of tolerance for light; unusual sensitivity to light that is characateristic of conditions that result from vitamin A deficiency.

photophosphorylation The synthesis of ATP that is coupled to the operation of an electron transport system in photosynthesis.

photopic vision Vision in bright light in which the cones of the retina function as light receptors.

photopolymerization A polymerization reaction that is induced by exposure to light.

photoprotection The protection of cells against the damaging effects of ultraviolet irradiation by prior exposure of the cells to light in the wavelength range of 310 to 370 nm; the effect may be due to an inhibition of cellular growth which allows more time for the repair of damaged DNA *Abbr* PP.

photoprotein One of a group of proteins that are responsible for luminescence in the jellyfish and related coelenterates. Light emission by photoproteins does not involve the luciferin-luciferase system.

photoreaction PHOTOCHEMICAL REACTION.

photoreactivating enzyme The enzyme that catalyzes photoreactivation. It binds to the thymine dimers that were produced in DNA as a result of ultraviolet irradiation. The DNA–enzyme complex then becomes activated by absorption of visible light and the enzyme cleaves the C—C bonds of the cyclobutyl rings of the thymine dimer, thereby converting it back to a normal pair of adjacent thymine residues. *Abbr* PR enzyme.

photoreactivation The recovery of cells from the damage caused by irradiation with ultraviolet light that is brought about by a photoreactivating enzyme when the damaged cells are exposed to visible light. *Abbr* PHR.

photoreactive center *See* reaction center.

photoreceptor A receptor that is stimulated by light; a photoreceptor may be either a photosensor or a photocoupler.

photorecovery PHOTOREACTIVATION.

photoreduction A reduction reaction that is caused by light. The conversion of $NADP^+$ to NADPH in association with photosystem I in photosynthesis is an example.

photorespiration The respiration of plant cells that occurs in the presence of light and while the cells are concurrently carrying on photosynthesis. Photorespiration utilizes reducing power generated by photosynthesis for the reduction of molecular oxygen; it does not involve mitochondria and does not yield ATP.

photorestoration PHOTOREACTIVATION.

photoselection The principle, in fluorescence polarization, that molecules having specific orientations of their dipoles will be preferentially excited.

photosensitive Sensitive to light; capable of being stimulated by light.

photosensitization The process whereby a substance, frequently a dye, sensitizes a biologically important molecule toward oxidation and achieves this by absorbing light energy and transferring this energy by means of various mechanisms to the target molecule. The oxidation of the target molecule may then occur either in the presence of molecular oxygen, resulting in a photodynamic action, or in the absence of oxygen, but in the presence of appropriate electron and/or hydrogen acceptors, resulting in a dye-sensitized photooxidation.

photosensitizer 1. A substance that, when added to a biological system, will increase the damage to the system when it is exposed to a subsequent dose of radiation. 2. A substance that brings about photosensitization. 3. PHOTODYNAMIC SUBSTANCE.

photosensor A photoreceptor that, upon stimulation by light, initiates a reaction that is driven by energy derived from sources other than the absorbed light energy. *See also* photocoupler.

photosynthate A product obtained as a result of photosynthesis.

photosynthesis The reaction whereby solar energy is captured by an organism and converted to chemical energy and which, in its most general form, can be written as $H_2D + A \xrightarrow{light} H_2A + D$, where D is an electron donor and A is an electron acceptor. Photosynthesis is carried out by a large number of organisms, both prokaryotic and eukaryotic, including plants, algae, and bacteria, and involves a variety of electron donors and electron acceptors. In green plants, photosynthesis takes place in chloroplasts and leads to the synthesis of carbohydrates from water and carbon dioxide and to the evolution of oxygen; it occurs in the presence of light and several chlorophyll pigments that are assembled in two photosystems (I and II) which are connected with electron transport systems.

photosynthesis cycle CALVIN CYCLE.

photosynthetic 1. Of, or pertaining to, photosynthesis. 2. PHOTOTROPHIC.

photosynthetic carboxylation The enzymatic fixation of carbon dioxide in photosynthesis.

photosynthetic cycle CALVIN CYCLE.

photosynthetic electron transport The flow of electrons through a chain of electron carriers that is induced by the light reaction of photosynthesis.

photosynthetic organism *See* photolithotroph; photoorganotroph.

photosynthetic phosphorylation PHOTOPHOSPHORYLATION.

photosynthetic pigment *See* primary pigment; accessory pigment.

photosynthetic quotient A measure of the photosynthetic activity of a system that is equal to the number of moles of oxygen evolved divided by the number of moles of carbon dioxide taken up.

photosynthetic unit The number of chlorophyll molecules that are required for the fixation of one molecule of carbon dioxide in photosynthesis.

photosystem A photosynthetic pigment or group of pigments that absorb light energy and that participate in the light reaction of photosynthesis. The photosynthetic reactions of chloroplasts are carried out by two such systems, designated I and II; each photosystem consists of a reaction center and associated antenna molecules and immediate electron carriers.

photosystem I The photosystem of chloroplasts that is based on the P_{700} pigment, requires light of longer wavelength, and is associated with the reduction of $NADP^+$ and with photophosphorylation.

photosystem II The photosystem of chloroplasts that is based on the P_{680} pigment, requires light of shorter wavelength, and is associated with the photolysis of water and the evolution of oxygen.

phototaxis A taxis in which the stimulus is light.

phototroph A cell or an organism that uses light as a source of energy.

phototropism A tropism in which light is the stimulus.

photovoltaic cell An electrical cell that operates on the principle that when light strikes certain metals or semiconductors, the flow of electrons produced is proportional to the intensity of the light; used in photometers and in some spectrophotometers.

pH paper Paper that is impregnated with indicator dyes and that is used for the approximate measurement of the pH values of solutions.

PHR Photoreactivation.

pH scale The range of pH values from 0 to 14 in which the value of 7 is that for pure water and the values of 0 and 14 represent approximately the hydrogen ion concentrations of 1.0 and 1.0×10^{-14} mol/L, respectively. *See also* pH.

pH-stat An instrument for maintaining a constant pH during the course of a chemical reaction. The acidic or basic groups released as the reaction proceeds signal the addition of titrant (base or acid) to the reaction mixture. The amount of titrant added as a function of time provides an assay of the kinetics of the reaction.

pH unit A change of 1.0 between two pH values on the pH scale.

phycobilin A linear tetrapyrrole derivative that occurs in conjugation with proteins and that functions in the form of a phycobiliprotein as an accessory pigment of photosynthesis in algal chloroplasts.

phycobiliprotein A conjugated protein that functions as an accessory pigment of photosynthesis in algal chloroplasts. Phycobiliproteins occur in phycobilisomes and include the phycocyanins, phycoerythrins, and allophycocyanins. *Aka* biliprotein.

phycobilisome A granule in red and blue-green algae that serves to harvest light energy and to transfer it to chlorophyll. Phycobilisomes are attached to the surface of the thylakoid membranes and contain the phycobiliprotein pigments; phycocyanin, allophycocyanin, and phycoerythrin.

phycocyanin A blue accessory pigment of algal chloroplasts that consists of a protein conjugated to a phycobilin.

phycoerythrin A red accessory pigment of algal chloroplasts that consits of a protein conjugated to a phycobilin.

phycology The study of algae.

phylloquinone Vitamin K_1.

phylogenesis PHYLOGENY.

phylogenetic Of, or pertaining to, phylogeny.

phylogenetic tree A diagrammatic representation of the development of species which indicates their interrelations and the times of their evolutionary divergence. Phylogenetic trees have been constructed on the basis of amino acid sequences in selected proteins, such as cytochrome c. *Aka* cladogram.

phylogeny The evolutionary development of a group of organisms, such as a species. *Aka* phylogenesis.

physical adsorption Adsorption that is brought about by physical forces, such as van der Waals forces.

physical biochemistry A branch of biochemistry that deals with the transformations of physical and chemical energies in biological systems, particularly as they relate to macromolecules.

physical half-life *See* half-life (1).

physical map A genetic map in which the distances between the genes have been determined by methods other than genetic recombination.

physicochemical Physical–chemical.

physics The science that deals with matter and energy, their interactions, and their changes.

physiological Of, or pertaining to, physiology.

physiological chemistry Biochemistry, particularly that of higher animal organisms.

physiological conditions The normal conditions pertaining to an organism such as, in the case of humans, a temperature of 37 °C and a pH of about 7.0.

physiological saline A 0.9% (w/v) solution of sodium chloride that is approximately isotonic to the blood and lymph of mammals and that is used for the temporary maintenance of living cells and tissues.

physiology The science that deals with the processes and the activities of living organisms or parts of organisms.

physisorption Sorption that is weaker than chemisorption and that is mainly the result of van der Waals interactions.

phytanic acid An oxidation product of phytol.

See also Refsum's disease.

phytanic acid storage syndrome REFSUM'S DIS-EASE.

phytic acid The hexaphosphate of myoinositol, an important phosphate storage compound in plants.

phytin The mixed calcium–magnesium salt of phytic acid.

phyto- Combining form meaning plant.

phytoagglutinin A lectin that is derived from plants and that agglutinates cells.

phytoalexins A group of stress metabolites produced by higher plants in response to fungal infection, physical damage, or exposure to certain chemicals.

phytochemistry The science that deals with the chemistry of plant materials.

phytochrome A chromophore–protein complex that occurs in plants and that is associated with photoperiodism; it functions as a switch, enabling a plant to sense its environment. Phytochrome is involved in the control of many types of plant growth including flowering, branching, leaf formation, and seed germination. The chromophore is an open chain tetrapyrrole, linked covalently to the protein. Phytochrome is believed to be a light-activated transcription factor.

phytoestrogen A compound, such as a specific flavonoid, that is derived from plants and that has estrogenic activity even though it is not a steroid.

phytohemagglutinin 1. A lectin that is derived from plants and that agglutinates red blood cells. 2. LECTIN.

phytohormone PLANT HORMONE.

phytokinin CYTOKININ.

phytol A long-chain alcohol that occurs in chlorophyll.

phytology Botany.

phytolysosome A lysosome of plant origin.

phytopathology The science that deals with the origin, the nature, and the course of plant diseases.

phytosterol A sterol that occurs in plants.

phytotoxin PLANT TOXIN.

phytylmenaquinone Vitamin K_1. *Aka* phylloquinone; phytonadione; phytylmenadione.

P_i Inorganic phosphate.

pi 1. A mathematical constant that is equal to 3.1416... and that expresses the ratio of the circumference of a circle to its diameter. 2. Osmotic pressure. *Sym* π.

pI Isoelectric point.

PI Phosphatidylinositol.

PIA Particle immunoassay.

pi bond A chemical bond formed by electrons that are in pi orbitals; a bond formed by the parallel overlap of p orbitals so that the electron density is concentrated above and below

the axis joining the two atomic nuclei.

picket fence porphyrin A synthetic porphyrin that contains bulky nonpolar peripheral substituents. The latter restrict access to the iron atom and provide a hydrophobic pocket that shields the bound oxygen molecule.

pico- Combining form meaning 10^{-12} and used with metric units of measurement. *Sym* p.

picornavirus A small, naked, icosahedral animal virus that contains single-stranded RNA; the poliovirus belongs to this group.

4-pi counter A geometrical arrangement for the standardization of beta radiation sources; produced by placing two windowless Geiger–Mueller detectors face to face and suspending the radiation source between them.

PI cycle Phosphatidylinositol cycle.

pi electron delocalization RESONANCE.

piericidin A An antibiotic, produced by *Streptomyces mobaraensis*, that resembles coenzyme Q in its structure and inhibits the electron transport system between NADH dehydrogenase and cytochrome b.

piezoelectric effect The generation of an electric current by application of pressure on certain crystals, such as quartz. The effect is reversible, that is, certain crystals can be made to oscillate by application of an electric field.

PIF *See* prolactin regulatory hormone.

pigeon crop sac assay A bioassay for prolactin; based on changes in the crop sac (proliferation and peeling of the epithelium) of pigeons in response to injected prolactin.

pigment 1. A naturally occurring coloring matter in an animal, a plant, or a microorganism; a biochrome. 2. A synthetic coloring matter; a dye.

PIH *See* prolactin regulatory hormone.

pi helix A variant of the alpha-helical structure of proteins that is occasionally seen in short stretches of the polypeptide chain. In this structure, the separation between successive hydrogen bonds is lengthened by one amino acid residue. *Sym* π helix.

pilus (*pl* pili) A small filamentous projection attached to the surface of a bacterium.

pinealectomy The surgical removal of the pineal gland.

pineal gland A small endocrine gland in the brain; it produces the hormone melatonin.

ping-pong chromatography A form of affinity chromatography that is useful for the purification of enzymes capable of forming covalent intermediates; involves attaching the substrate of the enzyme to the column, allowing the enzyme to become covalently linked to the substrate, and then eluting the enzyme by breaking the bond between it and the substrate.

ping-pong mechanism The mechanism of an

enzymatic reaction in which two or more substrates and two or more products participate, and the enzyme shuttles back and forth between its original and a modified form. According to this mechanism, after the binding of the first substrate by the enzyme, a product is released and the enzyme is converted to a modified form. The second substrate then binds to the modified form of the enzyme, and this is followed by the release of a second product and the regeneration of the original form of the enzyme.

pinocytosis The taking up of droplets of liquid. The uptake of small particles, solutes, and liquid droplets by a cell.

pinosome An endosome that contains nonparticulate soluble material destined for hydrolytic digestion.

pi orbital A molecular orbital that is a delocalized bond orbital, spread over two or more atoms, or over an entire molecule.

piperidine alkaloids *See* alkaloids.

PIPES Piperazine-*N*,*N'*-bis(2-ethanesulfonic acid); used for the preparation of biological buffers in the pH range of 6.1 to 7.5. *See also* biological buffers.

pipet A graduated open tube, usually made of glass, and used for measuring and transferring small and definite volumes of liquids. *Var sp* pipette.

pi–pi-star transition The excitation of an electron from a pi orbital to a pi-star orbital; such transitions are responsible for the most intense absorption bands of molecular spectra.

Pirani gauge A thermal conductivity vacuum gauge.

PI response *See* phosphatidylinositol cycle.

piscine Of, or pertaining to, fish.

pi-star orbital *See* antibonding orbital.

PITC Phenylisothiocyanate; *See* Edman degradation.

pitch The distance between two identical points along the axis of a helix; equal to the number of residues per turn of the helix, multiplied by the distance per residue along the axis of the helix.

pitocin OXYTOCIN.

Pitressin Trade name for vasopressin.

pituitary basophilism A tumor of the pituitary gland that is sometimes associated with Cushing's disease.

pituitary dwarfism Dwarfism that is caused by a deficiency in the secretion of growth hormone by the pituitary gland.

pituitary gland An endocrine gland, located below the brain, that regulates a large portion of the endocrine activity of vertebrates. The gland consists of an anterior lobe called adenohypophysis and a posterior lobe called

neurohypophysis. *See also* adenohypophysis; neurohypophysis.

pituitary lactogen PROLACTIN.

Pituitrin Trade name for hypophysin.

pK The negative logarithm, to the base 10, of an equilibrium constant based on activities. 2. The same as pK'.

pK' The negative logarithm, to the base 10, of an apparent equilibrium constant based on molar concentrations.

PK 1. Protein kinase. 2. Prekallikrein.

pK$_a$ The negative logarithm, to the base 10, of an acid dissociation constant.

pkat Picokatal.

pK$_b$ The negative logarithm, to the base 10, of a base dissociation constant.

P-K chain Porod–Kratky chain.

pk$_{int}$ The negative logarithm, to the base 10, of an intrinsic dissociation constant.

P-K reaction Prausnitz–Kuestner reaction.

PKU Phenylketonuria.

pK$_w$ The negative logarithm, to the base 10, of the ion product of water.

PL 1. Phospholipid. 2. Pyridoxal. 3. Placental lactogen.

placebo 1. An inactive substance that is identical in appearance to a biologically active one and that is given to a number of individuals out of a group while the remainder of the individuals receive the biologically active substance. The individuals receiving the placebo thus serve as controls which permit an evaluation of the effectiveness of the biologically active substance given to the other individuals. 2. An inert medication given to an individual for its suggestive and psychological effect.

placenta The structure by which the fetus is attached to the uterus and through which it exchanges materials with the maternal circulation, receiving nutrients and excreting waste products.

placental barrier A semipermeable membrane that restricts the type and quantity of material exchange between the fetus and the mother and that represents a partial block to the passage of antibodies from the mother to the fetus.

placental lactogen A single-chain polypeptide hormone that is synthesized by the placenta in increasing amounts during pregnancy and that has an amino acid sequence that is almost identical to that of growth hormone. It has lactogenic activity and some growth-promoting activity like growth hormone; it also stimulates the production of progesterone by the corpus luteum in rodents. *Abbr* PL. *Aka* chorionic somatomammotropin (CS); choriomammotropin.

plain dispersion SIMPLE DISPERSION.

planchet A thin disk, commonly of metal, used for the deposition and counting of radioactively labeled material.

planchet counter A radiation counter for radioactive samples deposited in planchets.

Planck's constant A universal constant that relates the energy of a photon to its frequency; equal to 6.625×10^{-27} erg-s or 1.58×10^{-34} cal-s. *Sym* h.

Planck's law The law that the energy of a photon is equal to the frequency of the radiation multiplied by Planck's constant.

plane of symmetry An imaginary plane that divides a symmetrical body into two mirror image halves.

plane-polarized light Light in which the electric field vectors oscillate in a plane that passes through the axis along which the light is being propagated.

planetesimals Small bodies of matter, believed to have been formed from primordial dust and gas, and to have consolidated subsequently to form the terrestrial planets and the asteroids.

planimeter A device for measuring the area under a curve; a graphical integrator that is used in chromatography and electrophoresis for estimating the relative amounts of separated components.

plant agglutinin A lectin extracted from plants. *See also* lectin.

plant bile pigments A group of open tetrapyrrole pigments, such as phycocyanins and phycoerythrins, that occur in plants; so called because of their chemical relation to the pigments of animal bile.

plant hormone *See* hormone.

plant lectin *See* lectin.

plant pigment A pigment of plant origin. *See also* carotenoid; chlorophyll; flavonoid.

plant sex hormone GAMONE.

plant sulfolipid A sulfonic acid derivative of a glycosyldiacylglycerol that occurs in plants.

plant toxin A toxin of plant origin such as the viscotoxins and the toxalbumins.

plant-type ferredoxin IRON–SULFUR PROTEIN (b).

plant virus A virus that infects plants and multiplies in them. *See also* virus.

plaque 1. A clear region in a culture plate that represents an area of cell lysis, devoid of intact cells. *See also* plaque assay; plaque technique. 2. An atheromatous deposit.

plaque assay An assay for counting the number of infections bacterial or animals viruses. The host cells are mixed with the virus in a gel, and the virus particles diffuse through the gel and infect and lyse the host cells. The viral progeny thus produced in turn infect and lyse adjacent host cells, resulting in the formation of a plaque or clear region that is devoid of intact cells at each site of infection by a viral particle.

plaque-forming cell *See* plaque assay; plaque technique.

plaque-reduction method An assay for interferon that is based on determining the dilution of interferon that will inhibit plaque formation in a plaque assay, using a tissue culture challenged with a virus.

plaque technique A technique, devised by Jerne, for counting antibody-producing lymphocytes. Lymphocytes from animals that have been immunized with red blood cells are mixed with some of the same red blood cells in a gel. The red blood cells become bound to the lymphocytes by means of the antibodies synthesized by the lymphocytes; the addition of complement then causes the complex to lyse and leads to the formation of a plaque around each antibody-producing cell.

plaque titer The number of virus particles, per unit volume of viral suspension, that are capable of forming a plaque under a given set of conditions.

plaque-type mutant A phage mutant that gives rise to a plaque of changed morphology.

plasma 1. The fluid obtained from blood by removal of the formed elements by means of centrifugation; serum plus fibrinogen. 2. A gas form present at very high temperatures (above $10^5\,°C$). At such high temperatures, atoms are stripped of their electrons and the plasma that results is a neutral gas mixture of nuclei and electrons.

plasmablast A proliferative cell that has developed from a small B lymphocyte but has not yet become a mature, immunoglobulin-secreting, plasma cell.

plasma cell A lymphocyte that is capable of synthesizing antibodies; a differentiated cell derived from a B lymphocyte.

plasma clearance CLEARANCE.

plasmacyte PLASMA CELL.

plasma-derived growth factor PLATELET-DERIVED GROWTH FACTOR.

plasma factors *See* extrinsic pathway; intrinsic pathway.

plasma gel ECTOPLASM.

plasmagene The smallest heritable unit in the plasmon; a self-replicating cytoplasmic gene such as the fertility factor.

plasmakinin *See* kinin.

plasmalemma CELL MEMBRANE.

plasmalemmasome MESOSOME.

plasmalemmosome MESOSOME.

plasma lipoproteins *See* lipoproteins.

plasmalogen A generic descriptor for glycerophospholipids in which the glycerol moiety bears a 1-alkenyl ether group. A phosphoglyceride that contains a phosphatidal group, such as phosphatidal choline, phosphatidal serine, or phosphatidal ethanolamine; plasmalogens are abundant in the membranes of muscle and nerve cells. *See also* phosphatidal group.

plasma membrane CELL MEMBRANE.

plasmapheresis A technique for decreasing the concentration of the plasma proteins of an animal; achieved by bleeding the animal repeatedly, collecting the blood cells, and reinjecting the blood cells, suspended in saline, into the animal.

plasma proteins A large group of proteins such as albumin, fibrinogen, prothrombin, and immunoglobulins, that are present in blood plasma. It is estimated that there are more than 100 plasma proteins, most of which are glycoproteins. Plasma proteins have a variety of functions such as the regulation of blood pH and osmotic pressure; the transport of ions, hormones, lipids, vitamins, and other metabolites; and the control of blood coagulation and immune reactions.

plasma sol ENDOPLASM.

plasma-specific enzyme An enzyme that is present in blood plasma and that has a specific function in plasma; an enzyme that functions in the reactions of blood clotting is an example.

plasma thromboplastic factor ANTIHEMOPHILIC FACTOR.

plasma thromboplastic factor B CHRISTMAS FACTOR.

plasma thromboplastin antecedent The factor that is activated by the Hageman factor in the intrinsic pathway of blood clotting. *Abbr* PTA.

plasma thromboplastin component CHRISTMAS FACTOR.

plasma transferrin *See* transferrin.

plasma transglutaminase Factor XIII. Actually, plasma protransglutaminase.

plasmenic acid Any derivative of *sn*-glycero-3-phosphate in which carbon 1 bears an *O*-(1-alkenyl) residue, and carbon 2 is esterified with a fatty acid.

plasmid An extrachromosomal genetic element in bacteria. A plasmid is a circular, double-stranded DNA molecule that usually confers some evolutionary advantage to the host organism such as resistance to antibiotics, production of colicins, and so on. Plasmids replicate independently of the bacterial chromosome and constitute a useful tool in recombinant DNA technology. *See also* episome.

plasmid amplification An increase in the number of plasmids per cell that occurs with high-

copy number plasmids when protein synthesis in the host cell is inhibited. The lack of proteins, involved in the initiation of DNA synthesis, inhibits initiation of replication for the host chromosomal DNA but not for the plasmid DNA.

plasmid cloning vector A plasmid that serves as a vector in recombinant DNA technology. *See* vector (3); recombinant DNA technology.

plasmid conduction *See* conduction (2).

plasmid copy number *See* copy number.

plasmid curing *See* curing.

plasmid donation *See* donation.

plasmid engineering RECOMBINANT DNA TECHNOLOGY.

plasmid fusion The linking together of two plasmids, one of which carries a transposon; a type of replicon fusion which involves formation of a cointegrate structure.

plasmid incompatibility *See* incompatible plasmids.

plasmid transfer *See* effective contact; mobilization; repliconation.

plasmin The proteolytic enzyme that catalyzes the hydrolysis of fibrin and thereby leads to the dissolution of intravascular blood clots.

plasminogen The inactive precursor of plasmin; it is converted to active plasmin by a number of serine proteases such as urokinase.

plasminogen activator One of a group of proteolytic enzymes that convert inactive plasminogen to active plasmin.

plasmodesmata (*sing* plasmodesma) Fine cytoplasmic channels that pierce cell walls and connect one cell to neighboring cells in higher plants. Each channel is lined with a plasma membrane, which is common to the two connected cells, and usually contains a fine tubular structure, called a desmotubule.

plasmodium A genus of parasitic protozoans which includes the organism that causes malaria in humans.

plasmogeny The artificial production of microscopic structures, the properties of which bear some resemblance to those of living cells.

plasmolysis The shrinking of cellular protoplasm that occurs when a cell is placed in a hypertonic solution so that water moves out of the cell.

plasmon A collective term for the total extrachromosomal hereditary complement of a cell.

plasmoptysis The swelling and rupturing of a cell and the escape of its protoplasm that occurs when a cell is placed in a hypotonic solution so that water moves into the cell.

plasmosome NUCLEOLUS.

plastic chlorophyll A chlorophyll molecule that undergoes somewhat different photochemical reactions from the normal ones because it differs from ordinary chlorophyll in its conformation, its packing within the chloroplast, or its environment.

plastid A DNA-containing cytoplasmic, self-replicating subcellular organelle of higher plants, some of which function in photosynthesis (chloroplasts) while others serve as storage vessels. Plastids that contain pigments are known as chromoplasts. These are responsible for plant and flower coloration. Plastids that are devoid of pigments are known as leukoplasts. These serve as storage sites for starch.

plastocyanin A copper-containing protein that serves as an electron carrier in chloroplast photosynthesis. *Abbr* PC. *Aka* blue protein.

plastogene The plasmagene of a plastid.

plastome The genetic complement of a plastid.

plastoquinone A compound that is closely related to coenzyme Q and that functions as a hydrogen donor and acceptor in the photosynthetic electron transport system. *Abbr* PQ.

plate 1. PETRI PLATE. 2. The tail plate of a T-even phage.

plateau 1. A region in a solution in which the concentration remains uniform but changes with time. This region is below the boundary in sedimentation, above the boundary in flotation, and between boundaries in moving boundary electrophoresis. A plateau is formed in these cases if the initial concentration was uniform throughout the solution. 2. That portion of the characteristic curve of a radiation detector in which the count rate is almost independent of the applied voltage.

plateaued rat A rat that has a slow rate of growth and that is used in the assay of growth hormone.

plateau phenomenon The progressive flattening of the pH gradient in the region of neutrality that occurs in isoelectric focusing during prolonged runs.

plate count A viable count of bacteria that is based on the number of colonies that develop on a solid nutrient medium when appropriately diluted aliquots of the original culture are plated out.

platelet A small, irregularly shaped disk that is present in the blood and that functions in blood clotting by releasing thromboplastin.

platelet-activating factor A phospholipid that, in vitro, causes platelets to change shape, aggregate, and release their contents. In vivo, it is present during anaphylactic shock and appears to mediate inflammation and allergic responses. The factor is released from IgE-sensitized leukocytes in the presence of antigen. *Abbr* PAF.

platelet cofactor I ANTIHEMOPHILIC FACTOR.

platelet cofactor II CHRISTMAS FACTOR.

platelet-derived growth factor A low molecular weight protein mitogen (MW 13,000), isolated from human platelets, that causes proliferation of mouse fibroblasts and a variety of other cells. *Abbr* PDGF.

plate theory The application of the theoretical plate concept, derived from countercurrent distribution, to chromatography and particularly to gas chromatography. *See also* theoretical plate.

plating The cultivation of microorganisms on a solid nutrient medium in a petri dish.

plating efficiency *See* absolute plating efficiency; relative plating efficiency.

platinosome An artificially induced, electron-dense, lysosomal body that contains platinum and that is formed in cultured animal cells after exposure to platinum complexes.

platysome NUCLEOSOME.

playback experiment An experiment in which a DNA strand, that has been used to form a DNA–RNA hybrid, is recovered and subjected to a reassociation kinetics experiment to show that it consists of nonrepetitive DNA. *See also* RNA-driven hybridization.

pleated sheet A configuration of protein molecules in which the polypeptide chains are partially extended. The chains are held together by means of interchain hydrogen bonds between the CO and NH groups of all the peptide bonds. The pleated sheet structure occurs predominantly in fibrous proteins; the pleated sheet is referred to as being parallel or antiparallel, depending on whether the polypeptide chains are parallel or antiparallel. *Aka* beta sheet.

plectonemic coiling The coiling of two threads in the same direction so that they cannot be separated except by uncoiling the threads; applies to the two strands in double-helical DNA. *See also* paranemic coiling.

pleiotropic mutation A single mutation that gives rise to multiple phenotypic effects.

pleiotropism The production of multiple, and apparently unrelated, phenotypic effects by a single gene. *Aka* pleiotropy.

pleiotypic response The coordinate control of the growth-related processes involved in the initiation of cell division and cell proliferation. These include membrane transport, RNA synthesis, DNA synthesis, protein synthesis, and protein degradation.

pleomorphism The occurrence of two or more forms, such as the different forms of an organism during its life cycle. *See also* doctrine of pleomorphism.

pleromer A component that can replace another component in a polymer with respect to the overall "balance" of components in that polymer. Thus, in a DNA molecule in which the mole percent of guanine equals that of cytosine plus 5-methylcytosine, the components cytosine and 5-methylcytosine are considered to be pleromers.

plexiglass Polymethylmethacrylate; a plastic.

PLI Pulsed-laser interferometry.

PLK Polylysine–kieselguhr.

-ploid Combining form indicating the multiple of the chromosome set in the nucleus, as in diploid and polyploid.

ploidy The degree of chromosome multiplicity; the chromosome state in which each chromosome is represented once, twice, etc. *See also* aneuploid state; euploid state; heteroploid state; polyploid state.

PLP Pyridoxal phosphate.

plus end *See* actin filament.

plus–minus method SANGER–COULSON METHOD.

plus strand That strand of viral nucleic acid that has the same (for RNA viruses) or essentially the same (for DNA viruses) base sequence as the corresponding viral mRNA. A minus strand is the corresponding complementary and antiparallel strand. (a) *RNA viruses*. In single-stranded RNA viruses containing a plus strand (classes IV and VI), the viral mRNA ultimately produced has the same base sequence as the original viral RNA; in single-stranded RNA viruses containing a minus strand (class V), the base sequence of the mRNA is complementary to the base sequence of the original viral RNA. In double-stranded RNA viruses (class III), the positive strand of the viral RNA has the same base sequence as that of the viral mRNA. (b) *DNA viruses*. In single-stranded DNA viruses containing a plus strand (class II), the base sequence of the viral DNA is the same as that of the viral mRNA (except that thymine in DNA is changed to uracil in RNA). In double-stranded DNA viruses (class I), the plus strand has the same base sequence as that of the viral mRNA (except that thymine in DNA is changed to uracil in RNA). *See also* virus; positive strand virus.

PM 1. Pyridoxamine. 2. Puromycin.

PMF Proton motive force.

PMP Pyridoxamine phosphate.

PM particle A ribosomal subparticle isolated from bacterial cells in which protein synthesis was inhibited by puromycin.

PMR Proton magnetic resonance.

PMS 1. Phenazine methosulfate; a reducible dye that can serve as an electron acceptor. 2. Pregnant mare's serum; *See* pregnant mare's serum gonadotropin.

PMSF Phenylmethylsulfonyl fluoride; an inhibitor of serine proteases.

PMSG Pregnant mare's serum gonadotropin.

PN 1. Pyridoxine. 2. Protease nexin.

PNA Pentose nucleic acid.

pneometer SPIROMETER.

Pneumococcus The organism, *Streptococcus pneumoniae*, which is the causal agent of pneumonia. A large number of strains are known; some are virulent, some are not. *See also* rough strain; smooth strain.

PNP 1. *p*-Nitrophenol. 2. Pyridoxine phosphate.

PNPP *p*-Nitrophenylphosphate.

PNS Peripheral nervous system.

P$_{O_2}$ 1. Oxygen tension. 2. Partial pressure of oxygen.

pocket ionization chamber A small dosimeter designed to be worn by an individual and used for monitoring the amount of radiation to which the individual has been exposed. *Aka* pocket dosimeter.

pock method A method for counting the number of infectious viral particles by counting the lesions produced in the chorioallantoic membrane of chick embryos following the infection of the membrane with the viral particles.

pOH The negative logarithm, to the base 10, of the hydroxyl ion concentration in moles per liter; $-\log[OH^-]$.

-poiesis Combining form meaning formation or production.

poikilocyte A red blood cell of irregular shape.

poikilothermic Descriptive of an organism whose temperature varies with the temperature of its environment. *Aka* cold-blooded.

pointed end *See* actin filament.

point group The combination of point symmetry operations that can be carried out for a given object; a symmetry class to which objects may belong by virtue of possessing elements of symmetry that pass through, or are arranged about, a single point which serves as the center of symmetry.

point mutation A mutation in which there is a change in only one nucleotide of a nucleic acid. *See also* transition (1); transversion.

point quenching The quenching that occurs in liquid scintillation counting when beta particles are absorbed by insoluble sample particulate matter.

point symmetry Any type of symmetry operation, such as rotation, rotoreflection, and inversion, about a point.

poise 1. *n* A unit of dynamic viscosity; one-hundredth of this unit is called the centipoise; the dimensions of poise are $g\ cm^{-1}s^{-1}$. 2. *v* To buffer an electrode potential. *See also* poising.

Poiseuille's law An equation for the volume rate of flow dV/dt of a liquid through a capillary; specifically, $dV/dt = \pi a^4 P/8\eta l$, where π is a constant equal to 3.1416..., a is the radius of the capillary, P is the pressure, η is the viscosity of the liquid, and l is the length of the capillary.

poising The resistance to change in electrode potential that is shown by an oxidation–reduction couple at and near the midpoint potential of the couple. At the midpoint potential the concentrations of the oxidant and the reductant are equal, and an oxidation–reduction couple acts as a potential buffer at and near this point, much as a weak acid and its conjugate base act as a pH buffer at and near the pK value where their two concentrations are equal.

poison A substance that alters the normal metabolism of an organism, is injurious to health, and may be lethal when a small amount of it is either taken into, or comes in contact with, the organism.

poisoning The inhibition of a catalyst by the product of the reaction. In enzymology this phenomenon is usually called feedback inhibition or end-product inhibition.

Poisson distribution A discontinuous probability distribution in which the variance equals the mean. If the total number of objects observed under certain conditions varies according to a Poisson distribution with a mean m, then the probability P_x of obtaining x objects is given by $P_x = e^{-m}m^x/x!$, where x is a whole number between zero and infinity, and e is the base of natural logarithms. The Poisson distribution is a limiting form of the binomial distribution for the case in which the probability of success for an individual trial is very small, the number of trials is very large, but the product of these two quantities is finite. The distribution applies to systems in which the number of events per unit of observation is being determined; the number of radioactive disintegrations in a fixed period of time or the number of bacterial colonies formed from a fixed volume of solution are two examples. *Aka* Poisson's law.

pokeweed mitogen A lectin, isolated from pokeweed (*Phytolocca americana*) that stimulates lymphocyte proliferation.

poky mutant A slow-growing mutant of the mold *Neurospora crassa*; the reduced growth rate is associated with a deficiency, or an absence, of certain components of the respiratory chain.

pol DNA polymerase. DNA polymerases I, II, and III of prokaryotes are designated as pol I, pol II, and pol III, respectively. DNA polymerases α, β and γ of eukaryotes are designated as pol α, pol β, and pol γ, respectively, *See* DNA polymerase.

polar Possessing polarity; having a permanent dipole moment.

polar amino acid An amino acid that has a polar side chain.

polar bond A covalent bond in which the elec-tron pair or pairs of the bond are held with unequal strength by the two bonded atoms.

polarimeter An instrument for measuring optical rotation. The instrument contains two nicol prisms (polarizer and analyzer) and is generally operated using light from a sodium lamp; the light passes in succession through the polarizer, the solution of the compound being studied, and the analyzer. *Aka* polariscope.

polarity 1. The property of having two poles, specifically in reference to a molecule in which the center of the positive charges does not coincide with the center of the negative charges; the degree to which a molecule is polar and possesses a permanent dipole moment. 2. The phenomenon in which a mutant gene leads to a decrease in the synthesis of proteins that are specified by genes that belong to the same operon as the mutant gene but are farther removed (more downstream) from the operator. 3. The sense in which a polynucleotide, or some other biopolymer, is transversed, synthesized, or functioning. Thus, the two strands in double-helical DNA are said to be antiparallel or to have opposite polarity; one strand runs $3' \rightarrow 5'$, the other runs $5' \rightarrow 3'$. 4. The existence of two mating types in a unicellular organism due to either the presence or the absence of a fertility factor.

polarity gradient The variation with distance along the operon of the effect of a polarity mutation in one gene on the expression of the remaining genes in the operon.

polarity mutation POLAR MUTATION.

polarity ratio The ratio of polar to nonpolar amino acid residues in a protein.

polarizability A measure of the tendency of a substance to have dipoles induced in it when placed in an electric field; equal to the magnitude of the induced dipole moment per unit strength of the applied electric field.

polarization 1. The state of charge separation, as that across a biological membrane, that results from the orientation and the distribution of ions and molecules and leads to establishment of a membrane potential. 2. The state of light, as that produced by passing light through certain substances, in which the electric and magnetic field vectors of the light oscillate only in specific directions.

polarization curve POLAROGRAPHIC WAVE.

polarization fluorescence *See* fluorescence polarization.

polarization microscope A microscope used for studying the anisotropic properties of objects and for visualizing objects by virtue of their anisotropic properties. *Aka* polarizing microscope.

polarized electrode An electrode, the potential of which varies with the current passing through it.

polarized light Light in which the electric and magnetic field vectors oscillate only in specific directions.

polarizer The nicol prism in a polarimeter that is used for producing plane-polarized light. *See also* analyzer.

polar lipid An amphipathic lipid.

polar mutation A mutation in a gene that reduces the rate of synthesis of proteins that are coded by genes that belong to the same operon as the mutant gene but are farther removed (more downstream) from the operator.

polar–nonpolar AMPHIPATHIC.

polarogram The record of a polarographic wave, either in the form of a direct visual display or in the form of a plot.

polarograph An instrument for conducting polarographic measurements.

polarographic wave The variation of electrode current as a function of potential that is produced in polarography.

polarography A method for electroanalytical studies of chemical substances, including the reduction of anions and cations, and for qualitative and quantitative microanalysis; based on measurements of the current produced at a microelectrode, such as at a dropping mercury electrode, as a function of the changing potential applied to an electric cell.

polaron A section of a chromosome within which gene conversion results in polarized genetic recombination.

polar requirement The slope of the line that is obtained by plotting the logarithm of the R_m value of an amino acid in a series of pyridine solvents as a function of the logarithm of the mole percent of water in the solvent. Amino acids having similar polar requirements are considered to be closely related. *See also* R_m value.

polar solvent A solvent that contains charged groups and/or dipoles.

Polenske number A measure of the volatile fatty acids in a fat that is equal to the number of milliliters of 0.1 N alkali which are required to neutralize the volatile, water-insoluble fatty acids in 5 g of fat. *Aka* Polenske value.

poliomyelitis The disease caused by the poliovirus; infantile paralysis. An acute infectious disease. Most cases involve respiratory and gastrointestinal infections but in some the disease develops with varying degrees of paralysis.

poliovirus A virus that causes poliomyelitis and that belongs to the enterovirus subgroup of picornaviruses.

poly- 1. Combining form meaning many. 2. Combining form meaning excessive.

poly(A) Polyadenylic acid.

polyacrylamide gel A cross-linked acrylamide gel prepared from the monomer acrylamide and the cross-linking compound, N,N'-methylenebisacrylamide, in the presence of a polymerizing agent such as ultraviolet light.

polyacrylamide gel electrophoresis A zone electrophoretic technique of high resolution in which a polyacrylamide gel is used as the supporting medium; important applications include SDS–PAGE and disc gel electrophoresis. *Abbr* PAGE.

polyadenylation The enzymatic reactions by which a sequence of adenylic acid residues (a poly(A) tail) is added to the 3'-end of most eukaryotic mRNAs.

polyaffinity theory The concept that there must be at least three points of contact (binding sites) between an enzyme and its substrate to account for the different reactivity of identical groups that are either part of a symmetrical substrate molecule, or are attached to a meso carbon of the substrate. *Aka* three-point landing; three-point attachment.

polyallomer A plastic that has a highly crystalline structure and that consists of polymerized crystalline segments of the constituent olefinic monomers.

polyamine A long-chain aliphatic compound that contains multiple amino and/or imino groups. Polyamines are widely distributed in nature and include compounds such as spermine, spermidine, cadaverine, and putrescine. Polyamines affect ribosomes, DNA, RNA, and other biological components, and their action is frequently attributed to an electrostatic interaction between the polyamine cation and a negatively charged molecule.

polyamino acid A synthetic, or naturally occurring, polymer of a given amino acid; polyglycine and polyglutamic acid are examples. *See also* sequence polymer.

polyampholyte A polyelectrolyte that can function as either a proton donor or a proton acceptor.

polyanion A molecule that possesses a large number of negative charges.

poly(A) tail A sequence of 20–200 adenylic acid residues that is added to the 3'-end of most eukaryotic mRNAs and that is believed to serve to increase the stability of mRNA by making it more resistant to nuclease digestion.

poly(C) Polycytidylic acid.

polycarbonate Polybisphenol-A-carbonate; a thermoplastic linear polyester of carbonic acid.

polycation A molecule that possesses a large number of positive charges.

polycephalic protein MULTIFUNCTIONAL PROTEIN.

polychlorinated biphenyls A group of industrial chemicals that are used as lubricants, heat-exchange fluids, insulators, and as plasticizers in paints, synthetic resins, and plastics. Polychlorinated biphenyls are of ecological interest, since residues of these compounds are found in a wide variety of tissues in fish, wildlife, and humans. Polychlorinated biphenyls have been shown to induce steroid hydroxylases, drug-metabolizing enzymes, and several cytochromes. *Abbr* PCB.

polycistronic messenger RNA A messenger RNA molecule that serves as a template for the translation of two or more polypeptide chains which are specified by adjacent cistrons in the DNA. *Aka* polycistronic message.

polyclonal antibodies Immunoglobulins derived from multiple clones of cells. The antibodies elicited in an individual organism by even a single pure antigen represent a mixture of molecules since different antibodies react with different parts of the same antigen molecule. Such an antibody preparation is, therefore, polyclonal; it contains antibodies synthesized by different clones of antibody-producing lymphocytes. *Abbr* PAbs.

polycloning site A short segment in a DNA vector that has been engineered to contain a number of cleavage sites for different restriction enzymes and thereby enhances the versatility of the vector.

polycythemia A condition characterized by the presence of abnormally large numbers of circulating red blood cells.

poly(dA) tail *See* homopolymer tail-joining.

polydeoxyribonucleotide A linear polymer of more than 10 deoxyribonucleotides that are linked by means of $3',5'$-phosphodiester bonds; a polynucleotide.

polydipsia Excessive thirst.

polydisperse Consisting of macromolecules that fall into a large number of classes with respect to their size.

poly(dT) tail *See* homopolymer tail-joining.

polyelectrolyte A linear polymer in which each monomer carries one or more ionic groups so that the polymer is a polyvalent ion with the charges distributed all along the chain.

polyene antibiotics A group of macrolide antibiotics, produced by *Streptomyces* species, that have 4–7 conjugated double bonds in the lactone ring.

polyenoic fatty acid A polyunsaturated fatty acid.

polyestrous Having more than one estrous cycle per year.

polyethylene A thermoplastic polymer of ethylene.

polyethylene glycol A chemical used for phase partitioning and as a fusogenic agent to pro-

mote the fusion of cells in tissue culture.

polyetiological theory A theory of cancer according to which cancer can be caused by a variety of chemical, physical, and biological agents.

polyfunctional *See* bifunctional.

polyfunctional protein POLYPROTEIN.

poly(G) Polyguanylic acid.

polygene One of a group of genes that control a quantitative trait such as size, weight, or pigmentation, and that are believed to function together and to have a cumulative effect. *See also* oligogene.

polygenic messenger POLYCISTRONIC MESSENGER RNA.

polyglucosan GLUCAN.

polygon A plane and closed figure that is bounded by many straight lines.

polyhead A long, hollow cylinder, produced by some phage mutants, that has the diameter of a normal phage head but differs from it in its properties.

polyhedron A solid that is bounded by many plane faces.

polyhydroxy Containing two or more hydroxyl groups.

polyisoprene A polymer of isoprene; the cis isomer is natural rubber and the trans isomer is gutta percha. *Aka* polyterpene.

polykaryocyte A multinucleated cell.

polyketide ACETOGENIN.

polylinker *See* linker DNA (2).

polylysine–kieselguhr An adsorbent that consists of polylysine bound to diatomaceous earth and that is used for column chromatographic fractionation of nucleic acids. *Abbr* PLK.

polymer A high molecular weight compound consisting of long chains that may be open, closed, linear, branched, or cross-linked. The chains are composed of repeating units, called monomers, which may be either identical or different.

polymerase *See* DNA-dependent DNA polymerase; DNA-dependent RNA polymerase; RNA-dependent DNA polymerase; RNA-dependent RNA polymerase.

polymerase chain reaction A technique for the synthesis of large quantities of specific DNA segments; consists of a series of repetitive cycles, one step of which involves a high temperature. The latter inactivates the DNA polymerase used originally, thus requiring the addition of fresh enzyme at each cycle.

polymerization The repetitive reactions whereby the repeating units of a polymer are linked together to form long chains; the formation of a polymer.

polymetaphosphate LINEAR POLYPHOSPHATE.

polymetaphosphate ethyl ester A polyphosphate compound, the hydrolysis of which has been used to drive the polymerization of amino acids in studies on the origin of life.

polymorph POLYMORPHONUCLEAR LEUKOCYTE.

polymorphic gene A gene that exists in the form of several prevalent alleles.

polymorphism 1. The occurrence of two or more forms, such as the different forms of a protein in individuals of the same species. 2. The occurrence of two or more genetically different individuals in the same breeding population.

polymorphonuclear leukocyte A white blood cell that contains a lobed nucleus and granular cytoplasm; includes neutrophils, basophils, and eosinophils, so named because of the affinity of their cytoplasmic granules for specific dyes. *Aka* granulocyte, polymorph.

polymyxin One of a group cyclic peptide antibiotics, produced by *Bacillus polymyxa*, that are surface active and that damage the bacterial cell membrane and increase its permeability to small molecules.

polyneme hypothesis The hypothesis that a newly formed chromatid contains more than one double-stranded DNA molecule.

polyneuritis A disease of birds caused by a deficiency of thiamine.

polynomial function A function that is given by an equation of the form $Y = b_0 + b_1x + b_2x^2 + \cdots + b_kx^k$. The highest power of x, having a nonzero coefficient, is the degree of the polynomial.

polynuclear complex A metal ion–ligand complex of the type —M—L—M—L—M— where the metal ions M are held together in chains by means of ligands L, each of which binds to two metal ions.

polynucleotidase A polynucleotide phosphatase.

polynucleotide A linear polymer of more than 10 nucleotides that are linked by means of 3′,5′-phosphodiester bonds. *See also* polydeoxyribonucleotide; polyribonucleotide.

polynucleotide kinase An enzyme that catalyzes the transfer of a phosphate group from ATP to the 5′-hydroxyl group of RNA or DNA; useful for the terminal labeling of a nucleic acid with ^{32}P.

polynucleotide ligase DNA LIGASE.

polynucleotide phosphorylase The enzyme that catalyzes the random polymerization of ribonucleoside diphosphates to polyribonucleotides, a reaction that is useful for the synthesis of synthetic mRNA molecules.

polyol A polyhydroxy alcohol.

polyoma virus A small, naked, icosahedral, oncogenic virus that contains double-stranded DNA and belongs to the group of papovaviruses; produces tumors in rodents.

polyoxin One of a group of antibiotics, produced by *Streptomyces cacaoi*, that are active against fungi.

polypeptide A linear polymer of more than 10 amino acids that are linked by means of peptide bonds.

polypheny PLEIOTROPISM.

polyphosphate granule VOLUTIN GRANULE.

polyploid state The chromosome state in which each type of chromosome is represented more than twice. *Aka* polyploidy.

polyprotein A polyfunctional protein; a large protein molecule that serves as a precursor for a number of biologically active peptides or proteins. Polyproteins are made from a polycistronic mRNA molecule and are cleaved by proteolytic enzymes after synthesis of the polypeptide chain has been completed.

polyprotic acid An acid that has several dissociable protons.

polyribonucleotide A linear polymer of more than 10 ribonucleotides that are linked by means of 3′,5′-phosphodiester bonds; a polynucleotide.

polyribonucleotide phosphorylase POLYNUCLEOTIDE PHOSPHORYLASE.

polyribosome POLYSOME.

polysaccharide A linear or branched polymer of more than 10 monosaccharides that are linked by means of glycosidic bonds.

polysaccharide phosphorylase *See* glycogen phosphorylase; starch phosphorylase.

polysheath A long phage sheath that is produced by some phage mutants in the absence of a phage tube.

polysome A strand of mRNA with two or more ribosomes attached to it.

polysome profile The tracing that shows the types and the relative amounts of different polysomes in a sample, and that is obtained by monitoring the sample material after it has been fractionated by density gradient centrifugation.

poly(T) Polythymidylic acid.

polytailtube A long fiber, produced by some phage mutants, that has the diameter of a normal phage tube.

polytene chromosome An exceptionally large chromosome that contains numerous strands of DNA attached side by side in the form of a giant cable. Polytene chromosomes are characterized by specific patterns of bands, perpendicular to the long axis of the chromosome, which result from the chromomeres being in register.

polyteny The replication of chromosomes that results in the formation of polytene chromosomes.

polyterpene *See* terpene.

poly(U) Polyuridylic acid.

polyunsaturated Highly unsaturated; containing many double and/or triple bonds between carbon atoms.

poly(U) paper A strong paper support to which polyuridylic acid has been linked covalently; serves to bind specifically the 3′-poly(A) tail of eukaryotic mRNA and hence is used for the isolation and purification of polyadenylated eukaryotic mRNA.

polyuria The excretion of excessive amounts of urine.

polyvalent 1. Having a high valence. 2. Having more than one valence.

polyvalent allosteric inhibition *See* multivalent allosteric inhibition.

polyvalent antiserum An antiserum that contains antibodies against many different kinds of antigens.

polyvalent vaccine A vaccine that contains antigens derived from two or more different types of bacteria or viruses.

polywater A liquid that has a density of about one and a half times that of ordinary water and is prepared by condensation of water in fine capillaries; originally described as a new, stable form of water, but now known to be an ordinary aqueous solution containing substances dissolved from the capillaries by the condensing vapor.

POMC Proopiomelanocortin.

Pompe's disease GLYCOGEN STORAGE DISEASE TYPE II.

pontal atom BRIDGING ATOM.

pool The total amount of a substance, or a group of similar substances in equilibrium with each other, that is not covalently bound and that is available for, participates in, the anabolic and the actabolic reactions of the steady state; may refer to substances in a cell, an organ, a tissue, or an organism.

POPOP 1,4-Bis-2-(5-phenyloxazolyl)benzene; a secondary fluor.

POPSO Piperazine-N,N'-bis(2-hydroxypropane sulfonic acid); used for the preparation of biological buffers in the pH range of 7.2 to 8.5. *See also* biological buffers.

population A collection of organisms, cells, or molecules that have some quality or characteristic in common; generally refers to a large collection contained within a particular space.

por The bare 16-membered ring structure of the porphyrin ring system.

P:O ratio A measure of oxidative phosphorylation in a system that is equal to the number of moles of ATP formed per gram-atom of oxygen taken up; also the number of molecules of ATP formed per oxygen atom taken up.

porcine Of, or pertaining to, swine.

pore A minute opening through a solid.

porin An integral protein of the outer membrane in gram-negative bacteria. Porins are arranged in a hexagonal lattice, with trimers at each corner, resulting in electron micrographs that have been interpreted as being indicative of pores; the latter are believed to be responsible for the permeability of the membrane to small polar (hydrophilic) molecules. *See also* osmoregulation.

Porod–Kratky chain A model for a stiff chain, used to describe the hydrodynamic behavior of DNA. The model envisions the polymer as a continuously curving chain, as opposed to the jagged, uneven contour of a highly flexible chain. The direction of curvature at any point is assumed to be random. *Abbr* P-K chain.

porosity The porous quality of a solid.

porous Having a large number of pores.

porous disk method A method for measuring the apparent translational diffusion coefficient of a macromolecule by an application of Fick's first law. The apparatus consists of two chambers connected by a porous disk across which diffusion takes place. The diffusion coefficient is calculated from the mass transfer across the disk. The method may be used for impure preparations provided an assay for the macromolecule of interest is available. *Aka* porous diaphragm method; porous plate method.

porphin The parent tetrapyrrole ring structure of the porphyrins.

porphobilinogen The monopyrrole that is formed by the condensation of two molecules of δ-aminolevulinic acid and that serves as an intermediate in the biosynthesis of the porphyrins. *Abbr* PBG.

porphyria One of a number of pathological conditions that are due to abnormalities in the metabolism of heme and porphyrins and that are characterized by the presence of excessive amounts of porphyrin in the urine. Some porphyrias are hereditary diseases; others are nongenetic in nature.

porphyrin The heterocyclic compound, present in hemoglobin, cytochromes, and other hemoproteins, that has a tetrapyrrole ring structure in which iron is chelated. Physiologically important types of porphyrins are (a) uroporphyrins—each pyrrole group carries an acetate and a propionate side chain; (b) coproporphyrins—each pyrrole group carries a methyl and a propionate side chain; (c) protoporphyrin—each of two pyrrole groups carries a methyl and a propionate side chain; each of the other two pyrrole groups carries a methyl and a vinyl side chain.

porphyrinuria The presence of excessive amounts of porphyrins, particularly of coproporphyrin, in the urine.

porphyropsin A visual pigment, present in freshwater fish, that consists of rod opsin and retinal$_2$ and that has an absorption maximum at 522 nm.

porter TRANSPORT AGENT.

Porter diagram The representation of an immunoglobulin molecule by means of straight and parallel lines for the light and heavy chains.

position isomer One of two or more isomers that differ from each other in the position of either substituents or functional groups on a chain or on a ring.

positive catalysis Catalysis that leads to an increase in the rate of a chemical reaction, as in the case where the binding of a substrate by an enzyme leads to a decrease of the activation energy for the conversion of the substrate to products.

positive control The initiation of a biological activity by the presence of a particular molecule; the initiation of gene expression in an induction–repression system by the presence of a particular regulatory protein is an example. *See also* positive regulation.

positive cooperativity Cooperative binding in which the binding of one ligand to one site on the molecule increases the affinity for the binding of subsequent ligands to other sites on the same molecule.

positive effector *See* effector.

positive electron POSITRON.

positive feedback A feedback mechanism, as in an autocatalytic reaction, in which there is a direct relation between the magnitude of the input into, and the output of, a system; a small input leads to a small increase in the subsequent output, while a large input leads to a large increase in the subsequent output.

positive gene control POSITIVE REGULATION.

positive hydration Hydration in which the water molecules in the primary hydration shell of an ion have lesser mobility than those in pure water.

positive polarity Descriptive of a single-stranded RNA or a single-stranded DNA molecule that has bases in the same sequence as that found in the corresponding mRNA molecule.

positive regulation The regulation of gene expression in which transcription is turned on by the binding of some effector (a protein, a small molecule, or a molecular complex) to the promoter. This activation of the promoter permits the binding of RNA polymerase and the subsequent synthesis of mRNA.

positive staining A staining technique, used in electron microscopy, in which components of the sample are visualized through their binding of an electron-dense material; the sample appears darker than the background. The staining of nucleic acids with uranyl acetate and the staining of antigens with ferritin-labeled antibodies are two examples.

positive strand virus 1. A virus containing single-stranded RNA that has positive polarity. The strand (called a plus strand) has the same base sequences as that found in the viral mRNA. The infecting single-stranded RNA could, therefore, function directly as viral mRNA; it contains sequence that can be translated into viral proteins after the RNA enters the cell. Actually, the infecting strand first gives rise to a negative strand which then leads to synthesis of a positive strand that function as mRNA. 2. A virus containing single-stranded DNA that has positive polarity. After infection, the positive strand gives rise to ±DNA which then is transcribed to yield viral mRNA having the same base sequence as that of the infecting single-stranded DNA (except that thymine in DNA is replaced by uracil in RNA). 3. A retrovirus containing single-stranded RNA that has positive polarity. The RNA is first copied by reverse transcriptase to yield a negative DNA strand which is converted to ±DNA and which then gives rise to viral mRNA having the same base sequence as the original viral single-stranded RNA. *See also* plus strand; virus.

positive supercoil POSITIVE SUPERHELIX. *See* superhelix.

positive superhelix *See* superhelix.

positron A positively charged electron. *See also* elementary particles.

positron emission tomography *See* PETT.

positron emission transaxial tomography *See* PETT.

postabsorptive state The state of a person or an animal after a fast that was of sufficient duration so that all of the last nutrients have been absorbed through the intestinal wall.

post dimer initiation The initiation of DNA synthesis that occurs by passing a thymine dimer and restarting chain growth beyond the thymine dimer block. This mechanism forms the basis of recombination repair.

posterior 1. Behind, or in the back part of, a structure. 2. After, in relation to time.

postmortem Of, or pertaining to, the period after death.

postnatal Of, or pertaining to, the period after birth.

postoperative Of, or pertaining to, the period following a surgical operation.

postpartum Of, or pertaining to, the period after childbirth.

postprandial Of, or pertaining to, the period following a meal.

postreplication repair 1. The repair of DNA that occurs after the replicating fork has moved some distance past the region requiring repair. 2. The repair of nonreplicating DNA.

postreplicative modification The series of chemical reactions whereby various purines and pyrimidines are occasionally modified after the DNA has been synthesized. This includes the formation of such derivatives as 5-hydroxymethyl cytosine in the T-even phages and the formation of 5-methyl cytosine and 6-methyl adenine in animal and bacterial cells.

post-transcriptional modification PROCESSING (1).

post-translational modification PROCESSING (2).

post-translational transport The transfer of a protein across a biological membrane that occurs after synthesis of the polypeptide chain has been completed. *Aka* post-translational transfer. *See also* cotranslational transport.

potassium An element that is essential to all plants and animals. Symbol, K; atomic number, 19; atomic weight, 39.102; oxidation state, +1; most abundant isotope, ^{39}K; a radioactive isotope, ^{42}K, half-life, 12.4 h, radiation emitted, beta particles and gamma rays.

potato spindle tuber viroid *See* viroid.

potency The degree of effectiveness of a drug in terms of the quantities required to produce certain effects.

potential A measure of the electrical energy of a half-cell in comparison to an arbitrary standard; the difference in electrical energy between an indicator and a reference electrode.

potential difference The difference in electrical potential between two points in an electrical circuit.

potential-drop method HIGH-RESISTANCE-LEAK METHOD.

potential energy barrier ENERGY BARRIER (1).

potential energy diagram A graphical representation of the potential energy barrier of a molecule in which the potential energy of the molecule is plotted as a function of the internuclear distance of its atoms.

potential energy well The ground state and the low energy levels of a molecule as represented by a potential energy diagram.

potential gradient The rate of change of potential with distance in a specified direction.

potential mediator An electromotively active system that is added during potentiometric titrations to accelerate the establishment of equilibrium.

potential well *See* potential energy well.

potentiation 1. The increase in the effectiveness of a drug, a hormone, or a carcinogen that is produced by either prior or simultaneous treatment of the organism with another agent. 2. The increase in the reaginic antibody response of an animal that is produced by injecting it with certain parasites. 3. SYNERGISM.

potentiometer An instrument for measuring electrical potentials.

potentiometric titration A titration in which the electrical potential is measured as a function of titrant added.

potentiometry The measurement of either electromotive force or electrical potential and the application of these measurements to the study of oxidation–reduction systems.

Potter–Elvehjem homogenizer A homogenizer that consists of a glass tube in which a tightly fitting pestle is rotated; the shear forces that develop between the pestle and the wall of the tube lead to the homogenization of materials introduced into the tube.

pour plate A solid medium that contains bacteria and that is prepared by adding a bacterial inoculum to melted nutrient agar, pouring the mixture into a petri dish, and allowing it to solidify.

powder method A method of x-ray diffraction in which a sample is used that is in powdered form.

powder pattern The x-ray diffraction pattern that is obtained from a sample in powdered form and that is equivalent to the aggregate pattern that would have been obtained from the same sample if it were present in the form of a large number of small, randomly oriented crystals. *Aka* powder diagram.

power stroke *See* rowboat model.

poxvirus The largest and most complex of the animal viruses that is brick-shaped and contains a double-stranded DNA core surrounded by membranes. Poxviruses infect mammals (including humans) and birds, and some are oncogenic.

PP 1. Inorganic pyrophosphate 2. Protoporphyrin. 3. Photoprotection.

PP-1 Phosphoprotein phosphatase.

ppb Parts per billion; a measure of concentration equal to the number of parts of a component per billion parts of the total sample, such as parts of solute per billion parts of solution.

pp60C-src *See* c-src gene.

PPD 1. Purified protein derivative. 2. *p*-Phenylenediamine.

PP factor Pellagra-preventative factor.

ppGpp *See* magic spot.

PP$_i$ Inorganic pyrophosphate.

ppm Parts per million; a measure of concentration equal to the number of parts of a component per million parts of the total sample, such as parts of solute per million parts of solution.

PPO 2,5-Diphenyloxazole; a primary fluor.

PPP Pentose phosphate pathway.

pppGpp *See* magic spot.

P protein A regulatory protein that functions in the reactivation of glutamine synthetase after its inactivation by adenylylation. *Aka* P_{II}.

ppt Precipitate.

PQ Plastoquinone.

PQQ *See* quinoprotein.

PRA 5-Phosphoribosyl-1-amine; an intermediate in the biosynthesis of purines.

practical Denoting a chemical that has not been rigorously purified.

Prausnitz–Kuestner reaction A skin test for the detection of human reagins in serum; performed by injecting the test serum intradermally into a healthy person and then eliciting a wheal and erythema response by injecting a dose of allergen into the same site 24 h later.

prealbumin 1. A plasma albumin that moves ahead of the major albumin fraction when subjected to electrophoresis under alkaline conditions. 2. *See* thyroxine-binding prealbumin.

pre-beta fraction VERY LOW-DENSITY LIPOPROTEIN.

prebiotic Pertaining to the period prior to the occurence of life on the earth.

prebiotic soup PRIMORDIAL SOUP.

Precambrian era The geologic time period that extended over about 1.6 billion years and that ended about 600 million years ago; it is divided into the Proterozoic and Archeozoic eras.

precancerous Descriptive of a cell or a tissue that is presently benign but from which a malignant tumor is expected to develop with a high degree of probability.

precession diagram An x-ray diffraction pattern obtained by means of a precession camera.

precipitant A substance that, when added to a solution, causes the formation of a precipitate.

precipitate The deposit of insoluble material that is obtained from a solution by an alteration of the conditions or by the addition of specific substances.

precipitating agent PRECIPITANT.

precipitating antibody PRECIPITIN.

precipitation membrane An artificially constructed membrane that, like a biological membrane, is specifically permeable to certain ions. An example is a $BaSO_4$ membrane, prepared by counterdiffusion of $Ba(OH)_2$ and H_2SO_4 across a sheet of cellophane.

precipitin 1. An antibody that forms a precipitate with an antigen in a precipitin reaction. 2.

The precipitate formed in a precipitin reaction.

precipitin curve A plot of the amount of antibody precipitated as a function of increasing amounts of antigen added to the solution.

precipitin reaction The formation of an insoluble precipitate by a reaction between antigens and antibodies.

precision A measure of the reproducibility of a measurement; the degree of agreement between two or more measurements made in an identical fashion.

precursor 1. A compound that precedes another compound in a metabolic pathway by one or more steps. 2. A simple low molecular weight molecule in the environment, such as carbon dioxide or nitrogen, that is used by living organisms for the synthesis of biomolecules.

precursor fragments OKAZAKI FRAGMENTS.

precursor messenger RNA A large RNA transcript that is a precursor of mRNA; contains both introns and exons and must be subjected to splicing to produce the final, functional mRNA molecule. *Abbr* Pre-mRNA.

precursor of serum prothrombin conversion accelerator PROCONVERTIN.

precursor ribosomal RNA The giant precursor RNA molecule from which the various species of rRNA are cleaved off by post-transcriptional processing. *Abbr* Pre-rRNA.

precursor transfer RNA A large RNA transcript that is a precursor of tRNA. In *E. coli*, this represents the actual tRNA molecule with two extra nucleotides at the 3′-end and 41 extra nucleotides at the 5′-end; in eukaryotes, precursor tRNA may contain one or more tRNA sequences. *Abbr* Pre-tRNA.

preearly RNA A virus-specific RNA that is synthesized by RNA polymerase very soon after the infection of the host cell by the virus (within 1 min in the case of phage infection). *Aka* immediate early RNA; prereplicative RNA; very early RNA.

preelectrophoresis The removal of ionic contaminants from an electrophoretic support, such as a gel, by subjecting it to an electric field prior to the electrophoresis of the sample.

preenzyme *See* preprotein.

pre-equilibrium approximation A method for deriving the rate equation of a chemical reaction that is based on two assumptions: (a) There exists a rate-determining step in the mechanism; (b) the concentrations of the intermediates, preceding this step, are governed by equilibrium conditions. The pre-equilibrium approximation is used in the Michaelis–Menten treatment of enzyme kinetics. *Aka* prior equilibrium approximation;

rapid equilibrium approximation.

preexponential factor *See* Arrhenius equation.

preferential association An immunological theory according to which specific viral antigens interact more strongly with certain allelic products of the host immunogenes than with others. As a result of this preferential association, or binding, the virus becomes more immunogenic, and hosts, having such allelic products, become more immune to viral infection.

prefolic A SERUM FOLATE.

preformation The concept that an organism develops through the appearance and growth of structures and functions that are already present in the egg. *See also* epigenesis.

preformed gradient isodensity centrifugation DENSITY GRADIENT SEDIMENTATION VELOCITY.

pregnancy hormone PROGESTERONE.

pregnane The parent ring system of the progestogens, mineralocorticoids, and glucocorticoids.

pregnanediol A major catabolite of progesterone that occurs as a glucuronide, especially in pregnancy urine.

pregnant mare's serum gonadotropin A gonadotropic hormone, present in the serum of pregnant mares, that is produced by the endometrium and that has similar biological effects to those of follicle-stimulating hormone. *Var sp* pregnant mare's serum gonadotrophin. *Abbr* PMSG.

pregnenolone A precursor of the steroid hormones. *See also* desmolase.

preincubation The incubation of a reaction mixture prior to the test incubation, the effect of which is being measured; preincubation may be for such purposes as the depletion of an endogenous component or the establishment of a binding equilibrium.

preinductive phase The time period that precedes the administration of an antigen to an animal.

prekallikrein A precursor of kallikrein; the conversion of prekallikrein to kallikrein can initiate the intrinsic pathway of blood coagulation. *Abbr* PK.

prelumirhodopsin BATHORHODOPSIN.

premature initiation The phenomenon, seen in some preparations of rapidly growing bacteria or rapidly replicating phages, in which a second initiation of DNA replication occurs before the first one is completed.

premelting The phenomenon that DNA, and other double-stranded nucleic acid molecules, begin to melt out much before the melting-out temperature (T_m) is reached. Due to the fact that some base pairs open up even at room temperature and the number of such base pairs increases gradually as the temperature is raised. *See also* breathing.

pre-messenger RNA 1. PRECURSOR MESSENGER RNA. 2. HETEROGENEOUS NUCLEAR RNA.

premix burner A burner used in atomic absorption spectrophotometry and designed so that the gases are mixed and the sample is atomized before entering the flame.

pre-mRNA 1. Precursor messenger RNA. 2. Heterogeneous nuclear RNA.

prenol A long-chain isoprenoid alcohol.

prenyl group The isoprene moiety.

PR enzyme Photoreactivating enzyme.

preparative method A method, such as ultracentrifugation, electrophoresis, or chromatography, that requires relatively large amounts of sample and that is used primarily for the isolation and purification of specific substances. *See also* analytical method.

preparative ultracentrifuge An ultracentrifuge, equipped with rotors of varying capacities, that is used for the preparative fractionation of macromolecules.

prephenic acid *See* chorismic acid.

prepriming protein One of a number of proteins that form part of the primosome complex; they enable the enzyme primase to synthesize RNA primers during DNA replication.

preprimosome The primosome complex without the enzyme primase.

preprohormone The initial ribosomal product of a peptide hormone. A molecule that is larger than the hormone and that undergoes post-translational proteolysis to form, first, the prohormone and, finally, the biologically active hormone. *See also* preproprotein.

preproinsulin *See* preprohormone; preproprotein.

preproopiomelanocortin The common precursor of several hormones (corticotropin, lipotropin, melanotropin) and endorphins (β-endorphin, γ-endorphin).

preproprotein The preprotein form of a proprotein; the inactive precursor of a secretory protein (a proprotein) that has the signal sequence still attached (a preprotein). *See also* preprotein; proprotein; signal hypothesis propeptides.

preprotein A secretory protein with the signal sequence still attached; the protein, according to the signal hypothesis, before the enzyme signal peptidase has cleaved off the signal sequence. *See also* signal hypothesis.

prereplicative RNA PREEARLY RNA.

pre-ribosomal RNA PRECURSOR RIBOSOMAL RNA.

pre-rRNA PRECURSOR RIBOSOMAL RNA.

pressor agent HYPERTENSIVE AGENT.

pressor amine An amine, such as vasopressin, that functions as a hypertensive agent.

pressor effect An increase in blood pressure.

pressor principle VASOPRESSIN.

pressure dialysis Dialysis in which there is either an application of pressure to the dialysis bag, or an application of a vacuum to the space surrounding the dialysis bag.

pressure-jump method A relaxation technique in which pressure is the variable that disturbs the equilibrium of a system. *See also* relaxation technique.

pre-steady-state kinetics The kinetics of an enzymatic reaction proceeding under conditions that precede the establishment of a steady state; generally investigated by means of rapid flow or relaxation techniques that permit a study of both the initial and the intermediate steps of the reaction. During the pre-steady-state period (induction period), the concentration of a given intermediate rises from zero to its steady-state level.

pre-transfer RNA PRECURSOR TRANSFER RNA.

pre-tRNA PRECURSOR TRANSFER RNA.

previtamin A precursor of a vitamin that is formed either during the in vivo conversion of a provitamin to the vitamin or during the in vitro conversion of a synthetic compound to the vitamin. *See also* provitamin.

PRF *See* prolactin regulatory hormone.

PRH *See* prolactin regulatory hormone.

Pribnow box A nearly universal sequence of nucleotides in prokaryotic DNA, about 10 base pairs upstream from the site at which transcription of structural genes starts. It has the consensus sequence TATAAT and is a site in the promoter to which the sigma subunit of RNA polymerase binds. *See also* TATA box.

PRIF *See* prolactin regulatory hormone.

PRIH *See* prolactin regulatory hormone.

primaquine An antimalarial drug.

primaquine sensitivity A genetically inherited metabolic defect in humans that is characterized by the tendency of erythrocytes to hemolyze upon the administration of a variety of compounds, including primaquine, to an individual; due, in most cases, to a deficiency of the enzyme glucose-6-phosphate dehydrogenase. *See also* glucose-6-phosphate dehydrogenase deficiency.

primary acidosis A deviation from the normal acid–base balance in the body that is due to overproduction, ingestion, or retention of acid. In the absence of compensatory mechanisms, such conditions lead to a lowering of the blood pH.

primary active transport *See* active transport.

primary alkali deficit PRIMARY ACIDOSIS.

primary alkali excess PRIMARY ALKALOSIS.

primary alkalosis A deviation from the normal acid–base balance in the body that is due to excessive loss of acid or an overdose of sodium bicarbonate. In the absence of compensatory mechanisms, such conditions lead to an increase in the blood pH.

primary amino acids STANDARD AMINO ACIDS.

primary bile acids *See* bile acids.

primary carbon dioxide deficit RESPIRATORY ALKALOSIS.

primary carbon dioxide excess RESPIRATORY ACIDOSIS.

primary charge effect The charge effect in a solution containing charged macromolecules that results from the differential movement of the charged macromolecules and their oppositely charged counterions. The primary charge effect leads to a decrease in the sedimentation rate and to an increase in the diffusion rate of a charged macromolecule.

primary culture A culture that is started from cells, tissues, or organs that are derived directly from an organism.

primary deficiency 1. A decreased hormone level that is due to impaired function or destruction of the endocrine gland that produces this hormone. 2. DIETARY DEFICIENCY.

primary derived protein *See* derived protein.

primary filament MYOSIN FILAMENT.

primary fluor A fluor that is excited by the radiation from a radioactive sample and that produces a flash of light during scintillation counting.

primary food producer A photosynthetic organism; a photolithotroph or a photoorganotroph.

primary hydration shell The layer of water molecules that are closest to an ion and which, in the case of a metal ion, are frequently considered to be molecules acting as ligands.

primary hypertension ESSENTIAL HYPERTENSION.

primary immune response *See* primary response.

primary ionization The ionization of matter that is produced by the ionizing radiation that impinges upon it or passes through it.

primary isotope effect An isotope effect in which the isotope itself is involved in bond cleavage, such that the bond to the isotope is broken in the transition state of the reaction.

primary lysosome *See* lysosome.

primary messenger HORMONE.

primary metabolite *See* metabolite.

primary oxaluria A genetically inherited metabolic defect in humans that is characterized by increased formation of oxalic acid and deposition of calcium oxalate in the kidneys and other tissues; involves a disorder in glycine metabolism such that virtually all of the synthesized glycine is oxidized via glyoxylic acid to oxalic acid as a result of a deficiency of the enzyme α-ketoglutarate:glyoxylate carboligase.

primary pigment The major photosynthetic pigment of an organism; the primary pigment of plants is chlorophyll and that of bacteria is bacteriochlorophyll. *See also* accessory pigment.

primary plot A direct plot of experimental enzyme kinetics data, such as a Lineweaver–Burk plot. *Aka* primary kinetic plot.

primary prostaglandin A prostaglandin of either type PGE or type PGF; so called, since prostaglandins of types PGA and PGB can be derived from them.

primary protein derivative *See* primary derived protein.

primary response 1. The immune response of an animal that is produced when the animal is first exposed to, or injected with, an antigen. 2. The direct induction of transcription of a few specific genes that is brought about by a steroid hormone.

primary solvent The solvent used in scintillation counting for the transfer of energy from the radioactive sample to the fluor.

primary standard 1. A purified chemical that can be weighed out and used for the preparation of solutions of known concentrations (standard solutions). 2. A reference, such as a source of radiation, that is used for purposes of calibration.

primary stimulus The immunogen that stimulates an animal to produce a primary immune response.

primary structure The basic structure of a polypeptide chain or a polynucleotide strand that is described by the type, the number, and the sequence of either the amino acids in the polypeptide chain or the nucleotides in the polynucleotide strand. The primary structure of proteins excludes the spatial arrangement of the atoms except for the configuration about the alpha carbon atoms of the amino acids; it likewise excludes disulfide bonds and, hence, is not equivalent to the covalent structure of the molecule.

primary tissue culture A short-term tissue culture.

primary transcript The initial RNA molecule transcribed from a section of DNA between a start and a stop signal for RNA polymerase; frequently, an RNA molecule that requires post-transcriptional modification (processing) to become a functional mRNA, rRNA, or tRNA molecule.

primary tumor The original tumor as contrasted with a secondary tumor, formed through metastasis.

primase The DNA-dependent RNA polymerase that functions in DNA replication by synthesizing the RNA primers which are then extended by DNA polymerase to yield newly synthesized DNA fragments. While being an RNA polymerase, primase is distinct from the RNA polymerase that functions in the transcription of DNA. *See also* DNA replication.

primate A mammal of the order Primates, which includes humans, apes, and monkeys.

primed cell A lymphocyte that has "recognized" an antigen; a Y cell according to the XYZ cell theory.

primed synthesis method SANGER–COULSON METHOD.

primer A macromolecule that stimulates the synthesis of another macromolecule by participating in the initiation of polymerization and that is linked covalently to the product of the reaction. In nucleic acid chemistry, a primer is a short, single-stranded RNA or DNA segment that functions as the starting point for the polymerization of nucleotides.

primer–template *See* template–primer.

prime strain A viral strain that is not well neutralized by antibodies that are specific for the wild-type strain.

primeval PRIMITIVE.

priming 1. The administration of antigens to an animal organism in such a fashion that the responsive immunocytes are activated. 2. The conversion of X cells to Y cells according to the XYZ cell theory.

priming dose The first dose of an antigen administered to an animal organism to produce an immune response.

primitive Of, or pertaining to, the early stages in the development of the earth and the evolution of life.

primitive atmosphere The atmosphere that surrounded the primitive earth and that is considered by many to have been a reducing atmosphere.

primitive earth The earth at an early stage of development at which time the first organic compounds are believed to have been formed.

primordial PRIMITIVE.

primordial soup The primitive oceans and other primitive bodies of water that contained a variety of organic compounds and that are thought to have been the site for the reactions in the primitive earth that led to the synthesis of the first macromolecules and to the assembly of the first living cells. *Aka* prebiotic soup.

primordium The early group of cells that subsequently undergoes mitosis to form a particular organ or structure.

primosome A multiprotein complex required for the priming action that precedes the synthesis of each Okazaki fragment during DNA replication; consists of prepriming proteins, proteins having ATPase activity, and the enzyme primase. Binding of the primosome to DNA is followed by synthesis of an RNA primer by primase and synthesis of a DNA fragment (Okazaki fragment) linked to the RNA primer. The primosome moves along the DNA strand that is synthesized discontinuously and moves in a direction that is opposite to that of RNA and DNA synthesis. This movement requires the input of energy that is obtained by the hydrolysis of ATP.

principle of evolutionary continuity The concept that the formation of biomolecules, subcellular structures, and ultimately living cells required a large number of small, but reasonably probable, steps.

principle of Le Chatelier *See* Le Chatelier's principle.

principle of maximum orbital overlap The principle, enunciated by Linus Pauling in 1931, that the strongest covalent bond between two atoms will be formed when the two atomic orbitals achieve maximum overlap.

principle of microscopic reversibility The thermodynamic principle which states that, at equilibrium, the forward rate must be equal to the reverse rate for every elementary step in the reaction mechanism. Applied to a simple reaction, consisting of a single elementary step, the principle leads to the conclusion that both the forward and the reverse reactions must proceed via the same activated complex. Applied to a complex reaction, consisting of several elementary steps, the principle leads to the conclusion that, if there are several paths whereby A can be converted to B, then each step in the mechanism must be reversible. Thus, the mechanism (a) is not possible while the mechanism (b) is possible

$$A \rightarrow B \qquad\qquad A \rightleftharpoons B$$
$$\nwarrow \swarrow \qquad\qquad \nwarrow\hspace{-0.3em}\nwarrow \; \swarrow\hspace{-0.3em}\swarrow$$
$$C \qquad\qquad\qquad C$$
$$\text{(a)} \qquad\qquad\quad \text{(b)}$$

principle of simplicity OCCAM'S RAZOR.

principle of unattainability of absolute zero An alternative statement of the third law of thermodynamics: it is impossible for any process, involving a finite number of steps, to reduce the temperature of any system to absolute zero.

prion A small, infectious protein that is believed to be the cause of a number of degenerative neurological diseases such as scrapie in sheep and goats and Creutzfeld–Jakob disease, kuru, and Gerstmann–Straussler syndrome in humans. Prions were formerly called "slow viruses" but are now known to be devoid of nucleic acids and are, therefore, neither viruses nor viroids. The name prion is a contraction of *pro*tein and infec*tion*. Prions are resistant to inactivation by procedures that modify nucleic acids. A link between prions and Alzheimer's disease has been postulated.

prior equilibrium approximation PRE-EQUILIBRIUM APPROXIMATION.

private antigen A rare blood group antigen that occurs only in one or in a few individuals.

privileged site A region in an organism that lacks normal lymphatic drainage and that constitutes a location where a tissue transplant may persist for extended periods of time without inducing transplantation immunity.

PRL Prolactin.

pro- Prefix indicating an inactive precursor, as that of an enzyme, a vitamin, or a hormone.

Pro 1. Proline. 2. Prolyl.

proaccelerin An accessory protein that participates in the activation of prothrombin to thrombin in both the extrinsic and intrinsic pathways of blood clotting.

proacrosome The acrosome at an early developmental stage.

pro-ACTH-endorphin PROOPIOMELANOCORTIN.

probability The relative frequency of occurrence of a specific type of event out of the total number of occurrences of this and other types of events, all equally likely to take place. *Sym* p.

probability curve NORMAL DISTRIBUTION.

probability distribution A frequency distribution in which a variable is divided into classes and the probabilities for these classes are indicated.

probability paper Graph paper in which one scale has been changed so that a plot of a normal distribution curve will yield a straight line.

probability value The probability expressed as a decimal fraction.

probable error An error, equal to 0.6745 times the standard deviation, such that there is a 50–50 chance that any other error will be larger than it.

probe 1. A group of atoms or a molecule that is attached to other molecules or cellular structures and that is used as an aid in studying the properties of these molecules and structures. *See also* reporter group; spin labeling. 2. A substance, frequently labeled with radioactive isotopes, that is used to identify or isolate a gene, a gene product, or a protein. The hybridization of mRNA with its DNA gene, the hybridization of chromosomal DNA with corresponding cDNA fragments, and the binding of specific protein molecules to monoclonal antibodies are some examples of the use of probes. *See also* northern blotting; Southern blotting; western blotting.

probiogenesis Primordial biosynthetic reactions; primordial biogenesis.

probit A statistical unit of measurement of probability based on deviations from the mean of a normal frequency distribution.

procarboxypeptidase The inactive precursor of carboxypeptidase that is synthesized by the pancreas. It is converted to the active enzyme by proteolytic action.

procarcinogen The inactive precursor of a carcinogen.

procaryon The nuclear region of a prokaryotic cell. *Var sp* prokaryon.

procaryote Variant spelling of prokaryote.

procaryotic Variant spelling of prokaryotic.

process An outgrowth or an extension, such as an axon or the dendrite of a neuron.

processed gene A eukaryotic gene that lacks introns but contains the 3'-poly(A) tail of the parental mRNA species; believed to arise in vivo by aberrant reverse transcription (using viral or cellular enzymes) of the mature mRNA or of nuclear RNA to yield a piece of double-stranded DNA which is then reintegrated into the host chromosome. *Aka* retrogene; processed retropseudogene.

processing 1. The series of chemical reactions whereby a primary RNA transcript is converted to a functional mRNA, rRNA, or tRNA molecule. These modifications include such reactions as removal of introns, splicing of exons, addition of a methylated cap, and addition of a poly(A) tail. *Aka* post-transcriptional modification. 2. The series of chemical reactions whereby a newly synthesized polypeptide chain is converted to a functional protein. These modifications include such reactions as removal of the *N*-formyl group from methionine, phosphorylation, acetylation, hydroxylation, formation of disulfide bonds from sulfhydryl groups, attachment of prosthetic groups, and cleavage of peptide bonds to convert a proprotein (proenzyme) to a protein (enzyme). *Aka* post-translational modification. 3. *See also* postreplicative modification.

processive enzyme An enzyme that remains bound to its substrate for a series of repeated catalytic events before being released; the continued polymerization of DNA by DNA polymerase I (or III) that occurs during DNA replication without release of the enzyme from the template, is an example of processivity.

processive exonuclease An exonuclease that, once bound to a polynucleotide strand, completely degrades the strand before it is released.

processivity 1. The repeated catalytic steps carried out by a processive enzyme. 2. CHANNELING.

prochiral compound An organic compound that has no chiral centers but has a potential for "handedness" (chirality); it is potentially able to react asymmetrically with an asymmetric site. Such a compound contains one or more carbon atoms to which are attached two identical and two different substituents (prochiral carbon; meso carbon). The two identical substituents have the potential of not reacting equally, for example, on the surface of an enzyme (at the active site). Replacement of one of the two identical ligands by a different one results in formation of a chiral center.

prochirality The potential for "handedness" (chirality) shown by prochiral compounds.

procollagen A high molecular weight (MW 150,000) precursor of collagen.

proconvertin The precursor of the active protease convertin which activates factor X in the extrinsic pathway of blood clotting.

proctodone An insect hormone that terminates diapause.

prodrug A drug that is inactive by itself but undergoes transformation to a reactive form as a result of metabolic reactions at some location in the body; a compound that is converted to an anticancer drug by a tumor-associated enzyme is an example.

product An atom, an ion, or a molecule that is produced in a chemical reaction. *Sym* P.

product inhibition The inhibition of an enzyme by a product of the reaction that the enzyme catalyzes.

production strain A strain of microorganisms that is used for the direct synthesis of an industrial product or for a step in the synthesis of such a product.

productive cell A cell that produces viral progeny when it is infected with a virus.

productive complex An enzyme–substrate complex in which the substrate is bound to the enzyme in such a fashion that catalysis is possible and that products can be formed.

productive infection A viral infection that leads to the formation of infectious viral progeny.

productive phase The stage in an immune response that follows the first appearance of antibodies in the serum and that corresponds to the time during which antibodies are synthesized vigorously.

prodynorphin The precursor from which dynorphin and rimorphin are derived.

proelastase The inactive precursor of elastase.

proenkephalin A The precursor from which the enkephalins are derived.

proenzyme ZYMOGEN.

profibrinolysin PLASMINOGEN.

profilactin A complex of profilin and actin molecules.

profile *See* elution profile; polysome profile; thermal denaturation profile.

profilin One of a number of small, cytoplasmic

proteins that bind primarily to actin monomers and retard actin polymerization.

proflavin A mutagenic acridine dye that causes frameshift mutations.

progenote The common ancestor of present day cells. The ancestor from which, it is believed, three lines of descent lead to prokaryotes, eukaryotes, and archaebacteria.

progeny 1. The offspring of an organism or of a cell. 2. The newly formed DNA molecules or viral particles.

progeria A genetically inherited metabolic defect in humans that involves defective DNA repair and that is associated with early aging and death.

progestational Preceding gestation.

progesterone The major female sex hormone required for the maintenance of pregnancy; a 21-carbon steroid that is secreted largely by the corpus luteum.

progestin Any substance with progesterone-like activity; includes both natural and synthetic compounds.

progestogen A substance that induces progestational changes in the uterus; progestin or related synthetic compounds.

program *See* computer program.

progress curve A plot of the concentration of either a reactant or a product of a chemical reaction as a function of the time that the reaction has been allowed to proceed.

progression *See* tumor progression.

prohead One of a number of precursor structures of the hexagonal, DNA-containing head of a T-even phage.

prohormone The inactive precursor of a peptide hormone. A molecule that is larger than the hormone and from which the active hormone is produced by proteolysis.

proinsulin The inactive, cyclic precursor of insulin that is converted to active insulin by hydrolytic removal of a peptide of 33 amino acids.

projection formula A two-dimensional representation of a molecule in which bonds projecting forward with respect to the plane of the page are indicated by horizontal lines, and bonds projecting backward are indicated by vertical lines.

prokaryon Variant spelling of procaryon.

prokaryote A simple, unicellular organism, such as a bacterium, that lacks a discrete nucleus surrounded by a nuclear membrane, and that contains its genetic material within a single chromosome. *See also* eukaryote.

prokaryotic Of, or pertaining to, prokaryotes.

prolactin A protein hormone, secreted by the anterior lobe of the pituitary gland, that is essential for the initiation of lactation in mammals. Prolactin also has a gonadotropic effect and stimulates progesterone secretion by the corpus luteum. *Abbr* PRL.

prolactin regulatory hormone One of two hypothalamic hormones (or factors) that, respectively, stimulate or inhibit the release of prolactin from the pituitary gland. The prolactin releasing hormone (or factor) is abbreviated PRH (PRF); the prolactin release-inhibiting hormone (or factor) is abbreviated as PIH (PIF) or PRIH (PRIF).

prolactin release-inhibiting hormone *See* prolactin regulatory hormone.

prolactin releasing hormone *See* prolactin regulatory hormone.

prolactoliberin PROLACTIN RELEASING HORMONE.

prolactostatin PROLACTIN RELEASE-INHIBITING HORMONE.

prolamin A simple, globular protein of plant origin that is insoluble in water but is soluble in 50 to 90% ethanol solutions. *Var sp* prolamine.

prolate ellipsoid of revolution An ellipsoid of revolution formed by rotation of an ellipse about its major axis.

prolidase An exopeptidase that is specific for N-terminal proline or hydroxyproline.

proline A heterocyclic, nonpolar alpha imino acid. Proline is a helix-breaking amino acid. *Abbr* Pro;P.

prolonged test A toxicity test that is performed on laboratory animals and that requires the administration of a chemical at least once daily for periods of about 1 to 3 months.

promiscuous DNA Sections of DNA that have been transferred as a result of transposition which took place during the early stages of evolution; the transfer can have occurred between organelles, such as mitochondria and chloroplasts, or between an organelle and a nucleus.

promitochondrion An abnormal mitochondrion in which all of the mitochondrially encoded gene products are missing. Such mitochondria are found in some types of petite mutants of yeast; they have a normal outer membrane but the inner membrane contains poorly developed cristae. Promitochondria are nonfunctional in oxidative phosphorylation but contain proteins specified by nuclear genes such as DNA and RNA polymerases, enzymes of the citric acid cycle, and inner membrane proteins. The term is also used to describe other degenerate or precursor mitochondria from which functional mitochondria are believed to develop.

promoter 1. The site on the DNA molecule to which RNA polymerase attaches and at which transcription is initiated. In an operon, the promoter is usually located adjacent to the operator and upstream from it (farther away from the point at which transcription begins) but other arrangements also occur. 2. A carcinogenic agent that brings about the second stage (promotion stage) in a two-stage or multistage mechanism of carcinogenesis; a cocarcinogen. A substance that, when applied repeatedly or continuously, can induce tumors in animals previously exposed to a tumor initiator. A promoter, by itself, is neither mutagenic nor carcinogenic. Several factors in the diet, such as pickled and salted foods and refined sugars, are suspected of being promoters. The dietary factor most clearly established as a promoter is dietary fat. For example, high-fat diets rich in linoleic acid (found in corn, safflower, and sunflower oils) act as promoters but similar diets rich in oleic acid (found in olive oil) do not act as promoters. *Var sp* promotor. *Aka* tumor promoter. *See also* antipromoter.

promoter-down mutation   *See* promoter mutation.

promoter gene   PROMOTER (1).

promoter mutation   One of three types of mutations involving the promoter region of DNA: (a) a mutation that decreases the efficiency of the initiation of transcription or inactivates a promoter and prevents mRNA synthesis (promoter-down mutation); (b) a mutation that increases the efficiency of the initiation of transcription or increases the binding of RNA polymerase to the promoter (promoter-up mutation); (c) a mutation that creates a new promoter sequence where one did not exist before.

promoter strength   The relative rate of synthesis of the full length RNA product from a given promoter; can also be expressed in terms of transcription initiation frequency (for example, 10 strands/min, once/generation).

promoter-up mutation   *See* promoter mutation.

promotion   The second stage in a two-stage or multistage mechanism of carcinogenesis, and the stage during which a precancerous cell is converted to a dependent cancer cell through the action of a promoter. Promotion causes transformed cells to proliferate and form a tumor. The promoting agent must be applied continuously; if it is removed, its effects are reversible. *See also* promoter (2).

promotor   Variant spelling of promoter.

pronase   A nonspecific proteolytic enzyme isolated from *Streptomyces griseus*.

pronucleus   The haploid nucleus of a gamete.

proofreading   Any mechanism that allows for the correction of errors in the process of replication, transcription, or translation; involves removal of an incorrectly incorporated unit (nucleotide or amino acid) in the growing polymer and its replacement by the correct

unit. *Aka* editing. *See also* proofreading function; kinetic proofreading; ribosome editing; double sieve mechanism.

proofreading function   The $3' \to 5'$ exonuclease activity of DNA polymerase. The catalytic site for this activity is believed to be separate from, but close to, the polymerization center on the enzyme molecule. The $3' \to 5'$ exonuclease activity removes a mismatched nucleotide from the $3'$-end of the growing strand; it thus moves in an opposite direction to that of the polymerase activity. It functions to ensure fidelity (lack of errors) in DNA replication and is found associated with DNA polymerases I, II and III. *Aka* editing function.

proopiocortin   PROOPIOMELANOCORTIN.

proopiomelanocortin   A large precursor molecule that is differentially processed in a tissue-specific manner to yield multiple active peptides such as corticotropin, β-lipotropin, and neuropeptides. *Abbr* POMC.

prooxidant   A substance that accelerates an autoxidation reaction.

propagation   ELONGATION.

propagation of errors   The carryover of an error from one experiment, measurement, or observation to another; the cumulative effect of an error is frequently much larger than the initial error.

propagule   Any disseminative or reproductive particle; gametes, spores, or mycelial fragments are examples.

propeptides   1. Large peptides, of the order of 100 amino acids, that occur at the N- and C-terminals of the precursor form of the tropocollagen molecule. 2. The peptides subsequently removed from a proprotein in its conversion to an active protein.

properdin   A normal serum globulin that plays a key role in the activation of complement (complement fixation) via the alternative pathway. A number of other factors, including magnesium ions, are required for this process. *See also* complement.

properdin pathway   The alternative pathway of complement fixation. *See* complement.

prophage   The stable, inherited, noninfectious, provirus form of a temperate phage in which the phage DNA has become incorporated into, and replicates with, the host bacterial DNA. The term also refers to the plasmid phage DNA for those cases in which the phage DNA does not become integrated into the host DNA.

prophage attachment sites   1. The base sequences in the bacterial chromosome at which the phage DNA can become integrated to form a prophage. 2. The base sequences in the phage DNA that attach to the bacterial chromosome during phage DNA integration. 3. The two base sequences flanking the integrated prophage in the bacterial chromosome.

prophage excision *See* excision (2).

prophage induction *See* induction (2).

prophage integration *See* integration (c).

prophage map The genetic map of a phage as determined from recombination studies between prophages.

prophage-mediated conversion *See* conversion.

prophase The first stage in mitosis during which the nuclear membrane breaks down.

propinquity effect PROXIMITY EFFECT.

propionate rule A rule that accounts for the number of methyl side chains in the aglycone portion of macrolide antibiotics which are produced by bacteria of the genus *Streptomyces*. The rule states that propionate may replace acetate in the building of the carbon skeleton of the antibiotic, and that one methyl side chain occurs every time that a propionate unit is incorporated.

propionibacteria *See* propionic fermentation.

propionic acidemia A genetically inherited metabolic defect in humans, characterized by massive ketosis, that is due to a deficiency of the enzyme propionyl CoA carboxylase.

propionic fermentation The fermentation of glucose, and generally also of lactic acid, that yields propionic acid and other products and that is characteristic of propionic acid bacteria (propionibacteria).

proplastid A colorless, and largely structureless, precursor of a plastid that reproduces by division.

proportional counter A radiation counter designed for operation in the proportional region.

proportional region That portion of the characteristic curve of an ionization chamber in which, during gas amplification, the chamber yields a charge that is proportional to the initial charge produced by the radiation.

proprotein An inactive precursor of a protein that is converted to the active form by removal of a peptide fragment; procarboxypeptidase and proinsulin are two examples. *See also* preproprotein.

prosome A ribonucleoprotein associated with repressed messenger ribonucleoproteins (mRNP) and believed to be involved in post-transcriptional cytoplasmic repression of mRNA translation.

ProSPCA Precursor of serum prothrombin conversion accelerator.

prostacyclin A compound derived from arachidonic acid and related to the prostaglandins. It contains a second 5-membered ring and is an inhibitor of platelet aggregation and a powerful vasodilator. These effects are oppo-

site those produced by thromboxanes. *Sym* PGI$_2$. *Aka* prostaglandin I$_2$.

prostaglandins A group of biologically active lipids that are derived from arachidonic acid and that are named by reference to a hypothetical compound, called prostanoic acid. Prostaglandins were first found in the prostate gland but are now known to occur in most, if not all, mammalian tissues. Prostaglandin effects appear to be hormonal in nature and include lowering of blood pressure, stimulating smooth muscle contraction, and the regulation of inflammatory reactions, blood coagulation, and the immune response. Prostaglandins are divided into groups, based on the structure of the substituted 5-membered ring:

prostanoic acid A hypothetical, 20-carbon fatty acid, that contains a 5-membered ring and that is considered to be the parent compound for naming the prostaglandins, prostacyclins, and thromboxanes.

prostanoids Collective term for the compounds derived from the hypothetical fatty acid, prostanoic acid; includes prostaglandins, prostacyclins, and thromboxanes. *See also* eicosanoids.

prosthetic group The cofactor of an enzyme or the nonprotein portion of a conjugated protein that is bound so tightly to either the enzyme or the protein that it cannot be removed by dialysis.

protamine A small, simple, and globular protein that is virtually devoid of sulfur but contains large amounts of arginine. Protamines are basic proteins that have a molecular weight of about 5000 and that are found in association with nucleic acids, primarily in sperm cells of fish.

protamine zinc insulin The zinc salt of a protamine–insulin complex which is less soluble than insulin. *Abbr* PZI. *See also* NPH insulin.

protean An insoluble, primary derived protein that is obtained by treatment of a protein with heat, acid, enzymes, or other agents.

protease PROTEOLYTIC ENZYME.

protease nexin One of a group of proteins, secreted by cultured normal human fibroblasts, that selectively form covalent linkages with certain serine proteases and thus modulate their activities. *Abbr* PN.

protecting group A chemical group that is reacted with, and bound to, a functional group in a molecule to prevent the functional group from participating in subsequent reactions of the molecule.

protective antigen An antigen that is derived from a pathogenic microorganism and that, when injected into an animal, will produce an immune response that will provide protection for the animal against infection by that microorganism.

protective colloid A colloid that is added to a food to prevent the separation of components in that food.

protective immunity Immunity that is produced in an organism to protect the latter against possible exposure to a pathogen or other harmful agent.

proteid Obsolete term for either protein or conjugated protein.

protein A high molecular weight polypeptide of L-amino acids that is synthesized by living cells. Proteins are biopolymers with a wide range of molecular weights, structural complexity, and functional properties. Proteins are variously classified on the basis of their (a) solubility (albumins, globulins, scleroproteins, etc.); (b) function (transport proteins, storage proteins, contractile proteins, enzymes, hormones, antibodies, etc.); (c) shape (globular proteins and fibrous proteins); (d) composition (simple proteins, conjugated proteins, and derived proteins).

protein A A cell wall protein of some strains of *Staphylococcus aureus* that combines with most human immunoglobulin molecules of the IgG type.

proteinaceous Consisting in part, or entirely, of protein.

proteinase 1. PROTEOLYTIC ENZYME. 2. A protease that shows specificity for intact (native) proteins.

protein biosynthesis *See* protein synthesis.

protein blotting A method for identifying and characterizing proteins in complex mixtures. Involves separating the protein mixture into its components by some gel electrophoretic technique, most commonly by SDS–PAGE. After electrophoresis, the proteins are eluted from the gel by a second electrophoresis, diffusion, or convection, and are adsorbed onto an immobilized matrix (nitrocellulose membrane filters, nylon-based membranes, diazotized paper, etc.) such that the original electrophoretic separation pattern is maintained. The immobilizing matrix, or blot, is reacted with an appropriate probe (antibody, lectin, etc.) so that the protein of interest can be detected.

protein-bound iodine The iodine in the blood that is conjugated to protein and that is a measure of the concentration of circulating thyroid hormone. *Abbr* PBI.

protein C FACTOR XIV. *See also* C-protein.

protein–calorie malnutrition The combined deficiency of both calories and proteins as it occurs during famine; a combination of the conditions of marasmus and kwashiorkor.

protein coat The protein shell that surrounds the nucleic acid of a virus. *See also* capsid.

protein conformation *See* chain conformation; primary structure; secondary structure; tertiary structure; quaternary structure; super secondary structure; domain.

protein domain *See* domain.

protein efficiency ratio A measure of the nutritive value of a protein defined as the gain in weight (in grams) per gram of protein consumed; eggs are considered to have a maximum protein efficiency ratio of about 4.4.

protein efficiency ratio method A method for determining the nutritive value of a protein by measuring the gain in weight of young rats that are fed a diet containing 10% of the particular protein. *Abbr* PER method.

protein engineering The design and construction of new proteins or enzymes, which have novel properties, by the methods of recombinant DNA technology.

protein error The change in the relative amounts of the undissociated and dissociated forms of an indicator that is brought about by the binding of one of these forms to a protein. The change in the relative amounts of indicator forms leads to a change in color; such a color change forms the basis of the albustix test.

protein evolution The molecular evolution of proteins. *See also* chemical evolution.

protein export The transport of a protein out of a cell; the secretion of an extracellular protein.

protein factor The factor 6.25 that, when multiplied by the weight of nitrogen (in grams) derived from a sample containing protein, gives the approximate weight (in grams) of the protein in the sample.

protein folding The processes involved in the conversion of an ensemble of newly synthesized (or denatured) polypeptide chain conformations to the unique, three-dimensional conformation of the native protein.

protein fractionation The separation of a mixture of different proteins for the purpose of isolating one particular type of protein; requires the use of one or more physical–chemical techniques such as precipitation, chromatography, centrifugation, or electrophoresis.

protein-free filtrate   A liquid, such as plasma or serum, from which protein has been removed by precipitation and filtration.

protein import   The transport of a protein within a cell; the movement of a protein, assembled on the ribosomes in the cytoplasm, to some other intracellular compartment.

protein index   The polarographic wave height, expressed in terms of current density, for a plasma filtrate divided by that for a plasma digest. The plasma filtrate is obtained by treating plasma with sulfosalicylic acid; the plasma digest is obtained by treating plasma with potassium hydroxide.

protein kinase   An enzyme that catalyzes the phosphorylation of a protein; a kinase acting on a protein.

protein machine   1. MULTIENZYME SYSTEM. 2. METABOLON (2).

protein modification   See processing (2).

proteinoid   A protein-like polymer formed by thermal polymerization of amino acids in the dry state.

proteinosis   The accumulation, in a tissue, of excessive amounts of normal proteins or of abnormal proteins.

protein overloading   IMMUNOLOGICAL UNRESPONSIVENESS (2).

proteinpolysaccharide   MUCOPROTEIN.

protein processing   See processing (2).

protein release factor   See release factor.

protein score   See chemical score.

protein sequencer   See sequenator.

protein sequencing   See overlap method; sequenator.

protein-sparing action   The decrease in protein catabolism that is produced by the intake of dietary carbohydrates or lipids.

protein structure   See chain conformation; primary structure; secondary structure; tertiary structure; quaternary structure; super secondary structure; domain.

protein synthesis   The process whereby proteins are synthesized on ribosomes according to the genetic information contained within mRNA. It is synonymous with protein biosynthesis and includes amino acid activation and the three stages of translation: chain initiation, chain elongation, and chain termination. The amino acids are brought to the ribosome in the form of aminoacyl-tRNA molecules and are then polymerized on the ribosome. The ribosome moves along the mRNA molecule and the amino acids are polymerized in the order dictated by the codons in the mRNA. See also translation; amino acid activation; initiation; elongation; termination.

protein synthesis factor   See initiation factor; elongation factor; termination factor.

protein-synthesizing system   CELL-FREE AMINO ACID INCORPORATING SYSTEM.

protein turnover   See turnover (2).

proteinuria   The presence of protein in the urine.

protein value   BIOLOGICAL VALUE.

proteo-   Combining form meaning protein.

proteoglycan   A high molecular weight substance that contains large amounts (95% or more) of heteropolysaccharide side chains linked covalently to a polypeptide chain backbone. Proteoglycans are polyanionic compounds that have properties that resemble those of polysaccharides more than those of proteins. They form the ground substance in the extracellular matrix of connective tissue and serve as lubricants and support elements. The carbohydrate portion of the proteoglycans was formerly called mucopolysaccharide but is now referred to as glycosaminoglycan. See also glycoprotein.

proteoglycan aggregates   The proteoglycan fraction extracted from cartilage. It can be fractionated into hyaluronic acid, disaggregated proteoglycans (proteoglycan subunits), and low molecular weight proteins, known as link proteins. The latter serve to connect the proteoglycan subunits to a long molecule of hyaluronic acid. The polypeptide backbone of each proteoglycan subunit is known as core protein.

proteoglycan subunits   See proteoglycan aggregates.

proteohormone   A hormone that is a protein.

proteolipid   1. A conjugated protein that contains a lipid component and that is soluble in some nonpolar solvents but is insoluble in aqueous solutions. See also lipoprotein. 2. A hydrophobic protein that may or may not contain a lipid component but that is soluble in some nonpolar solvents. Such proteins have a high content of hydrophobic amino acids, many of which are clustered at the surface of the protein. Some integral membrane proteins are proteolipids and interact strongly with the hydrocarbon core of the membrane bilayer.

proteoliposome   An artificial organelle, constructed by combining phospholipids with specialized proteins and enzymes to form a functional vesicle.

proteolysis   The hydrolysis of proteins, particularly that due to the action of proteolytic enzymes.

proteolytic   Of, or pertaining to, proteolysis.

proteolytic coefficient   A measure of peptidase activity that is equal to the unimolecular rate constant of the reaction catalyzed by a peptidase, divided by the peptidase concentration.

proteolytic enzyme   An enzyme that catalyzes the hydrolysis of peptide bonds.

proteolytic quotient   The ratio of two proteolytic coefficients that are determined with one enzyme and two different substrates.

proteose A partially hydrolyzed protein that is water soluble and precipitable with ammonium sulfate; an intermediate form between a protein and a peptone.

proteosomes Hydrophobic, membranous, multimolecular preparations of meningococcal outer membrane proteins that are also B cell mitogens and that are believed to serve as carrier proteins and as adjuvants to enhance peptide immunogenicity.

Proterozoic era The more recent of the two subdivisions of the Precambrian era that extended over about one billion years and ended about 600 million years ago; an era during which primitive invertebrates and algae evolved.

prothoracicotropic hormone A polypeptide hormone, produced by neurosecretory cells in the brain of insects, that stimulates the synthesis and secretion of ecdysone by the prothoracic gland. *Abbr* PTTH. *Aka* brain hormone.

prothrombin The inactive precursor of thrombin that is converted to thrombin by the action of accelerin.

prothrombin derivatives theory A theory of blood clotting according to which different factors, similar to those of the cascade mechanism, participate in the clotting reactions but are not present as such in the blood; the factors are newly made molecules, derived from prothrombin during the process of clot formation.

prothrombin factor Vitamin K_1.

prothrombokinase FACTOR X.

proti-, proto Proposed prefixes for the ^1H isotope. Proposed prefixes for the mixture of hydrogen isotopes that occur in nature are hydri- and hydro-. Thus, the term proton would mean the $^1H^+$ exclusively and not the H^+ in natural abundance; the latter would be called hydron. Likewise, the transfer of H^+ to a substrate would be called a hydronation reaction and not a protonation reaction.

proticity The flow of protons from a region of high protic potential (high $[H^+]$)to one of low protic potential (low $[H^+]$).

protic solvent 1. PROTOPHILIC SOLVENT. 2. PROTOGENIC SOLVENT.

protist 1. A unicellular or multicellular organism that lacks the tissue differentiation and the elaborate organization that is characteristic of plants and animals; some protists are plant-like, some are animal-like, and some have properties common to both kingdoms. The taxon protist includes algae, fungi, and protozoa (all of which are eukaryotic protists) and bacteria, blue-green algae, and prochlorophyta (all of which are prokaryotic protists).

2. Any unicellular organism.

protium The ordinary isotope of hydrogen that contains one proton and no neutrons in the nucleus. *Sym* H.

protoalkaloids The biogenic amines, so called because they can be precursors of alkaloids.

protobiochemistry The developments in the science of biochemistry that preceded the foundation of modern chemistry by Lavoisier and Dalton.

protobiont A primitive forerunner of a living organism.

protocell A primitive forerunner of a living cell.

protocollagen An experimentally produced form of procollagen that is deficient in hydroxyproline and hydroxylysine.

protoenzyme A primitive forerunner of an enzyme.

protofibril 1. MYOFILAMENT. 2. A small bundle of fibrous proteins such as a two-stranded fibrin molecule or a three-stranded keratin molecule.

protofilament *See* microtubules.

protogen LIPOIC ACID.

protogene A gene for a primitive protein such as the gene for a protoenzyme.

protogenic solvent An aqueous or nonaqueous, acidic solvent that has the capacity of donating protons to a solute; a hydroxylic solvent such as water, ethanol, or acetic acid.

protoheme HEME (2).

protolysis A reaction in which there is a transfer of a proton from an acid to a base; an acid–base reaction; the Bronsted concept of neutralization.

protolyte 1. ACID. 2. BASE.

protolytic Of, or pertaining to, protolysis.

protomer 1. The individual polypeptide chain in an oligomeric protein, such as the alpha or beta chain in hemoglobin. 2. One of the identical subunits, or monomers, of an allosteric enzyme, each of which has a catalytic site. 3. The basic building block of the capsomere of a virus. *See also* subunit; monomer.

proton An elementary particle of the atomic nucleus that has a charge of +1 and a mass of 1.0073 amu; identical to the nucleus of the hydrogen atom. *Sym* p.

proton abstraction The removal of a proton from a compound.

proton affinity A mass spetrometric term; the proton affinity of a molecule M is defined as the negative of the enthalpy change for the protonation reaction $M + H^+ \rightleftharpoons MH^+$. *Abbr* PA.

protonate To add protons to a group of atoms or to a compound, as in the conversion of an —NH_2 group to an —NH_3^+ group.

proton gradient The change in hydrogen ion

concentration with distance, particularly that across a biological membrane. The formation of an energy-rich proton gradient across the mitochondrial membrane constitutes a postulate of the chemiosmotic coupling hypothesis.

protonic acid BRONSTED ACID.

proton magnetic reasonance *See* nuclear magnetic resonance.

proton motive force The total electrochemical potential arising from an energy-rich gradient of protons; equal to the sum of the free energy changes due to the proton concentration gradient and due to the difference in electrical potential. *Abbr* PMF.

proton-motive hypothesis CHEMIOSMOTIC COUPLING HYPOTHESIS.

proton noise-decoupled mode A method of operating ^{13}C nuclear magnetic resonance instruments such that a single sharp resonance line is obtained for each unique (nonequivalent) kind of carbon atom present in the molecule being examined.

protonometry The determination of changes in proton concentration by measurements of the changes in the potential of a glass electrode.

protonophore A lipid-soluble compound that conducts protons across a membrane; a compound functioning like an ionophore for protons.

proton pump The structure and/or the mechanism that mediates the active transport of hydrogen ions across the inner mitochondrial membrane according to the chemiosmotic coupling hypothesis of oxidative phosphorylation.

proton pumping ATPase F_0F_1-ATPase.

proton transfer potential The free energy change per mole for the reaction $HA + H_2O \rightleftharpoons A^- + H_3O^+$.

proton translocating ATP synthase F_0F_1-ATPase.

proton translocation TRANSPROTONATION.

protooncogene A cellular gene that is a homologue of a retroviral oncogene and that controls the normal proliferation, and possibly also the differentiation, of cells. Protooncogenes can be converted to oncogenes (become activated) as a result of mutation or recombination with a viral genome. *See also* oncogene.

protoorganism The most recent common ancestor of all living things.

protopectin An insoluble form of pectin that occurs as the ground substance in the cell wall of plants; consists of pectin chains, linked covalently and noncovalently.

protophilic solvent A nonaqueous, basic solvent that has the capacity of accepting protons from a solute.

protophytolysosome A vesicle, derived from

the Golgi apparatus, that is rich in acid phosphatase and that is believed to function in the transport of acid phosphatase to intracellular sites.

protoplanet A body of matter considered to have been a forerunner of the planets.

protoplasm The living matter that forms the basis of animal, plant, and microbial cells; the substance of the cell that is surrounded by the cell membrane; the cytoplasm and the nucleus or the nuclear zone.

protoplast A bacterial cell that has been freed entirely of its cell wall. Protoplasts are prepared artificially by lysozyme digestion of gram-positive bacteria; they can survive only in hypertonic media and generally cannot multiply. *See also* spheroplast.

protoplast membrane CELL MEMBRANE.

protoporphyria A genetically inherited metabolic defect in humans that is due to a deficiency of heme synthetase (ferrochelatase).

protoporphyrin The biochemically most important porphyrin derivative that occurs in hemoglobin in the form of protoporphyrin IX. *Abbr* PP.

prototroph 1. An organism or a cell that is capable of synthesizing all of its metabolites from inorganic compounds and a carbon source; it can grow on a minimal medium. 2. A microorganism that has no nutritional requirements over and above those of the wild-type strain from which it is derived.

prototropic group A group capable of losing a proton; an enol group that tautomerizes to a keto group is an example.

prototropism TAUTOMERISM.

prototropy TAUTOMERISM.

protozoan (*pl* protozoa) A unicellular animal organism; a nonphotosynthetic, eukaryotic protist.

protransglutaminase FACTOR XIII.

provirus 1. A virus that is integrated into the chromosome of a host cell and is transmitted in that form from one host cell generation to another without leading to the lysis of the host cells. *See also* prophage. 2. A double-stranded DNA segment in a eukaryotic chromosome, equivalent to the genome of an oncogenic RNA virus, that is replicated with the cellular DNA and transmitted from one generation to another without causing cell lysis.

provirus hypothesis A hypothesis of cancer according to which infection of a eukaryotic cell with an oncogenic RNA virus results in the transcription of the viral RNA by reverse transcriptase to form a provirus. The provirus then becomes integrated into the host DNA and is replicated with it. The cell is thereby transformed and contains the information necessary for synthesis of new viral particles and for maintenance of the transformed state.

See also oncogene theory.

provitamin A naturally occurring precursor of a vitamin that is transformed to the vitamin in the animal body; beta carotene of plants, for example, is a provitamin of vitamin A. *See also* previtamin.

provitamin A BETA CAROTENE.

provitamin A carotenoid A generic descriptor for all carotenoids that exhibit qualitatively the biological activity of beta carotene.

provitamin D A sterol such as ergosterol that is converted to vitamin D upon irradiation with ultraviolet light.

Prower factor STUART FACTOR.

proximal Close to a particular location or to a point of attachment.

proximal carcinogen That form of a carcinogen in which it participates in the first chemical reaction of a multistep induction mechanism of cancer.

proximate analysis Quantitative analysis, rationally interpreted. Thus, the proximate analysis of a living tissue may consist of determining water by the weight lost upon drying at 100–105°C, determining protein by multiplying the amount of Kjeldahl nitrogen by the conversion factor of 6.25, determining lipids by the weight lost upon extracting the dry matter with nonpolar solvents, and so on.

proximity effect The contribution to the catalytic activity of an enzyme that results from the reactants being brought closer together on the surface of the enzyme which leads to a great increase in the effective concentrations of the reactants. *See also* orientation effect.

prozone 1. The concentration range in some agglutination reactions in which an undiluted, or slightly diluted, cell suspension fails to lead to agglutination, while a more dilute cell suspension produces a normal agglutination reaction. 2. AUTOINTERFERENCE.

PrP Prion protein; the scrapie agent.

PRPP 5-Phosphoribosyl-1-pyrophosphate.

PS 1. Peptide synthetase. 2. Phosphatidylserine.

pseudoacrosome A transient structure, formed in spermatozoa, that resembles the acrosome but disappears during development.

pseudoalleles Closely linked genes that behave in the complementation test as if they were alleles but that can separated by crossing over.

pseudocholinesterase CHOLINESTERASE.

pseudocyclic photophosphorylation A process that is similar to cyclic photophosphorylation but is dependent on the presence of oxygen; it leads to the reduction of an added hydrogen acceptor and the reoxidation of this acceptor by oxygen.

pseudofeedback inhibition Feedback inhibition that is caused by the analogue of a metabolite which is produced by, or participates in, a biosynthetic reaction. The feedback inhibition produced by a nucleotide that is formed by the incorporation of a base analogue is an example.

pseudo-first-order kinetics The kinetics of a chemical reaction in which more than one reactant is involved but which behaves kinetically as if only one reactant were present. Thus, in enzyme studies, when the [substrate]/[enzyme] is large (≥ 100), one can assume that the reaction is a pseudo-first-order reaction so that the velocity is directly proportional to the substrate concentration.

pseudogene A sequence in DNA that is nearly homologous to that of a functional gene but is unable to produce a functional product. A pseudogene is considered to be derived from a once functional gene by one or more mutations which resulted in the inactivation of this gene. *See also* processed gene.

pseudoglobulin A globulin that is sparingly soluble in water.

pseudohemoglobin An artificially prepared hemoglobin-like molecule in which the globin molecules are combined with iron porphyrins other than protoporphyrin IX. *See also* hybridization (2).

pseudoisoenzyme One of two or more different forms of an enzyme that catalyze the same reaction but are not true isoenzymes since they do not have genetically determined different primary structures. Pseudoisoenzymes consist of in vivo or in vitro modified enzymes, or enzymes in varying stages of aggregation, and are identical to isozyme (1).

pseudomessenger RNA *See* premessenger RNA.

Pseudomonas A genus of gram-negative bacteria that are widespread in soil and water; some species are capable of nitrate respiration.

pseudonucleoprotein Obsolete designation for a phosphorus-containing protein that is not a nucleoprotein.

pseudophotoaffinity labeling *See* photoaffinity labeling.

pseudoreversion The restoration of biological activity, lost as a result of a mutation, by means of a second mutation of the mutated base which leads to the formation of a codon that differs from the normal codon of the wild-type organism but constitutes an acceptable missense codon.

pseudo-U loop *See* arm.

pseudounimolecular reaction A bimolecular reaction in which A reacts with B but B is present in great excess so that its concentration effectively does not change during the reaction; as a result, the velocity of the reaction appears to be proportional only to the concentration of A.

pseudouridine An unusual nucleoside, occurring in tRNA, in which the normal base uracil is linked to ribose through carbon atom 5 of the uracil. *Sym* ψ; ψU; ψrd.

pseudouridylic acid The nucleotide formed from pseudouridine.

pseudoverification The deacylation of a correctly acylated, but noncognate, transfer RNA molecule by an aminoacyl-tRNA synthetase; can be brought about under special conditions in a mixed solvent.

pseudovirion An artificially prepared virus in which the DNA is derived from one virus and the protein coat from another virus.

pseudovitamin A compound, such as lipoic acid, that resembles a vitamin in its coenzyme function but does not constitute a dietary requirement.

psi The Greek letter ψ; *See* pseudouridine.

psi angle (ψ) The torsion angle that denotes the rotation about the C^α—C^1 bond of the peptide backbone in proteins.

psicofuranine A nucleoside antibiotic, produced by *Streptomyces hygroscopicus*, that has antitumor activity; it inhibits the last step in guanosine monophosphate (GMP) biosynthesis.

psi factor A protein believed to be responsible for the initiation of rRNA synthesis in bacteria. *Sym* ψ.

psilocin *See* psilocybin.

psilocybin A naturally occurring hallucinogenic drug that is an indolealkylamine; 4-phosphoryloxy-*N*,*N*-dimethyltryptamine. Psilocybin, and the related compound psilocin (4-hydroxy-*N*,*N*-dimethyltryptamine) occur in the fruiting body of the Mexican hallucinogenic fungus teonanacatl (*Psilocybe mexicana*). Both compounds are naturally occurring hallucinogenic drugs.

P site 1. PEPTIDYL SITE. 2. A cell surface receptor site on rat mast cells for adenosine and related compounds. The site is involved in the immunological release of adenyl cyclase and has an obligatory requirement for an intact purine ring. *See also* R site.

psoralen One of a group of photosensitive reagents that can cross-link base-paired regions in DNA.

PSP test Phenolsulfonphthalein test.

PSTV Potato spindle tuber viroid.

^{32}P suicide *See* suicide.

psychedelic drug HALLUCINOGENIC DRUG.

psycholytic drug HALLUCINOGENIC DRUG.

psychosine A generic descriptor for 1-monoglycosylsphingoids; a glycosphingolipid containing one monosaccharide unit.

psychotomimetic drug HALLUCINOGENIC DRUG.

psychotropic agents Compounds that affect the human psyche; narcotic drugs and hallucinogenic drugs.

psychrophile An organism that grows at low temperatures in the range of 0 to 25°C, and that has an optimum growth temperature in the range of 20 to 25°C.

psychrophilic Of, or pertaining to, psychrophiles; preferring low temperatures.

PTA Plasma thromboplastin antecedent.

PTC 1. Phenylthiocarbamyl group. 2. Plasma thromboplastin component.

PTC-amino acid Phenylthiocarbamyl amino acid.

pteridine A nitrogen-containing compound that consists of two fused, six-membered rings, and that is a structural component of biopterin, folic acid, and riboflavin.

pterin One of a group of widely distributed derivatives of the parent compound, 2-amino-4-hydroxypteridine which is structurally similar to guanine.

pteroic acid A structural component of folic acid that consists of a pterin attached to *p*-aminobenzoic acid.

pteroylglutamic acid FOLIC ACID.

PTH 1. Parathyroid hormone. 2. Phenylthiohydantoin.

PTH-amino acid Phenylthiohydantoin amino acid.

ptomaine A group of toxic substances, specifically amines, that are formed by microbial decomposition of proteins.

PTS Phosphotransferase system.

PTTH Prothoracicotropic hormone.

ptyalin Salivary alpha amylase.

Pu Purine.

PU Pregnancy urine; the urine of pregnant individuals that is used as a source of hormones.

public antigen A blood group antigen that occurs in a great number of individuals.

puckered conformation The conformation of ribose or deoxyribose in which 4 of the 5 atoms of the furanose ring lie in a plane, while the fifth atom (C-2' or C-3') protrudes out of the plane. The sugar ring conformation is denoted as endo if the protruding atom lies above the ring (on the same side as C-5') and is denoted as exo if the protruding atom lies below the ring (on the opposite side of C-5').

PUFA Polyunsaturated fatty acids.

puff *See* chromosome puff.

pulmonary Of, or pertaining to, the lungs.

pulsatile flow A flow that occurs in pulses rather than at a continuous pressure; sometimes used for perfusion of organs to more closely resemble the type of flow produced by the heart.

pulsating ribosome A model for the ribosome during protein synthesis according to which the two ribosomal subunits move apart and come together repeatedly as the amino acids are being polymerized.

pulse 1. A brief exposure to a radioactive isotope, as that used to label mRNA in bacterial cultures. 2. The amount of radioactive isotope used for the labeling of a substance by means of a brief exposure to the isotope. 3. The energy of the discharge in an ionization chamber. 4. The electric current produced by a discharge in an ionization chamber or by scintillations in a scintillation counter.

pulse-chase experiment An experiment in which a system is exposed very briefly to a radioactive substance and then to a large amount of the same, but unlabeled substance. *See also* pulse (1,2); chase; pulse-label experiment.

pulsed-field gel electrophoresis A gel electrophoretic method used for the separation of DNA molecules which can vary in size from several kilobase pairs to at least 4000 kbp. Involves changing the direction of the electric field repeatedly. This forces the molecules to change their direction of migration and to choose new pores in the gel. As a result both large and small DNA molecules can be made to move through the gel as a function of their size; small molecules migrate faster than larger ones. The technique eliminates the problem of reptation. *Abbr* PFG; PFGE. *See also* field inversion gel electrophoresis.

pulse discriminator *See* discriminator.

pulsed laser interferometry An optical technique that can be used in conjunction with the analytical ultracentrifuge, and in which intermittent laser illumination is used to produce interference patterns. *Abbr* PLI.

pulse height The intensity of the electric current produced in a scintillation counter.

pulse-height analyzer A device, consisting of two discriminators, that accepts pulses that have intensities that lie between the setting of the two discriminators. *See also* differential counting.

pulse-height shift method CHANNELS RATIO METHOD.

pulse-labeled RNA An RNA that is labeled by means of a brief exposure to a radioactive isotope and that is considered to represent largely, if not entirely, messenger RNA that has a short half-life.

pulse-label experiment An experiment in which a system is exposed very briefly to a radioactive isotope and then subjected to analysis. *See also* pulse-chase experiment; pulse (1,2).

pump The structure and/or the mechanism that mediates the primary active transport of a given substance across a biological membrane. *See also* active transport.

punctuation The elements of the genetic code that serve as initiation and termination signals of the messages, and that do not code for amino acids.

Puo Purine nucleoside.

Pur Purine.

pure Containing no contaminating material. *See also* purification; purity.

pure culture 1. A culture containing microorganisms from only one species. 2. A culture derived from a single cell.

pure line A strain of organisms that is homozygous as a result of continued inbreeding.

purging The replacement of one type of a gaseous environment by another; the flushing out of one atmosphere by another.

purification 1. The process whereby a preparation is freed of certain types of molecules to yield a sample that is enriched in, or consists solely of, molecules of a single type; the process whereby a specific enzyme, a nucleic acid, etc. is being isolated. 2. The ratio of the specific activity at a given step in the isolation of a substance divided by the specific activity at a reference step; applicable to the isolation of enzymes and other macromolecules, the activity of which can be measured. *See also* purity.

purified diet SYNTHETIC DIET.

purified parathyroid extract PARATHORMONE.

purified protein derivative A protein fraction obtained by ammonium sulfate precipitation of a culture of the tubercle bacillus, *Mycobacterium tuberculosis*, that had been grown in a synthetic medium. *Abbr* PPD.

purine 1. A basic, heterocyclic, nitrogen-containing compound that occurs in nucleic acids; common purines are adenine and guanine. *Abbr* Pu; Pur. *Aka* base; nitrogenous base. 2. The parent compound of adenine, guanine, and related compounds.

purine alkaloids The N-methylated xanthines: caffeine, theobromine, and theophylline.

purine antibiotics Structurally modified purines that have antibiotic activity; includes nucleosides, peptide-linked purines, and free bases.

purine cycle GLYCINE–ALLANTOIN CYCLE.

purine nucleotide cycle The group of reaction whereby AMP is deaminated to IMP and the latter is reaminated to form AMP.

purinergic nerves Motor neurons, innervating visceral organs of vertebrates, that use adenosine triphosphate (ATP) as a neurotransmitter.

purine salvage *See* salvage metabolic pathway.

purity 1. The state of a preparation in which all the molecules are those of a single type, as in a preparation containing only palmitic acid molecules. 2. The state of a preparation in which all the macromolecules are those of a single type, as in a preparation containing buffer ions and small molecules, but containing ribonuclease molecules as the sole macromolecules. 3. The degree to which a preparation consists of macromolecules of a single type. *See also* purification.

puromycin An antibiotic, produced by *Streptomyces alboniger*, the structure of which resembles the terminal grouping of the amino acid joined to adenosine in an aminoacyl-tRNA molecule. Because of this structural similarity, puromycin acts as an analogue of aminoacyl transfer RNA and inhibits protein synthesis in both prokaryotes and eukaryotes by binding to the growing polypeptide chain and causing its premature release from the ribosome in the form of peptidyl puromycin. *Abbr* PM.

purple membrane A specialized section of the cell membrane of the halophile *Halobacterium halobium* that converts energy from visible light into stored energy by pumping protons across the membrane. The purple membrane consists of lipid and a single kind of protein, bacteriorhodopsin, which functions as a light-driven transmembrane proton pump. *See also* bacteriorhodopsin.

purple sulfur bacteria A family of photosynthetic bacteria that can oxidize elemental sulfur to sulfate.

putrefaction The formation of foul-smelling products by microbial decomposition of high-protein materials such as meat and eggs.

putrescine A low molecular weight polyamine (1,4-diaminobutane) that contains two amino groups; it is formed by decarboxylation of ornithine and is found predominantly in prokaryotes.

putrescine cycle A proposed set of reactions in animal cells that involves the synthesis of spermine and spermidine, their conversion to *N*-acetyl derivatives, and oxidation of the latter to putrescine or to spermidine and monoacetylpropionaldehyde.

p value Probability value.

p50 value A measure of the affinity of hemoglobin or myoglobin for oxygen; defined as the partial pressure of oxygen at which 50% of the sites on the protein are oxygenated.

PVC Polyvinylcarbonate; a plastic.

PVN Phosvitin.

Py Pyrimidine.

pycnometer A small glass vessel that has a definite volume and that is used for the weighing of different liquids to determine specific gravities and densities. *Var sp* pyknometer.

pycnosis The shrinkage and condensation of the cell nucleus into a compact, densely staining structure that occurs when the cell dies.

pycocins Bacteriocins produced by strains of *Pseudomonas aeruginosa*.

Pyd Pyrimidine nucleoside.

pyknometer Variant spelling of pycnometer.

pyknosis Variant spelling of pycnosis.

Pyr Pyrimidine.

pyran A heterocyclic compound, the structure of which resembles the ring structure of the pyranoses.

pyranose A monosaccharide having a six-membered ring structure.

pyranoside A glycoside of a pyranose.

pyrenoid A proteinaceous granule around which starch may accumulate in chromatophores.

pyrenoid starch Starch that accumulates around pyrenoids in chromatophores.

pyridine alkaloids A group of alkaloids containing the pyridine ring system; nicotiana alkaloids, such as nicotine, are an example. Pyridine alkaloids occur in plants and as metabolic products of microorganisms.

pyridine-linked dehydrogenase A dehydrogenase that requires a pyridine nucleotide, NAD^+ or $NADP^+$, as a coenzyme.

pyridine nucleotide Any one of the oxidized or reduced forms of nicotinamide adenine dinucleotide or nicotinamide adenine dinucleotide phosphate; NAD^+(NADH) or $NADP^+$(NADPH).

pyridine nucleotide coenzymes The compounds NAD^+ and $NADP^+$, and their reduced forms. *Aka* pyridine coenzymes.

pyridine nucleotide cycle A salvage pathway in which nicotinamide, derived from the catabolism of NAD^+, is reused for the biosynthesis of NAD^+.

pyridoxal The aldehyde form of pyridoxine; a form of vitamin B_6. *Abbr* PL.

pyridoxal phosphate The coenzyme form of vitamin B_6 that functions in the metabolism of amino acids, as in the transamination reaction. *Abbr* PLP; PAL; PALP.

pyridoxamine The amine form of pyridoxine; a form of vitamin B_6. *Abbr* PM.

pyridoxine The alcohol form of vitamin B_6. *Abbr* PN.

pyridoxol PYRIDOXINE.

pyrimidine 1. A basic, heterocyclic, nitrogen-containing compound that occurs in nucleic acids; common pyrimidines are cytosine, thymine, and uracil. *Abbr* Py; Pyr. *Aka* base; nitrogenous base. 2. The parent compound of cytosine, uracil, thymine, and related compounds.

pyrimidine analogue ANTIPYRIMIDINE.

pyrimidine antibiotics Structurally modified pyrimidines that have antibiotic activity; includes nucleosides, peptide-linked pyrimidines, and free bases.

pyrimidine dimer The dimer formed by the linking of two adjacent pyrimidines, such as two thymines, in a nucleic acid strand as a result of ultraviolet irradiation.

pyrimidineless mutant *See* -less mutant.

pyrocondensation The condensation of molecules that is brought about by heat.

pyrogen Any fever-inducing substance.

pyroglutamic acid An internal lactam of glutamic acid formed by condensation of the α-amino group with the γ-carboxyl group; occurs in thyrotropin releasing hormone and in relaxin. *Aka* 5-oxoproline.

pyroglutamic aciduria OXOPROLINURIA.

pyrolysis The transformation of one substance into another that is brought about by heat alone; includes such processes as thermal decomposition, isomerization, and synthesis.

pyronin Y A basic dye used in cytochemistry for the staining of RNA.

pyrophosphatase The enzyme that catalyzes the hydrolysis of inorganic pyrophosphate to two molecules of orthophosphate.

pyrophosphate *See* inorganic pyrophosphate.

pyrophosphate cleavage The hydrolytic removal of a pyrophosphate group from either a nucleoside disphosphate or a nucleoside triphosphate.

pyrophosphate exchange A reaction, catalyzed by DNA polymerase, in which a DNA strand is shortened by one nucleotide in the presence of inorganic pyrophosphate; the nucleotide is cleaved off as a nucleoside triphosphate. The reaction may be represented as shown below. It involves phosphodiester bond formation in one direction and pyrophosphorolysis in the other direction. A similar reaction can also occur with RNA.

$$(dNMP)_n + P^*P^*_i \rightleftharpoons (dNMP)_{n-1} + dNPP^*P^*$$
$$(DNA) \qquad\qquad (DNA) \qquad (P^*P^*PN)$$

pyrophosphorolysis The reaction catalyzed by pyrophosphorylase.

pyrophosphorylase 1. The enzyme that catalyzes the formation of a nucleoside-5'-diphosphate sugar and pyrophosphate from a sugar-1-phosphate and a nucleoside-5'-triphosphate 2. A nucleotidyl transferase.

pyrrole A five-membered, heterocyclic, nitrogen-containing building block of porphyrins.

pyrrolidine alkaloids *See* alkaloids.

pyrrolizidine alkaloids *See* alkaloids.

pyrrolo quinoline quinone *See* quinoprotein.

pyruvate carboxylase The enzyme that catalyzes the anaplerotic reaction whereby pyruvic acid is carboxylated to oxaloacetic acid.

pyruvate dehydrogenase complex A multienzyme system, consisting of a large number of three different enzymes and five different coenzymes, that catalyzes the conversion of pyruvic acid to acetyl coenzyme A. *Abbr* PDH. *Aka* pyruvate dehydrogenase system.

pyruvate kinase The enzyme that catalyzes the phosphorylation of ADP to ATP by phosphoenolypyruvate.

pyruvate oxidation factor LIPOIC ACID.

pyruvic acid The three-carbon ketoacid that is the end product of glycolysis under aerobic conditions.

PZI Protamine zinc insulin.

Q

Q 1. Ubiquinone. 2. Glutamine. 3. *Q* value. 4. Metabolic quotient. 5. Queuine. 6. Queuosine. 7. Coenzyme Q.

Q_{10} The ratio of the velocity of a reaction at a particular temperature to the velocity at a temperature that is lower by 10°C; the Q_{10} is approximately 2 for chemical reactions.

Qβ *See* Q-beta.

QAE-Sephadex Quaternary aminoethyl Sephadex; diethyl-(2-hydroxypropyl) aminoethyl Sephadex, an anion exchanger. The ion exchanger contains the grouping $—C_2H_4N^+(C_2H_5)_2CH_2CHOHCH_3$ linked via ether bonds to the Sephadex.

Q-beta A small, tailless, icosahedral phage that contains single-stranded RNA and infects *E. coli*. *Abbr.* Qβ.

Q-beta replicase An RNA-dependent RNA polymerase of phage Qβ; a viral-encoded enzyme that is highly specific for the viral RNA. *Abbr* Qβ replicase.

QCD Quantum chromodynamics.

Q_{CO_2} *See* metabolic quotient.

Q cycle Quinone cycle.

Q-enzyme The enzyme that catalyzes the formation of $\alpha(1 \rightarrow 6)$ glycosidic bonds (branch linkages) in amylopectin.

Q gas A common gas mixture (1.3% isobutane in helium), used with Geiger–Mueller counters.

QH_2 Reduced coenzyme Q; reduced ubiquinone.

Q notation A method used in the past to denote enzyme activity, especially that of respiratory enzymes. The Q_s value of an enzyme was taken to be the number of microliters, at standard temperature and pressure, of the substrate used up per hour per milligram of enzyme. For nongaseous substrates, 1 μmol of substrate was considered to be equivalent to 22.4 μL.

Q_{O_2} *See* metabolic quotient.

quadri- Combining form meaning four.

quadrupole mass spectrometer A nonmagnetic mass spectrometer which employs a combination of dc and radio frequency potentials as a mass filter; consists of four parallel rods, arranged symmetrically.

Quaking mutation An autosomal, recessive mutation in mice that produces a myelin-deficient animal.

quantasome The smallest structural unit of photosynthesis; a membrane-enclosed vesicle containing 230 chlorophyll molecules, cytochromes, copper, and iron. Quantasomes are particulate subunits of the thylakoid membrane in chloroplasts. *Var sp* quantosome.

quantify 1. QUANTITATE. 2. To transform a relation from a qualitative into a quantitative form.

quantiles A class of values of a statistical variable that divide the total frequency distribution of a population into a given number of equal proportions. Percentiles and deciles are two examples.

quantitate 1. To measure the quantity of an item. 2. To express a relation in quantitative terms.

quantized In the form of discrete units, or quanta; used in reference to energy.

quantosome Variant spelling of quantasome.

quantum (*pl* quanta) The unit amount of energy that is released during the emission of radiation and that is taken up during the absorption of radiation; equal to *hv*, where *h* is Planck's constant, and *v* is the frequency of the radiation in cycles per second.

quantum chromodynamics A modern field theory of the strong force of particle physics. *See also* elementary particles.

quantum efficiency QUANTUM YIELD.

quantum flavordynamics WEINBERG–SALAM THEORY.

quantum mechanics The description of atomic and molecular phenomena in terms of energy quanta and quantized energy states rather than in terms of classical Newtonian mechanics; achieved by Heisenberg by using matrices and linear operators, and achieved by Schrödinger by considering the wave aspects associated with subatomic particles. The Schrödinger approach was originally called wave mechanics, but the term is now used as a synonymous expression for quantum mechanics.

quantum numbers A set of four numbers that describe an electron in an atom. The principal quantum number is the number of the main energy level. The other three numbers are the angular momentum, magnetic, and spin quantum numbers.

quantum requirement The reciprocal of the quantum yield.

quantum theory The theory that energy can be radiated and absorbed only in discrete packets, called quanta, the energy of which is proportional to the frequency of the radiation.

quantum yield The number of molecules that react chemically in a photochemical reaction, divided by the number of photons absorbed; the number of moles that react chemically in a photochemical reaction, divided by the number of einsteins absorbed.

quark *See* elementary particles.

quartet A quadruple peak, such as a nuclear magnetic resonance peak that has split into four peaks.

quartile A set of quantiles; specifically, the quartiles Q_1, Q_2, and Q_3 are values at or below which lie, respectively, the lowest 25%, 50% and 75% of a set of data.

quartz A glassy silicon dioxide that is used for the production of cuvettes that are utilized in absorbance measurements of ultraviolet light.

quasar Quasi-stellar radio source; a compact radio source with a star-like optical object; a "radio star."

quat Quaternary ammonium compound; one of a group of cationic detergents used as antiseptics and disinfectants. The compound cetyltrimethylammonium bromide (CTAB, cetavlon) is an example.

quaternary ammonium compound *See* quat.

quaternary nitrogen A positively charged nitrogen atom that is linked to other atoms or groups by means of four covalent bonds.

quaternary structure The structure of a protein that results from the interaction between individual polypeptide chains to yield larger aggregates; the arrangement in space of the subunits of a protein and the intersubunit contacts and interactions without regard to the internal structure of the subunits.

que Queuine.

queen substance Originally a term for the entire mandibular gland secretion of the queen bee which contains about 30 different substances; now a trivial name for the compound 9-oxo-*trans*-2-decenoic acid which serves as a pheromone for maintaining the division of labor in the beehive.

Quellung reaction The precipitin reaction that occurs between polysaccharides of bacterial capsules and antibodies to these polysaccharides; it results in an apparent swelling of the capsule.

quench correction curve A plot of counting efficiency versus the ratio of counts in two channels; used to correct the observed counts in liquid scintillation for quenching.

quenching 1. The process whereby secondary and subsequent ionizations in an ionization detector are stopped so that the detector becomes again sensitive to new, incoming ionizing radiation. 2. A decrease in the counting efficiency in liquid scintillation. 3. The de-

crease in fluorescence that results from an absorption of some or all of the emission energy. *See also* fluorescence quenching.

quetelet index A measure of obesity defined as the weight in kilograms divided by the square of the height in meters.

queuine A modified guanine found in tRNA. It differs from other modified (minor, rare) bases in that it is synthesized first as a base and then incorporated into mature tRNA by an enzyme-catalyzed exchange reaction in which guanine is removed from the tRNA and replaced by queuine. *Abbr* Q.

queuosine The ribonucleoside of queuine. *Abbr* Q; Quo.

quick-stop mutant A mutant of *E. coli* that immediately stops replication when the temperature is raised to 42°C.

quinacrine An acridine dye derivative that is a fluorochrome and that is used in the treatment of malaria and cancer.

quinary structure The group of macromolecular interactions between proteins that are transient in vivo.

quinine An alkaloid drug used in the treatment of malaria; a cinchona alkaloid.

quinoline alkaloids A group of alkaloids that contain the quinoline structure. They include the cinchona alkaloids, derived from the bark of tropical trees, especially *Cinchona succiruba*. The main alkaloid of the bark of *Cinchona* is quinine, a drug used for the treatment of malaria.

quinolizidine alkaloids *See* alkaloids.

quinone *p*-Dioxybenzene or a derivative of *p*-dioxybenzene. A particular quinone (coenzyme Q) serves as an electron carrier in the electron transport system.

quinone cycle A postulated mechanism for proton translocation according to the chemiosmotic coupling hypothesis; involves respiratory complex III and the participation of a semiquinone radical. A cyclic set of reactions in which coenzyme Q undergoes a two-stage reduction with the semiquinone as a stable intermediate.

quinoprotein One of a group of dehydrogenases that have the complex organic compound pyrrolo quinoline quinone (PQQ; methoxatin; 4,5-dihydro-4,5-dioxo-1*H*-pyrrolo-[2,3*f*]chinoline-2,7,9-tricarboxylic acid) as a prosthetic group. The latter serves as a coenzyme for the dehydrogenases that occur in methylotrophic and other types of bacteria.

quo Queuosine.

Q value The total energy per atom that is released in a nuclear reaction in which a nuclide is transformed into another, and ground-state, nuclide.

R

r 1. Ribo-, as in ribothymidine monophosphate (rTMP). 2. Ribosomal, as in ribosomal RNA (rRNA). 3. Roentgen 4. Correlation coefficient. 5. A rapid lysis mutant.

R 1. R group. 2. Gas constant. 3. R configuration. 4. Purine nucleoside. 5. Arginine. 6. Bacterial colony of rough morphology. 7. Resistance of a bacterial strain to an inhibitor or a phage. 8. Relaxed conformational form of an allosteric enzyme. 9. Roentgen. 10. A drug-resistant plasmid. 11. *See* RS system.

Ra Radium.

rabbit aorta contracting substance A substance, originally detected in the effluent from the lungs of anaphylactic guinea pigs, that is now known to consist primarily of a powerful vasoconstrictive thromboxane (TXA_2). *Abbr* RCS.

Rabin model A model, proposed by B.R. Rabin, according to which cooperative interactions can be ascribed to kinetic considerations. Specifically, the model applies to a single substrate enzyme reaction in which the enzyme can exist in two different conformational states, both of which can combine with the substrate.

racemase An enzyme that catalyzes the interconversion between two optical isomers, each of which has more than one asymmetric center. *See also* epimerase.

racemate RACEMIC MIXTURE.

racemic mixture An equimolar and optically inactive mixture of the two enantiomers of an optically active compound.

racemization The conversion of an optically active compound to a racemic mixture.

rachitic Of, or pertaining to, rickets.

rachitis RICKETS.

Racker band An absorption band that is produced by the binding of NAD^+ to a dehydrogenase and that is thought to be due to the formation of a charge transfer complex.

rack mechanism Originally, a version of enzyme catalysis according to which the enzyme literally tore its substrate apart by a rack mechanism. The term is now used in conjunction with the modern strain theory of catalysis. According to the latter, when the substrate binds to the enzyme, certain bonds in the substrate are distorted, resulting in an activated transition state.

rad Radiation absorbed dose.

radial chromatography CIRCULAR CHROMATOGRAPHY.

radial dilution The dilution of sedimenting components that is produced in the analytical ultracentrifuge due to the sectorial shape of the centrifuge cell and to the variation of the centrifugal force with distance from the center of rotation. *Aka* square dilution law.

radian The angle subtended by an arc that is equal in length to the radius of the circle.

radiation 1. The emission and propagation of waves of electromagnetic energy such as visible light, x rays, or gamma rays. 2. The emission and propagation of corpuscles such as alpha particles, beta particles, or electrons.

radiation absorbed dose The quantity of ionizing radiation that results in the absorption of 100 erg/g of irradiated material. *Abbr* rad. *See also* exposure dose.

radiation biochemistry An area of biochemistry that deals with the effects of radiation on biochemical compounds and biochemical systems.

radiation chimera A chimera produced experimentally by first irradiating an organism so as to destroy its antibody-producing cells, and then injecting it with antibody-producing cells from a different organism.

radiation curing *See* curing.

radiation dose The amount of radiation to which a specified tissue area or an entire organism is exposed.

radiationless transition A transition, involving an excited atom or molecule, in which no visible or ultraviolet radiation is emitted.

radiation sickness A pathological condition that results from exposure to x rays or other ionizing radiations; characterized in its mild form by nausea, vomiting, and weakness, and in its severe form by damage to blood-forming tissues and by loss of red and white blood cells.

radical A univalent group of atoms that acts as a unit in chemical reactions. *See also* carbon radical; free radical.

radical amino acid replacement RADICAL SUBSTITUTION.

radical anion An anion that is also a free radical.

radical cation A cation that is also a free radical.

radical ion An ion that is also a free radical.

radical scavenger A chemical compound that reacts readily with free radicals and that, when added to a biological system, provides protection against the indirect effects of radiation.

radical substitution The replacement in a protein of one amino acid by another, chemically different, amino acid such as the replacement of a polar amino acid by a nonpolar one or vice versa. A radical substitution is generally expected to lead to significant changes in the properties of the protein. *See also* conservative substitution.

radio- 1. Combining form meaning radiation. 2. Combining form meaning radioactive radiation.

radioactivation analysis ACTIVATION ANALYSIS.

radioactive antibody test A test that permits the identification of a bacterial colony that secretes a specific protein; involves pressing a plate, containing antibodies to the protein, onto bacterial colonies spread on an agar surface. If the protein is present, it will bind to the antibodies and stick to the plate. The latter is then placed in a solution containing labeled antibodies which will bind to the given protein and can then be located by autoradiography. *See also* colony hybridization.

radioactive contamination The deposition of radioactive material in preparations and/or in places where it was not intended to be deposited.

radioactive decay The changes occurring in the nucleus of a radioactive atom that lead to transformation of the nucleus into a different one and to the emission of ionizing radiation.

radioactive disintegration RADIOACTIVE DECAY.

radioactive half-life *See* half-life (1).

radioactive isotope An unstable isotope that undergoes radioactive decay.

radioactive radiation The electromagnetic or the corpuscular radiation emitted by radioactive isotopes.

radioactive series A succession of radioactive nuclides, each one decaying to the next by radioactive disintegration until a stable nuclide is formed.

radioactive suicide *See* suicide.

radioactive tracer *See* tracer (1).

radioactivity The spontaneous disintegration of certain unstable nuclides that is caused by changes in the atomic nucleus and that results in a transformation of the nucleus into a different one and in the emission of one or more types of ionizing radiation, such as alpha particles, beta particles, or gamma rays.

radioassay An assay in which radioactive isotopes are employed.

radioautograph *See* autoradiograph.

radioautographic efficiency *See* autoradiographic efficiency.

radioautography *See* autoradiography.

radiobiology A branch of biology that deals with the effects of radiation on biological systems.

radiocarbon dating A method for establishing the age of archeological, geological, or biological remains by determining the relative amounts of ^{12}C and ^{14}C in the specimen and by calculating its age from the known natural abundance and the known half-life of ^{14}C.

radiochemical 1. *n* A chemical that is labeled with a radioactive isotope. 2. *adj* Of, or pertaining to, radiochemistry.

radiochemical purity The degree of contamination of a radioactively labeled compound with other radioactive substances.

radiochemistry A branch of chemistry that deals with the chemistry of radioactive isotopes and their compounds, and with the applications of radioactive isotopes in other areas of chemistry.

radiochromatogram A chromatogram that contains substances labeled with radioactive isotopes.

radiochromatography Any chromatographic technique in which substances, labeled with radioactive isotopes, are separated.

radiocolloid An aggregate formed in solution by the clumping of molecules that contain radioactive isotopes.

radiodating The determination of the age of an object by measuring the extent of radioactive decay of a particular isotope contained within the object. *See also* radiocarbon dating.

radiogenic Produced by radioactivity, such as an element that is formed from another element by radioactive decay.

radiogram RADIOGRAPH.

radiograph The photographic record obtained in radiography.

radiography A photographic technique in which radiation other than light is passed through an object and a photograph is obtained that reflects the selective absorption of the radiation by various parts of the object.

radioimmunoassay The measurement of either antigen or antibody concentration that is based on the competitive inhibition of labeled antigens on the binding of unlabeled antigens, or vice versa, to specific antibodies. *Abbr* RIA. *Aka* displacement analysis; saturation analysis; competitive radioassay; competitive radioligand assay.

radioimmunochemistry The use of immunochemical techniques in which one or more of the components are radioactively labeled.

radioimmunoelectrophoresis Immunoelectrophoresis in which either the antigens or the antibodies used are radioactively labeled.

radioisotope RADIOACTIVE ISOTOPE.

radioisotopic enzyme assay An enzyme assay based on the measurement of radioactivity in a product of the enzyme-catalyzed reaction when one of the reactants is radioactively labeled.

radiolysis A chemical decomposition that is caused by radiation; the self-decomposition of aged tritium- or ^{14}C-labeled compounds are examples.

radiometer An instrument for measuring the intensity of radiation.

radiometric analysis The determination of an unknown compound by either reacting the unlabeled, unknown compound with a labeled reagent, or reacting the labeled, unknown compound with an unlabeled reagent; in either case, a radioactively labeled product formed in the reaction is then isolated and determined.

radiomimetic drug A chemical immunosuppressive agent, such as an alkylating agent, the effect of which on nucleic acids resembles that of ionizing radiation.

radionuclide A radioactive nuclide; a radioactive isotope.

radiopaque Describing material that does not transmit radioactive radiation.

radiophosphorus decay The radioactive disintegration of ^{32}P, particularly that in ^{32}P-labeled phage nucleic acid. The disintegrations lead to breaks in the sugar–phosphate backbone of the nucleic acid. Single-stranded nucleic acid is inactivated when a chain break occurs, but double-stranded nucleic acid is inactivated only when a break occurs in both strands.

radioresistance The relative resistance of cells, tissues, organs, or organisms to the harmful effects of radiation.

radiorespirometry The measurement of the kinetics of oxygen uptake and/or carbon dioxide evolution in a tissue or in an organism by means of radioactive isotopes.

radiosensitivity The relative sensitivity of cells, tissues, organs, or organisms to the harmful effects of radiation.

radiotherapy Therapy by means of x rays or other radioactive radiations.

radiotoxemia RADIATION SICKNESS.

radiotracer RADIOACTIVE TRACER.

radius of exclusion The distance of closest approach of a bound ion to a protein; equal to the sum of the radii of the protein and the bound ion.

radius of gyration A measure of the spatial extension of a polymer that is related to the distribution of mass in the polymer and to the

shape of the polymer. For a molecule that consists of an assembly of mass elements m_i, each located at a distance r_i from the center of mass, the radius of gyration R is given by $R^2 = \Sigma m_i r_i^2 / \Sigma m_i$. *See also* average radius of gyration.

radwaste Radioactive waste.

raffinose A nonreducing trisaccharide, composed of D-galactose, D-glucose, and D-fructose, that occurs in higher plants. *Aka* melitose.

RAIS Reflection–absorption infrared spectroscopy.

RAM Acronym for random access memory; the temporary memory in a computer that consists of specific instructions and data for a specific program and that can easily be altered. *Aka* read/write memory.

Ramachandran plot A plot of the degrees of rotation about the bond between the alpha carbon and the carbonyl carbon in the peptide bond versus the degrees of rotation about the bond between the alpha carbon and the nitrogen atom; constructed on the basis of van der Waals contact distances and the bond angles of the peptide bond, and used for indicating allowed and forbidden conformations of proteins.

Raman effect The light scattering that is produced when incident light leads to rotational and vibrational transitions of molecules; the scattered light has different frequencies from those of the incident light. *Aka* Raman scattering.

Raman optical activity A spectroscopic method for measuring the difference in the intensity of scattered right and left circularly polarized ultraviolet light by chiral compounds. *Abbr* ROA.

Raman spectrum The spectrum of the light that is emitted in the Raman effect.

Ramon method A method for determining the equivalence zone of a precipitin reaction by mixing a constant amount of antigen with varying dilutions of antibodies, and taking the tube in which precipitation occurs most rapidly to be indicative of the equivalence zone. *See also* Dean and Webb method; method of optimal proportions.

rancidity The development of unpleasant odors and tastes from fats and oils by the oxidation of the unsaturated fatty acid components and/or the hydrolysis of the triglycerides to diglycerides, monoglycerides, glycerol, and free fatty acids.

random coil A linear polymer in a relatively compact, irregular conformation in which there is little interaction between the side chains of the polymer. A random coil exhibits little resistance to rotation about single bonds and is continually contorted by impact of the solvent molecules. A random coil has no unique three-dimensional structure, only average dimensions, and its time-average shape is spherical.

random error INDETERMINATE ERROR.

random flight chain FREELY JOINTED CHAIN.

random genetic drift *See* genetic drift.

randomization The process in which a compound, labeled at a given position, gives rise to a product in which half of the molecules are labeled in one position, while the other half are labeled in another, symmetrical position.

randomly labeled GENERALLY LABELED.

random mechanism The mechanism of an enzymatic reaction in which two or more substrates participate, such that each substrate can readily undergo an association–dissociation reaction with the enzyme to form a binary complex prior to the formation of the ternary complex in which both substrates are associated with the enzyme.

random order RANDOM MECHANISM.

random primer A randomly polymerized oligodeoxyribonucleotide that may hydrogen bond to a complementary deoxyribonucleotide sequence in the template nucleic acid and then serve as a primer for the enzyme reverse transcriptase.

random process STOCHASTIC PROCESS.

random sample A sample of items that are selected from a population in such a fashion that all the items in the population have an equal chance of being included in the sample.

random variable VARIATE.

random walk The path in space taken by a molecule in which each step is uncorrelated with the preceding one; each step is being determined by chance, either in regard to direction or in regard to magnitude, or both. The path traced by a molecule in solution due to Brownian motion is an example; the conformation of an ideal, freely jointed random coil can likewise be treated as if the segments represented the random walk of a molecule.

random walk chain FREELY JOINTED CHAIN.

Raney nickel A preparation of finely divided nickel used as a catalyst for hydrogenation reactions and as a reactant for the desulfurization of sulfhydryl compounds.

range 1. The thickness of absorber required to absorb all of the radiation of a particular type. 2. The highest and lowest values for a set of results.

rank electrode A commonly used variation of the Clark electrode for measuring oxygen concentrations.

R antigen *See* rough strain.

Raoult's law The law that the lowering of the vapor pressure of the solvent by the solute is proportional to the mole fraction of the solute in the solution.

raphidosome One of a group of rod-shaped, intracellular particles found in bacteria and algae and believed to be associated with the nuclear region of the cell.

rapid equilibrium approximation PRE-EQUILIBRIUM APPROXIMATION.

rapid flow kinetics The kinetics of a chemical reaction that are determined by means of rapid flow techniques.

rapid flow technique A technique for studying fast chemical reactions in which the reactants are forced out of two syringes into a mixing chamber and the mixture is then allowed to flow through a tube for spectroscopic, or other, analysis. The distance along the tube is proportional to the reaction time. *See also* stopped flow technique.

rapidly labeled RNA PULSE-LABELED RNA.

rapidly reannealing DNA PEPETITIVE DNA.

rapidly reassociating DNA REPETITIVE DNA.

rapid lysis mutant A phage mutant that does not show lysis inhibition; such mutants (designated r) form larger plaques than those produced by wild-type phage particles.

rapid mixing technique RAPID FLOW TECHNIQUE.

rapid reaction *See* fast reaction.

rapid start complex OPEN PROMOTER COMPLEX.

rare amino acid An amino acid that occurs in only a few proteins, such as hydroxylysine or hydroxyproline, which are found in collagen and gelatin.

rare base MINOR BASE.

rare earth An element belonging to a group of 15 metals (atomic numbers 57 to 71) that have very similar properties.

ras genes Rat sarcoma genes; a ubiquitous eukaryotic gene family, first discovered as the transforming principle in retroviruses causing rat sarcomas.

RAST Acronym for radioallergosorbent test; an isotopic technique for the demonstration of reagins directed against specific allergens.

rat antidiuresis assay A bioassay for the activity of neurohypophyseal hormones in which the reduction of urine formation is measured in hydrated rats.

rat antispectacle eye factor INOSITOL.

rate *See* reaction rate.

rate constant A proportionality constant between the velocity of a chemical reaction and the concentrations of the reacting species; denoted k_{+n} for the forward, and k_{-n} for the reverse, reaction at the nth step of a reaction sequence. *Aka* rate coefficient.

rate-determining step The slowest step in a sequence of reactions; the step with the smallest rate constant. *Aka* rate-limiting step.

rate equation A mathematical expression for

the rate of a chemical reaction in terms of the rate constant of the various steps and the concentrations of the reactants and the products.

rate-limiting step   RATE-DETERMINING STEP.

ratemeter   A radiation detector that indicates the rate of emission of radioactive radiation.

rate-of-change method   A method of amplifying the ion current produced in an ionization chamber when high sensitivity is required.

rate of shear   The variation in the velocity of flow of a liquid flowing through a tube with the radial distance from the center of the tube; the velocity gradient, perpendicular to the direction of flow.

rate zonal centrifugation   DENSITY GRADIENT CENTRIFUGATION.

ratio   The supply of a nutrient to a tissue divided by the requirement of the tissue for the nutrient.

Rauscher leukemia virus   A mouse leukemia virus that belongs to the leukovirus group.

raw data   Data obtained directly from measurements (as opposed to those derived by calculations) and/or data that have not been subjected to statistical treatments (grouping, coding, etc.).

Rayleigh fringe   An interference fringe obtained with a Rayleigh interferometer.

Rayleigh interferometer   An interferometer in which constructive and destructive interference of light that has passed through one primary and two secondary vertical slits results in a series of light and dark fringes of fixed thickness.

Rayleigh quotient   RAYLEIGH RATIO.

Rayleigh ratio   A measure of the intensities of incident and scattered light in Rayleigh scattering; specifically, $R = (i_\theta/I_0)r^2$, where R is the Rayleigh ratio, i_θ is the scattered light intensity at angle θ, I_0 is the incident light intensity, and r is the distance from the observer to the source of the scattered light.

Rayleigh scattering   The light scattering that is produced by solutes in dilute solutions when the solute particles can be considered to be independent scatterers, when they are small compared to the wavelength of the incident light, and when the scattering is due to elastic collisions between photons and orbital electrons. The greatest dimension of the scattering particles must be less than about 0.05 times the wavelength of the incident light, and the scattered light is of the same frequency as that of the incident light.

Rb   Ribosome.

RBC   Red blood cell.

RBE   Relative biological effectiveness.

RBP   Retinol-binding protein.

R1,5BP   Ribulose-1,5-bisphosphate.

RBS   Rutherford backscattering.

RCF   Relative centrifugal force.

R configuration   *See* RS system.

RCS   Rabbit aorta contracting substance.

RDA   Recommended dietary (daily) allowance.

RDE   Receptor destroying enzyme.

r-determinant   *See* R plasmid.

rDNA   1. In general, any DNA segment that codes for rRNA; ribosomal DNA. 2. Specifically, a group of adjacent eukaryotic genes that code for rRNA and that differ sufficiently in their base composition from the bulk DNA to permit their easy separation from it.

rDNA amplification   The preferential replication of the genes that code for rRNA; occurs during oogenesis in amphibia and insects and also in the macronuclei of protozoa.

R-DNA polymerase   DNA POLYMERASE III.

R1,5DP   Ribulose-1,5-diphosphate; now designated as ribulose-1,5-bisphosphate (R1,5BP).

RDS   Respiratory distress syndrome.

RE   Retinol equivalent.

reactancy   The number of kinetically significant substrates or products in the Cleland convention. The reactancy is indicated by the syllables uni, bi, ter, and quad.

reactant   An atom, an ion, or a molecule that enters into a chemical reaction.

reaction center   The photochemically active complex that absorbs the excitation energy in photosynthesis from the antenna molecules. The complex consists of a few special chlorophyll and/or other pigment molecules that are bound to a few protein molecules and that becomes excited and initiate the reactions of the two photosystems.

reaction coordinate   The abscissa that represents the progress of a reaction as measured by some quantity indicated on the ordinate.

reaction kinetics   The rate behavior of a reaction.

reaction mixture   The mixture composed of sample material and reagents that is allowed to undergo a reaction under controlled conditions.

reaction of identity   The complete fusion of two precipitin bands in either two-dimensional double immunodiffusion or immunoelectrophoresis; obtained when two indistinguishable antigens react with an antibody in an adjacent field.

reaction of nonidentity   The complete crossing of two precipitin bands in either two-dimensional double immunodiffusion or immunoelectrophoresis; obtained when two unrelated antigens react with an antibody in an adjacent field.

reaction of partial identity   The partial fusion of, and spur formation by, two precipitin bands in either two-dimensional double immunodiffusion or immunoelectrophoresis;

obtained when two cross-reacting antigens react with an antibody in an adjacent field.

reaction order The sum of the powers of the reactant concentrations to which the reaction rate is proportional. *See also* reaction rate; order with respect to concentration; order with respect to time.

reaction paper chromatography A chromatographic technique for determining the number of various functional groups in a molecule, chiefly in an aromatic compound, on the basis of the chromatographic behavior of the unreacted molecule and of the molecule after it has reacted with appropriate reagents.

reaction rate The rate at which either a product is formed or a reactant is used up in a chemical reaction; for the first-order reaction $A \rightarrow B$, the rate is given by $v = -d[A]/dt = d[B]/dt = k[A]$, where brackets indicate molar concentrations, and k is the rate constant. For the second-order reaction $A + B \rightarrow C$, the rate is given by $v = k[A][B]$. *Aka* reaction velocity.

reactivation 1. The restoration of activity to cells or viruses that have suffered photochemical damage; photoreactivation, thermal reactivation, and multiplicity reactivation are examples. 2. The restoration of activity to an inactivated poxvirus, the protein coat of which has been denatured but the DNA of which has not been damaged; achieved by the presence of an infectious poxvirus, the enzymes of which lead to an uncoating of the inactivated virus. 3. The restoration of activity to an inhibited enzyme by removal of the inhibitor through a chemical reaction. The process has sometimes been termed reversible inhibition. *See also* irreversible inhibitor; reversible inhibitor.

reactive enzyme centrifugation *See* active enzyme centrifugation.

reactive hemolysis The reactive lysis of red blood cells.

reactive lysis The lysis of an unsensitized cell produced by the binding of complement to the cell surface.

reactive residue An amino acid residue in a protein that is accessible to and can undergo a reaction with a specific reagent.

reactor *See* nuclear reactor.

reading The process whereby the sequence information in one polymer is used to produce a defined sequence in another polymer; replication, transcription, and translation are examples of processes that entail reading.

reading frame The manner in which nucleotides in mRNA are grouped into codons for translation. The reading frame is one of three ways for translating a given nucleotide sequence, defined by the location of the initiation codon AUG. Thus, the possible reading frames of the mRNA sequence 5'-GCUAGCCUG····-3' are (a) 5'-(GCU)(AGC)···; (b) 5'-(CUA)(GCC)···; and (c) 5'-(UAG)(CCU)···; which sequence is used depends on the location of the initiation codon. Fixation of the starting point determines which groups of 3 bases in the mRNA sequence are interpreted as codons.

reading frameshift The shift in reading produced by a frameshift mutation.

reading mistake MISTRANSLATION.

readout The act of reading out. *See also* reading.

read-through 1. The transcription of DNA that proceeds past the normal termination signal in the DNA; may result from failure of RNA polymerase to recognize a terminator, temporary dissociation of a termination factor (such as rho in bacteria) from the terminator sequence, or the lack of attenuation by an attenuator. 2. The translation of mRNA that proceeds past the normal termination codon in the mRNA; may result from suppression of a termination codon by a nonsense suppressor tRNA.

read-through protein A protein that is produced as a result of a failure in the termination of translation of a polycistronic mRNA; such a protein consists of the regular amino acid sequence specified by its cistron plus a sequence of amino acids that corresponds to a translated intercistronic region.

reagent 1. A substance that participates in a chemical reaction. 2. A substance used for the detection or determination of another substance.

reagin A homocytotropic antibody of the IgE immunoglobulin class that is formed in response to an allergen and that, upon combination with the allergen, causes the release of histamine and other vasoactive agents of immediate-type hypersensitivity.

reaginic antibody REAGIN.

reannealing The renaturation of DNA; the joining of dissociated (denatured), complementary single strands to form duplex molecules. In reannealing, the single strands are from the same source while in annealing the single strands are from different sources, resulting in the formation of hybrid duplex molecules. *See also* annealing; reassociation.

rearrangement reaction A chemical reaction in which there is an alteration in the distribution of the atoms in a molecule.

reassociation The pairing of complementary DNA strands, or parts of DNA strands, that results in the formation of a duplex or sections of a duplex.

reassociation kinetics A technique for measuring the rate of reassociation of complementary DNA strands derived from the same source (as opposed to hybridization). The technique involves shearing the DNA into small fragments, denaturing these fragments by heating to form single-strand segments, and then allowing the latter to form double-stranded segments by slow cooling (annealing). The extent of reassociation is commonly followed by nuclease digestion, hypochromicity measurements, or hydroxyapatite chromatography. The data are analyzed by means of a cot curve. The DNA segments are usually classified into four categories: unique, slightly repetitive, middle repetitive, and highly repetitive DNA. *See also* cot curve; repetitive DNA.

reassortant virus A synthetically produced hybrid virus that contains DNA and protein from different species.

reassortment An exchange of genome segments as that which occurs with myxoviruses.

RECA RecA protein.

recapitulation theory The theory that an organism during its development passes through and recapitulates the stages that have occurred in the development of the species. *Aka* ontogeny recapitulates phylogeny.

RecA protein A protein (MW 38,000) that has several enzymatic activities, including a DNA-dependent ATPase activity. It plays a central role in genetic recombination and in SOS repair. *Aka* RecA protease. *See also* SOS repair.

recBCD enzyme EXONUCLEASE V.

receptor 1. A target site at the molecular level to which a substance becomes bound as a result of a specific interaction. As an example, the site may be on the cell wall, on the cell membrane, or on an intracellular enzyme, and the substance bound may be a virus, an antigen, a hormone, or a drug. The binding interaction might trigger a physiological or a pharmacological response. 2. A site in an organism that responds to specific stimuli such as a chemoreceptor, an osmoreceptor, or a photoreceptor.

receptor destroying enzyme NEURAMINIDASE.

receptor down regulation *See* down regulation.

receptor element *See* controlling element.

receptor gradient An arrangement of viruses in a series based on their reaction with, and their destruction of, receptor sites on red blood cells; any virus in the series will react with its own receptor sites and with those specific for viruses that precede it in the gradient, but will not react with receptor sites for viruses that follow it in the gradient.

receptor internalization *See* coated pit; receptosome.

receptor-mediated endocytosis LIGAND-INDUCED ENDOCYTOSIS.

receptosome A vesicular structure in animal cells formed during the down regulation by polypeptide hormones. It consists of a coated pit that has budded off from the cytoplasmic membrane and has entrapped receptor–ligand (hormone) complexes. Receptosomes are also formed in receptor-mediated endocytosis of other ligands such as low-density lipoproteins and lysosomal enzymes; they are probably identical to endosomes. *See also* coated pit.

recessive 1. RECESSIVE GENE. 2. The trait produced by a recessive gene in the homozygous state.

recessive gene A gene the expression of which is either partially or entirely suppressed when the dominant allelic gene is present. *Aka* recessive allele.

recessive lethal An allele that leads to the death of the cell or the organism that is either homozygous or heterozygous for the allele.

reciprocal activation The cyclic set of reactions in the intrinsic pathway of blood clotting whereby activated Factor XII converts prekallikrein to kallikrein which, in turn, generates more activated Factor XII.

reciprocal genes COMPLEMENTARY GENES.

reciprocal ion-atomsphere radius The term kappa(κ) of the Debye–Hueckel theory that is equal to the reciprocal of the thickness of the ion atmosphere; the reciprocal of the distance from the surface of the central ion to the outer edge of the ion atmosphere.

reciprocal lattice The three-dimensional crystal lattice deduced from a two-dimensional x-ray diffraction pattern; used to obtain the dimensions of the unit cell in the real crystal lattice and so called because the positions of the spots in the x-ray diffraction pattern are an inverse measure of the spacings in the real crystal. *Aka* reciprocal space.

reciprocal plot *See* single reciprocal plot; double reciprocal plot.

reciprocal recombination Recombination that involves a symmetrical exchange of genetic material by crossing over.

reciprocating shaker *See* shaker.

reciprocity The condition that exists when the product of dose rate, specifically that of radiation, and time of exposure is constant; thus, $(\text{dose rate})_1 \times \text{time}_1 = (\text{dose rate})_2 \times \text{time}_2$. *Aka* Bunsen–Roscoe law.

rec⁻ mutant Recombination-deficient mutant.

recognition A specific binding interaction occurring between macromolecules, as that between a tRNA molecule and an aminoacyl -tRNA synthetase, or that between an immunocyte and an antigen.

recognition site 1. tRNA SYNTHETASE RECOGNI-

TION SITE. 2. AMINOACYL SITE. 3. RECEPTOR.

recoil energy The energy of an atom after ejecting a subatomic particle; typically, the energy of the cation after the atom has lost a high-energy beta particle.

recombinant One of the progeny formed as a result of genetic recombination.

recombinant DNA technology The techniques by which genetic recombination is carried out in vitro. It entails the breakage and rejoining of DNA molecules from different organisms and the production and isolation of the modified DNA or fragments thereof. Thus, when a human gene, coding for insulin, is inserted into the DNA of a bacterial plasmid and the latter is cloned to produce many identical copies of the inserted gene, the methodology involved is referred to as recombinant DNA technology. The modified plasmid DNA is called a recombinant DNA molecule. Recombinant DNA technology generally involves the following: (a) using restriction enzymes to produce DNA fragments; (b) joining these DNA fragments (passengers) to other DNA molecules (vectors) by various techniques of ligation (splicing); (c) inserting the resultant recombinant DNA molecules into host cells where they self-replicate to produce multiple copies per cell of the inserted DNA fragments (cloning). *Aka* plasmid engineering; gene cloning; DNA cloning. *See also* cloning; genetic engineering; biotechnology; gene therapy.

recombinant joint The region of a heteroduplex that represents the linkage between two recombining DNA molecules.

recombinant RNA technology The techniques by which RNA molecules, from the same or from different species, are spliced together.

recombinase An enzyme functioning in genetic recombination; the recA protein is an example.

recombination The production of progeny that derives some of its genes from one parent and some from another, genetically different, parent; as a result, the combination of genes in the progeny is different from that of either of the parents. In higher organisms, recombination occurs by way of independent assortment or crossing over; in lower organisms, it occurs by way of transformation, conjugation, or transduction. *See also* general recombination; site-specific recombination; illegitimate recombination.

recombination-deficient mutant A mutant that is unable to produce recombinants. *Abbr* rec⁻ mutant.

recombination frequency *See* frequency of recombination.

recombinationless mutant RECOMBINATION-DEFICIENT MUTANT.

recombination nodule A protein-containing assembly, occurring at intervals on the synaptonemal complex, that is believed to play a role in crossing over.

recombination repair A postreplicative repair mechanism of DNA that involves an exchange of good for bad segments between two damaged, homologous, duplex molecules; the exchange is known as sister-strand exchange. Recombination repair is a dark (requiring no light) repair mechanism that serves to remove and repair lesions containing tymine dimers.

recombination value FREQUENCY OF RECOMBINATION.

recombinogenic Promoting recombination.

recommended dietary allowance A recommended quantity for the daily intake of calories, a food, or a vitamin that has been established by the Food and Nutrition Board of the National Research Council; recommended for a normal individual engaged in average activity and living in a temperate climate. *Abbr* RDA. *Aka* recommended daily allowance.

recon The unit of genetic recombination; the smallest section of a chromosome, which may be as small as a single nucleotide, that is capable of recombination and that cannot be divided by recombination.

reconstitute To reassemble a particle from its fragments or to reassemble a system from its fractions, as in the assembly of viruses, ribosomes, and protein-synthesizing systems.

reconstituted ghost An erythrocyte ghost that has been loaded with specific substances, and the membrane of which has been allowed to shrink back to its normal size and to return to its normal state of permeability.

recording spectrophotometer A spectrophotometer with an attached recorder for graphical representation of the data obtained.

recovery YIELD (2).

recovery heat The heat produced by a muscle that relaxes after a single contraction.

recovery time COINCIDENCE TIME.

recruitment factor A factor, believed to be involved in the regulation of mRNA translation in sea urchin eggs. *See also* masked mRNA.

recycling chromatography A column chromatographic technique in which resolution is improved by passing the column effluent back onto the same column; fractions may be collected and fresh solvent may be added during this process.

red cell agglutination HEMAGGLUTINATION.

red cell lysis HEMOLYSIS.

red drop The decrease in photosynthetic efficiency (the quantum yield) of chloroplasts that occurs at wavelengths longer than 680 nm.

red muscle A dark skeletal muscle that has a

relatively high content of myoglobin and cytochromes. It is rich in mitochondria, obtains most of its ATP through oxidative phosphorylation, and is capable of prolonged activity. *Aka* slow-twitch muscle.

redox Oxidation–reduction.

redox carrier ELECTRON CARRIER.

redox complex *See* complexes I–IV.

redox couple The electron donor and the electron acceptor species of a given half-reaction. *Aka* redox pair.

redox lipid A lipid, such as ubiquinone or tocopherol, that undergoes oxidation–reduction reactions and that contains polyisoprenoid chains.

redox loop mechanism A set of reactions, proposed as part of the chemiosmotic coupling hypothesis, according to which redox centers are arranged in the mitochondrial membrane in such a fashion that each center can simultaneously accept an electron and a proton from the matrix side of the membrane.

redox pair REDOX COUPLE.

redox potential OXIDATION–REDUCTION POTENTIAL.

red plaque A plaque that, in the plaque assay, is stained excessively with neutral red due to the increased binding of the dye by the lysosomes present in the virus-infected cells.

red shift BATHOCHROMIC SHIFT.

reduced hemoglobin Hemoglobin containing iron in the ferrous (+2) state.

reduced mean residue rotation The mean residue rotation corrected to that in a medium of unit refractive index; specifically, $[m'] = 3[m]/(n^2 + 2)$, where $[m']$ is the reduced mean residue rotation, $[m]$ is the mean residue rotation, and n is the refractive index of the medium.

reduced osmotic pressure The osmotic pressure (π) divided by the concentration of the solution (c); the term π/c.

reduced scattered light intensity RAYLEIGH RATIO.

reduced substrate concentration The substrate concentration [S] divided by the Michaelis constant K_m; the term $[S]/K_m$. Designated as $[S']$, σ, or α.

reduced viscosity The ratio of the specific viscosity of a solution to either the solute concentration or the volume fraction of the solute. *Aka* reduced specific viscosity.

reducing agent REDUCTANT.

reducing atmosphere An atmosphere that is rich in gases that are readily oxidized; a reducing atmosphere consisting of water, hydrogen, ammonia, nitrogen, methane, and hydrogen sulfide is believed by some to have been the primitive atmpophere of the earth about 4.5 billion years ago.

reducing end The end of an oligo- or a polysaccharide that carries the hemiacetal or the hemiketal grouping.

reducing equivalent A measure of reducing power equal to either one electron or one hydrogen atom.

reducing power The capacity of a substance to function as a reducing agent, that is, to provide hydrogen and/or electrons.

reducing sugar A sugar that will reduce certain inorganic ions in solution, such as the cupric ions of Fehling's or Benedict's reagent; the reducing property of the sugar is due to its aldehyde or potential aldehyde group.

reductant The electron donor species of a given half-reaction; the species that undergoes oxidation in an oxidation–reduction reaction.

reductase A dehydrogenase for which the transfer of hydrogen from the donor molecule is not readily demonstrated.

reduction The change in an atom, a group of atoms, or a molecule that involves one or more of the following: (a) loss of oxygen; (b) gain of hydrogen; (c) gain of electrons.

reduction division MEIOSIS.

reductionism The doctrine that a higher level of organization can be understood from a knowledge of lower levels. Thus, an organism or a natural phenomenon can be understood from a knowledge of the component parts; the entirety is equal to the sum of the parts. *See also* holism.

reduction phase The second stage of the Calvin cycle in which 3-phosphoglyceric acid is converted to glyceraldehyde-3-phosphate by means of ATP and NADPH.

reduction potential The electrode potential that is used in biochemistry and that measures the tendency of an oxidation–reduction half-reaction to occur by way of a gain of electrons. *Sym* E. *See also* electrode potential; standard electrode potential.

reductive carboxylic acid cycle A cyclic set of reactions that is essentially a reversal of the citric acid cycle and that involves the fixation of four molecules of carbon dioxide and the synthesis of one molecule of oxaloacetate per one turn of the cycle. The reactions occur in some photosynthetic bacteria. *Aka* reductive tricarboxylic acid cycle.

reductive pentose cycle CALVIN CYCLE.

reductive pentose phosphate cycle CALVIN CYCLE.

reductive tricarboxylic acid cycle REDUCTIVE CARBOXYLIC ACID CYCLE.

redundancy DEGENERACY (1). *See also* terminal redundancy; repetitive DNA.

redundant DEGENERATE.

redundant cistron A cistron that is present in multiple copies on the same chromosome.

redundant DNA REPETITIVE DNA.

reentrant surface The interior surface of a protein molecule, consisting of two or more atoms which are simultaneously in contact with a given probe.

reference BLANK.

reference electrode An electrode against which the potential of another electrode is being measured. *Aka* reference half-cell.

reflection The partial or comlete return of light waves or other types of radiation from a surface.

reflection–absorption infrared spectroscopy A technique for the study of surfaces in which infrared radiation, reflected from a suface, is used to measure the infrared spectrum of a surface component. *Abbr* RAIS.

reflection symmetry The symmetry of a body that exists if an identical structure of the body is produced when it is rotated about an axis and reflected through a plane perpendicular to that axis; the order in which these two processes are carried out is not significant. The rotation–reflection axis is denoted S_n, indicating that identical structures are produced by a rotation of $360°/n$.

refolding RENATURATION.

refractile Capable of refracting light.

refraction The change in the velocity and in the direction of light waves that pass obliquely from one medium into another.

refractive increment REFRACTIVE INDEX INCREMENT.

refractive index (*pl* refractive indices) A measure of the light-retarding property of a medium, equal to the ratio of the velocity of light in a vacuum to that in the medium; also equal to the ratio of the sine of the angle of incidence to the sine of the angle of refraction for light passing obliquely from a vacuum into the medium. *Sym* n. *Aka* index of refraction.

refractive index increment The rate of change of the refractive index of a solution with the concentration of the solution. *See also* specific refractive index increment.

refractometer An instrument for measuring refractive indices.

refractoriness DESENSITIZATION (3).

refractory Resistant to a given treatment or cure.

refractory period The period after the passage of an action potential during which a nerve axon or a muscle fiber is resistant to stimulation.

Refsum's disease A genetically inherited metabolic defect in humans, involving serious neurological problems (such as tremors, unsteady gait, and poor night vision), that is characterized by a large accumulation of phytanic acid in tissues and serum; due to a deficiency of the enzyme phytanate α-hydroxylase.

regeneration 1. The repair and replacement of damaged or lost tissue, as in the formation of liver tissue following partial hepatectomy. 2. The restoration of an ion-exchange resin to its original ionic form.

regeneration phase The third stage of the Calvin cycle in which ribulose-1,5 bisphosphate is regenerated by a series of reactions, beginning with glyceraldehyde-3-phosphate.

regiospecific Descriptive of a chemical reaction that proceeds in only one of two possible ways, such as a reaction that involves an addition to a double bond.

regression 1. A decrease in the size of a tumor or in the manifestations of a disease. 2. The relation between two statistical variables. *See* regression line.

regression coefficient A parameter that describes the rate of change of a dependent variable with respect to an independent variable; any coefficient in a regression equation, such as the parameters α and β in the linear regression equation $Y = α + βX$.

regression curve *See* regression line.

regression line A plot of the average of a dependent variable Y as a function of an independent variable X; a plot of $\bar{Y}_x$ versus X. The line defines the amount of change of one variable per unit change in the other; if the plot does not yield a straight line, it is referred to as a regression curve.

regulated secretory cell A cell, such as an endocrine cell or a neuron, that secretes large amounts of a protein at a rate that is much higher than that at which the protein is synthesized inside the cell. Such cells store large amounts of protein in secretory vesicles until the cell receives an appropriate stimulus for secretion. *See also* constitutive secretory cell.

regulator-constitutive mutant A mutant that results from a constitutive mutation in which the regulator gene has been mutated in such a way as to prevent the formation of a repressor or to produce a defective repressor; as a result, a previously inducible enzyme becomes a constitutive one.

regulator element *See* controlling element.

regulator gene A gene that is responsible for the synthesis of a repressor that, in turn, controls an operator. The regulator gene need not be adjacent to the operator. *Abbr* R gene. *Aka* regulatory gene. *See also* enzyme induction; enzyme repression.

regulatory enzyme An enzyme that has a regulatory function in metabolism and that has the capacity of having its catalytic activity modified. A regulatory enzyme is frequently the first enzyme in a reaction sequence or the enzyme at a branch point of metabolic pathways. Regulatory enzymes are of two kinds, allosteric enzymes or covalently modified enzymes. The activity of the former is modified through the binding of allosteric effectors to regulatory (allosteric) sites on the enzyme; the activity of the latter is modified as a result of a chemical alteration of the enzyme which, in turn, is catalyzed by other enzymes. *See also* allosteric enzymes; covalently modified enzymes.

regulatory factor HYPOTHALAMIC HORMONE.

regulatory hormone HYPOTHALAMIC HORMONE.

regulatory protein 1. ALLOSTERIC PROTEIN. 2. Any protein that has a regulatory function such as a protein that is produced by a regulatory gene, or the protein that functions in the bacterial phosphotransferase system (RPr).

regulatory sequence A segment of DNA that functions in regulating the expression of structural genes in an operon; operators, promoters, and attenuators are examples.

regulatory site A site on an allosteric enzyme to which an effector binds, as distinct from a catalytic site to which the substrate binds. *Aka* allosteric site.

regulatory subunit The subunit of the regulatory enzyme aspartate transcarbamylase that has no enzymatic activity but binds the negative effector CTP. *See also* catalytic subunit.

regulatory T cells *See* helper T cells; suppressor T cells.

regulon A set of nonadjacent structural genes that are under the control of a common regulatory gene. As opposed to an operon, in which different structural genes are adjacent, the different structural genes of a regulon are located at different sites on a chromosome, or are scattered over several chromosomes. Three examples of regulons are the HTP, Pho, and SOS regulons.

Reichert–Meissl number A measure of the volatile fatty acids in a fat; equal to the number of milliliters of 0.1 N alkali required to neutralize the volatile, water-soluble fatty acids in 5 g of fat. *Aka* Reichert–Meissl value.

Reichstein's compound A designation for some steroids; Reichstein's compounds H, F, and M refer, respectively, to corticosterone, cortisone, and cortisol.

Reid factor HIGH MOLECULAR WEIGHT KININOGEN.

reiterated DNA sequences REPETITIVE DNA.

reiterated genes *See* gene reiteration.

rejection *See* immunological rejection.

relative biological effectiveness The ratio of the biological effect produced by one ionizing radiation to that produced by an identical dose of a different ionizing radiation; also equal to the ratio of the doses of two different ionizing radiations that produce the same biological effect. For such calculations, the biological effect produced by x rays, gamma rays, or beta particles is generally assigned a value of unity. *Abbr* RBE.

relative centrifugal force The magnitude of the centrifugal force compared to the gravitational force; expressed in terms of multiples of g, as in $100,000 \times g$, where g is the gravitational acceleration. The relative centrifugal force (in multiples of g) is equal to $(1.12 \times 10^{-5})r(\text{rpm})^2$, where r is the distance from the center of rotation in centimeters, and rpm is the speed of the rotor in revolutions per minute. *Abbr* RCF.

relative configuration 1. The comparative spatial arrangement of the atoms about two or more asymmetric carbon atoms in one molecule. 2. The arrangement of the atoms in one molecule compared to that of the atoms in a different molecule.

relative counting The counting of radiation such that only a fraction of the actual radioactive disintegrations that occur in the sample are detected; consequently, the results are expressed as counts per minute rather than as disintegrations per minute.

relative deviation A deviation expressed in relative terms such as a percentage average deviation.

relative error 1. An error expressed in relative terms, such as in percentages. Thus, if a measurement has the value of 24.60 and the mean is 24.00, then the relative error is $(0.60/24.00)100 = 2.5\%$ 2. The number of standard deviations in the error.

relative infectivity The fraction of the initial infectivity of a virus preparation that remains when a virus is neutralized with antiviral antibodies and the reaction mixture is sampled as a function of time.

relative migration distance *See* retardation coefficient.

relative plating efficiency The percentage of cells that give rise to colonies when plated on a nutrient medium, compared to a control for which the absolute plating efficiency is arbitrarily taken as 100%.

relative retention The retention volume of a component that is separated by gas chromatography relative to the retention volume of a standard.

relative specific activity The ratio of the specific activity of the sample to that of a reference substance.

relative standard deviation COEFFICIENT OF VARIATION.

relative substrate concentration REDUCED SUB-

STRATE CONCENTRATION.

relative variance The square of the coefficient of variation.

relative viscosity The ratio of the viscosity of a solution to that of the solvent. *Sym* η_r.

relaxation 1. The transition of a system from a suddenly disturbed equilibrium position to a new equilibrium position. 2. The return of a muscle from its contracted to its resting state. 3. The conversion of a superhelical DNA molecule to one that has fewer, or no, superhelical turns.

relaxation complex An aggregate formed between some supercoiled DNA of *E. coli* plasmids and three tightly bound proteins. When the complex is heated, or treated with alkali, proteolytic enzymes, or detergents, one of these proteins (which is a nuclease) makes a site-specific nick in the DNA, thereby relaxing the supercoil to the nicked circular form. The nicking is believed to play a role in the transfer of the plasmid during conjugation by establishing the transfer origin.

relaxation effect The retardation of the electrophoretic mobility of a charged particle that results from the electric field set up by the differential movement of the charged particle and its ion atmopshere.

relaxation kinetics The kinetics of a system that undergoes relaxation.

relaxation protein SINGLE-STRAND BINDING PROTEIN.

relaxation technique A technique for studying either a rapid reaction or the intermediate steps in a complex reaction by means of relaxation; performed by allowing the system to come to equilibrium and then disturbing the system suddenly by means of a rapid change in one variable, following which the system is allowed to come to a new equilibrium position. Depending on the variable that is being altered, the technique is referred to as temperature jump, pressure jump, concentration jump, pH jump, etc.

relaxation time 1. A measure of relaxation that is equal to the time interval between the disturbance of the original equilibrium of the system and the achievement of the new equilibrium position. 2. A measure of relaxation that is equal to the time required for a system to change from its original equilibrium position to $1/e$ of this equilibrium value (e is the base of natural logarithms). *See also* rotational relaxation time. 3. The reciprocal of the rate constant for a relaxation experiment.

relaxed circle A double-stranded, circular DNA molecule in which supercoiling has been removed either by a single-strand break (nick)

or by the action of a topoisomerase. *Aka* relaxed DNA.

relaxed conformation *See* concerted model.

relaxed control 1. The continued synthesis of RNA that occurs in some bacterial mutants after removal of an essential amino acid from the medium. *See also* stringent control. 2. Plasmid replication that greatly exceeds that of the chromosome and that results in the production of 20 or more plasmids per cell.

relaxed DNA RELAXED CIRCLE.

relaxed helix A nontwisted helix; a helix that is not a superhelix; a relaxed circle.

relaxed muscle A muscle that has returned to is resting state following a contraction.

relaxed plasmid *See* copy number.

relaxed strain A bacterial strain that shows relaxed control.

relaxin A polypeptide hormone, produced by the corpus luteum, that causes relaxation of the symphyseal ligaments in mice and guinea pigs. Relaxin has been found in the blood of pregnant females of many species, including humans, but its role in the human female is unknown.

relaxing enzyme DNA-RELAXING ENZYME; *See* topoisomerase.

relaxing factor The calcium pump of the sarcoplasmic reticulum of muscle.

relaxing protein The complex formed between troponin and tropomyosin B.

release factor 1. One of a group of protein factors that function in the release of the polypeptide chain from the ribosome at the termination stage of translation. Release factors respond, in part, to termination codons. There are two such factors in *E. coli* and one in eukaryotes. Prokaryotic and eukaryotic release factors are designated RF and eRF, respectively. 2. RELEASING HORMONE.

releasing hormone A hormone that causes the release of another hormone. *Aka* releasing factor. *See also* hypothalamic hormone.

reliability The degree to which experimental data, or methods leading to such data, reflect both accuracy and precision.

relic model A model for the evolution of the genetic code according to which the early development of the code resulted from mechanistic processes while the later development resulted from stochastic processes.

REM 1. Roentgen equivalent man. 2. Roentgen equivalent mammal.

remission A temporary decrease in the size of a tumor or in the manifestations of a disease.

remodeling The continuous synthesis (formation) and degradation (resorption) of bone that proceeds throughout life; the steady state of bone.

renal Of, or pertaining to, the kidneys.

renal clearance *See* clearance.

renal compensation One of a number of mechanisms whereby the kidneys counteract the effects of either acidosis or alkalosis.

renal diabetes RENAL GLUCOSURIA.

renal glucosuria A pathological condition that is characterized by the recurrent excretion of glucose in the urine while the plasma concentration of glucose is either normal or slightly elevated; due to impaired reabsorption of glucose by the renal tubules.

renal hypertension Hypertension resulting from kidney disease or failure.

renal threshold *See* threshold.

renal tubular acidosis A systemic acidosis with an inappropriately high urinary pH as a result of congenital or acquired disorders of renal function.

renaturation The reformation of all, or part of, the native conformation of either a protein or a nucleic acid molecule after the molecule has undergone denaturation; a reversal of denaturation. *See also* annealing (1).

renatured Having undergone renaturation.

renin A proteolytic enzyme produced by the kidney that has hormone-like properties and catalyzes the conversion of angiotensinogen to angiotensin I.

renin substrate ANGIOTENSINOGEN.

rennet enzyme CHYMOSIN.

rennin CHYMOSIN.

renotropic Having a tendency or capacity to increase the activity of the kidneys.

reovirus A naked, icosahedral animal virus that contains double-stranded RNA. Reoviruses occur in the enteric and respiratory tracts but are not commonly associated with disease. Thus, as agents of disease, they are orphans, and hence the term reovirus (a contraction of respiratory enteric orphan virus). The virion of reoviruses consists of 10 double-stranded, linear RNA molecules.

REP Roentgen equivalent physical.

repair *See* DNA repair.

repair enzyme A DNA-dependent DNA polymerase that catalyzes the replacement of damaged and excised segments of single strands in double-stranded DNA. The enzyme uses the undamaged strand as a template and the repair is completed by a ligase that catalyzes the joining of the newly synthesized segments to the existing strand.

repairosome The complex of enzymes and other components that functions in the repair of DNA, damaged by ultraviolet light.

repair polymerase DNA POLYMERASE I.

repair replication The synthesis, by means of a repair enzyme, of single-stranded DNA segments to replace damaged segments that have been excised from double-stranded DNA; in this process, the undamaged strand serves as a template. *See also* repair enzyme; cut and patch repair; patch and cut repair.

repair synthesis The enzymatic filling of a gap in a DNA strand at a site at which a damaged segment (such as a thymine dimer) is excised.

reparase REPAIR ENZYME.

rep DNA Repetitive DNA.

repeated DNA sequences REPETITIVE DNA.

repeated gene family MULTIGENE FAMILY.

repeating polymer A polymer that consists of identical repeating units.

repeating unit The structural unit of a polymer, a large number of which are linked together to form the polymer; repeating units may be either identical or similar.

repeat pipet AUTOMATIC PIPET.

repeats Small tandem duplications. *See* tandem duplication.

repeat unit A major periodicity in molecular structure as deduced from x-ray diffraction patterns.

repetition frequency The number of copies of a given DNA sequence in the haploid genome. *Aka* repetition number.

repetition number REPETITION FREQUENCY.

repetitious DNA REPETITIVE DNA.

repetitive DNA A DNA that constitutes a significant fraction of the total DNA of eukaryotic cells and that is characterized by its content of a large number of copies of different nucleotide sequences. It is subdivided into slightly repetitive DNA (1 to 10 copies per haploid genome), middle (moderately) repetitive DNA (10 to several thousand copies per haploid genome), and highly repetitive DNA (several thousand to several million copies per haploid genome). Middle repetitive DNA contains the genes transcribed into rRNA, tRNA, and histones. Highly repetitive DNA frequently occurs as spacer DNA between structural genes. *Aka* repetitious DNA. *See also* reassociation kinetics; unique DNA.

repetitive genes MULTIGENE FAMILY.

replacement *See* amino acid replacement; conservative substitution; radical substitution.

replacement site A position in a gene at which a point mutation leads to an amino acid replacement in the protein coded for by the gene.

replacement vector *See* lambda cloning vector.

replica plating A method for producing a large number of identical patterns of bacterial colonies by pressing a cylinder, covered with sterile velvet, first against the bacterial colonies in a petri dish and then against a number of plain agar surfaces in other petri dishes.

replicase An RNA-dependent RNA polymerase that catalyzes the synthesis of RNA from the ribonucleoside-5'-triphosphates, using RNA as a template.

replicating fork The Y-shaped region of a replicating DNA molecule in which strand separation and synthesis of new strands takes place.

replicating form *See* replicative form.

replicating unit REPLICON.

replication 1. The process whereby a new daughter DNA molecule is synthesized from a parent DNA molecule which serves as a template for the synthesis; one or two daughter molecules will be synthesized depending on whether the parental DNA molecule was single- or double-stranded. 2. The process whereby a new daughter molecules is synthesized from either a parent DNA or a parent RNA molecule, with the parent molecule serving as a template for the synthesis. 3. A technique used in electron microscopy in which either a plastic or a carbon film is spread over the surface of the specimen, after which the specimen is removed and the remaining surface is subjected to shadowcasting. *See also* DNA replication.

replication bubble 1. EYE STRUCTURE. 2. D-LOOP.

replication-defective virus A virus that cannot complete its infective cycle due to the presence of one or more defective viral genes.

replication eye REPLICATION BUBBLE.

replication fork *See* replicating fork.

replication fragments OKAZAKI FRAGMENTS.

replication of DNA *See* DNA replication.

replication order The number of replicating Y-forks per DNA molecule that is undergoing replication.

replication origin A unique base sequence in double-stranded DNA at which replication begins. *Aka* Ori site.

replication polymerase DNA POLYMERASE III.

replication units Clusters of 20–80 replication origins in mammalian DNA that appear to be activated as a unit; within such units, each replication origin is spaced at intervals of 30–300 kbp from the next one.

replicative form A double-stranded intermediate that is formed during the replication of single-stranded DNA or RNA viruses; consists of the original viral nucleic acid strand which is hydrogen-bonded to a complementary strand. Abbr RF.

replicator A chromosome locus that forms part of the replicon and that, when acted upon by an initiator, initiates the replication of the DNA attached to it.

replicator locus *See* replicon.

replicon A functional unit of replication, analogous to the operon which is a functional unit of transcription. The replicon is a genetic element, of either DNA or RNA, that behaves as an autonomous unit during replication. In bacteria, plasmids, and viruses, the entire chromosome functions as a single replicon; in eukaryotes, each chromosome consists of many replicons. Each replicon contains two loci, an initiator locus and a replicator locus. For DNA replication, the initiator locus represents the site to which RNA polymerase binds and the replicator locus is the site at which DNA replication begins. The RNA primer synthesized by RNA polymerase is known as initiator.

repliconation The formation of a replicative functional plasmid in the recipient cell following plasmid transfer.

replicon fusion The linking together of two replicons, mediated by a transposon. If the two replicons are plasmids, the process is known as plasmid fusion. *Aka* cointegrate formation.

replicon misfiring Proposed term to describe an extra round of replication within any given chromosomal domain; an extra initiation at an origin of replication during one cell cycle. The probability for replicon misfiring is very low. *See also* firone.

replisome The large, multimolecular complex, consisting of DNA polymerase III, other enzymes, and proteins, that is assembled at the replicating fork of a bacterial chromosome and that carries out the various reactions in DNA replication. Commonly refers to all of the components except DNA polymerase I and DNA ligase.

replot SECONDARY PLOT.

reporter group A group of atoms or a molecule that can be introduced into a protein and that has a characteristic property, such as pK value, ultraviolet absorbance, or fluorescence, which is sensitive to changes in the polarity of the medium. The changes that occur in this characteristic property when the reporter group is attached to the protein are used to explore the nature of the immediate environment of the reporter group in the protein molecule.

rep protein The helicase that functions in DNA replication in *E. coli.*

representation ABUNDANCE.

repressible enzyme An enzyme, the synthesis of which is decreased when the intracellular concentration of specific metabolites reaches a certain level. *Aka* repressed enzyme. *See also* enzyme repression.

repressible system The regulatory system consisting of the components that function in enzyme repression. *See also* enzyme repression.

repressing metabolite COREPRESSOR.

repression *See* enzyme repression.

repressor 1. A protein molecule produced by a regulatory gene that either by itself, or in conjunction with a corepressor, prevents the synthesis of an enzyme by inhibiting the operator of the enzyme. *Aka* aporepressor. *See also* enzyme repression. 2. IMMUNITY SUBSTANCE.

reproducibility The degree to which an experimental measurement or result may be obtained repeatedly.

reproductive death The death of a cell that results from a failure of a DNA molecule to replicate.

reproductive mycelium AERIAL MYCELIUM.

reptation A "reptile-like," end-on, mode of migration, thought to account for the fact that large molecules exhibit nearly size-independent mobilities in high-voltage gel electrophoresis. Under these conditions, the sieving capacity of the gel becomes unimportant, and the molecules (typically, DNA molecules) are pulled through the gel like a train through a tunnel; all of the molecules migrate at the same velocity, independent of their size. *See also* field inversion gel electrophoresis.

repulsion 1. The repelling electrostatic force between two like charges. 2. The tendency of linked genes to be inherited separately on different chromosomes.

RER Rough endoplasmic reticulum.

RES Reticuloendothelial system.

resealed ghost RECONSTITUTED GHOST.

reserpine An alkaloid of the plant *Rauwolfia* that inhibits the packaging of norepinephrine in presynaptic vesicles; a drug that causes sedation and that is used in psychiatry.

residence time The average length of time that two molecules are close enough together to lead to a measurable result of their interaction. Thus, the residence time is the average length of time that a substrate and an enzyme must interact so that a product will be formed.

residual air The volume of air in the lungs that cannot be expelled voluntarily.

residual body The vacuole, with its contents, that remains in the lysosome after a primary lysosome has completed the digestion of a food vacuole.

residual index A measure of the agreement between a calculated structure and one obtained from x-ray diffraction data; essentially a measure of the difference between calculated structural factors and corresponding observed values.

residual relative infectivity The fraction of the initial infectivity of a virus preparation that remains after the neutralization reaction of the virus with antiviral antibodies has reached its equilibrium value.

residual variance In the analysis of variance, that part of the variability of the dependent variable that is attributable to chance or experimental error.

residue That portion of a monomer that is present in a polymer; the monomer minus the atoms removed from it in the process of polymerization. *See also* amino acid residue.

resilin An elastic protein in the exoskeleton of insects; resembles elastin in its properties and is rich in glycine and devoid of cysteine.

resin 1. A polymerized support used in chromatography. *See also* ion-exchange resin; electron-exchange resin. 2. A water-insoluble, heterogeneous plant material.

resin acid One of a group of acid constituents of natural resins; chemically, a resin acid is an aromatic diterpene.

resinate A salt or an ester of a resin acid.

resinoid 1. *n* A resin-like substance. 2. *adj* Resin-like.

resinous Of, or pertaining to, a resin.

resin soap An alkali metal salt of a resin acid.

resistance 1. The capacity of bacteria to resist infection by phage particles; results from the inability of the phage particles to adsorb to, and to inject their DNA into, the bacterial cells. 2. DRUG RESISTANCE.

resistance factor A bacterial episome that endows a recipient bacterium with resistance to an antibiotic; consists of a resistance-transfer factor (RTF) together with the genes for drug resistance (r-genes). *Abbr* R factor; RF. *Aka* R-plasmid. *See also* R-plasmid.

resistance-transfer factor *See* resistance factor.

resolution 1. The separation of enantiomers from a racemic mixture. 2. The minimum distance between two points in a microscopic specimen or in an x-ray diffraction pattern at which the points are seen as two distinct spots; resolution in which the minimum distance is small is referred to as high resolution, and resolution in which the minimum distance is large is referred to as low resolution. 3. The degree of separation between two components in a mixture as achieved by chromatographic, electrophoretic, or other separation techniques. 4. The cutting process that separates the two participating DNA molecules during genetic recombination, as in the Holliday model, the asymmetric transfer model, and replicon fusion.

resolvase An enzyme responsible for separation of the cointegrate structure into donor and recipient units.

resolve To achieve resolution

resolving gel *See* disc gel electrophoresis.

resolving power 1. The capacity of a magnifying system to reveal detail. *See also* resolution (2). 2. The capacity of a fractionating system to separate components. *See also* resolution (3).

resolving time   COINCIDENCE TIME.

resolving time loss   COINCIDENCE LOSS.

resonance   1. The phenomenon that a compound that can be represented by two or more equivalent, or nearly equivalent, electronic formulas has in reality a structure that is a composite of all the possible electronic formulas and that is more stable than any of the separate structures. 2. The phenomenon in which a system, subjected to impulses (such as sound or electromagnetic radiation) having the same frequency as one of the natural frequencies of the system, responds with this natural frequency or a multiple thereof.

resonance energy transfer   The stepwise and radiationless excitation of chromophores located near each other; light excites one chromophore which then excites an adjacent one with similar electronic properties, and so on. The transmission of the excitation energy by the antenna molecules in photosynthesis is an example.

resonance hybrid   The true structure of a compound for which resonance structures may be formulated; the composite of all the possible resonance structures of a compound.

resonance Raman spectroscopy   A form of Raman scattering in which one uses a laser beam of a wavelength that is absorbed in an electronic transition.

resonance stabilization   The stabilization of a compound as a result of resonance; due to the fact that in resonance the pi electrons of the compound are less localized and hence have a lower potential energy level than in the absence of resonance.

resorcinol test   SELIWANOFF TEST.

resorption   *See* remodeling.

respiration   1. The cellular oxidative reactions of metabolism, particularly the terminal steps, by which nutrients are broken down; the reactions, which require oxygen as the terminal electron acceptor, produce carbon dioxide as a waste product, and yield utilizable energy. The major pathway of respiration consists of (a) the formation of acetyl coenzyme A from carbohydrate, fatty acid, and amino acid metabolism, (b) the citric acid cycle, and (c) the electron transport system. 2. The physical and chemical processes by which an organism transports oxygen to the tissues and removes carbon dioxide from them. 3. The act of breathing; inhaling and exhaling; inspiration and expiration.

respiratory acidosis   A primary acidosis that is caused by a decrease in respiration which leads to an increase in the carbon dioxide and carbonic acid concentrations of the plasma.

respiratory alkalosis   A primary alkalosis that is caused by an increase in respiration which leads to a decrease in the carbon dioxide and carbonic acid concentrations of the plasma.

respiratory assembly   A self-contained unit in the inner mitochondrial membrane that consists of fixed numbers of molecules of the various electron carriers; the unit has been fractionated into four respiratory complexes designated complexes I–IV. *See also* complexes I–IV.

respiratory burst   An increase in the consumption of oxygen, with formation of hydrogen peroxide and superoxide anions, that accompanies (a) the engulfment of bacteria and other foreign objects by leukocytes, and (b) the destruction of tumor cells by macrophages.

respiratory chain   An electron transport system that functions in respiration; the mitochondrial electron transport system in which molecular oxygen is the terminal electron acceptor.

respiratory chain phosphorylation   OXIDATIVE PHOSPHORYLATION.

respiratory complex   *See* complexes I–IV.

respiratory control   ACCEPTOR CONTROL.

respiratory-control index   ACCEPTOR-CONTROL RATIO.

respiratory distress syndrome   A pathological condition, characterized by rapid and shallow breathing and by cyanosis, that results from a deficiency of lung surfactant and that is a major cause of death in premature babies. *Abbr* RDS.

respiratory enteric orphan virus   REOVIRUS.

respiratory enzyme   1. An enzyme that functions in cellular respiration. 2. CYTOCHROME OXIDASE.

respiratory enzyme complex   *See* complexes I–IV.

respiratory inhibitor   Any substance that inhibits the flow of electrons along the electron transport system and/or the synthesis of ATP in oxidative phosphorylation.

respiratory pigment   A protein pigment that functions in the reactions of cellular respiration; hemoglobin, myoglobin, and hemerythrin are examples.

respiratory poison   RESPIRATORY INHIBITOR.

respiratory protein   RESPIRATORY PIGMENT.

respiratory quotient   The number of moles of carbon dioxide produced by a tissue or an organism divided by the number of moles of oxygen consumed during the same time. *Abbr* RQ.

respiratory reduction   1. NITRATE RESPIRATION. 2. SULFATE RESPIRATION.

respiratory repression   The regulation of the synthesis of an enzyme by an exogenous electron acceptor that is independent of either specific or catabolite repression.

respiratory syncytial virus A virus belonging to the group of paramyxoviruses that, in humans, gives rise to localized diseases of the respiratory tract. *Abbr* RSV.

respiratory virus A virus that infects the respiratory system.

respirometer An instrument for measuring and/or recording respiratory movements.

respirometry The measurement of the kinetics of oxygen uptake and carbon dioxide evolution during respiration.

responder An animal that produces an immune response when challenged with a given antigen.

resting cell A metabolically active cell that is not in the process of dividing.

resting heat The heat produced by a resting muscle.

resting muscle A muscle that is not in the process of contracting or relaxing.

resting nucleus A metabolically active nucleus that is not in the process of dividing.

resting potential The membrane potential of an unstimulated membrane.

restitutive protection The protection of biomolecules against damage from an ionizing radiation by chemical substances that aid in the restoration of primary lesions to their original condition, but do not alter the number of these lesions. *See also* competitive protection.

restraining autacoid *See* autacoid.

restricted diffusion Diffusion through a porous medium in contrast to free diffusion which occurs in the gas phase or in solution.

restricted diffusion chromatography GEL FILTRATION CHROMATOGRAPHY.

restricted DNA A DNA from one cell that is prevented from replicating in a related, but not identical, cell because it is degraded by endogenous endonucleases of the related cell.

restricted rotation The limited rotation about a bond that can be attained by an atom or a group of atoms in a molecule; the rotation about a double bond and the rotation about the bond in a ring structure are examples.

restricted transduction SPECIALIZED TRANSDUCTION.

restricted virus A virus, the host range of which is limited as a result of a host-induced modification.

restriction The ability of a bacterial strain to degrade the DNA from a related, but not identical, strain. *See also* restricted DNA; restriction enzyme.

restriction allele *See* restriction gene.

restriction endonuclease RESTRICTION ENZYME.

restriction enzyme An endonuclease of prokaryotes that catalyzes the degradation of foreign DNA such as that of a different bacterium or a phage. Restriction enzymes recognize specific base sequences in DNA (generally palindromes or inverted repeats) and hence cut a DNA molecule into a relatively small number of fragments (restriction fragments). There are three major types of restriction enzymes. Type I (or class I) enzymes recognize specific base sequences but make cuts elsewhere in the DNA (at about 1000 bp from the recognition site); type II (or class II) enzymes recognize specific base sequences and make cuts only within or next to these sequences. A type II enzyme makes two single-strand breaks, one in each strand, at the recognition site. If the two breaks occur at the center of symmetry, the resulting strand ends are referred to as flush or blunt ends; if the two breaks are equidistant from, and on opposite sides of, the center of symmetry, the resulting single-strand ends are known as cohesive or sticky ends. Type III (or class III) enzymes recognize specific base sequences and make cuts at about 25 bp from the recognition site. Type I and III enzymes carry both endonuclease and methylase activity on a single protein molecule; type II enzymes have only endonuclease activity. *Aka* restriction endonuclease. *See also* isoschizomer; isocaudamer.

restriction fragment A segment of DNA produced from a larger DNA molecule by the action of a restriction enzyme. *See* restriction enzyme.

restriction fragment length polymorphism The variations that occur in the pattern of restriction fragments, produced by the same restriction enzyme, from the DNA of different individuals of the same species. Such variations will be observed if one or more mutations have occurred in the DNA of an individual, thereby either creating or abolishing restriction sites for the specific restriction enzyme. The analysis of restriction fragments is, therefore, useful in screening for mutations and genetic diseases. *Abbr* RFLP. *Aka* restriction polymorphism.

restriction gene A gene, the product of which is a restriction enzyme.

restriction map A diagrammatic representation of a linear or circular DNA molecule which shows the positions at which one or more restriction enzymes would make cuts.

restriction–modification system A system for the selective degradation of foreign DNA in prokaryotic cells that is based on the presence of two groups of enzymes, restriction enzymes (restriction endonucleases) and DNA methylases (modification methylases). The latter methylate the host DNA at restriction sites and thereby protect the host DNA against degradation by the host restriction enzymes. Foreign DNA, such as phage DNA, that enters the prokaryotic cell, is not protected by methylation at the same restriction sites and is, therefore, degraded by the host restriction enzymes.

restriction mutant A mutant that carries a restriction gene.

restriction point A point of no return in the G_1 phase of the cell cycle. Once this point has been passed, the cell will complete the rest of the cycle (S, G_2, and M phases) regardless of the external conditions. *Abbr* R point.

restriction polymorphism RESTRICTION FRAGMENT LENGTH POLYMORPHISM.

restriction site A site in DNA that contains a specific base sequence and that can be cleaved by a particular restriction enzyme.

restrictive cell A cell in which a conditional lethal mutant cannot grow.

restrictive conditions Conditions (such as the temperature or the type of host) under which a conditional mutant cannot either grow or express its mutant phenotype.

restrictive transduction SPECIALIZED TRANSDUCTION.

resultant spin The algebraic sum of the spin quantum numbers of all the electrons of the atom or the molecule; the resultant spin is zero if all of the electrons are paired.

retardation coefficient The slope of the line that is obtained when the relative migration distance of a protein in gel electrophoresis is plotted as a function of the reciprocal of the gel concentration. The relative migration distance is the ratio of the migration distance at a given gel concentration to that at a standard gel concentration.

retentate The material retained by a semipermeable membrane.

retention index *See* Kovats retention index system.

retention of configuration The maintenance of a given enantiomeric configuration about an asymmetric carbon atom in the course of a chemical reaction.

retention time The time between the injection of a sample into a gas chromatographic column and the appearance of the peak maximum.

retention volume The volume of gas in gas chromatography that is required to elute the compound of interest. *See also* holdup volume.

reticulate evolution An evolutionary pattern that, when diagrammed, resembles a net; such patterns are characteristic of the evolution of plant species.

reticulocyte An immature red blood cell that actively synthesizes hemoglobin and that possesses functioning pathways of glycolysis, the hexose monophosphate shunt, and the citric acid cycle.

reticuloendothelial system Collectively, the cells in spleen, liver, bone marrow, and other tissues that are involved primarily in phagocytosis and in the metabolism of hemoglobin. *Abbr* RES.

reticuloendothelium RETICULOENDOTHELIAL SYSTEM.

retinal The aldehyde form of vitamin A; $retinal_1$ is the aldehyde form of vitamin A_1 and $retinal_2$ is the aldehyde form of vitamin A_2. *See also* 11-*cis*-retinal; all-*trans*-retinal.

11-*cis*-retinal An isomeric form of retinal that is converted by the action of light to the all-trans isomer.

retinaldehyde RETINAL.

retinene RETINAL.

retinoid Collective term for vitamin A and its analogues.

retinol *See* vitamin A.

retinol-binding protein A plasma protein that binds and transports vitamin A, in the form of *trans*-retinol from the liver to extrahepatic tissues. The binding of vitamin A to the protein serves to solubilize vitamin A and to protect it against oxidation. *Abbr* RBP.

retinol equivalent A measure of vitamin A activity in foodstuffs equal to 1 μg of retinol or 6 μg of β-carotene.

retinyl ester An ester of all-*trans*-retinol.

retravirus Variant spelling of retrovirus.

retroconversion A biohydrogenation reaction, observed in liver and testes, whereby unsaturated fatty acids are converted to saturated ones.

retrogene PROCESSED GENE.

retrogradation The formation of microcrystals and precipitates that occurs in starch gels and in starch solutions upon standing; caused by the separation of amylose molecules that become aligned and hydrogen bonded.

retroinhibition FEEDBACK INHIBITION.

retroposition The RNA-mediated movement of genetic information from one locus to another; the movement of an RNA mobile genetic element; transposition that is RNA-mediated.

retroposon A mobile genetic element that transposes via an RNA intermediate; an RNA transposon. A retrovirus is a retroposon. *Aka* retrotransposon.

retroregulation The regulation of mRNA translation by DNA segments that lie downstream from the segment that codes for the mRNA.

retrosteroid A synthetic steroid that has a structure at carbon atoms 9 and 10 which is opposite to that in progesterone. Retrosteroids are made from lumisterols and are extremely active as progestogens.

retrotranscript A DNA segment formed via reverse transcription.

retrotransposon RETROPOSON.

retrovirus An RNA virus that contains the enzyme reverse transcriptase which allows the viral RNA genome to be transcribed into viral DNA. The name retrovirus refers to this "backward" transcription. The transcribed viral DNA is then integrated in, and replicated with, the host chromosome. Some retroviruses contain oncogenes and their infection of a host results in transformation of normal cells to cancer cells. Retroviruses contain two identical, single-stranded RNA molecules, held together by hydrogen bonding near their capped 5'-ends. *See also* oncornavirus.

reversal The change of cancer cells to normal cells or to benign tumor cells.

reverse bend BETA BEND.

reverse burst titration A titration procedure for the active site of an enzyme; based on measuring the decrease in absorbance resulting from the binding to the active site of a chromogenic compound that is not acted upon further by the enzyme. *See also* burst titration.

reverse dialysis A method for concentrating a solution of macromolecules. The sample is placed in a dialysis bag and the latter is packed in a dry, water-soluble, nondialyzable polymer such as polyethylene glycol. Water will leave the dialysis bag to equilibrate with the dry external phase and thus produce a more concentrated sample solution inside the bag. Sucrose can be used similarly but, since it is dialyzable, it must subsequently be removed from the bag by ordinary dialysis.

reversed micelle *See* micelle.

reversed phase chromatography Partition chromatography in which the mobile phase is more polar than the stationary phase. *Abbr* RPC. *Aka* reversed phase partition chromatography.

reverse electron transport The reduction of NAD^+ to NADH by mitochondria in the presence of ATP. *Aka* reverse electron flow.

reverse flow chromatography A column chromatographic technique in which the flow of solvent is reversed, proceeding from the bottom of the column to the top. This minimizes the decrease in flow rates that frequently occurs with conventional chromatography as a result of the packing of the stationary phase at the lower end of the column.

reverse isotope dilution analysis INVERSE ISOTOPE DILUTION ANALYSIS.

reverse isotopic trapping A technique for determining whether more than one metabolic pathway leads to the synthesis of compound A; performed by continuously administering a labeled precursor X, isolating compound A, and determining the specific activity of compound A. By comparing the specific activity of compound A with that of the administered precursor X, it is possible to decide whether A is synthesized solely from X, or also from other precursors by way of different pathways.

reverse mutation REVERSION (1).

reverse osmosis The movement of water or another solvent from a more concentrated to a more dilute solution across a semipermeable membrane; the desalination process in which seawater is forced across a semipermeable membrane is an example.

reverse passive anaphylaxis The anaphylactic reaction produced in an animal by injecting it first with an antigen that is itself an immunoglobulin and, following a latent period, by injecting it with an antiserum that is specific for the immunoglobulin used as an antigen.

reverse phase chromatography *See* reversed phase chromatography.

reverse transcriptase An RNA-dependent DNA polymerase that catalyzes the synthesis of DNA from deoxyribonucleoside-5'-triphosphates, using RNA as a template. The enzyme has been found in retroviruses and in some other viruses. The reaction catalyzed by the enzyme is in contradiction to the flow of genetic information described by the original central dogma of molecular biology: DNA → RNA → protein. The enzyme actually has three enzymatic activities: It copies a single-stranded RNA molecule to yield a double-stranded DNA–RNA hybrid; it copies a single-stranded DNA molecule to yield a double-stranded DNA molecule; and it degrades the RNA in a DNA–RNA hybrid (this is called its ribonuclease H, RNAase H, activity). In view of our current knowledge of this enzyme and of RNA replication, the original central dogma of molecular biology must now be represented as: DNA ⇌ RNA → protein. *Abbr* RT. *See also* retrovirus.

reverse transcription The reaction catalyzed by the enzyme reverse transcriptase.

reverse turn BETA BEND.

reversible boundary spreading test A test applied in moving boundary electrophoresis to assess the homogeneity of the sample. A boundary is first allowed to migrate for some distance and the electric field is then reversed; any sharpening of the boundary that occurs upon reversal of the electric field reflects heterogeneity in the sample and cannot be attributed to diffusion. The test can also be applied to zone electrophoresis.

reversible inhibitor   An inhibitor that binds to an enzyme in an equilibrium reaction so that the inhibition can be reversed by removal of the inhibitor from the enzyme by such processes as dialysis or ultrafiltration. *See also* irreversible inhibitor; reactivation.

reversible metabolic pathway   A metabolic pathway in which the equilibrium constants of the individual reactions are such that products can be converted to the reactants in significant amounts. The pathway is called directly reversible if the reaction sequence can be traversed in both directions: $A \rightleftharpoons B \rightleftharpoons C \rightleftharpoons D$; and is called indirectly reversible if it is traversed principally in one direction:

$$A \rightleftharpoons B \rightleftharpoons C \rightleftharpoons D$$
$$\diagdown\!\!\diagup$$
$$E$$

reversible process   A process in equilibrium thermodynamics in which the system goes from an initial to a final equilibrium state through a succession of equilibrium states. Thus, the system is always close to equilibrium and the direction of the process can be reversed by an infinitestimal change. The net entropy change for the system plus its surroundings is zero for such a process.

reversible reaction   A chemical reaction that (a) establishes an equilibrium and can be made to proceed in either direction by a change in the conditions, or (b) has an equilibrium constant of the order of 1.0, or (c) proceeds equally well in both directions because of other factors.

reversion   1. The change of a mutant gene to its state prior to mutation; a reverse mutation. Also referred to as true reversion as distinct from restoration of the genetic function by a suppressor mutation. 2. DEADAPTATION.

reversion index   The mutation index for a reverse mutation.

reversion spectroscope   A spectroscope that allows the formation of two spectra alongside each other, with the wavelength changing in one direction for one spectrum and in the opposite direction for the other spectrum.

revertant   1. A cell, a virus, or an organism that carries a gene that has undergone reversion. 2. A gene that has undergone reversion.

Reynold's number   A quantity that characterizes the flow of a liquid through a cylindrical tube; equal to $2\rho u a/\eta$, where u is the average velocity, a is the radius of the tube, ρ is the density, and η is the viscosity.

RF   1. Replicative form. 2. Resistance factor.

R factor   1. Resistance factor. 2. Release factor.

RFLP   Restriction fragment length polymorphism.

R_f value   The ratio of the distance traveled by a compound in flat-bed chromatography to that traveled by the solvent.

R_G   Average radius of gyration.

RGD   The tripeptide arginine-glycine-aspartic acid that occurs in many adhesive proteins, present in the extracellular matrix, such as fibronectin, collagen, and fibrinogen. The RGD sequence of the adhesive proteins is recognized by at least one member of a family of structually related receptors, called integrins, and this interaction is believed to play a major role in cell adhesion, cell migration, and cell differentiation.

R gene   1. Regulator gene. 2. Drug resistance gene; *see* resistance factor.

R group   1. That portion of an organic molecule that does not contain the functional group of interest; an organic radical. 2. The side chain of an amino acid.

R_g value   A value that is somewhat like an R_f value and that is used in the chromatography of carbohydrates; it is equal to the ratio of the distance traveled by a given carbohydrate to that traveled by glucose.

Rha   Rhamnose.

rhabdovirus   An enveloped, helical animal virus that contains single-stranded RNA. Rhabdoviruses contain lipoproteins in their envelope and include the causal agents of rabies and vesicular stomatitis.

rhamnose   A deoxysugar that occurs in some bacterial cell walls. *Abbr* Rha.

rhamsan gum   A water-soluble, heteropolysaccharide, produced by fermentation of some species of *Alcaligenes*, that yields solutions that have high viscosities at low concentrations.

rhapidosome   A rod-shaped or tubular nucleoprotein structure that occurs intracellularly in some bacterial species and that is believed to be derived from the cell membranes as the bacteria undergo lysis.

Rh blood group system   A human blood group system in which the Rh factor is the antigen and individuals are either Rh positive or Rh negative; so called since it was first discovered in the Rhesus monkey. An Rh positive baby of an Rh negative mother may be born with a hemolytic disease called erythroblastosis fetalis.

rheology   The science that deals with the deformation and flow of matter.

rheostat   A variable resistor.

Rhesus factor   Rh FACTOR.

rheumatoid factor   A specific serum γ-globulin, found in individuals afflicted with rheumatoid arthritis.

Rh factor   The antigen of the Rh blood group system.

rhinovirus A virus that belongs to a subgroup of picornaviruses and that causes respiratory infections.

Rhizobium A genus of gram-negative, aerobic, nitrogen-fixing bacteria that live as symbionts in the root nodules of leguminous plants. *Aka* nodule bacteria.

rhizopterin SLR FACTOR.

rhodanese The enzyme thiosulfate sulfur transferase that catalyzes the displacement of sulfur from thiosulfate by cyanide and thereby serves to detoxify cyanide.

rhodoplast A photosynthetic organelle of red algae; a red plastid that contains biliproteins.

rhodopsin A visual pigment, consisting of rod opsin plus retinal$_1$, that occurs in mammals and other vertebrates and that has an absorption maximum at 500 nm.

rhodopsin cycle VISUAL CYCLE.

Rhodospirillum rubrum A photosynthetic bacterium used for studies of photosynthesis.

rho factor A protein that is required for the correct termination of transcription. The *E. coli* protein is a hexamer of identical, 419 residue, subunits. The rho factor is an enzyme that catalyzes the unwinding of RNA–DNA and RNA–RNA double helices. *Sym* ρ.

rho-independent terminator A base sequence, in the DNA of *E. coli*, that signals the termination of transcription and that is recognized by DNA-dependent RNA polymerase in the absence of the rho factor.

RHP cytochrome A cytochrome that contains two hemes per molecule and that is considered to be a variant of cytochrome *c*; the heme groups are considered to be bound through thioether linkages as in cytochrome *c*, but to lack hemochrome linkages to extraplanar ligands of the protein at physiological pH.

rH scale A scale for evaluating oxidation–reduction reactions that is based on rH values rather than on electrode potentials. The rH value is the negative logarithm of the hydrogen pressure in atmospheres, and runs from zero for a hydrogen pressure of 1 atm to 41 for an oxygen pressure 1 atm. The rH value of a reaction is obtained by relating the reaction to the hydrogen half-reaction; rH values for most biochemical systems fall in the range of 0 to 25.

RIA Radioimmunoassay.

Rib Ribose.

α-ribazole The compound 1-α-D-ribofuranosyl-5,6-dimethylbenzimidazole, that forms part of the structure of vitamin B$_{12}$ and that has a nucleoside structure in which an uncommon nitrogenous base is linked to the sugar by means of an α-glycosidic bond.

ribitol A five-carbon sugar alcohol that forms part of the structure of riboflavin, FMN, and FAD.

ribodeoxyvirus RETROVIRUS.

riboflavin Vitamin B$_2$; a vitamin that is widely distributed in nature and the coenzyme forms of which are FMN and FAD.

riboflavin mononucleotide FLAVIN MONONUCLEOTIDE.

riboflavin phosphate FLAVIN MONONUCLEOTIDE.

riboflavin-5′-phosphate FLAVIN MONONUCLEOTIDE.

ribofuranose Ribose that has a 5-membered ring structure resembling that of the compound furan.

ribonuclease An endonuclease that catalyzes the hydrolysis of RNA; it cleaves 3′,5′-phosphodiester bonds in RNA.

ribonuclease III A ribonuclease that specifically hydrolyzes double-stranded RNA.

ribonuclease A A ribonuclease that leads to the production of mono- and oligonucleotides consisting of, or terminating in, a 3′-pyrimidine nucleotide. *Aka* pancreatic ribonuclease.

ribonuclease D A ribonuclease that removes a number of nucleotides from precursor tRNA thereby producing the 3′-terminus of mature tRNA.

ribonuclease H A ribonuclease that specifically degrades RNA in DNA–RNA hybrids. It is associated with reverse transcriptase and has also been isolated from other sources.

ribonuclease P A bacterial ribonuclease that cleaves an oligonucleotide from precursor tRNA, thereby producing the 5′-terminus of mature tRNA. The enzyme is unusual in that it contains 86% RNA and only 14% protein by weight. Moreover, the RNA possesses the catalytic activity, the protein does not. The protein merely serves to maintain the proper folding of the RNA to maximize its catalytic activity. Ribonuclease P is a ribozyme.

ribonuclease protection A method for determining the regions in RNA that interact and make contact with a given protein. The protein is allowed to bind to the RNA. The ribonucleoprotein complex (such as a tRNA molecule and its cognate aminoacyl-tRNA synthetase) is then treated with various endonucleases that digest away most of the RNA outside the regions protected by the bound protein. The remaining oligonucleotide fragments, representing the RNA regions that were in intimate contact with the protein, are then identified.

ribonuclease S An association of the two fragments (S peptide and S protein) that are formed from ribonuclease A by specific cleavage with subtilisin between amino acid residues 20 and 21. The two fragments are linked noncovalently and produce a fully active ribonuclease.

ribonuclease T1   A ribonuclease that leads to the production of mono- and oligonucleotides consisting of, or terminating in, a 3'-guanine nucleotide.

ribonuclease T2   A ribonuclease that leads to the production of mono- and oligonucleotides consisting of, or terminating in, a 3'-adenine nucleotide.

ribonuclease U$_2$A(U$_2$B)   *See* isoforms.

ribonucleic acid   The nucleic acid (*abbr* RNA) that occurs in three major forms as ribosomal, transfer, and messenger ribonucleic acid, all of which function in the biosynthesis of proteins. RNA is a polynucleotide that is characterized by its content of D-ribose and the pyrimidines uracil and cytosine.

ribonucleoprotein   A conjugated protein that contains RNA as the nonprotein portion. *Abbr* RNP.

ribonucleoside   A nucleoside of D-ribose.

ribonucleotide   A nucleotide of D-ribose.

ribonucleotide reductase   An enzyme that catalyzes the conversion of ribonucleoside diphosphates to deoxyribonucleoside diphosphates.

ribophorin   One of a number of glycoproteins that are an integral part of the membrane of the rough endoplasmic reticulum and that serve as binding sites for ribosomes. *See also* signal hypothesis.

ribose   The five-carbon aldose that is the carbohydrate component of ribonucleic acid. *Abbr* Rib.

ribose nucleic acid   *See* ribonucleic acid.

riboside   A glycoside of ribose.

ribosoid   Collective term for ribonucleic acid and ribonucleoprotein.

ribosomal   Of, or pertaining to. ribosomes.

ribosomal DNA   *See* rDNA.

ribosomal particle   A ribosome or any of its complete subunits.

ribosomal precursor RNA   PRECURSOR RIBOSOMAL RNA.

ribosomal protein   A protein that forms part of the ribosome. Ribosomal proteins are linked noncovalently to the ribosomal RNA and, together with the ribosomal RNA, form the two subunits of the ribosome. In bacterial ribosomes, there are 52 different ribosomal proteins per ribosome. The proteins of the large ribosomal subunit are designated by the letter L; those of the small subunit are designated by the letter S. *Abbr* r-protein.

ribosomal RNA   The RNA that is linked noncovalently to the ribosomal proteins in the two ribosomal subunits and that constitutes about 80% of the total cellular RNA. Three types of ribosomal RNA have been identified, having

sedimentation coefficients of about 5S, 16 to 19S, and 23 to 29S, depending on the source of the ribosomes. *Abbr* rRNA.

ribosomal sieve   *See* double-sieve mechanism.

ribosomal stalling   *See* attenuation.

ribosomal subparticle   1. A ribosomal subunit, such as the 30S or the 50S particle of bacterial ribosomes. 2. A particle that either is a precursor of a ribosome, or is prepared from a ribosome by the removal of some ribosomal proteins. *See also* intersome.

ribosomal subunit   One of the two ribonucleoprotein particles that make up the complete ribosome; the 30S or the 50S particle in bacteria, the 40S or the 60S particle in plant and animal cells.

ribosomal subunit exchange   *See* subunit exchange.

ribosome   1. One of a large number of subcellular, nucleoprotein particles that are composed of approximately equal amounts of RNA and protein and that are the sites of protein synthesis in the cell. Each ribosome is roughly spherical in shape, has a diameter of about 200 Å, and consists of two unequal subunits linked together noncovalently by means of magnesium ions and other bonds. Ribosomes occur free in the cytoplasm and attached to the endoplasmic reticulum. Four classes of ribosomes have been identified (bacterial, plant, animal, and mitochondrial) and they can be characterized by the sedimentation coefficients of the monomers, of the subunits, and of the ribosomal RNA. The bacterial ribosome contains 52 different protein molecules and 3 different RNA molecules; the smaller subunit contains 21 protein molecules and 1 RNA molecule; the larger one contains 31 protein molecules and 2 RNA molecules. The ribosome has two binding sites for transfer RNA (A site, P site) and can attach to messenger RNA as well. *Abbr* Rb. 2. RIBOSOMAL SUBUNIT.

ribosome binding site   SHINE–DALGARNO SEQUENCE.

ribosome binding technique   *See* binding assay (2).

ribosome crystal   An aggregate of ribosomes, packed in a regular array, that has been observed in several eukaryotic organisms.

ribosome cycle   The set of reactions whereby ribosomal subunits combine to form the intact ribosome during the initiation of translation, travel along the messenger RNA as intact ribosomes, and dissociate back to the subunits during the termination of translation. *See also* subunit exchange.

ribosome dimer   An aggregate of two ribosomes, consisting of two small and two large ribosomal subunits.

ribosome dissociating factor A factor that promotes the dissociation of the ribosome monomers into the two subunits during the termination of translation.

ribosome editing The hypothesis that errors in protein synthesis, resulting from the transfer of a peptide to an inappropriate aminoacyl-tRNA (thus producing an inappropriate peptidyl-tRNA), are detected by the ribosome; this detection then leads to the preferential dissociation of the inappropriate peptidyl-tRNA from the ribosome.

ribosome epicycle The steps through which the ribosome passes upon the addition of each amino acid during the elongation phase of protein synthesis; includes aminoacyl-tRNA binding to the ribosome, peptide bond formation, (transpeptidation), and translocation.

ribosome exchange SUBUNIT EXCHANGE.

ribosome monomer A complete ribosome that is composed of one small and one large subunit.

ribosome read-through See read-through.

ribosome receptor A receptor in the membrane of the rough endoplasmic reticulum to which a ribosome becomes bound during protein synthesis according to the signal hypothesis. The receptor is believed to be adjacent to a membrane pore and the ribosome becomes bound to the receptor after the signal sequence has become detached from the signal recognition protein.

ribosome runoff The loss from polysomes of ribosomes that have not completed the synthesis of the polypeptide chain.

ribosubstitution The replacement of some of the deoxyribonucleotides in a DNA by ribonucleotides; achieved by the in vitro synthesis of DNA under conditions that allow the incorporation of ribonucleotides from a mixture of ribo- and deoxyribonucleoside-5′-triphosphates. Ribosubstituted DNA can be used as an aid in the determination of the base sequence of the DNA.

ribosylthymine RIBOTHYMIDINE.

5-ribosyluracil PSEUDOURIDINE.

ribothymidine The ribonucleoside of thymine; an unusual nucleoside that does not, as a rule, occur in RNA. *Sym* Thd.

ribothymidylic acid The ribonucleotide of thymine.

ribotide A ribonucleotide.

ribotype The ribonucleoprotein complement of a cell.

ribovirus RNA VIRUS.

ribozyme A catalytic RNA segment that has the ability to break and form covalent bonds. Ribonuclease P and the intron in the rRNA precursor molecule of *Tetrahymena thermophila* are two examples.

ribulose A five-carbon ketose that is an intermediate in the hexose monophosphate shunt. *Abbr* Rul.

ribulose-1,5-bisphosphate A five-carbon ketose that is the acceptor of carbon dioxide in the Calvin cycle. Previously known as ribulose-1,5-diphosphate. *Abbr* R1,5BP; R1,5DP.

ribulose-1,5-bisphosphate carboxylase The enzyme that catalyzes the fixation of carbon dioxide by ribulose-1,5-bisphosphate in the Calvin cycle. *Abbr* Rubisco, RuBPCase; RuBP carboxylase.

ribulose diphosphate carboxydismutase RIBULOSE-1,5-BISPHOSPHATE CARBOXYLASE.

Rice test INDIRECT COMPLEMENT FIXATION TEST.

Richard's box A simple arrangement for constructing a model of a protein of known amino acid sequence from the electron density map of the protein; involves a half-silvered mirror which is used to superimpose a wire model of the structure onto a pile of lucite sections where the electron density map is plotted.

ricin A plant protein in the seeds of castor beans (*Ricinus communis*) that is toxic to animals and humans, inhibits protein synthesis, and has antitumor activity; a lectin that agglutinates red blood cells.

ricinin A toxic pyridine alkaloid from the seeds of castor beans (*Ricinus communis*); it is biosynthesized from nicotinic acid.

rickets A disease of children that is characterized by a softening and bending of the bones and that is caused by a deficiency of vitamin D. In rickets, as opposed to osteoporosis, the bone matrix remains intact and continues to be synthesized but formation of bone mineral is impaired. There is, therefore, a change in the ratio of bone mineral to bone matrix.

Rieske protein An iron–sulfur protein, containing a 2 Fe/S center, that occurs in association with cytochromes b and c_1 in complex III of the electron transport system.

rifampicin A semisynthetic antibiotic; the most commonly used form of the rifamycins.

rifamycins A group of antibiotics, produced by *Streptomyces mediterranei*, that inhibit the initiation of transcription (DNA-dependent RNA synthesis) in prokaryotes by binding to DNA-dependent RNA polymerase (RNA polymerase).

right splicing junction See splicing junctions.

rigor mortis The irreversible contraction of muscles upon the death of an animal.

rimorphin An opioid peptide, derived from prodynorphin.

ring A chain of atoms in which the ends of the chain are linked together covalently; a cyclic compound.

ring current The electric current set up in an aromatic ring when the latter is oriented perpendicularly to a strong magnetic field. Under these conditions, the pi (π) electrons circulate around the ring in a direction such that they induce a small local magnetic field which opposes the applied field in the middle of the ring but reinforces it outside the ring. The effect is important in nuclear magnetic resonance where it leads to a deshielding of aryl protons; the latter experience a magnetic field that is greater than the applied field and thus come into resonance at a lower applied field (downfield).

Ringer's solution A salt solution that is approximately isotonic to the blood and to the lymph of mammals; used for the temporary maintenance of living cells. As first proposed, it consisted of sodium chloride, potassium chloride, calcium chloride, and sodium bicarbonate; various modifications of this solution are now in use.

ring-flip The interconversion of chair conformations of cyclohexane rings that results in the interchange of axial and equatorial positions.

ring pucker *See* puckered conformation.

ring test A rapid and simple precipitin test in which a solution of antigen is layered over a solution of antibodies in either a test tube or in a capillary, and the presence or absence of a precipitate at the interface is determined.

RISA Acronym for radioimmunosorbent assay; an isotopic technique for the demonstration of minute amounts of immunoglobulins of the IgE type.

rise curve The increase in color intensity of a sample as a function of time; used in reference to determinations with an autoanalyzer.

rise period The time interval during which the extracellular titer of a phage multiplication cycle increases to a maximum.

ristocetin A glycopeptide antibiotic, produced by certain *Actinomycetes*, that prevents peptidoglycan synthesis.

R locus The position on a chromosome of a regulator gene.

rII locus A segment of the chromosome of T4 phage to which fine structure genetic mapping was first applied.

R-loop mapping An electron microscopic technique for visualizing complementary regions in DNA and the corresponding eukaryotic mRNA. The mRNA is hybridized with the DNA under certain annealing conditions (R-looping conditions) at which the DNA–RNA hybrid is more stable than the DNA–DNA duplex. As a result, the mRNA hybridizes with the DNA sense strand and this DNA–RNA hybrid displaces a segment of the DNA antisense strand which forms a loop (R-loop). One intron, present in the DNA but not in the mature mRNA, leads to formation of two R-loops, two introns yield three R-loops, and so on.

R-meter A radiation meter that is calibrated to indicate roentgens.

r mutant RAPID LYSIS MUTANT.

R mutant A bacterial mutant that gives rise to a rough colony. *See also* rough strain.

R_m value A chromatographic term that is related to the R_f value of a compound. It is defined in the literature both as $R_m = (1 - R_f)/R_f$ and as $R_m = \log [(1 - R_f)/R_f]$.

Rn Radon.

RNA Ribonucleic acid.

RNA-11 A double-helical conformation of RNA that resembles that of A-DNA. It has 11 base pairs per helical turn and has the base pairs inclined to the helical axis by about 14°. *Aka* A-RNA.

RNAase Ribonuclease.

RNA coding triplet CODON.

RNA-dependent (directed) DNA polymerase REVERSE TRANSCRIPTASE.

RNA-dependent (directed) RNA polymerase REPLICASE.

RNA dot blot *See* dot blot assay.

RNA-driven hybridization A DNA–RNA hybridization technique in which excess RNA is used so that all complementary sequences in the single-stranded DNA will form DNA–RNA hybrids. *See also* playback experiment.

RNA gene A segment of DNA that has the information for the synthesis of an RNA molecule other than mRNA. DNA segments that code for ribosomal or transfer RNA are examples.

RNA ligase An enzyme that catalyzes the formation of a phosphodiester bond in RNA between a 3'-hydroxyl group and a 5'-phosphate group, which may or may not be on the same polynucleotide strand. RNA ligases are involved in the splicing of RNA.

RNA maturation SPLICING (2).

RNA modification PROCESSING (1).

RNA nucleotidyltransferase RNA POLYMERASE.

RNA phage An RNA-containing phage.

RNA polymerase An enzyme that functions in the transcription of DNA and that catalyzes the synthesis of RNA from the ribonucleoside-5'-triphosphates using DNA as a template; referred to as DNA-dependent (directed) RNA polymerase to distinguish it from RNA-dependent (directed) RNA polymerase. *See also* RNA polymerase classes.

RNA polymerase classes Multiple forms of DNA-dependent RNA polymerase that occur in eukaryotes. All are high molecular weight proteins and all are found in the nucleus. Each form is responsible for the synthesis of a particular type of RNA and the various enzymes can be distinguished by their sensitivity to inhibition by α-amanitin; RNA pol I makes rRNA and is resistant to amanitin; RNA pol II makes mRNA and is highly sensitive to amanitin; RNA pol III makes tRNA and the 5S RNA of ribosomes and is somewhat sensitive to amanitin.

RNA primer A short RNA fragment onto which are added deoxyribonucleotides by DNA polymerase III during DNA replication. *See also* primer.

RNA processing PROCESSING (1).

RNA puff *See* chromosome puff.

RNA replicase *See* replicase.

RNase Ribonuclease.

RNA sequencing *See* Maxam–Gilbert method.

RNasin An inhibitor of ribonuclease.

RNA splicing The process whereby nontranslatable RNA sequences (introns) in the primary transcript of a split gene are excised, and the remaining translatable sequences (exons) are joined together to yield the functional gene product.

RNA synthesizer An automated setup for the chemical synthesis of oligonucleotide segments of RNA.

RNA synthetase RNA-DEPENDENT RNA POLYMERASE.

RNA virus An RNA-containing virus. *See also* oncornavirus.

RNP Ribonucleoprotein.

ROA Raman optical activity.

road map FLOW CHART.

Robertson model UNIT MEMBRANE HYPOTHESIS.

Robison ester Glucose-6-phosphate.

Rochelle salt Potassium sodium tartrate.

rocket electrophoresis A modification of immunoelectrophoresis in which varying amounts of antigen are placed in sample wells and electrophoresed into an agarose gel which contains antibodies. As the antigen moves into the gel it encounters antibody but does not precipitate out until the antigen and antibody concentrations are equivalent. The resulting precipitate pattern resembles a rocket and the distance traveled by the tip of the rocket is proportional to the amount of antigen in the sample well. *Aka* electroimmunodiffusion.

rod A light receptor in the retina of vertebrates that functions in night vision.

rodenticide A chemical that kills rodents.

rod threshold The lowest light intensity that the fully dark-adapted eye can detect.

roentgen A quantity of ionizing radiation that results in the formation of 2.083×10^9 ion pairs (one electrostatic unit of charge of either sign) per cubic centimeter of dry air at 0 °C and 760 mm Hg. *Sym* r; R.

roentgen equivalent man The product of the radiation absorbed dose (rad) and the relative biological effectiveness (RBE); the quantity of radiation that, when absorbed by man, produces an effect equivalent to the absorption of 1 R of x or gamma radiation. *Abbr* REM. *Aka* roentgen equivalent mammal.

roentgen equivalent physical The amount of ionizing radiation capable of producing 1.615 $\times$ 10^{12} ion pairs per gram of tissue; the amount of ionizing radiation that will result in the absorption by tissue of 93 erg/g of tissue. *Abbr* REP.

roentgen rays X-RAYS.

rohferment A crude enzyme preparation from almonds that is rich in glycosidases, particularly in β-D-glucosidase and β-D-galactosidase.

Rohrschneider constant A constant for relating the gas chromatographic retention behavior of a compound to the polarity of the liquid stationary phase.

rolling circle replication A model for the replication of duplex circular DNA. According to this model, a nuclease nick in one strand is followed by the addition of nucleotides to the 3'-end of the nicked strand, a reaction catalyzed by DNA polymerase. At the same time, the 5'-end of the strand is rolled out as a free tail of increasing length, resulting in intermediates larger than the original duplex. Small fragments are then synthesized complementary to the free tail and are eventually joined together through the action of a ligase. *Aka* sigma replication.

ROM Acronym for read only memory; the permanent memory in a computer that consists of basic instructions for processing information. This memory cannot be altered.

Romanowsky dyes A group of composite dyes used to stain blood cells and blood parasites. *Aka* Romanowsky stains.

Roman square LATIN SQUARE.

root-mean-square end-to-end distance A measure of the spatial extension of a polymer; equal to the square root of the average of the squares of the distances between the ends of the polymer, carried out over all possible conformations of the polymer.

Roseman hypothesis The hypothesis that cell adhesion is due to the noncovalent interaction of surface located glycosyl transferases of one cell with glycoproteins or glycolipids located on the surface of a second cell.

rosette technique IMMUNOCYTE ADHERENCE.

Rossman fold A characteristic mononucleotide binding domain in proteins for such compounds as ATP, ADP, AMP, and FMN. More complex dinucleotides, such as NAD^+, $NADP^+$, and FAD, may be bound to the protein by means of two Rossman folds.

Thus, in pyridine-linked dehydrogenases, the coenzyme binding site consists of two mononucleotide binding domains; one binds the AMP portion of NAD^+, the other binds the nicotinamide portion. *Aka* mononucleotide binding domain.

rotamer A rotational isomer; a conformational isomer resulting from a rotation about single bonds.

rotary *See* rotational.

rotary evaporator FLASH EVAPORATOR.

rotary shaker *See* shaker.

rotating crystal method A method for the analysis of a single crystal by means of x-ray diffraction. The crystal is mounted and rotated about an axis, thereby producing a large number of diffraction spots; rotation of the crystal about all of its axes produces the maximum number of diffraction spots.

rotation The turning about an axis.

rotational base substitution The process whereby a base in one DNA strand exchanges position with its complementary base in the second strand. This may occur if the glycosidic bonds between the bases and their sugar molecules are broken by irradiation, and the hydrogen-bonded base pair is then rotated prior to being reinserted into the strands.

rotational diffusion The rotation of molecules about their axes that results in their achieving a random orientation. *See also* rotational relaxation.

rotational diffusion coefficient A measure of rotational diffusion that depends on the size and shape of the diffusing particle; specifically, $\theta = RT/N\zeta$, where θ is the rotational diffusion coefficient, ζ is the rotational frictional coefficient, N is Avogaro's number, R is the gas constant, and T is the absolute temperature.

rotational frictional coefficient A measure of the frictional resistance experienced by a particle in solution that is equal to the frictional force divided by the angular velocity of the particle. *See also* rotational diffusion coefficient.

rotational relaxation The relaxation that takes place when a field that leads to the orientation of molecules, otherwise randomly oriented due to Brownian motion, is suddenly turned off. The return of the molecules to their random orientation is characterized by the rotational diffusion coefficient.

rotational relaxation time The time required, during rotational relaxation, for the average value of cost θ for all the solute molecules to fall to $1/e$ of its original value; e is the base of natural logarithms and $\emptyset$ is the angle through which the molecule has rotated away from its original direction of orientation.

rotational strength A quantity that is used in calculations of circular dichroism and that is related to the integrated value, over an absorption band, of the difference between the extinction coefficients for left and right circularly polarized light.

rotational symmetry The symmetry of a body that exists when identical structures are produced as the body is rotated about an axis. For a body containing subunits, all of the subunit centers can be set at the vertices of a regular polyhedron. *Aka* cyclic symmetry. *See also* axis of rotational symmetry.

rotational transition The transition of a molecule in which it rotates about an axis; rotational transitions require relatively small amounts of energy and are induced by infrared radiations of long wavelength.

rotation angle TORSION ANGLE.

rotation axis AXIS OF ROTATIONAL SYMMETRY.

rotation diagram An x-ray diffraction pattern obtained by the rotating crystal method. *Aka* rotation photograph.

rotation–reflection axis *See* reflection symmetry; rotoreflectional symmetry.

rotation technique *See* photographic rotation technique.

rotatory Of, or pertaining to, optical rotation.

rotatory dispersion *See* optical rotatory dispersion.

rotatory power SPECFIC ROTATION.

rotenone An insecticide, extracted from the roots of some tropical plants, that inhibits the electron transport system between the flavoproteins and coenzyme Q.

Rothera's test A qualitative test for ketone bodies in urine that is based on the production of a blue-purple color upon treatment of urine with sodium nitroprusside and ammonium hydroxide.

rotometer A device for measuring the rate of gas flow; used in gas chromatography.

rotor The container that is rotated in a centrifuge and that holds the tubes filled with the solution which is subjected to centrifugation. *Aka* head.

rotoreflectional symmetry The symmetry of a body that exists when identical structures are produced by a rotation about an axis, followed by a reflection through a mirror plane.

R_0t value A quantity used in RNA-driven hybridizations that is analogous to the C_0t value in DNA-driven hybridizations; equal to the product of the initial concentration of single-stranded RNA and the time allowed for reassociation. *See also* cot curve.

roughage DIETARY FIBER.

rough endoplasmic reticulum That portion of the endoplasmic reticulum to which a large number of ribosomes are attached; the site of membrane protein and lipid synthesis. *Abbr* RER. *Aka* rough-surfaced endoplasmic reticulum.

rough microsomes *See* microsomes.

rough-smooth variation *See* rough strain; smooth strain.

rough strain A bacterial strain, such as *Pneumococcus* (*Streptococcus pneumoniae*), that grows in the form of a colony that appears rough (has jagged edges). In some species, rough strains are nonvirulent or less virulent than the corresponding smooth strains, and they lack O antigens in the cell wall lipopolysaccharide; they have, instead, an R cell surface antigen. *Abbr* R strain.

round of replication A single transit of the replication system along a DNA molecule.

rounds of mating The number of matings in the line of ancestry of an average phage particle in the phage population.

Rous sarcoma A virus-induced malignant tumor in chickens.

Rous sarcoma virus A virus that contains single-stranded RNA and causes cancer in chickens; belongs to the group of leukoviruses. *Abbr* RSV.

routine A set of coded instructions that causes a computer to perform the various operations necessary for solving a given problem.

routine test dilution 1. The greatest dilution of virus particles that can produce confluent lysis of cells in monolayers. 2. The greatest dilution of phage particles that can produce confluent lysis of cells in a bacterial lawn. *Abbr* RTD.

rowboat model The current version of the sliding filament model of muscle contraction. According to this model, the myosin molecule "walks" along an actin filament, a process driven by ATP hydrolysis. Briefly, the model postulates the following: ATP binds to the head of a myosin molecule which leads to detachment of the myosin from an actin (thin) filament. The bound ATP is then hydrolyzed and the myosin head, carrying the products of ATP hydrolysis (ADP + P_i), moves close to a neighboring subunit on the actin filament; this occurs by diffusion and is made possible by the flexibility of the hinge regions in the myosin molecule. Next the myosin binds to the actin filament, releasing the ADP and P_i. This binding causes the myosin head to tilt or rotate and thereby exerts a pull on the rest of the myosin (thick) filament. At the end of this power stroke, a fresh molecule of ATP binds to the head and the cycle is repeated. *Aka* Huxley–Simmons model.

royal jelly A liquid nutrient produced by worker bees which, when fed to female larvae, leads to the production of queens.

R5P Ribose-5-phosphate.

RPC Reversed phase chromatography.

R plasmid A plasmid that carries drug resistance genes and that confers on the bacterial host resistance to one or more antibiotics. Most R plasmids consist of two contiguous DNA segments, known as resistance transfer factor (RTF) and r-determinant. The former carries genes for DNA replication and is required for transfer of the plasmid between bacteria; the latter carries genes for antibiotic resistance. *Aka* resistance factor.

rpm Revolutions per minute.

R point Restriction point.

RPP Reductive pentosephosphate cycle.

RPr A regulatory protein that functions in the bacterial phosphotransferase system.

r-protein Ribosomal protein.

RQ Respiratory quotient.

rRNA Ribosomal RNA.

rRNA transcription unit *See* Miller tree.

R_s A useful parameter for comparing regulatory enzymes; equal to the ratio of the substrate (or ligand) concentration at 0.9 of the maximum velocity to the substrate (or ligand) concentration at 0.1 of the maximum velocity. For an enzyme that follows Michaelis–Menten kinetics, $R_s = 81$; for an enzyme that yields a sigmoidal curve when the velocity is plotted as a function of the substrate concentration, R_s has different values. If $R_s < 81$, the enzyme shows positive cooperativity; if $R_s > 81$, the enzyme shows negative cooperativity.

RSD Relative standard deviation.

R site 1. A cell surface receptor site on rat mast cells for adenosine and related compounds. The site is involved in the immunological release of adenyl cyclase and has an obligatory requirement for an intact ribose moiety. *See also* P site. 2. Recognition site.

RS system A system for naming each of several chiral centers in a molecule in an absolute manner; based on ranking the substituents around each asymmetric (chiral) carbon atom in the order of decreasing atomic number. The molecule is viewed with the lowest ranking group pointing directly away from the viewer. If the sequence of the substituents in decreasing rank order is seen to be clockwise, the configuration around this chiral center is denoted R (Latin rectus, "right"); if the decreasing rank order is counterclockwise, the configuration is denoted S (Latin sinister, "left"). *Aka* Cahn–Ingold–Prelog sequence rules.

R strain Rough strain.

r-strand CRICK STRAND.

RSV 1. Rous sarcoma virus. 2. Respiratory syncytial virus.

RT Reverse transcriptase.

RTA Renal tubular acidosis.

RTD Routine test dilution.

RTF Resistance transfer factor.

rTU rRNA transcription unit.

rubber A natural high molecular weight polymer of isoprene units; a polyterpene.

rubella virus The virus that causes German measles and that belong to the paramyxovirus group.

Rubisco Ribulose-1,5-bisphosphate carboxylase.

Rubner's law The principle that the heat produced by the metabolism of animals is proportional to the surface area of the animals; the principle has been shown to be approximately correct.

RuBP Ribulose-1,5-bisphosphate.

RuBP carboxylase Ribulose-1,5-bisphosphate carboxylase.

RuBPCase Ribulose-1,5-bisphosphate carboxylase.

rubredoxin IRON–SULFUR PROTEIN (a).

Ruff degradation A degradative technique for aldonic acids whereby a carbon atom is removed by treatment with hydrogen peroxide in the presence of ferrous ions, and the sugar is converted to the next lower aldose.

ruffled edges Extensive cellular projections that function in the adhesion of eukaryotic cells to solid surfaces. *Aka* lamellipodia.

Ruhemann's purple An intensely blue-purple compound that is formed by the reaction of ninhydrin with amino acids; it consists of two ninhydrin moieties, linked via the nitrogen derived from the amino acid.

Rul Ribulose.

rule of mutual exclusion The principle that, for molecules with a center of symmetry, vibrational transitions that are allowed in the infrared are forbidden in the Raman effect, and vice versa.

rule of the ring The concept that apparently all viral DNAs are circular, or are converted into circles before replication, or appear to have been formed from circular molecules originally.

rumposome A structure consisting of two membrane-bound sheets that is frequently seen to connect the nuclear region with the dictyosome region in flagellated fungi.

running gel *See* disc gel electrophoresis.

run-off ribosome *See* ribosome runoff.

runt disease GRAFT-VERSUS-HOST REACTION.

runting syndrome GRAFT-VERSUS-HOST REACTION.

Rutaceae alkaloids *See* alkaloids.

rutherford The amount of radioactive substance that undergoes 10^6 disintegrations per second.

Rutherford backscattering A technique for the study of surfaces in which a high-energy beam of ions is directed at a surface. Measurements of the "shadows" cast by the surface atoms permit a pinpointing of the positions of surface atoms with great accuracy. *Abbr* RBS.

Rutherford scattering The scattering of radiation that results from elastic collisions of alpha or beta particles with atomic nuclei.

R value The fraction of the solute, in partition chromatography, that is present in the mobile phase.

S

s 1. Sedimentation coefficient. 2. Second. 3. Standard deviation.

$s_{20,w}^0$ Standard sedimentation coefficient.

S 1. Substrate. 2. Svedberg unit. 3. Sulfur. 4. Entropy. 5. Bacterial colony of smooth morphology. 6. Period of DNA synthesis in the cell cycle. 7. S configuration. 8. Sensitivity of a bacterial strain to a phage or an inhibitor (used as a superscript). 9. Serine. 10. Thiouridine. 11. Sphingomyelin. 12. *See* RS system. 13. Designating a ribosomal protein from the small ribosomal subunit.

^{35}S A radioactive isotope of sulfur that emits a weak beta particle and has a half-life of 87.2 days.

$[S]_{0.5}$ The substrate concentration at which an allosteric enzyme catalyzes a reaction at one-half the maximum velocity. *See also* K_m.

[S′] REDUCED SUBSTRATE CONCENTRATION.

SA Specific activity.

Sabin vaccine A poliomyelitis vaccine that is given orally and that consists of live, attenuated, poliovirus preparations.

saccharase SUCRASE.

saccharic acid ALDARIC ACID.

saccharide CARBOHYDRATE.

saccharifying amylase ALPHA AMYLASE.

saccharimeter An instrument, such as a polarimeter, or a device, such as a fermentation tube, for determining the amount of sugar in a solution. *Var sp* saccharometer.

saccharin *o*-Sulfobenzimide; an artificial sweetener.

saccharogenic amylase BETA AMYLASE.

saccharogenic method An assay of the enzyme amylase that is based on a determination of the amount of product formed.

Saccharomyces cerevisiae A species of yeast that includes the strains of baker's and brewer's yeast.

saccharopine The compound ε-*N*-(L-glutaryl-2)-L-lysine; an intermediate in the biosynthesis and degradation of lysine. The name derives from the fact that the compound functions in lysine synthesis in *Saccharomyces*. Hydrolytic and oxidation–reduction reacations convert saccharopine to lysine and α-ketoglutarate.

saccharopinuria A genetically inherited metabolic defect in humans, characterized by mental retardation, that is due to a deficiency of the enzyme saccharopine dehydrogenase.

saccharose SUCROSE.

sacculus The sack-like peptidoglycan structure of the bacterial cell wall.

sacrifice Euphemism for "to kill"; used in reference to experiments with animals.

Sagavac Trademark for a group of agarose gels used in gel filtration chromatography.

SAIDS Simian acquired immune deficiency syndrome.

Sakaguchi reaction A colorimetric reaction for arginine that is based on the production of a red color upon treatment of the sample with α-naphthol and sodium hypochlorite.

salimeter A hydrometer that is calibrated for the determination of either the specific gravity of a saline solution or the concentration of sodium chloride in the solution. *Aka* salinometer.

saline 1. *n* An aqueous solution of sodium chloride. 2. PHYSIOLOGICAL SALINE. 3. *adj* Of, or pertaining to, salt.

salivary juice The digestive juice, consisting of the saliva, that is secreted by the salivary glands into the mouth and that contains the enzyme ptyalin. *Aka* salivary fluid.

Salkowski reaction A modification of the Liebermann–Burchard reaction for cholesterol in which acetic anhydride is omitted.

Salk vaccine A poliomyelitis vaccine that is given by injection and that consists of formalin-killed poliovirus preparations.

salmine A protamine of 32 amino acids isolated from salmon sperm.

Salmonella A genus of gram-negative enteric bacteria that are widely used for genetic studies.

Salmonella test *See* Ames test.

saltatory conduction Descriptive of the conduction of a nerve impulse in myelinated axons. These axons are heavily insulated by the myelin sheath except for small bare regions (called Nodes of Ranvier), spaced about 1–2 mm apart. As a result, the nerve impulse effectively jumps from node to node; conduction is saltatory (dance-like). This type of conduction is much faster than that along unmyelinated axons and conserves metabolic energy since active excitation is confined to the small nodal regions.

saltatory motion Descriptive of the motion of lysosomes and other small vesicles that serves to move them from one place to another in a saltatory (dance-like) manner.

saltatory replication Descriptive of the lateral

replication of DNA during gene amplification in which large numbers of copies of a given segment of DNA are produced.

salt fractionation The fractional precipitation of proteins by means of inorganic salt solutions, frequently those of ammonium sulfate.

salt–gene theory A theory of hypertension according to which blood pressure is related to both salt intake and genetic background. The theory predicts that individuals, who are genetically susceptible, will develop high blood pressure on a high-salt diet.

salting in The increase in the solubility of a protein that is produced in solutions of low ionic strength by an increase of the concentrations of neutral salts; due to a stabilization of the charged groups on the protein as a result of a decrease in the activity coefficients of these groups.

salting out The decrease in the solubility of a protein that is produced in solutions of high ionic strength by an increase of the concentrations of neutral salts; due to a partial dehydration of the protein as a result of the competition between the protein and the salt ions for solvating water molecules.

salting-out chromatography A chromatographic technique in which water-soluble organic compounds are separated by ion-exchange chromatography using an aqueous salt solution for elution.

salting-out constant A constant characteristic of the solubility behavior of a protein; specifically, $\log S = \log S_0 - KI$, where S is the actual solubility of the protein, S_0 is the solubility in pure water, I is the ionic strength, and K is the salting-out constant.

salt link IONIC BOND.

salt precipitation SALT FRACTIONATION.

salt respiration ANION RESPIRATION.

salvage pathway A pathway that utilizes compounds formed in catabolism for biosynthetic purposes, even though these compounds are not true intermediates of the corresponding normal biosynthetic pathway. Thus, free purines may be salvaged from the hydrolysis of nucleotides and then used for the biosynthesis of nycleotides; likewise, free choline may be salvaged from the degradation of phosphatidyl choline and then used for the biosynthesis of phosphoglycerides.

SAM S-Adenosyl-L-methionine.

samesense mutation A mutational change in DNA in which a codon, coding for a given amino acid, is converted into a synonym codon for the same amino acid. As a result, the corresponding mRNA still directs the incorporation of the same amino acid into the protein; the protein is unchanged and the mutation is a silent one.

sample ampholyte An ampholyte of the mixture that is fractionated by isoelectric focusing, as distinct from a carrier ampholyte that is used to form the pH gradient.

sample gel *See* disc gel electrophoresis.

sampling error An error due to the method of obtaining samples, to the inadequacy of the number of samples, or to the inadequacy of the size of the sample.

Sandhoff's disease A genetically inherited metabolic defect in humans that is a rare form of Tay–Sachs disease and in which there is a deficiency of both hexosaminidase A and B.

sandwich technique INDIRECT FLUORESCENT ANTIBODY TECHNIQUE.

Sanfillipo's syndrome A mucopolysaccharidosis due to a deficiency of the enzyme heparan-N-sulfatase (syndrome of type A) or N-acetyl-α-glucosaminidase (syndrome of type B).

Sanger–Coulson method An enzymatic method of sequencing DNA. The method entails synthesis of radioactively labeled (^{32}P) fragments of various sizes using the DNA to be sequenced as a template. Further fragments are produced in a second (two-part) enzymatic stage in which either a single deoxyribonucleotide is present (plus system) or the same deoxyribonucleotide is lacking and only the remaining three deoxyribonucleotides are present (minus system). The two-part experiment is performed 4 times, with a specific deoxyribonucleotide present (or lacking) each time. Alternatively, $2',3'$-dideoxyribonucleoside triphosphates can be used as specific terminators of DNA synthesis in place of the minus system. The fragments are then separated by polyacrylamide gel electrophoresis and detected by autoradiography. *Aka* Sanger dideoxy method; plus-minus method.

Sanger reaction The reaction of the Sanger reagent, 1-fluoro-2,4-dinitrobenzene, with the free alpha amino group of amino acids, peptides, or proteins; the reaction yields a dinitrophenyl derivative that is useful for the chromatographic detection and quantitative estimation of amino acids, peptides, and proteins, as well as for endgroup analysis of N-terminal amino acids in peptides and proteins. The Sanger reagent reacts similarly with a number of other functional groups of the amino acids.

Sanger reagent *See* Sanger reaction.

S antigen Soluble antigen; an incomplete and noninfectious virus form that appears early in the course of certain viral infections.

sapogenin A steroid that occurs in plants in the form of glycosides known as saponins; the aglycone moiety of steroid saponins.

saponifiable fraction The fraction of total lipid

that, after saponification, is soluble in water and insoluble in ether.

saponifiable lipid A lipid that can be hydrolyzed with alkali to yield soap as one of the products; glycerolipids and sphingolipids are two major types of saponifiable lipids.

saponification The alkaline hydrolysis of a lipid, particularly a glyceride, that yields soap as one of the products.

saponification equivalent The number of grams of fat saponified by one mole of potassium hydroxide. *Abbr* SE.

saponification number A measure of the average chain length of the fatty acids in a fat; equal to the number of milligrams of potassium hydroxide required to saponify 1 g of fat. The greater the saponification number, the shorter the average chain length of the fatty acids in the fat. *Aka* saponification value.

saponin A water-soluble surface-active plant substance that forms soapy solutions even at high dilutions; saponins are powerful hemolytic agents that are glycosides and are classified according to the nature of the aglycone (also called genin) as steroid, triterpene, or steroid–alkaloid saponins.

saprophytic nutrition A mode of nutrition of "plant-like" organisms (such as bacteria and fungi) in which the organism derives its nutrients from dead or decaying plant or animal matter in the form of organic compounds in solution.

Sa protease A proteolytic enzyme, isolated from *Staphylococcus aureus*, that cleaves peptide bonds in which the carbonyl group is donated by aspartic or glutamic acid.

saprozoic nutrition A mode of nutrition of "animal-like" organisms (such as protozoa) in which the organism derives its nutrients from dead or decaying plant or animal matter in the form or organic compounds in solution.

sarcolemma The membrane surrounding the fiber of a striated muscle.

sarcoma A malignant tumor that arises from connective tissue.

sarcomere The longitudinal repeat unit of a myofibril; the segment that extends from one Z line to the next.

sarcoplasm The intracellular fluid that bathes the myofibrils of a muscle cell.

sarcoplasmic reticulum The smooth endoplasmic reticulum of a muscle cell.

sarcosine *N*-Methylglycine; an amino acid that is an intermediate in the metabolism of one-carbon fragments and that forms part of the structure of the antibiotic actinomycin D.

sarcosome A mitochondrion from a striated muscle.

sarcotubule A transverse tubule of the T-system of muscle.

sardinine A protamine isolated from sardines.

Sarkosyl Trademark for the detergent sodium *N*-lauryl sarcosinate.

Sarkosyl M-band technique A method for producing DNA–membrane complexes from bacterial lysates; based on the selective adsorption of membrane components to crystals of magnesium sarkosyl. These complexes can be recovered as discrete bands (M bands) by sucrose density gradient centrifugation. DNA is present in these bands by virtue of being bound to the cell membrane; it does not itself adhere to crystals of magnesium sarkosyl.

sat DNA Satellite DNA.

Satellite DNA A eukaryotic DNA fraction that differs sufficiently in its base composition from that of the bulk DNA (main band DNA) so that it can be separated from the latter by density gradient sedimentation equilibrium, using a cesium chloride gradient. Satellite DNAs are usually highly repetitive DNAs.

satellite DNA of mouse *See* mouse satellite DNA.

satellite phenomenon CROSS-FEEDING.

satellite RNA An RNA that has a sedimentation coefficient of 5 to 8S, small amounts of which are found in association with plant 26S ribosomal RNA.

satellite virus A small virus that occurs in association with another virus, such as the tobacco necrosis satellite virus or the adenovirus associated virus.

satellitism CROSS-FEEDING.

saturated fatty acid A fatty acid that contains a saturated alkyl chain. Most naturally occurring saturated fatty acids have an even number of carbon atoms. Even-numbered saturated fatty acids having less than 10 carbon atoms are liquid at room temperature; longer chain fatty acids are solids.

saturated solution *See* saturation (3).

saturating substrate concentration A substrate concentration that is, numerically, much larger than the Michaelis constant of the enzyme so that the velocity of the reaction is essentially equal to the maximum velocity.

saturation 1. The state of an organic compound in which it contains only single bonds between the carbon atoms. 2. The conversion of an unsaturated organic compound to a saturated one. 3. The state of a solution in which it contains the maximum amount of solute that can be dissolved permanently in that volume of solvent under specified conditions. 4. The state of a macromolecule in which it has bound the maximum number of ligands of a given type, as in the saturation of hemoglobin or myoglobin with oxygen, or in the saturation of an enzyme with its substrate.

saturation analysis RADIOIMMUNOASSAY.

saturation backscattering The maximum increase in counting rate that is observed when increasing thicknesses of backing material are placed under a radioactive sample.

saturation-backscattering thickness The thickness of backing material required to achieve saturation backscattering.

saturation current The current produced in an ionization chamber when the potential is of sufficient magnitude to permit the collection of all the ions; the saturation current is independent of the applied voltage.

saturation curve *See* oxygen saturation curve.

saturation fraction FRACTIONAL SATURATION.

saturation hybridization An in vitro hybridization experiment involving two types of polynucleotides, of which one is present in excess, so that all of the complementary sections of the other polynucleotide form hybrid duplex structures. *See also* DNA-driven hybridization; RNA-driven hybridization.

saturation kinetics The kinetics of a reaction in which the rate of the reaction levels off with increasing concentrations of a component, as is the case for a simple enzymatic reaction and for mediated transport.

saturation thickness The thickness of a radioactive sample such that any additional increase in its thickness will not increase the observed counts any further.

sawhorse projection A representation of the arrangement of the atoms in a molecule that provides a three-dimensional view of the molecule and resembles a sawhorse in outline.

SBP 1. Serum blocking power. 2. Sex steroid binding plasma protein.

sc 1. *adj* Subcutaneous. 2. *adv* Subcutaneously.

scaffold Chromosomal material whose exact nature and function are unclear; believed to consist of nonhistone proteins and RNA and to play a role in the compacting of the chromosome.

scalar 1. *n* A nondirectional quantity. 2. *adj* Of, or pertaining to, a nondirectional quantity.

scalar reaction A reaction that is nondirectional in which the components either are free to move at random or are fixed in a random order; an overall reaction in solution, but not the individual molecular event, is generally a scalar reaction.

scale method LAMM SCALE DISPLACEMENT METHOD.

scaler An electronic recording device that produces one output pulse for a given number of input pulses.

scanner 1. An instrument for measuring the distribution of either color intensity or radioactivity on a chromatogram or on an

electropherogram. 2. A photoelectric scanning attachment for the analytical ultracentrifuge that provides a plot of absorbance versus radial distance and that is used in conjunction with the ultraviolet absorption optical system.

scanning Any measurement performed systematically across an experimental pattern, such as the measurement of either color intensity or radioactivity as a function of distance across a radiochromatogram.

scanning electron microscope An electron microscope in which a three-dimensional view of the specimen is produced by the deflection of primary and secondary electrons and by the use of an electron beam to scan the specimen. *Abbr* SEM.

scanning hypothesis A hypothesis proposed to explain the initiation of translation in eukaryotic systems. According to this hypothesis, a 40S ribosomal subunit attaches at or near the 5'-cap of mRNA and then drifts along the mRNA, in the 3' direction, until it encounters an AUG initiation codon. At that point, the 40S subunit is joined by a 60S subunit to form an active 80S ribosome, and translation commences. The 80S ribosome dissociates into the two subunits upon reaching a termination codon.

scanning tunneling microscope A powerful microscope that produces high resolution, three-dimensional images of surfaces; it enables one to "see" surfaces, atom by atom, and it can resolve features that are a fraction of the size of an atom. The main difference between this and all other microscopes is that it uses no free particles of radiation and, hence, does not require special lenses and special light and electron sources. Instead, the bound electrons already existing in the sample under investigation serve as the exclusive source of radiation. Some of these electrons leak out from the sample's surface and form an electron cloud around the sample. The cloud is a result of the indeterminacy of the electron's location and the density of the cloud decreases with increasing distance from the surface. Since the electrons appear to be digging tunnels beyond the surface, the effect is known as tunneling. A scanning needle tip is swept across the sample surface and a voltage induced flow of electrons through the electron cloud (tunneling current) is measured. In this fashion, the tip follows the contours of the surface. The motion of the tip is then processed by a computer and displayed on a plotter. *Abbr* STM. *See also* field ion microscope.

scarce mRNA COMPLEX RNA.

Scatchard plot A graphical representation of binding data for the determination of intrinsic association constants and of the number of noninteracting binding sites per molecule; based on the Hill equation for the case of n_H = 1, and consists of a plot of $r/[S]$ versus r. *See also* Hill equation.

scatter diagram A plot of data as points in a plane rectangular coordinate system that is made to determine whether there is a correlation between the two variables indicated on the axes.

scattering The dispersion of radiation by matter in directions other than that of the incident beam as a result of collisions and/or interactions.

scDNA Single-copy DNA; *see* unique DNA.

Schardinger dextrin A group of oligosaccharides that are formed by the action on starch of amylase from *Bacillus macerans*; includes α-dextrins, which contain six glucose residues per molecule, and β-dextrins, which contain seven glucose residues per molecule.

Schardinger enzyme XANTHINE OXIDASE.

Schardinger reaction A test for oxidase activity in milk that is based on the incubation of milk with formaldehyde and methylene blue; the blue color of the solution disappears as the methylene blue is reduced in the process of serving as a hydrogen acceptor for the oxidation of formaldehyde.

schemochrome STRUCTURAL COLOR.

Scheraga–Mandelkern equation An expression for a function, denoted β, that is based on various physical parameters and that is used as an aid in determining the shape of a macromolecule in solution. For calculated values of β greater than 2.12×10^6, the molecule is best approximated by a prolate ellipsoid of revolution; for calculated values of β approximately equal to 2.12×10^6, the molecule is best approximated by an oblate ellipsoid of revolution.

Schiff base The condensation product formed between a primary amine and either an aldehyde or a ketone; a compound containing a carbon–nitrogen double bond (an imine group):

$$\diagdown \\ C = N{-} \\ \diagup$$

Schiff's reagent A reagent for the detection of aldehydes that consists of fuchsin, bleached with sulfurous acid; produces a red color upon reaction with an aldehyde.

Schiff's test 1. A test for uric acid that is based on the reduction of silver ions to metallic silver. 2. A test for urea that is based on the formation of a purple color upon treatment of the sample with furfural and hydrochloric acid. 3. A test for aldehydes that is based

on the reaction of aldehydes with Schiff's reagent.

Schilling test A test for measuring the rate of absorption of vitamin B_{12} from the intestine by administering vitamin B_{12}, labeled with radioactive cobalt, to an individual.

schistosome A parasitic worm in the blood vessels of birds and mammals.

schlepper A compound that, by combining with a weakly antigenic or nonantigenic substance, enhances the antigenicity of the latter.

schlieren optical system An optical system that focuses ultraviolet light passing through a solution in such a fashion that a photograph is obtained that represents a plot of refractive index gradient as a function of distance in the solution. A boundary in the solution appears as a peak on the photographic plate and measurements are made on this plate with a microcomparator. The optical system is used in the analytical ultracentrifuge and in the Tiselius electrophoresis apparatus.

Schmidt–Thannhauser procedure A procedure for the extraction of nucleic acids from tissue by digestion of the tissue with dilute alkali; DNA and RNA are obtained as separate fractions by this procedure, since the alkali hydrolyzes RNA but not DNA.

Schneider procedure A procedure for the extraction of nucleic acids from tissue by treatment of the tissue with either trichloroacetic acid or perchloric acid; DNA and RNA are not obtained as separate fractions by this procedure.

Schuetz–Borrisow rule An empirical rule that states that the velocity of an enzymatic reaction is proportional to the square root of the enzyme concentration; the rule was developed for pepsin and applies, under limited conditions, to pepsin and other proteolytic enzymes.

Schultz–Dale reaction An anaphylactic response that is produced in a sensitized animal by the introduction of antigens into its uterus.

Schwann cell The cell surrounding a myelinated nerve axon.

scintillation The emission of flashes of light by fluorescent substances subsequent to their excitation by means of radioactive or other radiation.

scintillation autography A modification of radioautography in which the labeled material is placed in a solution containing a scintillator, and the scintillations produced by the radioactive disintegrations expose a photographic film.

scintillation cocktail COCKTAIL (2).

scintillation counter A radiation counter in which incident ionizing particles or incident photons are counted by means of the scintillations that they induce in a fluor.

scintillation detector A solid or liquid fluor together with the electronic circuitry that converts the light flashes of the fluor to electrical pulses. *See also* external-sample scintillation counter; internal-sample scintillation counter.

scintillation spectrometer A scintillation counter designed to permit the measurement of the energy distribution of radiation.

scintillator FLUOR.

scintillon A subcellular, crystal-like structure that emits a flash of light upon acidification in the presence of oxygen; the structure has been isolated from some algae (dinoflagellates).

scissile Capable of being cut or broken; a scissile chemical bond, for example, is one that is readily cleaved.

scission The introduction of a break into a biopolymer, particularly a nucleic acid, which may be enzymatic or nonenzymatic.

SCK Serum creatine kinase.

scleroprotein A simple and generally fibrous protein that is insoluble in aqueous solvents and that serves as a structural component of tissues. Collagen and keratin, which occur in cartilage, connective tissue, hair, nails, and the like, are two examples.

sclerosis The pathological hardening of tissues.

S configuration *See* RS system.

scorbutic Of, or pertaining to, scurvy.

scotophobin A polypeptide of 15 amino acids which accumulates in the brains of rats trained to avoid darkness; claimed to be a memory-directing peptide which elicits a similar response in untrained individuals.

scotopic vision Vision in dim light in which the rods of the retina serve as light receptors.

scotopsin RHODOPSIN.

SCP Single-cell protein.

scrapie A transmissible, degenerative disease of the nervous system of sheep and goats. The disease has a long incubation period (months or years). The stricken animals develop an intense itch which is associated with emaciation and weakness until death ensues. The disease is believed to be caused by an infectious particle known as prion.

scrapie agent The agent causing scrapie, now believed to be identical to prion.

screw axis of symmetry An axis of symmetry such that a rotation about it, plus a translation parallel to the rotation axis, yields a structure that is identical to that before rotation. The screw axis is called an n-fold axis if the rotation involves $1/n$ of a turn, or $360°/n$.

screw symmetry The symmetry of a body that results from helical motion about an axis of symmetry.

scRNA Small cytoplasmic RNA.

scRNP Small cytoplasmic ribonucleoproteins; *see* small cytoplasmic RNA.

scurvy The disease caused by a deficiency of vitamin C.

SCYRP Small cytoplasmic ribonucleoproteins; *see* small cytoplasmic RNA.

SD Standard deviation.

SDA Specific dynamic action.

SDH Succinate dehydrogenase.

SDS Sodium dodecyl sulfate.

SD sequence Shine–Dalgarno sequence.

SDS gel electrophoresis *See* SDS–PAGE.

SDS–PAGE Sodium dodecyl sulfate polyacrylamide gel electrophoresis; an electrophoretic technique useful for the determination of the molecular weight of individual polypeptide chains. Based on the dissociation of oligomeric proteins with sodium dodecyl sulfate at neutral pH (to minimize the net charge of the protein) and in the presence of mercaptoethanol (to break disulfide bonds in the protein). The polypeptide chains form random coils and bind sodium dodecyl sulfate, yielding complexes that have essentially the same charge/mass ratios. These complexes are, therefore, separated by the sieving effect of the gel as a function of their molecular weight.

se Secretor gene.

SE 1. Standard error. 2. Saponification equivalent.

sebaceous gland A cutaneous gland that secretes sebum into a hair follicle.

sebum The oily secretion of a sebaceous gland that softens and lubricates the hair and the skin.

sec 1. *n* Second. 2. *adj* Secondary.

secondary acidosis COMPENSATED ACIDOSIS.

secondary active transport *See* active transport.

secondary alkalosis COMPENSATED ALKALOSIS.

secondary bile acids *See* bile acids.

secondary bonds WEAK INTERACTIONS.

secondary charge effect The charge effect in a solution of charged macromolecules that is due to the differential movement of the positive and the negative ions of the supporting electrolyte. The secondary charge effect may lead either to an increase or to a decrease in the sedimentation rate and in the diffusion rate of the charged macromolecule.

secondary culture A culture started from a primary culture.

secondary deficiency 1. The disorder that is caused by the inadequate intake of an essential nutrient in the diet, even though the dietary level of the nutrient is known to be adequate under normal conditions. Such exceptional conditions arise when there is an extra demand for the nutrient as may be the case during a disease or during pregnancy. *See also* conditioned vitamin deficiency. 2. A decreased hormone level that results from insufficient stimulation by the pituitary gland of the endocrine gland that produces the hormone. Thus, failure of the pituitary to secrete a trophic hormone (for example, as a result of a pituitary tumor) leads to a secondary deficiency of thyroid, adrenal, or gonadal hormones.

secondary derived protein *See* derived protein.

secondary electron An electron ejected from an atom or a molecule by collision with a charged particle or with a photon.

secondary filament ACTIN FILAMENT.

secondary fluor A fluor, such as POPOP, that is used in a scintillation counter to shift the wavelength of the light emitted by the primary fluor to longer wavelengths, to which the photomultiplier tube is more sensitive.

secondary hydration shell The layer of water molecules around an ion that is beyond the primary hydration shell, and that is not as firmly held as the primary hydration shell.

secondary immune response *See* secondary response.

secondary ionization The ionization of matter that is produced by the fragments that have resulted from a primary ionization.

secondary ion mass spectrometry A mass spectrometric technique that does not require heating of the smaple for its desorption and ionization. Instead, accelerated ion beams (argon, xenon, or cesium) bombard the sample, which is placed onto a metal, with or without a matrix material (such as glycerol). The technique is useful for desorption and ionization of nonvolatile biomolecules without prior derivatization. *Abbr* SIMS.

secondary isotope effect An isotope effect in which the isotope itself does not participate in bond cleavage; the bond to the isotope, though not broken, has a different energy during the transition state than it does in the ground state.

secondary lysosome *See* lysosome.

secondary malnutrition Malnutrition that results from diseases such as food allergies, ulcers, diabetes, or pernicious anemia; not a direct result of inadequate food intake.

secondary messenger *See* second messenger.

secondary metabolite *See* metabolite.

secondary plant constituents Nonnutritive components of plants such as indoles, flavones, and isothiocyanates. *See* also antipromoter.

secondary plot A plot of derived enzyme kinetics data that are obtained from a primary plot; a plot of intercepts (intercept replot) and a plot of slopes (slope replot) that are obtained from primary plots are examples.

Aka secondary kinetic plot.

secondary protein derivative SECONDARY DERIVED PROTEIN.

secondary response 1. The enhanced immune response of an animal organism to a second administration of the same antigen. 2. The delayed activation of some genes brought about by the gene products formed in the primary response to a steroid hormone.

secondary solvent The solvent used in scintillation counting for solubilizing a sample that is insoluble in the primary solvent.

secondary standard 1. A solution, the concentration of which is determined by means of titration against either a known weight of a primary standard, or a known volume of a standard solution of a primary standard. 2. A standard, such as a source of radiation, that has been calibrated against a primary standard.

secondary structure The regular folding of a polypeptide chain or of a polynucleotide strand along one axis of the molecule, as in a helix, that is due to the formation of intramolecular hydrogen bonds along the length of the chain or strand; the local spatial arrangement of segments of the polypeptide chain or of the polynucleotide strand without regard to the conformation of side chains or to the relation of one segment to other segments.

secondary tumor A tumor that is formed through metastasis of a primary tumor.

secondary valence bond VAN DER WAALS INTERACTION.

second critical concentration The concentration above the critical micelle concentration at which spherical micelles begin to undergo the structural changes that lead to the formation of liquid crystals.

second law of cancer biochemistry The principle that hormonal regulation of glycolysis at the initial hexokinase stage constitutes a major control mechanism of metabolism and of growth in both normal and malignant mammalian cells.

second law of photochemistry The law that absorbed light does not necessarily result in a photochemical reaction, and when it does, then only one photon is required for each molecule affected. *Aka* Stark–Einstein Law.

second law of thermodynamics The principle relating to the direction in which a process proceeds; all processes tend to proceed in such a direction that the entropy of the system plus its surroundings increases until an equilibrium is reached at which the entropy is at a maximum. The essence of this law lies in the definition of entropy and the prediction of the directionality of a process. Thus, the law can be stated in two parts: (a) The change in entropy of a system is equal to the heat absorbed or evolved by the system in a reversible process, divided by the absolute temperature at which the process occurs; (b) the change in entropy of the system and its surroundings is always positive for a spontaneous process. There are other, alternative formulations of this law. *See also* thermodynamics.

second messenger A substance that is released from a specific receptor in a target cell upon arrival of a chemical messenger such as a hormone, a neurotransmitter, or a prostaglandin; the compound $3',5'$-cyclic adenylic acid (cyclic AMP, cAMP) is an example of a second messenger. *See also* first messenger.

second moment The position in a sedimentation boundary that must be used for precise calculations; in the case of a symmetrical boundary, the second moment corresponds to a position that is farther away from the center of rotation than the position of the peak.

second-order reaction A chemical reaction in which the velocity of the reaction is proportional either to the product of the concentrations of two reactants, or to the square of the concentration of one reactant.

second-set rejection The accelerated sequence of events that leads to the rejection of an allograft by an individual who has already rejected a previous allograft at the same site. *Aka* second-set reaction.

second-site mutation SUPPRESSOR MUTATION.

second-site reversion SUPPRESSION.

secosteroid A steroid-like molecule; a modified steroid in which a ring has been broken. The D vitamins are secosteroids.

secretagogue One of a group of large polypeptides and other compounds that stimulate the secretion of gastric and pancreatic juices.

secretin A polypeptide hormone of 27 amino acids that is secreted by the duodenum and that stimulates the release of pancreatic juice. *Aka* nature's antacid.

secretion The movement of material, synthesized within a cell, across the cell membrane to the outside.

secretor An individual who secretes water-soluble forms of the blood group substances into his body fluids.

secretor gene A human gene that controls the secretion of water-soluble forms of A and B antigens of the ABO blood groups system into the saliva and other body fluids.

secretory component An additional polypeptide chain carried by a dimer of IgA molecules and found in secretions. The secretory component aids in the transport of IgA across epithelial cells and may also serve to protect

IgA against digestion by proteolytic enzymes in the secretions. *Aka* secretory piece.

secretory granule *See* vacuole.

secretory piece SECRETORY COMPONENT.

secretory protein A protein that is secreted from a cell to the outside.

secretory vesicle *See* vacuole.

secretosome A membranous, nerve-end preparation from the posterior lobe of the pituitary gland that secretes hormones. *Aka* neurosecretosome.

SECSY Spin-echo correlated spectroscopy; a two-dimensional nuclear magnetic resonance technique.

sector cell The standard cell used in the analytical ultracentrifuge that is designed in the shape of a sector to eliminate convection in the solution. Commonly used cells contain either a single or a double sector; the double-sector cell permits the simultaneous sedimentation of two samples or of a sample and a reference solution.

secular equilibrium The equilibrium that is established for a radioactive decay reaction if the half-life of the parent isotope is much greater than that of the daughter isotope.

sedation The calming and quieting of the nerves.

sediment 1. *n* The material removed from a solution by sedimentation. 2. *v* To subject a solution to sedimentation; to cause solute molecules of the solution to move by sedimentation.

sedimentation 1. The movement of molecules in solution under the influence of a centrifugal field and away from the center of rotation 2. The settling out of molecules from a solution under the influence of a gravitational field.

sedimentation coefficient A measure of the rate of sedimentation of a macromolecule that is equal to the velocity per unit centrifugal field; specifically $s = (dx/dt)/w^2x$, where s is the sedimentation coefficient, dx/dt is the velocity, w is the angular velocity, and x is the distance from the center of rotation. *Sym* s. *Aka* sedimentation constant. *See also* standard sedimentation coefficient.

sedimentation equilibrium Sedimentation, generally performed in an analytical ultracentrifuge, in which the centrifuge rotor is driven at relatively low speeds and for relatively long times so that an equilibrium is established in the solution between sedimentation and diffusion. No boundaries are formed in the solution during the sedimentation, and the photographic plate does not show peaks but only a curvature of the gradient curve. Sedimentation equilibrium is used for calculations of molecular weights. *See also* approach to sedimentation equilibrium;

density gradient sedimentation equilibrium; meniscus depletion sedimentation equilibrium; short column sedimentation equilibrium.

sedimentation equilibrium in a density gradient DENSITY GRADIENT SEDIMENTATION EQUILIBRIUM.

sedimentation field flow fractionation A separation technique, useful for the isolation and purification of macromolecules. In this technique, a sample is injected into, and moved along by, a flowing stsream of carrier liquid. The latter flows through a thin, open, ribbon-like channel and a centrifugal force is applied at right angles to the direction of flow. The centrifugal force pushes the macromolecules to the walls of the channel where they elute in order of increasing molecular weight. With appropriate detectors, the movement of the particle can be recorded as a "fractogram," which looks something like a chromatogram. *Abbr* SFFF; *See also* field flow fractionation.

sedimentation partition chromatography A technique for studying interactions between macromolecules. Involves a gradient that contains a narrow zone of one macromolecule. A layer of the second macromolecule is then placed on top of the gradient and the entire assembly is centrifuged.

sedimentation potential The electrical potential produced by the sedimentation of particles through a liquid as a result of gravity or the application of a centrifugal force.

sedimentation velocity Sedimentation, generally performed in an analytical ultracentrifuge, in which the centrifuge rotor is driven at high speeds and for relatively short times so that sedimentation exceeds diffusion and the macromolecules sediment through the cell. Boundaries are formed in the solution during the sedimentation and the photographic plate shows peaks. Sedimentation coefficients can be calculated from the movement of these peaks as a function of time. Sedimentation velocity is useful for studies of purity, homogeneity, association–dissociation equilibria, reaction kinetics, and other properties of macromolecules.

sedoheptulose A seven-carbon ketose that is an intermediate in the hexose monophosphate shunt.

segmented genome A viral genome that consists of two or more nonidentical nucleic acid molecules, each carrying separate genetic information. The genomes of influenza virus and of alfalfa mosaic virus are two examples. In the case of influenza virus, all the nucleic acid fragments are present in a single virion; such a virus is called isocapsidic. Alfalfa mosaic virus, on the other hand, contains 4

different RNA molecules, each packaged separately into a virion; such a virus is called heterocapsidic. Successful infection with alfalfa mosaic virus requires that a least one RNA molecule of each type enters the cell. A plant virus with a segmented genome is known as a multiple-component virus or a covirus.

segment long spacing An artificially prepared assembly of collagen molecules, aggregated side by side, in which the segments have the same length as the collagen molecule (2900 Å) and show a characteristic pattern of more then 40 cross-striations when examined with the electron microscope; produced from acidic solutions of collagen in the presence of polyanions such as ATP or chondroitin sulfate. *Abbr* SLS. *See also* fibrils long spacing.

segregation The separation of the two members of a pair of alleles during meiosis so that each gamete receives one of the alleles of the pair.

selected ion monitoring A mass spectrometric technique in which an ion current at a selected mass per charge value is used; a highly selective technique that permits the study of one, or a few, types of ions. *Abbr* SIM.

selected marker The desired gene, selected by means of the experimental conditions used in bacterial conjugation.

selection 1. The fourth stage in a multistage mechanism of carcinogenesis in which drug- and hormone-resistant cells are selected from a population of autonomous cancer cells. 2. The relative ability of different genotypes to survive and reproduce in the course of evolution. 3. The relative ability of different cells to survive and grow under experimental conditions, as in the selection of mutants for resistance to an antibiotic.

selection rules A set of quantum mechanical rules and formulas that are used to determine which transitions between energy states of a molecule are allowed and which are forbidden.

selective marker A marker, such as drug resistance or nutritional independence, that permits the selection of recombinants over the parental types.

selective medium A medium designed to allow the preferential growth of some organisms over that of others.

selective plating A method for isolating recombinants by plating two auxotrophic mutants on a minimal medium; only those recombinants that receive the normal allele of each mutant can grow under such conditions.

selective system Any experimental method designed to aid in the detection and isolation of a specific genotype.

selective theory 1. A theory of antibody formation according to which the information for antibody synthesis is genetically determined; the antigen is considered to react selectively with certain cells, receptors, or other biosynthetic units and to stimulate them to synthesize antibodies which they already were synthesizing at low levels or were potentially capable of synthesizing prior to immunization. *See also* instructive theory. 2. A theory according to which the evolution of the genetic code is due to natural selection, such that a large number of codes is theoretically possible, but one code has been selected over others because of its value for the survival of the organism; the code may, for example, be such as to minimize the effects of mutation or minimize the errors during translation. *See also* mechanistic theory.

selective toxicity The variable degree of harmfulness of a chemical such that it is toxic, at a given concentration, to one organism but not to another. A chemotherapeutic agent must show selective toxicity with espect to the host and the invading microorganism.

selective variant A mutated organism that can exist under conditions that are lethal to all like organisms not possessing the specific mutation.

selenium A element that is essential to humans and animals. Symbol, Se; atomic number, 34; atomic weight, 78.96; oxidation states, -2, $+4$, $+6$; most abundant isotope, ^{80}Se; a radioactive isotope, ^{75}Se, half-life, 120.4 days, radiation emitted, gamma rays. *See also* antipromoter.

self- An immunological term used in reference to the antigens and antibodies involved in autoimmunity.

self-absorption The absorption of radiation, particularly of radioactive radiation, by the material emitting the radiation.

self-assembly The spontaneous formation of a supramolecular structure from its component molecules or subunits, as in the assembly of ribosomes, viruses, membranes, or multienzyme systems; occurs without additional energy sources.

selfish DNA A segment of DNA that has no known function except to replicate itself. Spacer DNA and satellite DNA may be examples. *Aka* junk DNA.

self-marker theory A theory proposed by Burnet and Fenner in 1949 to explain an organism's capacity to distinguish between "self" and "nonself." According to this theory, all the antigens in an organism carry characteristic self-markers that can be recognized by the immunocompetent cells of that organism so that formation of antibodies to these antigens is prevented.

self-priming Not requiring a primer. The synthesis of RNA by DNA-dependent RNA polymerase is a self-priming reaction as opposed to the synthesis of DNA by DNA-dependent DNA polymerase which requires the 3′-OH end of a primer.

self-quenching INTERNAL QUENCHING.

self-recognition AUTOIMMUNITY.

self-scattering The scattering of radiation, particularly of radioactive radiation, by the material emitting the radiation.

self-transmissible plasmid A plasmid that is capable of transferring itself from one cell to another; a plasmid that is both conjugative and mobilizable.

Seliwanoff's test A colorimetric test for ketohexoses that is based on the production of a red color upon treatment of the sample with resorcinol and hydrochloric acid.

SEM 1. Scanning electron microscope. 2. Standard error of the mean.

semialdehyde A compound formed by the conversion of one of two carboxyl groups in a molecule to an aldehyde; the aldehydes formed from aspartic and glutamic acid are examples.

semiconservative replication A mode of replication for double-stranded DNA in which the parental strands separate and each daughter molecule consists of one parental strand and one newly synthesized strand. The usual model of DNA replication in which each parental strand serves as a template for the synthesis of a new, complementary strand. *See also* DNA replication.

semiconstitutive mutant A mutant that synthesizes an inducible enzyme at a greater rate than does the uninduced wild-type organism, and that can be induced to synthesize the enzyme at a level that is characteristic of the fully constitutive strain.

semidiscontinuous replication The overall synthesis of DNA according to the currently accepted mechanism in which one strand (leading strand) is synthesized essentially continuously while the other strand (lagging strand) is synthesized discontinuously. *See also* DNA replication.

semilethal mutation A mutation that results in the death of more than 50 but less than 100% of the mutants.

semilogarithmic paper Graph paper in which one axis has been scaled in terms of logarithms so that a plot of an exponential function will yield a straight line.

semiochemistry The study of substances that mediate interactions between organisms.

semipermeable membrane A membrane that allows the passage of only certain solutes but that is freely permeable to water.

semipolar bond COORDINATE COVALENT BOND.

semiquinone The half-reduced form of a quinone; the benzene ring carries one $>C=O$ and one $>CH—OH$ group.

semisynthetic Descriptive of a compound of which part of the structure has been isolated from natural sources and part of it has been synthesized. Semisynthetic penicillins and cephalosporins are examples of such synthetically modified antibiotics.

Semliki forest virus A virus that contains single-stranded RNA and that has a spherical nucleocapsid; it belongs to the group of arboviruses.

Sendai virus A virus that contains single-stranded RNA and that belongs to the paramyxovirus group of myxoviruses; used for cell fusion studies, since it modifies the surface of cells in such a manner that they tend to fuse.

senescence 1. The state of growing old; aging. 2. The phase of plant growth that extends from full maturity to death.

senility pigment AGE PIGMENT.

sense codon A normal, amino acid-specifying codon.

sense strand ANTICODING STRAND.

sensitive strain A bacterial strain that can be lysed by a temperate phage or by a lysogenic culture produced with that phage.

sensitive volume 1. That volume of a biological specimen in which an ionization must occur to produce a particular effect. 2. That volume of an ionization chamber through which the radiation must pass to be detected.

sensitivity The degree of responsiveness of an interconvertible enzyme to modification that is brought about by changes in the concentration of a given effector.

sensitivity spectrum The types of antimicrobial drugs that are effective against a given microorganism. *See also* antimicrobial spectrum.

sensitization The conditioning of an animal by administration of an allergen so that a second administration of the allergen will trigger an anaphylactic response. The sensitization may be active or passive depending on the type of anaphylaxis induced. *See also* active anaphylaxis; passive anaphylaxis.

sensitized erythrocyte An antibody-coated erythrocyte that is used in the detection of complement.

sensitized fluorescence The fluorescence that occurs when a photon excites a molecule, which then excites a second molecule by means of an energy transfer, and the second molecule dissipates the excitation energy by fluorescence.

sensitized phosphorescence The phosphorescence that occurs when a photon excites a molecule, which then excites a second molecule by means of an energy transfer, and the second molecule dissipates the exitation energy by phosphorescence.

sensitizer 1. PHOTOSENSITIZER. 2. ALLERGEN.

sensitizing agent SENSITIZER.

sensitizing injection The initial and harmless injection of an allergen into an animal which, if followed by a second injection, will trigger an anaphylactic response.

sensor gene *See* Britten–Davidson model.

sensor protein *See* Britten–Davidson model.

sentinel antibody An antibody-like receptor site on an antibody-producing cell by which the antigen is believed to stimulate the cell to produce antibodies.

separated plasma Plasma, obtained from whole blood, that is equilibrated with carbon dioxide at a given partial pressure.

separate package hypothesis The hypothesis according to which photosystems I and II of chloroplasts are two separate systems such that energy transfer is possible only between the pigments within each system, but not between the pigments from one system to those of the other. *See also* spillover hypothesis.

separation cell An analytical ultracentrifuge cell that allows for the separation and recovery from a mixture of the component having the smallest sedimentation coefficient.

separation factor The ratio of the retention times, or the ratio of the retention volumes, of two compounds that are separated by gas chromatography.

separation gel *See* disc gel electrophoresis.

separators Amphoteric substances that are added, in large amounts, to carrier ampholytes normally used in isoelectric focusing for the purpose of producing a relatively flat region in the pH gradient and thereby improving the separation of components.

Sephadex Trademark for a group of cross-linked dextran gels used in gel filtration.

Sepharose Trademark for a group of agarose gels used in gel filtration.

sepsis The presence of pathogenic microorganisms or their toxins in the blood or in the tissues.

septanose A monosaccharide that has a seven-membered ring structure.

septanoside A glycoside of a septanose.

septate junction An impermeable cell junction that is similar to a tight junction but occurs only in invertebrates; it differs from a tight junction in that the junctional proteins are arranged in a more regular fashion and that they form a seal without actually bringing the two plasma membranes into direct contact. *See also* cell junction.

septic Of, or pertaining to, sepsis.

septicemia The presence of pathogenic micro-organisms in the blood.

septum (*pl* septa; septums) A wall or a membrane that divides a cavity.

Sequenase Trademark for an enzyme preparation used in DNA sequencing. The enzyme is derived from bacteriophage T7 DNA polymerase and has been modified to improve its properties for sequencing.

sequenator An instrument for the automatic determination of amino acid sequences in a polypeptide chain; operation of the instrument is based on the repetitive application of the Edman degradation. *Aka* sequencer.

sequence 1. The linear order in which monomers occur in a polymer; the order of amino acids in a polypeptide chain, and the order of nucleotides in a polynucleotide strand are examples. 2. METABOLIC PATHWAY.

sequence complexity *See* complexity.

sequence gap A segment, consisting of one or more amino acids, that appears to be missing from a polypeptide chain when this chain is compared with others of the same protein but isolated from different sources, and when the chains are matched up so as to provide a maximum of sequence homology.

sequence homology The identity in sequence of either the amino acids in segments of two or more proteins, or the nucleotides in segments of two or more nucleic acids.

sequence hypothesis The hypothesis that the sequence of nucleotides in a nucleic acid specifies the sequence of amino acids in a protein.

sequence isomer One of two or more polymeric isomers that differ from each other in the sequence of the monomers in the chain.

sequence polymer A synthetic polypeptide consisting of identical repeating units, each of which is composed of more than one type of amino acid; the polymer (gly-ala-ser-val)$_n$ is an example. *See also* polyamino acid.

sequencer SEQUENATOR.

sequence rules RS SYSTEM.

sequence specificity The selectivity of a nuclease that accounts for its reaction being limited to specific base sequence in the nucleic acid.

sequencing The determination of the order of amino acids in a peptide, polypeptide chain, or protein, or the determination of the order of bases (nucleotides) in a nucleotide, polynucleotide strand, or nucleic acid.

sequencing gel A long, thin polyacrylamide gel slab used for nucleic acid sequencing.

sequential feedback inhibition The inhibition that is produced when one or more end products inhibit an enzyme in a metabolic pathway and the metabolite that accumulates as a result of this inhibition then inhibits the first enzyme in the sequence and thereby shuts off the entire pathway.

sequential induction Enzyme induction in which a single inducer brings about the synthesis of a number of inducible enzymes; the first enzyme induced acts on the inducer, thereby transforming it into an inducer for the second enzyme, which in turn acts on the second inducer, and so on. *See also* coordinate induction.

sequential mechanism The mechanism of an enzymatic reaction in which two or more substrates participate in such a fashion that all the substrates must become bound to the enzyme before any products can be released. The mechanism is ordered if the substrates add to, and the products leave, the enzyme in an obligatory sequence; the mechanism is random if the substrates add to, and the products leave, the enzyme in a nonobligatory sequence.

sequential model A model for allosteric enzymes, proposed by Koshland, Nemethy, and Filmer, according to which the enzyme undergoes a series of conformational changes as the various ligands become bound to the enzyme. Different types of site interactions may occur of which symmetry preservation, as in the concerted model, may be a special case. In general, however, the symmetry of the enzyme molecule is not conserved, since a subunit changes its conformation as a ligand becomes bound to it. The capacity of the enzyme to bind substrate, positive effectors, and negative effectors is altered by the conformational changes which the subunits undergo. *Abbr* KNF model.

sequential reactions CONSECUTIVE REACTIONS.

sequester To form a chelate.

sequestering agent CHELATING AGENT.

sequestration CHELATION.

sequestrene ETHYLENEDINITROLOTETRAACETIC ACID.

sequon An obligatory sequence of amino acids that is required for a specific reaction. The term has been used to describe the tripeptide asn-x-thr or asn-x-ser that must occur in a protein for the asparagine (asn) to be able to act as a site of attachment for a carbohydrate moiety, thereby giving rise to a glycoprotein. *Var sp* sequeon.

Ser 1. Serine. 2. Seryl.

SER Smooth endoplasmic reticulum.

serendipity The gift for discovering valuable or useful things not specifically sought but recognized in the process of dealing with other things.

serial dilution The systematic and progressive dilution that is frequently used in immunology, serology, and microbiology. A fixed volume of diluent is placed into a number of tubes and a given volume of sample, say 1.0 mL, is then added to the first tube. After mixing, 1.0 mL of solution is transferred to the second tube and, after mixing, 1.0 mL from this tube is transferred to the third tube, and so on.

serial passage 1. The repeated transfer of a pathogen (usually a virus) using a succession of animals, tissue cultures, or growth media with viral replication taking place after each transfer. Frequently used to attenuate the pathogen. 2. The transfer of an inoculum in tissue culture to a fresh medium; preparation of a subculture.

serial symbiosis theory The theory that eukaryotic cells developed by symbiotic associations with prokaryotic cells. Thus, mitochondria and microtubule organizing systems, present in modern eukaryotic cells, are believed to have evolved from various bacteria that lived as symbionts with primitive eukaryotic cells.

seriate In a series or a succession; serially.

seriatim In a series; serially.

sericate Of, or pertaining to, silk; silky.

sericin One of a group of proteins found in silk; silk gelatin.

sericulture The raising of silkworms (*Bombyx mori*) for the production of silk.

serine An aliphatic, polar alpha amino acid that contains an alcoholic hydroxyl group and that frequently occurs at or near the active site of an enzyme. *Abbr* Ser; S.

serine convention A method for assigning configurations to compounds by a comparison with the configuration of serine; thus (+) tartaric acid is designated as L_G when compared to glyceraldehyde, and as D_S when compared to serine.

serine esterase SERINE PROTEASE.

serine protease One of a group of proteolytic enzymes, including trypsin and chymotrypsin, that are similar in their three-dimensional structures and that contain a serine residue in the active site. Also called serine esterases because the mechanism of action involves formation of an ester between the hydroxyl group of the catalytically active serine and the carboxyl group of the cleaved peptide bond.

serologic Of, or pertaining to, serology. *Aka* serological.

serologic adhesion IMMUNE ADHERENCE.

serological pipet A measuring pipet in which the graduation marks extend to the tip of the pipet.

serology The science that deals with serums, particularly immune serums, and with the reactions and properties of antigens, haptens, antibodies, and complement.

serophyte Variant spelling of xerophyte.

serotonin 5-Hydroxytryptamine; a pharmacologically active mediator of immediate-type hypersensitivity. Serotonin is formed from tryptophan and is released from mast cells during the allergic response. Serotonin is a neurotransmitter that has hormone-like properties and causes vasodilation, increased capillary permeability, and contraction of smooth muscle.

serotonin hypothesis The hypothesis that schizophrenia is caused by an abnormality in the metabolism of serotonin in the brain, and that most hallucinogens act by either antagonizing or mimicking the functions of serotonin.

serotype A subdivision of a species or of a subspecies that is identifiable by serologic methods and that is distinguished by its antigenic character.

serotype transformation ANTIGENIC CONVERSION (2).

serous Of, or pertaining to, serum.

serous fluid A fluid resembling serum.

serous membrane One of various thin membranes that secrete a serous fluid, line cavities, or enclose the organs within them, and form the inner layer of a blood vessel.

SERS Surface-enhanced Raman spectroscopy.

serum (*pl* serums; sera). The fluid obtained from blood after it has been allowed to clot; plasma without fibrinogen.

serum Ac globulin ACCELERIN.

serum albumin The major protein in serum, the main function of which is the regulation of osmotic pressure.

serum blocking power The capacity of an immunoadsorbent to adsorb antibodies from a serum and to decrease the antibody titer of the serum. *Abbr* SBP.

serum converting enzyme The peptidase that catalyzes the conversion of angiotensin I to angiotensin II.

serum folate N^5-Methyltetrahydrofolic acid; a stable derivative of folic acid that is a major storage form of folate coenzymes in higher organisms and is the methyl group donor for the biosynthesis of methionine.

serum hepatitis *See* hepatitis.

serum L. caseii factor SERUM FOLATE.

serum proteins The proteins present in blood serum. *See also* plasma proteins.

serum prothrombin conversion accelerator PROCONVERTIN.

serum prothrombin converting factor PROCONVERTIN.

serum sickness A disease, characterized by fever and by local swelling at the injection site, that in humans may follow the injection of an immune serum prepared in an animal, and in animals may be initiated by the injection of large amounts of foreign protein. *Aka* serum disease.

serum sulfation factor SOMATOMEDIN.

serum thymic factor A nonapeptide in serum that has biological activity similar to that of thymic humoral factor.

servomechanism An automatic device for controlling the operation of a mechanism by having the output of the mechanism compared with its input, so that the error between the two quantities can be controlled in a prescribed manner.

sessile Attached directly to a base without a stalk; nonmotile; sedentary.

Sevag method A procedure for the deproteinization of nucleoprotein in which the protein is denatured by shaking a solution of the nucleoprotein with chloroform and isoamyl alcohol.

SEXAFS Surface-extended x-ray absorption fine structure spectroscopy.

sex chromatin BARR BODY.

sex chromosome A chromosome that is specifically connected with the determination of sex. *See also* X chromosome; Y chromosome.

sexduction The process whereby a segment of genetic material is transferred from one bacterium to another by attachment to the fertility factor; the formation of F' strains.

sex factor FERTILITY FACTOR.

sex hormone One of a group of hormones that are responsible for the development of secondary sex characteristics and that are capable of stimulating the development of accessory reproductive organs. Sex hormones are secreted principally by the gonads and consist of androgens, estrogens, and gestagens.

sex hormone binding globulin TESTOSTERONE–ESTRADIOL BINDING GLOBULIN.

sex linkage The linkage of genes located on a sex chromosome.

sex-linked gene A gene located on a sex chromosome.

sex pilus F-PILUS.

sex plasmid FERTILITY FACTOR.

sex steroid binding plasma protein TESTOSTERONE–ESTRADIOL BINDING GLOBULIN.

sexual conjugation CONJUGATION (3).

s_f Flotation coefficient.

s_f^0 Standard flotation coefficient.

SFC Supercritical fluid chromatography.

SFFF Sedimentation field flow fractionation.

SF$_1$ fragment One of two fragments of the head, or globular part, of the myosin molecule; produced when myosin is treated with the enzyme papain. *Aka* S1 fragment.

S-100 fraction A cell-free preparation obtained from a suspension of broken cells by first removing intact cells and cell debris by low-

speed centrifugation at $30,000 \times g$ (yielding an S-30 fraction), and then removing ribosomes by centrifugation at $100,000 \times g$. The fraction contains transfer RNA and amino acyl-tRNA synthetases, and is used in studies of cell-free amino acid incorporation. *See also* pH 5 fraction.

S1 fragment *See* SF$_1$ fragment.

SGF Skeletal growth factor.

SGOT Serum glutamate–oxaloacetate transaminase.

SGPT Serum glutamate–pyruvate transaminase.

SH Somatotropin.

shadowcasting A technique for preparing specimens for electron microscopy in which the specimen is covered by metal atoms deposited onto it at a fixed angle. Areas around particles on the far side of the metal source remain free of deposited metal atoms and form shadows that provide information about the size and the shape of the particle. *Aka* shadowing.

shake flask An Erlenmeyer flask, containing a liquid medium and a bacterial inoculum, that is placed on a shaker (or incubator–shaker) for aeration and growth.

shaker A laboratory device for the mechanical shaking of samples for purposes of mixing and aeration; the shaking action may involve circular motion (rotary shaker) or a back and forth motion (reciprocating shaker).

shallow groove MINOR GROOVE.

SH-antigen HEPATITIS B ANTIGEN.

shape factor *See* Oncley equation.

Sharples centrifuge An efficient, continuous, high-speed flow centrifuge. The rotor consists of a long, narrow cylinder which is rotated in an upright position, with slurry being fed in at the bottom and the supernatant flowing out at the top.

SHBG Sex hormone binding globulin.

shear The deformation experienced by a liquid as a result of the variations in the velocity of flow of different layers. Shear is associated with the flow of a liquid through a capillary, with the forcing of a liquid through a pipet, and with the homogenization of a suspension. *See also* rate of shear.

shear dichroism FLOW DICHROISM.

shear gradient RATE OF SHEAR.

shearing The process whereby particulate matter is degraded as a result of shear; the breaking up of DNA into fragments by treatment in a blender is an example.

shear stress The force per unit area resulting from solution flow.

sheath 1. A tubular structure that is formed around a chain of cells in some bacterial species. 2. The cell wall of bacteria belonging to the *Spirochetes*. *See also* capsule. 3. The covering of an axon; *see* myelin sheath.

sheath microfilaments *See* microfilaments.

Shemin cycle A complex set of reactions whereby the tetrapyrroles are synthesized from glycine and succinyl coenzyme A.

SH-enzyme An enzyme whose activity depends on the presence of one or more SH groups; an enzyme that has at least one cysteine residue at or near its active site.

SH group Sulfhydryl group.

shield A solid barrier for the protection of individuals from radiation or from potentially explosive laboratory setups.

shielded nucleus An atomic nucleus in a molecule that is surrounded by a relatively greater electron density than another nucleus. In nuclear magnetic resonance, such a nucleus will absorb radio frequencies of higher energy (upfield) than the second nucleus.

shift A chromosomal aberration in which a chromosome segment is removed from its normal place and is inserted elsewhere in the chromosome, with the original nucleotide sequence in the segment either maintained or reversed.

shift down A shift experiment in which the change in the medium leads to a decrease in the rate of growth of the cells.

shift experiment An interference with the balanced growth of cells in a culture by a precisely timed and defined change in the medium. *See also* shift down; shift up.

shift up A shift experiment in which the change in the medium leads to an increase in the rate of growth of the cells.

Shigella A bacterial genus that includes the causal agent of dysentery (*Shigella dysenteriae*). The latter is attacked by many *E. coli* phages.

shikimic acid A hydroxylated, unsaturated, acid derivative of cyclohexane that serves as a key intermediate in the biosynthesis of the aromatic amino acids.

shikimic acid pathway A pathway for the synthesis of shikimic acid in bacteria and for its conversion to tyrosine, tryptophan, and phenylalanine.

Shine–Dalgarno sequence A segment of 4–7 nucleotides that occurs in the leader section of mRNA. It base pairs with 16S rRNA and thereby places the AUG initiation codon of the mRNA in the proper orientation relative to the ribosome for initiation of translation to take place. The Shine–Dalgarno sequence is part or all of the following sequence: 5'-AGGAGGU-3'; it is named after the individuals who first recognized its significance. *Abbr*. SD sequence.

shock A circulatory failure due to loss of blood

from the vascular compartment by either hemorrhage or increased capillary permeability.

shocking dose The second injection of allergen that triggers the anaphylactic response in an animal.

shock-sensitive permeases PERIPLASMIC PERMEASES.

Shope papilloma virus A virus that produces papillomas in rabbits. *See* papilloma virus.

short column cell An analytical ultracentrifuge cell in which the sample solution constitutes a very short liquid column (about 1 to 3 mm); used for molecular weight determinations by sedimentation equilibrium.

short column sedimentation equilibrium A variation of the sedimentation equilibrium method for the determination of molecular weights in which short column cells are used so that the time required to reach equilibrium is greatly decreased; the method is especially useful for the simultaneous analysis of a large number of different samples.

short interspersed repeated sequences A family of short (70–300 bp), repetitive, segments in mammalian DNA that often occur in over 100,000 copies per genome; almost all appear to be retroposons. The Alu sequences constitute one such family. *Abbr* SINE.

short patch pathway A mechanism of DNA repair in both prokaryotes and eukaryotes in which the repaired DNA segments are relatively short.

short period interspersion A DNA structure characterized by relatively short segments of moderately repetitive DNA (about 300 bp each) alternating with relatively short segments of nonrepetitive DNA (about 1000 bp each).

short-range hydration The hydration by water molecules that are located in the primary hydration shell of an ion.

short-range interactions The attractive and repulsive forces between atoms and molecules that decrease rapidly with distance. *See also* weak interactions.

short trough technique An immunoelectrophoretic technique for the identification of a specific antigen in a mixture.

shotgun experiment The process of breaking up a large fraction of the DNA of an organism, collecting the fragments in a random manner, and cloning these DNA fragments in order to set up a clone library from which cloned fragments can later be selected.

SHRH *See* growth hormone regulatory hormone.

SHRIH *See* growth hormone regulatory hormone.

shufflon A clustered inversion; a section of

DNA in which multi-inversions of contiguous segments can occur.

shunt *See* metabolic shunt; hexose monophosphate shunt.

shuttle A mechanism whereby reducing equivalents are removed from cytoplasmic NADH and passed to the electron transport system in the mitochondrial membrane by way of intermediate compounds, since NADH itself cannot pass through the membrane.

shuttle vector A vector that can replicate in a number of different organisms (yeast and *E. coli*, for example).

Shwartzman reaction The production of an inflammatory lesion in an animal by the subcutaneous injection of an endotoxin from a gram-negative bacterium, followed by a second injection, about 24 h later, of the same endotoxin or of some other substance.

SHyp Mercaptopurine.

Si Silicon.

SI Système International d'Unités; an extension and a refinement of the metric system. The system is based on seven basic units from which other units are derived. The basic units (or base units), their symbols, and the quantities which they measure are meter, m, length; kilogram, kg, mass; second, s, time; ampere, A, electric current; Kelvin, K, thermodynamic temperature; candela, cd, luminous intensity; and mole, mol, amount of substance. Some derived units are the newton N for force, the joule J for energy, and the liter L for volume. *Aka* SI system.

Sia Sialic acid.

sial- *See* sialo-.

sialic acid *N*-Acylneuraminic acid or any of its esters or other derivatives of its alcoholic hydroxyl groups. Sialic acids are widely distributed in bacteria and animals; in most mammals, sialic acids occur in the form of *N*-acetyl (a) and *N*-glycolyl (b) derivatives of neuraminic acid:

$$\underset{\text{(a)}}{CH_3-\overset{\displaystyle\overset{O}{\|}}{C}-} \qquad \underset{\text{(b)}}{HOCH_2-\overset{\displaystyle\overset{O}{\|}}{C}-}$$

sialidase NEURAMINIDASE.

sialidosis A genetically inherited metabolic defect in humans due to a deficiency of the enzyme *N*-acetylneuraminidase.

sialo- 1. Combining form meaning of, or related to, saliva or the salivary glands (also used as sial-). 2. Combining form meaning of, or related to, sialic acid.

sialoglycosphingolipid GANGLIOSIDE.

sialogogue An agent that stimulates the secretion of saliva. *Var sp*. sialagogue.

sialosyl group The radical resulting from removal of a hydroxyl group from the anomeric

carbon of neuraminic acid or sialic acid.

sialoyl group The radical resulting from removal of a hydroxyl group from the carboxyl group of neuraminic acid or·sialic acid.

sickle cell An erythrocyte that has an abnormal crescent-like shape; such a cell is more fragile than the normal cell and tends to hemolyze in the blood capillaries.

sickle cell anemia A genetically inherited metabolic defect in humans that is characterized by the formation of an abnormal hemoglobin (sickle cell hemoglobin) which leads to sickling and hemolysis of erythrocytes. *See also* sickle cell disease; sickle cell trait.

sickle cell disease A condition in which an individual is homozygous (has 2 defective genes) for sickle cell anemia.

sickle cell hemoglobin The abnormal hemoglobin responsible for sickle cell anemia. It differs from normal adult hemoglobin in having valine in place of glutamic acid in the sixth position of the beta chain. *Sym* HbS. *See also* Murayama hypothesis.

sickle cell trait A condition in which an individual is heterozygous for sickle cell anemia.

sicklemia SICKLE CELL ANEMIA.

sickling The process whereby erythrocytes take on abnormal, crescent-shaped forms, as in sickle cell anemia.

side chain 1. A smaller chain attached laterally to a longer chain. 2. A chain attached to a ring. 3. AMINO ACID SIDE CHAIN.

side chain cleavage The removal of some or all of the carbon atoms from the aliphatic side chain attached to the steroid nucleus. Specifically, the set of enzymatic reactions whereby the aliphatic side chain of cholesterol is removed.

side chain theory EHRLICH'S RECEPTOR THEORY.

sidedness The vectorial character of either a molecule or a system.

sideramine *See* siderochrome.

side reaction A secondary reaction that occurs simultaneously with the major reaction.

siderochrome An early term for siderophore with the specific designations of sideramine and sideromycin applied to a siderophore that has growth supporting activity and antibiotic activity, respectively.

sideromycin *See* siderochrome.

siderophilin One of a group of nonheme, iron-binding, monomeric glycoproteins (MW about 80,000) that bind two Fe^{3+} ions per molecule concomitantly with two CO_3^{2-} (or HCO_3^-) ions. They are classified on the basis of their occurrence as transferrin (vertebrate blood, the major iron transport protein of plasma); lactoferrin (mammalian milk and other body secretions); and ovotransferrin (avian blood and egg white).

siderophore One of a number of low molecular weight, microbial, iron-containing or iron-binding organic compounds. Siderophores have a very strong affinity for Fe^{3+} (which they chelate) and function in the solubilization and transport of iron. Some also act as growth or germination factors, and some are potent antibiotics. Siderophores are classified as belonging to (a) the phenol–catechol type (such as enterobactin and agrobactin), or (b) the hydroxamic acid type (such as ferrichrome and mycobactin).

siderosis HEMOSIDEROSIS.

siderosome An artificially induced, electron-dense structure that contains iron particles; a lysosomal body that forms in cultured animal cells upon administration of iron complexes.

sidescattering The scattering of radiation in directions other than those of forward or backward scattering.

sieve tube A leaf capillary.

sievorptive chromatography A chromatographic technique that combines gel filtration and adsorption chromatography; ion filtration chromatography and intervent dilution chromatography are two examples.

SIF See growth hormone regulatory hormone.

sigma 1. The symbol Σ that indicates the summation of all the quantities that follow it. 2. Standard deviation (σ).

sigma bond A chemical bond formed by the electrons in sigma orbitals; a covalent bond of circular cross section in which orbital overlap is concentrated along the axis joining the two atomic nuclei.

sigma cycle The set of reactions composed of (a) the attachment of the sigma factor to the core enzyme of RNA polymerase at the initiation site of transcription; (b) the release of the sigma factor from the core enzyme during chain elongation; and (c) the subsequent binding of the sigma factor to the same, or to another, core enzyme at an initiation site.

sigma factor A protein subunit of *E. coli* RNA polymerase that functions in the recognition of the promoter and the initiation of transcription. The sigma factor, by itself, has no catalytic activity; it binds to the core enzyme prior to transcription. *Sym* σ. See also sigma cycle.

sigma orbital A molecular orbital that is a localized bond orbital, spread over two bonding atoms.

sigma replication ROLLING CIRCLE REPLICATION.

sigma star orbital See antibonding orbital.

sigma structure 1. The DNA structure formed during the replication of double-stranded circular DNA according to the rolling circle model. *Aka* lariat form. 2. The RNA structure formed during the splicing of pre-mRNA in eukaryotes.

sigma virus A virus, belonging to the group of rhabdoviruses, that infects the fruit fly, *Drosophila melanogaster*. Infected flies are sensitive to carbon dioxide; a brief exposure to CO_2 leads to paralysis and death.

sigmoid kinetics The kinetics of an enzymatic reaction that yield a sigmoid curve for a plot of reaction velocity versus substrate concentration; the sigmoid, or S-shaped, curve is characteristic of many allosteric enzymes as well as of other systems showing cooperative-type interactions.

signal amplification The production of a large amount of modified or unmodified interconvertible enzyme by a small amount of converter enzyme.

signal codons See signal hypothesis.

signal hypothesis The currently accepted model for the mechanism whereby secretory proteins are selected by the rough endoplasmic reticulum (RER) and exported through it (vectorially discharged). Key aspects of the signal hypothesis are as follows: (a) The mRNA that codes for a secretory protein contains a specific sequence of codons immediately following the initiation codon; this sequence is called the signal codons. (b) The translation of this mRNA is initiated by free (not membrane-bound) ribosomes in the cytosol which synthesize a peptide corresponding to the signal codons; this peptide consists of 15–30 amino acids, many of which are hydrophobic, and is known as the signal sequence or signal peptide. It is located at the N-terminal of the secretory protein. (c) As the signal sequence emerges from the ribosome, it becomes bound to a specific receptor in the RER which results in the binding of the free ribosome, bearing the signal sequence, to the RER. At the same time, a transient pore (transmembrane pore) is formed around the signal sequence. (d) As translation continues, the nascent polypeptide chain is extruded through this pore across the membrane of the RER; the signal sequence is cleaved off by a specific protease, called signal peptidase, before the polypeptide chain is completed and before the nascent protein is generally glycosylated. (e) Translation is terminated by complete deposition of the protein into the lumen of the RER, dissolution of the transient pore, and dissociation of the ribosome from the membrane of the RER. (f) The finished protein is then transported via the Golgi apparatus and guided to its particular cellular destination; this may involve additional glycosylation. *See also* vectorial discharge; glycosylation; signal recognition protein.

signal peptidase See signal hypothesis.

signal peptide See signal hypothesis.

signal recognition protein A multisubunit protein, present in the cytosol, that binds to ribosomes shortly after they have synthesized the signal sequence and serves to bind the ribosomes to the RER. It halts further translation until the ribosome has become bound to the RER. The protein is believed to recognize both the N-terminal of the nascent polypeptide chain and a receptor on the RER (called SRP receptor or docking protein). *Abbr* SRP. *Aka* signal recognition particle. *See also* signal hypothesis.

signal sequence *See* signal hypothesis.

signal-to-noise ratio The ratio of the electrical response of an instrument during the measurement of a sample to the response from random background electrical fluctuations.

significance of results A measure of the deviation of results from the mean. If the probability value of a result is equal to, or less than, 0.05 but greater than 0.01, the result is considered to be significantly different from the mean; if the probability value is less than, or equal to, 0.01 but greater than 0.001, the result is considered to be highly significant; and if the probability value is equal to, or less than, 0.001, the result is considered to be very highly significant. A probability value of 0.05 means that there is a 5% chance that the result will differ from the mean by ±2 standard deviations.

significance test A statistical test for accepting or rejecting a hypothesis. The test provides a criterion for deciding whether the difference between an assumed (expected) and an observed (measured) value of a parameter can reasonably be attributed to chance. If the difference is so small that it can be attributed to chance, then one has the option of accepting the hypothesis.

significant figures The digits in a number, the values of which are known with certainty, plus the first digit, the value of which is uncertain; the position of the decimal point is immaterial. Thus, the number 12.40 is considered to have four significant figures, and its true value lies between 12.395 and 12.405.

sign inversion mechanism A proposed mechanism for the action of DNA gyrase. It is based on changing a positive node in DNA to a negative one, thereby converting a positive superhelical segment to a negative one.

sign mutation FRAMESHIFT MUTATION.

SIH *See* growth hormone regulatory hormone.

SILA Suppressible insulin-like activity.

silage An animal feed prepared by the controlled fermentation (involving primarily *Lactobacillus* and *Streptococcus*) of various plant materials, packed firmly into silos.

silanizing The treatment of a chromatographic support, such as a diatomaceous earth, with trichloromethylsilane or a similar reagent that converts active silanol groups (a) to less polar silyl ethers (b), thereby changing the adsorptive properties of the support. Glass chromatographic columns can be treated similarly; this minimizes adsorption of sample material to the surface of the glass. *Aka* silanization.

silent allele *See* silent gene.

silent gene A gene that has no detectable product.

silent mutation A mutation that does not result in a detectable phenotypic effect. A silent mutation may be due to a transition or a transversion that leads to either a synonym codon or a codon which codes for an amino acid closely related to that coded for by the original codon. As a result, the polypeptide chain will be synthesized, either without a change in its amino acid sequence, or with replacement of an amino acid by a closely related one. In the latter case, the polypeptide, while altered, may still be fully functional so that the mutation produces no detectable effect. *See also* isocoding mutation.

silica gel An adsorbent used in column and thin-layer chromatography for the separation of nonionic organic compounds. *Aka* silica; silicic acid.

silicon An element that is essential to humans and several classes of animals and plants. Symbol, Si; atomic number, 14; atomic weight, 28.086; oxidation states, -4, $+2$, $+4$; most abundant isotope, ^{28}Si; a radioactive isotope, ^{31}Si, half-life, 2.6 h, radiation emitted, beta particles.'

siliconization The inactivation of a surface by physically coating it with a thin film of silicone oil; this contrasts with the chemical reaction of silanization.

silk fibroin A fibrous protein of silk that has an antiparallel pleated sheet structure.

silk gelatin SERICIN.

silver nitrate chromatography ARGENTATION CHROMATOGRAPHY.

silver stain A sensitive stain used for detecting proteins separated by gel electrophoresis.

silylation The introduction of a trimethylsilyl group $-Si(CH_3)_3$ into an organic compound; used for making volatile derivatives for gas chromatography.

SIM Selected ion monitoring.

Simha equation One of two equations that relate the viscosity increment to the axial ratio of either a prolate or an oblate ellipsoid of revolution.

simian Of, or pertaining to, monkeys.

simian acquired immune deficiency syndrome An infectious disease in Rhesus monkey that is strikingly similar to human AIDS and is also caused by a retrovirus. *Abbr* SAIDS.

simian virus 40 A small, naked, icosahedral, and oncogenic virus that contains double-stranded DNA and belongs to the group of papovaviruses. *Abbr* SV-40.

7S immunoglobulin A basic structural unit of the immunoglobulins. The IgG and IgD immunoglobulins as well as the monomeric units of IgA and IgM immunoglobulins all have sedimentation coefficients of 7S; the IgE immunoglobulins, however, have a sedimentation coefficient of 8S.

19S immunoglobulin IgM.

simple anhydride *See* acid anhydride.

simple diffusion 1. The movement of a solute across a biological membrane that does not require the participation of transport agents; unmediated transport. *Aka* passive transport. 2. FREE DIFFUSION. 3. SINGLE DIFFUSION.

simple dispersion *See* simple optical rotatory dispersion.

simple enzyme 1. An enzyme that consists only of protein. 2. An enzyme that contains a non-protein component which does not participate in the catalytic process.

simple goiter A thyroid enlargement that is unaccompanied by either hyper- or hypothyroidism.

simple hapten A low molecular weight hapten that constitutes a separate part of a complete antigen but that does not give a visible precipitin reaction with the appropriate antibody.

simple lipid NEUTRAL LIPID.

simple microscope A microscope having one lens.

simple optical rotatory dispersion An optical rotatory dispersion that can be described by a one-term Drude equation.

simple protein A protein that is composed only of amino acids.

simple-sequence DNA SATELLITE DNA.

Simplesse Trademark for a low-calorie, cholesterol-free fat substitute that is made from milk and egg white proteins; involves a heating and blending process (called microparticulation) that shapes the proteins into small spherical particles (0.1–2.0 μm in diameter). In this shape and size range, the particles roll smoothly over one another so that the tongue perceives them as fluid rather than as individual particles. This creates the sensation of smoothness, richness, and creaminess normally associated with fat.

simple sugar MONOSACCHARIDE.

simple triglyceride A triglyceride containing only one type of fatty acid.

SIMS Secondary ion mass spectrometry.

simultaneous reactions A mechanism in which one reactant can give rise to either of two different products by either of two different reactions.

SINE Short interspersed repeated sequences.

single-burst experiment A technique for studying viral multiplication by isolating single infected cells.

single carbon unit ONE-CARBON FRAGMENT.

single-cell protein Protein derived from single-cell organisms such as bacteria, yeasts, and fungi; of interest as a potential food for man and animals. *Abbr* SCP.

single-copy DNA UNIQUE DNA.

single-copy plasmid A plasmid having a copy number of one; a plasmid present in bacterial cells in the ratio of one plasmid per host chromosome.

single diffusion Immunodiffusion in which either the antigen diffuses through a gel containing antibody, or the antibody diffuses through a gel containing antigen.

single-displacement mechanism SEQUENTIAL MECHANISM.

single-event curve SINGLE-HIT SURVIVAL CURVE.

single-hit survival curve A survival curve in which the absorption of one photon leads to loss of viability of the active unit; such a survival curve reflects simple exponential inactivation kinetics.

single-ion monitoring MASS FRAGMENTOGRPHY.

single reciprocal plot A plot of $1/Y$ or X/Y versus X, where X and Y are two variables; the Scatchard plot and the Eadie-Hofstee plot are two examples.

single-site mutation POINT MUTATION.

single-strand assimilation The displacement of one DNA strand in a duplex by the homologous strand, leading to formation of a D-loop. The reaction occurs during recombination, and formation of the heteroduplex involves the RecA protein.

single-strand binding protein One of a group of proteins that bind to the single strands of DNA at the Y-fork during DNA replication; they allow replication to proceed by preventing the unwound strands from coming back together and annealing. *Abbr* SSB. *Aka* unwinding protein.

single-strand break NICK.

single-stranded Descriptive of a nucleic acid molecule that consists of only one polynucleotide chain. *Abbr* ss.

single-stranded DNA binding protein *See* single-strand binding protein.

single-strand exchange The pairing of a DNA

strand from one duplex molecule with a complementary strand in another duplex DNA, thereby displacing the homologous strand from the second duplex.

single-substrate enzyme An enzyme that catalyzes a reaction involving only one substrate.

singlet A single peak, as that obtained in nuclear magnetic resonance.

single-tailed test ONE-SIDED TEST.

singlet state The electronic state of an atom or a molecule in which two single electrons in two orbitals (two unpaired electrons) have their spin in opposite directions and are spin-paired so that $S = 0$ and $2S + 1 = 1$, where S is the resultant spin and $2S + 1$ is the spin multiplicity. Such an atom or molecule has no net magnetic moment. *See also* triplet state.

SIno Thioinosine.

sintered Descriptive of porous material that is formed by the partial fusion of a substance by heat.

siphon An inverted U-shaped tube with legs of unequal length that is used for the delivery of liquid from one container to a second one at a lower level.

Sips distribution A frequency distribution that closely resembles a Gaussian distribution and that is used for the treatment of data from equilibrium dialysis measurements of the antigen–antibody reaction.

siroheme An iron tetrahydroporphyrin of the isobacteriochlorin type that is the prosthetic group of nitrite reductase from *Neurospora crassa* and of sulfite reductase from *E. coli*.

sirohydrochlorin A siroheme from which the iron has been removed.

sister chromatids The two identical nucleoprotein molecules held together by a centromere.

sister-strand exchange *See* recombination repair.

sistrand The translational unit of mRNA as opposed to the genetic unit, the cistron. The sistrand is that unit of RNA that lies between an initiation and a termination signal. The term is derived from single initiation single termination srand.

site 1. A specific region of a macromolecule at which a binding interaction with another molecule, or with an atom or an ion, takes place; the active site of an enzyme and the antigen binding site of an antibody are two examples. 2. The position of a mutation in a gene.

site heterogeneity The heterogeneity of antibodies that results from the antigen molecule possessing several antigenic determinants; consequently, one antigen leads to the production of several types of antibodies.

site-specific endonuclease RESTRICTION ENZYME.

site-specific inversion *See* inversion (2).

site-specific recombination Genetic recombination in which the exchange of genetic material takes place at specific sites at one or both of the participating DNA segments. The segments need not be homologous and the recombination results in an altered nucleotide sequence. The exchange is catalyzed by an enzyme (site-specific recombinase) that requires specific nucleotide sequences on one or both of the recombining DNA molecules. *Aka* site-specific exchange. *See also* general recombination.

skeletal band An infrared absorption band that is characteristic of the entire molecule rather than of a specific group in the molecule. *See also* group frequency band.

skeletal growth factor A large protein that stimulates the growth of bone cells. *Abbr* SGF.

skeletal muscle A striated muscle that is attached to, or that moves parts of, the skeleton.

skew conformation Any conformation partway between a staggered and an eclipsed conformation.

skewed distribution A distribution that is not symmetrical; it is known as a positively skewed distribution if the tail is to the right and the mean exceeds the median; it is known as a negatively skewed distribution if the tail is to the left and the mean is less than the median.

skin-sensitizing antibody REAGIN.

slant culture A bacterial culture grown in a tube that contains a solid nutrient medium which was solidified while the tube was kept in a slanted position.

slice technique *See* tissue slice.

sliding filament model A model proposed by Huxley and Hanson to explain the changes in the length of a striated muscle upon contraction and stretching. According to this model, thick and thin muscle filaments slide alongside each other, thereby leading to varying degrees of interpenetration without changes in the lengths of the filaments themselves. *See also* rowboat model.

sliding microtubule mechanism A proposed mechanism according to which the bending movements of cilia and flagella are brought about by the sliding of adjacent doublet microtubules past each other within the axoneme.

slightly repetitive DNA *See* repetitive DNA.

slime *See* capsule.

slope replot *See* secondary plot.

sloppy agar Semisolid agar.

slow-assembly end *See* actin filament.

slow component That fraction of DNA (usually unique DNA) that reassociates last in a reassociation kinetics experiment.

slow hemoglobin A hemoglobin which, after electrophoresis, is located closer toward the cathode than normal adult hemoglobin.

slow-reacting substance *See* SRS; SRS-A.

slow-reacting substance of anaphylaxis *See* SRS-A.

slow stop mutant A temperature-sensitive mutant of *E. coli* that can complete its round of replication when placed at the restrictive temperature but stops at that point; it cannot initiate another round of replication.

slow-twitch muscle RED MUSCLE.

slow virus *See* prion.

SLR factor N^{10}-Formylpteroic acid; a factor that stimulates the growth of *Streptococcus lactus* R; rhizopterin.

SLS Segment long spacing.

small-angle x-ray diffraction A method of x-ray diffraction in which the scattering of the x rays is measured at small angles; used for the analysis of large molecular spacings, as those between monomeric groups in a polymer.

small calorie CALORIE.

small cytoplasmic ribonucleoproteins *See* small cytoplasmic RNA.

small cytoplasmic RNA A class of small RNA molecules, consisting of 100–300 nucleotides, that are located in the cytoplasm of eukaryotic cells. Their distinguishing feature is their association with specific protein molecules to form small cytoplasmic ribonucleoproteins (scRNP); little is known about the function of the latter. *Abbr* scRNA.

small lymphocyte A term used to describe both B cells and T cells.

small nuclear RNA A class of small RNA molecules, consisting of 100–300 nucleotides, that are located in the nucleus of eukaryotic cells. Their distinguishing feature is their association with specific protein molecules to form small nuclear ribonucleoproteins (snRNP or snurp); some of the latter participate in RNA processing. *Abbr* snRNA.

small nuclear ribonucleoprotein *See* small nuclear RNA.

smallpox virus VARIOLA VIRUS.

S1 mapping A technique for locating the 5′-end of mRNA in vivo by the use of S1 nuclease.

smectic liquid crystal *See* liquid crystal.

Smith degradation The degradation of a polysaccharide by means of periodate oxidation, followed by reduction with sodium borohydride and acid hydrolysis; the method yields fragments that indicate the mode of linkage of the monosaccharides in the original polysaccharide.

Smithie's theory A somatic recombination theory of antibody formation.

smooth endoplasmic reticulum That portion of the endoplasmic reticulum to which few or no ribosomes are attached; contains a variety of enzymes used to detoxify drugs and other toxic compounds. *Abbr* SER. *Aka* smooth-surfaced endoplasmic reticulum.

smooth microsomes *See* microsomes.

smooth muscle An involuntary muscle such as a muscle of an internal organ or a muscle of a blood vessel; so called because it does not appear striated.

smooth–rough variation *See* smooth strain; rough strain.

smooth strain A bacterial strain, such as pneumococcus (*Streptococcus pneumoniae*) that grows in the form of a colony that appears smooth (has no jagged edges). In some species, smooth strains are virulent and have complete cell wall lipopolysaccharides (O antigens). *Abbr* S strain.

SMP Submitochondrial particle.

S mutant A bacterial mutant that gives rise to a smooth colony. *See also* smooth strain.

sn Stereospecific numbering.

SN Steroid number.

snake venom A mixture of toxins produced in the venom glands of poisonous snakes; consists of various toxic proteins (such as neurotoxins, cardiotoxins, and protease inhibitors) and a number of enzymes (such as hyaluronidase, acetylcholinesterase, and L-amino acid oxidase).

snapback The rapid renaturation of heat-denatured double-stranded DNA or RNA that occurs when the nucleic acid contains a short segment in which the nucleotide pairs are intact, with the base pairing in perfect register. The minimum size of this segment varies with the composition of the nucleic acid; the segment generally consists of between 12 and 20 nucleotide pairs. *See also* minimal stable length.

snapback DNA FOLDBACK DNA.

sneak synthesis BACKGROUND CONSTITUTIVE SYNTHESIS.

Snell's law The law governing the refraction of light as it passes from medium 1 to medium 2; specifically, $n_1 \sin i_1 = n_2 \sin i_2$, where n is the index of refraction, i_1 is the angle of incidence, and i_2 is the angle of refraction.

S_N1 mechanism A unimolecular, nucleophilic substitution mechanism that can be formulated as $RX \rightarrow R^+ + X^-$; $N^- + R^+ \rightarrow RN$; $v = k[RX]$; v is the velocity of the reaction, and k is the rate constant. Also designated SN_1 mechanism.

S_N2 mechanism A bimolecular, nucleophilic substitution mechanism that proceeds with an inversion of configuration and that can be formulated as $N^- + RX \rightarrow NR + X^-$; $v =$

$k[N][RS]$; v is the velocity of the reaction, and k is the rate constant. Also designated SN_2 mechanism.

Sno Thioinosine.

snRNA Small nuclear RNA.

snRNP Small nuclear ribonucleoprotein.

S1 nuclease An endonuclease, from *Aspergillus oryzae*, that catalyzes the hydrolysis of single-stranded DNA or single-stranded regions in double stranded DNA, to yield 5′-phosphoryl mono- and oligonucleotides.

snurp Small nuclear ribonucleoprotein.

soap The salt of a long-chain fatty acid; commonly refers to either the sodium or the potassium salt.

soap bubble meter A device for measuring low rates of gas flow that is used in gas chromatography.

SOD Superoxide dismutase.

sodium An element that is essential to humans and several classes of animals and plants. Symbol, Na; atomic number, 11; atomic weight, 22.9898; oxidation state, +1; most abundant isotope, ^{23}Na; a radioactive isotope, ^{22}Na, half-life, 2.60 years, radiation emitted, positrons and gamma rays.

sodium azide *See* azide.

sodium dodecyl sulfate A detergent used in the solubilization of membrane fractions. *Abbr* SDS; *Aka* sodium lauryl sulfate.

sodium dodecyl sulfate polyacrylamide gel electrophoresis *See* SDS–PAGE.

sodium error The apparent decrease in the pH of a solution that is measured with most glass electrodes when they are used at high pH values; the effect is due to the fact that most glasses that are responsive to protons are also somewhat responsive to sodium ions.

sodium lauryl sulfate SODIUM DODECYL SULFATE.

sodium pentobarbital *See* Nembutal.

sodium pump The structure and/or the mechanism that mediates the active transport of sodium and potassium ions across a biological membrane in higher animals; the operation of the pump requires cellular ATP and an ATPase. *See also* Na^+,K^+-ATPase.

sofa conformation A half-chair conformation.

soft center An electrophilic center in a molecule which, when attacked by a nucleophile, leads to a transition state that involves the formation of an unusual bond. *See also* hard center.

soft clot The blood clot formed by the aggregation of fibrin molecules in the absence of calcium ions and fibrin stabilizing factor.

soft ice Water that is oriented, bound, and frequently compressed by polar interactions, as distinct from water that is in the form of icebergs, in which the water is stabilized by nonpolar interactions.

soft ligand A large atom, or group of atoms, of high polarizability; sulfur and iodine atoms are examples. *See also* hard ligand.

soft soap A potassium salt of a long-chain fatty acid; soft soaps are more water-soluble than hard soaps.

software The procedural specifications required for the operation of computers; includes programs, routines, translators, assemblers, etc.

soft water Water that either contains low concentrations of calcium, magnesium, and iron ions, or is devoid of these ions; as a result, the surface-active action of ordinary soap molecules is not appreciably interfered with.

soft x rays Low-frequency x rays that have long wavelengths and small penetrating power.

sol A liquid colloidal dispersion.

solation The transition from a gel to a sol.

solenoid structure A supercoiled form of DNA that is formed during condensation of the chromosomes in the nuclei of eukaryotic cells.

solid medium A gel that contains nutrients.

solid phase synthesis An automated technique for the synthesis of polypeptides in which the chain grows while it is attached to a solid support, and excess reagents as well as by-products are washed away. The chain grows from the C-terminal to the N-terminal, and the process is initiated by attaching the C-terminal amino acid through its carboxyl group to an insoluble resin. This amino acid is then reacted with the next amino acid, the amino group of which has been blocked. After peptide bond formation, the amino-blocking group is removed and the dipeptide is then reacted with the next amino acid, and so on.

solid scintillation counter An external-sample scintillation counter in which a solid fluor is used as a detector.

solid scintillation fluorography A method for visualizing labeled compounds in a specimen by covering the specimen with solid fluors and allowing the photons, which are emitted by the scintillating fluors, to expose a photographic emulsion; used to locate labeled compounds in a chromatogram.

solid substrate room temperature phosphorescence An analytical technique based on the enhancement of phosphorescence of analytes adsorbed on solid substrates such as cellulose, inorganic crystals, polymer salts, or other matrix materials. The technique permits a determination of analyte concentration in a sample in the subnanogram range and is frequently carried out in the presence of heavy metal atoms; particularly useful for single-component analysis in a mixture without prior separation. *Abbr* SSRTP. *Aka* solid surface (matrix) room temperature phosphorescence.

soliton A solitary wave; the collective amide I vibrations of successive peptide groups in a polypeptide chain which propagate themselves in wave-like fashion along the chain.

solubility The amount of a substance that will dissolve in a given volume of solvent under specified conditions.

solubility curve A plot of the amount of solute in solution as a function of the amount of solute added.

solubility product The product of the concentrations of the ions of a sparingly soluble electrolyte in a saturated solution of the electrolyte; the concentration of each ion is raised to a power equal to the number of ions of that type per molecule of electrolyte. When the solubility product is exceeded, some of the electrolyte is precipitated out.

solubilization chromatography A chromatographic technique in which compounds of insufficient solubility in water are separated by ion-exchange chromatography by using an aqueous solution of an organic solvent for elution.

solubilize To disperse in a solution; used to describe both the dispersion of membranes by treatment with detergents, and the release of an enzyme from particulate matter.

soluble Capable of going into solution; capable of dissolving.

soluble antigen An antigen that is not present on a cell, a virus, or some other particle in contrast to a particulate antigen.

soluble enzyme An enzyme that exists in the free state in the cytoplasm (in contrast to a particulate enzyme), is readily extracted from cells, and can be purified by the general methods of protein fractionation.

soluble fibrin SOFT CLOT.

soluble fraction A subcellular fraction that contains material of the intracellular fluids but that is devoid of cellular or intracellular particulate structures.

soluble RNA TRANSFER RNA.

soluble starch A partially hydrolyzed starch.

solute That component of a solution, consisting of two components, that is present in the smaller amount; for ordinary solutions of solids in liquids, it is the solid.

solution A homogeneous mixture of two or more substances; ordinary solutions are mixtures of one or more solutes and a solvent.

solution hybridization LIQUID HYBRIDIZATION.

solvated electron An electron surrounded by solvent molecules, commonly one surrounded by water molecules (hydrated electron). A hydrated electron is a strong reductant that is unstable in aqueous media but may be stabilized in nonaqueous media. Hydrated electrons can be produced by pulse radiolysis, that is, by brief irradiation of aqueous solutions with high-energy electrons or gamma rays.

solvation The process whereby solvent molecules surround, and bind to, solute ions and molecules. Solvation may enhance or diminish the asymmetry of the solute particles in solution.

solvent That component of a solution, consisting of two components, that is present in the greater amount; for ordinary solutions of solids in liquids, it is the liquid.

solvent demixing The separation of a solvent system into its constituent components.

solvent drag The increased rate of movement of a solute over that attributable to simple diffusion, and caused by the flow of the solvent; the transfer of solutes across membranes by water which moves in response to an osmotic gradient is an example.

solvent extraction The removal of a wanted or an unwanted substance from a liquid by mixing the liquid with an immiscible, or a partially miscible, solvent, and by separating the phases.

solvent fermentation ACETONE–BUTANOL FERMENTATION.

solvent front The line of advancing solvent in chromatography.

solvent perturbation A technique for studying the stereochemistry and the location of chromophoric groups in a protein by measuring the changes in a property of the protein, such as absorbance or fluorescence, upon the addition of perturbing solutes to a solution of the protein. A chromophore on the surface of the protein molecule is expected to be sensitive to added perturbing solutes, while a chromophore in the interior of the molecule is expected to be insensitive to them; a chromophore in a crevice is expected to be sensitive to perturbing solutes of low molecular weight but not to those of high molecular weight.

solvent regain *See* water regain.

solvolysis A generalized concept for the reaction between a solvent and a solute by which new substances are formed. In most cases the solvent donates a proton to, and/or accepts a proton from, the solute. When the solvent is water, the reaction is referred to as hydrolysis.

soma The totality of the somatic cells of an organism.

somatic Of, or pertaining to, soma.

somatic antigen 1. An antigen of a somatic cell in higher organisms. 2. A cell surface antigen of bacteria as distinct from flagellar or capsular antigens.

somatic cell Any cell of a multicellular organism other than the mature gametes and the germ cells from which they develop.

somatic cell genetic engineering The modification of somatic cells to correct for genetic defects; such corrections are not hereditary. The insertion of genes for insulin production into defective pancreatic cells is an example.

somatic cell hybrid A hybrid cell produced by fusion of two somatic cells.

somatic crossing over MITOTIC RECOMBINATION.

somatic mutation A mutation occurring in a cell that is not destined to become a germ cell.

somatic mutation theory 1. A selective theory of antibody formation according to which antibody formation is based either on the hypermutation of specific genes or on the preferential expansion of clones in which advantageous mutations have occurred. *See also* somatic theory. 2. A theory of carcinogenesis according to which a tumor results from either a spontaneous or an induced mutation, or mutations, of the somatic cells of an organism.

somatic recombination theory A selective theory of antibody formation according to which antibody formation is based on somatic recombinations that occur between genes that are responsible for the synthesis of antibodies. *See also* somatic theory.

somatic theory A theory of the origin of the genes that code for the variable regions of antibody molecules and hence allow for the great diversity of antibodies. According to this theory, these genes have arisen through modifications of a smaller number of germline genes. Such modifications involve mutations as in the somatic mutation theory, or recombinations as in the somatic recombination theory.

somatocrinin SOMATOTROPIN-RELEASING HORMONE.

somatoliberin SOMATOTROPIN-RELEASING HORMONE.

somatomedin One of a group of low molecular weight polypeptides (MW 7000–10,000) that mediate the action of growth hormone on skeletal tissue and produce insulin-like effects in various target tissues. Somatomedins are released by the liver and/or the kidneys by the action of growth hormone. They lead to increased incorporation of sulfate into cartilage and, hence, are also known as sulfation factor. Several of the somatomedins are known as insulin-like growth factors.

somatostatin GROWTH HORMONE RELEASE-INHIBITING HORMONE.

somatotrophin Variant spelling of somatotropin.

somatotropin GROWTH HORMONE.

somatotropin regulatory factor SOMATOTROPIN REGULATORY HORMONE.

somatotropin regulatory hormone GROWTH HORMONE REGULATORY HORMONE.

somatotropin release-inhibiting hormone GROWTH HORMONE RELEASE-INHIBITING HORMONE.

somatotropin releasing hormone GROWTH HORMONE RELEASING HORMONE.

Somogyi–Nelson method An analytical procedure for the determination of glucose in blood in which proteins are precipitated with zinc sulfate and barium hydroxide, and a blue color is produced by treatment of the protein-free filtrate with copper sulfate and an arsenomolybdate reagent.

Somogyi unit The quantity of amylase that liberates sugars with a reducing value equivalent to 1 mg of glucose during the course of a 30-min reaction at $40\,°C$ and at a pH of 6.9 to 7.0.

sonication ULTRASONICATION.

sonicator An instrument for the generation of ultrasonic sound waves and for the rupture of cells by means of these sound waves.

sonic oscillation ULTRASONICATION.

sonic oscillator SONICATOR.

sonification ULTRASONICATION.

sonifier SONICATOR.

sorbate 1. ABSORBATE 2. ADSORBATE. 3. The anion of sorbic acid.

sorbent 1. ABSORBENT (1). 2. ADSORBENT.

sorbic acid The compound $CH_3(CH)_4COOH$; it is inhibitory to fungi and some bacteria and, hence, is used as a food preservative.

sorbitol A sugar alcohol derived from glucose; a 6-carbon sugar in which each carbon carries a hydroxyl group.

sorbose A 2-ketohexose that is biosynthesized from sorbitol; an intermediate in the commercial synthesis of ascorbic acid (vitamin C).

Sørensen buffer 1. A phosphate buffer prepared by mixing x mL of M/15 KH_2PO_4 and $(100 - x)$ mL of M/15 Na_2HPO_4; pH range 5.0 to 8.2. 2. A glycine buffer prepared by mixing x mL of 0.1 M glycine in 0.1 M NaCl and $(100 - x)$ mL of either 0.1 N HCl or 0.1 N NaOH; pH range 1.2 to 3.6 or 8.4 to 13.0. 3. A citrate buffer prepared by mixing x mL of 0.1 M disodium citrate and $(100 - x)$ mL of either 0.1 N HCl or 0.1 N NaOH; pH range 2.2 to 4.8 or 5.0 to 6.8.

Sørensen titration FORMOL TITRATION.

Soret band A characteristic absorption band of the cytochromes around 400 nm.

sorption 1. ABSORPTION (1). 2. ADSORPTION (1).

SOS box One of several operator sequences in *E. coli* (about 20 nucleotides long) to which the LexA repressor binds. An SOS box occurs, for example, adjacent to the lexA and recA genes.

SOS functions SOS RESPONSE.

SOS regulon The regulon responsible for SOS repair; includes over 10 genes, among them those coding for the recA protein and the lexA repressor. *See also* SOS repair.

SOS repair An error-prone repair mechanism of DNA in *E. coli* that is induced as a result of damage to the DNA; DNA strands are formed even though they contain incorrect bases. The frequency of replication errors is allowed to increase when necessary; survival with mutation is apparently better than no survival at all. SOS repair appears to be the major cause of UV-induced mutations, such as the formation of thymine dimers. The repair mechanism involves the coordinate induction of several enzymes, the key one being an enzyme called RecA protein (RecA protease; RECA). The RecA protein is formed in sufficient amounts only after damage to the DNA has occurred. It then cleaves a protein called the LexA repressor (LexA protein). Cleavage of this repressor results in the activation of many genes involved in the repair. *Aka* error-prone repair; mutagenic repair. *See also* transdimer synthesis.

SOS response A complex group of inducible responses in *E. coli* to conditions that damage DNA or inhibit DNA replication. Includes the induction of synthesis of large amounts of RecA protein, enhanced DNA repair capacity, induced mutagenesis, prophage induction, and UV reactivation. *Aka* SOS functions; SOS system. *See also* SOS repair.

source The material that emits the radiation.

Southern blotting A blotting technique, developed by E.M. Southern, for the transfer of DNA fragments, separated by agarose gel electrophoresis, to a sheet of nitrocellulose. The DNA fragments in the gel are denatured by soaking with NaOH. The gel is then placed on a sheet of nitrocellulose and a weight is placed on the gel to squeeze out the liquid. The denatured, single-stranded, DNA fragments bind to the nitrocellulose sheet and are then located by hybridization with a radioactively labeled DNA or RNA probe. *Aka* Southern transfer; Southern hybridization. *See also* blotting.

Sowden–Fischer synthesis A reaction whereby a one-carbon fragment is added to a carbonyl group by means of nitromethane, as in the extension of a monosaccharide from a pentose to a hexose.

Soxhlet extractor A device that is interposed between a flask of boiling solvent and a condenser and that allows for the continuous extraction of a specimen with the solvent.

soybean trypsin inhibitor A protein, isolated from soybeans, that is an inhibitor of the enzyme trypsin; the Kunitz and the Bowman–Birk inhibitors are two such proteins. *Abbr* STI.

SP 1. Split proteins. 2. Secretory piece. 3. Substance P.

space-filling model A compact molecular model that shows the full bulk of each atom and the effective shape of the molecule. In this model, the bond angles are correct and the distances between the atoms are to scale based on their van der Waals radii.

space group A symmetry class to which objects may belong by virtue of possessing elements of symmetry over and above those of a point group; the additional elements of symmetry may be symmetry operations such as rotation and translation.

spacer A chemical grouping inserted between the ligand and the polymer matrix in affinity chromatography. The spacer ensures that the ligand is far enough removed from the column matrix so that the specific binding of protein to ligand can proceed without stearic hindrance. *Aka* spacer arm.

spacer DNA 1. Untranscribed sections of DNA, located between structural genes. It occurs in eukaryotic and in some viral genomes and usually consists of highly repetitive DNA. 2. LINKER DNA.

spacer gel *See* disc gel electrophoresis.

spacing *See* nucleosome spacing.

spare receptors Receptors that are unoccupied by bound ligand under conditions where the ligand produces a full biological repsonse. These receptors are not really "spare" or "superfluous" since they serve to increase cellular sensitivity to low concentrations of the ligand.

sparing effect The decrease in dietary requirement of one substance that is occasioned by the presence of a metabolically related substance in the diet; tyrosine, for example, exerts a sparing effect on phenylalanine. *See also* brain sparing.

sparking The phenomenon that a catalytic amount of a certain di- or tricarboxylic acid, such as oxaloacetic acid, must be present for initiation of the beta oxidation of fatty acids. The acid is oxidized, thereby yielding ATP which is required for fatty acid activation; the acid also combines with the acetyl coenzyme A formed during beta oxidation.

sparsomycin An antibiotic, produced by *Streptomyces sparsogenes*, that inhibits protein synthesis in eukaryotes by preventing peptide bond formation.

SPCA Serum prothrombin conversion accelerator.

specialized transduction Transduction in which

only certain bacterial genes of the donor bacterium may become transduced, namely those genes close to the site of phage integration into the host chromosome. *Aka* restrictive transduction.

special structure The structural elements of the bacterial cell wall that are additional to the peptidoglycan framework and that include polysaccharides, teichoic acids, polypeptides, proteins lipopolysaccharides, and lipoproteins. The nature of the special structure varies greatly from organism to organism and contributes to the antigenic properties of the bacterial cells and to their acceptor specificity for viruses and bacteriocins.

speciation 1. DENDRITIC EVOLUTION. 2. The transformation of one species into another.

species (*pl* species) A taxon that forms a division of a genus and that consists of a group of individuals of common ancestry that closely resemble each other structurally and physiologically and that, in the case of sexual forms, are capable of interbreeding.

specific 1. Of, or pertaining to, species. 2. Designating a physical or a chemical property per unit amount of substance. 3. Of, or pertaining to, specificity. 4. Constituting or falling into a category.

specific acid–base catalysis The catalysis in solution in which the catalysts are free protons H^+, H_3O^+, and/or free hydroxyl ions; it is not affected by other acidic or basic species present. *See also* general acid–base catalysis.

specific activity The number of activity units per unit of mass; the number of enzyme units per milligram of protein, the number of katals per kilogram of protein, and the number of microcuries per micromole are examples. *Abbr* SA.

specific adsorption The adsorption of specific cations to a surface that results from the formation of coordination complexes between groups of atoms on the surface and the cations. *Aka* ion pairing.

specifically labeled Designating a compound in which one or more known atoms contain all of the label, not necessarily in an even distribution; the positions of the labeled atoms are included in the name of the compound.

specific dynamic action The extra heat, over and above that due to the basal metabolism, that is produced by an animal upon the ingestion of a food. The heat represents the extra energy released as a result of the metabolism of the food, and amounts to approximately 30% for proteins, 6% for carbohydrates, and 4% for lipids. *Abbr* SDA.

specific extinction coefficient The extinction coefficient that attains when the concentration of the solution is expressed in terms of grams per liter (mg/mL).

specific gravity The ratio of the weight of a given volume of a substance to the weight of an equal volume of water.

specific growth rate The rate of growth of a bacterial population, either per cell or per unit mass of cells; equal to $(1/x)(dx/dt)$, where t is the time and x represents either the number or the mass of the cells.

specific heat The quantity of heat required to raise the temperature of 1 g of a substance by 1 °C. *Aka* specific heat capacity.

specific immune suppression The loss of the ability of an organism to respond to a particular antigen after exposure to that antigen; the ability to respond to different antigens is not affected.

specific immunity The immunity that is due to the formation of antibodies in response to the recognition of specific antigens, as contrasted with nonspecific immunity which is due to nonimmunological mechanisms.

specific interaction theory A theory of the evolution of the genetic code according to which the code evolved as a result of specific physical–chemical relations between the amino acids and their codons or anticodons. Support for this theory comes from the finding that amino acid polarity, amino acid hydrophobicity, and bulkiness of amino acid side chains have all been shown to correlate with several properties of the corresponding amino acid anticodons.

specific ionization The number of ion pairs formed per unit distance along the path of an ionizing particle.

specificity 1. The degree of selectivity shown by an enzyme with respect to the number and types of substrates with which the enzyme combines, as well as with respect to the rates and the extents of these reactions. *See also* polyaffinity theory. 2. The degree of selectivity shown by an antibody with respect to the number and types of antigens with which the antibody combines, as well as with respect to the rates and the extents of these reactions. 3. The degree of selectivity shown by a membrane, or a membrane component, with respect to the type and the degree of permeability to substances transported across the membrane in mediated transport.

specificity constant A measure of the effectiveness of a substrate; defined as the ratio k_{cat}/K_m, where k_{cat} is the catalytic rate constant and K_m is the Michaelis constant.

specificity factor A protein that associates temporarily with the core component of RNA polymerase and thereby determines to which promoter the enzyme will bind and which genes will be transcribed. The sigma subunit of the *E. coli* RNA polymerase is an example.

specific osmotic pressure REDUCED OSMOTIC PRESSURE.

specific radioactivity The specific activity of radioactive material, frequently expressed in terms of the number of microcuries per micromole.

specific rate constant RATE CONSTANT.

specific reaction rate RATE CONSTANT.

specific refractive index increment The contribution to the refractive index per gram of solute; equal to $(n - n_0)/c$, where n is the refractive index of the solution, n_0 is the refractive index of the solvent, and c is the concentration of the solution in grams per cubic centimeter. *Aka* specific refractive increment.

specific retention volume The volume of liquid, per gram of adsorbent, that passes through a column in displacement chromatography before a particular substance is eluted from the column.

specific rotation A measure of the optical rotation at a particular wavelength per unit amount of a substance; specifically, $[\alpha]_D^t = \alpha/dc$, where $[\alpha]_D^t$ is the specific rotation at a temperature of $t°C$ for the sodium D line, α is the observed rotation in degrees, d is the optical path length in decimeters, and c is the concentration in grams per milliliter.

specific rotatory power SPECIFIC ROTATION.

specific substrate concentration REDUCED SUBSTRATE CONCENTRATION.

specific viscosity A measure of the fractional change in viscosity that is produced by the solute; equal to the relative viscosity minus one. *Sym* η_{sp}.

specific volume The volume occupied by 1 g of material; the reciprocal of the density.

specimen screen A screen made of copper or gold and used as a support for samples to be examined with the electron microscope.

spectinomycin An antibiotic, produced by *Streptomyces spectabilis*, that inhibits protein synthesis but does not cause misreading of the genetic code; it acts on 30S ribosomes and the inhibition can be reversed by washing the ribosomes.

spectral Of, or pertaining to, a spectrum.

spectral bandpass *See* bandpass.

spectral shift The change of an absorption spectrum or an absorption band to either longer or shorter wavelengths.

spectrin A peripheral protein, located at the cytoplasmic side of the red blood cell membrane. Spectrin is a long, fibrous protein that occurs as a mixture of dimers and tetramers. These are arranged in a filamentous network, held together by actin and other protein molecules. The network is believed to be responsible for maintaining the biconcave shape of red blood cells.

spectrofluorometer A fluorometer in which the desired excitation and emission wavelengths are selected by means of a monochromator. Spectrofluorometers may be of a corrected or an uncorrected type depending on whether the intensity of the exciting light is controlled as a function of the wavelength, and whether the response of the detector is controlled as a function of the emission wavelength. *Var sp* spectrofluorimeter.

spectrogram The photographic record of a spectrum.

spectrograph An instrument for separating light into its component wavelengths and for obtaining a photographic record of the spectrum thus produced.

spectrometer 1. An instrument for measuring either the wavelengths or the frequencies of a spectrum. 2. A liquid scintillation assembly that includes a detector, scaler, sample changer, print out, and electronic circuitry.

spectrophotometer A photometer in which a monochromator, composed of prisms or diffraction gratings, is used for the isolation of narrow bandwidths. Spectrophotometers allow the measurement of the selective absorption of light; they are used for both qualitative and quantitative analysis of chemical substances, and cover the ultraviolet, visible, and infrared ranges of the electromagnetic spectrum. *See also* double beam in space spectrophotometer; double beam in time spectrophotometer; recording spectrophotometer.

spectropolarimeter An instrument that is a combined spectroscope and polarimeter and that is used for measurements of optical rotation as a function of wavelength.

spectroscope An instrument for separating light into its component wavelengths and for examining the spectrum thus produced.

spectroscopic splitting factor g VALUE.

spectroscopy The production and study of spectra.

spectrum (*pl* spectra; spectrums) 1. The variation of radiation intensity as a function of either wavelength or frequency that is generally represented in graphical form. 2. A range of either wavelengths or frequencies of a radiation. 3. ELECTROMAGNETIC SPECTRUM. 4. A specific type of radiation wavelengths or frequencies, such as an absorption or an emission spectrum.

S peptide A peptide formed by cleavage of ribonuclease with subtilisin; it consists of amino acid residues 1 through 20. The remaining, larger fragment (residues 21–124) is called S protein. Subtilisin hydrolyzes only this one peptide bond in ribonuclease (between residues 20 and 21).

spermaceti A solid wax obtained from the sperm oil secreted in the jaw region of the sperm whale; used commercially as a basis for pharmaceutical and cosmetic creams.

spermatocyte A cell that develops into a mature sperm upon meiosis.

spermatogenesis The formation and development of spermatozoa.

spermatozoon (*pl* spermatozoa) The male reproductive cell of animals; the male gamete.

spermidine A low molecular weight polyamine (*N*-[3-aminopropyl]-1,4-butanediamine) that occurs in both prokaryotes and eukaryotes; contains two amino groups and one imino group and its synthesis involves *S*-adenosyl methionine.

spermine A low molecular weight polyamine (*N*,*N*′-bis[3-aminopropyl]-1,4-butanediamine) that occurs in eukaryotes but not in prokaryotes; contains two amino groups and two imino groups and its synthesis involves *S*-adenosyl methionine.

spherical phage *See* minute phage.

spherocyte A spherical red blood cell that has a smaller diameter than the normal erythrocyte.

spherocytosis A genetically inherited metabolic defect in humans, characterized by erythrocytes that look like bags rather than disks; due to a deficiency in the synthesis of one of the polypeptide chains in spectrin.

spheroplast A bacterial cell that is largely, but not entirely, freed of its cell wall. Spheroplasts are prepared artificially from gram-negative bacteria by either lysozyme digestion of the cells or by growing them in the presence of penicillin. Spheroplasts are osmotically sensitive but may, at times, convert to an L form. *See also* protoplast.

spherosome A lysosome-like organelle of plants. It is derived from the endoplasmic reticulum and serves as a major site of lipid storage.

spherule LIPOSOME.

sphinganine 2-Amino-1,3-octadecanediol; dihydrosphingosine.

4-sphingenine SPHINGOSINE.

sphingoid A generic descriptor for sphinganine, its homologues and stereoisomers, and the hydroxy and unsaturated derivatives of these compounds.

sphingolipid Any lipid containing a sphingoid; a lipid that contains either sphinganine, its homologue, its isomer, or its derivative, and that is especially predominant in brain and nervous tissue.

sphingolipidosis One of a number of genetically inherited metabolic defects in humans that are characterized by the accumulation of various sphingolipids and that are due to deficiencies of lysosomal enzymes. Gaucher's disease, Tay–Sachs disease, Niemann–Pick disease, and Krabbe's disease are some examples.

sphingolipodystrophy SPHINGOLIPIDOSIS.

sphingomyelin A phosphosphingolipid that consists of sphingosine, a fatty acid, a phosphate group, and choline, and that is predominant in the myelin sheath of nerves; a ceramide-1-phosphocholine.

sphingomyelinosis NIEMANN–PICK DISEASE.

sphingophospholipid Any phospholipid derived from sphingosine or related compounds.

sphingosine A long-chain, unsaturated amino alcohol that is the parent compound of the sphingolipids; 2-amino-4-octadecene-1,3-diol; *trans*-4-sphingenine. *Aka* 4-sphingenine.

spike A characteristic protrusion on viral envelopes.

spike potential ACTION POTENTIAL.

spiking INTERNAL STANDARDIZATION.

spillover hypothesis The hypothesis according to which photosystems I and II of chloroplasts have specific points of contact between them so that energy transfer is possible not only between the pigments within each photosystem, but also between the pigments of one photosystem and those of the other. *See also* separate package hypothesis.

spin The rotation about an axis, as the rotation of an electron or an atomic nucleus.

spinach ferredoxin A ferredoxin that contains two iron atoms per molecule, has a molecular weight of about 12,000, has a standard reduction potential (E_0') of −0.42 V, and serves as an early acceptor for electrons from the excited P_{700} pigment in photosystem I of chloroplasts.

spin adduct *See* spin trapping.

spin coupling *See* spin–spin coupling.

spin decoupling A technique used in nuclear magnetic resonance in which the effect of spin coupling with nucleus A on the spectrum of nucleus B is eliminated by irradiation with the resonance frequency of nucleus A. *Aka* double irradiation; spin–spin decoupling.

spindle A structure, composed of microtubules, which is responsible for eukaryotic chromosome alignment and movement during nuclear division.

spindle poison *See* colchicine.

spin flip The change, by an oriented nucleus in nuclear magnetic resonance, from a lower energy state to a higher energy state upon absorption of incident radiation.

spin imaging A nuclear magnetic resonance technique that involves the use of magnetic field gradients to provide information about the spatial distribution of molecules within a sample; based on obtaining two- or three-dimensional images of the proton signals from water within animals and human beings.

spin labeling The introduction into a protein of a substituent, the electron paramagnetic resonance of which is sensitive to changes in the environment of the substituent. Measurements of the electron paramagnetic resonance of the substituent in the protein can be used to explore the environment of the substituent in the protein molecule. *See also* reporter group; spin probe.

spin–lattice relaxation A radiationless process whereby the energy of a nucleus in a high spin state is dissipated to the lattice of surrounding nuclei by means of oscillating magnetic fields.

spin multiplicity The term $2S + 1$, where S is the resultant spin.

spin probe A spin label that is bound noncovalently to a protein.

spin quantum number The value of either $+\frac{1}{2}$ or $-\frac{1}{2}$ that is arbitrarily assigned to one of the two directions of spin of an orbital electron.

spin–spin coupling The interaction between the magnetic moment of a proton, or some other nuclear dipole, and those of neighboring dipoles in nuclear magnetic resonance; this interaction leads to the splitting of a single peak into multiple peaks. *Aka* spin–spin interaction.

spin-spin decoupling *See* spin decoupling.

spin–spin relaxation A measure of the energy exchange between nuclei in nuclear magnetic resonance; a radiationless process in which the energy of one nucleus is transferred to another.

spin–spin splitting The production of multiple peaks from a single peak in nuclear magnetic resonance as a result of spin–spin coupling.

spin trapping A technique of electron paramagnetic resonance that permits the identification of transient free radicals; based on the ability of certain compounds (spin traps) to react with highly unstable radicals to yield long-lived, stable products, termed spin adducts. The latter yield electron spin resonance parameters that reflect the nature of the trapped free radicals.

spirillum (*pl* spirilla) A bacterium having a helically shaped cell; spirilla represent one of the three major forms of bacteria. *See also* coccus; bacillus.

spirometer An instrument for measuring the volume of air inhaled and exhaled; used in measurements of the basal metabolic rate.

SPK Synthase phosphorylase kinase.

spleen A large organ in the abdominal cavity that functions in the destruction of erythrocytes and in the production of antibodies.

splenectomy The surgical removal of the spleen.

splenomegaly An enlargement of the spleen.

spliceosome A multicomponent complex that carries out the splicing of RNA; evidence for, and characterization of, such a complex has been based on experiments involving both yeast and mammalian systems.

splice sites The cutting sites for removal of an intron; the 5′-splice site is adjacent to the GU-end of the donor junction, the 3′-splice site is adjacent to the AG-end of the acceptor junction. *See also* splicing junctions.

splicing 1. Gene splicing. The process whereby DNA fragments from different sources are joined covalently to form a recombinant DNA molecule. *See also* recombinant DNA technology. 2. RNA splicing. The process whereby introns are excised, and exons are joined, in the conversion of a primary RNA transcript to the finished (mature) mRNA molecule in eukaryotes.

splicing homeostasis *See* maturase.

splicing junctions Segments of nucleotides at the ends of introns that function in RNA splicing; the sequence at the 3′-end of an intron is called the acceptor (right) junction, and that at the 5′-end is called the donor (left) junction. *See also* GU-AG rule.

split gene A gene that occurs in pieces; it consists of DNA sequences that are transcribed and translated (exons) and of DNA sequences that are transcribed but not translated (introns). The latter are excised from the primary RNA transcript and the exons are then spliced together to yield the functional gene product.

split proteins The proteins that, when removed from ribosomes, leave behind core particles. Core particles and split proteins are inactive in protein synthesis but can be reassociated into functionally active ribosomes. *Abbr* SP.

splitter A device for decreasing the amount of sample that is introduced into a gas chromatographic column from the inlet.

splitting The separation, in terms of energy, between previously degenerate energy levels in an atom or a molecule.

spontaneous generation The doctrine that living things can come from nonliving matter; abiogenesis.

spontaneous hypersensitivity Hypersensitivity that is produced in an animal in the absence of any known contact of the animal with an antigen.

spontaneous induction The induction of a prophage that is caused by the random interaction between the immunity substance and the prophage.

spontaneous mutation A naturally occurring

mutation for which there is no observable cause; a mutation that results from a chance exposure of an organism to a mutagen in the environment.

spontaneous process A process that is accompanied by an increase in entropy.

spontaneous reaction A chemical reaction that is accompanied by a negative free energy change; an exergonic reaction.

spontaneous tumor A tumor that arises without known exposure of the organism to a carcinogen, such as one formed in certain inbred strains of mice that are genetically prone to tumor development.

sporangium (*pl* sporangia) A special structure housing spores.

spore A dormant cellular form, derived from a bacterial, fungal or plant cell, that is devoid of metabolic activity and that can give rise to a vegetative cell upon germination; it is dehydrated and can survive for prolonged periods of time under drastic environmental conditions.

spore coat One of two outer layers that surround the spore cortex and that consist largely or entirely of cross-linked proteins having a high cystine content.

spore core The central body of a spore; the cytoplasm of a spore.

spore cortex A thick layer that contains glycopeptides and that is located between the spore coats and the spore membrane.

spore membrane The membrane that surrounds the spore core.

spore peptide The glycopeptide material released from germinating spores.

spore tip mucilage A naturally occurring strong adhesive that is stored in a dehydrated form in a specialized compartment at the apex of rice blast fungus spores (*Magnaporthe grisea*). Upon hydration of the spore, the adhesive is also hydrated and released; it then effects attachment of the phytopathogenic spores to the host plant. *Abbr* STM.

spore wall SPORE MEMBRANE.

sporicide An agent that kills spores.

sporogenesis The production of spores.

sporont A cell from which a spore may subsequently develop.

sporophyte The spore-producing, diploid individual in the life cycle of an organism exhibiting alternation of generations.

sporulation 1. The differentiation of a cell into a spore. 2. The discharge, or liberation, of spores.

spot 1. *n* The location of a separated and visualized component in either chromatography or electrophoresis. 2. *v* To apply material in small amounts to either a chromatographic or an electrophoretic support.

spot desmosome A cell junction that acts like a rivet holding epithelial cells together at specific points of contact. *See also* cell junction.

spray-freezing A modification of the freeze-fracturing technique in which the cooling rate is increased by spraying the sample into a liquid at very low temperatures; the frozen droplets are then collected, "glued" together with butyl benzene, and subjected to freeze-fracturing.

spreading factor HYALURONIDASE.

spreading position effect The phenomenon in which several genes, located near a chromosomal aberration of inversion or translocation, appear to be activated simultaneously.

spreading reaction The change in permeability and other properties of the cell membrane that occurs in a phage-infected cell in the vicinity of each bound phage particle.

spread plate A petri dish, containing solid medium, in which the inoculum has been spread over the entire surface.

S-protein A large peptide formed by cleavage of ribonuclease with subtilisin; it consists of amino acid residues 21–124. The other, smaller fragment (residues 1–20) is called S-peptide. Subtilisin hydrolyzes only this one peptide bond in ribonuclease (between residues 20 and 21).

S-100 protein 1. An acidic, brain-specific protein that has been implicated in neurophysiological functions and in the process of learning; the protein has a molecular weight of 20,000 and is rich in aspartic and glutamic acids. 2. One of a group of low molecular weight, Ca^{2+}-binding, modulator proteins that function in cell cycle progression, cell differentiation, and cytoskeletal-membrane interactions.

SP-Sephadex Sulfopropyl Sephadex; a cation exchanger. The ion exchanger contains the grouping $—C_3H_6SO_3^-$, linked via ether bonds to the Sephadex.

spur The overlapping portion of two precipitin arcs in immunodiffusion; one such spur is formed in a reaction of partial identity, and two such spurs are formed in a reaction of nonidentity.

spurious counts Counts of radioactivity that are not caused by the sample but that result from outside factors, such as malfunctioning of the apparatus.

squalene A terpenoid that serves as the immediate precursor of sterols in their biosynthesis; it gives rise to lanosterol which is then converted to cholesterol.

square bacteria Halophilic bacteria that are found in hypersaline environments and that have square structures with a cell wall comprised of regularly arranged subunits.

square dilution law RADIAL DILUTION.

square wave polarography A polarographic technique in which a square-wave alternating potential is superimposed on the normal dc applied potential and the alternating component of the current is measured.

squiggle The symbol ~ used to designate a high-energy bond, specifically one involving the phosphate group.

^{90}**Sr** Strontium-90.

src gene An oncogene of Rous sarcoma virus; its product (src protein) is a protein kinase that can modify a large number of proteins. *See also* c-src gene.

Srd Thiouridine.

S region EXTRA ARM.

SRF *See* growth hormone regulatory hormone.

SRH *See* growth hormone regulatory hormone.

SRIF *See* growth hormone regulatory hormone.

SRIH *See* growth hormone regulatory hormone.

SR mutant Smooth–rough mutant; *see* smooth strain; rough strain.

sRNA Soluble RNA.

4S RNA TRANSFER RNA.

5S RNA An RNA molecule that has a sedimentation coefficient of 5S and that forms part of the ribosomal RNA in both prokaryotes and eukaryotes; it is similar to transfer RNA in its base composition and sedimentation coefficient, but it does not contain minor bases.

5.8S RNA An RNA molecule that has a sedimentation coefficient of 5.8S and that forms part of the ribosomal RNA in eukaryotes.

SRP Signal recognition protein.

SRP receptor DOCKING PROTEIN.

S-rRNA RNA of the small ribosomal subunit.

SRS Slow-reacting substance; a substance that causes slow contractions of bronchial smooth muscles.

SRS-A Slow-reacting substance of anaphylaxis. A substance that causes slow contractions of bronchial smooth muscles and that is released upon exposure to specific antigens. SRS-A consists of a mixture of leukotrienes LTC$_4$ and LTD$_4$; the ratio of these two leukotrienes varies with the source of the SRS-A, the stimulus, and the time elapsed after stimulation.

SRS technique Separation–reaction–separation technique; a separation technique for steroids in which the steroids are first separated in one dimension by thin-layer chromatography, are then exposed to radiation or are reacted with chemical reagents, and are then separated in the second dimension by thin-layer chromatography with the same solvent used for the first dimension.

S → R variation Smooth–rough variation.

ss single-stranded.

SSA test Sulfosalicylic acid test.

SSB Single-strand binding protein.

SSC Standard saline citrate.

ssDNA Single-stranded DNA.

S-shaped curve *See* sigmoid kinetics.

ssRNA Single-stranded RNA.

SSRTP Solid substrate room temperature phosphorescence.

SSS Steady-state stacking.

S strain Smooth strain.

stab culture A bacterial culture made by plunging an inoculating needle into a solid medium.

stability constant FORMATION CONSTANT.

stabilizer EMULSIFYING AGENT.

stable factor PROCONVERTIN.

stable isotope An isotope that is not radioactive.

stacking 1. BASE STACKING. 2. ISOTACHOPHORESIS.

stacking energy The free energy of the stacking interactions between two base pairs in a double-helical nucleic acid structure.

stacking gel *See* disc gel electrophoresis.

stacking interactions The hydrophobic interactions between the base pairs that are arranged in parallel planes in the interior of a double-helical nucleic acid structure.

staggered conformation The conformation of a molecule in which, in a Newman projection, few atoms are either partially or fully concealed from view by other atoms. In such a conformation, interatomic distances are relatively great and interatomic interactions are minimized; as a result, staggered conformations are more stable than eclipsed ones.

staggered cuts Two cuts in duplex DNA, one in each strand, that are removed from each other by one or more base pairs; the type of cuts carried out by many restriction enzymes.

stains all A complex cationic carbocyanine dye that stains most macromolecules. It stains RNA blue-purple, DNA blue, protein red, phosphoproteins blue, and mucopolysaccharides blue to purple.

staircase reaction A stepwise chemical reaction; the hydrolysis of starch, which proceeds through the formation of intermediate dextrins and oligosaccharides to the formation of glucose, is an example.

staircase response The gradual increase in steroid excretion upon the administration of adrenocorticotropic hormone that occurs in individuals who suffer from adrenal insufficiency as a result of either pituitary or hypothalamic dysfunction.

stalk *See* supermolecule.

standard *See* standard solution.

standard amino acids The 20 amino acids that occur in proteins as distinct from other amino acids (nonprotein amino acids) that function only in various areas of metabolism. *Aka* primary amino acids; normal amino acids; standard set.

standard conditions *See* standard temperature and pressure.

standard curve A plot of a physical property, commonly absorbance, of a compound or of its derivative as a function of the concentration of the compound; used for the determination of unknown concentrations of the compound from measurements of the physical property.

standard deviation A measure of the scatter of values about the mean of a set of measurements; for a normal distribution curve, it is the range about the mean within which 68.27% of all the measurements will fall. The standard deviation is equal to $[\Sigma(X - \bar{X})^2/(N - 1)]^{1/2}$, where N is the number of measurements, and $\Sigma(X - \bar{X})^2$ is the sum of the squared deviations from the mean $\bar{X}$. It is denoted s for a sample and σ (sigma) for a population; it is equal to the positive square root of the variance. *Abbr* SD.

standard deviation of the mean STANDARD ERROR.

standard diffusion coefficient The value of a translational diffusion coefficient that is calculated from data which have been extrapolated to zero concentration, and that is corrected to a diffusion coefficient under standard conditions which are taken as the diffusion in a medium of water at 20°C. *Sym* $D^0_{20,w}$.

standard electrode potential The electrode potential for a half reaction at 25°C and 1 atm of pressure in which all reactants and products are present in their standard states; denoted E^0. It is usually approximated as the midpoint potential for a system in which all reactants and products are present at 1 M concentrations and is then denoted E_0. The biochemical standard electrode potential is a reduction potential for a half-reaction at 25°C and 1 atm of pressure in which all reactants and products are present at 1 M concentrations except protons which, unless otherwise specified, are present at a concentration of 10^{-7} M (pH 7.0); this potential is measured as a midpoint potential and is denoted E'_0.

standard enthalpy change *See* standard state.

standard entropy change *See* standard state.

standard error A measure of the reliability of the mean, expressed as SE = $\sigma/\sqrt{N}$, where SE is the standard error, σ is the standard deviation, and N is the number of individual results. *Abbr* SE.

standard error of estimate The square root of the residual variance in regression analysis; the standard deviation of the differences between the observed and the calculated values. It is equal to $[\Sigma(Y - \hat{Y})^2/(n - 2)]^{1/2}$ where Y is the observed value of Y for a given value of X, $\hat{Y}$ is the calculated value of Y (using a regression equation) for a given value of X, and n is the total number of observations.

standard error of the mean STANDARD ERROR.

standard flotation coefficient The value of a flotation coefficient that is calculated from data that have been extrapolated to zero concentration, and that is corrected to a flotation coefficient under standard conditions which, for many lipoproteins, are taken as the flotation in a sodium chloride solution having a density of 1.063 g/mL at 26°C. *Sym* s^0_f.

standard free energy change The free energy change of a reaction at 25°C and 1 atm of pressure in which all reactants and products are present in their standard states; denoted ΔG^0. The biochemical standard free energy change is the free energy change of a reaction at 25°C and 1 atm of pressure in which all reactants and products are present in their standard states except protons which, unless otherwise specified, are present at a concentration of 10^{-7} M (pH 7.0); this free energy change is denoted $\Delta G^{0'}$.

standardization The determination of the concentration of a solution by titration against either a primary or a secondary standard.

standard oxidation potential The standard electrode potential for an oxidation half-reaction.

standard potential *See* standard electrode potential.

standard reaction conditions The theoretical thermodynamic standard reaction conditions are a temperature of 25°C, a pressure of 1 atm, and a chemical composition in which all components are in their defined standard states; that is all reactants and products have an activity of 1.0. In practical terms, for general chemical reactions, the latter requirement is taken to mean that all reactants and products are present at a 1.0 M concentration. For biochemical systems, the theoretical conditions are as above, but, in practical terms, the chemical composition aspect is modified as follows: (a) the proton concentration is 10^{-7}M (pH 7.0); (b) the total concentration of all species (ionized forms, metal chelates, etc.) of a given compound is 1.0 M; that is, the sum of the concentrations of all forms for each reactant and product is 1.0 M. *See also* standard state.

standard reduction potential The standard electrode potential for a reduction half-reaction.

standard saline citrate A buffer composed of

0.15 M sodium chloride and 0.015 M triso-dium citrate, pH 7.0; used in studies of DNA in solution. *Abbr* SSC.

standard sedimentation coefficient   The value of a sedimentation coefficient that is calculated from data that have been extrapolated to zero concentration, and that is corrected to a sedi-mentation coefficient under standard condi-tions which are taken as the sedimentation in a medium of water at 20 °C. *Sym* $s_{20,w}^0$.

standard set   The group of amino acids that commonly occur in proteins. *Aka* standard amino acids.

standard solution   A solution of known concen-tration.

standard state   A thermodynamic reference state in which the temperature is 25 °C, the pressure is 1 atm, and the composition is such that all components are present in their de-fined reference states. The standard state of pure substances is the state of the pure liquid or solid at 25 °C and 1 atm of pressure. The standard state for components in solution is usually taken as that in which the activities of both solutes and solvent are equal to unity; for solutes, an activity of one is approximated by a 1 M concentration. In biochemical sys-tems, the standard state of hydrogen ions is taken to be the hydrogen ion concentration which corresponds to pH 7.0. *See also* stand-ard reaction conditions.

standard temperature and pressure   A tempera-ture of 0 °C and a pressure of 1 atm (760 mm Hg). *Abbr* STP.

Staphylococcus   A genus of gram-positive bac-teria that occur commonly as parasites and pathogens of humans and animals; includes *Staphylococcus aureus*, the causal agent of staph infections.

staphylokinase   The enzyme that is present in filtrates of *Staphylococcus* cultures and that promotes the lysis of human blood clots by catalyzing the activation of plasminogen to plasmin.

star   *See* star gazing.

starch   The major form of storage carbohy-drates in plants. It is a homopolysaccharide, composed of D-glucose units, that occurs in two forms; amylose, which consists of straight chains, and in which the glucose residues are linked by means of $\alpha(1 \rightarrow 4)$ glycosidic bonds; and amylopectin, which consists of branched chains, and in which the glucose residues are linked by means of both $\alpha(1 \rightarrow 4)$ and $\alpha(1 \rightarrow 6)$ glycosidic bonds.

starch gel electrophoresis   A zone electrophore-tic technique of high resolving power in which a gel of partially hydrolyzed starch is used as the supporting medium.

starch granule   A storage particle of starch that

occurs in the cytoplasm of plant cells and that also contains proteins and enzymes that func-tion in the synthesis and breakdown of starch. The granules vary in size and shape, and may be used to classify the plants from which they are derived.

starch indicator   A starch solution used as an indicator in iodometric titrations; starch gives a dark blue color with iodine.

starch phosphorylase   The enzyme that cata-lyzes the successive hydrolytic removal of glucose residues in the form of glucose-1-phosphate from the nonreducing end of starch; this reaction is the first step for the utilization of starch in glycolysis. *See also* phosphorylase.

starch synthase   The enzyme that catalyzes the synthesis of straight chains of starch (amylose) from ADP-glucose.

star gazing   A technique for studying the radioactive decay of phosphorus in nucleic acids by labeling DNA or a virus with radioac-tive phosphorus (^{32}P) and then embedding it in a sensitive photographic emulsion. Star-shaped patterns of beta particle tracks result from the decay of the radioactive phosphorus atoms and a count of these tracks can be used to calculate the molecular weight of the nuc-leic acid.

Stark effect   The change in the absorption spec-trum of a pigment that is placed in a strong electric field.

Stark–Einstein law   SECOND LAW OF PHOTO-CHEMISTRY.

Starling's hypothesis   The hypothesis that the rate of exchange of fluid between the plasma in the blood capillaries and the interstitial fluid is governed by a balance of hydrostatic pressure and osmotic pressure, primarily the hydrostatic pressure due to the blood and the osmotic pressure due to the plasma proteins.

start codon   INITIATION CODON.

starter   1. PRIMER. 2. INOCULUM.

starter tRNA   INITIATOR TRANSFER RNA.

starting potential   The potential in that portion of the characteristic curve of a Geiger Mueller counter that precedes the plateau and at which there is a sharp rise in the curve.

startpoint   The base pair in DNA that codes for the first nucleotide incorporated into the RNA molecule (primary RNA transcript), synthesized by RNA polymerase. *Aka* start-site.

starvation   1. An extreme case of undernutri-tion due to severe and prolonged inadequate intake of most of the required nutrients. 2. Undernutrition with respect to one or more nutrients, as in the withholding of an amino acid from a growing bacterial culture.

starvation diabetes   A condition of temporary

carbohydrate intolerance that is characterized by glucosuria; follows the ingestion of carbohydrate after prolonged starvation and is due to either the decreased output of insulin or the decreased ability to synthesize glycogen.

-stat Combining form meaning constant. *See also* cryostat; pH stat; thermostat.

state function Any one of the four thermodynamic parameters: internal energy, enthalpy, entropy, and free energy. The values of these parameters for a given process depend on the initial and the final state of the system and not on the path taken to proceed from the initial to the final state.

state of a system The description of a system in terms of the thermodynamic state functions. The state before a process occurs is known as the initial state, and the state after the process has occurred is known as the final state.

static osmometer An osmometer in which osmosis is allowed to take place so that a pressure difference develops.

static quenching Quenching in which the fluorophore is in the ground state.

statin A release-inhibiting hormone; *see* regulatory hormone.

stationary electrolysis ISOELECTRIC FOCUSING.

stationary phase 1. The phase of growth of a bacterial culture that follows the exponential phase and in which there is little or no growth. 2. The liquid, solid, or solid plus adsorbed liquid that serves as a support in chromatography.

stationary state STEADY STATE.

stationary-state approximation STEADY-STATE APPROXIMATION.

stationary substrate STATIONARY PHASE (2).

statistic A quantity (such as a mean or a standard deviation) that is calculated from a sample of measurements. *See also* parameter.

statistical Of, or pertaining to, statistics.

statistical error An error in the counting of radioactive materials that is due to the random nature of radioactive decay.

statistical factor The ratio between the binding (association) constant for the first ligand and the binding (association) constant for the second ligand for a macromolecule that has two binding sites for a given ligand and that exhibits independent binding. For such systems, the statistical factor is equal to 4.

statistical mechanics A description of the equilibrium properties of macroscopic systems that is based on the application of statistics to atomic and molecular energy states.

statistical segment A theoretical building block of a real polymer chain. The latter is approximated as a polymer of freely jointed statistical segments, each of which is randomly oriented

with respect to all other segments. The segment is chosen to be long enough to assure its random orientation.

statistics The science that deals with the collection, analysis, interpretation, and presentation of masses of numerical data.

status quo hormone ALLATUM HORMONE.

Staudinger equation An equation relating intrinsic viscosity $[\eta]$ to molecular weight M by means of two empirical constants, A and α, such that $[\eta] = AM^{\alpha}$.

Staudinger index INTRINSIC VISCOSITY.

steady state 1. The nonequilibrium state of a system in which matter flows in and out at equal rates so that all of the components remain at constant concentrations. In a chemical reaction sequence, a component is in a steady state if the rate at which the component is being synthesized (produced) is equal to the rate at which it is being degraded (used) 2. The maximum color intensity of a sample that is obtained as a function of time; used in reference to determinations with an autoanalyzer.

steady-state approximation A method for deriving the rate equation of a chemical reaction that is based on two assumptions: (a) There exists a rate-determining step in the mechanism; (b) the concentrations of the intermediates preceding this step are governed by steady-state conditions. The steady-state approximation is used in the Briggs–Haldane treatment of enzyme kinetics. *Aka* stationary-state approximation.

steady-state electrolysis ISOELECTRIC FOCUSING.

steady-state kinetics The kinetics of an enzymatic reaction proceeding under steady-state conditions; essentially the kinetics of the rate-determining step of the reaction.

steady-state stacking ISOTACHOPHORESIS.

steapsin The lipase present in the pancreatic juice.

stearic acid A saturated fatty acid that contains 18 carbon atoms and that occurs in animal fat.

steatolysis The hydrolysis and the emulsification of fats during digestion.

steatorrhea Failure to digest and/or absorb lipid through the intestinal wall and consequent excretion of these lipids with the stool.

steatosis 1. ADIPOSIS. 2. FATTY DEGENERATION.

stefin *See* cystatin.

stem 1. A structural part of transfer RNA; *see* arm. 2. STALK. 3. The base paired region of single-stranded DNA, folded back upon itself in a hairpin structure.

stem-and-loop DNA *See* foldback DNA.

stem cell An undifferentiated cell from which specialized cells are subsequently derived.

step-down A decrease in a metabolic pathway or in the growth of an organism that is brought about by a decrease in the concentration of nutrients.

step-growth polymer A polymer formed by the linking of monomers with the elimination of either water molecules or other small molecules. *Aka* condensation polymer.

ST-EPR Saturation transfer electron paramagnetic resonance; *see* electron–electron double resonance.

stepwise development A chromatographic technique, used particularly with paper and thin-layer chromatography, in which the sample is developed repeatedly with different solvents.

stepwise elution The chromatographic elution in which two or more solutions that differ in composition are added to a chromatographic column by abruptly changing from one solution to the next.

sterane GONANE.

stereo carbon *See* stereogenicity.

stereochemical Pertaining to the three-dimensional arrangement of atoms in a molecule.

stereochemical theory A theory according to which the evolution of the genetic code is due to a stereochemical relation between an amino acid and either its codons or its anticodons.

stereochemistry The branch of chemistry that deals with the spatial arrangement of atoms in a molecule.

stereoelectronic Pertaining to the spatial aspects of atomic and molecular orbitals.

stereogenicity The characteristic that, in a chiral molecule, the transposition of two bonded atoms produces a new stereoisomer. An atom that displays this property is termed a stereogenic atom or a stereocenter. Thus, in the compound CHBrClF, only the carbon atom is stereogenic (a stereo carbon). Stereogenicity deals with stereoisomerism as opposed to chirotopicity which deals with the local geometry of compounds.

stereoisomer One of two or more isomers that have the same molecular composition and the same atom-to-atom sequence, but that differ from each other in the spatial arrangement of the atoms in the molecule; stereoisomers cannot be distinguished on the basis of their two-dimensional structures. *See also* structural isomer.

stereology A body of mathematical methods that relates three-dimensional parameters of a structure to two-dimensional measurements obtainable from sections of the structure. The sampling of the sections is statistical and measurements of the sections may involve electron microscopy.

stereomer STEREOISOMER.

stereomutation The conversion of one stereoisomer into another; particularly the interconversion of geometrical isomers, as in the conversion of oleic acid (cis) to elaidic acid (trans).

stereopopulation control The concept that the combined effect of the multipoint attachment of the substrate to, and its precise fit into, the active site of an enzyme would tend to restrict the rotational freedom of the substrate and "freeze" it into a unique conformation, suitable for entering the transition state.

stereoselective reaction A chemical reaction in which one stereoisomer is generated or destroyed preferentially over another. Enzymes are stereoselective; as an example, an L-amino acid oxidase will catalyze a reaction with an L-amino acid but not with a D-amino acid. *See also* stereospecific reaction.

stereospecific numbering A system for numbering derivatives of symmetrical compounds, such as glycerol, that is based on representing the parent molecule always in one specific stereochemical configuration. *Sym* sn.

stereospecific reaction A chemical reaction in which there is a relation between the configurations of the reactants and those of the products. Enzymes are stereospecific; as an example, pyridine-linked dehydrogenases catalyze the addition of a hydrogen atom to, or its removal from, a particular side of the pyridine ring in the coenzyme part of these enzymes. *See also* stereoselective reaction.

steric STEREOCHEMICAL.

steric contour diagram RAMACHANDRAN PLOT.

steric factor A mathematical factor, used in the collision theory of chemical kinetics, that allows for the fact that collisions of sufficient energy may take place between molecules without leading to a chemical reaction because of an improper steric orientation of the colliding molecules.

steric hindrance The prevention of a molecule from undergoing a chemical reaction or from achieving a particular conformation or configuration that is ascribed to the stereochemical arrangement of the atoms and groups of atoms in the molecule; bulky substituents in the molecule exercise steric hindrance by tending to compress reactive groups and forcing them too close to their unbonded neighbors.

steric strain A strain in a ring structure that is due to two nonbonded groups repelling each other if they approach too closely and attempt to occupy the same point in space.

sterile Free from viable microorganisms.

sterilization The complete destruction of all viable microorganisms in a material by physical and/or chemical means.

sterilizer AUTOCLAVE.

Stern potential The potential across the double layer of a molecule or a particle; consists of a layer of fixed surface charges plus a layer of immobile counterions. The large diffuse layer, consisting mostly of mobile counterions, that extends beyond the Stern potential represents the zeta potential.

Stern–Volmer equation An equation that permits the calculation of the ratio for the efficiency of fluorescence in the absence and presence of a quencher. *Aka* Stern–Volmer relation.

steroid A cyclic compound of animal or plant origin, the basic nucleus of which consists of three 6-membered rings and one 5-membered ring, fused together to yield perhydrocyclopentanophenanthrene. The steroids represent a wide variety of compounds, including sterols, bile acids, adrenocortical hormones, and sex hormones.

steroid alkaloids A group of nitrogen-containing steroids that occur in plants, often in the form of carbohydrate-linked compounds (glycoalkaloids).

steroid alkaloid saponin *See* saponin.

steroid conjugate A steroid breakdown product that is conjugated to either glucuronic acid or sulfuric acid and that is excreted in this form.

steroid diabetes A condition produced by the prolonged administration of glucocorticoids and characterized by glucose production in the liver and by suppression of the action of insulin.

steroid glycoside A carbohydrate derivative of a steroid, such as a cardiac glycoside or a saponin.

steroid hormone A collective term for androgens, estrogens, and corticoids.

steroid number A number assigned to a steroid on the basis of the number of carbon atoms and the types of functional groups that it contains; specifically, $SN = S + F_1 + F_2 + \cdots + F_n$, where SN is the steroid number, S is the number of carbon atoms in the parent steroid, and $F_1, F_2, \cdots, F_n$ are arbitrary values, characteristic of the functional groups of the steroid.

steroidogenesis The biosynthesis of steroids.

steroid receptor One of a group of cytoplasmic proteins that bind to specific steroid hormones. After activation of the hormone–receptor complex, the latter moves into the cell nucleus where it binds to a particular site on the DNA to regulate the expression of specific genes.

steroid saponin *See* saponin.

sterol A steroid in which an alcoholic hydroxyl group is attached to position 3, and an aliphatic side chain of eight or more carbon atoms is attached to position 17 of the steroid nucleus;

a steroid alcohol.

sterol carrier protein One of a group of cytoplasmic proteins that bind to specific sterols and transport them from one endoplasmic reticulum bound enzyme to another in the course of metabolism.

Sterzl theory XYZ CELL THEORY.

STH somatotropic hormone.

STI Soybean trypsin inhibitor.

Stickland reaction The coupled decomposition of two amino acids such that one amino acid is oxidized while the other is reduced. The reaction occurs in some species of the genus *Clostridium*, which cannot degrade single amino acids but can degrade suitable pairs of amino acids.

sticky end COHESIVE END.

sticky region A region in a nucleic acid molecule that is rich in guanine and cytosine.

stilbestrol 1. 4,4′-Dihydroxystilbene. 2. Diethylstilbestrol.

still An apparatus for the purification of liquids by distillation.

stimulant amine *See* antidepressant.

stimulin INSULIN-LIKE ACTIVITY.

STM 1. Scanning tunneling microscope. 2. Spore tip mucilage.

stochastic Referring to the presence of a random variable. Thus, stochastic variation is a variation in which at least one of the elements is a random variable (variate), and a stochastic process is one that incorporates an element of randomness.

stochastic process 1. A process of problem solving that provides a solution that is close to the best; the steps in the process are based on an uncertainty which is indicated by the laws of probability and the movement at each step is random. Conjecture and speculation are used to select a possible solution which is then tested against known evidence, observation, and measurement. 2. A process of problem solving that consists of random trial and error steps. *See also* algorithm; heuristic process; stochastic.

stochastic variable VARIATE.

stoichiometric amount The amount of a substance that is used in a chemical reaction either as a reactant or as a product.

stoichiometric model The representation of an enzymatic reaction in a manner that is analogous to that of a typical chemical reaction. Thus, the formulation $E + S \rightleftharpoons ES \rightleftharpoons E + P$ represents the simplest stoichiometric model of an enzymatic reaction. *Aka* stoichiometric reaction scheme.

stoichiometry The quantitative relations between the elements in a compound or between the reactants and the products in a chemical reaction.

stoke A unit of kinematic viscosity; one-hundredth of this unit is the centistoke. The dimensions of stoke are cm^2s^{-1}.

Stokes–Einstein equation An equation that relates the diffusion coefficient (D) to the radius (r) of a spherical particle. Specifically, $D = kT/6\pi\eta r$ where k is the Boltzmann constant, η is the viscosity of the solution, and T is the absolute temperature.

Stokes' law An expression for the frictional coefficient of a sphere; specifically, $f = 6\pi\eta r$, where f is the frictional coefficient, π is a constant equal to $3.1416\ldots$, η is the viscosity of the solvent, and r is the radius of the sphere (Stokes' radius).

Stokes' radius The radius of a perfect anhydrous sphere that has the same rate of passage through a gel filtration column as the protein under study. The Stokes radius is formally defined by Stokes' law.

Stokes–Raman scattering A type of Raman scattering in which a vibrating molecule gains a quantum of energy; in anti-Stokes–Raman scattering, the molecule loses a quantum of energy.

Stokes' reagent A solution of ferrous sulfate, tartaric acid, and ammonia that is used in testing for hemoglobin.

Stokes' shift The change in the wavelength of light in fluorescence from that of the exciting light to that of the emitted light. The emitted light is usually of longer wavelength than the exciting light, and the change in wavelength is attributable to energy lost as vibrational energy.

stoma (*pl* stomata) A minute opening in a leaf through which gases pass; a "breathing tube" of leaves.

stone CALCULUS.

stop codon TERMINATION CODON.

stopped flow technique A technique for studying fast chemical reactions in which the reactants are forced out of two syringes into a mixing chamber, and the mixture is then allowed to flow into an observation cell where the flow is halted abruptly and the mixture is analyzed spectroscopically or by other means. *See also* rapid flow technique.

storage disease *See* glycogen storage disease; lipid storage disease.

storage mRNA MASKED mRNA.

STP Standard temperature and pressure.

straggling The variation in the range of particles of a specific radiation in which initially all of the particles had the same energy.

straight chain OPEN CHAIN.

strain 1. *n* A subdivision of a species that possesses distinguishing characteristics. 2. *n* A deformation in a molecule. 3. *v* To pass through a coarse filter such as cheesecloth. *See also* cell strain.

strand 1. A polynucleotide chain. 2. A polypeptide chain.

strand displacement The synthesis of a new strand (or segment) of duplex DNA that proceeds while an old strand (or segment) is being pushed out of the way. The in vitro activity of DNA polymerase I proceeds with strand displacement.

strand exchange BRANCH MIGRATION.

strand polarity *See* polarity (3).

strand selection The ability of DNA-dependent RNA polymerases to choose one strand (the transcribed strand) over the other for transcription.

streak 1. To apply material as a strip to either a chromatographic or an electrophoretic support, usually for preparative purposes. 2. To apply a bacterial inoculum as a strip to the surface of a solid nutrient medium.

streak plating A method for isolating bacterial strains by drawing an inoculation needle, containing a culture, lightly over the surface of a solid nutrient medium.

stream birefringence FLOW BIREFRINGENCE.

streamer sedimentation DROPLET SEDIMENTATION.

streaming birefringence FLOW BIREFRINGENCE.

streaming current The electrical current produced by the movement of an electrolyte through a tube.

streaming potential The electrical potential produced by the movement of a conducting liquid through a porous medium as a result of hydrostatic pressure; an electrokinetic phenomenon that is the reverse of electroosmosis.

streamline flow LAMINAR FLOW.

stream potential STREAMING POTENTIAL.

Strecker synthesis The synthesis of amino acids from aldehydes, ammonia, and hydrogen cyanide; proposed as a possible mechanism for the synthesis of amino acids under conditions existing on the primitive earth.

street virus A virus obtained from a naturally infected animal. *See also* fixed virus.

streptococcal fibrinolysin STREPTOKINASE.

Streptococcus A genus of gram-positive bacteria that includes the causal agents of pneumonia (*Streptococcus pneumoniae* or *pneumococcus*) and strep throat (*Streptococcus pyogenes*).

streptodornase One of a number of extracellular nucleases produced by species of the genus *Streptococcus*.

streptogenin peptides A group of natural products, such as liver extracts or partial hydrolysates of proteins, or of synthetic peptides that stimulate the growth of microorganisms, especially that of lactic acid bacteria. The unknown growth stimulant is called streptogenin.

streptokinase A protein, produced by β-hemolytic *Streptococci*, that is devoid of protease activity but is a potent activator of plasminogen and thereby promotes the lysis of human blood clots. Streptokinase binds tightly to plasminogen, alters the conformation of plasminogen in this complex, and thereby renders it catalytically active.

streptolydigin One of a group of antibiotics that bind to bacterial RNA polymerase and prevent RNA elongation during transcription.

streptolysin A toxin, produced by the genus *Streptococcus*, that causes hemolysis. Streptolysin-S is a leucocidin.

Streptomyces A genus of gram-positive, aerobic soil bacteria, many species of which produce antibiotics.

streptomycin An aminoglycoside antibiotic, produced by *Streptomyces griseus*, that inhibits proteins synthesis by binding to the 30S ribosomal subunit, and that causes misreading of the genetic code.

streptomycin suppression The suppression of the effects of streptomycin (inhibition of protein synthesis and misreading of the genetic code) that is shown by some mutants having an altered ribosomal protein of the small ribosomal subunit (S12). Such mutants can initiate protein synthesis in the presence of streptomycin and show decreased misreading of the genetic code.

streptonigrin An antibiotic, produced by *Streptomyces flocculus*, that leads to chromosome breaks.

streptovaricin One of a group of antibiotics, produced by species of *Streptomyces*, that are closely related to the rifamycins in both mechanism of action and antimicrobial spectrum.

stress fibers Bundles of actin filaments and associated proteins that are attached to the plasma membrane of nonmuscle cells, underlie some coated pits, and are characteristic of the cytoskeleton of eukaryotic cells in tissue culture; they function in cell motility.

stress metabolites *See* phytoalexins.

stress proteins HEAT SHOCK PROTEINS.

stretched muscle A muscle that has been lengthened by stretching.

stretching vibration A vibration in a molecule that results in a lengthening of an interatomic bond distance.

striated muscle A muscle, such as a skeletal or a cardiac muscle, that is characterized by transverse striations.

strict aerobe OBLIGATE AEROBE.

strict anaerobe 1. OBLIGATE ANAEROBE. 2. An organism or a cell that cannot tolerate oxygen and is inhibited or killed by oxygen or by oxidized components of the medium.

striction The decrease in the volume of a solution, compared to the sum of the volumes of the separate solute and solvent, as a result of solute–solvent interactions. *See also* covolume; electrostriction.

stringent control 1. The control of the rate of RNA synthesis by $_{pp}G_{pp}$ and $_{ppp}G_{pp}$ (magic spots) in the stringent response. 2. Plasmid replication that is synchronized with that of the chromosome; that is, the plasmid:chromosome ratio remains constant (usually, 1:1). *See also* superinfection immunity.

stringent factor The enzyme that catalyzes the synthesis of $_{pp}G_{pp}$ and $_{ppp}G_{pp}$ (magic spots); the latter are the factors whose concentration regulates the rate of RNA synthesis in the stringent response. Synthesis of the magic spots involves pyrophosphate transfer from ATP to GDP or GTP. The enzyme is located exclusively in the 50S ribosome but only in about 1 in 200 ribosomes. It is activated only when two conditions are met: (a) the 50S subunit is part of a 70S ribosome that is bound to mRNA and engaged in translation; (b) the A site of the ribosome is occupied by an uncharged tRNA molecule.

stringent plasmid *See* copy number.

stringent response The decrease of RNA synthesis that occurs normally in wild-type bacteria after removal of an essential amino acid from the medium. *See also* relaxed control.

stringent strain A bacterial strain that shows stringent control.

stripped particle CORE PARTICLE.

stripped transfer RNA An aminoacyl transfer RNA molecule from which the amino acid has been detached by hydrolysis.

stripping 1. The hydrolytic removal of an amino acid from an aminoacyl-tRNA molecule. 2. The removal of ribosomal proteins from ribosomes, resulting in the production of ribosomal subparticles.

stroke A sudden neurological afflication caused by a loss of blood to the brain as a result of embolus formation, thrombus formation, or cerebrovascular hemorrhage. A stroke may lead to paralysis, weakness, speech defects, or death.

stroma 1. The matrix material between grana in a chloroplast. 2. The connective tissue framework of a gland or an organ as distinct from the parenchyma.

stroma starch The starch of chromatophores that is not clustered around pyrenoids.

stromatin A structural protein in the cell membrane of erythrocytes.

stromatolite A macroscopic structure composed of fossilized bacterial mats; colonies of bacteria embedded with minerals.

strong electrolyte An electrolyte that is completely dissociated into ions in water.

strong interactions The attractive forces between atoms and molecules that result in the formation of covalent and ionic bonds, and the repulsive forces of ion–ion interactions.

strong promoter HIGH-LEVEL PROMOTER.

strontium An element that is essential to a few species of organisms. Symbol, Sr; atomic number, 38; atomic weight, 87.62; oxidation state, +2; most abundant isotope, ^{88}Sr; a radioactive isotope, ^{85}Sr, half-life, 64.7 days, radiation emitted, gamma rays.

strontium-90 The heavy radioactive isotope of strontium that occurs in the fallout from the explosion of nuclear weapons; it has a half-life of 25 years and is incorporated into biological systems.

strophantidin OUABAIN.

strophantin G OUABAIN.

structon A proposed palindromic DNA sequence in the promoter region of the rRNA cistron that is believed to function in switching on the synthesis of rRNA.

structural color A color created by optical effects (interference, refraction, etc.) resulting from the physical nature of surfaces.

structural formula A two-dimensional representation of the structure of a molecule in which the atoms are connected by one or more lines, with each line indicating a pair of shared electrons.

structural gene 1. A gene, the nucleotide sequence of which determines the amino acid sequence of a polypeptide chain; a cistron. 2. A gene, the nucleotide sequence of which determines the nucleotide sequence of a polynucleotide strand.

structural isomer One of two or mroe isomers that have the same molecular compositions but that differ from each other in their atom-to-atom sequence; they can be distinguished on the basis of their two-dimensional structures. *See also* stereoisomer.

structural protein 1. A protein that functions primarily as a structural component of cells and tissues. 2. A noncatalytic protein of mitochondrial membranes, the existence of which is in doubt. 3. SCLEROPROTEIN.

structural unit The monomer of the viral capsomer that consists of one or more polypeptide chains. *See also* capsid; capsomer.

structured smectic *See* liquid crystal.

strychnine An indole alkaloid from tropical plants of the genus *Strychnos*; a highly toxic compound (a neurotoxin) used as a rodent poison.

Stuart factor The first factor that is common to both the intrinsic and the extrinsic pathways of blood clotting. The Stuart factor is activated by the Christmas factor in the intrinsic pathway and by proconvertin in the extrinsic pathway. *Aka* Stuart–Prower factor.

Student's t test A statistical test for evaluating the difference between two sample means; that is, the same measurement is made on two separate groups (frequently a control and a test group) and the means of the two groups are compared. The one-sample t test is a test of the null hypothesis that a random sample comes from a normal population; the two-sample t test is a test concerning the difference between the means of two normal populations having the same standard deviation; the paired sample t test is an application of the one-sample t test to the difference between paired data.

sturine A protamine isolated from sturgeon.

SU Thiouridine.

sub- 1. Prefix meaning below or under. 2. Prefix meaning fraction of or less than.

subacute test PROLONGED TEST.

subcellular fraction A preparation containing either one or more specific components of the cell, or a portion of the total cellular material; a preparation of an enzyme, or of mitochondria, or of cell membranes, etc.

subcloning A variation of the molecular cloning technique in which a DNA segment is cloned, isolated, and then cut into successively smaller fragments by the use of several restriction enzymes. A specific fragment is then recloned in a suitable plasmid.

subculture A culture that is produced by transferring a portion of a stock culture to fresh medium.

subcutaneous injection An injection under the skin. *Abbr* sc.

subfragment 1 SF$_1$ FRAGMENT.

sublethal gene A gene that, when mutated, gives rise to a subvital mutation.

sublimation The transition of a solid to a vapor without passage through the intermediate liquid state.

submitochondrial particles INSIDE-OUT PARTICLES.

subnatant The liquid below another liquid or below solid material.

subribosomal particle RIBOSOMAL SUBPARTICLE.

subribosome RIBOSOMAL SUBPARTICLE.

subroutine A portion of a complete computer routine, consisting of a set of instructions that cause a computer to carry out well-defined mathematical or logical operations.

substance P A peptide neurotransmitter found in brain and in the digestive tract. It consists of 11 amino acids, is a powerful promoter of muscular contractions in the intestinal tract, and is a potent vasodilator. *Abbr* SP.

substituent An atom or a group of atoms that is introduced into a molecule by replacement of another atom or group of atoms.

substituent constant *See* Hammett equation.

substituted enzyme 1. The modified enzyme that is produced by the reaction of the original enzyme molecule with the first substrate in a ping-pong mechanism. 2. ENZYME–SUBSTRATE INTERMEDIATE.

substitution *See* base substitution; conservative substitution; radical substitution; substitution reaction.

substitution loop The loop formed by two non-complementary DNA segments when two, otherwise complementary, DNA strands are hybridized; a region of noncomplementarity in both strands. *See also* deletion loop.

substitution reaction A chemical reaction in which an atom or a group of atoms attached to a carbon atom is removed and replaced by another atom or group of atoms.

substitution vector *See* lambda cloning vector.

substrate The compound acted upon by an enzyme; the molecule or structure, the transformation of which is catalyzed by an enzyme. *Sym* S.

substrate anchoring The concept that substrates, confined to the active site of an enzyme, have relatively long residence time compared to the time interval that the same substrates would be within striking distance of each other if they were in random motion in solution.

substrate-assisted catalysis An approach to the engineering of enzyme specificity. The method involves removal of part of the catalytic structure of the enzyme (for example, by site-directed mutagensis). The enzyme activity is then partially restored by supplying the missing functionality from a bound substrate. In this way substrates are distinguished primarily by their ability to actively participate in the catalytic mechanism, thereby permitting the design of extremely specific enzymes.

substrate-binding site ACTIVE SITE.

substrate–bridge complex *See* bridge complex.

substrate constant The equilibrium (dissociation) constant of the reaction ES $\rightleftharpoons$ E + S, where E is the enzyme and S is the substrate. *Sym* K_S.

substrate cycle The combination of a forward reaction, catalyzed by one enzyme, with the reverse reaction, catalyzed by a different enzyme. When these two reactions occur at comparable rates, they may result in a futile cycle.

substrate elution chromatography A type of column affinity chromatography in which an enzyme is bound nonspecifically to a polymer and the binding is then specifically overcome by formation of an enzyme–substrate complex when the column is eluted with a solution containing the substrate of the enzyme. The method can also be used for other protein–ligand systems.

substrate inhibition The inhibition of the activity of an enzyme by the substrate of the enzyme; generally most pronounced at high substrate concentrations.

substrate phosphorylation The synthesis of ATP that is coupled to the exergonic hydrolysis of a high-energy compound and that is not linked to an electron transport system; the synthesis of ATP in glycolysis is an example. *Aka* substrate level phosphorylation.

substrate synergism The acceleration by a substrate, or substrates, of a reaction undergone by other substrates of a multisubstrate enzyme system.

subtilin A peptide antibiotic produced by *Bacillus subtilis* that consists of 32 amino acids.

subtilisin A nonspecific proteolytic enzyme derived from *Bacillus subtilis*. *See also* S peptide; S protein.

subtractive Edman degradation A modification of the Edman degradation in which the N-terminal amino acid, removed as the phenylthiocarbamyl derivative, is identified by determining the amino acid composition of the peptide before and after treatment with the reagent.

subunit 1. The smallest covalent unit of a protein; it may consist of one polypeptide chain or of several chains linked together covalently, as by means of disulfide bonds. 2. The functional unit of an oligomeric protein, such as the structure in hemoglobin that consist of one alpha and one beta chain. 3. A definite substructure of a macromolecule such as the 30S or the 50S unit of the 70S bacterial ribosome. *See also* monomer; protomer.

subunit exchange The association of the small ribosomal subunit from one ribosome with the large subunit from a different ribosome, and vice versa, as a result of the operation of the ribosome cycle.

subunit model *See* concerted model; sequential model.

subvital mutation A mutation that decreases viability but results in the death of less than 50% of the organisms carrying the mutation.

succinate–coenzyme Q reductase COMPLEX II.

succinate–glycine cycle SHEMIN CYCLE.

succinate pathway A catabolic pathway whereby methionine, isoleucine, and valine are converted to succinic acid which then enters the citric acid cycle.

succinate:ubiquinone oxidoreductase COMPLEX II.

succinic acid A symmetrical, four-carbon, dicarboxylic acid that is an intermediate in the citric acid cycle where it is formed from α-ketoglutaric acid.

succinylation The modification of a protein by reaction with succinic anhydride; introduction of a 3-carboxypropionyl group ($-COCH_2CH_2CO_2^-$) into a protein. Succinylation involves primarily the amino groups (α and ε) of amino acids but may also involve the phenolic group of tyrosine. *Aka* 3-carboxypropionylation.

succinyl–coenzyme A A high-energy compound that is an intermediate in the citric acid cycle.

succotash A mixture of corn (low in lysine) and beans (low in tryptophan) that provides a nutritionally adequate supply of the essential amino acids and that was used by New World Indians.

sucrase The enzyme that catalyzes the hydrolysis of sucrose to glucose and fructose.

sucrose A disaccharide of glucose and fructose that is abundant in the plant world and that is the sugar used in the home.

sucrose density gradient A density gradient prepared by using sucrose solutions of different concentrations, frequently covering a range of 5 to 25% (w/w).

sucrose gradient centrifugation Density gradient sedimentation velocity in which a sucrose density gradient is used.

sucrose intolerance A genetically inherited metabolic defect in humans due to a deficiency of either sucrase or isomaltase.

sucrose polyester One of a group of synthetic compounds, consisting of polyacylated sucrose in which the acyl groups are those of long-chain fatty acids. These esters resemble cooking oils but are not digestible; they have been used in experimental diets.

Sudan black B A lysochrome.

sugar CARBOHYDRATE.

sugar acid An acid derivative of a mono- or an oligosaccharide.

sugar alcohol An alcohol derivative of a mono- or an oligosaccharide.

sugar nucleotide NUCLEOSIDE DIPHOSPHATE SUGAR.

sugar pucker *See* puckered conformation.

suicide The loss of infectivity or other biological activity by either a molecule or a particle as a result of radioactive decay of an incorporated radioisotope. *See also* radiophosphorus decay; star gazing; suicide substrate.

suicide inhibition *See* suicide substrate.

suicide substrate A substrate that normally does not become linked covalently to an enzyme but is so altered that a reactive group is generated which then reacts with a nearby group on the enzyme, thereby linking the substrate covalently to the enzyme and inactivating the latter.

sulfa drug One of a class of antibacterial drugs that are derivatives of sulfonamide RSO_2NH_2 and that inhibit the growth of many bacteria. Sulfa drugs function by being competitive inhibitors of *p*-aminobenzoic acid, which is required by these bacteria for the synthesis of folic acid. Their toxicity to humans is relatively low.

sulfanilamide Benzene sulfonamide; a sulfa drug.

sulfate assimilation The reduction of sulfate, carried out by plants and bacteria, that leads to the biosynthesis of cysteine.

sulfate dissimilation *See* dissimilatory reduction; sulfate respiration.

sulfate reduction *See* sulfate assimilation.

sulfate respiration The dissimilatory reduction of sulfate in the course of cellular respiration in which sulfate replaces oxygen as the terminal electron acceptor; the sulfate is reduced to hydrogen sulfide which is excreted.

sulfatide A glycosphingolipid containing a sulfate group; a sulfoglycosylsphingolipid; a ceramide monosaccharide sulfate.

sulfating agent ACTIVE SULFATE.

sulfation The incorporation of sulfate into a compound.

sulfation factor SOMATOMEDIN.

sulfhydryl group The radical —SH.

sulfhydryl reagents Reagents, such as iodoacetate and *p*-hydroxymercuribenzoate, that react specifically with the sulfhydryl group (SH— group) of cysteine in peptides and proteins.

sulfite oxidase deficiency A genetically inherited metabolic defect in humans that is associated with mental retardation and that is due to a deficiency of the enzyme sulfite oxidase.

sulfitolysis The cleavage of a covalent bond by reaction with sulfite (SO_3^{2-}). In the case of peptides containing a disulfide bond (R_1—S—S—R_2), the latter can be broken by reaction with sulfite to yield two sulfonic acid molecules (R_1—S—SO_3H and R_2—S—SO_3H).

sulfolipid 1. Any sulfur-containing lipid. 2. PLANT SULFOLIPID.

sulfonamide *See* sulfa drug.

sulfone One of a group of compounds that are antibacterial agents and appear to function much like the sulfa drugs but with greater toxicity. Sulfones have the structure R_2SO_2. An example is 4,4'-diaminodiphenyl sulfone (DDS, dapsone) which is an important drug in the treatment of leprosy.

sulfonic acid An organic acid containing the radical —SO_3H.

sulfosalicylic acid test A test for protein in urine and in other biological fluids that is based on the turbidity formed upon addition of sulfosalicylic acid to the sample. *Abbr* SSA test.

sulfur An element that is essential to all plants and animals. Symbol, S; atomic number, 16; atomic weight, 32.964; oxidation states, -2, $+4$, $+6$; most abundant isotope, ^{32}S; a radioactive isotope, ^{35}S, half-life, 87.2 days, radiation emitted, beta particles.

sulfur amino acid An amino acid that contains sulfur, such as cysteine or methionine.

sulfur bacteria A group of bacteria that are chemolithotrophs and that use carbon dioxide as their carbon source, oxidation–reduction reactions as their energy source, and inorganic sulfur-containing compounds as electron donors.

sulfur dioxide The gas SO_2, used as a food preservative. Its bactericidal and fungicidal properties are believed to be due to its ability to reduce the disulfide bonds of enzymes as it undergoes oxidation from SO_3^{2-} to SO_4^{2-}.

sulfur mustard Di(2-chloroethyl)sulfide; a chemical mutagen and alkylating agent that is structurally related to nitrogen mustards but contains sulfur instead of nitrogen. *See also* alkylating agent.

Sulkowitch test A test used for the semiquantitative determination of calcium in urine; based on measurements of the turbidity that is produced upon addition of oxalate, buffered with acetate, to urine.

Sullivan reaction A colorimetric reaction for cysteine that is based on the production of a red color upon treatment of the sample with sodium 1,2-naphthoquinone-4-sulfonate and sodium sulfite.

SUN Serum urea nitrogen; *see* blood urea nitrogen.

super- 1. Prefix meaning above or in the upper part of. 2. Prefix meaning excessive.

superacid catalysis Catalysis due to a positively charged ion, particularly a metal ion, that acts as a Lewis acid.

superagonist A compound that strongly mimics the positive biological activity of a naturally occurring molecule.

supercoil SUPERHELIX.

superconductor A substance that conducts electricity at greatly reduced resistance; ceramic copper oxide materials, for example, can conduct electricity without resistance at the temperature of liquid nitrogen.

supercool To cool a liquid below its freezing point without the separation of solid matter.

supercritical fluid chromatography A chromatographic technique in which compressed fluids, at temperatures slightly above their critical temperatures, are used as mobile phases; useful for the separation of substances that are not readily separated by gas or liquid chromatography. *Abbr* SFC.

superficial Near the surface.

superfusion A technique in which blood, plasma, or some other fluid is allowed to drip onto, or to flow over, the surface of a perfused organ.

supergene A segment of linked genes that is protected from crossing-over and that is transmitted intact from one generation to another.

superhelix 1. The structure formed when double-stranded, circular DNA is further twisted; one such twist converts the circle into a figure eight; additional twists result in multiple figure eights, linked together. The superhelix can be formed by twisting the DNA in an opposite sense to that of the double helix. Such a superhelix is termed left-handed, negative, or underwound. All naturally occurring superhelical DNA shows negative supercoiling or underwinding. If the superhelix is formed by twisting the DNA in the same sense as that of the double helix, then the resulting superhelix is called right-handed, positive, or overwound. 2. A helix composed of two or more component chains and produced by the winding of two or more helical chains, either polynucleotide or polypeptide chains, about each other.

superhelix density The ratio of the writhing number (W) to the twisting number (T); that is, W/T. It has a value of about 0.05 for all naturally occurring superhelical DNA (negatively supercoiled DNA). In other words, there is one negative twist produced per 200 base pairs or 0.05 negative twists per turn of the double helix.

superinducible mutant A mutant that synthesizes an inducible enzyme at concentrations of inducer that are lower than those required by the wild-type organism.

superinfection 1. An extensive invasion of an organism by pathogenic microorganisms, as that which may arise from its infection by drug-resistant microorganisms. 2. A repeated phage infection of a bacterial culture that is already infected with phage. Such a culture is characterized by possessing bacterial cells that contain more than one phage particle per cell.

superinfection curing *See* curing.

superinfection exclusion The phenomenon that superinfection of a bacterial culture with phage at times leads to failure of the DNA of the superinfecting phage to enter the host cell; a change in cell permeability (a "sealing reaction") has been invoked to explain this phenomenon.

superinfection immunity The phenomenon that two identical, or very similar, plasmids cannot coexist in the same cell; this is usually the case when plasmid replication is subject to stringent control.

superior Above, or higher than, a given structure.

supermolecule The polymeric protein system that functions as the energy transducing unit and constitutes the structural unit of the inner mitochondrial membrane according to the conformational coupling hypothesis. The supermolecule consists of seven complexes, four of which are electron transfer complexes, and the other three are the ATP synthetase, the transprotonase, and the transhydrogenase complexes. The supermolecule is in the shape of a knoblike structure composed of a basepiece, a stalk, and a headpiece; this structure is formed by arranging six of the complexes around a central unit referred to as a tripartite repeating unit (TRU). The headpiece of the TRU is the site for synthesis or hydrolysis of ATP; the stalk of the TRU, which connects the headpiece and the basepiece, is a regulatory device which determines whether ATP will be synthesized or hydrolyzed; and the basepiece of the TRU is the membrane-forming element that functions as a linkage system around which are arranged the four electron transfer complexes, the transhydrogenase, and the transprotonase. *Aka* elementary particle.

supernatant The liquid above sedimented material or above a precipitate. *Aka* supernate.

superoxide anion A very active radical ($O_2^- \cdot$) that is formed when a molecule of oxygen gains an electron. The radial reacts readily with protons, other superoxide anions, and hydrogen peroxide to produce a variety of reactive and toxic species.

superoxide dismutase The enzyme that catalyzes the reaction $O_2^- \cdot + O_2^- \cdot + 2H^+ = H_2O_2 + O_2$. The enzyme appears to be ubiquitous among aerobic organisms and protects the organism against action by the superoxide radical $O_2^- \cdot$, which is believed to be a mutagenic radical, produced in an organism by ionizing radiation. *Abbr* SOD.

superprecipitation The contraction of an actomyosin gel to a small plug that occurs in the presence of ATP; the process is used for in vitro studies of muscle contraction.

superrepressed mutant A mutant that synthesizes a repressible enzyme at a rate that is characteristic of the uninduced wild-type organism, regardless of the presence or absence of the inducer. The gene, responsible for synthesis of the enzyme, is in an uninducible state.

superrepressor A repressor that remains permanently bound to the operator.

supersaturated solution A solution that holds temporarily more solute than will be contained by an equal volume of a saturated solution under specified conditions.

super-secondary structure A recurring pattern of protein structure that is at a higher level than secondary structure but does not constitute an entire domain. Greek-key structure, beta barrel, and beta meander are some examples.

supersonic 1. ULTRASONIC. 2. Having a velocity exceeding that of sound.

supersonic oscillation ULTRASONICATION.

supersuppression A mutation that can partially or completely restore a number of genetic functions by suppressing the expression of several mutations (usually nonsense mutations) at several, different, sites on the chromosome.

supertwist SUPERHELIX.

supplemental air The volume of air that can be forcibly expired from the lungs after the normal tidal air has been expired.

supplementary action The increase in the biological value of two proteins when they are fed together, over the sum of their biological values when they are fed separately.

support 1. The solid material that either is or holds the stationary phase in chromatography. 2. The solid material in which electrophoresis is performed.

suppressible insulin-like activity Insulin-like activity that is present in the serum and that is lost upon treatment with anti-insulin antibodies. *Abbr* SILA.

suppressible mutation A mutation such that the genetic function, the loss of which is caused by the mutation, can be restored by means of either an intragenic or an intergenic suppressor mutation.

suppression The partial or complete restoration of a genetic function, lost as a result of a mutation, by means of a second mutation, occurring at a different site on the chromosome.

suppression test A clinical test for differentiating between hyperadrenalism due to adrenocortical hyperplasia and that due to an adrenal carcinoma. The test is based on the administration of a cortisol analogue which, in normal individuals but not in those with an adrenal carcinoma, has a feedback effect and decreases the pituitary output of adrenocorticotropic hormone; this, in turn, leads to a suppression of the output of steroids by the adrenal glands.

suppressor 1. SUPPRESSOR GENE. 2. The product of a suppressor gene.

suppressor factors Protein regulatory factors isolated from suppressor T cells.

suppressor gene A gene that can partially or completely reverse the effect of a mutation in another gene.

suppressor mutation A mutation that partially or completely restores a genetic function, lost as a result of another mutation, and that is located at a site other than that which sustained the primary mutation. The suppressor mutation may occur either in the gene that was originally mutated (intragenic suppressor mutation) or in a different gene (intergenic suppressor mutation).

suppressor-sensitive mutant A conditional mutant that behaves normally when grown in some hosts but exhibits its mutant phenotype when grown in other hosts. Normal growth occurs in hosts having suppressor genes. The products of these genes either compensate for the defect in the mutant or enable the altered gene to produce a functional gene product. *Abbr* sus.

suppressor T cells A group of T cells that suppress the response of specific T or B lymphocytes to an antigen. *Aka* suppressor T lymphocytes.

suppressor transfer RNA An unusual tRNA molecule, produced by a suppressor gene, that brings about missense or nonsense suppression. The tRNA molecule leads to the incorporation of the correct amino acid that would have been specified by the codon prior to its being changed as a result of either a missense or a nonsense mutation. The term is usually applied specifically to a tRNA molecule that can pair with a nonsense (termination) codon.

supra- Prefix meaning above.

supramolecular complex 1. A complex composed of several or many molecules, such as a ribosome, a biological membrane, or a virus. 2. MULTIENZYME SYSTEM. 3. METABOLON (2).

suprarenal gland ADRENAL GLAND.

suprarenin EPINEPHRINE.

supravital staining The staining of living cells after their removal from the host.

Sur Thiouracil.

SUra Thiouracil.

surface-active agent A substance that alters the surface tension of a liquid, generally lowering it; detergents and soaps are typical examples.

surface activity The strong adsorption of surface-active agents at surfaces or interfaces of liquids in the form of oriented monomolecular layers.

surface antigen An antigen that is present on the surface of a cell, such as the H, O, or Vi antigen of bacteria.

surface balance An instrument for measuring surface pressure.

surface-enhanced Raman spectroscopy A technique for identifying molecules adsorbed on a surface by the enhancement of Raman scattering produced by these molecules. *Abbr* SERS.

surface enzyme An exoenzyme that remains attached to the cell surface.

surface extended x-ray absorption fine structure spectroscopy A technique for the study of surfaces in which an x-ray source, aimed at the surface, scans through a range of x-ray energies. At characteristic energies, associated with certain atomic transitions, the radiation is absorbed, causing the atoms in the sample to emit electrons. Measurements of x-ray absorption and of emitted Auger electrons provides information about the arrangement of atoms on the surface. *Abbr* SEXAFS.

surface factor HAGEMAN FACTOR.

surface membrane proteins Proteins that are associated with the hydrophilic surfaces of the lipid bilayer of biological membranes.

surface potential The difference between the electrical potential of the clean surface of a liquid and that of a monolayer on the surface of the same liquid. *Aka* surface film potential.

surface pressure The lowering of the surface tension of a liquid by a monolayer that is present on the surface of that liquid.

surface tension The tension exerted by the surface of a liquid as a result of the intermolecular attractive forces between the molecules of the liquid.

surfactant SURFACE-ACTIVE AGENT.

surfactant vesicle *See* vesicle.

surfactin An extracellular product of *Bacillus subtilis* that causes lysis of red blood cells (hemolysis) and lysis of some protoplasts. It consists of an oligopeptide linked to a fatty acid derivative and has, therefore, detergent-like properties.

surroundings That part of the universe, in the thermodynamic sense, that is not under study; the part of the universe that is being studied is known as the system.

survival curve A dose–response curve that indicates the surviving fraction of active units as a function of the radiation dose.

survival of the fittest *See* natural selection.

surviving slice TISSUE SLICE.

sus Suppressor-sensitive mutant.

suspension A colloidal dispersion in which the particles are so large that they will settle out of solution. *See also* colloidal dispersion; colloidal solution.

suspension culture A cell culture, used in virology, in which animal cells are kept in suspension in a medium that is low in divalent ions.

SUV Small unilamellar vesicle; *see* vesicle.

SV Satellite virus.

SV-40 Simian virus 40.

Svedberg A unit of the sedimentation coefficient equal to 10^{-13} s. *Sym* S.

Svedberg equation An equation used for calculating moleculear weights from sedimentation and diffusion data; specifically, $M = RTs/D(1 - \bar{v}\rho)$, where M is the molecular weight, T is the absolute temperature, s is the sedimentation coefficient, D is the diffusion coefficient, $\bar{v}$ is the partial specific volume of the solute, and ρ is the density of the solution.

swelling The uptake of water by a gel or by a tissue.

swinging bucket rotor A centrifuge rotor used in a preparative analytical ultracentrifuge for density gradient centrifugation. During the centrifugation the buckets, which hold tubes filled with solution, swing out to a horizontal position that is at right angles to the axis of rotation; this eliminates the extensive convection produced by centrifugation in a fixed angle rotor.

switchgene A gene that causes an organism to follow a different developmental pathway.

switching sites Points in a chromosome at which breaks occur and at which gene segments combine during gene rearrangements.

switch regions The regions between the constant and the variable portions in both the light and the heavy immunoglobulin chains.

swivel A region in double-stranded DNA in which rotation of one strand around the other takes place.

swivelase *See* DNA gyrase; topoisomerase.

symbiont An organism that lives in symbiosis with another organism.

symbiosis The living together in close association, and for their mutual benefit, of two organisms from different species.

symbiotic Of, or pertaining to, symbiosis.

symbiotic nitrogen fixation The conversion of atmospheric nitrogen to ammonia by the combined action of plants and bacteria; applicable primarily to leguminous plants and to bacteria of the genus *Rhizobium*.

symmetrical Possessing symmetry. *Aka* symmetric.

symmetric transcription The synthesis of RNA from complementary segments of the two strands of DNA; this occurs, for example, when the core enzyme of RNA polymerase is used in vitro.

symmetry The geometrical regularity in the structure of a body such that the sizes, shapes, and relative positions of structural parts are distributed equally about dividing planes, axes, and centers in the body. *See also* specific types.

symmetry-breaking model SEQUENTIAL MODEL.

symmetry-conserving model CONCERTED MODEL.

symmetry model CONCERTED MODEL.

sympathoadrenal system The adrenal medulla and the sympathetic nervous system which function in concert in response to stress.

sympathomimetic amine A substance that acts like epinephrine and that produces effects that are similar to those brought about by a stimulation of the sympathetic nervous system.

symplast The intracellular compartment of a plant consisting of the living cells (including the phloem tubes), linked via plasmodesmata, and bounded by the combined plasma membranes. *See also* apoplast.

symport The linked transport in the same direction of two substances across a membrane.

syn 1. Referring to a nucleoside conformation in which the base has been rotated around the sugar, using the C—N glycosidic bond as a pivot, so that the sugar is placed directly below the base. This represents a sterically more hindered conformation than the anti conformation; in polynucleotides, it leads to the bulky portions of the bases being pointed toward the sugar–phosphate backbone of the chain. 2. Referring to a cis configuration for certain compounds containing double bonds, such as the oximes which contain the grouping $>C=N—OH$. 3. Referring to the position occupied by two radicals of a stereoisomer in which the radicals are closer together as opposed to the anti position in which they are farther apart. *See also* anti.

synalbumin An insulin-inhibitory polypeptide in the blood of some diabetics; may be identical or similar to the B chain of insulin which binds to albumin under certain conditions and which acts as an inhibitor of insulin in that form.

synapse The area of functional contact between two nerve cells; consists of the nerve terminals, the specialized regions of the two nerve cells in the immediate vicinity of the nerve terminals, and the gap (synaptic cleft) between the two cells. A synapse is a communicating junction between two cells but the communication is indirect even though the cells are in physical contact. The "sending" cell (presynaptic cell) secretes a chemical (neurotransmitter) that diffuses across the synaptic cleft and signals the "receiving" cell (postsynaptic cell). The majority of synapses are such chemical synapses but some synapses are electrical; in the latter, the signal passes directly from one neuron to another through a gap junction. *See also* cell junction; gap junction.

synapsin A protein substrate of $Ca^{2+}/$ calmodulin-dependent protein kinase.

synapsis The pairing of homologous chromosomes during meiosis.

synaptic cleft *See* synapse.

synaptic vesicle One of a group small vesicles, located in the presynaptic cell of a synapse, that store acetylcholine and play a role in the regulation of acetylcholine within the nerve cell.

synaptonemal complex A complex protein structure that forms between, and parallel to, the two paired homologous chromosomes during the early stages of meiosis.

synaptosome A largely artificial structure that is produced by disruption of the nerve endings in a synapse. Typically formed by homogenization of brain or spinal cord which results in the snapping off of nerve terminals. The membranes of the latter then reseal to form artifactual, osmotically active organelles that are separable by centrifugation. These structures contain acetylcholine and acetylcholinesterase.

synarchy The working together of two interrelated intracellular messengers in regulating various biological functions. The coupling of the actions of cyclic AMP and calcium, that appears to be shared by nearly all differentiated cells of higher organisms, is an example. *Aka* synarchic regulation.

syncarcinogenesis Synergistic carcinogenesis.

syncatalytic process A process that is synchronous with the catalytic action of an enzyme. A substrate-dependent increase in the reactivity of a functional group of the enzyme or the inactivation of an enzyme by the product of the enzymatic reaction, are two examples.

synchronous growth Growth in which all of the cells are at the same stage in cell division at any given time. *Aka* synchronized growth.

synchronous muscle A muscle that yields a single contraction for every motor nerve impulse that it receives.

synchronous reaction CONCERTED REACTION.

synchrotron An accelerator designed to impart high kinetic energy to charged particles by means of a high-frequency electric field and a low-frequency magnetic field.

syncytium A group of cells, joined by cytoplasmic bridges and not separated by cell membranes; an aggregate that contains many nuclei and maintains cytoplasmic continuity.

syndein ANKYRIN.

syndesine An amino acid that has been isolated from cross-linked collagen chains and that represents the product of an aldol condensation between a molecule of hydroxyallysine and a molecule of allysine.

syndet Synthetic detergent.

syndiotactic polymer A polymer in which the R groups of the monomers alternate regularly on both sides of the plane that contains the main chain.

syndrome A group of symptoms that occur at the same time and that characterize a disease.

syneresis The shrinkage of a gel with the expulsion of liquid. *See also* clot retraction.

synergism The phenomenon in which two or more agents work together cooperatively such that their combined effect is greater than the sum of the effects when either agent is acting alone. *See also* substrate synergism.

synergistic Of, or pertaining to, synergism.

synergy 1. SYNTROPY. 2. SYNERGISM.

synexin A protein that occurs in several tissues and causes the Ca^{2+}-dependent aggregation of isolated chromaffin granules; believed to promote fusion of the granules with the plasma membrane during exocytosis.

syngeneic Referring to genetically identical individuals of the same species; used in reference to tissue transplants.

syn genes Mitochondrial genes of yeast that code for mitochondrial protein synthesizing machinery such as tRNA and rRNA.

syngraft A transplant from one individual to a genetically identical individual of the same species.

synhibin *See* calelectrin.

synomone *See* allomone.

synonym codon One of several codons that code for the same amino acid, such as the codons UUU and UUC, both of which code for the amino acid phenylalanine.

synovial fluid The fluid present at the joints of vertebrates.

syntenic genes Genes that are believed to be located on the same chromosome because of their behavior during cell hybridization.

synthase 1. LYASE. 2. An enzyme that is not a lyase but for which it is desirable to stress the synthetic aspects of the reaction.

synthase-phosphorylase kinase A converter enzyme that catalyzes the interconversion of the two allosteric forms of phosphorylase (a and b); it catalyzes the ATP-dependent phosphorylation of phosphorylase b to a. *Abbr* SPK.

synthesis The process whereby a more complex substance is produced from simpler substances by a reaction or a series of reactions; the simpler substances, or portions thereof, are combined to form the more complex substance.

synthetase 1. LIGASE. 2. An enzyme that is not a ligase and that catalyzes the formation of a compound by some other mechanism; an example is the enzyme thymidylate synthetase.

synthetic 1. Of, or pertaining to, synthesis. 2. Man-made; synthesized in vitro; prepared artificially as opposed to being isolated from natural sources.

synthetic auxin A synthetic organic compound

that has auxin activity.

synthetic boundary cell An analytical ultracentrifuge cell in which solvent or a less concentrated solution is layered at low rotor speeds over the sample solution, thereby establishing a "synthetic boundary." The cell is useful for measurements of small sedimentation coefficients and for determinations of apparent diffusion coefficients.

synthetic diet A diet composed only of known chemical ingredients.

synthetic linkers LINKER DNA (2).

synthetic medium A medium composed only of known chemical ingredients.

synthetic messenger RNA A synthetic polyribonucleotide that is used as messenger RNA in a cell-free amino acid incorporating system.

synthetic polyribonucleotide An RNA molecule made in vitro in the absence of a nucleic acid template; this may be accomplished either by using the enzyme polynucleotide phosphorylase or by chemical synthesis.

synthon A structural unit within a molecule that can be formed and/or assembled by known or conceivable synthetic operations; the intermediates prepared during the synthesis of genes or proteins are examples.

syntoxic Descriptive of a substance or a condition that leads an animal to live with and tolerate, rather than fight and resist, an irritation, a toxin, or similar factors.

syntrophy The nutritional and metabolic interactions between organisms that are placed in the same environment. The phenomena of cross-feeding and of the nutritional interdependence of plants, animals, and microorganisms with respect to the carbon, oxygen, and other cycles are examples. *Aka* syntrophism.

synzyme A synthetic enzyme; a synthetic macromolecule that has enzymatic activity; a synthetic enzyme analogue.

Syp Mercaptopurine.

system 1. A set of subcellular components that perform one main function, such as an amino acid incorporating system or a transport system. 2. That part of the universe, in the thermodynamic sense, that is under study; the rest of the universe is known as the surroundings. 3. A set of organs performing one main function, such as the digestive system, the central nervous system, or the endocrine system.

systematic error DETERMINATE ERROR.

systematic name A name created on the basis of definite rules of nomenclature to distinguish clearly one item from related ones; the name of a compound based on its structure or the name of an enzyme based on the reaction that it catalyzes are examples.

systemic 1. Relating to a system. 2. Relating to the entire organism and not just to one of its parts.

Szilard–Chalmers reaction A reaction in which a chemical compound is altered by bombardment with neutrons in such a way as to allow chemical separation of the reaction products.

T

t Student's statistic; *See* Student's t test.

t₁/₂ Half-life.

T 1. Thymine. 2. Thymidine. 3. Tritium. 4. Transmittance. 5. Absolute temperature. 6. T-even phage. 7. T-odd phage. 8. Threonine. 9. Tensed conformational form of an allosteric enzyme. 10. Tera 11. Tesla. 12. Tocopherol. 13. Ribothymidine. 14. Twisting number.

T₁; T₂; T₃ *See* glucose-6-phosphatase.

T₃ Triiodothyronine.

T₄ Thyroxine.

T₁/₂ Melting out temperature.

tachometer A device for measuring the angular velocity of a rotating shaft. *Var sp* tachymeter.

tachykinins A group of neuropeptides that includes substance P and substance K.

tachyphylaxis A pharmacological phenomenon in which the first, or the first few, doses of a drug produce a response and lead to establishment of resistance so that subsequent doses of the drug fail to elicit any further response.

tactic Of, or pertaining to, taxis.

tactoid A paracrystalline aggregate of molecules which resembles that in a crystal but is limited to one dimension; the resulting linear arrangements of molecules are packed parallel to each other. Descriptive, for example, of the aggregation of sickle cell hemoglobin in sickle cell anemia.

TAF Tumor angiogenesis factor.

tag LABEL.

tail 1. The long fibrous portion of the myosine molecule. 2. The 3′-hydroxyl end of an oligo- or a polynucleotide strand. 3. The elongated, cylindrical structure attached to the head of a T-even phage. 4. The passive portion of a condensing unit. 5. Poly(A) tail. *See also* terminal transferase. 6. The nonpolar, hydrocarbon portion of a fatty acid or a phospholipid molecule. 7. The region at the beginning and at the end of a statistical distribution. 8. The long narrow portion of a sperm that contains many mitochondria and a flagellum and that serves to propel the sperm to the egg.

tail growth *See* tailward growth.

tailing The chromatographic and electrophoretic phenomenon in which a peak appears lopsided and a band, or a spot, appears ill-defined; due to the fact that in the support the front edge of the region that contains the component is sharp while the back edge is diffuse. *See also* leading; trailing.

tail-to-tail condensation The condensation of two molecules by way of their passive ends, as in the condensation of isoprene units to form carotenoids.

tailward growth A polymerization mechanism in which the activated head of a monomer adds to the passive tail of a chain, thereby making its own tail the receptor for the next addition of monomer. *See also* head-to-tail condensation; headward growth. *Aka* tail polymerization.

Taka amylase *See* Taka diastase.

Taka diastase Trademark for an amylase (diastase) preparation from the mold *Aspergillus oryzae* The preparation contains an alpha amylase that consists of a single polypeptide chain (MW 50,000) and that has calcium as a cofactor; the preparation also contains an adenosine deaminase which converts adenosine to inosine and is not involved in the degradation of starch.

Tamm–Horsfall glycoprotein *See* uromodulin.

tandem One behind the other.

tandem duplication A chromosomal aberration in which two identical chromosomal segments lie one behind the other on the same chromosome; the order of genes is the same in both segments.

tandem mass spectrometry A technique in which two or more independently operable mass analyzers are used in sequence. As the ion beam traverses the instrument, it is subjected to a series of operations which involve changes in mass, charge, or reactivity of the ions. *Abbr* MS/MS.

tandem repeats 1. TANDEM DUPLICATION. 2. Multiple copies of a gene that lie on the same chromosome and are separated by spacers; components of repetitive DNA.

tangent 1. A straight line that touches a curve at only one point. 2. The ratio of the length of the side facing an acute angle in a right triangle to the length of the side facing the other acute angle.

Tangier disease A genetically inherited metabolic defect in humans that is characterized by an almost complete absence of plasma HDL as well as by excessive deposition of cholesterol esters in many tissues. *Aka* familial HDL deficiency.

tannic acid *See* tannins.

tannins A group of complex phenolic, non-nitrogenous compounds in the bark, fruits, and leaves of many plants. Tannins can be divided into two groups: (a) Hydrolyzable tannins consist of a sugar (usually glucose), esterified with one or more polyhydric phenols such as gallic acid (3,4,5-trihydroxybenzoic acid). Tannic acid is a hydrolyzable tannin; it has the empirical formula $C_{76}H_{52}O_{46}$. (b) Condensed (nonhydrolyzable) tannins are derivatives of bioflavonoids. Tannins can cross-link proteins, a property made use of in leather tanning; they also frequently have astringent properties.

Tanret's reagent A reagent that contains potassium iodide, mercuric chloride, and acetic acid, and that forms a white precipitate when added to urine containing albumin.

t antigen An antigen that is related to, but smaller than, the T antigen of SV-40 virus. The two proteins have the same N-terminal amino acid sequences but different C-terminal sequences. *Aka* little t.

T antigen 1. An antigen occurring in the nuclei of cells that are infected with, or transformed by, certain oncogenic viruses, such as polyoma virus or SV-40 virus; a virus-encoded protein that is required for viral DNA replication and that has several other functions. *Aka* big T. 2. An antigen that is present on normal human red blood cells but that is unreactive unless the cells are first treated with the enzyme neuraminidase.

tape An informational molecule that functions in replication, transcription, or translation; used specifically for mRNA.

tape theory The theory according to which replication, transcription, and translation proceed by the currently accepted template-type mechanisms.

TAPS Tris(hydroxymethyl) methylaminopropane sulfonic acid; used for the preparation of biological buffers in the pH range of 7.7 to 9.1. *See also* biological buffers.

TAPSO 3-[*N*-Tris(hydroxymethyl)methylamino]-2-hydroxypropane sulfonic acid; used for the preparation of biological buffers in the pH range of 7.0 to 8.2. *See also* biological buffers.

tare A counterweight for balancing a container during weighing.

target cell 1. A receptor cell that binds cytotropic antibodies in anaphylaxis. 2. A receptor cell that is acted upon by a hormone.

targeted mutagenesis The production of mutations that is limited to the actual sites of damage in DNA. The mutations arising at the thymine dimer sites produced by ultraviolet irradiation are an example.

target molecular weight The molecular weight calculated from the dose of radiation that re-

sults in the survival of 37% of the irradiated units. *See also* thirty-seven percent survival dose.

target organ 1. The receptor organ that is acted upon by a hormone. 2. The organ, the cells of which are damaged by the multiplication of an infecting virus.

target sequence A short nucleotide sequence, in the recipient DNA molecule, that functions in transposition. As the transposon is being inserted into the recipient DNA, the target sequence is duplicated and the transposon is sandwiched in between two identical copies of the target sequence.

target theory The theory that a cell will be killed, or an enzyme molecule, virus particle, etc., will be inactivated if struck by radiation in a small, sensitive target volume in which one or more hits (ionizing events) will bring about the specific effect.

T arm *See* arm.

TATA box A nearly universal sequence of nucleotides in eukaryotic DNA, about 25 bp upstream from the site at which transcription of structural genes by RNA polymerase II starts. It has the consensus sequence TATAAAT and is believed to be a site to which transcription factors, rather than RNA polymerase II, bind; hence, it is not considered to be part of the promoter proper. *See also* Pribnow box.

tau *See* elementary particles.

tau particles Intracellular particles of phage-infected bacteria that have been detected by electron microscopy and that are believed to be phage precursors.

tau proteins Proteins that have molecular weights of about 60,000–70,000 and that are associated with, and enhance the polymerization of, microtubules. *See also* MAP.

taurine An aminosulfonic acid, derived from cysteine, that forms a bile salt when it is conjugated to a bile acid; it may also function as a general detoxifier and remover of xenobiotics.

taurocholic acid A compound formed by the conjugation of taurine and cholic acid; one of the bile salts.

tautomer One of the two isomers that exhibit tautomerism.

tautomeric shift A reversible change in the location of a hydrogen atom in a molecule that occurs in enol–keto tautomerism and that converts one tautomer into the other.

tautomerism A rapid equilibrium between two isomeric forms of a molecule that differ significantly in both the relative positions of the atoms and in the bonds between the atoms; the molecule may react in either of the two isomeric forms depending on the conditions.

taxis The movement of a cell or an organism in response to a stimulus such that the latter controls the direction of movement. The movement may be toward, or away from, the stimulus and is then referred to as either positive or negative taxis.

taxon (*pl* taxa) A taxonomic group of any rank or size.

taxonomic Of, or pertaining to, taxonomy.

taxonomy The scientific classification of plants and animals that is based on their natural relationships; includes the systematic grouping, ordering, and naming of the organisms.

Tay–Sachs disease A genetically inherited metabolic defect in humans that is characterized by an accumulation of gangliosides in the brain and that leads to a progressive and fatal disease associated with blindness and brain deterioration; caused by a deficiency of the enzyme hexosaminidase A.

TBG Thyroxine-binding globulin.

t-Boc group Tertiary butyloxycarbonyl group; the grouping $(CH_3)_3C—O—CO—$, used as a blocking agent of the amino group of amino acids in solid phase protein synthesis.

TBPA Thyroxine-binding prealbumin.

TC Transcarboxylase.

TC$_{50}$ Median tissue culture dose.

TCA 1. Tricarboxylic acid. 2. Trichloroacetic acid.

TCA cycle Tricarboxylic acid cycle.

TψC arm *See* arm.

TCD$_{50}$ Median tissue culture dose.

TCDD *See* dioxin.

TC detector Thermal conductivity detector.

T cell A thymus-derived lymphocyte. In both mammals and birds, hemopoietic stem cells (bone marrow cells) migrate via the blood to the thymus where they differentiate into thymus lymphocytes. These then migrate to peripheral lymphoid tissues to become T cells. T cells function in cell-mediated immune responses against invading microorganisms, including fungi, parasites, intracellular viruses, cancer cells, and foreign tissues. *See also* cytotoxic T cells; regulatory T cells.

T cell growth factor INTERLEUKIN 2.

T cell helper *See* helper T cell.

T cell line A line of antigen-specific T cells that can be produced in tissue culture by activating T cells with a specific antigen and then having them proliferate indefinitely in the presence of interleukin 2.

TCID$_{50}$ Tissue culture infectious dose.

TψC loop *See* arm.

TCT Thyrocalcitonin.

t$_D$ 1. Doubling time. 2. Thermal death time.

TD$_{50}$ Median toxic dose.

T-DNA *See* crown gall tumor.

TDP 1. Ribosylthymine diphosphate. 2. Ribosylthymine-5′-diphosphate.

TEAE-cellulose *O*-(Triethylaminoethyl)cellulose; an anion exchanger.

TEBG Testosterone–estradiol–binding globulin.

Teflon Trademark for polytetrafluoroethylene; a plastic.

Teichmann's crystals Rhombic crystals, produced by heating hemoglobin in the presence of NaCl and glacial acetic acid; used for the microscopic detection of blood. *Aka* chlorhemin crystals.

teichoic acid An accessory polymer in the cell wall of gram-positive bacteria that consists of long chains of either glycerol or ribitol which are linked by means of phosphodiester bonds and which carry various substituents, including both amino acids and monosaccharide residues.

teichuronic acid A teichoic acid-type polymer that contains uronic acid substituents.

tektin SPECTRIN.

telecommunication The transmission of data between a computer and another computer, or a terminal, in a different location by telephone, radio, or other means.

teleology The doctrine that the occurrence of any structure or function in a living organism implies that the structure or function is valuable, has a purpose, and has conferred an advantage on the organism in the course of evolution. As originally formulated by Aristotle, the concept invoked a supernatural agent that has foreseen the final value of the structure or of the function. To circumvent this connotation, the term teleonomy has been introduced. *See also* teleonomy.

teleolysosome A cytoplasmic particle containing acid hydrolases.

teleonomy The doctrine that the occurrence of any structure or function in a living organism implies that the structure or function has conferred an advantage on the organism in the course of evolution; the genetic character of an organism is considered to have become adapted to its environment through the process of evolution rather than through theurgic forces. *See also* teleology.

telestability The degree to which the stability of one region of the DNA double helix is affected by an adjacent region; specifically the stabilization of a dA-dT segment by an adjacent dG-dC segment.

TELISA Thermometric enzyme linked immunosorbent assay; a variation of the ELISA technique. Involves measurement of heat formation, as a result of enzyme action, by means of an enzyme thermistor unit.

telomere One of the two terminal chromomeres of a chromosome; a DNA sequence required for the stability of the ends of eukaryotic chromosomes.

telopeptides Sequences of about 12–25 amino acids that occur at the N- and C-terminals of the tropocollagen molecule.

telophase The final stage in mitosis during which the two new nuclear membranes are formed and which is followed by cytokinesis to produce the two daughter cells.

TEM 1. Tetramine. 2. Transmission electron microscope.

TEMED N,N,N,N-Tetramethylethylenediamine; an initiator of acrylamide polymerization.

temperate cycle LYSOGENIC CYCLE.

temperate phage A phage that is incorpoated as a prophage into the host chromosome and allows the host cell to survive in the form of a lysogenic bacterium; it generally does not cause lysis of the bacterial cell that it has infected but may cause lysis (undergo a lytic cycle) under certain conditions.

temperature The degree of hotness measured on one of several arbitrary scales; a measure of the kinetic energy of molecules. *See also* temperature scale.

temperature coefficient The rate of a reaction at one temperature divided by the rate at a second temperature; usually expressed as the Q_{10} value.

temperature jump method A relaxation technique in which temperature is the variable that disturbs the equilibrium of a system. *See also* relaxation technique.

temperature programming The controlled increase in temperature at a predetermined rate; used in the construction of thermal denaturation profiles and in the operation of gas chromatography columns.

temperature scale *See* absolute temperature scale; Celsius temperature scale; Fahrenheit temperature scale; Kelvin temperature scale.

temperature-sensitive mutant A conditional mutant that behaves normally at a lower (permissive) temperature range but displays its mutant phenotype at a higher (restrictive) temperature range. *Abbr* ts.

template A macromolecule that functions as a mold or pattern for the synthesis of another macromolecule. A template determines the composition of the product and directs its synthesis during the process of polymerization but does not have to be linked covalently to the product. In nucleic acid chemistry, a template is the polynucleotide strand whose base sequence determines the base sequence of the newly synthesized strand.

template chromatography A column chromatographic technique for the fractionation of oligonucleotides. Based on linking defined oligonucleotides covalently to soluble polymers and then fixing the complex onto a column

(such as DEAE-cellulose) in a noncovalent manner. This leaves the nucleotide groups on the column free for interaction with complementary oligonucleotides in the mobile phase. The complementary oligonucleotides become adsorbed by base pairing and can be eluted by means of a temperature gradient.

template–primer A macromolecule that functions as both a template and a primer. A double-stranded DNA molecule that is replicated in vitro with DNA polymerase I is an example; one strand functions as a template, the other is used as both a template and a primer.

template RNA MESSENGER RNA.

template switching The alternate replication of the two strands of DNA in the Kornberg mechanism; DNA polymerase I uses one strand as a template and then switches (jumps) to use the second, displaced strand as a template.

template theory *See* antigen template theory.

TEMPO 2,2,6,6-Tetramethylpiperidine-1-oxyl; used as a spin label.

tensed conformation *See* concerted model.

tensiometer An instrument for measuring surface and interfacial tensions.

tension The concentration of a gas in a solution; expressed in terms of the partial pressure of the gas which is in equilibrium with the solution.

TEPA Generic name for Tris(1-aziridinyl)phosphine oxide; triethylenephosphoramide; a phosphine oxide containing three aziridine (R) groups (a). An aziridine mutagen.

$$O{=}P{-}R$$

with R groups above (R), to the right (R), and below (R)

(a)

teprotide A nonapeptide that is an antagonist of serum converting enzyme which catalyzes the conversion of angiotensin I to angiotensin II.

ter- 1. Combining form meaning three or thrice. 2. Referring to three kinetically important substrates and/or products of an enzymatic reaction; thus a ter bi reaction is a reaction with three substrates and two products.

tera- Combining form meaning 10^{12} and used with metric units of measurement. *Sym* T.

teratogen An agent that causes birth defects.

teratogeny The origin and production of congenital malformations. *Aka* teratogenesis.

teratology The study of malformations and of the abnormal developments of organisms and parts of organisms.

teratoma A tumor derived from embryonic tissue and consisting of a disorganized mass of many tissue types. Many cells represent a variety of differentiated tissues; others are undifferentiated stem cells that continue to divide to generate yet more differentiated tissues.

ter cutting *See* terminase.

terminal 1. The end of a polymeric chain, such as the C- or N-terminal of a polypeptide chain, or the 3'- or 5'-terminal of a polynucleotide strand. *Aka* terminus. 2. A piece of equipment that has a keyboard for input and a device, such as a printer, for output; used for communicating with a computer.

3'-terminal That end of an oligo- or polynucleotide strand which carries either a free or a phosphorylated hydroxyl group at the 3'-position of the terminal ribose or deoxyribose.

5'-terminal That end of an oligo- or polynucleotide strand that carries either a free or a phosphorylated hydroxyl group at the 5'-position of the terminal ribose or deoxyribose.

terminal cisterna (*pl* terminal cisternae) A transverse, connecting channel of the sarcoplasmic reticulum.

terminal deletion A chromosomal aberration in which genetic material is lost from the end of a chromosome. A deletion occurring elsewhere on the chromosome is called an intercalary deletion.

terminal deoxynucleotidyl transferase TERMINAL TRANSFERASE.

terminal enzyme An enzyme that catalyzes the addition of nucleotides to the terminal of a nucleic acid strand in the absence of a template, with nucleoside triphosphates serving as substrates.

terminal redundancy The occurrence of identical sequences at both ends of a DNA molecule. *Aka* terminal repetition. *See also* long terminal repeat.

terminal repetition TERMINAL REDUNDANCY.

terminal transferase An enzyme that catalyzes the sequential addition of deoxyribonucleotides to the 3'-OH end of a DNA strand without any requirement for a template. The enzyme is used to add homopolymer tails. Any two populations of DNA fragments can be made complementary by appropriate homopolymer extension of their 3'-ends, followed by annealing and ligation; this procedure is used in recombinant DNA technology. *Aka* terminal deoxynucleotidyl transferase.

terminase An enzyme that functions in the rolling circle (sigma) replication of phage lambda. The enzyme cuts the phage DNA at cos sequences (ter cutting) and thereby initiates the packaging of the DNA chromosome into new capsids. *Aka* ter system.

termination 1. The final stage in protein synthesis during which the completed polypeptide chain is released from the ribosome, and the ribosome is released from the mRNA. 2. The third step in a chain reaction. 3. The final stage in the transcription of DNA by RNA polymerase.

termination codon A codon that codes for termination of the translation of an mRNA molecule; it signals the termination of a growing polypeptide chain. Termination codons do not code for amino acids and are the codons UAA, UAG, and UGA. *Aka* stop codon; nonsense codon.

termination factor RELEASE FACTOR.

terminator A sequence of nucleotides in DNA that provides a signal for the termination of transcription by RNA polymerase; it signals the termination of a growing RNA stand. *Aka* terminator sequence.

terminology NOMENCLATURE.

terminus (*pl* termini) TERMINAL.

termolecular reaction A chemical reaction in which three molecules (or other entities) of reactants interact to form products.

ternary Consisting of three parts.

ternary complex mechanism SEQUENTIAL MECHANISM.

ternary complex model MOBILE RECEPTOR MODEL.

terpene A hydrocarbon terpenoid. Terpenes are classified according to the number of isoprene units that they contain: hemiterpene (1); monoterpene (2); sesquiterpene (3); diterpene (4); triterpene (6); tetraterpene (8); pentaterpene (10); and polyterpene (large number).

terpene alkaloids *See* alkaloids.

terpeneless oil An essential oil from which less odorous terpenes have been removed by deterpenation.

terpenoid A polyisoprenoid compound that may be linear or cyclic and in which the isoprene units are usually linked in a head-to-tail manner. *See also* terpene.

terramycin *See* tetracycline.

terreactant reaction A termolecular reaction in which three different reactants interact to form products.

ter system TERMINASE.

tert Tertiary.

tertiary base pairs A group of nine base pairs that occur in the folded cloverleaf structure (L type) of tRNA and that are responsible for maintaining the tRNA in its three-dimensional configuration. Eight of the base pairs are linked via tertiary hydrogen bonds; one is a standard Watson–Crick type base pair (a G-C base pair).

tertiary coiling *See* superhelix.

tertiary hydrogen bonds Unusual hydrogen bonds that are formed between various hydrogen bond donor and acceptor groups in the folded cloverleaf structure (L type) of tRNA. These are not the ordinary hydrogen bonds of base pairs in double-helical RNA segments; rather, these are bonds that serve to maintain the tRNA in its folded tertiary structure.

tertiary interactions *See* tertiary hydrogen bonds.

tertiary response The immune response to a third dose of antigen that is essentially similar in its characteristics to those of the secondary response.

tertiary structure The irregular three-dimensional folding of the polypeptide chain upon itself, as in a globular protein, that results from the interaction of amino acid side chains which are either close or far apart along the chain; the arrangement in space of all the atoms of a protein or of a subunit without regard to the relation of the atoms to neighboring molecules or subunits. The term may likewise be applied to the three-dimensional structure of a polynucleotide strand.

TES *N*-Tris(hydroxymethyl)methyl-2-amino-ethanesulfonic acid; used for the preparation of biological buffers in the pH range of 6.8 to 8.2. *See also* biological buffers.

tesla A unit of magnetic field strength; one tesla equals 10^4 gauss. *Sym* T.

testectomy The surgical removal of a testis.

test meal A food that consists of one or more selected items and that, after it has been eaten, is followed by the removal of the gastric contents for analysis.

test of significance *See* significance of results.

testosterone A steroid hormone that is the major male sex hormone in humans; it is secreted by the testes and is responsible for the development of secondary male characteristics. Testosterone functions as a prohormone when it is converted to 17β-estradiol by the enzyme aromatase.

testosterone–estradiol–binding globulin A specific carrier for testosterone in the blood. *Abbr* TEBG. *Aka* sex hormone binding globulin (SHBG).

tetra- Combining form meaning four.

tetraantennary *See* high mannose oligosaccharides; complex oligosaccharides.

tetracyclines A group of broad-spectrum antibiotics obtained from various species of *Streptomyces*. Tetracyclines contain four, linearly fused, six-membered rings; they inhibit protein synthesis in prokaryotes by binding to the ribosomes and preventing the normal binding of aminoacyl-tRNA to the A site. Chlortetracycline (aureomycin) is produced by *Streptomyces aureofaciens*; oxytetracycline (ter-ramycin) is produced by *Streptomyces rimosus*; and tetracycline (achromycin; tetracyn) is obtained by reductive dehalogenation of chlortetacycline.

tetracyn *See* tetracyclines.

tetrahedral intermediate A transition state in enzyme catalyzed reactions in which a specific carbon has the four bonds in their usual tetrahedral arrangement.

tetrahedron A polyhedron with four equal faces; a pyramid composed of four triangles. A tetrahedron is used for representation of the carbon atom, with the nucleus of the atom inside the pyramid, and with the four single bonds extending to the corners of the pyramid.

tetrahydrocannabinol The psychotropic principle of hashish and marijuana. A narcotic that occurs in several isomeric forms; Δ'-tetrahydrocannabinol is abbreviated as THC.

tetrahydrofolic acid The reduced form of the vitamin folic acid and the parent compound of the coenzyme forms of this vitamin; the coenzymes function in the metabolism of one-carbon fragments. *Abbr* FH_4; THFA; THF.

tetrahydropteroylglutamic acid TETRAHYDROFOLIC ACID.

tetrahymena A genus of ciliate protozoans widely used for genetic studies and found in a variety of freshwater habitats.

tetramer An oligomer that consists of four monomers.

tetramine An aziridine mutagen that contains three aziridine groups attached to the three carbon atoms in a 6-membered ring in which carbon and nitrogen alternate; triethylene-melamine; 2,4,6-tris (1-aziridinyl)-*s*-triazine. *Abbr* TEM.

tetranucleotide hypothesis An early hypothesis of the structure of DNA according to which DNA has a uniform structure that is produced by the polymerization of one basic repeating unit, namely a tetranucleotide containing one residue each of adenine, cytosine, guanine, and thymine.

tetrose A monosaccharide that has four carbon atoms.

T-even phage A large phage that infects the bacterium *E. coli* and that has a complex tadpole-like structure consisting of a head, tail, and tail fibers. The head is icosahedral and contains double-stranded DNA; to the head is attached the tail through which the DNA is ejected into the host cell. The tail terminates in tail fibers by means of which the phage adsorbs to the bacterium. The tail core, or tube, is surrounded by a sheath and is attached to the head by means of a collar and to the tail fibers by means of a plate. Tail pins are attached to the tail plate.

text 1. The sequence of nucleotides in a nucleic acid. 2. The sequence of amino acids in a protein.

TF Transfer factor.

T factor ELONGATION FACTOR T.

Tg Generation time.

TGF Transforming growth factor.

TGF-β Transforming growth factor β. A widely distributed peptide growth factor that can stimulate or inhibit the growth of cells in vitro. The specific physiological function of TGF-β is unknown but it appears to be involved in the development of mammary glands.

TGFA Triglyceride fatty acids.

TGN Trans-Golgi network.

TH Transhydrogenase.

thalassemia A heritable disorder characterized by a reduced rate of synthesis of one or more of the globin chains of hemoglobin. The imbalance in globin chain production leads to precipitation of the excess chains, lowered hemoglobin levels (anemia), and reduced red blood cell survival. In α or β thalassemia there is a deficiency of the α or β globin chains, respectively. Absence of a chain is designated as α^0 or β^0, respectively; decreased synthesis of a chain is designated as α^+ or β^+, respectively. Homozygotes, with α^0 thalassemia exhibit a syndrome called Hydrops fetalis in which death occurs prior to or shortly after birth. The homozygous condition is known as thalassemia major and the heterozygous condition is known as thalassemia minor. *Aka* Cooley's anemia.

thalidomide A drug (2,6-dioxo-3-phthalimido-piperidine), formerly used as a sedative but since withdrawn from the market since it led to fetal abnormalities.

thallus The undifferentiated growth of a plant body as that of a fungus or a mold.

THAM TRIS.

thanatochemistry The chemical reactions that occur in a tissue or an organism after death. *Aka* mortichemistry.

thaumatin A basic and carbohydrate-free protein, consisting of a single polypeptide chain of 207 amino acids. The protein has very high specificity for sweet taste receptors so that it is approximately 100,000 times sweeter than sugar on a molar basis.

THC Tetrahydrocannabinol.

Thd Ribothymidine.

theine The name frequently given to the stimulatory substance in tea; it is chemically identical to the caffeine in coffee beans.

theobromine A methylated xanthine that occurs in tea and cacao beans. A purine alkaloid that acts as a diuretic, smooth muscle relaxant, cardiac stimulant, and vasodilator.

theophylline A methylated xanthine that is present in tea. A purine alkaloid that has a stimulatory effect and acts like caffeine; it also inhibits the phosphodiesterase that converts cyclic AMP to inactive 5'-AMP and thereby prolongs the adrenalin-producing effect of cyclic AMP.

Theorell–Chance mechanism An ordered sequential mechanism of an enzymatic reaction in which two substrates and two products participate and in which the steady-state concentration of the ternary complexes is very low. The two ternary complexes in this mechanism are the enzyme plus the two substrates, and the enzyme plus the two products.

theoretical plate The theoretical stage in countercurrent distribution at which perfect equilibrium is established between the two phases. The theoretical plate concept has been adapted for use in distillation and chromatographic columns where perfect equilibrium between the phases is not established, since the phases are constantly in motion relative to each other. Hence, it is customary to refer to the length of column over which the separation effected is equivalent to that of a theoretical plate; this length of column is known as a height equivalent to a theoretical plate (HETP). The shorter the HETP, the more efficient is the column. *Aka* theoretical stage.

theory A confirmed explanation of observed phenomena; a scientific doctrine; a hypothesis that has been tested and confirmed with facts not known when the hypothesis was first proposed.

theory of absolute reaction rates The theory of chemical kinetics according to which the velocity of a chemical reaction is proportional to the concentration of an activated complex that is formed from the reactants; the reactants must first be activated by means of an activation energy to form the activated complex before they can be converted into products. The activated complex is a transient phase; an unstable complex held together by weak bonds.

theory of antibody formation *See* clonal selection theory; Ehrlich's receptor theory; germline theory; instructive theory; selective theory; side chain theory; Smithie's theory; somatic mutation theory; somatic recombination theory.

theory of cancer *See* autocrine hypothesis; Busch theory; catabolic deletion hypothesis; chromosome theory of cancer; deletion hypothesis; feedback deletion hypothesis; Greenstein hypothesis; imbalance theory; Mason's theory; membron theory of cancer; multistep induction theory; oncogene theory;

polyetiological theory; provirus hypothesis; somatic mutation theory; virogene theory; Warburg theory. *See also* initiation (2); promotion.

therapeutic index A measure of the safety of a drug that is equal to the ratio of the median lethal dose to the median effective dose.

thermal Of, or pertaining to, either heat or temperature.

thermal chromatography A column chromatographic technique in which elution is carried out with one eluent but at increasing temperatures.

thermal conductivity detector A detector in which the basic component is a thermal conductivity cell of either the thermistor or the katharometer type; used in gas chromatography, primarily for detecting inorganic gases or large amounts of organic compounds. *Abbr* TC detector.

thermal death point The lowest temperature required for the sterilization of a standard suspension of bacteria in 10 min.

thermal death time The minimum time required for the sterilization of a standard suspension of bacteria at a given temperature. Sym t_D.

thermal denaturation The denaturation of macromolecules by heat, particularly the heat-induced breaking of the hydrogen bonds of the base pairs in double-helical nucleic acid segments which leads to the formation of random coils.

thermal denaturation profile The plot of a hydrodynamic property, such as viscosity, or of an optical property, such as absorbance, for a nucleic acid solution as a function of temperature. The curve describes the degree of separation between the two strands of double-stranded molecules, or between parts of the same strand of folded single-stranded molecules, that results from the breaking of hydrogen bonds as the temperature is increased. The temperature at which one-half of the maximum change in the measured property is observed is referred to as the melting out temperature and is denoted T_m.

thermal inactivation point The temperature at which a suspension of virus particles must be kept for 10 min to inactivate all of the virus particles; used specifically for plant viruses such as tobacco mosaic virus. *Abbr* TIP.

thermalite A subtransition in the thermal denaturation profile of DNA.

thermal neutron A neutron that has a kinetic energy that is equivalent to the energy of a gas molecule at room temperature.

thermal noise DARK CURRENT.

thermal polymer A high molecular weight compound produced by the heat-induced

polymerization of monomers.

thermal quenching CHEMICAL QUENCHING.

thermal reactivation The increase in the survival of ultraviolet-irradiated bacteria that is manifested if the bacteria are incubated at an elevated temperature immediately following their irradiation and are then allowed to grow at a lower temperature; this is in comparison to those bacteria that, following their irradiation, are allowed to grow only at the lower temperature.

thermine A polyamine, isolated from the extreme thermophile *Thermus thermophilus*, that has the chemical structure $NH_2—(CH_2)_3—NH—(CH_2)_3—NH—(CH_2)_3—NH_2$.

thermistor A semiconductor device, the resistance of which changes with temperature; used for measuring temperatures.

thermo- Combining form meaning heat.

thermoacidophiles A group of archaebacteria that grow in hot, acidic environments. They have been isolated from hot sulfur springs and smoldering piles of coal tailings. *See also* archaebacteria.

thermobarometer A control manometer used in conjunction with the Warburg apparatus to correct for changes in room temperature and barometric pressure during the experiment.

thermochemistry A branch of thermodynamics that deals specifically with enthalpy changes of chemical reactions. Based on two statements that follow from the first law of thermodynamics and that are known as the laws of thermochemistry. (a) Law of Lavoisier and Laplace: The quantity of heat required to decompose a compound into its elements is equal to the quantity of heat evolved when the compound is formed from its elements. (b) Hess's law (law of constant heat summation): The total heat change of a chemical reaction, at either constant volume or constant pressure, is the same whether this reaction takes place in one or several steps.

thermochromism The reversible change of color by a compound as a function of temperature.

thermocouple A device for measuring temperatures by the production of an electromotive force between the two junctions of two different metals in a closed circuit; the electromotive force is proportional to the temperature difference between the two junctions.

thermoduric Heat-enduring; capable of surviving high temperatures.

thermodynamic property STATE FUNCTION.

thermodynamics The science that deals with the interconversion of different forms of energy and with the spontaneous direction of processes; involves the study of heat, work, and energy, their interconversions and the changes that they bring about. Thermodynamics is based upon fundamental laws that are generalizations derived from human experience. These laws cannot be proven in an exact or direct way; their truth is ascertained by inference, that is, from the fact that, so far, all of the consequences derived from these laws, have been verified experimentally. Classical thermodynamics (also called energetics or equilibrium thermodynamics) deals with the bulk properties of macroscopic systems at equilibrium; it considers the initial and final states of a system and its surroundings. Statistical thermodynamics deals with individual atoms and molecules and sums up their collective behavior and properties. Irreversible thermodynamics deals with nonequilibrium systems and with rate processes. *See also* first law of thermodynamics; second law of thermodynamics; third law of thermodynamics; zeroth law of thermodynamics.

thermodynamics of irreversible processes *See* irreversible thermodynamics.

thermodynamic variable STATE FUNCTION.

thermogenic Heat-producing.

thermolabile Inactivated by treatment with heat; sensitive to high temperatures.

thermolysin A zinc-containing, heat-stable, proteolytic enzyme from the thermophilic bacterium *Bacillus thermoproteolyticus* that catalyzes the cleavage of peptide bonds in which the NH-group is contributed by one of several nonpolar amino acids. *Aka* thermophilic bacterial protease.

thermolysis PYROLYSIS.

thermometer A device for measuring temperatures; usually implies either a mercury-in-glass or an alcohol-in-glass device.

thermonuclear reaction A nuclear reaction that is induced by a thermal activation of the reacting nuclei.

thermophile An organism that grows at high temperatures in the range of 45 to 70°C (or higher temperatures) and that has an optimum growth temperature in the range of 50 to 55°C.

thermophilic Of, or pertaining to, thermophiles; preferring high temperatures.

thermopile An instrument for measuring the total amount of radiant energy irrespective of its wavelengths; based on the principle that a truly black surface absorbs all of the incident light energy and converts all of it into heat.

thermoplastic polymer A polymer that, when heated to its melting point, softens and flows and can be cooled and remelted many times without undergoing any change. *See also* thermosetting polymer.

thermosetting polymer A polymer that, when heated to its melting point, undergoes a permanent change and sets to a solid which cannot be remelted. *See also* thermoplastic polymer.

thermostable Stable to treatment by heat; not inactivated by high temperatures.

thermostable enzyme An enzyme that has an unusually high optimum temperature.

thermostat A device for maintaining a constant temperature by means of the automatic regulation of the source of heat.

thermotolerance Resistance to heat; the ability of an organism to withstand high temperatures after exposure to lower, but still elevated, temperatures which induces the formation of heat-shock proteins.

thermotropic liquid crystal *See* liquid crystal.

theta replication The bidirectional replication of double-stranded, circular DNA; so called because the structure resembles the Greek letter theta (θ).

theta structure The DNA structure formed in circular double-stranded DNA, by two replicating forks moving in opposite directions (bidirectional replication). This is the case for bacterial DNA. *Aka* Cairns molecule.

THF 1. Thymic humoral factor. 2. Tetrahydrofolic acid.

THFA Tetrahydrofolic acid.

thiamine One of the B vitamins (vitamin B_1), the coenzyme form of which is thiamine pyrophosphate, and a deficiency of which causes beriberi. *Var sp* thiamin.

thiamine pyrophosphate The coenzyme form of the vitamin thiamine that functions in the decarboxylation of α-ketoacids. *Abbr* TPP; ThPP; DPT.

thiazole ring The ring structure occurring in thiamine and thiamine pyrophosphate.

thick filament MYOSIN FILAMENT.

thin filament ACTIN FILAMENT.

thin-layer chromatography A chromatographic technique in which the stationary phase is a thin layer of solid, such as silica gel, spread on a flat glass or plastic plate; the technique allows for rapid analysis of very small amounts of sample. *Abbr* TLC.

thin-layer electrophoresis An electrophoretic technique in which the supporting medium is a thin layer of silica gel, alumina, or other adsorbent spread on a flat glass or plastic plate; the technique allows for rapid analysis of very small amounts of sample. *Abbr* TLE.

thin-layer gel filtration A separation technique in which gel filtration is carried out on thin layers of gel spread on a flat glass or plastic plate.

thio- Combining form meaning sulfur.

thioalcohol THIOL.

thiochrome A blue, fluorescent compound that is produced by mild alkaline oxidation of thiamine and that serves for the quantitative determination of thiamine.

thioclastic reaction A phosphoroclastic reaction in which lipoic acid participates. *Aka* thioclastic split.

thioctic acid LIPOIC ACID.

thioester A compound formed by the joining of a carboxyl group to a sulfhydryl group through the elimination of a molecule of water; a compound having the general formula R—CO—S—R'.

thioether A compound in which two radicals are linked by means of a sulfur atom; a compound of the general formula R—S—R'.

thioglycollic acid treatment A procedure for breaking disulfide bonds in proteins.

thiokinase The enzyme that catalyzes the formation of a fatty acyl coenzyme A ester from the fatty acid and coenzyme A in fatty acid activation. *Aka* acyl-CoA synthetase.

thiol A compound containing a sulfhydryl group; a compound of the general formula R—SH.

thiolase The enzyme that catalyzes the thiolytic cleavage reaction in beta oxidation.

thiolate The anion of a thiol; a nucleophile that has the structure R—S$^{\bar{\cdot}}$.

thiolation The introduction of a sulfhydryl group into an organic compound.

thiol enzyme SH-ENZYME.

thiol ester THIOESTER.

thiol ether THIOETHER.

thiol group SULFHYDRYL GROUP.

thiolysis THIOLYTIC CLEAVAGE.

thiolytic cleavage The last step, catalyzed by the enzyme thiolase, in the cyclic reaction sequence of beta oxidation; in this step, a fatty acyl coenzyme A derivative, in the presence of coenzyme A, is cleaved to produce acetyl coenzyme A and a fatty acyl coenzyme A of a fatty acid that contains two carbon atoms less than the fatty acid that entered the cycle.

thiomethylgalactoside *See* TMG.

thiophorase The enzyme that catalyzes the conversion of an acyl coenzyme A ester and a free fatty acid to the opposite pair of acyl coenzyme A ester and free fatty acid; this reaction represents an alternative reaction for fatty acid activation to that catalyzed by fatty acid thiokinase.

thioredoxin A heat-stable protein that can exist as a dithiol, thioredoxin-$(SH)_2$, or as a disulfide, thioredoxin-S_2, and that serves to reduce ribonucleoside diphosphates to deoxyribonucleoside diphosphates.

thioredoxin reductase The enzyme that catalyzes the reduction of thioredoxin-S_2 to thioredoxin-$(SH)_2$ with the simultaneous oxidation of NADPH to NADP$^+$; the enzyme is an FAD-containing flavoprotein.

thiostrepton An antibiotic that prevents protein synthesis in prokaryotes by inhibiting the enzyme translocase.

thiotemplate mechanism A mechanism for the synthesis of naturally occurring, but unusual, peptides (such as cyclic peptides, those containing D-amino acids or uncommon amino acids, or those containing covalent bonds other than peptide bonds). The synthesis involves activation of an amino acid by ATP, yielding an aminoacyl adenylate. The amino acid is then bound to an enzyme complex through the sulfhydryl group of the cofactor 4'-phosphopantetheine, rather than being linked to tRNA. Covalent linkage of the amino acid residues occurs on this enzyme complex and the sequence of the amino acids is determined by the enzyme.

Thio-TEPA Tradename for triethylenethiophosphoramide; a mutagenic alkylating agent. The compound contains three aziridine groups (R) and has the structure

$$S=P-R$$

with R groups attached.

thioracil A minor base occurring in tRNA in which a sulfur atom has replaced an oxygen atom at position 2 or 4 in uracil.

third factor NATRIURETIC HORMONE.

third law of thermodynamics The principle relating to the magnitude of entropy: All substances have finite positive entropies and all simple, crystalline substances at a temperature of absolute zero have equal entropies which are assigned a value of zero. *See also* thermodynamics; principle of unattainability of absolute zero.

third-order reaction A chemical reaction in which the velocity is proportional to the product of three concentration terms involving one, two, or three different reactants.

thirty-seven percent survival dose The dose of radiation that results in 37% survival of the irradiated units. The 37% survival value corresponds to an average of one lethal hit per sensitive target and can be used for calculations of target molecular weights. The number 37 is based on exponential inactivation kinetics, since $e^{-1} = 0.37$.

thixotropy The property of some gels of undergoing a reversible, isothermal, gel–sol transformation upon agitation; the solid gel becomes a liquid sol as a result of the agitation.

ThPP Thiamine pyrophosphate.

Thr 1. Threonine. 2. Threonyl.

three-carbon plants *See* C$_3$ plants.

three-factor cross A series of genetic crosses, involving three nonallelic linked genes, that is used primarily for purposes of chromosomal mapping.

three-point attachment POLYAFFINITY THEORY.

three-point cross THREE-FACTOR CROSS.

three-point landing POLYAFFINITY THEORY.

threo configuration The configuration of a compound in which two asymmetric carbon atoms have identical substituents on opposite sides, as in the configuration of threose.

threonine An aliphatic, polar alpha amino acid that contains an alcoholic hydroxyl group. *Abbr* Thr;T.

threonine dehydratase An enzyme that catalyzes the deamination of threonine to α-ketobutyric acid; an inducible enzyme in the liver of higher organisms. *Aka* threonine dehydrase; threonine deaminase.

threshold 1. A measure of the ability of the kidney to absorb a substance from the blood. *See also* threshold of appearance; threshold of retention. 2. A measure of the sensitivity of the eye to light. *See also* cone threshold; rod threshold; visual threshold.

threshold dose The smallest dose of a radiation above which the radiation produces a detectable effect.

threshold limit value The airborne concentration of a chemical to which humans may be exposed day after day in their working environment without suffering adverse effects. *Abbr* TLV.

threshold of appearance The plasma concentration of a substance above which the substance cannot be fully absorbed by the kidney and appears in the urine. *Aka* threshold of excretion.

threshold of retention The plasma concentration of a substance when it is identical to the concentration of the substance in the urine; at higher plasma concentration values the urine will be more concentrated in the substance than the plasma, whereas at lower plasma concentration values the plasma will be more concentrated in the substance than the urine.

threshold potential That portion of the characteristic curve of a Geiger–Mueller counter that follows the starting potential and at which the curve begins to level off to a plateau.

threshold substance A substance that appears in the urine in substantial amounts only when its concentration in the plasma exceeds a certain value.

thrombin The proteolytic enzyme that functions in blood clotting by catalyzing the hydrolytic cleavage of fibrinopeptides A and B from fibrinogen, thereby converting fibrinogen to fibrin.

thrombocyte PLATELET.

thrombogen PRCTHROMBIN.

thrombokinase THROMBOPLASTIN.

thromboplastin An accessory lipoprotein factor that is released from blood platelets in injured tissue and that is involved in initiating the sequence of reactions leading to the formation of a blood clot in the extrinsic pathway of blood clotting. *Aka* tissue thromboplastin.

thromboplastinogen ANTIHEMOPHILIC FACTOR.

thrombosis The occlusion of a blood vessel by formation of a blood clot (thrombus).

thrombosthenin A contractile protein in blood platelets that is similar to G-actin.

thromboxanes A group of substances, derived from arachidonic acid, that differ from prostaglandins in their ring structure; thromboxanes contain a 6-membered, cyclic ether (oxane ring). Thromboxanes were first isolated from thrombocytes (blood platelets) and stimulate platelet aggregation and smooth muscle contraction. These effects are opposite those produced by prostacyclin. *Abbr* TX.

thrombus (*pl* thrombi) A blood clot formed on site within a blood vessel or within the heart. *See also* embolus.

Thunberg technique A technique for the estimation of dehydrogenase activity by a photometric measurement of the reduction of methylene blue.

Thunberg tube A tube, used in the Thunberg technique, that can be evacuated and that has a side arm for the addition of reagents.

Thx Thyroxine.

Thy Thymine.

Thy-1-antigen A cell membrane antigen of thymus lymphocytes that can be used to distinguish them from other lymphocytes.

thylakentrin FOLLICE-STIMULATING HORMONE.

thylakoid disk One of a large number of flattened vesicles that contain the photosynthetic pigments and the electron carriers of chloroplasts and that are stacked to form the grana of the chloroplasts.

thymectomy The surgical removal of the thymus.

thymic Of, or pertaining to, the thymus.

thymic humoral factor A heat-labile polypeptide from the thymus that may be a hormone; it restores several T-cell immunological responses in in vivo and in vitro assays. *Abbr* THF.

thymidine The deoxyribonucleoside of thymine. Thymidine mono-, di-, and triphosphate are abbreviated, respectively, as dTMP, dTDP, and dTTP. The abbreviations refer to the 5'-nucleoside phosphates unless otherwise indicated. *Abbr* Thd;T.

thymidine factor SOMATOMEDIN.

thymidine kinase The enzyme that catalyzes the phosphorylation of thymidine to thymidine monophosphate and that is believed to play a major role in the control of DNA synthesis. The level of the enzyme increases markedly during infection with certain viruses.

thymidylate kinase The enzyme that catalyzes the phosphorylation of thymidine monophosphate to thymidine diphosphate and the phosphorylation of thymidine diphosphate to thymidine triphosphate.

thymidylate synthetase The enzyme that catalyzes the methylation of deoxyuridine-5'-monophosphate (dUMP) to thymidine-5'-monophosphate (dTMP). *Aka* thymidylate synthase.

thymidylic acid The deoxyribonucleotide of thymine.

thymin THYMOSIN.

thymine The pyrimidine, 5-methyl-2,3-dioxypyrimidine, that occurs in DNA. *Abbr* T;Thy.

thymine dimer A pyrimidine dimer, produced by ultraviolet irradiation of DNA, in which two adjacent thymines in a DNA strand form a dimer and block the replication of the DNA at that point.

thymineless death The death of bacteria that results from a lack of thymine; the absence of thymine leads to breaks in single strands of DNA and thereby blocks the synthesis of DNA. Thymineless death can be produced by depriving a thymine auxotroph of thymine or by exposing a bacterial culture to 5-fluorouracil which prevents the synthesis of deoxythymidylic acid. *Abbr* TLD.

thymineless mutant *See* -less mutant.

thymine starvation *See* starvation (2).

thymocyte A lymphocyte that occurs within the thymus gland.

thymol A bactericidal compound; 5-methyl-2-isopropylphenol.

thymol turbidity test A liver function test that is based on the production of turbidity when serum from individuals with one of several forms of hepatitis is treated with a barbiturate buffer that contains thymol.

thymonucleic acid Obsolete term for DNA.

thymopoietin 1. A polypeptide of 49 amino acids that is a component of thymosin 2. THYMOSIN.

thymosin A heterogeneous group of mitogenic polypeptides that stimulate the growth and proliferation of T lymphocytes. *Aka* thymin; thymopoietin.

thymus A gland, located in the lower part of the neck, that produces lymphocytes; the thymus is large in young animals but atrophies after the animals attains sexual maturity.

thymus nucleic acid An early designation for DNA.

thyrocalcitonin That fraction of the hormone

calcitonin that is secreted by the thyroid gland. *Abbr* TCT.

thyroglobulin A large, iodinated, glycoprotein that represents the form in which iodine is stored in the thyroid colloid and from which thyroxine and triiodothyronine are formed by proteolysis.

thyroid colloid The gelatinous material in the follicles of the thyroid gland in which the iodine is stored, principally in the form of thyroglobulin.

thyroid crisis THYROID STORM.

thyroidectomy The surgical removal of the thyroid gland.

thyroid gland An endocrine gland, located in the neck, that produces the hormones thyroxine and triiodothyronine.

thyroid hormones The hormones thyroxine and triiodothyronine.

thyroid hyperfunction HYPERTHYROIDISM.

thyroid hypofunction HYPOTHYROIDISM.

thyroid-stimulating hormone THYROTROPIN.

thyroid-stimulating hormone releasing hormone THYROTROPIN RELEASING HORMONE.

thyroid storm A severe clinical state due to overactivity of the thyroid gland and excessive secretion of thyroid hormones into the blood stream. A thyroid storm may occur spontaneously or be precipitated by infection or stress; it is characterized by high fever, rapid pulse, and acute respiratory distress. *Aka* thyroid crisis.

thyroliberin THYROTROPIN RELEASING HORMONE.

thyrotoxic effect A toxic condition produced by certain forms of excessive activity of the thyroid gland or by excessive doses of thyroid hormones.

thyrotoxicosis GRAVE'S DISEASE.

thyrotrophic hormone THYROTROPIN.

thyrotropic hormone THYROTROPIN.

thyrotropic hormone releasing factor THYROTROPIN RELEASING HORMONE.

thyrotropic hormone releasing hormone THYROTROPIN RELEASING HORMONE.

thyrotropin A protein hormone, secreted by the anterior lobe of the pituitary gland, that stimulates the synthesis of thyroid hormones and the release of thyroxine by the thyroid gland. *Var sp* thyrotrophin.

thyrotropin releasing hormone A hypothalamic hormone that controls the secretion of thyrotropin from the pituitary gland. *Var sp* thyrotrophin releasing hormone. *Abbr* TRH. *Aka* thyrotropin releasing factor (TRF).

thyroxine An iodinated aromatic amino acid that is the major hormone of the thyroid gland and that controls the rate of oxygen consumption and the overall metabolic rate. It is formed by the coupling of two molecules of 3,5-diiodotyrosine. *Abbr* Thx; T_4.

thyroxine-binding globulin A glycoprotein that serves as the major and specific carrier of thyroxine in plasma. It has a molecular weight of 58,000 and has one thyroid hormone binding site (either thyroxine or triiodothyronine will bind). Also known as inter-alpha-globulin because it has an electrophoretic mobility between that of α_1- and α_2-globulin. *Abbr* TBG.

thyroxine-binding prealbumin An albumin that serves as a secondary carrier of thyroxine in plasma. A tetramer (MW 55,000) with four binding sites for thyroid hormones (either thyroxine or triiodothyronine will bind). *Abbr* TBPA. *Aka* transthyretin.

TIBC Total iron-binding capacity.

tidal air The volume of air that enters and leaves the body with each normal respiratory movement.

tiered metabolic pathway A metabolic pathway of the type

$$A \rightarrow B$$
$$\downarrow$$
$$C \rightarrow D$$
$$\downarrow$$
$$E \rightarrow F.$$

tight coupling The state of cellular respiration in which the mitochondria are characterized by having a high acceptor control ratio.

tight junction A cell junction in which the interacting plasma membranes of two cells are so close together that there is no intercellular space between them; an impermeable junction that seals the space between cells. *See also* cell junction.

tilde SQUIGGLE.

tilt angle The angle between the perpendicular to the axis of a helical nucleic acid and the plane of an individual base.

time average The average value of a quantity that is measured over a number of time intervals; used for quantities that vary as a function of time, such as atmospheric temperature, blood flow, and caloric intake.

time constant RELAXATION TIME.

time constant of a reaction A measure of the duration of a reaction; for an exponential decay, the time constant is the time required for 63% (the fraction $1 - 1/e$) of the total change to occur.

time factor effect The effect of exposure time on the results produced by irradiation when reciprocity does not hold, so that the product of dose rate and exposure time is not constant; under these conditions, a short-time irradiation at a high intensity is frequently more effective than a long-time irradiation at a low intensity.

time-lapse microcinematography *See* microcinematography.

time-lapse photography A photographic technique in which exposures are taken on a photographic film at selected time intervals and the film is then projected at a faster speed so that the actions shown are speeded up.

time-of-flight mass spectrometer A mass spectrometer in which one measures the time required by an ion to travel from the ion source to the detector. All the ions receive the same kinetic energy during acceleration but since they have different masses, they separate into groups according to velocity and, hence, mass. *Abbr* TOF-MS.

times gravity convention *See* relative centrifugal force.

tin An element that is essential for humans and animals. Symbol, Sn; atomic number, 50; atomic weight, 118.69; oxidation states, +2, +4; most abundant isotope, ^{120}Sn; a radioactive isotope, ^{119}Sn, half-life, 250 days, radiation emitted, gamma rays.

tincture An alcoholic solution of a chemical or a drug.

TIP 1. Thermal inactivation point. 2. Translation inhibitory protein.

Ti plasmid *See* crown gall tumor.

Tiselius apparatus An apparatus that can be used for either diffusion or electrophoresis measurements; consists of a three-piece U tube that allows the formation of sharp boundaries between solvent and solution. The spreading of these boundaries, or their movement under the influence of an electric field, is then analyzed by means of either a schlieren or an interferometric optical system.

tissue An aggregate of cells and intercellular material that form a definite structure; the cells are generally of similar structure and function.

tissue culture The maintenance of living and metabolizing cells, tissues, or organs in artificial media.

tissue culture infectious dose MEDIAN TISSUE CULTURE DOSE.

tissue factor 1. THROMBOPLASTIN. 2. AUTACOID.

tissue hormones Hormones produced by individual specialized cells, scattered throughout a tissue that is specialized for some other function.

tissue mince Tissue that is cut or chopped into small pieces.

tissue plasminogen activator A protease that converts plasminogen to plasmin and thereby causes blood clots to dissolve. Both the enzyme and plasminogen bind to fibrin in close proximity, but the enzyme has low affinity for plasminogen and great affinity for fibrin. As a result, the enzyme promotes fibrinolysis at the site of a blood clot while having little effect on circulating plasminogen or fibrinogen. *Abbr*

TPA. *Aka* tissue-type plasminogen activator.

tissue polypeptide antigen A polypeptide in human blood; there appears to be a significant correlation between the presence of this substance and the incidence of cancer. *Abbr* TPA.

tissue slice A thin slice of tissue that can carry out metabolic reactions and that freely exchanges gases and metabolites with the suspending medium. The metabolic reactions in such a slice are frequently studied in a Warburg apparatus and monitored by manometric techniques.

tissue-specific enzyme An enzyme that is present in appreciable concentrations in only one tissue or in one organ; the plasma concentration of such an enzyme is of clinical value in assessing the functional state of that tissue or of that organ.

tissue thromboplastin THROMBOPLASTIN.

titer 1. The amount of a standard solution that is required for a defined titration of a given volume of a second solution. 2. The greatest dilution of a sample solution that gives a positive test under defined conditions; the greatest dilution of a virus sample that gives a positive hemagglutination test, and the greatest dilution of an antiserum that gives a positive precipitin test are two examples. *See also* plaque titer.

titin One of three unusually large myofibrillar proteins (MW 1 to 5×10^6) that occur in several vertebrate and invertebrate striated muscles. Titins are found in M lines, Z lines, and at the junctions of A and I bands. They may help to keep the myosin thick filaments centered within the sarcomere during force generation.

titrant The solution that is added to a given volume of a second solution in the course of a titration.

titratable Capable of being titrated. *Aka* titrable.

titratable acidity A measure of the acidity of urine, particularly that due to $H_2PO_4^-$, expressed in terms of the number milliliters of 0.1 N sodium hydroxide required to neutralize a 24-h volume of urine.

titrate To carry out a titration.

titration A method of volumetric analysis in which a test solution is added from a buret to a known volume of a standard solution, or a standard solution is added from a buret to a known volume of a test solution.

titration curve A plot of the amount of acid or base added during a titration as a function of the pH of the solution.

titration equivalent weight The average molecular weight of a fatty acid in a mixture of fatty acids; equal to the number of milli-grams of fatty acids in the sample, divided by the number of milliequivalents of alkali used in the titration of the sample to a phenol-phthalein end point.

titrimetry Chemical analysis by means of titration.

TLC Thin-layer chromatography.

TLD Thymineless death.

TLE Thin-layer electrophoresis.

t-like RNA 5S RNA.

TLV Threshold limit value.

T lymphocyte T CELL.

T_m 1. Melting out temperature. 2. Transport maximum.

TMG Thiomethyl-β-D-galactopyranoside; a nonmetabolizable analogue of lactose that has been used in studies of the lactose transport system in *E. coli*. A gratuitous inducer of β-galactosidase. *Aka* thiomethylgalactoside.

TMP 1. Ribosylthymine monophosphate (ribothymidylic acid). 2. Ribosyl-5'-monophosphate (5'-ribothymidylic acid).

TMS Tetramethylsilane; *see* chemical shift.

TMV Tobacco mosaic virus.

Tn Transposon.

TN Troponin.

TNBS 2,4,6-Trinitrobenzene sulfonic acid; used to examine the orientation of some phospholipids in biological membranes.

TNF Tumor necrosis factor.

Tn3 transposon A large transposon that contains several genes. It differs from a composite transposon in that it is not flanked by insertion sequences.

TNV Tobacco necrosis virus.

tobacco alkaloids NICOTIANA ALKALOIDS.

tobacco mosaic virus A helical plant virus that contains single-stranded RNA and infects tobacco leaves. *Abbr* TMV.

tobacco necrosis virus A spherical plant virus that contains single-stranded RNA and infects seed plants. *Abbr* TNV.

tocol The parent substance of the tocopherols. Tocol is the trivial name for 2-methyl-2(4',8',12'-trimethyl-tridecyl) chroman-6-ol.

tocopherol A generic descriptor for all mono-, di-, and trimethyl tocols. Tocopherol is not synonymous with vitamin E.

T-odd phage A phage that infects the bacterium *E. coli*. A T-odd phage is smaller than a T-even phage and contains DNA that has the same base composition as that of the host cell.

Toepfer's reagent An alcoholic solution of *p*-dimethylaminobenzene that is used in testing for free hydrochloric acid in gastric juice.

TOF-MS Time-of-flight mass spectrometer.

togavirus ARBOVIRUS.

tolerance 1. IMMUNOLOGICAL TOLERANCE. 2. The decrease in, or the loss of, the response of an animal to a dose of a chemical to which it responded on a prior occasion. 3. The limit of error permitted in the graduation of measuring instruments, standarized products, or analytical evulations. 4. DESENSITIZATION (3).

tolerance test   *See* epinephrine tolerance test; galactose tolerance test; glucose tolerance test; lactose tolerance test.

tolerogenic antigen   An antigen that produces immunological tolerance. *Aka* tolerogen.

Tol G protein   A protein in the outer membrane of gram-negative bacteria that appears to function in F-pilus-mediated conjugation and to serve as a receptor for certain phages.

Tollen's reagent   A dilute solution of silver oxide (Ag_2O) in ammonia.

Tollen's test   A test for reducing sugars that is based on the reduction of silver ions to metallic silver in an alkaline solution.

toluenized cells   Bacterial cells that have been treated with toluence to increase their permeability or promote lysis.

tomato bushy stunt virus   A plant virus that contains single-stranded RNA.

tomography   An x-ray technique that provides a picture of a given plane in a specimen; used medically to scan a planar section of a patient. The measurements are then analyzed mathematically to create a cross-sectional map representing a reconstructed image of bone and tissue structure. *See also* PETT.

tonfilaments   KERATIN FILAMENTS.

tonometry   The measurement of tension or pressure such as the measurement of blood pressure or the measurement of the partial pressure of CO_2, in equilibrium with the blood.

tonoplast   The membrane surrounding an intracellular vacuole. *Aka* vacuolar membrane.

tophus (*pl* tophi). A deposit of sodium urate in cartilage that occurs in gout.

topography   The description and the mapping of the features of a surface; the delineation of the configurations and the structural relationships of a surface.

topoisomer   A topological isomer.

topoisomerase   One of a group of enzymes that control the topological state of DNA and catalyze the interconversion of topological isomers of DNA. These reactions include changes in superhelicity and the formation of knotted and catenated structures. All topological interconversions of DNA require the transient breakage and rejoining of DNA strands. Accordingly, topoisomerases are divided into two groups: Type I enzymes produce a transient break in a single strand of the duplex, while Type II enzymes produce a transient double-strand break in the duplex. Some of the Type I enzymes (originally named DNA relaxing enzymes, swivelases,

untwisting enzymes, untwistases, or nicking-closing enzymes) carry out the relaxation of superhelical DNA. The Type II enzymes include the DNA gyrases, enzymes that convert relaxed, closed circular DNA to a superhelical form.

topological isomers   Isomers that differ in their topological structure; duplex DNA molecules that are relaxed, have positive superhelicity, or have negative superhelicity, are some examples. *Aka* topoisomer.

topology   1. The study of the properties of geometric shapes and figures that are subjected to deformations and transformations such as those produced by twisting or bending. *See also* topoisomerase. 2. A branch of mathematics that deals with those properties of shapes and figures that remain unchanged if the shape or the figure is subjected to one-to-one continuous transformations.

torand   One of a group of synthetic macrocyclic compounds, so called because of their toroidal shape. They are rigid, doughnut-shaped, and appear to have useful metal-binding properties; they are unsaturated analogues of 18-crown-6-ethers.

toroidal supercoil   One of a class of possible superhelical structures for circular double-stranded DNA. A toroid is a surface generated by a plane closed curve, rotated about a line in its plane that does not intersect the curve. A supercoiled duplex in which the axis is wound as if about a cylinder.

Torr   A unit of pressure equal to 1/760 atm; a pressure of 1 mm mercury.

torsional strain   A strain in a molecule, believed to be due to the slight repulsion between electron clouds in the C—H bonds as they pass each other in close proximity when the molecule is in the eclipsed conformation. *Aka* eclipsing strain.

torsion angle   A rotation angle that can be defined as the angle between the plane containing atoms A, B, and C and the plane containing atoms B, C, and D for a system that consists of four atoms linked in the sequence A-B-C-D. The torsion angle can also be described as the angle between the projection of A-B and that of C-D when the system of four atoms is projected onto a plane normal to the bond B-C. *See also* phi angle; psi angle; omega angle.

Tos   Tosyl group.

tosyl group   The *p*-toluenesulfonyl grouping; a compound containing this group is known as a tosylate. The tosyl group is used as a protecting agent for blocking the amino group in peptide synthesis.

total activity   The total number of activity units in a sample, such as the total number of enzyme units or the total number of microcuries.

total cell count   The total number of bacteria, both viable and noviable, in a sample.

total consumption burner   A burner used in atomic absorption spectrophotometry and designed so that the gases and the sample are not mixed before entering the flame.

total inhibition   The inhibition of the activity of an enzyme in which a saturating concentration of inhibitor can cause the reaction velocity to become zero.

total iron-binding capacity   The concentration of iron that is equal to the sum of the actual iron concentration in plasma and the unsaturated iron-binding capacity; generally expressed in terms of micrograms per 100 mL of plasma. *Abbr* TIBC.

totally labeled   GENERALLY LABELED.

total osmotic pressure   The osmotic pressure produced when a membrane separating two solutions is impermeable to all solutes regardless of their size.

total titer   The titer of phage particles that is obtained after disruption of the infected cells and that is a measure of the maturation of phage particles in the host cells. *See also* extracellular titer; intracellular titer.

totipotence   The capacity of a cell to express all of its genetic information under appropriate conditions and to develop into a complete and fully differentiated organism. *Var sp* totipotency.

Townsend avalanche   The flood of ions produced by gas amplification in an ionization chamber.

toxalbumin   One of a group of proteins that are toxic to plants; a phytotoxin. Ricin and abrin are two examples.

toxemia   A pathological condition characterized by the presence of toxins in the blood.

toxic   1. Of, or pertaining to, a poison. 2. Poisonous.

toxic goiter   1. A thyroid enlargement accompanied by hyperthyroidism. 2. GRAVE'S DISEASE.

toxicity   The degree of harmfulness of a substance for an organism; the capacity of a substance to produce injury.

toxicity index   The ratio of the highest dilution of a germicide that kills the cells of chick heart tissue in 10 min to the highest dilution of the germicide that kills the test microbial organisms in the same time and under the same conditions.

toxicology   The science that deals with the harmful effects of chemicals on biological systems.

toxicosis   A pathological condition caused by the action of a poison or a toxin; systemic poisoning; toxemia. *See also* thyrotoxicosis.

toxigenicity   The production of a toxin, particularly the production of a toxin by bacteria. *Aka* toxicogenicity.

toxin   A high molecular weight compound of plant, animal, or bacterial origin that is toxic and generally antigenic in animal species.

toxohormone   A toxic substance that inhibits the activity of the enzyme catalase; apparently a polypeptide that can be extracted from cancer cells and that may also occur in normal cells.

toxoid   A toxin that has lost its toxic properties as a result of denaturation or chemical modification but that has retained its antigenic properties.

toxophore   The group of atoms in a toxin molecule that is responsible for the toxic properties of the molecule.

TP   1. Transport piece. 2. Transprotonase.

TPA   1. Tissue plasminogen activator. 2. Tissue polypeptide antigen. 3. *See* phorbol esters.

TPCK N-Tosyl-L-phenylalanine chloromethylketone; an irreversible inhibitor of chymotrypsin that acts by alkylating a histidine residue in the enzyme.

T phage   *See* T-even phage; T-odd phage.

TPN$^+$ Triphosphopyridine nucleotide. *Aka* NADP$^+$.

TPNH   Reduced triphosphopyridine nucleotide. *Aka* NADPH.

TPP   Thiamine pyrophosphate.

trabecular bone   *See* lamellar bone.

trace element   An element, such as cobalt, copper, manganese, or zinc, that is an essential nutrient for an organism but that is required only in minute amounts of the order of milligrams or micrograms per day for humans and animals. *See also* bulk elements; macronutrients; micronutrients.

tracer   1. An isotope, either radioactive or stable, that is used to label a compound. 2. A compound labeled with either a radioactive or a stable isotope.

tracking dye   A dye used as a marker in gel electrophoresis.

track radioautography   The radioautography of beta particle tracks that are produced by decaying radiophosphorus atoms incorporated into nucleic acids. *See also* radiophosphorus decay; star gazing.

trailer   A nontranslated sequence of nucleotides, excluding the poly(A) tail, that occurs after the termination signal at the 3'-end of mRNA. *Aka* trailer sequence.

trailing   The chromatographic and electrophoretic phenomenon in which the material that is separated not only appears in peaks, spots, or bands, but also appears along part of, or the entire, migration path. *See also* tailing; leading.

trans 1. Referring to the configuration of a geometrical isomer in which two groups, attached to two carbon atoms linked by a double bond, lie on opposite sides with respect to the plane of the double bond. 2. Referring to two mutations, particularly those of pseudoalleles, that lie on different chromosomes.

transacetylase ACETYLTRANSFERASE.,

trans-acting locus A region on a DNA molecule that affects the activity of genes that are located on a different molecule. *See also* cis-acting locus.

transacylase An enzyme that catalyzes the transfer of an acyl group from one compound to another.

transaldolase An enzyme that catalyzes the transfer of a dihydroxyacetone group from a 2-ketosugar to carbon number one of various aldoses.

transamidation The reaction in which the amide nitrogen of glutamine is transferred as an amino group to another compound.

transamidinase AMIDINOTRANSFERASE.

transamidination The reaction in which the guanido group of arginine is transferred to glycine, thereby forming guanidinoacetic acid.

transaminase An enzyme that catalyzes the transfer of an amino group from an amino acid to a ketoacid, thereby giving rise to the opposite pair of ketoacid and amino acid. Transaminases require a derivative of vitamin B_6 as coenzyme.

transaminase-type mechanism *See* ping-pong mechanism.

transamination The reaction catalyzed by the enzyme transaminase.

transcapsidation The formation of a hybrid virus during phenotypic mixing; the genome of one virus ends up being contained within the capsid of another virus.

transcarboxylase *See* acetyl-CoA carboxylase.

transcarboxylation A reaction in which a carboxyl group is transferred from one compound to another; the reaction is catalyzed by a biotin-requiring enzyme.

transcellular transport The transport across a cell or a layer of cells that moves material both into and out of a cell; the transport across the kidney tubules and the transport across the gastric mucosa are two examples. *See also* intracellular transport; homocellular transport.

transcortin An α_1-globulin that binds and transports both cortisol and corticosterone in the blood.

transcribed spacer A segment of RNA that occurs in the primary RNA transcript (as that of rRNA) but is subsequently removed during post-transcriptional processing when a secondary, smaller, but functional RNA transcript is formed from the original transcript.

transcript A transcribed sequence; a nucleic acid molecule formed during transcription.

transcriptase DNA-DEPENDENT RNA POLYMERASE.

transcription The process whereby the genetic information of DNA is copied in the form of RNA; a sequence of deoxyribonucleotides in a strand of DNA gives rise to a complementary sequence of ribonucleotides in a strand of RNA. *See also* reverse transcription.

transcriptional control The regulation of protein synthesis at the level of transcription. *See* operon hypothesis.

transcription enhancer *See* enhancer.

transcription initiation frequency *See* promoter strength.

transcription unit A stretch of DNA that is transcribed as a single, continuous mRNA strand by RNA polymerase; includes the signals for initiation and termination of this transcription. A simple transcription unit is one that carries information for the synthesis of only one protein; a complex transcription unit carries information for the synthesis of more than one protein molecule. *Aka* operon.

transdeamination The deamination of amino acids that involves the coupling of a transamination reaction and the reversal of an ammonia fixing reaction. In the former, the amino acid is deaminated to a ketoacid and glutamate is formed from α-ketoglutarate. In the second reaction, glutamate is converted to α-ketoglutarate and NH_3 is released.

transdetermination A switch from one heritable state to another; a change in developmental fate. The differentiation of a group of cells in tissue culture into structures that are inappropriate with respect to the cells from which the culture was derived, is an example.

transdimer synthesis The synthesis of DNA that occurs across a thymine dimer block despite the distortion of the helix caused by the dimer. The process forms the basis of the SOS repair.

transduce 1. To transfer genetic material from one bacterium to another by transduction. 2. To transform one form of energy into another.

transduced element The chromosomal fragment that is transferred from one bacterium to another during transduction.

transducer A device that transforms one form of energy into another; a photocell that converts light energy into electrical energy is an example.

transducer cell One of a number of neurosecretory cells in the hypothalamus that are stimulated by the central nervous system to secrete the hypothalamic releasing hormones which act on the adenohypophysis.

transducin A GTP-binding protein (G protein) that has a key role in the visual excitation process; it mediates the light activation signal from photolyzed rhodopsin to cGMP phosphodiesterase. *See also* G protein.

transducing particle A phage particle containing host (bacterial) DNA.

transducing phage A phage that brings about transduction.

transductant A bacterial cell that has received DNA by transduction.

transduction 1. The genetic recombination in bacteria in which DNA from a donor cell is transferred to a recipient cell by means of a phage. A segment of the donor DNA is first incorporated into the phage DNA and is then incorporated, by recombination, into the recipient DNA. 2. The transformation of one form of energy to another. The stimulation of a flow of matter by a gradient in electrical potential (electrophoresis) and the production of ATP through dissipation of a proton gradient (oxidative phosphorylation) are two examples.

trans effect The influence of one gene on the expression of another gene that is located on a different chromosome.

transeliminase PECTATE LYASE.

transesterification INTERESTERIFICATION.

transfection 1. The transformation of competent bacterial cells by infection with naked phage DNA to produce infectious phage particles. Involves the viral DNA alone, without the protein capsid, and the bacterial cells must first be rendered sufficiently permeable to the DNA (that is, one uses protoplasts or spheroplasts). 2. The introduction of foreign DNA into eukaryotic cells in tissue culture; this can be viral DNA or some other type of DNA.

transfectoma a transfected cell; a cell altered by transfection.

transfer BLOTTING.

transferase 1. An enzyme that catalyzes a reaction in which there is a transfer of a functional group from one substrate to another. *See also* enzyme classification. 2. ELONGATION FACTOR.

transferase I ELONGATION FACTOR T.

transferase II TRANSLOCASE (2).

transfer factor 1. ELONGATION FACTOR. 2. A factor that can be extracted from lymphocytes and that can be used to transfer delayed hypersensitivity in humans. *Abbr* TF.

Transfer origin A unique base sequence in plasmid DNA at which a nick is made during transfer of the plasmid DNA to the recipient bacterium during conjugation. *See also* relaxation complex.

transfer pipet VOLUMETRIC PIPET.

transfer potential *See* group transfer potential.

transfer rate coefficient The permeability coefficient divided by the volume from which diffusion takes place.

transferred DNA *See* crown gall tumor.

transferred immunity *See* adoptive immunity.

transferred tolerance *See* adoptive tolerance.

tranferrin *See* siderophilin.

transfer RNA A low molecular weight RNA molecule, containing about 70 to 80 nucleotides, that binds an amino acid and transfers it to the ribosomes for incorporation into a polypeptide chain during translation. Transfer RNA has a sedimentation coefficient of about 4S, is characterized by having a high content of minor bases, and binds to the codon in messenger RNA by way of a complementary anticodon that is present in the transfer RNA. The secondary structure of transfer RNA consists of a clover leaf configuration, with the chain folded back upon itself and held together by means of hydrogen bonding. The clover leaf is folded once more to yield an L-shaped tertiary structure in which portions of the RNA strand are held together by means of tertiary hydrogen bonding. *Abbr* tRNA. *See also* L-type structure.

transfer RNA multiplicity The occurrence of two or more forms of the same transfer RNA molecule, all of which can be charged with the same amino acid.

transformant A bacterial cell that has undergone transformation.

transformasome An extension of the membrane on the surface of transformation-competent bacteria (such as *Hemophilus influenza*) that is responsible for DNA binding and uptake during transformation.

transformation 1. The genetic recombination in bacteria in which a DNA fragment of a purified DNA preparation is incorporated into the chromosome of a recipient bacterial cell. 2. The change of a normal cell to a malignant one, as that resulting from infection of normal cells by oncogenic viruses. 3. The morphological and other changes that occur in both B and T lymphocytes upon exposure to an antigen. *Aka* blast transformation. 4. The mathematical change of a variable to simplify calculations or for some other purpose.

transforming growth factor One of two polypeptides that alter the phenotype of some normal cells to that of transformed cells; the peptides also promote angiogenesis. *Abbr* TGF.

transforming principle The purified DNA that is incorporated into the recipient bacterial cell during transformation; first described for the transformation of *Pneumococcus. Aka* transforming factor.

transgenic Descriptive of an organism that contains some genetic material that has been experimentally transferred into it from some other source.

transgenome A genome that contains some genes that have been experimentally transferred into it from some other source.

transgenosis The overall phenomenon of the transfer of bacterial genes to plants and their subsequent expression within the plant cells; includes gene transfer, gene maintenance, transcription, translation, and metabolic function.

transglycosidation The reaction catalyzed by a glycosyl transferase.

trans-Golgi network *See* GERL.

trans-Golgi reticulum *See* GERL.

transhydrogenase The mitochondrial enzyme that catalyzes the interconversion of NAD^+ and NADPH to NADH and $NADP^+$. *Abbr* TH. *See also* supermolecule.

transhydrogenation The reaction $NAD^+ +$ NADPH $\rightleftharpoons$ NADH $+ NADP^+$ that is catalyzed by a transhydrogenase and that occurs in mitochondria.

transient Temporary; transitory; short-lived.

transient dipole-induced dipole interactions DISPERSION FORCES.

transient equilibrium The equilibrium that is established for a radioactive decay reaction if the half-life of the parent isotope is not much greater than that of the daughter isotope.

transient phase A temporary phase in enzyme reactions which precedes the steady-state phase. *See also* pre-steady-state kinetics.

transient state isoelectric focusing A variation of isoelectric focusing that entails a study of the kinetic behavior of the charged amphoteric molecules during their transport by the electric field; achieved by repetitive optical scanning (absorbance, fluorescence, refraction, etc.) of the concentration distribution of the migrating solutes, followed by computer analysis of the peaks. The method can be used for an isoelectric focusing experiments using either a gel or a density gradient as the supporting medium. *Abbr* TRANS-IF.

transient state kinetics PRE-STEADY-STATE KINETICS.

transient time The time required to reach equilibrium in a sedimentation equilibrium experiment.

TRANS-IF Transient state isoelectric focusing.

transinhibition *See* transstimulation.

transinteractions Noncovalent interactions between segments, located on different polypeptide chains of the immunoglobulin molecule.

trans isomer *See* trans (1).

transition 1. A point mutation in either DNA or RNA in which there is a replacement of one purine by another or a replacement of one pyrimidine by another. In double-stranded nucleic acids, a complementary base is then inserted into the second strand so that a new base pair is produced. 2. A change from one state to another, such as a change in the electronic configuration of an atom or a molecule upon excitation, or a change in the conformation of a macromolecule upon denaturation. *See also* electronic transition, rotational transition, vibrational transition.

transitional mutant A mutant that differs from the wild-type organism by a transition.

transition dipole moment The dipole moment induced in a molecule as a result of the displacement of charge during the absorption of a photon by the molecule.

transition element One of a group of metal elements, including iron and cobalt, in which filling of the outermost shell to 8 electrons within a period is interrupted to bring the penultimate shell from 8 to either 18 or 32 electrons. These elements can use outermost, as well as penultimate, shell orbitals for bonding, and they form chelates of importance in biochemistry. *Aka* transition metal.

transition moment *See* magnetic transition moment; transition dipole moment.

transition probability The probability that a molecule will undergo a transition from one energy state to another if it is supplied with the energy required for this transition.

transition state ACTIVATED COMPLEX.

transition state inhibitor A compound that is structurally similar to the transition state and that acts as an inhibitor by binding to the surface of an enzyme in place of the substrate.

transition state theory THEORY OF ABSOLUTE REACTION RATES.

transition temperature MELTING OUT TEMPERATURE.

transitory complex An intermediate in an enzyme-catalyzed reaction that can undergo a unimolecular reaction with the release of a substrate or a product, or that is capable of isomerizing into such an intermediate. *See also* central complex.

transketolase An enzyme that catalyzes the transfer of a glycolaldehyde group from a 2-ketosugar to carbon number one of various aldoses.

translation 1. The process whereby the genetic information of messenger RNA is used to specify and direct the synthesis of a polypeptide chain on a ribosome; the sequence of codons in messenger RNA gives rise to a sequence of amino acids in a polypeptide chain. Synthesis of the polypeptide chain includes chain initiation, chain elongation, and chain termination.

2. A motion along a line without rotation.

translational amplification A mechanism for the production of large amounts of a specific protein by prolongation of the lifetime of the corresponding mRNA.

translational control The regulation of protein synthesis at the level of translation. The regulation of the expression of a gene that is achieved by a control of the rate at which the mRNA, specified by that gene, is being translated.

translational coupling The regulation of the translation of a distal gene in a polycistronic mRNA by a proximal gene. This occurs when the termination codon of the proximal gene overlaps the initiator region of the distal gene and forms a stem-and-loop structure with it. As a result, translation of the distal gene depends on completion of translation of the proximal gene and the opening up of the stem-and-loop structure.

translational diffusion The macroscopic flow of material from a region of high concentration to a region of lower concentration that results from the random Brownian motion of the molecules.

translational diffusion coefficient A measure of translational diffusion that depends on the size and the shape of the diffusing particle; specifically, $D = RT/Nf$, where D is the translational diffusion coefficient, f is the translational frictional coefficient, N is Avogadro's number, R is the gas constant, and T is the absolute temperature. *Aka* diffusion coefficient. *See also* standard diffusion coefficient.

translational frictional coefficient A measure of the frictional resistance experienced by a particle in solution that is equal to the frictional force divided by the translational velocity of the particle. *Aka* frictional coefficient. *See also* translational diffusion coefficient.

translational symmetry The symmetry of a body that exists when idential structures are produced by moving the body along an axis of symmetry.

translation error model A model for the evolution of the genetic code according to which the code evolved so as to minimize errors during translation.

translation inhibitory protein ANTIVIRAL PROTEIN.

translocase 1. TRANSPORT AGENT. 2. The enzyme that catalyzes the GTP-dependent translocation reaction in protein synthesis in which the peptidyl-tRNA is shifted from the A site to the P site in the ribosome, and in which the mRNA is shifted simultaneously by one codon.

translocation 1. TRANSPORT. 2. The reaction catalyzed by the enzyme translocase. 3. An interchromosomal aberration in which a chromosome fragment becomes inserted into another, nonhomologous chromosome. 4. The movement of an activted hormone–receptor complex from the cytoplasm to the nucleus. Steroid hormone–receptor complexes redistribute themselves in this fashion, thereby leading to activation or inhibition of the transcription of specific genes. 5. INVERSION(2).

translocation factor TRANSLOCASE.

translocator TRANSPORT AGENT.

translocon A cluster of genes that code for the variable and constant regions of the immunoglobulins.

translucent Descriptive of a substance that permits only a partial passage of the light rays striking it; partially transparent.

transmembrane potential MEMBRANE POTENTIAL.

transmembrane protein A protein that spans the entire width (thickness) of a membrane; it extends from one side of the membrane to the other.

transmethylation The reaction whereby a labile methyl group is transferred from one compound to another.

transmissible plasmid *See* self-transmissible plasmid.

transmission 1. TRANSMITTANCE. 2. NERVE IMPULSE TRANSMISSION.

transmission electron microscope *See* electron microscope.

transmittance The ratio of the intensity of the transmitted light to that of the incident light. *Sym* T.

transmittancy The ratio of the transmittance of the solution to that of the solvent.

transmitter *See* chemical transmitter.

transmutation The transformation of one nuclide into the nuclide of a different element, as that which occurs during radioactive decay.

transparent Descriptive of a substance that permits the passage of light rays striking it and that can be seen through.

transpeptidase PEPTIDYL TRANSFERASE.

transpeptidation The reaction catalyzed by peptidyl transferase; peptide bond formation.

transphoresis ISOTACHOPHORESIS.

transphosphorylation The transfer of a phosphoryl group, a pyrophosphoryl group, or an adenylyl group from one molecule to another.

transpiration The passage of water vapor through the surface of an organism, as the passage through a leaf or the skin.

transplant The part of an animal that is transferred within the same, or to a different, animal by transplantation.

transplantation The transfer of a part of an animal, such as a tissue or an organ, to another site of the same animal or to a site of a different animal.

transplantation antigen   HISTOCOMPATIBILITY ANTIGEN.

transplantation immunity   The reactions of the immune response that occur in the recipient of a transplant and that are caused by the antigens of the transplant.

transplant rejection   The immunological reactions by which the recipient of a transplant brings about the destruction of the cells of the transplant.

transport   The movement of material from one place to another, particularly in reference to the movement of material within a biological fluid or across a biological membrane. *See also* active transport; mediated transport, passive transport.

transport agent   A substance, generally a protein or an enzyme, that is instrumental in transporting material across a biological membrane or within a biological fluid. A transport agent may act as an actual carrier, affect a translocation, or in some other way affect the transport process. *Aka* transporter; carrier; porter; translocase; permease.

transporter   TRANSPORT AGENT.

transport maximum   1. The maximum rate at which the kidney can either absorb or secrete a particular substance. *Sym* T_m. 2. The maximum rate of flux in facilitated diffusion that occurs when all membrane-bound transport sites are occupied by solute.

transport-negative mutant   A mutant that has lost the ability to synthesize an enzyme, a group of enzymes, or some other protein that is required for a transport system.

transport piece   SECRETORY PIECE.

transport process   A physical property that is based on the transport of macromolecules in solution; sedimentation, diffusion, electrophoresis, and viscosity are examples.

transport protein   TRANSPORT AGENT.

transport system   The various components, including transport agents, that function in a given type of transport.

transposable code   An early version of the genetic code according to which the DNA strand that is not transcribed consists either of all nonsense codons, or of codons that code for the same amino acids as those coded for by the complementary codons in the transcribed strand. In this fashion each genetic locus would give rise to only one type of protein, regardless of whether one or both of the DNA strands were being transcribed.

transposable element   A mobile genetic segment; a section of DNA that can move from one chromosomal site to another. Transposable elements occur in both prokaryotes and eukaryotes. Prokaryotic transposable elements are usually called transposons. *Aka* transposable genetic element. *See also* transposition.

transposable genetic element   TRANSPOSABLE ELEMENT.

transposase   An enzyme required for formation of the cointegrate structure which is believed to be an obligatory intermediate in transposition.

transposition   The movement of genetic information from one locus to another; the movement of a mobile genetic element. Transposition occurs in both prokaryotes and eukaryotes. The term is used specifically for DNA-mediated events as opposed to retroposition which refers to RNA-mediated events.

transposon   A transposable genetic element in bacteria. Three types have been identified: insertion sequences, composite transposons, and Tn3 transposons. The term is also used as a synonym for any transposable element more complex than an insertion sequence.

transprotonase   An enzyme that catalyzes a transprotonation reaction. *Abbr* TP. *See also* supermolecule.

transprotonation   The transport of protons across a membrane, as that postulated by the chemiosmotic hypothesis.

transsplicing   Intermolecular splicing; the splicing of two exons from two separate precursor molecules.

transstimulation   A phenomenon observed with some reversible transport systems, that is, those that can operate in either direction. In these cases, extracellular addition of a given transport substance may increase (transstimulate) or decrease (transinhibit) the movement of the same transport substance from the inside of the cells to the outside.

transsulfuration   The biosynthesis of cysteine from serine via a pathway that passes through homocysteine (which can arise from methionine).

transthyretin   A protein that binds the retinol-binding protein/vitamin A complex in the plasma and serves to carry vitamin A to the eye. The protein also binds the thyroid hormones (thyroxine and triiodothyronine). *Aka* thyroxine-binding prealbumin.

transtubular network   *See* GERL.

transudate   The fluid exuded during transudation.

transudation   The movement of solvent plus solutes through the pores of a membrane or through the interstices of a tissue.

transverse diffusion   The movement of proteins and lipids from one side of a biological membrane to the other; flip-flop.

transverse mutation   TRANSVERSION.

transverse relaxation   SPIN–SPIN RELAXATION.

transverse system T SYSTEM.

transverse tubule *See* T system.

transversion A point mutation in either DNA or RNA in which there is a replacement of a purine by a pyrimidine or vice versa. In double-stranded nucleic acids, a complementary base is then inserted into the second strand so that a new base pair is produced.

transversional mutant A mutant that differs from the wild-type organism by a transversion.

Trappe's eluotropic series *See* eluotropic series.

Traube's covolume *See* covolume.

Traube's rule The rule that, for a homologous series of surfactants, the concentration of surfactant required to produce an equal lowering of surface tension in a dilute solution decreases by about a factor of three for each additional CH_2 group in the surfactant molecule.

Trautman plot A graphical method for the representation of data obtained by the Archibald method that is particularly useful for the comparison of a number of different experiments.

treadmilling A steady-state phenomenon of microtubules and actin filaments in which the length of the polymeric structure remains constant despite the fact that monomeric molecules constantly add to, and come off from, both ends of the microtubule or the actin filament. In the case of microtubules, tubulin molecules come off predominantly (but not exclusively) at one end, and are added on predominantly (but not exclusively) at the other end. The net rate of addition at one end equals the net rate of loss at the other end, so that the net rate of growth of the polymer is zero. A similar situation exists in the case of actin filaments. For both types of polymers, the process is driven by nucleoside triphosphate hydrolysis; if this were not the case, then treadmilling would represent a type of perpetual motion machine in violation of the second law of thermodynamics.

tree diagram A graphical representation of nuclear magnetic resonance data; shows the individual effects of each coupling constant on the overall nuclear magnetic resonance pattern.

trehalose A nonreducing disaccharide of glucose that occurs in the hemolymph of many insects (hence, also called "blood sugar") and also in bacteria and fungi (hence, also called "mushroom sugar").

TRF Thyrotropin releasing factor; *See* thyrotropin releasing hormone.

TR factor Transfer RNA releasing factor; an enzyme, postulated to be involved in the release of tRNA from ribosomes during protein synthesis.

TRH Thyrotropin releasing hormone.

tri- Combining form meaning three or thrice.

triacylglycerol A compound formed by the esterification of one glycerol molecule with three fatty acid molecules. *Aka* triglyceride.

trial and error A method for obtaining a desired result by trying various ways and/or values, noting the errors and causes of failure, and eliminating them in subsequent steps.

triangulation A method for approximating the area under a symmetrical peak by assuming the peak to be an equilateral triangle. The area is then given by the width of the peak, at one-half the maximum height, multiplied by the maximum height.

triangulation number 1. The total number of small equilateral triangles into which the surfaces of an icosadeltahedron have been divided. 2. The number of 60-unit sets in the capsid of icosahedral viruses. Thus, a virus having 240 subunits is said to have a triangulation number of 4.

triantennary *See* high-mannose oligosaccharides; complex oligosaccharides.

tribasic Descriptive of a compound that contains three hydrogen atoms replaceable by a metal (such as $Na_2H_3IO_6$) or an acid that can furnish three hydrogens ions (such as H_3PO_4).

tribology The science and technology of interacting surfaces in relative motion.

tricarboxylic acid cycle CITRIC ACID CYCLE.

trichloroacetic acid A halogenated derivative of acetic acid, used to precipitate proteins; the compound Cl_3C—COOH. *Abbr* TCA.

trichothecenes A family of sesquiterpenoids that constitute a large group of mycotoxins.

tricine *N*-Tris(hydroxymethyl)methylglycine; used for the preparation of biological buffers in the pH range of 7.4 to 8.8 *See also* biological buffers.

tridentate Describing a ligand that is chelated to a metal ion by means of three donor atoms.

triethylenephosphoramide *See* TEPA.

triethylenethiophosphoramide *See* thio-TEPA.

trigger protein A hypothetical protein that may have to be accmulated by a cell up to threshold levels before the cell can proceed beyond the restriction point (R point) in the G_1 phase of the cell cycle and thus complete the latter. *Aka* unstable protein (U protein).

trigger reaction A reaction that initiates another reaction or a sequence of reactions.

triglyceride TRIACYLGLYCEROL.

triiodothyronine A minor hormone of the thyroid gland that has the same functions as thyroxine but is present in much smaller amounts. *Sym* T_3.

triketohydrindene hydrate Ninhydrin. *See also*

ninhydrin reaction.

trimer An oligomer that consists of three monomers.

trimethylpsoralen A low molecular weight, planar molecule that intercalates with double-stranded DNA and, upon exposure to ultraviolet light, becomes linked covalently to pyrimidines; it can link to a single strand or cross-link two strands.

triose A monosaccharide that has three carbon atoms.

triparental recombinant A progeny phage that contains genes, derived from three different phages which had simultaneously infected the same bacterial cell. *See also* Visconti–Delbrueck hypothesis.

tripartite repeating unit *See* supermolecule.

triphosphopyridine nucleotide NICOTINAMIDE ADENINE DINUCLEOTIDE PHOSPHATE.

triple bond A covalent bond that consists of three pairs of shared electrons.

triple-chain length A crystalline form of glycerides in which three acyl groups form a unit structure.

triple helix Three intertwined helical chains such as the three left-handed polypeptide chains of tropocollagen that are wound together to form a right-handed triple helix.

triplet 1. CODON. 2. A triple peak, as that obtained in nuclear magnetic resonance.

triplet code A genetic code in which an amino acid is specified by a codon that is composed of a sequence of three adjacent nucleotides.

triplet state The excited electronic state of an atom or a molecule in which two single electrons in two orbitals (two unpaired electrons) have their spin in the same direction and are not spin-paired so that $S = 1$ and $2S + 1 = 3$, where S is the resultant spin and $2S + 1$ is the spin multiplicity. A triplet state is a long-lived, metastable state with a lifetime of about 10^{-3} to 1 s; a triplet state atom or molecule has a net magnetic moment. *See also* singlet state.

triplex A triple helix.

triploid state The chromosome state in which each of the various chromosomes, except the sex chromosome, is represented three times. *Aka* triploidy.

Tris Tris(hydroxymethyl)aminomethane; used for the preparation of biological buffers in the pH range of 7.2 to 9.2. *Aka* THAM. *See also* biological buffers.

tris- Prefix indicating three phosphate groups linked at three different positions. Thus, adenosine-2′,3′,5′-trisphosphate is now preferred to adenosine-2′,3′,5′-triphosphate.

triskelion *See* clathrin.

triti-, trito- Proposed prefixes for the ^{3}H isotope. *See also* proti-, proto-.

tritiated Labeled with tritium.

tritium The heavy radioactive isotope of hydrogen that contains one proton and two neutrons in the nucleus; a weak beta emitter having a half-life of 12.26 years. *Sym* T.

tritium gel filtration A method for studying the hydrogen exchange of macromolecules by gel filtration chromatography of a tritiated sample and analysis of the eluate for tritium content.

triton 1. The tritium nucleus that consists of one proton and two neutrons. 2. Trademark for a series of organic, nonionic surface-active agents.

tritosome A lysosome that contains large amounts of the nonionic detergent Triton WR-1339. Such secondary lysosomes have lower densities than primary lysosomes and, hence, are readily isolated by differential centrifugation.

triturate To grind or rub to a powder, usually with the aid of a liquid.

trityl group A triphenylmethyl group; used to block the amino group of amino acids. *Abbr* Trt.

trivial name A working or common name that is not based on rules of nomenclature, such as the name of an enzyme that is not based on the classification rules of the Enzyme Commission. *See also* systematic name.

Trizma Trademark for a group of Tris buffers.

tRNA Transfer RNA.

tRNA^{AA} Transfer RNA that is specific for the amino acid AA.

tRNA arm *See* arm.

tRNA^{fmet} Methionine transfer RNA that allows the methionine, after if becomes attached, to be enzymatically formylated; also abbreviated tRNA$_f^{met}$.

T4 RNA ligase A ligase from phage T4 that catalyzes the ATP-dependent formation of 3′,5′-phosphodiester bonds between 5′-phosphate ends and 3′-hydroxyl ends of oligoribonucleotides.

tRNA^{met} Methionine transfer RNA that does not allow the methionine, after it becomes attached, to be enzymatically formylated.

tRNA nucleotidyltransferase *See* CCA enzyme.

tRNA stem *See* stem; arm.

tRNA synthetase AMINOACYL-tRNA SYNTHETASE.

tRNA synthetase recognition site The site on the transfer RNA molecule to which the aminoacyl-tRNA synthetase becomes bound.

Trojan horse inhibitor An inhibitory compound introduced into an active site of an enzyme or a protein by the technique of affinity labeling.

tropane alkaloids *See* alkaloids.

-trophic Combining form meaning related to nutrition; a common ending for the names of many hormones where it is used interchangeably with -tropic.

trophic hormone A hormone, the main function of which is to stimulate the secretion of another hormone from an endocrine gland.

trophic value The value of a nutrient in terms of the raw material that it furnishes for the building and the maintenance of the metabolic machinery; it is greater than that described by the caloric value of the nutrient.

trophosome A structure in the hydrothermal worm *Riftia pachyptila* that contains chemoautotrophic bacteria as suppliers of nutrients.

-tropic Combining form meaning a turning; a common ending for the names of many hormones where it is used interchangeably with -trophic.

tropism An involuntary response of an organism to a stimulus. The response, such as bending, turning, or directional growth, may be either toward, or away from, the stimulus and is then referred to as positive or negative tropism.

tropocollagen The basic structural unit of collagen that consists of a triple helix having a molecular weight of about 300,000. Tropocollagen units associate to form collagen fibrils and the fibrils associate to form larger fibers. The term is also used to describe the collagen of embryonic or fetal tissue which is much more soluble than the collagen of adult tissue.

tropomyosin A minor protein of the myofilaments of striated muscle; a rigid, rod-shaped protein.

tropomyosin A A water-insoluble form of tropomyosin that is present in the catch muscles of mollusks.

tropomyosin B A water-soluble form of tropomyosin that is present in the thin filaments of the I bands of striated muscle; consists of two nonidential subunits (each having a molecular weight of 34,000), has no ATPase activity, and has a helical content of about 90%.

troponin A minor protein component of the thin filaments of striated muscle. Troponin (MW 76,000) consists of three different subunits, each named according to its function: TN-T (MW 37,000) contains the binding site for tropomyosin; TN-I (MW 21,000) is an inhibitor of actomyosin ATPase; it inhibits the interaction between myosin and actin; TN-C (MW 18,000) binds calcium ions and regulates the interactions among TN-T, TN-I, and other components of the contractile system. *Abbr* TN.

troponin C-like protein CALMODULIN.

Trp 1. Tryptophan. 2. Tryptophanyl.

Trt Trityl group.

TRU Tripartite repeating unit.

true fat NEUTRAL FAT.

true order ORDER WITH RESPECT TO CONCENTRATION.

true plasma Plasma that is obtained from whole blood after the blood has been equilibrated with carbon dioxide at a specific partial pressure.

true toxin An exotoxin against which an antitoxin can be produced.

trypanosome An intracellular protozoan parasite; a unicellular eukaryotic flagellate and the causative agent of several diseases in humans and livestock, including sleeping sickness and Chagas' disease.

trypsin An endopeptidase that acts primarily on peptide bonds in which the carbonyl group is contributed by either arginine or lysine.

tryptamine A biogenic amine derived from tryptophan by decarboxylation. It stimulates the contraction of smooth muscle and occurs in both plants and animals.

tryptase A trypsin-like mast cell proteinase, present in fairly large concentrations in skeletal muscle, lung, and skin, and believed to be involved in inflammatory reactions.

tryptic Of, or pertaining to, trypsin.

trypticase A tryptic digest of casein.

tryptic peptide A peptide obtained by digestion of a protein with trypsin.

tryptone A peptone produced by the proteolytic action of trypsin.

tryptophan An aromatic, heterocyclic, and nonpolar alpha amino acid. *Abbr* Trp; W.

tryptophan dioxygenase A widely distributed enzyme that utilizes a heme prosthetic group to cleave the pyrrole portion of the indole ring of tryptophan to yield *N*-formylkynurenine. The enzyme is inducible in the liver of higher organisms. *Aka* tryptophan pyrrolase.

tryptophan load test *See* load test.

tryptophan pyrrolase TRYPTOPHAN DIOXYGENASE.

tryptophan synthetase The enzyme that catalyzes the terminal step in the biosynthesis of tryptophan; it consists of four polypeptide chains, one of which, the A chain, has been used for studies of amino acid replacements.

T$_s$ *See* elongation factor.

ts Temperature-sensitive mutant.

TSH Thyroid-stimulating hormone.

ts mutant Temperature-sensitive mutant.

T suppressor cell *See* suppressor T cell.

T system A system of tubules that interconnects the sarcolemma and the myofibrils of striated muscle; it is instrumental in communicating the depolarization of the sarcolemma almost simultaneously to all the myofibrils of the muscle fiber. *Aka* transverse system.

t test *See* Student's *t* test.

TTP 1. Ribosylthymine triphosphate. 2. Ribosylthymine-5'-triphosphate.

T tubule *See* T system.

T$_u$ *See* elongation factor.

tube dilution method A method for determining the sensitivity of a microorganism to an antimicrobial drug; based on measuring the concentration of drug that prevents microbial growth in one or more tubes of a series of tubes that contain identical microbial inocula but different concentrations of the drug.

tuberculin A protein preparation obtained from the tubercle bacillus *Mycobacterium tuberculosis* that is used in the tuberculin test for delayed-type hypersensitivity. *See also* old tuberculin; purified protein derivative.

tuberculin hypersensitivity Delayed-type hypersensitivity to inoculation with tuberculin following an infection by the tubercle bacillus *Mycobacterium tuberculosis*.

tuberculin PPD Tuberculin purified protein derivative; *See* purified protein derivative.

tuberculin test A test, performed in humans or animals, for delayed-type hypersensitivity to tuberculin.

d-tubocurarine An active component of the neurotoxin curare that acts like α-bungarotoxin.

tubular ion exchange A set of ion exchange reactions that proceed across the kidney tubules and that represent the major mechanism for the acidification of urine by the kidney.

tubulin A dimeric protein, composed of two closely related subunits (α and β), that is the principal component of microtubules. *See also* treadmilling.

tumor NEOPLASM.

tumor angiogenesis factor A factor, released from tumors capable of unlimited growth, that causes the growth of blood vessels toward the tumors. *Abbr* TAF.

tumor antigen T ANTIGEN (1).

tumorigenesis The formation of a tumor.

tumorigenic Capable of causing tumor formation.

tumor initiator *See* initiator (4).

tumor necrosis factor A cytokine, produced naturally by macrophages in response to bacterial infection and other challenges. It appears to work synergistically with interferon and results in the killing of tumor cells. *Abbr* TNF. *See also* cytokines.

tumor progression 1. The series of changes whereby a cancerous lesion becomes more and more malignant with time. 2. The third stage in a multistage mechanism of carcinogenesis in which a dependent cancer cell is converted to an autonomous cancer cell.

tumor promoter *See* promoter (2).

tumor-specific transplantation antigen An antigen that is present on tumor cells but not on the normal cells of the organism in which the tumor developed. The presence of this antigen can induce an immune response, resulting in rejection of an situ or a transplanted tumor.

tumor virus ONCOGENIC VIRUS.

tunicamycin One of a group of homologous nucleoside antibiotics produced by *Streptomyces lysosuperificus*. Tunicamycins consist of uracil, *N*-acetylglucosamine, an aminodeoxydialdose (tunicamine), and a fatty acid of variable chain length. Tunicamycins are active against viruses, gram-positive bacteria, yeast, and fungi.

tunneling The transfer of a particle, such as an electron or a proton, across a potential energy barrier without the particle acquiring sufficient energy to surmount the energy barrier. The transfer occurs as a result of the probability distribution of the particle on both sides of the barrier. *See also* scanning tunneling microscope.

tunneling electron microscope *See* scanning tunneling microscope.

turbidimetry The quantitative determination of a substance in suspension that is based on measurements of the decrease in light transmission by the suspension due to the scattering of light by the suspended particles. *See also* nephelometry.

turbidity 1. A measure of the scattering of light by a solution; the apparent absorbance caused by scattering. Defined as $(1/l)\ln(I_0/I)$, where I is the intensity of the transmitted light, I_0 is the intensity of the incident light, ln is the natural logarithm, and l is the length of the light path; sometimes defined as $\ln(I_0/I)$. 2. The cloudiness of a solution that is caused by fine suspended particles.

turbidostat A type of chemostat in which light is passed through the growing culture and onto a photocell. When the intensity of the light striking the photocell falls below a preset value, the culture is diluted with fresh medium.

turbid plaque A plaque produced in the plaque assay or in the plaque technique when a mixture of cells is present so that not all of the cells in the area of the plaque are lysed.

turbid plaque mutant A phage mutant (plaque-type mutant) that produces turbid plaques.

turbulent flow The disturbed flow of a liquid along a tube; produced by obstacles and/or high rates of shear. *See also* laminar flow.

turgor The outward pressure of a cell caused by the osmotic imbalance between the intracellular and extracellular fluids.

turnip yellow mosaic virus A plant virus, containing single-stranded RNA, that infects turnips and other vegetables. *Abbr* TYMV.

turnover 1. The rate at which a substrate is acted upon by an enzyme and measured by the turnover number. 2. The rate at which an intracellular protein is renewed; the balance between the rate of degradation and the rate of synthesis. 3. TURNOVER TIME.

turnover number A measure of enzymatic activity, expressed either as molecular activity or as catalytic center activity. *See also* molar activity.

turnover time 1. The time required for the transfer of a substance into or out of a pool under steady-state conditions such that the amount of the substance transferred is equal to the amount of the substance in the pool. 2. The half-life of an intracellular protein that results from its intracellular degradation.

Tween Trademark for a series of nonionic detergents that consist of fatty acid esters of polyoxyethylene sorbitan.

twin ion technique A mass spectrometric technique in which a substance is labeled with two isotopes and the substance and its metabolic transformations are recognized and identified by the characteristic twin ions in their mass spectra.

twist *See* twisting number; twist conformation.

twist conformation The conformation of furanoses in which the ring is not planar; three adjacent atoms of the ring are coplanar while the other two atoms lie above or below the plane of the ring. *See also* envelope conformation.

twisting number The total number of turns of the double helix of DNA. The twisting number (T) is equal to the number of base pairs in duplex DNA divided by the number of base pairs per turn of the double helix. It is a measure of the pitch of the double helix and is related to the linking number (L) as follows: $L = W + T$, where W is the writhing number. *Sym* T.

two-carbon fragment 1. A group of atoms or a compound containing only two carbon atoms. 2. An early designation of acetyl coenzyme A.

two-dimensional chromatography A flat-bed chromatographic technique in which the compounds are first separated in one direction and, after rotation of the chromatogram by 90°, are then separated in a second direction.

two-dimensional electrophoresis A flat-bed electrophoretic technique in which the compounds are first separated in one direction and, after rotation of the electropherogram by 90°, are then separated in a second direction.

two-factor cross A genetic recombination experiment involving two nonallelic genes.

two-five A One of a group of unusual nucleotides having the formula $pppA(2'p5'A)_n$ where $n = 1$ to 10; they are produced as part of the mechanism by which interferon enables mammalian cells to resist attack by RNA viruses. The oligonucleotides are formed from ATP after a particular synthase is activated by interferon; they activate an endonuclease which degrades viral mRNA and thus inhibits viral protein synthesis.

twofold rotational symmetry The type of symmetry present in a palindrome; the sequence of bases is identical in both strands when each strand is traversed in the same sense of polarity. Thus, the double-stranded segment (a) shows twofold rotational symmetry.

$$3'-AGCT-5'$$
$$5'-TCGA-3'$$
$$(a)$$

two-genes–one-polypeptide chain *See* one-gene–one-polypeptide chain hypothesis.

two-messenger hypothesis The hypothesis that hormonal action can be interpreted by invoking a two-messenger mechanism. *See also* first messenger; second messenger.

two out of three method A proposed alternative method of codon reading that might apply in some cases; according to this method, only the first two nucleotides of a codon are recognized by the anticodon.

two-point cross A genetic recombination experiment involving two linked genes.

two-sided test A test for which the region of rejection comprises areas at both extremes of the distribution of the test statistic. *aka* double-tailed test; two-tail test.

two-state model A proposed model to explain the functioning of receptors. According to this model, a receptor exists in two states, an active and an inactive one, which are in equilibrium. Agonists bind preferentially to the active form of the receptor, thus shifting the equilibrium between the two forms toward the active form. *See also* mobile receptor model.

two-tail test TWO-SIDED TEST.

TX Thromboxane.

TXA A thromboxane that contains the following oxane ring structure:

TXB A thromboxane that contains the following oxane ring structure:

ty elements Transposable elements of yeast; the term is derived from transposon-yeast.

tygon A copolymer of vinyl chloride and vinyl acetate.

TYMV Turnip yellow mosaic virus.

Tyndall effect The phenomenon that the path of a beam of light through a solution containing colloidal particles becomes visible as a result of the scattering of light by the particles.

Tyndallization FRACTIONAL STERILIZATION.

type A hepatitis *See* hepatitis.

type A RNA virus *See* oncornavirus.

type A virus *See* oncornavirus.

type B hepatitis *See* hepatitis.

type B RNA virus *See* oncornavirus.

type B virus *See* oncornavirus.

type C RNA virus *See* oncornavirus.

type C virus *See* oncornavirus.

type I error In hypothesis testing, the erroneous rejection of a null hypothesis. *Aka* α error; error of the first kind.

type II error In hypothesis testing, the erroneous acceptance of a null hypothesis. *Aka* β error; error of the second kind.

type I immunoglobulin TYPE K IMMUNOGLOBULIN.

type II immunoglobulin TYPE L IMMUNOGLOBULIN.

type K immunoglobulin An immunoglobulin containing light chains of the kappa type.

type L immunoglobulin An immunoglobulin containing light chains of the lambda type.

type 1 reaction ANAPHYLAXIS.

type 2 reaction A form of immediate-type hypersensitivity in which antibodies combine with cell surface antigens or with cell-bound antigens.

type 3 reaction A form of immediate-type hypersensitivity that involves complement fixation.

type 4 reaction CELLULAR IMMUNITY.

type-specific antigen An antigen that is found in only one subdivision of a family of viruses.

Tyr 1. Tyrosine. 2. Tyrosyl.

tyrocidin One of a group of cyclic polypeptide antibiotics, produced by *Bacillus brevis*; gramicidin S (misnamed) is an example.

tyrosinase The enzyme that catalyzes the oxidation of tyrosine to dopa and the oxidation of dopa to dopa quinone which is then converted to melanin.

tyrosine An aromatic, polar alpha amino acid that contains a phenolic hydroxyl group. *Abbr* Tyr;Y.

tyrosinemia I TYROSINOSIS.

tyrosinemia II A genetically inherited metabolic defect in humans that is due to a deficiency of the enzyme tyrosine aminotransferase.

tyrosinosis A genetically inherited metabolic defect in humans that is characterized by excessive excretion of *p*-hydroxyphenylpyruvic acid and excretion of hawkinsin (an unusual, sulfur-containing, dicarboxylic amino acid); due to a deficiency of the enzyme *p*-hydroxyphenylpyruvic acid oxidase.

U

U 1. Uracil. 2. Uridine. 3. Enzyme unit. 4. Uranium. 5. Uniformly labeled.

U1 One of a group of small nuclear ribonucleoproteins (snRNP) that function in the splicing of nuclear pre-mRNA. U1 consists of several polypeptides and an RNA molecule (165 nucleotides) that contains a segment that is complementary to the acceptor and donor junctions of introns. Other members of the group that have been identified include U2, U4, U5, and U6.

Ubbelohde viscometer A capillary-type viscometer that permits the measurements of viscosity as a function of concentration, since the solution can be diluted directly in the viscometer.

ubiquinol Reduced coenzyme Q; $CoQH_2$.

ubiquinol : cytochrome C oxidoreductase Complex III.

ubiquinone Coenzyme Q.

ubiquitin A small, heat-stable, nonenzymatic polypeptide of 76 amino acids that is widely distributed (as its name implies) in eukaryotic cells. ATP hydrolysis drives a reaction whereby ubiquitin becomes covalently linked, via its carboxyl group, to the amino group of lysine in proteins that are substrates for proteolysis. It is believed that the attachment of ubiquitin marks a given protein for rapid intracellular degradation; it may serve to tag damaged or abnormal proteins. Ubiquitin is also involved in histone modification and appears to be part of certain receptor sites on lymphocyte surfaces.

ubiquitination The covalent linking of ubiquitin to a protein.

ubiquitous RNA A class of RNA molecules that form part of the small nuclear RNA complement and that have wide occurrence. Ribosomal 5S RNA and transfer RNAs are examples.

UDP 1. Uridine diphosphate. 2. Uridine-5'-diphosphate.

UDPG Uridine diphosphate glucose; see UDP-glucose.

UDP-Gal Uridine diphosphate galactose.

UDP-Glc Uridine diphosphate glucose; see UDP-glucose.

UDP-glucose Uridine diphosphate glucose; a nucleoside diphosphate sugar that serves as the donor of a glucose residue in the biosynthesis of glycogen. *Abbr* UDPG; UDP-Glc.

UEP Unit evolutionary period.

UFA Unesterified fatty acids.

UIBC Unsaturated iron-binding capacity.

ultimate carcinogen The form of a carcinogen in which it ultimately reacts with cellular macromolecules; the reactive intermediate formed from a carcinogen that is inacitve per se. The ultimate carcinogen is believed to be a form that contains an electrophilic center (such as a carbonium ion, free radical, epoxide, or metal cation) that attacks electron-rich centers in proteins and nucleic acids.

ultimate precision *See* method of ultimate precision.

ultra- 1. Prefix meaning excessive. 2. Prefix meaning beyond the range of.

ultracentrifugation Centrifugation, performed in an ultracentrifuge, in which high centrifugal forces are used.

ultracentrifuge A high-speed centrifuge capable of generating speeds of approximately 60,000 rpm and centrifugal forces of approximately $500,000 \times g$.

ultracryotomy The preparation of thin sections from frozen tissue specimens.

ultrafiltration The filtration of a solution through a filter or a membrane that will retain macromolecules.

ultramicrotome A microtome for cutting very thin sections of tissues, about 0.05 to 0.10 μm in thickness, for electron microscopic examination.

ultrasonic Of, or pertaining to, ultrasound.

ultrasonication The exposure of material to ultrasound; used for the rupture of cells and the denaturation of proteins.

ultrasound Sound waves that have frequencies greater than those of audible sound; refers particularly to sound waves that have frequencies of 500,000 cps or higher.

ultrastructure The fine structure of tissues, cells, and subcellular particles beyond that revealed by the light microscope.

ultraviolet dichroism The dichroism produced when ultraviolet light is absorbed by oriented samples such as nucleic acid preparations.

ultraviolet microscope A microscope that utilizes ultraviolet light for illumination of the specimen and for image formation. It has twice the resolving power of an ordinary light microscope and may be used for estimating the amount of nucleic acid in a sample.

ultraviolet optics ABSORPTION OPTICAL SYSTEM.

ultraviolet spectrum That part of the electromagnetic spectrum that covers the wavelength range of about 1.3×10^{-6} to 4×10^{-5} cm and that includes photons which are emitted or absorbed during electronic transitions; radiation of the shorter wavelengths is known as far ultraviolet, and that of the longer wavelengths is known as near ultraviolet. *Abbr* UV spectrum.

umber codon The codon UGA; one of the three termination codons. *Aka* opal codon.

umber mutation A mutation in which a codon is mutated to the umber codon, thereby causing the premature termination of the synthesis of a polypeptide chain.

UMP 1. Uridine monophosphate (uridylic acid). 2. Uridine-5'-monophosphate (5'-uridylic acid).

unbalanced growth The growth of cells such that not all of the cellular components increase by the same factor and, hence, the overall composition of the cells does not remain constant. A given increase in the concentration of protein, DNA, RNA, lipid, or some other macromolecule is not coincident with the same increase in the mass and number of cells.

uncertainty principle The principle, enunciated by Heisenberg, that it is impossible to know simultaneously, with absolute precision, both the position and the speed of a small particle, such as an electron. Specifically, the product of the uncertainty in position Δx and the uncertainty in momentum Δp can be no smaller than Planck's constant h. Thus, $\Delta x \cdot \Delta p \geq h$. Depending on the precise definition of Δx and Δp, the right side of this expression can be h, $h/2\pi$, or $h/4\pi$. *Aka* Heisenberg uncertainty principle.

uncharged polar amino acid A polar amino acid that carries no charge in the intracellular pH range of about 6 to 7.

uncharged tRNA A transfer RNA molecule that does not have an amino acid covalently linked to it.

uncoating The removal of the protein coat from a virus that occurs extracellularly in the case of bacterial viruses and, apparently, intracellularly in the case of animal viruses.

uncoating enzyme An enzyme that functions in the removal of the protein coat from a virus.

uncoded amino acid An amino acid, such as hydroxyproline or hydroxylysine, for which no codon exists. Such an amino acid is derived by enzymatic modification of the parent amino acid (e.g., proline or lysine) after the parent amino acid has become incorporated into a polypeptide chain in response to its codon.

uncompensated acidosis An acidosis in which the blood pH falls due to insufficient compensatory mechanisms.

uncompensated alkalosis An alkalosis in which the blood pH rises due to insufficient compensatory mechanisms.

uncompetitive inhibition The inhibition of the activity of an enzyme that is characterized by a decrease in the maximum velocity compared to that of the uninhibited reaction and by a reciprocal plot (1/velocity versus 1/substrate concentration) which is parallel to that of the uninhibited reaction. *See also* degree of inhibition.

uncompetitive inhibitor An inhibitor that binds to the enzyme–substrate complex but does not bind to the free enzyme.

uncoupler Uncoupling agent.

uncoupling 1. The separation of the two processes that constitute oxidative phosphorylation such that ATP synthesis is dissociated fromt he electron transport system at one or more of the phosphorylation sites. As a result of uncoupling, ATP synthesis is inhibited but electron transport and respiration are allowed to proceed and may even be stimulated. 2. The decreased sensitivity of target cells to stimulation by a ligand that is due to the fact that specific receptors, while still present, are unable to activate a particular enzyme. The two processes of ligand binding to a receptor and enzyme activation by the ligand–receptor complex have been separated. *See also* desensitization (3).

uncoupling agent An inhibitor that brings about the uncoupling of ATP synthesis from the electron transport system at one or more of the phosphorylation sites.

undecaprenol A long-chain lipid, containing 11 isoprene units, that serves as a carrier (in the form of undecaprenol phosphate) of carbohydrates in the biosynthesis of peptidoglycan in bacteria. *See also* glycosylation.

undecaprenol phosphate *See* undecaprenol.

undernutrition Malnutrition resulting from the consumption of inadequate amounts of food so that one or more of the essential nutrients are lacking in the diet. *Aka* undernourishment.

underwinding Negative supercoiling; see superhelix.

uneconomic species An undesirable species; the species that is eliminated by another, desirable, species through the use of chemicals.

uni- 1. Referring to one kinetically important substrate and/or product of a reaction; thus a uni bi reaction is a reaction with one substrate and two products. 2. Prefix meaning one.

unidirectional replication DNA replication that

proceeds in one direction; one replicating fork moving in a given direction, as in the rolling circle model.

uniformitarianism *See* doctrine of uniformitarianism.

uniformly labeled Designating a compound that is labeled in all of the positions in the molecule in a statistically uniform or nearly uniform manner; generally refers to ^{14}C-labeled compounds. *Sym* U.

unifunctional feedback A feedback mechanism that affords control in only one direction so that the input of a system is affected by either an increase or a decrease of the output; a system in which the pH will be adjusted if it rises above the normal value is an example.

unilamellar vesicle *See* vesicle.

unimolecular reaction A chemical reaction in which one molecule (or other entity) of a single reactant is transformed into products.

unineme hypothesis The hypothesis that a newly formed chromatid contains a single double-stranded DNA molecule.

unionized calcium The physiologically inactive form of calcium that cannot diffuse through a semipermeable membrane through which calcium ions (Ca^{2+}) can pass; represents largely that fraction of calcium that is bound to plasma proteins, particularly albumin.

uniport The transport of a substance across a membrane that is not linked to the transport of another substance across the same membrane. *See also* antiport; symport.

unique DNA DNA sequences that occur only once in the haploid genome. Most structural genes and their introns are unique DNA. *Aka* nonrepetitive DNA. *See also* repetitive DNA.

unitarian hypothesis The hypothesis that the injection of an antigen results in the formation of a single type of antibody that has multiple functions and that can react with antigens under all conditions, such as those of agglutination, precipitation, and complement fixation.

unitary rate The rate constant of an elementary step.

unit cell The smallest portion of a crystal that embodies the structural characteristics of the crystal and that, when repeated indefinitely, will generate the entire structure.

unit evolutionary period The time required for the amino acid sequence of a protein from two species to change by 1% (that is, develop a difference of 1 amino acid residue per 100 residues) after the two species have diverged in the course of evolution. *Abbr* UEP.

unit mass The quantity m/z, used in mass spectrometry, where m and z are the mass and charge of the fragment, respectively.

unit membrane hypothesis The hypothesis that the structure of biological membranes can be described in terms of a unit membrane, about 90 Å thick. The unit membrane is considered to be a bilayer of polar lipids, such as phospholipids, surrounded on both sides by protein. The lipids are arranged with their nonpolar portions inward and their polar portions outward where they are coated with protein molecules that are in their beta configuration. *Aka* Danielli–Davson–Robertson model; Robertson model. *See also* Danielli–Davson model; fluid mosaic model.

unit mitochondrion hypothesis A controversial idea, proposed by some biologists, according to which there exists only one mitochondrion per cell in vivo. The hundreds and thousands of mitochondria seen in electron microscope preparations are considered to be artifacts, arising from disruption of massive in vivo mitochondria as the sample is prepared for electron microscopy.

unit of complement The amount of complement that lyses 50% of the sensitized erythrocytes. *Abbr* $C'H_{50}$ unit.

unit of enzyme *See* enzyme unit.

unit of information *See* bit.

unit of transcription *See* transcription unit.

unity of biochemistry The phenomenon that widely different microbial, plant, and animal cells show a high degree of similarity with respect to both the types of molecules that they contain and the metabolic reactions that these molecules undergo.

universal buffer A mixture of several buffers that can be used over a relatively wide pH range.

universal code A code in which the same codons code for the same amino acids in all types of organisms.

universal donor An individual of the O type in the ABO blood group system, who can donate blood to any recipient.

universal recipient An individual of the AB type in the ABO blood group system, who can receive blood from any donor.

universal red shift The phenomenon that the wavelength of maximum absorption for a compound in solution generally occurs at a longer wavelength than that for the compound in the gas phase.

universe The thermodynamic system plus its surroundings.

unmasking The conversion of an unreactive amino acid residue or an unreactive site on a protein molecule to a reactive one that is accessible to specific reagents and that can participate in specific binding or other reactions. The process may involve conformational changes in the molecule and/or proteolytic removal of blocking groups.

unmixing 1. The sorting out of solute molecules from solvent molecules in a solution, a process that is accompanied by a decrease in entropy. 2. The formation of a single complex from separate particles, a process that is accompanied by a decrease in entropy.

unpaired electron A single electron, instead of the usual two, in an orbital.

unresponsiveness IMMUNOLOGICAL UNRESPONSIVENESS.

unrestricted Descriptive of a restricted mutant that has undergone the opposite modification. *See also* restricted virus.

unsaponifiable lipid The fraction of lipid in a sample that remains insoluble after saponification of the sample with alkali; it consists principally of steroids and terpenes.

unsaturated fatty acid A fatty acid that contains one or more double bonds in the alkyl chain. Unsaturated fatty acids have lower melting points than corresponding saturated fatty acids; all of the common unsaturated fatty acids are liquid at room temperature. Some unsaturated fatty acids are essential in animals but they are probably obtained in sufficient amounts in the typical human diet.

unsaturated iron-binding capacity The difference in the concentration of iron between the maximum potential concentration (transferrin fully saturated with iron) and the actual concentration (transferrin only about 25 to 30% saturated with iron); generally expressed in terms of micrograms per 100 mL of plasma. *Abbr* UIBC.

unscheduled DNA synthesis DNA synthesis that occurs during some stage of the cell cycle other than the S phase. *See also* cell cycle.

unselective marker A marker that does not affect the growth of the organism on a selective medium.

unstable mutation A mutation that has a high frequency of back mutation to the original, nonmutated state.

unstable protein TRIGGER PROTEIN.

unstructured smectic *See* liquid crystal.

untargeted mutagenesis The production of mutations that occur at locations other than the actual sites of damage in DNA. Mutations arising in ultraviolet irradiated DNA that do not occur at thymine dimer sites, are an example.

untwistase *See* topoisomerase.

untwisting enzyme *See* topoisomerase.

unwindase SINGLE-STRAND BINDING PROTEIN.

unwinding protein SINGLE-STRAND BINDING PROTEIN.

unzippering *See* zippering.

u orientation The sense of insertion of a DNA fragment into a vector such that the genetic maps of the fragment and the vector have different orientations. *See also* n orientation.

UP Uroporphyrin.

upfield Describing a peak in nuclear magnetic resonance that is on the high magnetic field side of the spectrum. *See also* shielded nucleus.

UPG Uroporphyrinogen.

uphill reaction ENDERGONIC REACTION.

up promoter mutation *See* promoter up mutation.

U protein Unstable protein; *see* trigger protein.

upstream Describing a location or a sequence of units in the opposite direction to that in which a process occurs: (a) during transcription, the location or sequence of deoxyribonucleotides on the transcribed strand of DNA from the 5′- to the 3′-end; (b) during translation, the location or the sequence of ribonucleotides on the mRNA from the 3′- to the 5′-end; (c) during replication, the location in a direction that is opposite to that of the replicating fork movement; (d) in a polypeptide chain, the sequence of amino acids in which they are linked together, from the C- to the N-terminal; (e) in the electron transport system, the sequence of electron carriers in a direction opposite to the flow of electrons, from oxygen to metabolite. *See also* downstream.

upstream activation sites Sites in eukaryotic DNA that are usually 50–300 bp upstream from the promoter and that have a regulatory role in transcription; they are believed to be binding sites for transcription factors.

upward flow A column chromatographic technique, frequently used in gel filtration, in which the eluent is passed upward through the column. This minimizes compression of the gel bed and allows for finer control of the flow rate.

UQ Ubiquinone.

UQH$_2$ Dihydroubiquinone; reduced coenzyme Q.

Ura Uracil.

uracil The pyrimidine 2,4-dioxypyrimidine that occurs in RNA. *Abbr* U; Ura.

uracil-DNA glycosidase *See* uracil-N-glycosylase.

uracil fragments Fragments that are excised during DNA replication as a result of the misincorporation of uracil in place of thymine. The fragments consists of uracil and nucleotides from which uracil has been removed. The fragments are formed by the combined action of several enzymes, including uracil-*N*-glycosylase and AP endonuclease.

uracil-N-glycosylase An enzyme that functions in DNA repair by removing a uracil that was either erroneously incorporated or formed by mutation. The action of this enzyme is then followed by that of AP endonuclease and DNA polymerase. *Aka* uracil-DNA glycosidase.

uracil-rich code An early version of the genetic code according to which the codons of all the amino acids contain at least one uracil nucleotide.

urate A salt of uric acid.

urate oxidase URICASE.

Urd Uridine.

urea A compound that is formed in the urea cycle during amino acid catabolism and that is one of the major forms in which nitrogen is excreted in the urine; urea is also the end product of purine catabolism in most fishes and amphibia. *Aka* carbamide.

urea cycle The cyclic set of reactions whereby two amino groups, derived from the catabolism of two amino acids, and a molecule of carbon dioxide are converted to urea which is then excreted.

urease An enzyme that appears to have an absolute specificity and that catalyzes only the single reaction whereby urea is hydrolyzed to carbon dioxide and ammonia.

urenzyme A rudimentary form of an enzyme in a primitive cell.

ureotelic organism An organism, such as a mammal, the excretes the nitrogen from amino acid and purine catabolism primarily in the form of urea.

urethane Ethyl carbamate; a carcinogenic agent.

Urey equilibrium The equilibrium established in the reaction in which the reactants are carbon dioxide and calcium silicate, and the products are calcium carbonate and silica. The reaction has been proposed by Urey to be the one by means of which the carbon dioxide pressure was maintained at low levels in the primitive atmosphere of the earth.

ur genes The genes coding for a "minimal" biological system that is essential for all organisms; the genes coding for rRNA, ribosomal proteins, and the enzymes and other proteins involved in DNA replication, DNA transcription, and RNA translation.

-uria 1. Combining form meaning the presence of a substance in the urine. 2. Combining form meaning the presence of excessive amounts of a substance in the urine.

uric acid A purine that does not occur in nucleic acids but that is an intermediate of purine catabolism in some organisms, and an end product of purine catabolism in humans and other organisms.

uricase A copper-containing enzyme that catalyzes the oxidation of uric acid by molecular oxygen to allantoin and hydrogen peroxide.

uricolysis The reaction catalyzed by the enzyme uricase.

uricosome An early term for peroxisome; so coined, because of the presence of urate oxidase in these organelles.

uricosuria The presence of excessive amounts of uric acid in the urine.

uricosuric drug A drug that tends to promote the excretion of uric acid.

uricotelic organism An organism, such as a bird, that excretes the nitrogen from amino acid and purine catabolism primarily in the form of uric acid.

uridine The ribonucleoside of uracil. Uridine mono-, di-, and triphosphate are abbreviated, respectively, as UMP, UDP, and UTP. The abbreviations refer to the 5'-nucleoside phosphates unless otherwise indicated. *Abbr* Urd; U.

uridine diphosphate glucose *See* UDP-glucose.

uridine nucleotide coenzyme *See* nucleotide coenzyme.

uridylic acid The ribonucleotide of uracil.

uridylylation The transfer of a 5'-UMP group (5'-uridylyl group) from UTP; specifically, the transfer of a uridylyl group from UTP, catalyzed by the enzyme uridylyl-transferase. This reaction is part of the complex regulation of the activity of the enzyme glutamine synthetase in *E. coli*.

urinalysis The chemical and physical analysis of urine.

urinometer A hydrometer designed for measuring the specific gravity of urine.

urkaryote The original eukaryotic cell from which present day eukaryotes evolved.

U1 RNA *See* U1.

urochrome The principal pigment of urine that is composed of a peptide of unknown structure and either urobilin or urobilinogen.

urogastrone A gastrointestinal hormone that inhibits the secretion of HCl by the stomach; appears to be identical to human epidermal growth factor.

urokinase A serine protease, synthesized in the kidney and found, as its name implies, in the urine; a plasminogen activator that converts plasminogen to plasmin.

uromodulin A glycoprotein of 616 amino acids that has immunosuppressive properties in vitro. It contains about 30% carbohydrate and is synthesized by the kidney; it appears to be identical to the Tamm–Horsfall glycoprotein which is the most abundant protein of renal origin in normal human urine.

uronic acid A monocarboxylic sugar acid of an aldose in which the primary alcohol group has been oxidized to a carboxyl group.

uroporphyrin The urinary pigment derived

from uroporphyrinogen. *Abbr* UP. *See also* porphyrin.

uroporphyrinogen An intermediate in the biosynthesis of the porphyrins that is formed by the linking of four molecules of porphobilinogen. *Abbr* UPG.

ursolic acid An unsaturated, pentacyclic, triterpene carboxylic acid that occurs widely in plants as the free acid or as the aglycone of triterpene saponins.

usnic acid A antibiotic produced by some lichens that is active against many gram-positive bacteria (including *Mycobacterium tuberculosis*) and some fungi.

USP Unites States Pharmacopeia; denotes a chemical that meets specifications set out in the U.S. Pharmacopeia.

USPHS United States Public Health Service.

Ussing chamber technique A technique for evaluating active transport by mounting a piece of tissue in a Lucite chamber in such a way that both electrical and chemical gradients are eliminated.

Ussing's short circuit An experimental setup for the measurement of membrane potentials and of concentration gradients across a membrane.

ustilagic acids A group of extracellular glycolipids, produced by *Ustilago* species (fungi that are plant parasites).

utile cycle A cyclic set of reactions that does accomplish something in metabolism and that, therefore, does not constitute a futile cycle.

UTP 1. Uridine triphosphate. 2. Uridine-5'-triphosphate.

UV Ultraviolet.

UV endonuclease An *E. coli* enzyme (composed of three subunits) that functions in the repair of DNA which contains thymine dimers that were produced by ultraviolet irradiation. The enzyme makes a cut at or near the 5'-end of the thymine dimer. *Aka* UVR⁺ endonuclease.

UV-induced dimer A pyrimidine dimer that is formed in DNA by exposure of the DNA to ultraviolet light. *See also* pyrimidine dimer; thymine dimer.

UV monitor A monitor, the operation of which is based on the measurement of the absorption of ultraviolet light by a solution.

UV reactivation The phenomenon that the survival of UV-irradiated lambda phage particles is higher on an irradiated host than on an unirradiated host; represents a DNA repair mechanism that involves the SOS repair system. *Aka* Weigle reactivation; W reactivation.

UVR⁺ endonuclease UV ENDONUCLEASE.

V

v Reaction rate.

v₀ 1. Initial velocity. 2. Control velocity; initial velocity in the absence of an inhibitor.

v̄ Partial specific volume.

V 1. Maximum velocity. 2. Volume. 3. Volt. 4. Valine. 5. Vanadium.

vacant lattice point model A model for the structure of water according to which the water structure is an essentially crystalline system that is closely related to an open, expanded, ice-like structure into which free and nonassociated water molecules can easily fit.

vaccenic acid An unsaturated fatty acid that contains 18 carbon atoms and one double bond; it is the major unsaturated fatty acid in *E. coli*.

vaccination An immunization in which a vaccine is administered to humans or to animals for the purpose of establishing resistance to an infectious disease.

vaccine A suspension of antigens that are derived from infectious bacteria or viruses and that, upon administration to humans or to animals, will produce active immunity and will provide protection against infection by these, or by related, bacteria or viruses.

vaccinia virus A virus of the poxvirus group that infects humans and many animals.

vacuole An intracellular structure, surrounded by a single membrane (tonoplast) and filled with fluid. Plant vacuoles are usually large and occupy a major portion of the cell volume. Animal vacuoles are much smaller and are also known as secretory vesicles or secretory granules.

vacuolysosome A lysosome that has fused with a vacuole.

vacutainer An evacuated tube used in the drawing of blood.

vacuum evaporator A vacuum chamber in which metal atoms are evaporated in the process of shadowcasting specimens for electron microscopy.

vacuum ultraviolet The range of the ultraviolet spectrum that encompasses wavelengths less than 1.8×10^{-5} cm.

Val 1. Valine. 2. Valyl.

valence 1. The number of electrons of an atom or a group of atoms that participate in the formation of chemical bonds. 2. ANTIGEN VALENCE. 3. ANTIBODY VALENCE.

valence bond theory The theory of chemical bonding that is developed by considering the atoms to have intact atomic orbitals, and then moving the atoms closer to each other with a resultant overlap of the atomic orbitals. According to this theory, a covalent bond is formed by the overlap of two orbitals, one from each bonding atom, and with each orbital holding one electron. A coordinate covalent bond is formed by the overlap of an orbital of one atom, holding two electrons, with an unoccupied orbital of a second atom.

valence electron An electron that is located in the outermost energy shell of an atom and that participates in the formation of chemical bonds.

valency Variant spelling of valence.

-valent Combining form meaning valence; used either with mono, di, ..., poly to indicate the chemical valence of atoms and ions, or with uni, bi, ..., multi to indicate the immunological valence of antigens or antibodies.

valine An aliphatic, branched, nonpolar alpha amino acid that contains five carbon atoms. *Abbr* Val; V.

valinemia A genetically inherited metabolic defect in humans due to a deficiency of the enzyme valine aminotransferase.

valinomycin An ionophorous antibiotic, produced by *Streptomyces fulvissimus*, that acts as an uncoupler of oxidative phosphorylation; a depsipeptide antibiotic.

vanadium An element that is essential to humans and animals. Symbol, V; atomic number, 23; atomic weight, 50.942; oxidation states, +2, +3, +4, +5; most abundant isotope, ^{51}V; a radioactive isotope, ^{48}V, half-life, 16 days, radiation emitted, positrons and gamma rays.

vancomycin A glycopeptide antibiotic, produced by some species of *Actinomycetes*, that inhibits peptidoglycan biosynthesis.

van Deemter equation An equation that expresses the height equivalent of a theoretical plate in chromatography as the sum of three terms: an eddy diffusion term, a molecular diffusion term, and a mass transfer term.

van Deemter plot A plot of height equivalent of a theoretical plate as a function of average gas flow rate in a gas chromatographic column.

van den Bergh reaction A colorimetric reaction for bilirubin that is based on the formation of a diazo dye from bilirubin and dia-zotized sulfanilic acid. The direct van den Bergh reaction measures direct-acting bilirubin, and the indirect van den Bergh reaction measures indirect-acting bilirubin.

van der Waals compound A compound in which neutral atoms or molecules are held together by van der Waals interactions; the dimers formed from noble gas atoms are an example.

van der Waals distance The distance between two atoms at which the attractive van der Waals force is just balanced by the repulsive force due to the orbital electrons of the two atoms; the distance is equal to the sum of the van der Waals radii of the two atoms. *Aka* van der Waals contact distance; van der Waals separation.

van der Waals forces VAN DER WAALS INTERACTIONS.

van der Waals interactions The attractive and repulsive forces between atoms and molecules that involve various dipole–dipole interactions and that consist of three components: a dispersion effect, an induction effect, and an orientation effect. The three components are classified as weak interactions, since the energy of these interactions is proportional to r^{-6} where r is the distance between the interacting species. *See also* dispersion effect; induction effect; orientation effect; weak interactions.

van der Waals radius One-half of the van der Waals distance between two like atoms; one-half of the equilibrium internuclear distance between two non-bonded atoms when the attractive and repulsive forces between the two atoms are exactly balanced. *Aka* van der Waals contact radius.

van der Waals repulsive force A repulsive force that develops when noncovalently bonded atoms or molecules approach each other closer than their van der Waals distance; due to the repulsion of the overlapping electron clouds.

van der Waals shell The space surrounding the nucleus of an atom that is described by the van der Waals radius.

van der Waals surface The complex surface of a folded protein that results when each atom is depicted as a sphere of its van der Waals radius and overlapping spheres, where the atoms are covalently bonded, are truncated. This surface has a strictly defined area and encloses a definite volume.

vanilmandelic acid 3-Methoxy-4-hydroxy-mandelic acid; a compound that is quantitatively the most important metabolite of the catecholamines and that is used to assess the endogenous production of catecholamines. *Abbr* VMA.

van Slyke method A method in which a che-mical reaction is measured by either the volume or the pressure of a gas released during the reaction.

van't Hoff complex A catalyst–reactant complex for which the conversion to free catalyst plus product proceeds at a much faster rate than the reversion to free catalyst plus reactant. *See also* Arrhenius complex.

van't Hoff equation An equation that describes the variation of the equilibrium constant K with the absolute temperature T at constant pressure; specifically, $d\ln K/dt = \Delta H^0/RT^2$, where ΔH^0 is the standard enthalpy change and R is the gas constant.

van't Hoff factor A measure of the deviation of a solution from ideality; equal to the ratio of the measured value of a colligative property of the solution to the value calculated on a molar basis.

van't Hoff isobar VAN'T HOFF EQUATION.

van't Hoff isochore VAN'T HOFF EQUATION.

van't Hoff limiting law An expression for the limiting value of the osmotic pressure of an ideal solution; specifically, $\lim_{c \to 0} \Pi = RT/M$, where Π is the osmotic pressure, c is the concentration of the solute, M is the molecular weight of the solute, R is the gas constant, and T is the absolute temperature.

van't Hoff plot A plot of the logarithm of the equilibrium constant as a function of the reciprocal of the absolute temperature. *See also* van't Hoff equation.

V antigen A surface antigen of some viruses.

vapor diffusion method A method for the slow crystallization of a substance; used in the preparation of transfer RNA crystals.

vapor–liquid partition chromatography GAS CHROMATOGRAPHY (2).

vapor phase chromatography GAS CHROMATOGRAPHY.

vapor pressure osmometer An osmometer, the operation of which is based on the lowering of the vapor pressure of a solvent by the addition of a solute.

variable Generally, any quantity that varies. More precisely, a variable in the mathematical sense; that is, a quantity that may take on any one of a specified set of values. *See also* dependent variable; independent variable; variate.

variable arm *See* arm.

variable genes *See* V genes.

variable of state STATE FUNCTION.

variable region That part of the immunoglobulin molecule in which differences in amino acid sequences are found when immunoglobulins from different sources are compared. The region comprises portions of both the light and the heavy chains and includes the two antigen binding sites. Variable regions consists of hypervariable regions, where the bulk of amino acid sequence variability occurs, and of framework regions, in which the amino acid sequence is relatively constant. *Aka* variable domain. *See also* constant region.

variable substrate The substrate the concentration of which is varied while the concentrations of other substrates are maintained at fixed values; used in kinetic studies of multisubstrate enzyme systems.

variance 1. The square of the standard deviation. 2. DEGREES OF FREEDOM (1).

variance analysis *See* analysis of variance.

variance-ratio distribution An *F* distribution; *See* F test.

variance-ratio test F TEST.

variant 1. *n* One of several forms of a protein, occurring within one species or distributed among several species, that differ from each other either in their state of aggregation, as in the case of isozymes, or in their amino acid sequence, as in the case of abnormal hemoglobins. *See also* multiple forms of an enzyme. 2. A strain that differs in some aspects from the given microorganism. 3. *adj* Dissimilar; showing variation. Said of nonidentical amino acid residues that occupy similar positions in the same polypeptide chain which is isolated from different sources.

variant repetition The occurrence of nonidentical, but related, genes per nucleus; gives rise to a family of related proteins (variants).

variate A random variable; a variable that can take on any of the values of a specified set with specified relative frequency or probability. *Aka* random variable; chance variable; stochastic variable.

varicella The primary form of the disease produced by varicella virus in a host that was not previously infected by the virus.

varicella virus A virus of the herpesvirus group that leads to varicella in hosts without immunity to the disease and to herpes zoster in hosts with immunity. It is the causal agent of chicken pox and shingles. *Aka* varicella–zoster virus.

variola virus The virus of the poxvirus group that causes smallpox.

varion A variable codon; the number of codons for a pair of homologous proteins that have been free to fix mutations over some part of the period during which the two proteins have diverged from their common ancestor; it is a joint property of the pair of nucleotide sequences. *See also* covarion.

vascular Of, or pertaining to, vessels that conduct a biological fluid; such vessels include those that carry the blood and those that conduct the sap of plants.

vasocative Having an effect on blood vessels.

vasoactive intestinal peptide A peptide hormone of 28 amino acids that is found in the gastrointestinal tract and that possesses a wide variety of biological functions when infused into an animal. These functions include vasodilation, increase of cardiac output, increase of glycogenolysis, inhibition of gastric secretion, stimulation of intestinal secretion, and stimulation of lipolysis. *Abbr* VIP.

vasoconstriction A decrease in the diameter of the blood vessels.

vasoconstrictor An agent that causes vasoconstriction.

vasodepressor An agent that brings about a lowering of the blood pressure.

vasodilation An increase in the diameter of the blood vessels.

vasodilator An agent that causes vasodilation.

vasopressin A cyclic peptide hormone, consisting of nine amino acids, that increases the blood pressure and increases the absorption of water by the kidneys; it is secreted by the posterior lobe of the pituitary gland and occurs in mammals. *Abbr* VP.

vasopressor principle VASOPRESSIN.

vasotocin A cyclic peptide hormone, consisting of nine amino acids, that is related to vasopressin in its structure and in its function; it is secreted by the posterior lobe of the pituitary gland and occurs in birds, reptiles, and some amphibians.

VCD Vibrational circular dichroism.

vector 1. A directional quantity. 2. An organism that serves to transfer a parasite from one host to another. 3. A self-replicating DNA molecule that serves to transfer a DNA segment (passenger) into a host cell in recombinant DNA technology. The three most common types of vectors used are bacterial plasmids, phages, and other viruses.

vectorial Of, or pertaining to, a vector.

vectorial discharge Descriptive of the transmembrane insertion of proteins during protein synthesis according to the signal hypothesis; the term vectorial indicates directionality of transport from the cytosol to the lumen of the rough endoplasmic reticulum and directionality with respect to the protein (N-terminal end inserted into the membrane first); the term discharge emphasizes that this is an active, energy-requiring process. *See also* signal hypothesis.

vectorial enzyme An enzyme that is directional in its action, such as one that is fixed in a membrane.

vectorial phosphorylation The coupled processes of transport and phosphorylation of monosaccharides in the bacterial phosphotransferase system.

vectorial reaction A reaction that is directional

and in which the components either are not free to move at random or are fixed in a nonrandom order; a reaction across a membrane is generally a vectorial reaction.

vectorial thioesterification The coupled processes of transport and thioesterification whereby a fatty acid is converted to fatty acyl coenzyme A in *E. coli.*

vectorial transfer CHANNELING.

vegan A strict vegetarian; an individual who eats only plant food products. *See also* lacto vegetarian; lacto–ovo vegetarian.

vegetative Pertaining to growth, particularly of plants, as opposed to reproduction.

vegetative cell An actively growing cell, as distinct from one forming spores.

vegetative DNA The pool of genetically competent DNA of a phage that is produced during the vegetative state of the phage but that has not yet been assembled into complete phage particles.

vegetative map The genetic map of a phage that is deduced from its vegetative replication.

vegetative mycelium That portion of the mycelium of a fungus that penetrates into the medium and absorbs nutrients.

vegetative nucleus 1. The macronucleus of a ciliate. 2. The tube nucleus of a pollen grain. 3. The nucleus of a cell during interphase.

vegetative phage A phage in its vegetative state.

vegetative replication The replication of vegetative DNA.

vegetative reproduction 1. Asexual reproduction. 2. The reproduction in plants without true seeds.

vegetative state The state of a phage during which it replicates actively and autonomously within the host cell; a noninfective state during which infective phage particles have not yet been assembled. During this state the phage controls the synthesis by the host of components that are necessary for the production of infective phage particles.

vehicle VECTOR (3).

vein A blood vessel that transports blood from the tissues to the heart.

velocity REACTION RATE.

velocity constant RATE CONSTANT.

venom substrate STUART FACTOR.

venous Of, or pertaining to, veins.

ventilation The process whereby oxygen is supplied to the lungs and to the blood in the capillaries of the lungs; aeration of the blood in the lungs.

venule A small vein or veinlet.

verification The deacylation of a misacylated transfer RNA molecule by its aminoacyl-tRNA synthetase, thereby regenerating the uncharged tRNA molecule.

Verner–Morrison syndrome PANCREATIC CHOLERA.

vernier A small, movable, auxiliary scale that is attached to a larger scale and that has divisions of slightly different width than those of the larger scale; used for increasing the accuracy of a measurement.

Veronal Trademark for barbital.

Versene Trademark for ethylene dinitrolotetraacetic acid.

vertebrate 1. *n* An animal that has a backbone; includes fishes, amphibia, reptiles, birds, and mammals. 2. *Adj* Of, or pertaining to, an animal that has a backbone.

vertical evolution The process wereby a species changes over time, without splitting, and evolves into a new and distinct species.

vertical transmission The transmission of viruses from one generation of hosts to the next. *Aka* vertical infection.

very early RNA PREEARLY RNA.

very high-density lipoprotein A plasma lipoprotein that has a density above 1.21 g/mL. An increase in the concentration of very high-density lipoproteins (VHDL) is believed to be linked to a decrease in the incidence of atherosclerosis. VHDL contain about 57% protein, 21% phospholipid, 17% cholesterol and cholesterol esters, and 5% triacylglycerols (triglycerides). VHDL have molecular weights of about $1.5-2.8 \times 10^5$, a sedimentation coefficient of 2–10S, and are classified as the α-fraction on the basis of electrophoresis. *Abbr* VHDL. *See also* lipoprotein.

very high-lipid lipoprotein VERY LOW-DENSITY LIPOPROTEIN.

very long-chain fatty acids *See* adrenoleukodystrophy.

very low-density lipoprotein A plasma lipoprotein that has a density of 0.95–1.006 g/mL. An increase in the concentration of very low-density lipoproteins (VLDL) is believed to be linked to an increase in the incidence of atherosclerosis. VLDL contain about 10% protein, 18% phospholipid, 22% cholesterol and cholesterol esters, and 50% triacylglycerols (triglycerides). VLDL have molecular weights of about $5.0-20 \times 10^6$, a flotation coefficient of 12–400S, and are classified as the pre-beta fraction on the basis of electrophoresis. *Abbr* VLDL. *See also* lipoprotein.

very low-lipid lipoprotein VERY HIGH-DENSITY LIPOPROTEIN.

vesicle 1. A small sac; a membranous cavity. 2. A closed bilayer structure. The structure may consist of a single bilayer (unilamellar vesicle) or of several bilayers (multilamellar vesicle). Unilamellar vesicles may be small (SUV; 0.02–0.05 μm), or large (LUV; ≥ 0.06 μm); multilamellar vesicles (MLV) range in size from 0.1 to 5.0 μm. Vesicles formed from synthetic surfactants are called surfactant vesicles; those composed of phospholipids are called liposomes.

vesicular stomatitis virus A virus of the genus rhabdovirus that infects cattle, pigs, and horses, producing symptoms as in foot and mouth disease. *Abbr* VSV.

vesiculin A highly acidic, low molecular weight (10,000) protein in synaptosome vesicles.

V genes Genes that code for segments of the variable region of immunoglobulin molecules.

v_H The variable region of the heavy chains of the immunoglobulins.

VHDL Very high-density lipoprotein.

v_i Initial velocity in the presence of an inhibitor.

viable Describing a cell or an organism that is alive and capable of reproduction.

viable count The number of viable cells in a bacterial culture.

Vi antigen A surface antigen of bacteria that is different from the O antigen.

vibrating reed electrometer A sensitive electrometer that is used for measuring the small currents produced in an ionization chamber.

vibrational circular dichroism A spectroscopic method in which one measures the difference in absorption of right- and left-circularly polarized light by chiral compounds. *Abbr* VCD.

vibrational optical activity A group of related spectroscopic methods which includes vibrational circular dichroism and Raman optical activity. *Abbr* VOA.

vibrational transition The transition of a molecule in which, as a result of the stretching or the bending of a bond, the molecule undergoes a vibration so that the atomic nuclei are temporarily displaced, but their equilibrium positions are not changed. Vibrational transitions require energies that are intermediate between those of electronic and those of rotational transitions, and they are induced by short wavelength infrared radiation.

vibronic transition A molecular transition in which the energy of vibration is added to that of the electronic transition; a transition in which both the electronic and the vibrational quantum numbers change.

vic- Combining form meaning vicinal.

vicinal Referring to two substituents on adjacent carbon atoms. *Abbr* vic.

villi Large, fringe-like projections of an epithelium. *See also* microvilli.

villikinin A gastrointestinal hormone that controls the movement of the villi.

villin A protein, isolated from intestinal microvilli, that binds to and fragments actin filaments.

vimentin A protein (MW 55,000) that is a component of intermediate filaments; vimentin filaments are found in fibroblasts.

vinblastine An indole alkaloid used as an antitumor agent in the treatment of Hodgkin's disease and other cancers. Vinblastine inhibits microtubule formation; it disrupts microtubules of the mitotic spindle and preferentially kills rapidly dividing cells.

vinca alkaloids A group of indole alkaloids that includes mitotic poisons, hypertensive agents, and anticancer compounds.

vincaleukoblastine VINBLASTINE.

vincristine An indole alkaloid, related to vinblastine and used in the treatment of leukemia and various other types of cancer.

vinculin A fibrous protein, believed to be involved in anchoring actin filaments to the inner side of the cell membrane. Vinculin is also present at patches on the cell surface, called adhesion plaques, believed to be responsible for intercellular adhesion. Cells infected with Rous sarcoma virus produce a kinase (src protein) that modifies vinculin, a reaction that may play a role in cytoskeletal changes following cell transformation by the virus.

vinegar souring ACETIFICATION.

vinyl The radical $-CH=CH_2$ that is derived from ethylene.

viologen A 4,4'-bipyridinium salt. Compounds of this type can be polymerized to yield redox polymers (polyviologens or polymeric viologens) that can serve as electron donors for some dehydrogenases.

viosterol CALCIFEROL.

VIP Vasoactive intestinal peptide.

viral Of, or pertaining to, viruses.

viral coat CAPSID.

viral hepatitis *See* hepatitis.

viral interference The inhibition of viral multiplication that may occur upon multiple infection of cells with the same virus or upon mixed infection with two or more different viruses; due to a large extent to the action of interferon that is produced in response to the infecting viruses.

viral multiplication cycle *See* multiplication cycle.

viral particle VIRUS.

viral-specific VIRUS-SPECIFIC.

viral yield The average number of infectious viral particles obtained per productive cell.

viremia The presence of virus particles in the blood. Viremia is said to be primary or secondary depending on whether it is due to progeny virions produced principally at the site of the initial infection, or whether it is due to those produced in other organs.

virgin cell A lymphocyte in peripheral lymphoid tissue that has not yet encountered an antigen. When such a cell encounters an antigen, it may be stimulated to multiply and to become an effector cell or a memory cell.

virial coefficient *See* virial expansion.

virial expansion A mathematical expansion of the general form $Y = A + BX + CX^2 + \cdots$, where X is a variable and A, B, C, etc. are the first, second, third, etc., virial coefficients. *Aka* virial equation.

viricide A chemical or a physical agent that inactivates viruses.

virion A complete viral particle consisting of a nucleocapsid and any additional structural proteins and/or envelopes; a virus.

virogene theory ONCOGENE THEORY.

viroid A virus-like infectious particle that consists of a single-stranded, covalently closed, circular RNA molecule, does not have a protein coat, and has a molecular weight of about 100,000 (about 300 nucleotides). Viroids are the causal agents of some plant diseases, such as potato spindle tuber disease, and are resistant to a wide range of treatments to which viruses are usually sensitive. Viroids appear to be related structurally to some introns and are believed by some to represent escaped introns. *Aka* infectious RNA; pathogene; pathogenic RNA.

virology The science that deals with viruses and viral diseases.

viroplasms Electron-dense aggregates of viral particles or viral components that occur in the cytoplasm of some virus-infected cells.

viroplast One of a group of spherical or elongated bodies that occur in the cytoplasm of some virus-infected plant cells; consists of an accumulation of viral particles with or without cytoplasmic components.

virosome An artificial macromolecular complex that is formed when animal viral membrane proteins and lipid films (prepared from lecithin and diacetyl phosphate) are mixed and dialyzed free of nonionic detergents.

virulence 1. PATHOGENICITY. 2. The multiplication of virulent viruses.

virulent Of, or pertaining to, virulence.

virulent virus A virus that causes lysis of the host cell that it infects; such a virus multiples in the host cell and forms progeny viral particles which are released into the medium when the cell bursts. In the case of bacteria, a virulent phage cannot become a prophage.

virus An infectious agent that consists of protein and either DNA or RNA, both of which are arranged in an ordered array and are sometimes surrounded by a membrane. A virus is generally smaller than a bacterium and is an obligate intracellular parasite at the genetic level; it uses the cell machinery to produce viral products specified by the viral nucleic acid. Viruses can be classified into 6 classes on the basis of the type of nucleic acid

that they contain and the mechanism involved in the production of viral mRNA (± represents double-stranded nucleic acid; +DNA has the same polarity as the mRNA; +RNA is identical to the mRNA; and −RNA is complementary to the mRNA):

Class I   (±DNA) → +mRNA → Protein

Class II (+DNA) → ±DNA → +mRNA → Protein

Class III (±RNA) → +mRNA → Protein

Class IV (+RNA) → −RNA → +mRNA → Protein

Class V (−RNA) → + mRNA → Protein

Class VI (+RNA) → −DNA → ±DNA → +mRNA → Protein

See also plus strand.

virus antigen A surface antigen of some viruses.

virusoid Plant satellite RNA.

virus receptor A cell membrane receptor to which a virus attaches.

virus-specific Designating a protein or a nucleic acid molecule that is coded for by a viral gene rather than by a host cell gene, and that is produced in the host cell after its infection by the virus.

virus theory of cancer *See* oncogene theory.

viscid VISCOUS.

viscoelastic Descriptive of a substance, such as a concentrated polymer solution, that exhibits properties of both liquids and solids.

viscometer An instrument for measuring the viscosity of a liquid.

Visconti–Delbrueck hypothesis The hypothesis according to which the types and frequencies of phage recombinants that are obtained upon phage infection of bacterial cells are due to the multiplication and the repeated mating of the phages in the host cells. Mating occurs in pairs and at random, and each mating produces phage recombinants as a result of the exchange of genetic material by one or more crossovers between the parental phages. The total intracellular collection of vegetative phage genomes is called a mating pool.

viscose rayon Fibers of regenerated cellulose.

viscosimeter VISCOMETER.

viscosity The resistance of a fluid to flow; the internal friction of a fluid. For Newtonian fluids, the viscosity is the force required to maintain unit velocity for a fluid flowing between two parallel plates of unit area each and a unit distance apart. *Sym* η. *Aka* coefficient of viscosity.

viscosity-average molecular weight An average molecular weight that is obtained from viscosity measurements. It is given by $\bar{M}_v = (\Sigma n_i M_i^{\alpha+1}/\Sigma n_i M_i)^{1/\alpha}$ where n_i is the number of moles of component i, M_i is the molecular weight of component i, and α is a quantity that varies with the system being studied. When $\alpha = 1$, $\bar{M}_v$ becomes equal to the weight-average molecular weight ($\bar{M}_w$); when $\alpha < 1$, $\bar{M}_v$ falls between the number-average ($\bar{M}_n$) and the weight-average molecular weight; and when $\alpha > 1$, $\bar{M}_v$ falls between the weight-average molecular weight and the z-average molecular weight ($\bar{M}_z$). *Sym* $\bar{M}_v$.

viscosity increment A measure of the asymmetry of a solute molecule that is equal to the ratio of the intrinsic viscosity of the solution to the partial specific volume of the solute.

viscosity number REDUCED VISCOSITY.

viscosity ratio RELATIVE VISCOSITY.

viscotoxins A group of homologous proteins, containing 46 amino acid residues and 3 disulfide bonds; the viscotoxins are plant toxins (phytotoxins) that act as hypotensive agents, slowing the rate of heart beat.

viscous 1. Possessing viscosity. 2. Thick; sticky; glutinous.

viscous drag The frictional force that counteracts and balances either the electrial driving force in electrophoresis or the centrifugal force in sedimentation.

visible dichroism The dichroism that is produced when visible polarized light is absorbed by oriented samples.

visible mutation A mutation that results in some alteration of the morphology of an organism.

visible spectrum That part of the electromagnetic spectrum that covers the wavelength range of about 4×10^{-5} to 7.5×10^{-5} cm and that includes photons that are emitted or absorbed during electronic transitions.

visual cycle A cyclic set of reactions that occur in both the rods and the cones of the retina whereby (a) light leads to the isomerization of 11-*cis*-retinal to the all-*trans*-retinal and to its dissociation from the appropriate opsin, and (b) the all-trans isomer is reconverted enzymatically to the 11-cis isomer which recombines with the opsin.

visual pigment One of several conjugated proteins that consist of an opsin and a form of vitamin A aldehyde and that function in the biochemical reactions that pertain to vision.

visual purple RHODOPSIN.

visual threshold The minimum light intensity required to produce a visual sensation.

vital capacity The greatest volume of air that can be expired after a forced inspiration; includes the tidal, supplemental, and complemental airs.

vitalism The doctrine that life and its phenomena are not fully explicable in terms of the laws and processes of chemistry and physics, and that they require special vital forces that are found only in living organisms. *See*

also mechanistic philosophy.

vital stain A stain that can penetrate the cell membrane of a living cell and that can stain the contents without injury to the cell.

vitamer One of two or more forms of a vitamin; vitamins A$_1$ and A$_2$ are examples of vitamers.

vitamin An organic compound that (a) occurs in natural food in extremely small concentrations and is distinct from carbohydrates, lipids, proteins, and nucleic acids; (b) is required by the organism (generally restricted to animals) in minute amounts for normal health and growth, and generally functions as a component of a coenzyme; (c) when absent from the diet, or improperly absorbed from the food, leads to the development of a specific deficiency disease; (d) cannot be synthesized by the organism and must, therefore, be obtained exclusively through the diet.

vitamin A A generic descriptor of all β-ionone derivatives, other than provitamin A carotenoids, that exhibit qualitatively the biological activity of all-*trans*-retinol. Vitamin A is a fat-soluble vitamin that is structurally related to the carotenes and that is required for certain aspects of metabolism, particularly the biochemistry of vision. Vitamin A$_1$ (retinol$_1$) predominates in higher animals and marine fish, and vitamin A$_2$ (retinol$_2$) predominates in freshwater fish; the two forms differ by one double bond in the molecule. A deficiency of vitamin A causes night blindness and xerophthalmia. The recommended names for vitamin A derivatives are as follows: retinol (vitamin A$_1$ alcohol; axerophthol); retinal or retinaldehyde (vitamin A$_1$ aldehyde; retinene); retinoic acid (vitamin A$_1$ acid); and 3-dehydroretinol (vitamin A$_2$). *See also* antipromoter.

vitamin A$_1$ *See* vitamin A.

vitamin A$_2$ *See* vitamin A.

vitamin A$_1$ acid *See* vitamin A.

vitamin A$_1$ alcohol *See* vitamin A.

vitamin A$_1$ aldehyde *See* vitamin A.

vitamin B 1. VITAMIN B COMPLEX. 2. The original antiberiberi activity.

vitamin B$_1$ THIAMINE.

vitamin B$_2$ RIBOFLAVIN.

vitamin B$_3$ PANTOTHENIC ACID.

vitamin B$_4$ An activity, isolated from yeast or liver, that could alleviate muscular weakness in rats and chicks. The existence of vitamin B$_4$ has not been confirmed since all purported vitamin B$_4$ deficiency symptoms could be alleviated by known nutritional factors such as thiamine, glycine, arginine, and cystine.

vitamin B$_5$ A growth-stimulating activity in pigeons that is probably identical with nicotinic acid.

vitamin B$_6$ A generic descriptor for all 3-hydroxyl-2-methyl pyridine derivatives that exhibit qualitatively the biological activity of pyridoxine. Major forms of the vitamin are pyridoxine, pyridoxamine, and pyridoxal. Vitamin B$_6$ is widely distributed in nature, and its phosphorylated forms function as coenzymes in amino acid metabolism, as in the transamination reaction.

vitamin B$_7$ CARNITINE.

vitamin B$_8$ The nucleotide adenylic acid that is no longer classified as a vitamin.

vitamin B$_9$ A designation that has not been used for a B vitamin.

vitamin B$_{10}$ An activity that promotes growth and feathering in the chick and that is a mixture of folic acid and vitamin B$_{12}$.

vitamin B$_{11}$ An activity that promotes growth and feathering in the chick and that is a mixture of folic acid and vitamin B$_{12}$.

vitamin B$_{12}$ A generic descriptor for all corrinoids that exhibit qualitatively the biological activity of cyanocobalamin. Vitamin B$_{12}$ is a cobalt-containing vitamin and its inadequate absorption from the intestine causes pernicious anemia. Coenzyme forms of vitamin B$_{12}$ are termed cobamides. Different forms of vitamin B$_{12}$ are named by reference to cobalamin, a derivative of cobamide. *See also* cobamide; cobalamin.

vitamin B$_{13}$ A compound, isolated from distillers dried solubles, that was provisionally called vitamin B$_{13}$ and later shown to be orotic acid, an intermediate in pyrimidine metabolism. Orotic acid is not recognized as a vitamin by United States and Canadian drug authorities.

vitamin B$_{14}$ An unconfirmed vitamin.

vitamin B$_{15}$ Preparations marketed in the United States as vitamin B$_{15}$ or pangamate (pangamic acid) have been shown to contain one or more of the following: calcium gluconate, glycine, *N,N*-dimethylglycine, and *N,N*-diisopropylamine dichloroacetate. Only the latter compound has pharmacological activity leading to decreased blood pressure and body temperature in rats. Since the compositions of these preparations are undefined, adequate clinical studies to test the claims for pangamate have not been made. At present, there is no scientific evidence that pangamate preparations have vitamin activity.

vitamin B$_{17}$ A term used to describe laetrile and/or amygdalin. Neither laetrile nor amygdalin (nor any similar cyanogenic glucosides) are recognized by United States or Canadian drug authorities as vitamins; nor is the term vitamin B$_{17}$ recognized by these authorities. *See also* laetrile; amygdalin.

vitamin B$_c$ An activity that prevents nutritional anemia in the chick and that is known to be folic acid.

vitamin B complex A group of diverse water-soluble vitamins that are classified as a group primarily for historical reasons though, to some extent, they are found together in nature. The complex includes niacin, riboflavin, folic acid, thiamine, pyridoxine, pantothenic acid, biotin, and cobalamine. The compounds choline, lipoic acid, inositol, and *p*-aminobenzoic acid are usually also classified as B vitamins. Most, if not all, of the B vitamins function as components of coenzymes.

vitamin B group VITAMIN B COMPLEX.

vitamin B$_p$ An activity that prevents perosis in the chick and that is replaceable by a mixture of choline and manganese.

vitamin B$_t$ An activity that promotes insect growth and that contains carnitine as the active component.

vitamin B$_w$ BIOTIN.

vitamin B$_x$ *p*-AMINOBENZOIC ACID.

vitamin C A generic descriptor for all compounds that exhibit qualitatively the biological activity of ascorbic acid. Vitamin C is a water-soluble vitamin that occurs in fruits and vegetables. Vitamin C functions in the regulation of oxidation–reduction reactions in metabolism; a deficiency of vitamin C causes scurvy. *See also* antipromoter.

vitamin C$_2$ *See* bioflavonoids.

vitamin D A generic descriptor for all steroids that exhibit qualitatively the biological activity of cholecalciferol; a group of fat-soluble vitamins, structurally related to the sterols, that are active in the prevention and cure of rickets. Since they can be derived from sterols by ultraviolet irradiation, vitamin D is not required in the diet if the organism has adequate access to ultraviolet light (present in sunlight). Vitamin D affects the absorption and deposition of calcium phosphate. *See also* 1,25-dihydroxycholecalciferol.

vitamin D$_1$ Originally considered to be a pure vitamin but later shown to be a mixture of vitamin D$_2$ (ergocalciferol) and lumisterol.

vitamin D$_2$ ERGOCALCIFEROL.

vitamin D$_3$ CHOLECALCIFEROL.

vitamin D-resistant rickets A form of rickets that is indistinguishable in its symptoms from ordinary rickets but differs from the latter in being resistant to treatment with vitamin D. It can be due to a number of unrelated causes. The most common form is a genetically inherited metabolic defect involving the renal transport mechanism for phosphorus.

vitamin E A generic descriptor for all tocol and tocotrienol derivatives that exhibit qualitatively the biological activity of α-tocopherol; a group of fat-soluble vitamins that are required for normal growth and fertility of animals. Tocopherols are widely distributed in nature and function primarily as antioxidants. *See also* antipromoter.

vitamin F 1. Obsolete designation for the activity of the essential fatty acids as reflected in the prevention of atherosclerosis in animals; confirmed to be the compound *cis,cis*-linoleic acid. 2. Obsolete designation for thiamine.

vitamin G Obsolete designation for riboflavin.

vitamin GH₃ *See* gerovital.

vitamin H Obsolete designation for biotin.

vitamin H₃ *See* gerovital.

vitamin hypothesis The hypothesis that certain compounds having the properties of vitamins constitute an essential dietary requirement of an organism.

vitamin I Obsolete designation for vitamin B_7.

vitamin J Obsolete designation for vitamin C_2.

vitamin K A generic descriptor for 2-methyl-1,4-naphthoquinone and all of its derivatives that exhibit an antihemorrhagic activity in animals fed a vitamin K-deficient diet. The compound 2-methyl-3-phytyl-1,4-naphthoquinone is generally called vitamin K or phylloquinone (phytylmenaquinone; phytonadione). The vitamin K derivatives represent a group of widely distributed fat-soluble vitamins that have quinone-type structures and that are required for the formation of prothrombin. A deficiency of vitamin K causes prolonged clotting times and hemorrhagic disease.

vitamin L₁ A liver filtrate activity believed to be necessary for lactation and shown to be anthranilic acid.

vitamin L₂ A yeast filtrate activity believed to be necessary for lactation and shown to be adenylthiomethylpentose.

vitamin M An activity that prevents nutritional anemia and leucopenia in the monkey; the compound is an active form of folic acid.

vitamin N Obsolete designation for a mixture obtained from either brain or stomach and believed to inhibit cancer.

vitamin P *See* bioflavonoids.

vitamin PP Obsolete designation for nicotinic acid.

vitamin R An activity that promotes bacterial growth and that is apparently related to folic acid.

vitamin S An activity that promotes the growth of chicks and that is related to the peptide streptogenin.

vitamin T A group of activities isolated from termites, yeasts, or molds and reported to increase growth rates, improve wound healing, alleviate skin disorders, and accelerate insect development; a varied mixture of folacin,

vitamin B_{12}, deoxyribosides, and amino acids.

vitamin U The methylsulfonium salts of methionine, which occur naturally in cabbage and other green vegetables, and which have been claimed to alleviate peptic ulcers in guinea pigs. These claims were not supported by later clinical trials and the compounds are not recognized as vitamins by United States and Canadian drug authorities.

vitamin V An activity from tissue that promotes bacterial growth and that has been shown to be a mixture of NAD^+ and NADH.

vitellenin The lipid-free protein of lipovitellenin.

vitellin The principal protein in hens' egg yolk; the lipid-free protein of lipovitellin. A glycoprotein that contains about 1% phosphate and that is derived from vitellogenin.

vitelline membrane A membrane that surrounds the ovum.

vitellogenesis The formation of egg yolk.

vitellogenic hormone ALLATUM HORMONE.

vitellogenin An egg yolk glycoprotein of amphibia and birds. It is the precursor of the proteins phosvitin and lipovitellin. The synthesis of vitellogenin is stimulated by female sex hormones.

vitreous Glass-like.

vitreous humor The gel-like material that fills the posterior cavity of the eye.

vivisection The cutting of, or operating on, a living animal for purposes of experimentation.

v_L The variable region of the light chains of the immunoglobulins.

VLCFA Very long-chain fatty acids.

VLDL Very low-density lipoprotein.

VMA Vanilmandelic acid.

V_{max} Maximum velocity.

VOA Vibrational optical activity.

Voges–Proskauer test A test for organisms that carry out butylene glycol fermentation; based on the production of a pink color by the reaction of acetoin, formed during the fermentation, with creatine in an alkaline solution.

void volume 1. The volume of solvent in column chromatography that is equal to the total bed volume of the column minus the volume occupied by the particles of the support; the volume of solvent that is external to the support particles. 2. OUTER VOLUME.

vol Volume.

volatile buffer A buffer that can be evaporated without leaving a residue, as one consisting of ammonium formate and ammonium hydroxide.

volt A unit of electrical potential which is equal to the potential required to make a current of 1 A flow through a resistance of 1 Ω. *Sym* V.

voltage clamp technique A technique for studying the effects of membrane potentials that are intermediate between the resting potential and the peak of the action potential; based on the production of a sudden displacement of the membrane potential from its resting value by means of a pair of electrodes, and on holding the potential across the membrane at this new level by means of a feedback amplifier. The current that flows through a definite area of the membrane, maintained by a space clamp, under the influence of this applied voltage (called command voltage) is then measured with a separate pair of electrodes and a separate amplifier.

voltage-gated channel *See* gated channel.

voltammetry The electroanalytical study of chemical rections by the measurement of the currents produced by the electrolysis of oxidizable or reducible substances as a function of the applied voltage.

voltammogram *See* cyclic voltammetry.

volume fraction The fraction of the solution volume that is occupied by the solute.

volume receptor A receptor in the central nervous system that responds to changes in the volume of the blood.

volumetric analysis A method of chemical analysis that is based on the measurement of the volume of a standard solution which is required to react completely with a sample of the substance that is being determined.

volumetric pipet A pipet that is enlarged to a bulb in its center and that is used for the transfer of a fixed volume of liquid.

volumetric technique A gasometric technique in which gas volumes are measured.

volutin granule A microbial storage granule of polymetaphosphate that stains metachromatically; found in cyanobacteria, yeast, and other microorganisms.

von Gierke's disease GLYCOGEN STORAGE DISEASE TYPE I.

von Willebrand factor FACTOR VIII.

von Willebrand's disease A genetically inherited metabolic defect in humans, characterized by blood coagulation disorder and slow aggregation of platelets, and due to a deficiency of Factor VIII.

vortex stirrer A device for mixing a solution in a test tube by forming a vortex in the solution.

VP Vasopressin.

VPC Vapor phase chromatography.

VPg Genome-linked viral protein; a small protein that is linked covalently to the 5'-end of the RNA of poliovirus. Its function is to prime viral replication.

V region Variable region.

VSV Vesicular stomatitis virus.

V system A system composed of a regulatory enzyme and an effector, such that the effector alters the maximum velocity of the reaction but does not alter the substrate concentration at which one half of the maximum velocity is obtained.

vulcanization The cross-linking of natural rubber chains by means of sulfur.

v/v The concentration of a solution that is expressed in terms of volume per unit volume.

vWF von Willebrand factor.

W

W 1. Tryptophan. 2. Watt. 3. Wyosine. 4. Writhing number.

Walden inversion The alteration of the configuration of an asymmetric center in a molecule as a result of a bimolecular displacement reaction.

walking *See* chromosome walking.

walking down the helix The movement of the replisome along the DNA during replication.

wall effect The spreading and curving of a zone as it migrates downward in a chromatographic column as a result of the increased flow rate at the wall of the tube compared to that at the center.

walling-off effect AFFERENT INHIBITION.

wandering spot procedure A method for sequencing synthetic oligonucleotides. Involves labeling the oligonucleotide at the 5'-end with ^{32}P, using polynucleotide kinase, and subsequent partial digestion of the oligonucleotide, using snake venom phosphodiesterase. The resulting labeled fragments are fractionated, first by electrophoresis on cellulose acetate paper at pH 3.5, and then by homochromatography on DEAE-cellulose. The sequence is determined by the characteristic mobility shifts of the labeled fragments.

Warburg apparatus A manometer that is used for studying the cellular respiration of tissue slices or cells by measuring oxygen uptake and/or carbon dioxide evolution. *Aka* Warburg manometer; Warburg–Barcroft apparatus.

Warburg coefficient METABOLIC QUOTIENT.

Warburg–Dickens–Horecker cycle HEXOSE MONO-PHOSPHATE SHUNT.

Warburg–Dickens pathway HEXOSE MONOPHOS-PHATE SHUNT.

Warburg effect 1. The inhibition of photosynthesis by high concentrations of oxygen. 2. The overproduction of lactic acid that occurs in many tumors. *See also* Warburg theory.

Warburg method A manometric method for studying the cellular respiration of tissue slices or cells by means of a Warburg apparatus; the tissue slices or cells are maintained at a constant temperature and the changes in gas pressure are measured with a constant volume manometer. *See also* Warburg's direct method.

Warburg's atmungsferment CYTOCHROME OX-IDASE.

Warburg's direct method A manometric

method for studying the cellular respiration of tissue slices or cells in which the only gases exchanged are oxygen and carbon dioxide, and in which respiration is measured by the absorption of carbon dioxide in alkali.

Warburg's respiratory enzyme CYTOCHROME OX-IDASE.

Warburg's yellow enzyme OLD YELLOW EN-ZYME.

Warburg theory A theory of cancer that is based on the universal occurrence and importance of glycolysis in the metabolism of cells. According to this theory, cancer results from an irreversible injury to respiration, specifically to the electron transport system, which is followed by a changeover in the cell to an anaerobic, glycolytic, and fermentative-type metabolism for energy production; the main difference between a normal and a tumor cell is, therefore, a shift in metabolism toward that of an anaerobic state, in which glycolysis is emphasized.

Warburg vessel A receptacle for the tissue slices or the cells, the cellular respiration of which is being measured in a Warburg apparatus.

warfarin A synthetic compound, closely related to dicumarol, that is used as a rat poison; it blocks prothrombin activation in blood clotting.

Waring Blendor Trademark for a blender used in the preparation of tissue homogenates.

warm antibody An antibody that has a higher titer at elevated temperatures.

warm-blooded HOMOIOTHERMIC.

Wasserman test A complement fixation test that is used in the diagnosis of syphilis and that is based on the reaction of cardiolipin with the Wasserman antibody; cardiolipin serves as an antigen and is mixed with lecithin and cholesterol to form micelle-like structures that enhance its reactivity.

wasting disease A disease that is characterized by a decrease in the amount of certain tissues or organs; a pathological condition in which there is atrophy.

water balance The reactions and factors involved in maintaining a constant internal environment in the body with respect to the distribution of water between the various fluid compartments and with respect to the establishment of an equilibrium between the intake of water and its output.

water bath A water-filled container for maintaining immersed tubes, flasks, etc., at a constant temperature.

water compartment FLUID COMPARTMENT.

waterfall sequence CASCADE MECHANISM.

water hydrate model A model for the structure of water according to which the water structure results from the formation of clathrate-type structures which consist of ordered, but highly random, labile frameworks in which the cages are occupied by unbonded water molecules.

water intoxication An extreme case of hypotonic expansion that may lead to convulsions and death.

water of hydration Water that surrounds and binds to solute molecules and/or forms hydrates.

water regain The uptake of water by a dry gel expressed as grams of water taken up per gram of gel when the latter is suspended in water and allowed to swell.

water softening The conversion of hard to soft water by ion exchange chromatography or by some other method.

water-soluble B An early designation for a fraction of water-soluble vitamins prepared from egg yolk.

water-soluble vitamin One of a group of vitamins that are soluble in water and that include those of the vitamin B complex as well as some other vitamins such as vitamin C and vitamin P; most, if not all, of the water-soluble vitamins function by being components of coenzymes.

water structure See distorted bond model; flickering cluster model; vacant lattice point model; water hydrate model.

Watson–Crick model The model proposed by Watson and Crick in 1953 for the structure of DNA. According to this model, DNA consists of two right-handed helical polynucleotide chains coiled around the same axis to form a double helix. The chains are antiparallel, with the deoxyribose–phosphate backbone on the outside and the purines and pyrimidines stacked perpendicularly to the fiber axis on the inside. The bases are held together by hydrogen bonds and they are specifically paired: adenine and thymine by means of two hydrogen bonds, and guanine and cytosine by means of three hydrogen bonds; the two chains are, therefore, complementary.

Watson–Crick-type DNA A DNA molecule that can be described by the Watson–Crick model of DNA.

Watson strand The DNA strand of Watson–Crick-type DNA that is transcribed in vivo; the sense strand. *Abbr* W strand.

watt unit of power equal to 1 J/s. *Sym* W.

wavelength The distance, along the axis of propagation, between two identical points of a wave.

wave mechanics quantum mechanics.

wave number The reciprocal of the wavelength; the reciprocal of waves per unit length.

wave shifter SECONDARY FLuorer.

wax A neutral lipid consisting wavelength. from fatty acids and long-chain other formed than glycerol.

wax acids Fatty acids that occur in wax.

wax alcohols Long-chain, monohydroxy, phatic alcohols that form waxes by esterification to fatty acids.

WBC White blood cell.

weak electrolyte An electrolyte that is only partially dissociated into ions in water.

weak interactions The attractive and repulsive forces between atoms and molecules that are less strong than those of covalent bonds, ionic bonds, and ion–ion interactions. Weak interactions include the forces of hydrogen bonds, hydrophobic bonds, van der Waals interactions, and charge fluctuation interactions.

weak promoter LOW-LEVEL PROMOTER.

weight-average molecular weight An average molecular weight that is weighted toward the heavier molecules in a mixture of molecules; specifically, $\bar{M}_w = \Sigma n_i M_i^2 / \Sigma n_i M_i$, where n_i is the number of moles of component i, and M_i is the molecular weight of component i. *Sym* $\bar{M}_w$. *See also* average molecular weight.

weighted average The average of a set of numbers obtained by multiplying each number by a weight (factor) expressing its relative importance and then dividing the sum of these products by the sum of the weights.

weighted mean WEIGHTED AVERAGE.

Weigle reactivation UV REACTIVATION.

Weinberg–Salam theory A unified field theory of the weak and electromagnetic forces of particle physics. *See also* elementary particles.

Welan gum A water-soluble, heteropolysaccharide, produced by fermentation of some species of *Alcaligenes*, that yields solutions of high viscosity at low concentrations.

western blotting A variation of the Southern blotting technique in which proteins are separated electrophoretically, transferred to a special paper which binds them covalently, and are then located by reaction with a radioactive probe (for example, radioactive double-stranded DNA for DNA binding proteins or radioactive antibodies for protein antigens). *Aka* western transfer. *See also* blotting.

wet application A method of sample application in electrophoresis in which the sample is

applied to a previously wetted support (~
as cellulose acetate, for example). ...at en-

wetting agent A surface-active age~ a solid
hances the spreading of a lio~·
~sue, or of col-
surface.

wet weight The weight o~ater has not been
lected cells, from w~·
~cous solution from the
removed.

Wharton's jelly ~at is frequently used as a
umbilical c~ for the preparation of hyalur-
starting~Trademark for a group of filter pap-
onic

W~ **and** **erythema** **response** A local,
cutaneous anaphylactic reaction in humans
that is produced by the intracutaneous injection of an allergen. *Aka* wheal and flare
response.

wheat germ agglutinin A lectin, isolated from
wheat germ, that binds to *N*-acetyl-
glucosamine on cell surfaces.

whey proteins The milk proteins that are
obtained by acidifying skim milk to pH 4.7
and removing the precipitated casein. Whey
proteins account for about 20% of the total
milk proteins.

white adipose tissue WHITE FAT.

white fat 1. Ordinary adipose tissue as opposed
to brown fat. 2. Adipose tissue in which the
fat is present in a single droplet within the fat
cell. *Aka* white adipose tissue. *See also* brown
fat.

white muscle A pale skeletal muscle that has a
relatively low content of myoglobin and
cytochromes; it is nearly devoid of mitochondria and obtains almost all of its ATP from
glycolysis. It is capable of short bursts of
activity. *Aka* fast-twitch muscle.

white plaque A plaque that does not stain with
neutral red in the plaque assay; the lack of
staining is due to a destruction of lysosomes in
the virus-infected cells.

White's solution A synthetic medium for the
growth of plant cells.

whole blood Blood that has not been fractionated in any way.

whole body counter A large, external-sample-
type liquid scintillation counter that is designed for the determination of total body
radioactivity in humans or in animals.

whole plasma Plasma that has not been fractionated in any way.

whole serum Serum that has not been fractionated in any way.

wiggle SQUIGGLE.

Wijs iodine number An iodine number determined by the use of a solution of iodine in
glacial acetic acid, with iodine chloride serving
as an accelerator of the reaction.

wild-type 1. The typical, most frequently encountered phenotype of an organism in nature. 2. The phenotype of an organism that is
used as a standard of comparison for mutants
of the same organism.

wild-type allele *See* wild-type gene.

wild-type gene 1. The normal, most frequently
encountered allele of a given gene of an
organism. 2. An allele of a given gene that is
arbitrarily selected as a standard of comparison for mutant alleles of the same gene.

Willebrand factor *See* von Willebrand factor.

Wilson chamber CLOUD CHAMBER.

Wilson's disease A genetically inherited metabolic defect in humans that is due to a deficiency of ceruloplasmin, resulting in an increase in the level of copper in the brain and
in the liver.

Wilzbach method A method for the random
labeling of a compound with tritium by exposing the compound to tritium gas in a sealed
container for several weeks. A modification
of this method entails the exposure of the
compound to tritium gas in the presence of
a silent electric discharge. *Aka* Wilzbach gas
exposure.

winding number LINKING NUMBER.

window CHANNEL.

windowless counter A radiation counter in
which the sample is not separated from the
ionization detector by either a window or a
membrane.

windowless gas flow counter A radiation counter that incorporates the characteristics of
both a gas flow and a windowless counter.

wobble base The third base (the 5'-end) in the
anticodon of transfer RNA; the wobble base
can bind with one of several possible bases at
the 3'-end of the codon.

wobble hypothesis A hypothesis proposed by
Crick to explain how one tRNA molecule can
"recognize" more than one codon. According
to this hypothesis, the first two bases (the
3'-end) of the anticodon in tRNA bind to the
first two bases (the 5'-end) of a codon in regular base-pairing fashion. The third base,
however, of the tRNA anticodon (the 5'-end),
while hydrogen bonding, has a certain amount
of play or wobble that permits it to bind to
one of several possible bases at the 3'-end of
the codon.

wobble pairing The base pairing that is allowed
according to the wobble hypothesis.

Wohl–Zemplen degradation A degradative
technique for aldoses whereby the carbon of
the reducing group is removed by treatment
with hydroxylamine, and the sugar is converted to the next lower aldose.

Wolfson's method An analytical procedure for
determining gamma globulins in serum by

precipitating them with ammonium sulfate and sodium chloride.

Wolman's disease A genetically inherited metabolic defect in humans that is due to a deficiency of a lysosomal lipase (cholesterol ester hydrolase) that normaly catalyzes the hydrolysis of cholesterol esters carried by low-density lipoproteins; one of the lysosomal diseases.

wood alcohol Methanol; methyl alcohol.

wood sugar D-XYLOSE.

wool fat LANOLIN.

Woolf–Augustinsson plot A single-reciprocal plot of the Michaelis–Menten equation; a plot of v as a function of $v/[S]$, where v is the velocity of the reaction and $[S]$ is the substrate concentration. *Aka* Woolf–Augustinsson–Hofstee plot.

wool wax LANOLIN.

word CODON.

working hypothesis A hypothesis the formulation of which provides a basis for further experiments.

wormlike coil model A model that is frequently invoked to describe the DNA molecule in solution. The model is characterized by a contour length and a persistence length; the latter increases with increasing stiffness of the molecule but is independent of the former. According to this model, the DNA molecule is best approximated as a rod having a continuum of flexible distortions; such a structure is intermediate between that of a rigid rod and that of a random coil. The wormlike coil model can be considered to be a limiting case of a freely jointed chain. *See also* freely jointed chain.

wound tumor virus A plant virus, containing double-stranded RNA, that infects a large number of unrelated plants.

woven bone Bone that consists of random, nonparallel collagen fibers; found in embryonic life and during bond repair in adults.

W reactivation Weigle reactivation; *see* UV reactivation.

writhe WRITHING NUMBER.

writhing number The number of superhelical turns in DNA; the number of turns that the axis of the double-stranded helix makes in space in the process of forming the superhelix. The writhing number (W) is a measure of the extent of supercoiling of the DNA. It is related to the linking number (L) as follows: $L = W + T$, where T is the twisting number. *Sym* W.

W strand Watson strand.

wt Weight.

w/v The concentration of a solution that is expressed in terms of weight per unit volume.

w/w The concentration of a solution that is expressed in terms of weight per unit weight.

wyosine A hypermodified nucleoside, related to guanosine, that is found in tRNA; the base consists of three nitrogen-containing, fused rings. *Abbr* W.

X

X 1. Xanthine. 2. Xanthosine. 3. An uniden-
tified amino acid in an amino acid sequence.

Xan Xanthine.

xanthine A purine that is formed in catabolism
by the deamination of guanine, and that does
not occur in nucleic acids. *Abbr* Xan; X.

xanthine oxidase An enzyme of purine catabolism that catalyzes the oxidation of xanthine to
uric acid and the oxidation of hypoxanthine
to xanthine; a molybdenum-containing flavoprotein that also catalyzes the oxidation of
aldehydes. *Abbr* XO.

xanthine oxidase factor Obsolete designation
for inorganic molybdate.

xanthinuria A genetically inherited metabolic
defect in humans that is characterized by the
presence of excessive amounts of xanthine in
the urine and that is due to a deficiency of the
enzyme xanthine oxidase.

xanthoma A deposit of lipid in the skin; usually yellow-orange in color due to the presence
of lipid-soluble pigments such as carotenes.

xanthophyll An oxygenated carotenoid.

xanthoproteic reaction A reaction for the qualitative determination of proteins that is based
on the successive production of a white, yellow, and orange precipitate upon treatment of
the sample with nitric acid, followed by heating and the addition of alkali.

xanthopterin A pterin that functions as a pigment in insects.

xanthosine The ribonucleoside of xanthine.
Xanthosine mono-, di-, and triphosphate are
abbreviated, respectively, as XMP, XDP, and
XTP. The abbreviations refer to the 5'-
nucleoside phosphates unless otherwise indicated. *Abbr* Xao; X.

xanthurenic aciduria A genetically inherited
metabolic defect in humans due to a deficiency of the enzyme kynureninase, an enzyme
of tryptophan metabolism.

xanthylic acid The ribonucleotide of xanthine.

Xao Xanthosine.

X body VIROPLAST.

X cell *See* XYZ cell theory.

X chromosome A sex chromosome of which
the female generally carries two and the male
carries one per cell.

XDP 1. Xanthosine diphosphate. 2. Xan-
thosine-5'-diphosphate.

xeno- Prefix meaning foreign.

xenobiotic 1. *n* A synthetic compound that is
foreign to living systems; drugs, insecticides,

anesthetics, and petroleum products are some
examples. 2. *adj* Of, or pertaining to, a compound that is foreign to living systems.

xenogeneic Referring to a transplant from one
species to another.

xenograft HETEROGRAFT.

xenoplastic Referring to a transplant from one
genus to another.

xenosome Proposed term for intracytoplasmic
symbiotic bacteria found in the marine protozoan *Parauronema acutum*. These symbionts
are obligate parasites that grow and divide in
synchrony with the host and cannot be cultured outside the protozoan.

xenotropic virus An endogenous retrovirus
that does not replicate in animal cells in which
it is endogenous but can grow in cells of other
species.

xero- Prefix meaning dry.

xeroderma pigmentosum A genetically inherited metabolic defect in humans that is due to
a deficiency in the excision–repair of pyrimidine dimers in DNA. Patients with this disease are abnormally sensitive to sunlight and
show a high incidence of skin cancer.

xerogel A gel in which the removal of the dispersing agent (the solvent) results in the structure shrinking to an unswollen state.

xerophthalmia A pathological change in the
eye that results from a deficiency of vitamin
A; involves an extreme dryness of the con-
junctiva which loses its luster and becomes
skin-like from a lack of intrinsic secretion.

xerophthol VITAMIN A.

xerophyte A plant growing in an arid environment; a drought-resisting plant. *Var sp*
serophyte.

Xis Excisionase.

X linkage SEX LINKAGE.

XMDIC Xylylene-*m*-diisocyanate; a bifunctional reagent used to label antibodies with
ferritin.

XMP 1. Xanthosine monophosphate (xanthylic
acid). 2. Xanthosine-5'-monophosphate (5'-
xanthylic acid).

XO Xanthine oxidase.

x ray An electromagnetic radiation that is produced when high-speed electrons are suddenly
stopped by a metal target. The impinging
electrons raise target electrons to higher energy levels, and when these electrons fall back
to lower energy levels the excitation energy
is given off in the form of x rays. The

wavelength of x rays extends from about 10^{-10} to 2.5×10^{-6} cm; x rays are a form of ionizing radiation. *See also* hard x rays; soft x rays.

x-ray analysis X-RAY CRYSTALLOGRAPHY.

x-ray crystallography The study of the three-dimensional structure of molecules in a crystal by means of x-ray diffraction.

x-ray diffraction The diffraction patterns obtained when x rays are reflected from the atoms in a crystal; a useful method for determining the structure of macromolecules. The x rays are scattered by the extranuclear electrons of the atoms as opposed to neutrons which are scattered by the atomic nuclei. *Abbr* XRD. *Aka* x-ray scattering.

x-ray diffraction pattern The pattern of spots, arcs, and rings that is produced by x-ray diffraction.

x-ray film A photographic film that is coated with a sensitive emulsion on both sides and that is used in x-ray crystallography.

x-ray microanalysis A technique for determining the relative and absolute concentrations of elements within cells and tissues. Based on bombarding freeze-dried specimen sections with electrons (as in the electron microscope) and then analyzing the x rays produced by various metallic and nonmetallic elements in the specimen.

x-ray structure The structural arrangement of the atoms in a molecule or in a crystal as deduced from x-ray diffraction patterns.

XRD X-ray diffraction.

XRF X-ray fluorescence; a spectrophotometric method in which fluorescence is produced by means of x rays.

X^2 test Chi-square test.

XTP 1. Xanthosine triphosphate. 2. Xanthosine-5′-triphosphate.

Xul Xylulose.

Xyl Xylose.

xylan A homopolysaccharide of xylose that occurs in plants.

xylem A group of specialized plant cells that are dead and that have thickened cell walls. They are aligned to form tubes and function in the transport of water and inorganic ions in the plant. *See also* phloem.

xylitol A sugar alcohol derived from xylose that can be used as a sugar substitute in the treatment of human diabetes.

xylose A five-carbon aldose. *Abbr* Xyl. *Aka* wood sugar.

xylulose A five-carbon ketose that is an intermediate in the hexose monophosphate shunt. *Abbr* Xul.

X-Y recorder A recorder in which two signals are recorded simultaneously by one pen; the pen is driven in one direction, the X axis, in response to one signal, and it is driven in the other direction, the Y axis, in response to the second signal.

XYZ cell theory A theory according to which immunocytes are classified into three categories: X cells are immunologically competent cells that are not yet engaged in any specific immunological response; Y cells are immunologically activated, or primed, cells as a result of an interaction between X cells and antigen; and Z cells are antibody-producing cells, formed as a result of a second stimulation of the Y cells by antigen. *Aka* Sterzl theory.

Y

Y 1. Pyrimidine nucleoside. 2. Tyrosine. 3. Yttrium.

Y base A highly modified guanine residue in tRNA that exhibits a characteristic fluorescence.

Y cell *See* XYZ cell theory.

Y chromosome A sex chromosome that is generally the mate of the X chromosome in the male.

yeast A lower fungus that reproduces by budding and that is characterized by either short or nonexistent mycelia; refers particularly to fungi of the genus *Saccharomyces*.

yeast eluate factor Obsolete designation of vitamin B_6.

yeast filtrate factor Obsolete designation for pantothenic acid.

yeast nucleic acid An early designation for RNA.

yellow enzyme One of a group of flavoprotein dehydrogenases that contain a yellow flavin prosthetic group. *See also* old yellow enzyme.

yellow protein reaction XANTHOPROTEIC REACTION.

Y fork REPLICATING FORK.

yield 1. For a general chemical reaction: the weight of product obtained divided by the theoretical yield of product. 2. For the isolation of an enzyme: the total activity at a given step in the isolation divided by the total activity at a reference step.

yield coefficient The weight of bacteria obtained from a culture divided by the weight of a limiting material that was utilized by the bacteria during their growth. *Aka* yield constant.

yin–yang hypothesis The hypothesis that cyclic AMP (cAMP) and cyclic GMP (cGMP) are the opposing arms of a bidirectional intracellular control system. This is based on the finding that these two compounds often undergo inverse changes in concentration (one increase while the other decreases) in cellular events in which opposing or bidirectional processes are being regulated.

ylide A dipolar compound with adjacent positive and negative charges; a dipolar carbanion. The dipolar form of thiamine pyrophosphate is an example:

$$\overset{+}{N}\underset{}{R},\ CH_3,\ (CH_2)_2-O-\overset{O}{\underset{OH}{P}}-O-\overset{O}{\underset{OH}{P}}-OH$$

yogurt A fermented milk product, generally made by adding a culture of *Lactobacillus bulgaricus* to milk and incubating the mixture. *Var sp* yoghurt; yohourt.

Young–Helmholtz trichromatic theory The theory according to which color vision is due to a set of at least three pigments in the cones for the perception of red, green, and violet, respectively, and the perception of other colors results from the combined stimulation of two or more of these pigments.

Yphantis method MENISCUS DEPLETION SEDIMENTATION EQUILIBRIUM.

Z

Z 1. Average net charge of an ion. 2. Atomic number. 3. The sum of glutamic acid and glutamine when the amide content is either unknown or unspecified. 4. Impedance.

z-average molecular weight An average molecular weight that is weighted toward the heavier molecules in a mixture of molecules; specifically, $\bar{M}_z = \Sigma n_i M_i^3 / \Sigma n_i M_i^2$, where n_i is the number of moles of component i, and M_i is the molecular weight of component i. *Sym* $\bar{M}_z$. *See also* average molecular weight.

Z cell *See* XYZ cell theory.

Z DNA *See* DNA forms.

zeatin A purine derivative that occurs free in maize and many other plants; a naturally occurring cytokinin. *Aka* maize factor.

Zeeman effect The splitting of the degeneracies of the excited states of a chromophore by an external magnetic field.

zein A seed protein of corn.

zeolite A naturally occurring alkali metal– or alkaline earth–aluminum silicate that has a network structure and ion-exchange capacity; used as an ion exchange resin for water softening and as a molecular sieve. *See also* permutite.

zero layer line EQUATOR.

zero meniscus concentration method MENISCUS DEPLETION SEDIMENTATION EQUILIBRIUM.

zero mobility position The position occupied by an uncharged substance in an electrophoresis experiment.

zero-order kinetics The kinetics of a zero-order reaction.

zero-order reaction A chemical reaction in which the velocity of the reaction is independent of the concentrations of the reactants.

zero point The point on an x-ray diffraction pattern where the incident beam strikes the photographic film.

zero-point mutation A mutation that is expressed immediately following the irradiation of cells with a mutagenic radiation.

zeroth law of thermodynamics The law that establishes a quantitative concept of temperature so that the state of every thermodynamic system includes temperature either explicitly or implicitly. The law can be phrased as follows: If body A is in temperature equilibrium with body C, and body B is in temperature equilibrium with body C, then bodies A and B are in temperature equilibrium with each other. This principle is assumed whenever a thermometer is used to compare the temperature of two systems. *See also* thermodynamics.

zero time binding DNA That fraction of the DNA, in a reassociation kinetics experiment, that contains repeating sequences and, therefore, forms duplexes at the very start of the reaction.

zero time control A control used in enzyme studies in which the enzyme is inactivated prior to addition of, and incubation with, the substrate.

zeta potential The potential difference across the plane of motion between two phases, particularly the potential across the surface of shear between a charged particle and its surrounding ion atmosphere in electrophoresis. *See also* Stern potential.

zeugmatography SPIN-IMAGING.

Z form *See* DNA forms.

zigzag scheme Z-SCHEME.

Zimm–Crothers viscometer A Couette-type viscometer in which the inner cylinder is a self-centering float; used for viscosity studies of DNA.

Zimmerman method *See* electrofusion.

Zimm plot A double extrapolation used in the analysis of light scattering data when the scattering particles are larger than those involved in Rayleigh scattering. A plot of KC/R_θ versus $\sin^2(\theta/2) + kc$ is extrapolated both to $c = 0$ and to $\theta = 0$, where K is an optical constant, c is the concentration, R_θ is the Rayleigh ratio, θ is the angle at which scattering is observed, and k is an arbitrary constant chosen to provide a convenient spread of the data. Both the molecular weight and the radius of gyration can be obtained from the plot. *Aka* Zimm grid.

zinc An element that is essential to all plants and animals. Symbol, Zn; atomic number, 30; atomic weight, 65.37; oxidation state, +2; most abundant isotope, ^{64}Zn; a radioactive isotope, ^{65}Zn, half-life, 243.7 days, radiation emitted, positrons and gamma rays.

zinc finger A structural domain in proteins, formed by folding of a polypeptide chain about a zinc atom. Such domains have been found in nucleic acid binding proteins and are believed to be involved in the binding of these proteins to the nucleic acid.

zinc sulfate turbidity test A liver function test

that is based on the production of turbidity when serum from individuals with one of several forms of hepatitis is treated with a barbiturate buffer containing zinc sulfate.

zippering The rapid formation of a helical DNA or RNA duplex from the two separated and complementary strands; it follows a slower helix nucleation step in which a short stretch of double helix is formed. The reverse process, separation of the two strands, is known as unzippering (unzipping). *Aka* zipping; zippering up.

Z line The dark line that bisects the I band of the myofibrils of striated muscle.

Zn Zinc.

Zollinger–Ellison syndrome A clinical syndrome, characterized by excessive secretion of gastric juice, hyperplasia of gastric mucosa, and severe peptic ulcer disease; due to the presence of gastrin-producing tumors (gastrinomas).

zonal centrifugation DENSITY GRADIENT CENTRIFUGATION.

zonal centrifuge A specially designed centrifuge that allows large scale and continuous fractionation of material by density gradient centrifugation.

zonal centrifuge A specially designed centrifuge that allows large scale and continuous fractionation of material by density gradient centrifugation.

zonal diffusion A method for determining diffusion coefficients from the diffusion profile that is produced by the diffusion of a thin layer of macromolecules which have been placed in a shallow density gradient in a zonal centrifuge rotor.

zonal electrophoresis ZONE ELECTROPHORESIS.

zonal rotor A high-capacity rotor used for zonal centrifugation in a preparative ultracentrifuge.

zone centrifugation DENSITY GRADIENT CENTRIFUGATION.

zone convection electrofocusing A technique for conducting horizontal isoelectric focusing in free solution rather than in a density gradient.

zone electromigration ZONE ELECTROPHORESIS.

zone electrophoresis An electrophoretic technique in which components are separated into zones or bands in a buffer that is generally stabilized by a solid, porous, supporting material such as filter paper, starch gel, agar gel, or polyacrylamide gel.

zone precipitation The precipitation of a protein as a zone in a gel filtration column that is brought about by using a gradient of a protein precipitating agent.

zone spreading The broadening of a zone in either chromatography or electrophoresis as a result of processes such as eddy diffusion and eddy migration.

zoopherin Obsolete designation for vitamin B_{12}.

zoosterol A sterol that occurs in animals.

zootoxin ANIMAL TOXIN.

Z scheme The series formulation for the photosynthetic reactions of photosystems I and II of chloroplasts.

ZTP 5-Amino-4-imidazole carboxamide riboside 5′-triphosphate; a compound believed to function as an alarmone in response to shortages of folate coenzymes.

zwischenferment GLUCOSE-6-PHOSPHATE DEHYDROGENASE.

zwitterion DIPOLAR ION.

zwitterion-pair chromatography A form of ion-pair chromatography in which a zwitterion-pairing agent (hetacron; a zwitterionic compound that forms preferentially quadrupolar ion pairs with zwitterionic solutes) is added to the predominantly aqueous eluent in reversed phase liquid chromatography. The method is useful for the separation of bases, nucleosides, and nucleotides.

zygospore A spore formed by the conjugation of two other spores.

zygote 1. The cell produced by the union of the male and female gametes in reproduction. 2. The organism that develops from a zygotic cell.

zygotic induction The induction of a prophage that is transferred during conjugation from a lysogenic *Hfr* cell to a nonlysogenic F^- cell.

zymase A heat-labile enzyme fraction that is obtained from yeast and that catalyzes the reactions of alcoholic fermentation.

zymogen The inactive precursor form of an enzyme that is generally converted to the active form by limited proteolysis.

zymogen granule A membrane-surrounded, cytoplasmic, secretory vesicle that is formed by the Golgi apparatus. Zymogen granules serve to store, and subsequently to secrete, the zymogens synthesized by the ribosomes of the endoplasmic reticulum.

zymogram The record of a zone electrophoresis experiment in which the enzymes in a sample, particularly esterases, are separated according to their charge and molecular dimensions, and in which the activity of these enzymes is indicated by specific staining reactions. A zymogram thus provides a measure of the types and the relative amounts of various enzymes in the sample; zymograms prepared from bacterial samples may be used as an aid in bacterial taxonomy.

zymology The science that deals with fermentations.

zymolyase An enzyme preparation from

Arthrobacter luteus that is useful for lysing yeast cells and producing spheroplasts.

zymosan A polysaccharide derived from yeast cells that inactivates complement.

zymosis FERMENTATION.

zymosterol An intermediate in the biosynthesis of cholesterol from lanosterol.

zymurgy The application of fermentation to the manufacture of alcoholic beverages.